AF441820

DICTIONARY OF ALTERNATIVE DEFENSE

DICTIONARY OF ALTERNATIVE DEFENSE

BJØRN MØLLER

Lynne Rienner Publishers • Boulder

Adamantine Press • London

Published in the United States of America in 1995 by
Lynne Rienner Publishers, Inc.
1800 30th Street, Boulder, Colorado 80301

Published in the United Kingdom by
Adamantine Press
3 Henrietta Street, Covent Garden, London WC2E 8LU

Library of Congress Cataloging-in-Publication Data
Møller, Bjørn.
    Dictionary of alternative defense / by Bjørn Møller.
    Includes bibliographical references.
    ISBN 1-55587-386-3 (alk. paper)
    1. Europe—Military policy—Dictionaries. 2. Offensive (Military
strategy)—Dictionaries. 3. Deterrence (Military strategy)—
Dictionaries. I. Title.
    UA646.M73 1994
    355.4—dc20   92-27347
                                                            CIP

British Cataloguing in Publication Data
A Cataloguing in Publication record for this book
is available from the British Library.
ISBN 0-7449-0048-4

Printed and bound in the United States of America

The paper used in this publication meets the requirements
of the American National Standard for Permanence of
Paper for Printed Library Materials Z39.48-1984.

5 4 3 2 1

To my family

Contents

Preface:
Why Alternatives?

"Alternative" has a certain oppositional and not-quite-serious ring to it. This is at least the impression that the "establishment" seeks to convey, for the obvious reason that alternatives are invariably alternatives to what they stand for. Because what makes an establishment established is precisely its holding of the reins of power, alternatives are always disadvantaged by being in fact oppositional.

Nevertheless, it has to be acknowledged that alternatives are good because to have them at hand means being able to choose. With no alternatives available, conclusions become foregone and the liberty of decision-makers shrinks. Furthermore, unless one is prepared to postulate that alternatives are always wrong, one must admit that such foregone conclusions will often be suboptimal in the sense of being made on a deficient basis of information. There is thus a genuine need for alternative ideas.

Alternatives are, moreover, one of the most decisive engines driving historical development. Most of the values and "eternal verities" to which we have now grown accustomed originated as alternatives. Democracy began as an alternative to the absolute monarchy (and in 1989–1991 experienced a renaissance as an alternative to communist dictatorship), liberalism emerged as an alternative to mercantilism, and so on. Without alternatives and, by implication, without alternative thinkers, historical development would soon come to a standstill. Unless one is prepared to agree with Fukuyama's postulate that we have reached "the end of history,"† one must acknowledge such an outcome as most unfortunate.

What is alternative is subject to change. All innovations, for instance, begin life in the alternative stage. Some ideas remain alternatives forever; others are gradually being implemented (either wholesale or in a piecemeal fashion), whereby they cease to be alternative and become part and parcel of the status quo, often to such an extent that their alternative pasts recede into oblivion. This is, indeed (if somewhat paradoxically), the whole purpose of alternatives: to change the status quo, that is, to create a new status quo.

---

†Francis Fukuyama, "The End of History," *National Interest,* no. 16 (Summer 1989): 3–18.

Alternatives might be ordered along a spectrum according to degree of radicality. At one end of the spectrum we have mere modifications, the advocates whereof would often resent the label "alternative." At the other end we find revolutionaries who envisage a complete topsy-turvy and who accept the heritage from the past only as something deplorably inevitable. In between we have various degrees of alternative thinking, differing according to scope as well as intensity: conceptions of incremental changes to be made across the entire universe alongside proposals for drastic changes in certain corners of the universe (whatever this may be).

Whereas most would therefore have to agree (at least after a moment's reflection) that alternatives are per se desirable, there is a tendency to reject each concrete alternative; indeed, this is the very essence of conservatism. In many cases, conservatives may be right because being alternative is far from tantamount to being advisable. National socialism, fascism, and Stalinism were, for example, alternatives in the 1930s, to the temptation of which fortunately only certain nations succumbed. In other instances, however, conservatism is an obstacle to necessary innovation.

Matters of national defense have not generally been regarded (at least by the establishment) as a field particularly in need of alternative thinking. On the contrary, because of the extreme importance of the matter, the status quo has been portrayed as almost sacrosanct, and alternatives have been regarded as dangerous. Even to discuss the wisdom of the prevailing arrangements of security and defense policy would allegedly undermine the credibility of a state's defense and hence jeopardize national security. Such a don't-rock-the-boat attitude, for example, characterized NATO's discussions in the 1980s, perhaps paradoxically reflecting a deeply felt uncertainty about the viability of NATO strategy rather than an unswerving belief in its value:

- The unidirectional orientation of NATO against the USSR and the Warsaw Pact seemed obvious, the facts notwithstanding that certain member states (Greece and Turkey, for instance) feared each other more than they did the Eastern foe, and that not all NATO states really believed the Soviet Union to be expansionist and aggressive.
- "Flexible response" was a nice concept, even though opinions differed widely between Europeans and Americans (as well as among Europeans themselves) about what it meant, and whether the envisioned deliberate nuclear escalation should ever be carried out.
- "Forward defense" was equally desirable, but nobody really believed it to be feasible.

The foundations of the two sacrosanct pillars of NATO strategy, flexible response and forward defense, have been seriously undermined since the mid-1980s by developments in both East and West. The INF controversy undermined popular, and to some extent elite, support for flexible response in the West; the "Gorbachev revolution" in the East undermined the enemy image held in the West of the Eastern superpower; and German unification in 1990 finally removed all justification for (as well as the practical option of implementing) forward defense.

The Western alliance is now, for the first time in its history, beginning seriously to contemplate alternatives. Even though it would be unrealistic to expect radical changes to be made overnight, it does seem likely that at least some of the formerly alternative conceptions will be incorporated into NATO strategy, if only in truncated versions. By implication, some of the previous alternatives will therefore change status to become, at most, proposals for modifications of NATO strategy. Some of them (probably the least extreme) may, as a consequence, come under attack from more radical alternative thinkers for their having been adopted by the establishment, perhaps even for exacerbating the underlying problems, because piecemeal reform may give the prevailing strategy a longer lease on life.

Compared with the relatively stable NATO, everything in the former East is in flux: the most basic assumptions of security policy are being revised; realignment is being actively pursued by several countries (whereas others are left in limbo); military doctrines are being developed; and armies are being built from scratch. Were it not for the apparent quest for NATO compatibility, the various alternative defense models might be ideal guidelines in these endeavors, and in some countries they have actually served as such. In light of political developments in the former USSR (for example, the electoral feat of the semifascist and Great Russia imperialist Zhirinovsky in the Russian elections), defensive restructuring of the large armed forces inherited from the Soviet Union may be the best chance of preventing war between Russia and its neighbors within the next decade or so. Defensive restructuring might also recommend itself for (contributing to) solving some of the outstanding conflicts in the Third World that may otherwise lead to nuclear proliferation and perhaps war: between the two Koreas, India and Pakistan, China and its neighbors, and the like.

Alternative defense concepts have thus become topical to a far greater degree than ever before in the postwar period. The obvious and urgent need for alternative thinking is, however, far from matched by any widespread awareness of the actual span of defense alternatives. On the contrary, the field seems characterized by national as well as disciplinary parochialism.

- The Anglo-Saxon audience is only very marginally aware of what is, beyond comparison, the richest body of alternative defense literature, namely, that of Germany. The Germans, in their turn, know the U.S. and British literature insufficiently. The French seem unaware of any literature except that written in French, in which, on the other hand, nobody else seems to take much interest. Finally, practically nobody outside the countries in question is aware of alternative defense writing in small language areas such those of the Nordic countries, Spain, Italy, the Netherlands, or the Slavonic states (to say nothing of countries with "odd" languages such as Finnish or Hungarian).

- Alternative defense designers (many of whom are active or, more often, retired military officers) tend to disregard the wider security political implications of their proposals, whereas political scientists advocating alternative defense tend to be either uninterested in or incompetent with regard to the nuts-and-bolts aspects of alternative defense, or both. The debate hence tends to become bifurcated, whereas any feasible alternative would have to take both sides of the matter into due consideration.

The present volume is intended as a contribution to rectifying these deficiencies. First of all, it is about alternative defense, broadly conceived in a dual sense:

- "Alternative" has been defined permissively to include proposals both for drastic transformations and for incremental reforms. Furthermore, the category includes both good and bad (in the opinion of the author), viable and infeasible, offensive and defensive alternatives, albeit with a certain preference for the (presumably) good, feasible, and defensive alternatives.
- Defense has been conceived of to encompass not only military defense but also nonviolent defense and various forms of prophylactic security policies that might, it is hoped, largely eliminate the need for an actual defense. The focus, however, is placed on military alternatives.

Second, the volume is global in its scope, and more language areas have been covered than is usually the case: the Anglo-Saxon languages, German, French, Italian, Spanish, Dutch, Danish, Norwegian, and Swedish. It was, however, impossible to cover the Slavonic languages as well as those of Finland and Hungary. This would not have been a very serious omission a few years ago, before glasnost, perestroika, and East European emancipation and democratization (because defense alternatives were not discussed, much less written about, in the communist countries), but it may well be a serious limitation under the present circumstances. There is actually a debate in progress on alternative defense, both in what used to be the Soviet Union (which had committed itself to nonoffensive defense but not yet decided upon the configuration of the new armed forces) and in the former Warsaw Pact member states, which are now reconsidering their entire security policies.

The author's background for accepting the monumental task of writing a dictionary of alternative defense is (besides an access to the aforementioned languages) that he has, from 1985 until the present day, been attached to (much of the time has de facto constituted) the research project Non-Offensive Defense in Europe, conducted under the auspices of the Centre for Peace and Conflict Research at the University of Copenhagen.

In this capacity the author has, inter alia, edited an international research newsletter, *NOD: Non-Offensive Defence* (since 1991, *NOD and Conversion*), which has functioned as a node in the emerging all-European, and to a certain extent global, network of alternative defense thinking. Also, he has been a frequent participant in the numerous conferences and symposia held on the subject, particularly in the period 1987–1990, which has allowed him to acquaint himself personally with a number of the most prominent thinkers in the field. Finally, his own research has allowed him to cover quite a good part of the alternative defense field, as will emerge from the entry on the author, which has (perhaps immodestly) been included alongside those on other authors in the field, as well as from the bibliography. In 1993 he received a major grant from the Ford Foundation for a project to create a "Non-offensive Defence Network." In his capacity as project director, he will be visiting a great number of countries, especially in the former communist bloc and in the Third World, seeking to disseminate information about NOD and assessing the opportunities for defensive restructuring.

The *Dictionary of Alternative Defense* is intended as an all-around vade mecum on the entire field of alternative defense. For the benefit of specialists and other academics, the text is thoroughly annotated to comply with the relevant scholarly criteria. For the benefit of the general reader, the entries have been written so as to require no prior expertise in the field, only a very general knowledge about security policy. Were it not that the term *encyclopedia* makes readers expect a multivolume work covering a wide range of subjects, the term might have been more appropriate for the present work than *dictionary*. The form of many entries is actually closer to what one would expect from an encyclopedia than from a dictionary.

The first part is a short introduction to the subject of alternative defense, dealing, inter alia, with the following subjects:

- Why there is a need for alternative thinking.
- The history of the alternative defense debate.
- The purposes of alternative defense (war prevention, disarmament, damage limitation, détente, and so on).
- The present status of the idea(s).
- How the different problems link together (for example, the link between nuclear no-first-use and conventional stability).
- How the subject matter has been defined (Where do modifications end and alternatives begin?).

The second part constitutes by far the major component of the volume and is organized alphabetically. The entries fall into the following categories:

- Persons (politicians, authors)
- Groups and organizations
- Countries and regions
- Events and periods
- Technology
- Concepts
- Full-fledged theoretical constructions
- Problem areas
- Cross-references

The third part is a comprehensive bibliography, covering the aforementioned languages as well as a small assortment of works in other languages, based on references in works directly accessible to the author. The bibliography also serves as the database to which the citations in parts 1 and 2 refer.

Asterisks preceding italicized words refer the reader to other entries. References in square brackets are to relevant items in the Bibliography. Foreign-language book titles and quotations have been translated by the author.

For inadvertently omitting an entry that should rightly have been included, I apologize.

*   *   *

I want to thank the Centre for Peace and Conflict Research, where I have been employed since 1985, for providing an inspiring environment.

I would also like to thank Lynne Rienner Publishers for shouldering the task of seeing this project through to final publication.

To more than anybody else, however, I wish to express my gratitude to my family—my wife, Ulla, and my two children, Hans and Ditte—for tolerating an excessive obsession with work, not only during office hours but late into the night, on weekends as well as workdays. Without their support, this dictionary could never have been written.

Bjørn Møller
Copenhagen, February 1995

Introduction

This introduction is intended to set the information contained in the following dictionary of alternative defense (AD) somewhat into context. It should thus be viewed as an "AD primer," outlining the main themes of the AD discourse, with a special focus on nonoffensive defense (NOD). For documentation and detail, the reader is referred to the main body of the dictionary.†

Historical Background

Alternative defense and security have been debated for ages but have enjoyed a special prominence in the aftermath of major wars as well as in periods of high international tension and widespread fear of war. In the following four periods in the twentieth century, interest in AD peaked:

1. In the 1920s and 1930s, that is, in the aftermath of *WWI*;
2. During the first Cold War (from the late 1940s to the early 1960s), with a special peak in the mid-1950s, when German rearmament was on the agenda;
3. In the late 1960s and early 1970s, when antiauthoritarian moods prevailed in large parts of the younger generation;
4. During the second Cold War (from the late 1970s to the late 1980s); and
5. During the second détente (from the late 1980s to the present).

The driving forces were not entirely the same in the four periods, nor did the debate proceed along similar lines.

The Interwar Years

Interest in AD in the interwar years was spurred above all by the desire to avoid a repetition of the unprecedented disaster of World War I. There was a rather widespread recognition that this disaster had been brought about by a combination of factors:

---

†As in the main body of the dictionary, words preceded by asterisks refer to entries.

- The (as it turned out erroneous) belief in the strength of the offensive, manifested, inter alia, in the *Schlieffen Plan* and its French counterpart, the *Plan 17*;
- The *arms race* that preceded it and the resultant extraordinarily high level of armaments; and
- The last-minute *mobilization* race, which acquired an unstoppable momentum because no state dared be caught off-guard by an attacker.

From this recognition sprang, on the one hand, a quest for viable supranational or international institutions, that is, for a form of *collective security*. The *League of Nations* wherein this conception came to be materialized did, however, prove inadequate for both small- and large-scale threats to peace. It was able to prevent neither the Spanish Civil War nor the Italian attack against Ethiopia (both of which could, in hindsight, be seen as dress rehearsals for World War II), and it failed completely to muster an adequate response to the combined German-Italian-Japanese aggressions in 1939–1940.

The league did play a role in the disarmament negotiations in the 1930s, particularly in the 1932 *World Disarmament Conference*, where the focus was placed on reducing or abolishing so-called offensive weapons. It proved impossible to find any definition of such weapons that was acceptable to all participating states, and the negotiations were exploited for launching allegations against adversaries. One of the few partly successful sets of disarmament negotiations was that on naval armaments, especially the 1922 *Washington Naval Conference;* the 1930 and 1932–1936 conferences in London were largely unsuccessful.

There was also a debate in certain countries on the prevailing trends in weapons technology and their effects on the offense/defense balance, as well as concrete suggestions for strengthening the defensive. In the U.K., the belief in the supremacy of the defense (advocated above all by *Liddell Hart*) supported the policy of continental disengagement and appeasement associated with the Chamberlain government, which, in their turn, facilitated Hitler's expansionist drive. In France, similar beliefs led to the construction of the *Maginot Line*, a comprehensive system of fortifications along the northeastern border of France. Although this defensive line was probably quite effective, successive French governments failed to complement it with mobile reserves capable of dealing with breakthroughs and circumventions.

The (in hindsight at least partly erroneous) belief in the supremacy of the defensive left the European democracies vulnerable to German attacks that exploited the emerging offensive potentials of *tanks* and aircraft, working in combination, for *blitzkrieg* strategies.

The First Cold War

After the collapse of the anti-German coalition by the late 1940s, a bipolar system developed in Europe, which was divided by an "iron curtain" between the two opposing military alliances: *NATO and (after 1955) the *Warsaw Pact. Bipolarity was underpinned by nuclear weapons fielded in ever-increasing numbers. The concurrence of a debate on German rearmament since the early 1950s with the deployment of tactical and theater nuclear weapons in Europe spurred a growing interest in AD.

With a view to preventing German rearmament, within the framework of NATO, from spoiling the last chances of German reunification, as well as from leading directly to war, a number of proposals were made, both by Germans (*Bogislav von Bonin, the *SPD, the *FDP, and others) and foreigners (*George Kennan, Hugh Gaitskill, and others) for various forms of *disengagement. Proposals were put forward for *neutralization and for the establishment of *nuclear-weapons-free zones, which coincided with (and may have been partly inspired by) Soviet and East European proposals to the same effect (above all the *Rapacki Plan). Even though the German question, and thereby Central Europe as a whole, was at the heart of the disengagement idea, similar proposals were made for other parts of Europe, for example, for a Nordic nuclear-weapons-free zone (the *Kekkonen Plan).

The disengagement proposals were accompanied by some thinking on alternative means of defense. Some of the aforementioned disengagement advocates accompanied their proposals with sketchy suggestions of other ways of defending Germany and Europe. Whereas Bonin proposed a genuine military NOD-type defense, Kennan came close to advocacy of *nonviolent defense. Others (*Steven King-Hall among them) arrived at similar conclusions and recommended a shift to nonviolent defense, either from religious or ethical motives, or from other angles.

The Antiauthoritarian Revolt

The first *détente period, from the early 1960s to the mid-1970s, did not produce much alternative defense or security thinking, but the tumultuous years around 1968 did spur an interest in particular forms of AD. These were the heyday of *civilian-based defense, interest in which was inspired by the apparent success with similar methods of struggle for various domestic purposes: as a means of antiauthoritarian struggle in the students' revolt; as a means of civil rights struggle in the U.S. (led by *Martin Luther King); and as a suitable method for the women's liberation movement, the emerging *Greens, and so on.

Although the interest was, above all, directed to the methods as such, some thought was also given to their potential for replacing military forms of defense. Most of the classic works on civilian-based defense were thus written in a short span of time, the late 1960s to the early 1970s, by *Adam Roberts, *Gene Sharp, *Theodor Ebert, *Anders Boserup, Andrew Mack, and others.

The interest in AD was also spurred by the new sense of solidarity with the Third World, in its turn partly reflecting the *Vietnam War, which was by then in progress. Whence sprang not only an interest in *guerrilla strategy (as exemplified in the Vietnamese guerrillas, as well as in the theory and practice of *Mao Tse-Tung) but also an interest in the thoughts of *Mahatma Gandhi on nonviolence, which fitted well with the aforementioned interest in nonviolent defense.

The Second Cold War

The debate in the late 1960s reflected a general sense of security; the renewed controversy in the late 1970s and early 1980s reflected the exact opposite, namely, a prevalent fear of war.

Nuclear strategy appeared to many observers to be heading in a dangerous direction, which was made to look even more frightening because the arsenals on both sides had by now reached the overkill level. Around 1976, this led a few academics (*Horst Afheldt and others) to put forward worked-out proposals for an alternative and strictly defensive, but still military, defense of Central Europe.

Although *arms control achieved some apparent successes from the early 1960s onward, a decade later disillusionment with its prospects was spreading. This spurred both more-hawkish attitudes toward arms control and détente (particularly in the U.S.) and a number of critical theoretical analyses of the logic of arms control, with *Robert Jervis standing out for his proposals for distinguishing between offensive and *defensive weapons.

The most blatant failure of arms control was the *INF debacle, running parallel to a highly conspicuous new and frightening development in nuclear strategy. This directly affected the Europeans in general, and the Germans in particular, not just because of the immediate implications in terms of creating high-priority targets on European soil but also because of the political implications of deployment, which threatened to jeopardize the accomplishments of détente in Europe.

The direct opposition of the *peace movements to the deployment of Pershing-2 ballistic missiles and the Tomahawk ground-launched cruise missiles (GLCM) was accompanied by a growing interest in all sorts of alternatives to the nuclear-infected strategy of NATO. Opposition was far from limited to what the establishment might regard as "the loony left" but included numerous members of the establishment itself. Proposals were thus made by former high-ranking NATO officials for a shift to a *no-first-use strategy, and old plans for nuclear-weapons-free zones and the like were resurfaced.

The antinuclear sentiments became so widespread in the population as to change the rules of the game according to which politics was played. Subsequent changes in U.S. and NATO strategy were therefore couched in terms of raising the nuclear *threshold* (as was the case with the *FOFA* plan) or even of making nuclear weapons "impotent and obsolete," as it was claimed that the *SDI* plan of the Reagan administration could accomplish.

The nuclear debate had matured considerably in comparison with that of previous decades. The general level of knowledge was higher, and a large body of alternative expertise had emerged that was capable of attacking established strategic thinking head-on. One effect thereof was to increase the interest in nonnuclear alternatives because it was realized that nuclear weapons were there for a purpose and hence had to be rendered superfluous before their abolition would become a realistic goal.

This was where AD proponents in general, and NOD advocates in particular, made one of their most important contributions. Concrete proposals were made for a restructuring of the armed forces, intended to ensure that NATO would possess a conventional posture that would be adequate for deterring (or better: dissuading) a Soviet attack, thus rendering nuclear deterrence redundant. A few alleged NOD proponents (*Albrecht von Müller*) along with establishment strategists (*Samuel Huntington* and the *ESECS* group) envisioned accomplishing this through what might be called *nuclear equivalence*, that is, by doing the same by conventional means that had previously been done by nuclear means, inter alia retaliation and long-range interdiction.

The majority of AD proponents, however, including all genuine NOD advocates, proposed the exact opposite: a build-down of offensive capabilities in the conventional field, that is, a shift to a strictly *defensive defense* (also called *nonprovocative defense* or *nonoffensive defense*). This should not merely be more effective for actual defense purposes (thus diminishing the need for nuclear back-up) but also remove lucrative targets for the enemy's nuclear weapons: the so-called *no-target principle*. In addition, a shift to NOD would presumably facilitate disarmament and solidify détente.

Although the NOD proponents remained within the military paradigm, antimilitarism also flourished in these years, both in the peace movements and in the (partly overlapping) "green" movements and parties that enjoyed considerable support in the populations and electorates of most West European countries.

The Second Détente

The AD proposals were not really taken seriously by the establishment before the late 1980s but were either ignored or rejected as incompatible with NATO strategy and ipso facto unacceptable. In the latter half of the

1980s this situation changed when, quite unexpectedly, the *Soviet Union* expressed interest in the concept. Previously, nearly all NOD proposals had been addressed to NATO, both because their proponents were Westerners and because it was regarded as inconceivable that the East might be persuaded to change its strategy. At best, Western NOD proponents had hoped for inducing a change indirectly: by making NATO reform its strategy first and thereby motivating the USSR to change course.

The Soviet sea change came as a complete surprise, inter alia because the strategy and the operational and tactical conceptions of the Red Army had thitherto been highly offensive. In the course of a couple of years, however, the Soviet (Communist) leadership had committed itself politically to a complete shift to NOD, accompanied by an equally wholesale commitment on the part of the Warsaw Pact.

This was reflected not only in a certain (albeit rather slow and, from a Western point of view, unsatisfactory) unilateral restructuring and build-down of the armed forces but also in numerous concrete proposals for negotiations concerning both nuclear and conventional forces. One result was the start of a new set of negotiations on conventional armed forces in Europe to supersede the unsuccessful *MBFR* negotiations. The mandate of the new *CFE* talks was initially likewise limited to conventional ground forces in Europe. It was, however, somewhat broader, both geographically (by covering the entire *ATTU* area, from the Atlantic to the Urals) and thematically (by dealing with equipment, that is, major army weapons systems as well as combat aircraft and helicopters). The mandate was also more focused and relevant because the proclaimed intention was to reduce offensive capabilities in general, and those of *surprise attacks* in particular.

The talks were, in one sense, remarkably successful, and a treaty was signed in November 1990. In another sense, they had clearly been overtaken by events: the division of the participating states into two alliances had been rendered obsolete by the effective collapse (and subsequent formal dissolution) of the Warsaw Pact; one participating state (the *GDR*, that is, East Germany) had shifted sides and subsequently ceased to be, through a merger with the FRG; and, finally, the pace and magnitude of the unilateral reductions exceeded those stipulated in the CFE treaty.

By the early 1990s, many NOD conceptions had been incorporated into the establishment strategic discourse, and many of the goals set by NOD proponents had at least been partly achieved. By now, though, new problems were appearing on the horizon, with which AD proponents are currently struggling:

- What to do about genuine *multipolarity*, when *alliances* change and small-scale interstate and civil wars in the periphery of Europe have become more likely than a large-scale war in the center?

- What to do (militarily as well as politically) with the "German problem" lurking beneath the surface of the apparently tranquil center of Europe?
- What to do about the rapidly collapsing Soviet Union and empire so as to prevent a Soviet or Russian threat from ever reappearing?
- How to prevent the (it is hoped) pacific Europe from being merely a tranquil island in a sea of turmoil and war, that is, how to make the rest of the globe, and in particular the Third World, peaceful?
- Whether, and if so how, to combine a shift to alternative defense with an enhanced role for the United Nations, both militarily and otherwise?

The prevailing mood in the AD community seems to be one of optimism and a new belief in the prospects of institution-building. *Regimes* tend to be regarded as a natural reflection of a growing and multifaceted *interdependence* that should be further promoted. Collective security might thus be the appropriate security arrangement for such a European continent (and ultimately: world) of *integration* and federalization.

Attention thus seems to be drifting toward broader forms of alternative *security*. Even so, it has to be acknowledged that some sort of defense must remain part of such arrangements, and that the present defense postures and strategies are quite unsuitable for this purpose. There thus remains a need for alternative defense.

The Goals and Principles of Alternative Defense

Let us now proceed with presenting (in skeleton form) the theoretical foundations of AD in general and NOD in particular.

Alternative defense proponents have tended to agree on a set of goals while disagreeing about how to prioritize them. The following are the primary goals about which all AD proponents would tend to agree:

- Disarmament
- War prevention
- Defensive strength
- Damage limitation
- Détente, *entente,* and democracy

Disarmament

AD advocates strive for an abolition, or at least a build-down, of the huge armed forces now fielded, for several reasons:

1. The very existence of (standing) armed forces is regarded by many as compounding the risk of war (cf. below);
2. The greater the magnitude of the military in general and of its nuclear element in particular, the greater will be the devastation in the case of war (cf. below);
3. AD proponents tend to view military expenditures as wasteful consumption that they would like redirected to socially useful purposes. This points directly to the need for combining AD thinking with plans for *conversion* from military to civilian production and consumption.

AD advocates tend to have little faith in the traditional (at least in the post-war period) approach to disarmament, namely, *arms control*: its goals are too modest; its objectives are often irrelevant, sometimes even harmful; and it tends to fail.

First, the relevant goal should not be merely to control the growth of the arsenals through ceilings and the like but to reduce their size. Still, most AD (and especially NOD) advocates tend to agree with the arms control school that the focus should not primarily (or at least not merely) be placed on the size of the arsenals but also on their composition.

Second, arms control is criticized for focusing on *balance* and parity between contestants, concepts that are notoriously difficult to define and even harder to accomplish in practice. The proclivity for worst-case analysis is strong, just as is each side's desire for a margin of superiority over its opponent(s). Furthermore, even if balance should be attainable (which is unlikely), it would not really matter much because neither of two adversaries would feel comfortable or secure in a situation of parity. Strategic surprise might still allow one side to defeat the other in that it would allow the attacker to concentrate overwhelming forces at decisive points.

Third, and inter alia because of the aforementioned defects, the cumulative score of arms control endeavors over the past decades has been far from impressive. With the partial exception of the 1972 *ABM Treaty*, the few negotiations that have been successful have yielded only irrelevant results (such as the SALT-1 ceilings on launchers, when warheads were what mattered), whereas negotiations on relevant matters have usually failed.

Certain AD proponents have suggested abandoning arms control entirely in favor of *unilateralism* and/or *gradualism* (more on which below). Others have merely demanded a refocusing of arms control endeavors.

To be relevant and to stand a reasonable chance of success, arms control should focus on the element of mutual threats, that is, on offensive capabilities. The theoretical underpinning of this recommendation is the notion of the *security dilemma*, elaborated upon by *Robert Jervis*, *John Herz*,

Barry Buzan, and others, according to whom states tend to threaten each other inadvertently. When one state acquires weaponry or fields armed forces to safeguard against what it perceives as a threat from other states, the threat it then constitutes to the other states likewise grows, and they are likely to reciprocate, whence may well spring a spiral of arms buildup, even if neither side has aggressive intentions.

The solution proposed by NOD advocates is to build down offensive capabilities while maintaining (or even strengthening) defensive capabilities, which cannot be perceived as threatening by adversarial states and therefore do not spur any arms buildup on the part of the latter. A stable state would be attainable in the form of *mutual defensive superiority*, that is, a situation wherein neither side would be capable of winning a war of aggression, and both would be sure to prevail in a defensive war against an aggressor. A certain margin of defensive superiority would even allow either side to reduce its forces unilaterally without thereby incurring any significant risks, whence might spring an inverted arms race, that is, a spiral of disarmament.

The above perspectives do, however, presuppose that meaningful distinctions can be made between offensive and defensive armaments. A few NOD proponents speak of defensive and offensive weapons; others prudently relegate the distinction to higher levels of analysis and speak of defensive and offensive formations, or entire military postures, as well as of defensive versus offensive strategies.

War Prevention

Even though war prevention is today nearly universally acknowledged as a high-priority goal, AD proponents tend to place even greater emphasis on the goal than does the establishment and to regard war as permissible only for strictly defensive purposes. Neither wars of *intervention* nor *preventive wars* are thus deemed acceptable under any circumstances.

With regard to the means of war prevention, AD proponents tend to take a broader approach than the adherents of deterrence. Although they sometimes acknowledge that deterrence through retaliation (be it nuclear or conventional) may be an appropriate counter to premeditated aggression, they tend to regard other types of wars as more likely, against which deterrence will be, at best, of limited utility and, at worst, positively counterproductive.

Preventive wars and *preemptive attacks* are motivated by the fear of war itself, that is, they are launched by a state fearing that its adversary may soon attack (or has already embarked on aggression) and that time is of the essence. Waiting will mean only that chances of prevailing will deteriorate; striking first may improve the chances. The choice facing a state in this situation is thus not one of war or no war but, rather, of war now or war later. Even a politically defensive state may choose to attack, dictated by military logic, and more in sorrow than in anger. According to this reasoning, a state cannot guard against preventive or preemptive war through measures of deterrence, say, by mobilizing or raising the state of alert of its armed forces; this would merely confirm its opponent's fears of an imminent attack. On the contrary, appeasement is called for in order to allay the fears on the part of the opponent, so as to dissuade it from attacking.

The problem is, of course, that it may be impossible to distinguish between defensive precautions and attack preparations, whence might spring mutually reinforcing (mis)perceptions and reactions. It is therefore imperative to devise means of simultaneously hedging against aggression and appeasing possibly peaceful adversaries. Here, too, the distinction between offensive and defensive military means provides a solution: strictly defensive armed forces might safely be mobilized and alerted in the event of an expected attack but could not trigger such an attack, unless the opponent was determined to attack for positive purposes. Preventive and preemptive wars would thus become well nigh inconceivable.

NOD advocates have, furthermore, recommended the rendering of preemptive attacks ineffective as well as unnecessary by removing the targets against which such attacks might be directed: airfields, missile sites, tank assembly areas, and so on. Such removal would be an instance of the aforementioned *no-target principle.*

Defensive Strength

Because it is usually impossible to ascertain the intentions of one's opponent beyond any doubt, it has to be acknowledged that genuine aggressive wars are conceivable, even if not particularly likely. For it to be feasible, any alternative defense therefore has to maintain an undiminished strength vis-à-vis a determined aggressor, that is, it must be effective, even according to traditional military criteria, regardless of the abstention from offensive capabilities.

Advocates of *civilian-based defense* assert that their recommended *nonviolent resistance* would suffice for preventing an aggressor from establishing control over a country after invasion, that is, that it would prevent successful occupation from which an aggressor might benefit. Accordingly, it would presumably dissuade an attack in the first place.

The majority of AD proponents (as well as, of course, the entire establishment) remain unconvinced that civilian-based defense (its numerous merits notwithstanding) would suffice and envision the maintenance of military means of defense. The efficiency of such a strictly *defensive defense* is not believed to suffer as a result of the abstention from offensive capabilities; on the contrary, to prepare merely for waging war on his own territory allows the defender to reap all the fruits of specialization. The defender is able to construct *barriers and fortifications; to fight almost entirely from shielded positions; to streamline his equipment and training practices for the relevant type of terrain; to operate on *interior lines; to integrate civilian means of communication into defense planning, and so on.

NOD proponents thus subscribe to *Clausewitz's thesis about the inherent superiority of the defensive form of combat, which they seek to amplify through strategic, operational, and tactical innovations; a pointed deployment of forces; and arms procurement policies focusing on weapons of particular suitability for defensive purposes. NOD proposals differ widely with regard to detail, but the majority have the following principles in common.

- Strategically, the defender should exploit the time factor through the ability to wage a protracted war of resistance.
- Operationally, the forces should be largely stationary, that is, possess only limited long-range mobility, and ideally no border-crossing potential at all.
- Tactically, units should be largely self-contained and operate in great *dispersion. No decisive battles should be fought, but defensive strength should, rather, be maintained with a view to protracted resistance. To improve survivability, units should preferably fight only from shielded positions and not move about under enemy fire.
- The deployment of forces should be rather dispersed, hence invulnerable to massive enemy strikes. Preferably, the entire territory should be covered, for example, through a *web-like mode of deployment.
- The typical weapons would be rather short-range and specialized as counters to the enemy's weapons platforms: antitank weapons, air defense weapons, antiship missiles, and so on. The defender's own mobile weapons platforms should be reduced to the greatest possible extent.

Whether nuclear weapons would have any role to play remains disputed. At the very least, an NOD-type defense should suffice for defending against a conventional attack, but some NOD proponents feel a need for a *minimum deterrence nuclear force as a hedge against nuclear attack, or threats thereof, that is, nuclear blackmail. Others, however, dispute this assumption and regard their conventional defenses as entirely self-sufficient. Still others (only a small minority) recommend research into means of defense against nuclear attack, something that the great majority, however, regard as utterly infeasible.

Damage Limitation

Damage limitation has always ranked high on the list of AD goals. In fact, the first AD proposals were motivated by the desire to devise less-than-suicidal means of defense as an alternative to *nuclear deterrence*, which may well be painless for as long as it lasts, but whose breakdown would result in complete annihilation of the defended country, along with, perhaps, the rest of the globe.

The answer of civilian-based defense advocates to this problem has been complete and, if need be, unilateral disarmament. General and complete disarmament (also pertaining to would-be aggressors) would obviously render war impossible and thus secure absolute damage limitation. Moreover, unilateral disarmament would presumably reduce the damage endured by a country under attack because it would dissuade at least the use of massive weapons on the part of the aggressor.

NOD's answer to the same problem has been, first, to improve the strength of conventional defenses and thus to contribute to a build-down (it is hoped, abolition) of nuclear weapons. Second, to render nuclear use on the part of an aggressor less militarily rewarding through a less inviting target profile (the no-target principle, once again). Third, various proposals have been made for diminishing the destruction wreaked by conventional warfare, which would (as a result of the shift to NOD) largely take place on one's own territory. For example, it has been proposed to declare cities *open towns* and, as a corollary, to leave them militarily undefended.

NOD advocacy has sometimes been combined with the (increasingly widespread) belief that a conventional war fought in Europe will inevitably be totally destructive. This *Totally Destructive Conventional War* hypothesis is, however, logically compatible only with civilian-based defense (and perhaps nuclear deterrence), but certainly not with NOD.

Détente, Entente, and Democracy

The immediate purpose of AD has been to solve military problems, which has given rise to two critiques: that there are other (presumably better) solutions to the same problems; and that there are other (perhaps more serious) problems that AD could not possibly solve. On inspection, neither critique seems appropriate: AD proponents have been quite aware that there might be entirely different solutions to the same problems, just as they have acknowledged these problems constitute merely part of the general security problem.

First, *democracy* is generally believed to serve as a powerful inhibition against aggressive war. Democratic states tend to get involved in nearly as many wars as totalitarian states, but they seldom (if ever) fight one another. Democratization may thus serve the goal of war prevention as well as constitute a *confidence-building measure*. No AD proponents would likely deny this. Various forms of AD may, in fact, promote such democratization, at least in the negative sense of removing obstacles to democratic development. A complete abolition of the armed forces, such as envisioned by civilian-based defense proponents, would remove the risk of military takeovers, and the methods of nonviolent resistance have proved quite effective counters to attempted coups (most recently in the USSR in August 1991).

Less radical transformations of the defense posture may likewise further democracy: *militia structures tend to make the armed forces a more representative sample of the general population than is the case of professional forces; and the envisioned *decentralization of the armed forces would tend to curtail the power of the officer corps, that is, the most likely instigators of military coups.

Second, a general relaxation of tension, that is, détente, would go a long way toward making war between adversarial states unlikely and thus reduce the salience of military matters. Ultimately, détente might be superseded by a true entente, under which a group of (formerly perhaps adversarial) states might come to constitute a *security community, between whose members war would become not just unlikely but inconceivable, from which point on military factors (force ratios, deployment modes, strategy, and so on) would cease to matter at all.

AD proponents would willingly concede as much but maintain that until that point is reached, military matters will continue to play a role. In this context the purpose of AD would be gradually to downplay the military element. It might thus be called "a military solution to render military solutions superfluous," which is only seemingly a contradiction.

AD can presumably promote détente in several ways:

- It can help dismantle enemy images by removing their foundations and by enhancing transparency;
- It can serve as a *CBM (confidence-building measure) or *CSBM (confidence-and security-building measure) in its own right; and
- It has the potential of neutralizing the military logic and thus allowing political relations to be guided by political considerations.

By promoting détente and entente, AD might also indirectly contribute to solving other problems on which it has no direct impact: by means of its contribution to disarmament, it could, for example, help make a peace dividend available for environmental protection, development aid, or other worthy nonmilitary causes.

The Span of Alternatives

Defense proposals in general, and AD proposals in particular, differ according to several criteria:

- The envisioned role (if any) of nuclear weapons;
- The envisioned role (if any) of the armed forces;
- The envisioned role (if any) of nonviolent forms of defense;
- The proposed strategy for, and configuration of, the nuclear forces, conventional armed forces, and nonviolent means of defense; and
- The context of the defense proposal: geographical, temporal, alliance-wise, scenario-wise, and so on.

Applying these criteria allows us to divide the field into several, partly overlapping, categories and subcategories. The main categories are shown in Table 1.

Table 1.   Traditional and Alternative Defense

	Conventional Military		Nonmilitary	Mixed
	Offensive	Defensive		
Nuclear	Soviet strategy	(Flexible response)		
Nonnuclear		NOD	Civilian-based defense	Total defense

The strategies described as nuclear are, of course, not based entirely on nuclear weapons; the term merely signifies that nuclear weapons play a significant role in the strategies. This has been the case for Soviet and NATO strategy, both of which have, however, undergone transformation. Pre-Gorbachev Soviet strategy was (and Soviet strategy to some extent remained until the late 1980s) very offensive-minded conventionally. NATO's strategy of flexible response and forward defense has all along been largely defensive as far as the conventional realm has been concerned, even though *counter-offensive* operations have certainly been envisioned.

Among the nonnuclear strategies, we find (the broad variety of) NOD proposals, which may envision maintenance of some nuclear weapons, yet only as a temporary measure and merely as a counter to an enemy's nuclear weapons. Conventionally, these strategies are strictly defensive, but they still rely primarily on military means of defense. This sets them apart from civilian-based defense, which is also both nonnuclear and defensive, but which rejects the use of military means altogether, as a matter of principle.

The two types of defensive, nonnuclear strategies may be combined into a *total defense*, just as nonmilitary means of defense might, in principle at least, be combined with offensive and nuclear strategies. In real life, however, they tend not to be.

The category of nonnuclear, defensive military options, that is, NOD, might be further subdivided according to criteria defining the concrete deployment and configuration of the proposed forces. The competing NOD models differ, inter alia, with regard to the envisioned degree of mobility and mode of deployment. Some models maintain forward defense as, in principle at least, strictly defensive, as well as ideal from a damage-limitation point of view: for a state to defend its own territory immediately at the border spares its population the horrors of war yet threatens nobody, if only the state abstains from border-crossing capabilities.

Other NOD proponents have preferred an in-depth and *territorial defense*-like deployment that would presumably be more *tous azimuts* (and thereby more confidence-inspiring) as well as less vulnerable to surprise attack. Still others have sought the best of both worlds through some combination of successive zones with different types of defenses. Typically, it has been recommended to place the more stationary, and thus defensive, elements up front and to withdraw mobile and potentially offensive formations to rearward positions, an arrangement that would also reduce any capabilities of surprise attack.

Most NOD models are considerably less mobile than the present postures, especially since they deliberately eschew border-crossing capabilities in order to minimize offensive potentials. However, whereas some models envision genuinely stationary forces, others place great emphasis on tactical (that is, short-range) mobility and agility with a view to diminishing vulnerability. The majority of models envision some combination of stationary and mobile formations, even though the concrete mix differs.

In Table 2, some typical exponents of the different combinations are offered as examples. The combination of mobility and in-depth deployment is exemplified with guerrilla forces, tactics derived from which are used to a large extent by the neutral and non-aligned countries in Europe.

Table 2.   NOD

	In-depth	Forward	Zonal
Mobile	*Guerrilla strategy*		
Stationary	*Horst Afheldt*	*N. Hanning*	*Eckhard Afheldt*
Mixed	*SAS*		*J. Löser*

The category of mixed concepts, that is, those that combine military (perhaps even nuclear) and nonmilitary means of defense, might be subdivided according to mode of combination and what is being combined. On the one hand, we have genuine NOD models that envision a civilian-based defense element, which might complement the military defense in time (by providing a fallback option), or in certain places (for example, in cities, particularly *open towns*), or for special circumstances. On the other hand, we have logically possible, yet distinctly rare, proposals for combining civilian-based defense with nuclear deterrence, usually in a considerably scaled-back version (see Table 3). There are many additional ways of subdividing AD proposals, but for further detail the reader is referred to the main body of the dictionary.

Table 3. Mixed Concepts

	Nonnuclear	Nuclear
Temporal	*ADC*	
Spatial	H-H. & *W. Nolte*	
Ad hoc	*A. Roberts*	*F. Dyson*

Modes of Implementation

AD proposals are nearly always blueprints, that is, descriptions of a different defense or security arrangement, as it might look at some point in the future. Even though the road-map task, that is, to devise ways of getting from here to there, does not follow logically from the blueprint task, most AD proponents have also devoted considerable attention to it. In these respects as well, opinions differ among AD proponents.

Some have favored *unilateralism* in order to circumvent the numerous obstacles to traditional arms control. According to this school (*Robert Neild* and others), states ought simply to shift to NOD (or civilian-based defense) regardless of whether other states would follow suit because to do so would simply be advantageous.

Others have conceded as much yet maintained that a shift on both sides would be far preferable, and quite realistic. Various methods for inducing reciprocation have been devised, among which one could mention *Tit-for-Tat* and various versions of *GRIT*, all of which have been conceived of as forms of *gradualism*, coordinated unilateralism, or informal bilateralism.

Still others have all along adhered to, or have gradually become convinced of the merits of, negotiations as the appropriate method for implementing the envisioned strategic transformation. Numerous concrete proposals for alternative *CFE* mandates and outcomes have thus been put forward by NOD proponents, many of whom have also suggested an expanded agenda, to include, inter alia, maritime and nuclear forces as well.

The Present Status of NOD

As mentioned in the section on the history of AD, interest in AD in general and NOD in particular has grown considerably in recent years. In the early 1980s, AD was a preoccupation of a few academics and some rather marginalized political groupings, but by the early 1990s support for AD had grown considerably. The basic conception had been adopted wholesale by most of the socialist parties in Western Europe, by the Soviet leadership (and apparently by its successors), and by the former Warsaw Pact countries. Furthermore, the notion of building down offensive capabilities had been the subject of arms control negotiations (the *CFE*), and military doctrines had become a topic for international high-level discussion (at the 1990 *Vienna Seminar* and its successors).

More and more prominent politicians were beginning to include concepts and terminology with origins in the AD debate and even top-brass military officials had begun to acknowledge openly the need for reform of strategy. The stage thus seemed set for proceeding with realizing the AD ideas, to which the following alphabetical dictionary is devoted.

A

ABM TREATY. Treaty signed 26 May 1972 between the U.S. and the USSR that prohibited, with certain exceptions, the deployment of anti-ballistic missile (ABM) systems. The ABMT stipulated that each state was entitled to two ABM systems (one protecting an ICBM field, another defending the national command authority, i.e., the capital), the subsequent Vladivostok agreement of 1974 reduced this to one system, i.e., to a choice between defending one or the other [in Carter & Schwartz (eds.) 1984: 427–439; cf. Schwartz 1984; Schneiter 1984; Garthoff 1984; 1985: 127-198, 426]. The U.S. came to abandon ABM entirely; the USSR maintained its ABM system around Moscow until the present day [Stevens 1984; Pike 1985].

The ABMT was based on an understanding of the ambivalence inherent in the *offense/defense distinction* as applied to *strategic defense*: if combined with offensive nuclear weapons, the ability to defend against a (first or retaliatory) nuclear attack would not diminish but, rather, strengthen offensive capabilities, perhaps to the point of constituting a disarming first-strike capability [cf. Sloss 1984; Glaser 1986; Stockton 1986; Freedman 1987; Drell & al. 1986].

The *SDI* program of the Reagan administration would have violated the ABMT, hence SDI advocates have urged an abrogation of the treaty, yet to no avail [Schneider 1993]. However, certain other weapons programs that might, according to some interpretations, also infringe upon the letter, if not the spirit, of the ABMT, have nevertheless been allowed to proceed: the plans for an *extended air defense* and the Global Protection Against Limited Strikes (*GPALS*) of the Bush administration [Brauch 1987b; Pike & Stambler 1992; Pike & al. 1992; Wright 1992]. Because of the disappearance of the East-West conflict, however, it seems unlikely that Russia will object to such minor reinterpretations, especially in light of the prospects of partaking in GPALS [Savelyev 1993; 'Documentation,' *Comparative Strategy,* 12:1 (January–March 1993): 109–113].

ACKER, ALEXANDER. Lieutenant colonel of the Bundeswehr, collaborator on the research project Alternative Security Policy led by *Horst Afheldt*. His main contribution was elaboration of the details on the rocket artillery component of the model [Acker 1984; cf. Afheldt 1983: 91–96].

ACTION-REACTION PHENOMENON. See *Armaments Dynamics.*

ACTIVE DEFENSE. Common term for the largely defensive Western operational doctrine in the 1970s and early 1980s.

1. **U.S. Army.** The operational doctrine of the U.S. Army found in the 1976 version of the *Field Manual 100-5: Operations,* formulated in light of the U.S. experience in the *Vietnam War*, as well as of the Yom Kippur War [FM 100–5 1976; cf. Czege 1984: 105–108; Bellamy 1987: 132–133; Mallin 1990: 200–205].

AD placed the main emphasis on defense and sought to capitalize on the inherent defense superiority (cf. *three-to-one rule*), inter alia through full use of cover and concealment, mine laying, and obstacle construction. Counterattack was to be conducted "only when the gains to be achieved are worth the risks involved in surrendering the innate advantages of the defender" [FM 100–5 1976: 5.14].

As early as 1979, it was decided to embark on a revision, which was completed in 1982, when AD was superseded by the *AirLand Battle* doctrine. The reason for the presumed inadequacy of AD was the Soviet move toward a greater emphasis on fluid maneuver warfare (by means of, e.g., OMG, Operational Maneuver Groups), against which a positional operational doctrine was ineffective. Furthermore, AD's emphasis on *forward defense* necessitated large-scale lateral movements along the front of formations and supplies that were presumably vulnerable to enemy interdiction. (See *Land Forces*).

2. **NATO.** NATO's operational doctrine, formulated in the manual *ATP-35* (Allied Tactical Publication-35): *Land Force Tactical Doctrine,* inspired by and largely identical with the U.S. doctrine. When the AD of the U.S. Army was replaced with AirLand Battle, a revision was likewise undertaken in NATO's AD, leading to the issue of *ATP-35(A),* heralding a greater emphasis on maneuver warfare [Garrett 1989: 43-45; Bellamy 1987: 124-129].

The least radical *NOD* advocates might actually be seen as simply opposing these more recent, offensive innovations, and as recommending essentially a return to AD.

ADC.   Alternative Defence Commission, established in 1980 by a number of British academics, many of whom were close to the *peace movement.* The ADC included, among others, Frank Blackaby, April Carter, Malcolm Dando, Owen Greene, *Mary Kaldor*, Michael Randle, Paul Rogers, Josef Rotblat, *Dan Smith*, and Jonathan Steele, as well as members of the *Labour Party* and the Liberal Party.

Before its dissolution, the ADC published two reports, *Defence Without the Bomb* [1983] and *The Politics of Alternative Defence* [1987], and a selection of contributions to the international consultation held in Bradford in 1986 [Randle & Rogers (eds.) 1990].

In the 1983 report, various ways of making the British nuclear deterrent superfluous (thus allowing for its unilateral abolition) were explored, including *territorial defense* and *civilian-based defense.* The ADC recommended a shift to *defensive deterrence*, based on "the capacity to inflict heavy losses on any invading force, but at most only a limited capacity to mount offensive operations in the opponent's territory" [1983: 9]. Protracted *guerilla warfare* and civilian resistance were taken into consideration as viable fallback strategies to which the state might resort in case an aggressor should succeed in occupying the country. Three force structures were considered in detail, all of which envisioned a reduced emphasis on offensive-capable weapon systems and military formations and were estimated to allow for saving on the defense budget [1983: 282–284]. (See Table 4.)

In the 1987 report, the focus was on the alliance-wide and pan-European aspects of alternative defense and security. A key concept was that of *dealignment*, envisioned as a form of *disengagement*, and thus as a step in the direction of a dissolution of the opposing alliances. Britain's contribution to this development might be one of conditional membership in NATO, i.e., a continued membership combined with a quest for changing the policies and strategy of the alliance. Should these attempts prove unsuccessful, Britain should withdraw from NATO and establish a national defense based on the principles of defensive deterrence while relentlessly working for its emulation on the part of other states [ADC 1987: 182–237; cf. Dan Smith 1990; Kaldor & Falk 1987].

Table 4. Alternative Defense Postures

		Force Structures		
		I Year 5	II Year 10	III Year 20
Army	Regular	160,000	80,000	50,000
	Reserves	350,000	70,000	70,000
Air Force	Personnel	60,000	60,000	48,000
	Aircraft	389	389	293
	SAM sqdr.	12	12	6
Navy	Personnel	35,000	35,000	28,000
	Major ships	101	101	104
Cost	Year 20, £Mill.	10,610	8,140	5,950

AFES PRESS. Arbeitsgruppe Friedensforschung und Europäische Sicherheitspolitik (Working Group Peace Research and European Security Studies). Small international study group dealing, among other topics, with *NOD*. The chairman of AP is *Hans Günter Brauch*; the deputy chairmen are *John Grin* and *Bjørn Møller*.

AFGHANISTAN, SOVIET WAR IN.   From the Soviet 1979 invasion until the withdrawal of the last Soviet forces in 1988 the theater of intense *guerrilla warfare*, subsequent to a civil war with guerilla features.

Even though they succeeded to a greater extent in keeping their involvement limited, the Soviet forces nevertheless found themselves in a quagmire very similar to the one in which the U.S. had been in Vietnam. The original plan for a swift intervention by means of *desant* forces and a small contingent of ground forces was rapidly upset and the invaders had to augment their numbers, reaching a plateau of 100,000–125,000 men [Bradsher 1985: 169–188; Gibson 1986; Dunnigan & Bay 1985: 99–100]. All attempts at defeating the *mujahedeen,* however, failed miserably [Chaliand 1981; McCormick 1987; Urban 1988; Cordesman & Wagner 1990(3): 3–237].

1. **First Stage.** In 1980–1981, the Soviets apparently hoped to be able only to reinforce the indigenous government forces, whose dissolution simply accelerated, leaving the Soviet forces practically without local support [Bradsher 1985: 205-207]. Due to the *mujahedeen*'s lack of an adequate supply of modern antitank weapons, however, the Soviet army units were still relatively invulnerable in their armored vehicles, obliging the guerrillas to focus on "soft-skin" targets such as trucks. Mines and avalanches exacted a significant toll on the Soviet convoys passing through valleys in guerrilla-controlled territory. The establishment of scattered garrisons provided the guerrillas with targets for nocturnal raids and made the invaders increasingly dependent on logistical chains [Victor 1985: 589–590; McCormick 1987: 61–64; Chaliand 1981: 95–105]. Soviet air raids against villages presumed to be *mujahedeen* strongholds and against transit routes from Pakistan only contributed to support for the resistance, as well as to a surge of refugees into the neighboring countries. In addition, the raids gradually depopulated certain regions, making it all but impossible for the *mujahedeen* to live off the country in those regions, an offensive use of a *scorched earth strategy* [Dunnigan & Bay 1985: 101].

2. **Second Stage.** In 1982–1983, the Soviets introduced special forces (paratroopers, *speznaz* units, etc.), by means of which they attempted sweep operations intended to free certain areas of guerillas and to establish security zones. After each initial success, however, the *mujahedeen* were soon able to reestablish a foothold. Furthermore, the Soviet forces experienced increasing difficulties with road support: lorry convoys were constantly attacked by the *mujahedeen.* Mine-clearing operations enjoyed some success, but safe transport still depended on the availability of helicopter support. Furthermore, the regime took advantage of the political divisions within the resistance to try to rally loyal forces in local *militias,* etc. [Victor 1985: 590–594; McCormick 1987: 64–67].

Even though many of the accounts of Soviet use of chemical weapons ("yellow rain") and "toy mines" have been revealed as unfounded, there is no disputing the fact that highly lethal antipersonnel weapons were used quite extensively, and that the civilian population suffered severely [Dunnigan & Bay 1985: 111–112; Bradsher 1985: 210–211]. Nevertheless, even these methods failed to break the resistance, and only flooded the refugee camps in Pakistan and Iran [Bradsher 1985: 279–280; Karp 1986: 1044–1045].

3. **Third Stage.** Thenceforth, the Soviet forces launched major combined-arms offensives, facilitated to a certain extent by the introduction of new weaponry that had been designed with the Afghan war in mind; the helicopters proved an especial success. Subsequently, the Soviet forces experienced increasing difficulties through being deprived of their unchallenged air superiority after the introduction of highly effective air-defense systems (above all the man-portable *SAM Stinger*) [Victor 1985: 594–595; *Jane's Defence Weekly,* 29 November 1986; 10 October 1987]. The morale of the Soviet forces was gradually undermined as the war became increasingly "dirty," with widespread drug abuse and occasional desertions as a result [Bradsher 1985: 272–273; Jukes 1989].

4. **Disengagement and Withdrawal.** As it became increasingly obvious that the war could not be won without disproportionate effort, and because the war in progress seriously damaged the Soviet image, particularly in the Third World [R.F. Miller 1989; Holmes 1989; Saikal 1989], thus hampering the Soviet peace offensive under *Gorbachev,* the USSR began to disengage. Initially it seems to have placed its hopes in an "Afghanization" of the war, a (partly successful) attempt at building up the Afghan army [Karp 1986: 1039–1040], simultaneously with entering into multinational negotiations in Geneva toward settlement of the conflict. As stipulated in the 1988 Geneva Accords, it simply withdrew, without much hope for its indigenous allies [Dupree 1989; Maley 1989]. In early 1992 the Afghan government finally succumbed to the *mujahedeen* offensive.

5. **Lessons.** Among the reasons for the Soviet defeat in A were the following. The guerillas enjoyed the support of the majority of the population, even though they were severely hampered by their factionalism and lack of a social and political program [Chaliand 1981: 77–91]. Second, they enjoyed widespread international support, to a considerable degree materialized in arms provisions from Arab countries, China, and the U.S. [Bradsher 1985: 276–279]. Third, they took advantage of the highly inaccessible mountainous terrain, in which the invader's heavy equipment was of little use, thus forcing it to place the main emphasis on *air power.* Moreover, they gradually made this air power ineffectual, and were thus able to maintain relatively sheltered enclaves similar to base areas.

AFHELDT, ECKART.   Brigadier general (ret.) in the Bundeswehr; cousin of *Horst Afheldt* and collaborator on his research project, to which EA contributed a proposal, for a "Defense Without Suicide: Proposal for an Employment of Light Infantry" [1984; 1984a; H. Afheldt 1983: 66–80].

The gist of EA's proposal was to substitute, to a considerable extent, *infantry* for armored forces, especially in forward areas. The proposal would presumably lend itself to implementation in a step-by-step manner, just as it, in turn, would merely constitute a step toward the all-encompassing *territorial defense* scheme of Horst Afheldt.

EA proposed the creation of a 70–100-km-wide forward zone along the eastern border of West Germany, defended by means of a *web* of around 100,000 light infantry troops (*Jägers*) without assistance of *tanks*, i.e., a tank-free *zone*. The *Jägers* were to be deployed in-depth and in great *dispersion* (an average of two *Jägers* per 24 square km). The typical equipment would include ordinary light infantry weapons, obstacle construction equipment, and antitank missiles, as well as artillery rockets, inter alia for the laying of landmines, rocket launchers for close combat (with maximum ranges of 2 km), multiple rocket launchers (with ranges up to 8 km), various types of *PGMs*, etc. The forces were to be stationary, to exploit terrain advantages, and to fight only from full cover.

AFHELDT, HORST.   German peace researcher who in 1976 first introduced the idea of *NOD* (*defensive defense*) [1976; 1978a].

The point of departure was HA's involvement with the research team that in 1970 published *War Consequences and War Prevention,* proving that nuclear war on German soil would completely devastate the country [Weizsäcker (ed.) 1971; Afheldt 1971; 1971a; idem & Roth 1971: 296–297], thus pointing toward the need for an alternative to *flexible response*. The theory was further developed in the framework of two consecutive research projects at the Max Planck Institute (since 1980, Max Planck Society) in Starnberg, FRG. From 1970 to 1983 he developed the details of his NOD model in two books: *Defense and Peace: Politics by Military Means and Defensive Defense* [H. Afheldt 1976; 1983 (French translation: 1985b); Weizsäcker (ed.) 1984; cf. Carton 1984; Trempnau 1983; Møller 1991: 63–75]. From 1984 until his retirement in 1989, he served as codirector (with *Hans-Peter Dürr* and *Albrecht A.C. von Müller*) of the Working Group Afheldt, which was devoted to a project on stability-oriented security politics [Dürr & al. 1984]. During this period, he published *Nuclear War: The Curse of Politics by Military Means* [Afheldt 1987b] and *The Consensus: Arguments for the Policy of Reunification of Europe* [1989a; extracts in idem 1992].

HA's theoretical background was the theory of stable deterrence, on which he based his own "peace policy with military means" [Afheldt 1976 (summed up in 1978a)]: a defense posture that would be stable in the sense of rewarding neither *preemption* or escalation, nor arms racing. Hence the *no-target principle* and the principle of abstaining from offensive capabilities, in order to escape the *security dilemma*: "In the case of multipurpose armies of the same structure, i.e., armies with which one can also attack, one is secure only in the worst of cases if one is sufficiently superior. But then it is urgent for the adversary to rearm" [Afheldt 1983: 37].

The defender should renounce symmetry and opt instead for an asymmetrical response (cf. the *indirect approach*) because specializing in defense means that the demands on *logistics*, armor protection, and the command system would be reduced [1983: 47-49]. In his model of *defensive defense*, HA envisioned such specialization by replacing *forward defense* with an area-covering *web* of small light *infantry* units, i.e., *Jägers* or *techno-commandoes* (inspired by *Brossolet* and *Spannocchi*), which should be stationary, in the sense of possessing neither strategic nor operational mobility. All heavy armored, and thus mobile, force elements were to be abolished, along with aircraft, above all in order to avoid being lucrative targets. The techno-commandoes should, at most, be tactically mobile but have several alternative prepared fighting positions at their disposal. In the absence of operational mobility, area coverage should be assured by a superimposed, somewhat coarser, web of missile commands, equipped with short-range missiles, by means of which firepower could be concentrated against enemy strongholds and attack columns [1983: 57–58, 89–91; Acker 1984]. The two overlapping webs would be interconnected by a *C³I* web [Freundl 1984], which should emphasize redundancy, e.g., by integrating the civilian communications arrays into military planning.

The lack of mobility further implied that an invader could not be expelled by force. In 1976 HA envisioned a restoration of the status quo ante bellum by means of the threat of nuclear first-use, but he later came to support *no-first-use* [1984b; cf. 1976: 86, 233]. The imperative of damage limitation ruled out a military defense of cities, hence HA recommended declaring them *open towns* and using various means of *civilian-based defense* [1983: 133–143; cf. Ebert 1984b; Nolte & Nolte 1984]. Regarding implementation, HA consistently favored *unilateralism*, albeit in a step-by-step mode, where reciprocation on the part of the Warsaw Pact should be invited [1987a: 7; 1989; 1989a: 31–40].

AGENDA FOR PEACE.   Title of the report by *U.N.* Secretary-General Boutros Boutros-Ghali released in June 1992 [Boutros-Ghali 1992; cf. idem 1992a; 1993], i.e., in light of the U.N. experience with the *Gulf War* and of the disappearance of the East-West conflict that had thitherto hampered the organization [Howard 1993; Roberts & Kingsbury 1993; Gati 1991]. The rationale for the AFP was to explore ways of exploiting "an opportunity [that] has been regained to achieve the great objectives of the Charter—a United Nations capable of maintaining international *peace* and *security*', by creating a modicum of a *collective security* system [see, e.g., Weiss (ed.) 1993]. Besides the well-known concept of *peacekeeping*, the report mentioned three new concepts:

- *Preventive diplomacy*, defined as "action to prevent disputes from arising between parties, to prevent existing disputes from escalating into conflicts and to limit the spread of the latter when they occur," inter alia through the good offices of the U.N., including the secretary-general [see Franck & Nolte 1993].
- *Peacemaking*, defined as "action to bring hostile parties to agreement," including what has been called *peace enforcement* in accordance with Article 42 of the *U.N.* Charter: "The Security Council . . . may take such action by air, sea or land forces as may be necessary to maintain or restore international peace and security. Such action may include demonstrations, blockade, and other operations by air sea or land forces . . ." [in Roberts & Kingsbury (eds.) 1993: 511].
- *Peacebuilding*, defined as "action to identify and support structures which will tend to strengthen and solidify peace in order to avoid a relapse into conflict" [1992: 475].

The terminology and taxonomy have not been entirely clear [Hill 1993; cf. Cox 1993; Flory 1993; Lodgaard 1993; Ruggie 1993; Brock 1993], and some analysts have feared a gradual erosion of the boundaries between traditional UN missions, such as peacekeeping with the consent of the parties involved [Ansprenger 1993; cf. Rikhye & Skjelsbaek (eds.) 1990; Goulding 1993; Morphet 1993; Roper & al. 1993; Berdal 1993], and more controversial missions, such as the restoration of peace by military means, almost inevitably directed against one side to a conflict—operations that might also be very demanding militarily, and that could well exact a toll in human lives [cf. Mackinley 1993; Connaughton 1992; 1992a; Urquhart 1991; 1993;

1993a; Allen & al. 1993; Isaacs 1993; Berdal 1993; Kühne 1993; Renner 1993]. Not least, German authors have been concerned about this, which would have implications for the specific German constraints on the use of military force, codified inter alia in the *FRG*'s constitution [Lutz 1987d; 1989a; 1990a; 1992; 1993; 1993a-b]. In Table 5 a clarification is attempted.

Table 5. An Agenda for Peace

	Preventive Diplomacy ("Prevent conflict, prevent spread")	Peacemaking ("Restore peace, resolve conflict")	Peacekeeping ("Maintain peace or armistice")	Peacebuilding ("Create lasting peace structure")
Nonmilitary means		Mediation Negotiation Arbitration Adjudication Amelioration of problems Sanctions	Humanitarian aid Monitoring of elections Repatriation of refugees	Strengthen democracy Protect human rights Cooperative projects
Military-related means	Confidence building • CBMs • Risk-reduction centers Fact-finding Early warning	Enforcement of sanctions	Military observation Police duties	Disarming Weapons destruction De-mining Demilitarization
Military instruments	Preventive deployment Demilitarized zones • with the consent of both sides • with the consent of only one side	Peace enforcement: Military action (Article 42) Restore cease-fire ("peace-keeping plus")	Interposition between potential belligerents "Robust peacekeeping"	

Another controversial point in the AFP was whether to change the very basis of the U.N. as an organization comprising sovereign states in favor of something else. The secretary-general was somewhat ambiguous on this point, stating, on the one hand, that "the foundation-stone in this work is and must remain the State," and, on the other hand, acknowledging that "'the time of absolute and exclusive sovereignty, however, has passed" [1992: 474]. This has spurred a debate on the pros and cons of humanitarian intervention, i.e., a forceful enforcement on individual states of universal human rights by the international community, or states acting on its behalf [Farer & Gaer 1993; Rochester 1993; Rodley (ed.) 1992; Deng 1993. For a theoretical analysis: Camilieri & Falk 1992.] Some headway, at least according to some analyses, was made in this direction by the operations to protect the Kurds in Iraq during and after the Gulf War and by the operations in Somalia. The lack of U.N. action in the former *Yugoslavia* might seem to point in the opposite direction [Freedman & Boren 1992; Eikenberg 1993].

In its turn, such a radical expansion of the U.N.'s prerogatives was almost bound to raise a new debate on the structure of, and distribution of influence within, the organization; radical calls for *democratization* [Singer & Wildawsky 1993; cf. Archibugi 1993; Volger 1991] have been voiced alongside more modest proposals for reform, e.g., by the incorporation of new permanent members into the Security Council [see e.g., Kaiser 1993; Wagner 1993].

On the basis of his enumeration of urgent new tasks, Boutros-Ghali also suggested several organizational and financial innovations. The Military Staff Committee should be reinvigorated and enabled to take charge of the planning of operations [cf. Boulden 1993; Grove 1993]; member countries should assign forces specifically to the U.N.; U.N. operations should be an item in the defense rather than the foreign policy budget; etc. [Boutros-Ghali 1992; cf. Bertrand 1993; Wilenski 1993].

AGRELL, WILHELM.   Swedish peace researcher and military analyst who for a number of years has advocated *NOD.

Even though *Sweden has occasionally been described as a model of NOD by virtue of its *territorial defense strategy [e.g., Roberts 1976: 84–123; ADC 1983: 113-114], WA's attitude toward his country's security and defense policy has consistently been critical. In *If War Does Not Come* [Agrell 1979; cf. 1984a; idem & Wiberg 1980], he criticized the very core conceptions of Swedish defense planning: those of "marginal effect" (implying that Sweden would be only a secondary target of aggression) and of "perimeter (or peripheral) defense," tantamount to a special form of *forward defense. According to WA, the repeated Soviet submarine incursions into the Swedish archipelago were harbingers of a more immediate Soviet threat to Sweden as well as the rest of Scandinavia, necessitating an adaptation of Swedish defense to the new threat [Agrell 1986; 1986a; 1987a].

WA recommended a shift to a *militia-type defense that should capitalize on the technological innovations in antitank and in coastal and air defense, particularly through a skilful application of *PGM-type weapons [1979: 125–156; cf. 1983; 1984; 1987; 1989]. Even though he elaborated upon the military element of national defense in great detail, WA also underlined the importance of nonmilitary safeguards, inter alia, in the form of an enhanced civilian *invulnerability and preventive conflict resolution, as well as various *disengagement efforts, and included therein the creation of a Nordic *nuclear-weapons-free zone [1979: 157-197; 1982: 144-152; 1984b].

AIR COMMAND.   See *Air Power.

AIR CONTROL.   See *Air Power.

AIR DEFENSE.   Defense of civilian or military targets against aerial attack. Occasionally, defense against missiles is also counted as AD or *Extended AD. In principle, there are three forms of AD:

1. **Offensive Counter-Air** (*OCA) implies destroying the enemy's aircraft on the ground, a method that has occasionally achieved decisive results, at least in situations when the side launching the attack could benefit from the surprise factor. Against forewarned opponents equipped with effective point air defense systems, and with dispersed (hence less vulnerable) defense postures, however, the case for OCA seems less convincing.

OCA tends to damage *crisis stability, particularly if both sides to a conflict rely on it for AD, because this is tantamount to poising two offensive-capable air forces against each other, thus giving either strong incentives to preempt against the other. On the other hand, there are powerful counters to OCA, such as *dispersion, hardening, point AD, etc.

2. **Active AD** implies the interception of enemy aircraft in the air, traditionally the task of fighter interceptors, which are, however, by their very nature ambivalent.

In principle, the fighters would be targeted, above all, against the enemy's bombers and other ground-attack aircraft because these constitute the most serious offensive threat. However, because the latter are usually accompanied by fighter escorts, the first round in an aerial battle is almost bound to be a dogfight between fighters, a fact that further illustrates the intrinsic ambivalence of the fighter aircraft. The ensuing, perhaps decisive, fighter battle may be fought either in a defensive context, i.e., in one's own airspace, or in the context of an offensive, above enemy territory and in conjunction with, or as preparation for, bombing raids or ground assaults. The fighters employed for defensive and offensive tasks are nearly always indistinguishable, even though range requirements tend to be higher for the offensive craft.

In an *NOD* posture, it would therefore be preferable with weapons that were effective only above one's own territory but powerless above enemy territory. A close approximation to this ideal seems to be *SAM*s, which may, of course, be mobile in principle but not under wartime conditions, due to lack of armor protection. Most NOD proponents therefore recommend a replacement (complete or partial) of fighters with SAMs.

3. **Passive AD** is a pejorative term for modifications of the "target set," whereby a reduction of its vulnerability is sought. Such modifications may include the active protection of targets by means of hardening (point AD, silo basing, concrete shelters, and so on). However, in addition to such "active passive AD" precautions, states may also protect their assets through what might be called "passive passive AD": by dispersing them, e.g., by having secondary, dispersal airfields and the like available and/or by enhancing their mobility and hence elusiveness. Finally, and most radically, it is in certain instances possible to eliminate targets altogether.

The shift to a NOD posture would remove a wide range of otherwise lucrative targets (nuclear storage sites, tank assembly areas, and so on), thus lessening the need for active AD measures, perhaps to such an extent that the manned fighters might eventually become superfluous [cf. Møller 1989; Hagena 1990b; 1994; SAS 1984b; Unterseher 1989g; Wissdorf 1992].

AIR POWER.   Generic term for a range of military capabilities, signifying "the ability to project military force by or from a platform in the third dimension above the earth or the sea" [cf. Armitage & Mason 1983: 2]. For all its versatility and effectiveness in armed combat, however, control of the aerial dimension possesses no intrinsic value but merely facilitates the performance of military functions invested with such intrinsic value. AP is thus, at most, a means to an end.

1. **Air Control** refers to the control of the aerial dimension exercised by a party to an armed conflict and may be subdivided according to degree and exclusivity. "Air command" signifies an absolute and exclusive form of control; "air supremacy" (or superiority) is somewhat less exclusive, referring merely to "the ability to exercise sufficient control over a particular airspace at a particular time as to be able to carry out one's own air operations effectively with little or no enemy interference, while, at the same time, denying the same opportunity to the enemy" [Armitage & Mason 1983: xi]. Whereas such exclusive forms of AP are incompatible with *Common Security* principles and *NOD* conceptions, air control may be compatible, especially if the notion is applied merely to a certain airspace. If, for instance, circumstances could be created under which each side would enjoy air supremacy above its own territory, this would be tantamount to an aerial manifestation of the principle of *mutual defensive superiority* and might indeed nearly suffice for bringing about such a stabilizing force ratio as far as the *land forces* were concerned [cf. Møller 1989; 1992: 183-187].

2. **Mission Structure.** AP refers to the ability simultaneously to perform a

range of missions. Certain of these functional capabilities would be indispensable even in a strictly defensive posture; others would need modification in order not to violate the NOD principles; others again might be superfluous or downright incompatible with NOD. The last category includes the following:

- Strategic bombardment of an adversary's nonmilitary assets, i.e., civilian population, industrial structure, or perhaps political leadership: so-called countervalue warfare, in the postwar period usually (albeit not logically) associated with *nuclear deterrence* [cf. on *massive retaliation:* Freedman 1989a: 76–90]. The questionable efficiency of such warfare aside [cf. Quester 1989; Gibson 1988: 319–334; Armitage & Mason 1983: 43–45, 106–111; Clodfelter 1989] (not to speak of the legal and ethical problems associated with targeting civilians [Best 1980: 262–285]), it is also incompatible with NOD, with one possible exception: in the form of *minimum deterrence* with the sole purpose of deterring nuclear attack and accompanied by a *no-first-use* policy.
- Interdiction of enemy forces is a way of preventing (some of) them from ever reaching the FEBA (forward edge of battle area), rather than engaging them there, and is thus a form of indirect support for the land battle. Whereas shallow interdiction (to a depth of, say, 50 km) is unproblematic from an NOD point of view, deep interdiction, such as envisioned in NATO's *FOFA* program is more problematic: it may furnish the adversary with both incentives for, and options of, *preemptive attack* against the aircraft assigned to deep interdiction missions, thus damaging *crisis stability* [cf. Wijk 1986; 1986a; Møller 1986a; 1987a; 1988a; 1989b]. Furthermore, it inevitably diverts resources from *forward defense*.
- Counterindustrial warfare (i.e., the targeting of an enemy's war-making potential) is comparable to interdiction in also constituting an attempt at preventing enemy forces from reaching the FEBA, through preventing them from coming into being in the first place.

The category of aerial missions that, duly modified, might be both compatible with, or even indispensable for, NOD includes the following:

- Close (or tactical) air support, which is ambivalent in the sense of having both offensive and defensive utility. It encompasses, inter alia, tactically defensive missions such as battlefield surveillance, *air defense* of army formations against enemy ground-attack aircraft, and internal reinforcement of isolated forces in terms of firepower, supplies, and evacuation support. However, such tactically defensive functions may also form part of offensive operations, e.g., in *offensive-turned-defensive* situations [cf. Mearsheimer 1983: 134–164; Freedman 1987: 12–21]. Aircraft may, of course, also be used in a directly (tactically) offensive mode: by attacking ground targets. In terms of gross firepower, aircraft are inferior to the artillery; in terms of effective firepower, their mobility and versatility make them highly effective, provided that a suitable target set is available (which is precisely not the case in *guerrilla warfare* and/or against certain types of NOD postures).
- Aerial surveillance is useful for early *warning* of impending attack, thereby minimizing the surprise factor and enhancing stability. It can also be used for tactical surveillance of the battlefield, thus contributing to the "transparency revolution" (see *C^3I*). Both forms of surveillance have obvious potentials for target acquisition, whereby they contribute to

expanding the battlefield, which is, on balance, an undesirable development from an NOD point of view. If possible, a distinction between surveillance and target acquisition would therefore seem desirable [Herolf 1988].

- Naval aviation, i.e., carrier-based or land-based aircraft assigned to maritime roles, is likewise ambivalent. As a general rule, land-based aircraft hamper seaward invasions (by their ability to sink surface ships and conduct *ASW). They thereby strengthen the defensive side, without providing significant offensive capabilities because of range limitations (except in very narrow waters). *Aircraft carriers, on the other hand, tend to extend the inland reach of naval power, thus strengthening the offensive side (cf. *Maritime Strategy).
- Air defense, i.e., the defence of civilian as well as military objects against air attack is *prima facie* a clearly defensive task. It may, however, have offensive implications if combined with the means of performing offensive missions.

3. **NOD and AP.** One might almost divide the NOD constituency into four groups:

- Those who disregard the topic almost entirely
- Avowed air force opponents, such as *Horst Afheldt* [1983], *Dieter Lutz* [1989c; 1989e; 1990d] and the late *Anders Boserup* [idem & Graabaek 1990]
- Equally avowed AP supporters, such as *Ludger Dünne* [1988; 1988a] and *Carlo Jean* [1990]
- Those who envision a limited yet still important role for AP within a multiservice NOD scheme, such as the *SAS* [1984b; Unterseher 1988c; 1989g; Møller 1989; Hagena 1990b; 1994; Wissdorf 1992]

4. **Nonoffensive Air Power.** One might sum up what would seem to be the approximate consensus among NOD supporters with regard to an aerial posture in Table 6.

Table 6.   Nonoffensive Air Power

Criteria		Posture
Minimum deterrence	→	Few long-range bombers
No long-range, ground-attack capability	→	No long-range cruise missiles
No deep interdiction or OCA capability	→	No fighter-bombers (AS-weapons), no long-range target acquisition
No-target principle	→	Dispersed air bases and/or STOVL aircraft
No long-range or mobile naval aviation	→	No aircraft carriers
Tactical concentration, close air support	→	Short-range aircraft and helicopters with AS weapons
Air superiority in national airspace	→	Interceptors with AA weapons
Air defense capability, including defense against helicopters	→	SAMs, interceptors (AA weapons), low-level air defense systems
Strategic reconnaissance	→	Satellites, airships
Tactical reconnaissance	→	RPVs, scout helicopters, sensor arrays, infantry

Note: AA = air-to-air; AS = air-to-surface; RPV = remotely piloted vehicle; SAM = surface-to-air missile; STOVL = short take-off, vertical-landing

AIR SUPREMACY.   See *Air Power*.

AIRCRAFT CARRIER.   Large surface ship carrying conventional (but not necessarily conventionally armed), fixed-wing aircraft.

Since WWII, ACs have been the most important means of power projection and *intervention* by sea, as well as of gunboat diplomacy, and thus the manifestation par excellence of *sea power* [cf. Cable 1981]. Their high offensive potential is due to their possession of onboard *air power* that allows them to operate effectively far from friendly shores, and even in the vicinity of the enemy coast. Even though AC have an impressive striking power, particularly if the aircraft are equipped with nuclear weapons (as is still the case), the net striking power is smaller than it might appear because of the need for air defense when operating within reach of enemy land-based aviation and coastal artillery. Out of, say, a *Nimitz* class AC's ninety aircraft, only thirty-four represent genuine striking power; the rest are devoted primarily to self-protection [Luttwak 1984: 220–222].

Even though AC also have a certain defensive utility, most of their defensive task might be performed by other means. For *SLOC* protection, smaller ships optimized for *ASW* tasks would be more cost-effective; for the air defense of allies (such as *Norway*, to be defended by placing U.S. ACs in Norwegian fiords), land-based fighters would be preferable; and so on. A defensive naval posture would therefore have to dispense with ACs in favor of a combination of land-based naval aviation, smaller surface vessels (including helicopter carriers), and submarines [cf. Møller 1987; 1989e; 1992: 187–194].

AIRCRAFT.   See *Air Power*.

AIRLAND BATTLE.   Operational doctrine of the U.S. Army, described in the 1982 version of the *Field Manual on Operations,* with some slight modifications in the 1986 version. It superseded the *Active Defense* doctrine of 1976, which was believed to be ill-adapted to the type of fluid maneuver warfare apparently planned for by the *Soviet Union* [FM 100–5 1982; cf. Czege 1984: 105–108; Lindner 1984; Mallin 1990: 234–239; Garrett 1989: 45–47; Bellamy 1990: 131–138; Sutton & al. 1984; Strachan 1984: 34–38; Richardson 1986].

ALB emphasizes maneuver warfare over the positional and firepower-oriented form that characterized active defense. It is envisioned to conduct deep penetrations beyond the FEBA (forward edge of battle area), and to prepare these with *deep strikes* (the "extended battlefield"). Such deep penetration presupposes ample air support, wherefore the ALB requires a high degree of collaboration between the U.S. Army and Air Force, with the latter performing both *air defense* (including *OCA*, i.e., offensive counter-air) and ground-attack missions.

ALB is very offensive-minded and obliges field commanders always to capture the initiative through offensive operations. Nevertheless, in the field manual, the envisioned offensives constituted merely tactical and operational-scale, but not strategic, *counteroffensives*; no invasion capability vis-à-vis the Soviet Union was thus sought [cf. for the distinctions: Mackenzie 1991; Reid 1991]. On the other hand, the 1982 version of ALB envisioned an integrated use of both battlefield nuclear weapons and chemical weapons, which would further magnify the war damage likely to result from an ALB-type war.

ALB has implications for weapons procurement, inter alia, because of its greater emphasis on highly capable *tanks*, self-propelled artillery (e.g., the MLRS, Multiple-Launch Rocket System) and ground-attack aircraft. However, seemingly defensive types of forces and armaments would also need to be procured in great numbers: light *infantry* units, battlefield *air defense*, antitank weaponry, etc., because of the need to operate under intense enemy fire and in all types of terrain, including conurbations [Mazarr 1990: 32–33, 106–109; Downing 1986; W. Dupuy 1985; Gates 1989; Kafkalas 1986; W. J. Olson 1985; Petraeus 1984; Wickham 1985].

According to U.S. authorities, the ALB was never applicable to Europe, where *NATO doctrine took precedence (cf. below) [Rogers 1985; cf. *Military Review*, 65:12 (Dec. 1985): 4–11]. A number of ALB critics (including almost the entire *NOD community) found this unbelievable, e.g., because the same U.S. Army formations would indisputably be fighting according to the ALB guidelines elsewhere. This was the case of the U.S. Army forces dispatched to Saudi Arabia for Operation Desert Storm, which became the first large-scale war fought according to the ALB doctrine [Friedman 1991: 129–130]. (See *Gulf War.*)

The U.S. Army's shift to ALB was followed by NATO's revision of its land-force doctrine, heralded by the issuing of *ATP-35(A)* in 1983, as well as by a new NorthAG (Northern Army Group) operational concept, both of which reflected a greater emphasis on maneuver warfare. The NATO document thus underlined that the defensive battle should be fought "with imagination, energy and aggressive spirit and with the aim to seize the initiative wherever and whenever possible" [Bellamy 1987: 124–129; Pollard 1991: 123–124].

ALB has attracted the almost unanimous condemnation of the entire NOD community (with the exception of *Jochen Löser* [Löser & Anderson 1984: 69–70, 75–76, 143–144, 151]) for being excessively offensive-minded, jeopardizing *crisis stability* and spoiling the chances of *arms control* and *disarmament*, as well as for being unaffordable [cf. e.g., Møller 1987a–b, 1988a, 1989b; Unterseher 1987; Wijk 1986; Webber 1990: 26–33; for a "Green" assessment: die Grünen im Bundestag 1984; and for the social democratic view: Voigt 1984; 1987a]. More mainstream military analysts have tended to agree with the latter point of criticism, in addition to which they have pointed to the opportunity costs of ALB in terms of an inevitable weakening of *forward defense*, and/or they have questioned the potential of the emerging technologies (see *E.T.*) envisioned under ALB auspices [Garrett 1989: 45–47; 1990: 111–112; Mearsheimer 1981; Canby 1983; 1984; 1985; 1985a; Brown & Leney 1984; Papp 1984; Willliams & Wallace 1984; Gormley 1986; Brown 1986; Flanagan 1986; 1987; Kipp 1987].

AIRLAND BATTLE 2000.   The name of two different, albeit closely connected, long-term military planning documents from the early 1980s:

1. A planning document of the U.S. Army's Training and Doctrine Command (TRADOC) from 1982, which was later superseded by the *Army-21* conception. Neither ALB-2000 nor its successor ever became actual doctrine but merely served as a theoretical background for the *AirLand Battle* doctrine, and as a rationale for weapons procurement and R&D (research and development) ventures, particularly in the field of *ET.

The ALB-2000 envisioned a distinctly hi-tech, extremely lethal (with extensive use of nuclear, chemical, and biological weapons), and highly mobile and fluid battlefield in the twenty-first century: an extended, integrated, and automated battlefield. Protracted fighting would allegedly be out of the question, hence the strategy should not rely on attrition but seek a swift annihilation of enemy forces through an offensive orientation [in die Grünen im Bundestag 1984: D27-D30; Horlohe 1984: 24–36; Bellamy 1987: 136–138; *Jane's Defence Weekly*, 06.12.1986: 1355].

2. A long-term planning document formulated jointly by the U.S. Army and the Bundeswehr in August 1982, known as the *Glanz-Meyer Paper* after the two signatories, U.S. Army Chief of Staff Edward Meyer and Army Inspector Meinhard Glanz [die Grünen im Bundestag 1984: D31–D36; Horlohe 1984: 37–49].

Highly mobile units and deep strikes against the enemy's rear were envisioned on a battlefield resembling that described by TRADOC (albeit with no mention of biological weapons), and a general offensive posture was recommended (albeit only on the tactical and operational levels). Critics therefore alleged that the paper violated the *FRG's Constitution, particularly the article about the prohibition against preparations for attack [cf. Lutz 1987d; 1989a; 1993; 1993a-b].

ALBANIA.   See *Balkans.

ALBRECHT, ULRICH.   Started his career as an engineer in the aerospace industry but was later appointed professor of international relations at the Free University in Berlin. UA has been one of the most prominent German proponents of political *disengagement. During the interregnum period in the postrevolutionary *GDR, UA served as a consultant in the Foreign Ministry.

UA advocated a combined military and political disengagement, including a "nonnuclear neutral security zone" in Central Europe, from which not only nuclear weapons but also heavily armored shock divisions and various offensive-capable weapons systems should be removed. The political element would consist of a deliberate decoupling (see *dealignment), implying maintenance (but preferably no actual exercise) of a neutralist option with a view to enhancing flexibility. The purpose should be to opt out of the bloc confrontation, driven primarily by the superpowers. The Scandinavian model (cf. *Nordic Balance) might be applied to Central Europe, where a "Swedenization" of the FRG and a "Finlandization" of the GDR might be combined with a scaling down of the alliance commitments of the Benelux states to the level of Norway and Denmark. The resultant differentiated security zone might presumably serve as a *buffer zone. At its heart would be West Germany, which would have to be "militarily strong enough to ensure that its territory could not be used for aggressive purposes by either side in the event of a European war, yet not so strong as to pose a threat to its neighbors. German defenses would have to be organized in a benign, nonprovocative manner." Even though these ideas were developed prior to the renewed debate on German unification, UA regarded the guidelines as even more imperative under these new conditions [1982; 1985; cf. 1983a; 1990].

ALGERIA, LIBERATION WAR OF.   Guerilla war for national independence waged in the French colony Algeria from 1954 to 1962 by the FNL (*Front Nationale de Liberation*).

The war was both protracted and bloody, featuring, e.g., indiscriminate attacks on civilians on both sides. Its principled rejection of terrorist methods notwithstanding, the FNL gradually came to wage war against the *pied noirs* (French civilians), which in turn spurred counterterrorism (most atrociously conducted by the Organisation de l'Armé Secret, OAS), and vice versa. The war was further exacerbated by the intense internecine strife between competing factions and personalities within the FNL.

Apart from this, the FNL's struggle followed the traditional lines of *guerrilla warfare. *Decentralization was emphasized through a compartmentalization of the FNL's armed forces into six regional *willayas,* which were largely autonomous in terms of operational responsibility. The profoundly political nature of the war allowed for a coordination of the guerilla struggle with strikes, etc. Self-sufficiency was attempted by establishing munitions factories and by attacking enemy encampments in order to lay hands on their weapons. The FNL thus avoided becoming entirely dependent on foreign aid, although in due course substantial amounts did materialize, most of it coming from other Arab countries, Algerian residents in France, and (to a limited extent) the socialist world. Finally, the FNL benefited from its superior, albeit extremely low-tech, intelligence and surveillance network (including smoke signals and dogs). Fighting was concentrated in the most inaccessible terrain: mountainous regions and the *maqui* rather than the open desert. Furthermore, the guerillas employed base areas, although the most important of these was across the border in Tunisia.

The French were not very successful in their counterinsurgency warfare, in spite of the fact that many of the officers serving in Algeria had prior experience in Vietnam. The first phases of the French struggle for the maintenance of colonial rule has been likened to killing fleas with a sledgehammer. In the second phase, the French, on the one hand, introduced more specialized units (the paratroopers and the Foreign Legion) and fighting techniques (semiguerilla detachments) designed for the job. On the other hand, the government gradually lost control over the military, a process culminating in the attempted military coup d'état in 1961. Nor were the attempts at depriving the guerillas of their rural basis by means of resettlement of peasants in barbed-wire encampments successful; rather, they increased political support for the FNL.

Eventually the French invented a fairly effective means of counterinsurgency: the so-called Morice Line along the border with Tunisia. It consisted of huge amounts of barbed wire, high voltage wires, etc., supported by heavy artillery, and provided with target coordinates from a continuous surveillance array. The FNL's initial response was costly and largely futile attempts at breaking through the line. Eventually, FNL leader Boumedienne succeeded in enforcing a new strategy that was more effective: nominal assaults against the line should freeze French troops at the border so as to allow the resistance a freer hand in the interior. Furthermore, the French eventually became entrapped in the self-defeating military logic of *escalation*. Initial prohibitions against indiscriminate bombing, etc. were gradually eroded under the spell of, e.g., the doctrine of collective responsibility. Moreover, outright illegal practices such as systematic torture flourished, without, however, breaking the resistance. The escalation (and especially the torture) contributed to undermining the colonial cause by spurring a widespread antiwar movement in France [Horne 1977; Carver 1981: 121–150].

ALLIANCE, THE. Political coalition between the SDP and the Liberal Party. See *U.K.*

ALLIANCE. Generic term that might be defined as a durable agreement on collaboration between two or several actors for the attainment of some goal [cf. Møller 1992: 49–77].

1. **Taxonomy of International A.** Leaving aside domestic political and other A, security political alliances between states may be subdivided into species according to various characteristics, including those shown in Table 7.

Table 7. Alliance Dichotomies and Trichotomies

Formal	Informal	
Permanent	Durable	Temporary
Voluntary	Enforced	
Egalitarian	Hegemonic	Imperial
Collaborative	Adversarial	
For something	*Against* someone	

Needless to say, the dichotomous distinctions constitute simplifications. First, it is often a question of more or less rather than of either/or. Second, each dimension might be further subdivided, implying, e.g., that A may be directed against someone in one respect, yet also be based on a communality of values in another respect; and that what are voluntary A for some members might be enforced, as far as others are concerned, etc.

2. **Pros and Cons of A.** Whether A are good (i.e., further *peace* and other values) or bad remains disputed, both *in abstracto* and as far as concrete A are concerned.

In traditional theory of international politics, particularly that inspired by *Realism*, A play an important role, e.g., as a manifestation of the *balance*-of-power mechanisms, and as means to the end of "national interest defined in terms of power." Competitive alliance building might thus serve as a functional substitute for war [Morgenthau 1960: 181–194; Liska 1962: 12, 26–27, 108–115, 168–201]. These functions did, however, presuppose a propensity for states to balance rather than to "bandwagon," i.e., it was assumed that states tended to respond to the growth of a potentially hostile alliance by counteralignment rather than by joining it [Walt 1987: 17–33, 147–180; Evera 1990; Christensen & Snyder 1990; Jervis & Snyder (eds.) 1991].

Whether A actually further peace and *stability* remains disputed, and the available historical evidence is contradictory. On the one hand, the pre-*WWI* alliances may have made war less avoidable as well as widened its scope; on the other hand, the absence of binding A may have facilitated the combined *blitzkrieg* and "salami" strategy of Nazi Germany and thus have contributed to bringing about *WWII*.

A are, furthermore, fraught with dilemmas, including what has been called "the alliance *security dilemma*," manifesting itself, inter alia, in the abandonment-versus-entrapment dilemma: on the one hand, states fear abandonment unless they comply with the alliance norms (say, by opting for a partial *dealignment*); on the other hand they also fear entrapment as a result of too-tight alliance bonds that might draw them into a conflict from which they might otherwise stay aloof [G. Snyder 1984; cf. J.M.O. Sharp 1987].

3. **A and Alternative Defense.** Proponents of alternative defense have differed in their attitudes toward A. Western *NOD* advocates might, e.g., be subdivided into the following categories, according to their attitudes toward *NATO* (during the Cold War):

- Neutralists, urging unilateral *neutralization* and often viewing NOD primarily as a contribution to this end [e.g., Mechtersheimer 1984; 1986; 1987; Löser & Schilling 1984; P. Johnson 1985].
- Advocates of various forms of *disengagement*, intending to enlarge the scope for autonomous action, as well as to avoid entrapment [e.g., Afheldt 1983; 1989a; Albrecht 1982; 1983; 1983a; idem & al. 1988; Lutz 1982; 1985; Møller 1987c].
- Proponents of conditional alliance membership, acknowledging the desirability of reforming NATO as a whole but seeing unilateral action as the most promising way to accomplish this. They might thus recommend a unilateral shift to NOD, combined with efforts to persuade the alliance to do likewise, yet leave open the option of withdrawal if these attempts should fail [ADC 1987: 214–238].
- Supporters of essentially unchanged NATO membership, who merely work for an alliance-wide shift to NOD [SAS (ed.) 1984; 1989].

4. **Neutrality-Alignment Continuum.** The majority of NOD proponents have belonged to the second and third categories, envisioning a certain loosening but no

complete severance of alliance bonds. It may thus be helpful to conceive of alignment as a continuum ranging from complete *neutrality* at one pole to complete alignment at the other, but with an intermediate zone of varying degrees of semi-neutrality and semialignment [cf. Ørvik 1986a]. The variables determining each state's position would include the following (which are not really comparable):

- The state's contribution to the alliance's common defense;
- The integration of its national armed forces in alliance command structures;
- The deployment of foreign troops;
- Significant reservations with regard to alliance strategy.

The most aligned country of all, was accordingly, the *FRG*, contributing quite substantially to NATO defense, integrating its armed forces in their entirety into NATO, hosting both foreign troops and nuclear weapons, and having no significant strategic reservations. Countries such as the U.K. and the U.S. have contributed substantially in absolute terms, but a good part of their forces have all along remained outside NATO integration. The *U.S.* has, moreover, hosted neither foreign troops nor weapons systems and has for a quite long period departed significantly from NATO strategy. Countries such as *Denmark* and *Norway* have had noteworthy reservations about strategy and have rejected any peacetime stationing of forces as well as nuclear weapons on their territory, but their forces would have come under NATO command in war. *France* has withheld its forces entirely from NATO integration, a position emulated by *Spain* upon its entry into NATO. Finally, Iceland has never fielded any armed forces at all yet has hosted a U.S. base. Because the neutrals are, in their turn, equally heterogeneous, the total picture ever since 1945 has been one of different shades of alignment.

5. **Post–Cold War A Patterns.** With the waning of the East-West conflict, some of the hitherto neutrals (e.g., *Sweden*) have reassessed their options and moved closer to the Western alliance, or what used to be regarded as sister organizations (the EC/EU and the *WEU*). On the other hand, the NATO integration of, e.g., the FRG has been loosened somewhat with unification and the cessation of four-power rights. Additional shades have been added with the dissolution of the *Warsaw Pact*, whose former members are apparently eager either to join or at least establish cooperative relations with NATO (see *NACC*, *Partnership for Peace*). The emerging picture may be one of varying shades of alignment (in contrast to the Cold War: in or around the same alliance), whereas the extremes of complete neutrality and complete alignment seem to be receding [cf. Møller 1991b].

ALT, FRANZ.   See *Christian Democrats in Favor of Disarmament*.

ALTERNATIVE DEFENCE COMMISSION.   See *ADC*.

ALTERNATIVE DEFENSE PROJECT.   See *U.S.*

ALTERNATIVE DEFENSE WORKING GROUP.   See *IDDS*.

ALTERNATIVE SECURITY.   See *Security*.

ALTERNATIVE SECURITY WORKING GROUP.   See *U.K.*

AMPHIBIOUS FORCES.   See *Sea Power*.

ANTISHIP MISSILE.   See *Sea Power*.

ANTISUBMARINE WARFARE.   See *ASW*.

ANTITANK WEAPONS.   See *PGM*.

ARBATOV, ALEXEI G.   Head of department at the Soviet (subsequently Russian) research institution *IMEMO*, since 1990 director of the independent Center for Arms Control and Strategic Stability. AGA is the son of *Georgy A.*, one of the first Soviet advocates of *NOD*. His contributions to the alternative defense debate are primarily found in three fields: NOD strategy in general, conventional arms control, and nuclear strategy.

AGA's point of departure was acknowledgment of the need for a distinction between offensive and defensive postures, on several levels of analysis: "If the forces are stationed along the frontier, this points to a defensive strategy. If, however, the troops are arranged in striking 'fists' at specific points along the line, this could indicate offensive intentions" [1990: 15; idem & Savelyev 1989]. He expressed concern over the prevailing trends in the development of *NATO* and U.S. strategy, especially as far as *AirLand Battle* and *FOFA* were concerned [1988: 204–207], but he was equally critical toward Soviet strategy: it was defensive in regard to political aspects, but its military-technical aspects were clearly offensive [1988: 207–214;].

Regarding implementation of the recommended defensive restructuring, AGA emphasized the importance of reciprocity, best attainable through *arms control*. As the appropriate guideline for the *CFE* negotiations, he proposed a partial *disengagement* of forces in three *zones* along the Central Front. This arrangement would be tantamount to a partial withdrawal of offensive-capable forces (including aircraft) from the forward line [1989c-d; idem, Kishilov, Amirov & Streltsov 1989] (see Table 8).

Table 8.   CFE Reductions

| | Proposed Ceilings | | |
	Jaruzelski Zone	Intermediate Zone	ATTU Zone
Tanks	4,700	2,500	4,600
Artillery	2,000	800	5,600
Tactical-strike aircraft	400	600	1,000

Source: A. G. Arbatov 1989c: 274–279.

AGA also contributed significantly to the development of Soviet (and subsequently Russian) thinking about nuclear strategy. Although in favor of deep cuts in the nuclear inventories, he was particularly concerned with *stability*, i.e., the prevention of first-strike and *preemption* incentives, in which sense he might be seen as representing the U.S. approach to nuclear arms control [1988a-c; 1989e-f; 1990b; idem & Cochran 1991; idem & Lednev 1988; idem & Savelyev 1988]. He also differed from most Soviet analysts and policymakers (as well as from the majority of NOD proponents) in advising against a complete abolition of tactical and theater nuclear weapons, recommending instead a build-down to "minimum (finite) tactical nuclear deterrent capabilities," in the range of around 500 warheads for each side [Arbatov 1990c].

ARBATOV, GEORGY.   Director of the *ISKAN* in Moscow, member of the *Palme Commission*, former member of the Communist Party Central Committee, prominent advisor to successive general secretaries, and father of *Alexei A.*

In 1983 the views of GA on security and defense policy were elaborated upon in a series of interviews [Arbatov & Oltmans 1983], wherein the focus was on Soviet-U.S. relations. Making sure to convey official Soviet viewpoints, GA placed greater emphasis on *détente* and subscribed to a somewhat broader conception thereof than was usually the case for Soviet representatives, who often regarded it as simply "a continuation of the class struggle by other means," just like *peaceful coexistence*. According to GA, however, "It is a process that . . . is aimed not at gaining victory in conflicts by means short of nuclear war, but at the settlement and prevention of conflicts, at lowering the level of military confrontation, and at the development of international cooperation" [ibid.: 13]. Through the *Palme Commission*, GA was also instrumental in bringing about the Soviet endorsement (under *Gorbachev*) of the concept of *Common Security*.

ARCHIPELAGO DEFENSE.   Term used by *J. F. C. Fuller* to describe a special form of antitank defense based on a system of dispersed defensive strongpoints whence fire could be directed against approaching tank columns [Fuller 1944] (see also *checkerboard defense*).

AREA DEFENSE.   Term (*Raumverteidigung*) used by *Horst Afheldt* and others for a *territorial defense*. It was a means to highlight the feature that there would be no undefended rear because the entire territory was to be covered by military defense, almost inevitably implying an abandonment (or at least a substantial weakening) of *forward defense* [Afheldt 1976; 1983].

AREA-COVERING DEFENSE.   Term (*Raumdeckende Verteidigung*) used by, e.g., *Jochen Löser* for a differentiated *area defense* configuration of the armed forces. This would allow for a significant *forward defense* capability simultaneous with a complete coverage of the country to be defended, albeit almost inevitably with rather weak (and/or cadred) covering forces [Löser 1981].

ARMAMENTS DYNAMICS.   Generic term for the propellants of the arms buildup [for a general overview, see Gleditsch & Njølstad (eds.) 1989].

A common distinction is between external and internal dynamics, or exogenous and endogenous causal factors. Only as far as the external (exogenous) dynamics are concerned, may the term "arms race" be appropriate, and it even remains disputed whether to reserve the term for an intense and accelerating armaments competition, or to apply it to the everyday competition [Buzan 1987a: 69–75; Albrecht 1990a: 89–90; cf. Gleditsch 1990: 11].

1. **External Dynamics.** *NOD* advocates tend to focus on the external dynamics, where the competition manifests itself in the so-called actions-reactions phenomenon (ARP) [Rathjens 1973]. It may be found on several levels of analysis. Its simplest manifestation is found on the level of individual weapons systems, the development of which may constitute a sequence of measures, countermeasures, counter-countermeasures, etc. More complex ARPs transcend the boundaries between weapons categories and occasionally even that between the conventional and nuclear spheres [Forsberg 1985].

A phenomenological explanation of the ARP may be the proclivity for "fallacies of the last step." Decisionmakers tend to take into account merely the immediate gains to be reaped from a particular step, while disregarding the likely countermeasures on the part of their adversaries. An additional factor is the proclivity for worst-case analysis, leading to systematically exaggerated threat assessments, in their turn driving (or at least legitimizing) arms buildup.

A more fundamental cause of the ARP is the *security dilemma. *Robert Jervis described what might be called the armaments security dilemma: "When states seek the ability to defend themselves, they get too much and too little: too much, because they gain the ability to carry out aggression; too little because others, being menaced, will increase their own arms and so reduce the first state's security" [Jervis 1976: 64].

NOD advocates have sought an escape from this dilemma through a distinction between offensive and defensive postures [Jervis 1978: 197; cf. Afheldt 1976: 324–325; Hannig 1984: 15–16; Galtung 1988; Møller 1990f; 1992: 79–104; idem & Wiberg 1989]. If defensive measures and armament programs are clearly recognizable as defensive (with regard to both underlying intentions and the options they provide), they will call for no balancing steps. Other NOD advocates questioned the security dilemma element (regarding alleged Soviet fears as deceptive), yet still urged a Western shift to NOD because this would have deprived the Soviet leaders of the legitimation for their arms buildup [Löser 1981: 241–242]. Finally, NOD would presumably facilitate *arms control by modifying the concept of *balance or by rendering it irrelevant.

A mixture between the above external and the internal armaments dynamics is found in what might be called the "internalized ARP," i.e., the phenomenon of "anticipatory reaction" [cf. Buzan 1987a: 86–89]. With the present long (and growing) lead time for major weapons systems, such anticipation may well be a precondition for not falling behind in the arms competition. Due to the uncertainties surrounding threat assessment, however, such anticipatory reaction tends to become entirely self-propelling [Allison 1983; idem & Morris 1975], thus nearly indistinguishable from the genuine internal dynamics.

2. **Internal Dynamics.** The internal dynamics (also referred to as autism) refer to societal forces seeking to maintain at least a minimum (but preferably a maximum) of armed force, regardless of external threats. These forces include the military apparatus, the arms industry, parts of the technological community, and the bureaucracy, sometimes referred to as the *MIC, i.e., military-industrial complex (in capitalist states) or the military-bureaucratic complex (in the USSR) [Pursell (ed.) 1972; Senghaas 1972; 1990f; Prins (ed.) 1983: 133–168; Albrecht 1990a; Thee 1990; Evangelista 1990; cf. for a critique: Sarkesian (ed.) 1972]. The impact of a shift to NOD on the endogenous factors remains an almost entirely unexplored topic. Still, NOD might in fact have a certain impact, as a side effect of its modification of the weapons profile of the armed forces; it would be tantamount to a shift of the industrial emphasis from the current main contractors (airframe plants, shipyards, etc.) toward other companies. This would, at least, weaken the present MIC, while not inevitably creating a new MIC [Møller 1990f; 1992: 93–95].

ARMENGOL, VERENC FISAS.   See *Spain.

ARMS CONTROL.   New approach to the regulation of the *armaments dynamics that in 1960 nearly replaced the quest for *disarmament, especially in the U.S.

1. **Rationale of AC.** AC was more in line with the dominating paradigm of international politics, namely, *Realism, and appeared more realistic than disarmament by virtue of its modest goals: merely to regulate (rather than halt) the armaments dynamics with a view toward enhanced *stability—both *crisis stability and *arms race stability. Whereas these objectives in some instances might require a quantitative build-down, in other instances they might call for a partial expansion

and, above all, for a change in the composition of the arsenals [Bull 1987; Schelling 1966: 256–259; 1986; idem & Halperin 1961: 56–58; Blechman 1983; Kissinger 1960: 210–229].

2. **Pitfalls.** An explanation of the rather unimpressive record of AC may be the several pitfalls in the AC approach, either as originally conceived or as actually practiced, among which are the following [Møller 1992: 79–84; cf. 1990f. For a right-wing critique to almost the same effect, see Gray 1992; 1993]:

- The focus on *balance* in the sense of parity, which was both ill-defined and politically unobtainable, and which would be unrecognizable as such even if attained;
- The incentives for "loophole armament" that were created through negotiations on specific types of weaponry that rewarded a diversion of resources into unconstrained fields;
- The temptation of "bargaining-chip armaments," i.e., the search for a prenegotiation position of strength, either through the actual production of weapons or through plans for production, which might then be bartered for actual reductions on the part of the adversary;
- The Western insistence on adequate verification (including on-site inspections), which went against the Soviet propensity for secrecy (until the *INF* treaty) and thereby blocked many possible agreements [cf. Schelling & Halperin 1961: 91–106; Myrdal 1976: 293–316; Heckrotte 1986; cf. Altmann & Rotblat (eds.) 1989; Grin & Graaf (eds.) 1990; Kokoski & Koulik (eds.) 1990; Kokoyev & Androsov 1990];
- The focus on already-deployed weapons to the near exclusion of what might have been a fruitful approach, namely, "prophylactic AC" [cf. Kissinger 1960: 280]; and
- An inbuilt propensity for misplaced emphasis because neither side had any strong incentives to limit, much less to renounce entirely, high-priority weapons systems, hence negotiations have tended to deal with irrelevant topics: conceivable activities that were never seriously contemplated by either side, potential weapons systems that neither side really considered for development or deployment, etc. [cf. A. Carter 1989].

3. **Alternatives to AC.** Limitations to the AC approach made many *NOD* advocates dismiss the very notion of negotiated arms regulations out-of-hand [Neild 1989] in favor of *unilateralism* or *gradualism*; others have merely suggested relegating AC to one among other viable means of military restructuring and disarmament.

ARMS RACE.   See *Armaments Dynamics*.

ARMS RACE STABILITY.   Propensity of a system to avoid a spiraling *armaments dynamics*. The lower the degree of ARS, the higher the probability that the states involved will arms race against one another, with the amount of available resources constituting the only limit to their military expenditures. The higher the ARS, the greater margin will they experience for unilateral arms build-down, and the higher the probability of successful *disarmament*. The ARS is a function of the possibility of making meaningful *offense/defense distinctions* and of a situation of *defense dominance* [cf. Jervis 1978; Møller 1992: 79–103].

ARMS SALES.   See *Arms Trade Regulations*.

ARMS TRADE REGULATIONS.   Regulation of the international arms trade.

1. **Taxonomy.** One might distinguish between at least four relevant dimensions of ATR, as well as any combination thereof:

- Quantitative ATR, i.e., attempts to limit total size of arms sales;
- Geographical ATR, i.e., attempts to prevent an arms buildup through arms imports in particular regions, especially in the Third World;
- Temporary ATR, i.e., the limitations of arms deliveries to countries actually at war or apparently preparing for war; and
- Qualitative ATR, i.e., attempts at preventing the proliferation of special types of weapons.

A further categorization might distinguish between the parties to an ATR agreement:

- Agreements between arms-exporting countries;
- Agreements between importing countries; and
- Multilateral agreements between suppliers and recipients.

2. **Rationale for ATR.** There are many reasons for seeking ATR. First, the global scale of arms transfers leads to a proliferation of sophisticated, and often very lethal, weapons throughout the entire world, including regions where they may well be used for actual war, hence the growing intensity and destructiveness of Third World wars. The trend of the 1970s toward expanding arms sales abated in the 1980s and 1990s, but the explanation for this seems to be a shortage of funds rather than a diminished desire to purchase arms [cf. Pierre 1982; Klare 1984; Sampson 1977; Allebeck 1989; Anthony & al. 1991]. Second, when operating *out-of-area*, as during the *Gulf War*, the major arms exporting countries tend to find themselves fighting against forces equipped with "their own" weapons. Third, it is increasingly being acknowledged that militarization and excessive arms procurement in Third World countries hampers their economic, hence also political, development, thereby often preventing a stabilization of entire regions [Ball 1988; Catrina 1988; 1990; Weiss & Kessler (eds.) 1991; cf. Brandt Commission 1980].

3. **Achievements.** Few attempts have been made to regulate the global arms trade, and none of them have been really successful:

- The *NPT* of 1968 prohibited the transfer of nuclear weapons to other states.
- In the years 1977–1978, the U.S. and the Soviet Union conducted Conventional Arms Transfer Talks, CAT (occasionally CATT) without any results [Husbands & Kahn 1988; Nolan 1988].
- The *U.N.* has on several occasions debated the issue and passed a number of resolutions without any tangible effects on the global arms trade [Corradini 1990; Wulf 1991b]. In 1991, however, the Conventional Arms Register was opened, which will, at least, give greater visibility to arms transfers, and thereby perhaps indirectly lead to some restraint [Anthony 1993; Chalmers & Greene 1993; Laurence & al. 1993; Wulf 1993].
- On a regional scale, certain agreements have been reached between recipient states not to introduce qualitatively new types of armaments into a region (see *Latin America*);
- In 1987 a number of the main arms-exporting nations signed the Missile Technology Control guidelines, thus inaugurating the *MTCR*, intended to limit the proliferation of long-range ballistic missiles [Anthony 1991a; Bailey 1991a; Carus 1990; A. Karp 1991a; 1992; Kniest 1992; Mack 1991b; Nolan 1992; Shuey 1992; Zimmerman 1992; Findlay (ed.) 1991; Neuneck & Ischebeck (eds.) 1992].

- The major arms-exporting nations have since 1991 made repeated statements about the need for multinational ATR, especially pertaining to the Middle East and Persian Gulf region. The "Perm Five" (the permanent members of the U.N. Security Council), for example, in October 1991 agreed on some "lines of conduct," including to "consider carefully whether proposed transfers will: promote the capabilities of the recipient to meet needs for legitimate self-defence." Furthermore, they would "avoid transfers which would be likely to prolong or aggravate an existing armed conflict; increase tension in a region or contribute to regional instability; or introduce destabilizing military capabilities in a region" [*Atlantic News*, 2362 (23 October 91): 3]. However, by early 1994 no binding regulations had yet been agreed upon.

4. **Obstacles.** There are numerous possible explanations for the poor results achieved with ATR so far, including

- What might be called "the arms exports' *prisoner's dilemma*": even though, in the long run, all states might benefit from a curtailment of the global arms trade, a single state would stand to benefit even more if all but it limited their arms sales, leaving it as the sole defector in control of the market. Furthermore, all exporters would suffer from compliance with an ATR agreement unless all other states complied as well.
- Arms transfers were an element in the superpower competition, i.e., a functional substitute for direct involvement or alliance formation, hence the reluctance of both superpowers to limits arms sales [Pierre 1982; Spear & Croft 1990].
- The continued demand for weapons in many regions of the Third World, where states are likewise victims to the prisoner's dilemma: compliance with an agreement to curtail arms imports may jeopardize a state's security unless potential adversaries comply as well.
- Finally, arms exports are profitable, both for the companies producing the weapons and in certain cases for the respective states, especially if long and cost-effective production lines presuppose exports. Were states to buy weapons built exclusively for domestic consumption, unit procurement costs would be higher.

The final cause may well become even more decisive in the years ahead, also because of the reductions in permitted weapons holding following from the *CFE* Treaty, as well as the general trend toward arms build-down as a corollary of the disappearance of the East-West conflict. Many weapons will become dispensable (hence exportable), and many arms factories will have to limit production or close down entirely unless they exploit export opportunities—in the West, but perhaps even more so in the East, above all in the former USSR.

5. **Proposals.** Numerous proposals have been made over the years for improved ATR, ranging from suggestions to prohibit the arms trade completely to more modest recommendations for strengthening legal constraints [Klare 1984: 247–250; Pierre 1982: 291–296; Kozyrev 1990]. The idea most in line with alternative defense thinking may be the proposal by Alva Myrdal in 1976 for "the forbidding of, or at the very least restraint of the production and transfer of, arms that are typically offensive in character while leaving purely defensive capacities less restrained" [Myrdal 1976: 148–149; Klare 1993].

Even though a definition with universal validity of "offensive weapons" would be impossible to find, it might well be feasible to single out, for particular regions, the types of weaponry that each state would prefer its enemies not to possess as the types of military hardware eligible for ATR (cf. *offense/defense distinction*). For them to be really effective, any such ATR would have to be combined with other measures intended to contain both the "supplier-push" and "recipient-pull" propellants of the global arms trade, including

- *Conversion* of military industries from military to civilian production, thereby relieving the pressure for arms exports for economic and employment reasons; and
- Regional arms control negotiations with the purpose of devising stable, i.e., less offensive, military postures for all states involved, thereby diminishing the pressure for arms imports [Kemp & Stahl 1991; Stahl & Kemp (eds.) 1992] (cf. *Armaments Dynamics*).

ASPIN, LES.   See *U.S.*

ASW.   Antisubmarine warfare, which may be subdivided according to target: (1) strategic ASW is directed against the opponent's strategic submarines, i.e., nuclear missile platforms (SSBNs: Strategic Submarine, Ballistic, Nuclear), and (2) tactical ASW is directed against the opponent's attack submarines.

Strategic ASW (such as that envisioned to be waged against the *Soviet Union* under the *Maritime Strategy* of the U.S. Navy) is usually regarded as destabilizing because it endangers the second-strike capability of the adversary and thereby provides it with an incentive to preempt (thus damaging *crisis stability*). It also spurs a buildup of the opponent's SSBN fleet as a hedge against strategic ASW, thus hampering nuclear *arms control*.

Tactical ASW may be stabilizing, neutral, or destabilizing, depending on circumstances. As an element in NATO's *SLOC* defense (the defense of sea lines of communication), it was stabilizing and defensive; as an element in anti-SLOC operations or a *guerre de course*, it tends to be destabilizing [Mearsheimer 1986; Møller 1987; 1989b; 1989e].

A distinction between strategic and tactical ASW may also be made geographically and/or in terms of equipment. ASW operations on the part of NATO tended to be strategic if conducted in the Barents and White Seas (i.e., within the Soviet SSBN *bastions*), but tactical if conducted south of the GIUK (Greenland-Iceland-U.K.) or GIN (Greenland-Iceland-Norway) perimeters. Capabilities for under-ice ASW, likewise, pointed to strategic ASW as the main mission, whereas diesel-electric propulsion or land-based (and fairly short-range) ASW aircraft or helicopters pointed to tactical ASW. Most combinations are, however, dual-capable, such as ordinary nuclear-powered attack submarines (SSN).

ATBM.   Antitactical Ballistic Missile (defense). See *Extended Air Defense, *GPALS, *Strategic Defense.*

ATLANTIC-TO-THE-URALS.   See *CFE.*

ATM.   Antitactical Missile (defense). See *Extended Air Defense, *GPALS, *Strategic Defense.*

ATTRITION.   Gradual wearing down of one or both sides to an armed conflict.

In pure wars of A, the outcome depends on two variables: the initial force-to-force ratio and the relative attrition rates. According to classical strategic theory (e.g., the *Lanchester Laws*), the A rates in each engagement depend on the initial force ratios, i.e., in part on *concentration*. Presumably, the side with the greatest force concentration at the moment of contact invariably wins since it is able to inflict disproportionate attrition on the weaker side: the traditional argument in favor of mobile forces and thus (paradoxically) also of maneuver warfare. Even though this is the antithesis of A warfare, it is at the same time the logical conclusion toward which the classical approach to A seems to lead.

However, it has been demonstrated that many variables other than the degree of concentration determine the A rate, some of which militate against maneuver and mobile forces. First, the inversely interconnected factors of surveillance (including target acquisition) and concealment are of growing importance because of the increasing accuracy of both direct and indirect fire systems. Whoever pinpoints the other side first, by virtue of an advantage in terms of surveillance over concealment, tends to win the engagement. Second, the increasing ranges over which targets can be engaged imply that mobility and physical concentration become superfluous in many instances (where directed firepower is what matters) and positively harmful to the extent that they reveal the location of the weapons platform, thus facilitating the opponent's target acquisition. *Decentralization*, combined with adequate battlefield surveillance and sufficiently accurate weapons, tends to minimize A [Neild 1986].

Most *NOD* proponents have placed their faith in A, according to the logic that the capability of waging protracted war would deter (or dissuade) attack by virtue of the prospects it opens of a steadily accumulating A to which the attacker is bound to succumb, deprived of the option of enforcing a decisive battle [Afheldt 1983; cf. Mearsheimer 1983].

In addition to actual (i.e., material) A, one might conceive of "virtual A." Material A is the result of actual fighting. Virtual A signifies the A that a prospective attacker must anticipate as a result of the defender's deployment and force configuration: if the opposing defense posture includes submarines, an invasion fleet is, for instance, virtually attrited by its need to bring along tactical *ASW* forces (which may never go into action), at the expense of genuine attack forces. This conception of virtual A also amounts to a partial rebuttal of the critique occasionally voiced against certain NOD models (particularly those wherein an *attrition net* plays a central role): that most of the defense *web* would remain inactive because the attacking forces would concentrate along narrow axes. Even though they remain inactive, these parts of the attrition net may nevertheless inflict a virtual attrition on the attacker.

ATTRITION NET.   Description of the *web*-like posture advocated by many *NOD* proponents (such as *Brossollet* [1975], *Horst Afheldt* [1976, 1983], *Löser* [1981], the *SAS* [1984; 1989], and others.

ATTU.   Atlantic to the Urals. Scope proposed by France, and agreed upon for the *CFE* negotiations.

AUGMENTATION FORCES.   *NATO*'s concept for forces kept at a low state of readiness, i.e., reserve forces that can be mobilized in an emergency.

1. **NATO Decisions.** By 1990–1991 practically everybody in the West concurred that whatever "standing-start" attack options the *Soviet Union* might have possessed in the past had disappeared because of the dissolution of the *Warsaw Pact*, the accompanying Soviet withdrawal from Eastern Europe, and the negotiated as well as unilateral reductions of the Soviet armed forces. Hence the availability of a longer strategic *warning*, which allowed NATO to shift partially from in-place-stationed to home-based forces, and from large to smaller standing armies, yet with an enhanced growth potential.

After investigations by NATO's military staffs, decisions were made by the alliance's political authorities (NATO's Defense Planning Committee and the Nuclear Planning Group in ministerial sessions in Brussels on 28–29 May 1991) on a three-category division of NATO's military potential into (in-place) "main defensive forces"; (immediately available) "rapid reaction units"; and AF that were to be mobilized and sea- or airlifted in an emergency [*Jane's Defence Weekly,* 15:16 (20 April 91): 623; ibid. 15:24 (15 June 91): 1076]. At the summit meeting of the North Atlantic Council in Rome on 7–8 November 1991, the document *The Alliance's New Strategic Concept* was adopted, wherein it was stated that "the overall size of the Allies' forces, and in many cases their readiness, will be reduced," but that they "will be structured so as to permit their military capability to be built up when necessary" [*Atlantic Documents:* 75 (09 November 91)].

2. **NOD Proposals.** Many *NOD* advocates have for years recommended such an expanded role for reserve forces because they have assumed that a lower state of readiness would render otherwise offensive-capable forces less threatening. Before launching an attack, a state would have to go through a fairly lengthy process of bringing its forces to combat readiness (calling up reserves, conducting refresher courses, matching personnel with equipment, moving units into position, etc.), which would *ipso facto* provide the prospective defender with ample warning, and thus time for preparing its defense. AF would thus seem to enhance *crisis stability,* as well as be economically advantageous [Löser 1981; Bülow 1988c; cf. for an early precursor: Kant 1795: 17–18].

Other NOD proponents have, however, pointed to certain flaws in this logic. *WWI* demonstrated how *mobilization* processes can start malign interactions between adversaries, leading to mounting war scares and ultimately, through a process of *escalation,* to the point of war [cf. Tuchman 1962; Taylor 1969; Evera 1985; Snyder 1985; cf. for a critique: Trachtenberg 1990]. The advantages accruing from a lower state of readiness may therefore presuppose that the reserve forces intended for possible mobilization are deprived, as far as possible, of offensive capabilities, so that their mobilization would constitute no threat to an adversary that might provoke a possibly avoidable war [Unterseher 1987a; cf. Møller 1989b; 1992: 122–126].

AUSTRALIA.   The alternative defense debate in A is of a fairly recent vintage and spurred by several factors, including

- Concern about nuclear strategies, highlighted, e.g., by the dispute between the U.S. and *New Zealand*;
- Apparently destabilizing trends in the naval confrontation in the Pacific region; and
- Direct influence from the European debate.

The center of the debate has been the Peace Research Centre (PRC) at the Research School of Pacific Studies at the Australian National University (ANU) in Canberra.

Until 1991 its leader was Andrew Mack (subsequently professor of international relations at the ANU). In the early 1970s, Mack had worked at a peace research institute in Denmark, one result of which was a modern classic on *civilian-based defense, War Without Weapons,* authored jointly with *Anders Boserup* [Boserup & Mack 1974]. Since his return to A Mack has been influential in disseminating the European *NOD* ideas [Mack 1984; 1989; 1990b]. Furthermore, he has been active in working out their possible application to Australian defense policy [1986; 1990e] and to *disarmament* and *détente* in the Asia-Pacific area [1990c; 1991], including the opportunities for various forms of *naval arms control* and *maritime C(S)BMs,* and for applying the *MTCR* to the region [1990; 1990a; 1990d].

Another researcher at the PRC has been Geoffrey Wiseman (part of whose academic career has been at Oxford University), who in 1989 published a major work, *Common Security and Non-Provocative Defence,* an exposition of the theory of *Common Security* and NOD [1989]. He subsequently worked with the application of these principles to A and the entire Asia-Pacific region [1994], until his assignment as program officer for the Ford Foundation in the U.S.

The most concrete proposals for an NOD-type defense posture for A have been published by Graeme Cheeseman, likewise from the PRC but with a background in the Australian regular army and the Department of Defence. In proposing an alternative posture, GC emphasized the need for a scale-down of political ambitions to the objectives of defense of Australian territory and vital interests; the need for integrating the army, navy, and air force into a coherent whole; and the need for a less expensive defense [1988; 1989; 1990; 1990a; 1991]. In 1993 GC published a set of recommendations for a nonoffensive defense of A—pointing, as a preliminary, to an inescapable problem: that forces intended to defend only A might still constitute a threat to its neighbors because the distance to, say, Indonesia is shorter than that across A itself. He recommended that A assume greater responsibility for its own defense, yet by strictly defensive means: a maritime command, responsible for "high-level threats" within the "area of direct military interest," which should essentially consist of Australian territory and territorial waters; and a home defense command, based to a large extent on reserve forces. By thus opting for defense specialization, A would send the right political signals to its neighbors and might set in motion a process of disarmament that could maintain the existing balance on a lower, hence less costly, level of armed forces [1993].

In 1990 GC coedited with St. John Kettle the anthology *The New Australian Militarism,* which attracted considerable attention, to the extent that Defense Minister Kim Beazly expressly asked to have a rejoinder to the proposals included in the volume. The editors had concluded with a "Statement of Concern," stating, "Our future security depends . . . on:

- an affordable, strong and non-provocative defence system;
- the solution of regional conflicts by political means; and
- initiatives to create new arms control machinery in the region."

In his response, the minister emphasized that "our strategy is in the broadest possible sense defensive," while declining to renounce "the use of offensive tactics to achieve a defensive goal" [Cheeseman & Kettle (eds.) 1990: 207-208; Beazley 1990].

Another politician taking an interest in NOD has been Foreign Minister Gareth Evans, who has likewise endorsed the ideas of common security and *collective security* as a guideline for foreign and defense policy [1993].

AUSTRIA.   Small neutral state in Central Europe, and perhaps the country in the world whose defense posture is most in line with the recommendations of *NOD* advocates.

1. **Postwar Settlement.** In 1955 a four-power settlement (U.S., USSR, U.K., France) was reached concerning the German WWII ally, which terminated the Allied occupation and granted full sovereignty to A. In the Moscow Memorandum, A committed itself to permanent *neutrality,* and in the "State Treaty," it renounced its right to acquire, among other things, nuclear weapons; other potential weapons of mass destruction; "self-propelled or guided munitions"; torpedoes or "devices which serve for the launch and control hereof"; sea mines; manned torpedoes; submarines; motor torpedo boats; specialized attack vehicles; artillery pieces with

ranges exceeding 30 km; and certain specified chemical substances (similar to the prohibitions stipulated in the peace treaties with *Finland*, Italy, Bulgaria, *Romania*, and Hungary) [Tanner (ed.) 1992]. The four powers, furthermore, reserved to themselves the right to add further prohibitions [Rotter 1984; Vetschera 1985; 1985a; 1992; idem & Rocca 1985; Aichinger & Maiwald 1989; Steiner 1988].

Contrary to the other formerly occupied countries (which soon had their bans lifted or reinterpreted), A remained debarred from the acquisition of guided missiles until 1988, with an ineffective *air defense* posture as the inevitable result. In 1989 the Austrian government did, however, decide upon the purchase of air defense missiles, encountering no protests [*Jane's Defence Weekly,* 22 April 1989: 695].

2. **Defense Strategy.** Whether in spite of or because of these constraints, A developed a genuine *territorial defense* strategy, designed by *Emilio Spannocchi* [1976]. It was finalized in 1985 when the government published an elaborate defense plan [Bundeskanzleramt 1985; cf. Spannochi 1990; Danzmayr 1990; 1991a; Fuhrer 1987; Schneider 1987; Tauschitz 1990; Mayer 1992]. It was conceived as a *total defense* consisting of four main elements: military, spiritual, civil, and economic. The guidelines for the military element were entirely in line with the recommendations of NOD proponents:

- It was conceived of as a form of *deterrence by denial*, renouncing, as a matter of principle, the option of retaliation by "carrying the war to the enemy."
- All fighting was intended to take place on A's own territory, i.e., in a territorial defense mode.
- The goal should be to preserve as large a part of the national territory as possible under A's control, but if need be part of the territory might be ceded temporarily in order to preserve the ability to continue armed resistance from a heavily defended *bastion*.
- To allow for a smooth transition from the peacetime exercise of sovereignty to actual defense, it was envisioned to combine *forward defense* with an in-depth defense.
- As far as the latter was concerned, the armed forces should to a large extent consist of reserves to be mobilized and fight locally, i.e., on familiar terrain and able to take advantage of prepared fighting positions.
- No decisive battle should be sought, but the aggressor should rather be gradually attrited through a sequence of small encounters (cf. *nonbattle*).
- Forces should be deployed in great *dispersion* (cf. *no-target principle*) [Bundeskanzleramt 1985: 54–66].

3. **Alternative Defense Debate.** There has been little debate about neutrality as the appropriate status for A, but some disagreement on its modalities and limits. Shifting governments have found closer ties to the West permissible, but the *peace movements* have favored a policy of equidistance between the blocs, e.g., with a view to letting A play the role of mediator and bridge builder between East and West [Binter 1986; 1986a; 1991; W. Graf 1986; Hinteregger 1989; Marolz 1987; Skuhra 1986a; 1987]

Because of the predominantly nonoffensive defense policy of A, a popular debate on NOD has been largely missing, even though certain peace researchers and others have criticized the government, the armed forces, and the foreign NOD proponents who point to A as a model [cf. Löser & Schilling 1984], for excessive militarism [cf. Binter 1987; 1989; Marolz 1987a; Semlitsch 1987; Skuhra 1986; Wörister 1986].

In the same vein, A has all along had its advocates of complete and unilateral military disarmament in favor of *social defense*, just as foreign *civilian-based defense* proponents have urged A to take advantage of its presumably wider scope for national action (as compared with alliance members) by abolishing military defense [Ebert 1987b; Herz 1971; G. Jordan 1989; Stock & Krottmayer 1986].

AUTONOMOUS DISSUASION.   Term employed by Hans-Heinrich and *Wilhelm Nolte* for their combination of *civilian-based defense* and military *NOD* [Nolte & Nolte 1984].

AXELROD, ROBERT.   Professor of political science and public policy at the University of Minnesota and author of *The Evolution of Cooperation* [1984; cf. idem & Keohane 1985], wherein he analyzed the cooperative strategy of *Tit-for-Tat*, and found it to be capable of bringing about cooperation between egoists ("from nations to bacteria") without central authority. The strategy would thus presumably represent a solution to the *security dilemma*.

More recently, RA has analyzed the preconditions of *stability* with regard to conventional forces in Europe [1990]. He found *crisis stability* to be much lower on the conventional than on the nuclear level because of the weaker inhibitions against the use of conventional weapons. At the same time, however, a conventional war, once started, would considerably compound the risk of nuclear war, hence the importance of *conventional stability*, in its turn determined, e.g., by the prevailing beliefs about, as well as the actual relative strength of, offense and defense.

B

BÄCHLER, GÜNTHER. See *Switzerland.

BAHR, EGON. A former journalist, ambassador, state secretary, member of the *Palme Commission*, longtime MP for the *SPD* and director of the *IFSH* until 1994 [see Lutz (ed.) 1992].

Having been (as Willy Brandt's chief advisor) the main architect of Ostpolitik in general and *change through rapprochement* in particular [1963; cf. 1985; 1988a: 95; 1991], EB was also one of the first to formulate (perhaps indeed the inventor of) the concept *common security* (CS), above all under the auspices of the Palme Commission [Palme Commission 1982; Bahr 1985b: 31; cf. Lutz 1986a; 1992a; Kelstrup 1987].

In EB's view, it was necessary to clarify the relationship between CS and *nuclear deterrence* by distinguishing between deterrence as a fact (compatible with, perhaps indeed the central premise of CS) and deterrence as a doctrine, the antagonism inherent in which was incompatible with CS [Bahr 1985b: 33; 1986: 18–19]. To abandon the doctrine of deterrence, however, presupposed *conventional stability*, as a guideline for which EB as early as 1983 suggested the absence of capabilities for attack, i.e., *NOD* [Bahr 1983: 144–145]. EB also served as the first chairman of the SPD's working group on new strategies, tasked with developing the modalities of, inter alia, conventional restructuring and nuclear build-down, in which capacity EB was instrumental in bringing about the SPD's commitment to NOD [1984]. Unlike other leading party members, however, EB never lent his support to any specific NOD model but merely advocated the establishment of a *zone* without offensive weapons [1987a–b; idem & al. 1988].

In 1988 EB published a small book, *European Peace: A Response to Gorbachev*, wherein he predicted within a short time "the elimination of the Soviet threat plain and simple" [Bahr 1988a: 24, 30–33, 37]. Because *Gorbachev*'s panoply of peace initiatives created unique opportunities for conventional stability in Europe, it was no longer necessary to approach the problems through *gradualism* or similar cautious strategies. "The direct road to common security in Europe is open." The implied European "CS regime" would be one of *defense dominance* (i.e., *mutual defensive superiority*), which might be brought about, e.g., through an immediate realization of the old notion of a "nonoffensive corridor." Gradually, a situation of robust conventional stability would emerge, which would render nuclear weapons on European soil entirely superfluous, thus allowing for a "third zero solution," i.e., elimination of short-range nuclear weapons [1988a: 52, 61–68, 75–77].

EB, furthermore, in 1989 coauthored (with *Karsten Voigt* and *Andreas von Bülow*) a document on *European Security 2000*. Its forward-looking title notwithstanding, it consisted mostly of concrete and short-term *arms control* proposals, albeit couched in terms of promoting NOD. The authors envisioned a reduction of WTO and NATO forces to 50 percent of current NATO levels by the year 2000, combined with some restructuring of remaining forces and the creation of 100-km-deep "restricted military areas" along the border from which offensive-capable forces would be banned: battle *tanks*, armored personnel carriers, infantry fighting vehicles, artillery, combat helicopters and aircraft, ballistic missiles, and various forms of equipment suitable for offensive operations (such as bridge building) [Bahr & al. 1989].

EB did, however, also speculate more freely about future European security structures, e.g., creation of a genuine European security community, the activities of which should be unconstrained by veto powers and have armed forces at its disposal, i.e., a *collective security* system ["Sicherheit durch Annäherung," *Die Zeit,* 29 June 1990; 1991a].

BALANCE.   A central concept in both traditional and alternative defense and security policy, invariably invested with positive connotations, yet with several different meanings, among which the following:

1. **Balance of Power** is a central concept in *Realism*, presumably serving as a stabilizer in the anarchical international system. However, confusion surrounds both elements of the concept. First, B may signify either any temporarily stable state in a given system to which it presumably automatically reverts after each crisis or a special form of equilibrium, the maintenance and restoration of which presupposes "balancing behavior" [Morgenthau 1960: 167–223; Kissinger 1957: 1–6; Bull 1977: 101–126; Aron 1984: 137–144; Wolfers 1965; Haas 1972]. Second, B may be either static, permitting no change at all, or dynamic, in the sense of allowing for change within the system, yet leaving it unclear at what point the system as such is replaced by another. Third, "power" may refer to either influence or potentials, and may rest on military or nonmilitary foundations. Finally, "power" may carry either assertive ("offensive") or protective ("defensive") connotations, i.e., it may either refer to the ability to force other actors to comply with one's wishes or to the ability to withstand such attempts on the part of others. In the latter sense, power becomes nearly synonymous with *security*.

2. **Parity.** A special form of B has attracted the attention of decisionmakers, particularly in the era of *arms control* negotiations, namely, that of equality of military potentials. The premise of most arms control negotiations (most recently *START* and *CFE*) has been that equal ceilings were an appropriate goal at which to aim. Upon closer scrutiny, the very notion of parity reveals itself as flawed [cf. Møller 1986a; 1990f; 1992: 81–84; 1992a]:

First, it is unclear between whom B should obtain, especially under conditions of *multipolarity* or even loose bipolarity. If all states are of equal strength, then each state is facing a potential hostile *alliance* of overwhelming superiority; as a matter of historical fact, such was the way the *Soviet Union* used to view itself. If all states were to strive for B with the totality of their potential adversaries, this would be tantamount to a quest for superiority, hence the propensity for arms racing.

Second, even in bipolar relations a quantitative comparison between differently configured military potentials is difficult, not to say impossible: How do, for instance, *aircraft carriers* compare with *tanks*, or tanks with fighter aircraft? Even between, say, tanks there may be vast differences in quality rendering a mere counting close to meaningless [cf. Chalmers & Unterseher 1987; 1988; 1989].

Third, reciprocal perceptions imply that even a perfect numerical balance would be unlikely to be recognized as such by either side. States tend systematically to underestimate their own military capabilities and to exaggerate those of other states because they prudently allow for a certain probability of malfunction of their own weapons but (equally prudently) assume a near-perfect performance of those of a potential attacker.

Fourth, static and dynamic *force comparisons* tend to get confused [see, e.g., Epstein 1985; 1988; 1990: Posen 1984a; 1988; Mearsheimer 1988; Kupchan 1989]. On the plausible assumption that an aggressor will always enjoy a certain *mobilization* lead over the defender, states (prudently) tend to compare their own standing forces with an opponent's mobilizable potential.The combined effect of these factors is that even states seeking nothing but B tend to strive for what is actually superiority, something that obviously only one side can achieve, hence the inbuilt *armament dynamic.*

Ironically, B is not merely undefinable, unobtainable, and unrecognizable but also close to irrelevant because parity is neither a necessary nor a sufficient condition for war prevention or successful defense. On the one hand, states may well defend themselves conventionally against numerically stronger attackers (cf. *three-to-one rule*), just as they may annihilate an attacker through nuclear retaliation, regardless of initial force ratios (cf. *Minimum Deterrence*). On the other hand, under less favorable circumstances, particularly in a complete *surprise attack*, states may well be defeated by numerically inferior attackers, both conventionally and through a disarming nuclear first strike.

3. **Balance of Options.** More relevant, as well as more in line with the alternative defense logic, would be a B of options, of which several varieties are conceivable.

A B of "potential next steps" was the guiding principle of the *Nordic balance* that contributed to low tension and peaceful relations in the Nordic region during the Cold War. More generally, *security regimes* are based on a B of unexercised options, i.e., on reciprocal restraint, as would be a *common security* regime under which all parties would refrain from placing the security of their respective adversaries at risk.

The particular type of B recommended by most *NOD* proponents, namely, *mutual defensive superiority*, is merely a special case thereof: a B based on the superiority of either side over the other, depending on who is acting as aggressor and who as defender. This may be expressed in the formula $D^A > O^B$ and $O^A > D^B$, where a and b are two states and O and D stand for offensive and defensive strength [cf. Boserup 1985]. In contrast to parity, this form of B would further both *arms race stability* and *crisis stability*, hence also war prevention because neither side would have anything to gain from launching a war [Møller 1992: 84–89].

BALD, DETLEV. Scientific director of the Social Science Institute of the Bundeswehr in Munich and author of several books on military sociology and related topics.

DB has primarily contributed to the German *NOD* debate with his studies on *militias*, in the context of an *FRG* of the 1990s facing mounting problems with filling the ranks because of an unfavorable age pyramid [Bald (ed.) 1987; idem & Klein (eds.) 1988; cf. Grass 1984; 1990]. DB has, however, also approached the topic from a historical angle. In the interwar period as well as in the 1950s, serious consideration was given to a shift to militia-like armed forces as the most appropriate form of *conscription* for numerically inferior countries with defensive ambitions (cf. *Stülpnagel*) [Bald 1987; 1987a; 1989].

BALKANS. Region in southeastern Europe including Bulgaria, Romania, *Yugoslavia*, Albania, Greece, and (at least in certain respects) Turkey.

1. **Past Conflicts.** The security problems in the B have always been more complex than in the rest of Europe, inter alia, because of the combinations of three dimensions of conflict, some of which have, furthermore, transcended the

regional boundaries [cf. Chipman 1988; Brzezinski 1989; J. Snyder 1990; Larrabee 1990]:

- The East-West conflict, which pitted Greece and Turkey against Bulgaria and to some extent Romania, with Yugoslavia and Albania being neutral;
- Traditional state-to-state conflicts (e.g., between Greece and Turkey, or Yugoslavia and Albania) over territorial issues, ethnic problems, and the like; and
- Intrastate conflicts (ethnic, religious, political, economic) in nearly all countries (with Greece as a partial exception).

2. **Cold War Alternatives.** The most important potential contribution to solving (or at least defusing) the regional repercussions of the East-West conflict in the past was the proposals for a *nuclear-weapons-free zone* in the B. For all their merits, the proposals suffered from the same weakness as those of the proposed Nordic zone, namely, that only NATO actually deployed nuclear weapons in the region, in Greece and Turkey [Behar & Nedev 1983; Drougos 1988; Platias & Rydell 1982; cf. Arkin & Fieldhouse 1985: 219–220, 232–233, 264; Lilow 1989]. Furthermore, there were strong neutralist (i.e., anti-NATO) sentiments in Greece, caused primarily by a feeling of being betrayed by the alliance in the conflict with Turkey [Veremis 1988; 1990; cf. Karaosmanogulo 1988].

Neutralism was also a feature of Romanian politics, manifested both in an abstention since 1963 from participation in *Warsaw Pact* maneuvers and other military cooperation [Holden 1989: 21-24, 56-59; Dunay 1990: 63-65], and in the development of an independent military strategy. The latter was inspired by the *territorial defense* strategy of Yugoslavia (but also bore a certain resemblance to that of *Austria*): a "homeland defense by the entire people," specifically designed with an "exclusively defensive character." Guerilla-like operations were envisioned for a protracted in-depth defense, which should transform a prospective *blitzkrieg* into a protracted defensive struggle, in which the *attrition* rates would presumably benefit the defender because of its advantages of *interior lines*, prepositioned *barriers*, and adaptation to the terrain. Romania's defense strategy was thus the closest real-life approximation to *NOD* within (albeit on the periphery of) the *Warsaw Pact* [Cernat & Stanislav (eds.) 1976: 74–99, 120–137 & passim; Coman & al. 1984; cf. Burke 1982].

3. **Post–Cold War Developments.** With the Soviet abandonment of the Brezhnev Doctrine, the democratic revolutions in Bulgaria, Romania, and Albania, and the subsequent collapse of the Warsaw Pact, the framework of security policy in the B changed drastically, and the resultant uncertainties were further complicated by two new developments: the outbreak of civil war in Yugoslavia and the *Gulf War* against Iraq, each of which had repercussions in the region. This situation presented both risks and opportunities.

Taking advantage of the removal of "artificial" conflicts, Greece in 1991 proposed to the governments of Bulgaria and Turkey that they remove all "*offensive weapons*" from their common borders, i.e., that *tanks*, armored vehicles, artillery pieces, combat aircraft, and helicopters be withdrawn from Thracia, the European part of Turkey, and the southern parts of Bulgaria bordering on Greece or Turkey. The Bulgarian government was immediately favorably inclined towards the proposal; Turkey did not at first react [*Neue Stuttgarter Zeitung,* 05 July 91]; and nothing had by early 1994 come of it.

The dissolution of the Warsaw Pact has also faced Bulgaria especially with the task of developing an indigenous military doctrine to replace the tight integration with the Soviet Union [Schulz (ed.) 1993b]. In this endeavor, it seems that the formerly loyal WP member has been drawing closer to NATO, e.g., by signing a cooperation agreement with Greece, according to which the latter would help Bulgaria reorganize its armed forces [*Jane's Defence Weekly*, 16:23 (07 December 91): 1083]. Romania is likewise reorganizing its armed forces, albeit less profoundly than Bulgaria and with an emphasis on depolitization [Schulz (ed.) 1993c].

BALLISTIC MISSILE DEFENSE. See *Strategic Defense, *Extended Air Defense, *GPALS.

BALLISTIC MISSILE PROLIFERATION. See *MTCR.

BALTIC REGION. Subregion in Europe around the Baltic Sea, consisting of (parts of) other subregions such as

- The constituent parts of the *Nordic Balance* system: *Denmark, *Sweden, and *Finland, with an occasional, albeit not quite logical, inclusion of *Norway [cf. Holst 1991a];
- Central European states: the *FRG (and until 1990 also the *GDR) and *Poland;
- The Baltic states (previously the Baltic republics), i.e., the former republics in the *Soviet Union (Estonia, Latvia, and Lithuania); and
- A small part of the Russian Federation around the former German port city Kaliningrad (Königsberg), and St. Petersburg (formerly Leningrad).

1. **Cold War Patterns.** During the Cold War, the BR was rather fragmented with a low level of regional institutionalization [cf. Westing (ed.) 1989; Joenniemi & Vesa (eds.) 1988; Joenniemi (ed.) 1991]. For example, because of the presence of both alliance systems, Denmark, Norway, and the FRG belonged to *NATO; the Soviet Union, Poland, and the GDR were members of the *Warsaw Pact; and Sweden and Finland were neutral. Furthermore, the region was characterized by a rather high level of militarization: sizable standing forces (including navies), and a high frequency of large-scale military exercises and other peacetime military activities (including submarine "incidents" in the Swedish archipelago) [Agrell 1986; 1986a; 1987a].

Various proposals were made during this period for lowering tension and promoting bloc-transcending cooperation, none of which were implemented:

- Some of the proposals for a Nordic *nuclear-weapons-free zone* envisioned the inclusion of the Baltic Sea and parts of the USSR in the zone, even though the modalities of such an inclusion were unclear [cf. SNU 1982; Lodgaard 1982; Møller 1985; Bomsdorf 1989a: 229–284; Krohn 1991c].
- Proposals were made for the introduction of *maritime C(S)BMs* to the navies operating in the Baltic [SPD & PVAP 1985; 1988; Lodgaard 1989; Møller 1990b].
- *NOD advocates recommended a shift to less-offensive naval postures on a regional scale, including elimination of naval nuclear weapons, and a reduction of large surface warships in general and amphibious craft in particular, in favor of maritime defense postures relying on small, missile-carrying patrol craft, land-based coastal artillery, and submarines [Boserup 1990a; Møller 1990b].

2. **Post–Cold War Developments.** The collapse of the Warsaw Pact and the Soviet Union have transformed the security political framework of the BR profoundly in several respects. Above all, the three Baltic states have achieved independence, after what they (correctly) regard as having been subject to Soviet occupation. However, they were left with the problem of a continued Soviet

(subsequently Russian) military presence. At a meeting of the Baltic Council (5 October 1991), the three states unanimously demanded the immediate withdrawal of forces and (even more urgently) nuclear weapons [*Atlantic News* 2357 (09 October 1991): 3]. Both the then-Soviet and subsequently *CIS* political leaders, and their counterparts in Russia, promised a withdrawal as soon as possible, and no later than 1994. However, by the end of 1993 only the withdrawal from Lithuania had been completed.

In parallel with this withdrawal, however, the Baltic states will have to devise national military doctrines and strategic conceptions, as well as raise armed forces to meet the requirements implied thereby. Even though the level of ambition has been modest (Estonia, 5,000; Latvia, 9,000; Lithuania, 20,000 troops), it has proved very difficult to procure arms for these forces (because of lack of funds) and to develop a suitable strategy (because of the near total absence of military expertise) [Vare 1991; Jundzis 1991; Petersen 1992; Vares & Haab 1993]. In addition to collaborating with one another (inter alia by forming a joint batallion for U.N. peacekeeping missions [*Jane's Defence Weekly,* 20:22 (27 November 93): 5]), all three states have sought as close as possible links with the West, including NATO and the EC.

The BR has been further transformed by German unification, making the united Germany beyond comparison the strongest regional state, economically, politically, and militarily. Furthermore, the emancipation of Poland and its development of an indigenous military doctrine was also bound to affect military relations in the BR; however, apparently in the direction of a lessened emphasis on offensive operations and force postures [Schulz (ed.) 1992].

There thus seemed to be unprecedented opportunities for a growing interest in, and a plethora of proposals for, increased cooperation in the BR in the 1990s. An institutional framework toward that end was established in 1991: the Council of the Baltic Sea States [cf. Ilnicki & al. 1991; Joenniemi (ed.) 1991; 1993; Krohn 1991; 1993; Visuri 1991a; Møller 1991b; Vares & Haab 1993].

BALTIC STATES.   See *Baltic Region.*

BAR-LEV LINE.   See *Israel.*

BARNABY, FRANK.   Nuclear physicist, director of *SIPRI* from 1971 to 1981, and chairman of the British *Just Defence* [Barbaby & Windass 1983].

FB's main preoccupation has been with military technology, including proliferation issues [1992; 1993]. His points of departure for the study of *NOD* have been, first, an acknowledgement of *NATO*'s need for credible conventional defenses against a Soviet attack as a precondition for a removal of nuclear weapons, and second, a belief in the potentials of modern weapons, particularly when used for defensive purposes [1984; 1986].

In 1982 FB published a concrete NOD scheme jointly with *Egbert Boeker* (inspired by, among others, *Jochen Löser* and *Norbert Hannig*) that envisioned the creation of a 50-km-deep defense *zone* along the East-West border, to be defended by means of highly (but only tactically) mobile light forces equipped with sophisticated antitank and *air defense* weapons of both gun and missile types. Weapons would, as a general rule, be short-range (maximum range around 80 km) and as light as possible. The combat efficiency of the forces would be increased by their access to a decentralized, computerized, and redundant *C^3I* network based on prepositioned sensors linked with the command centers by underground fiberglass cables. A further means of strictly *defensive defense* would be various forms of *tank* *barriers*, such as artificial tank ditches, to be created by the demolition of preemplaced explosives (of a binary type, making them harmless in peacetime) [Barnaby & Boeker 1982; 1988; 1989; Barnaby 1986: 162–167].

BARRIERS.   Generic term for various obstacles intended to prevent or hinder the mobility of an opponent [Messenger 1987]. Even though the focus has (rightly) been placed on one particular species of B, namely, artificial *tank* B, other types of militarily significant B are conceivable.

1. **Taxonomy.** Table 9 presents a number of examples, divided into two dimensions (terrestrial and maritime) and three subcategories: natural, "coincidental" (i.e., man-made, albeit for other than defense purposes), and specific (i.e., man-made with the explicit purpose of creating B).

Table 9.   Barriers

Land Barriers			Sea Barriers	
Natural	Coincidental	Specific	Natural	Specific
Woods, jungles	Canals, parks, plantations, etc.	Minefields	Ice, straits, archipelagos	Minefields
Rivers, lakes	Railways	Tank ditches		
Mountains	Roads	Walls		
Marshes	Cities	Barbed wire		

It is apparent that some countries are more fortunate than others with regard to B: e.g., *Switzerland* and *Afghanistan* are protected by their mountains; *Vietnam* by its jungles; and *Finland* by its lakes and forests. It also is apparent that much can be accomplished by less fortunate countries.

2. **Efficiency.** Numerous studies have shown how B may serve as force multipliers, i.e., enhance the combat strength of a given formation or weapons system. According to (vintage 1975) U.S. Army estimates, the efficiency of a U.S. M60 tank, e.g., rose by 165 percent through protection with B; that of a TOW antitank missile increased by as much as 1,097 percent [Epstein 1990: 67–72; cf. Kaufmann 1983: 68; Simpkin 1986: 57–77; Gupta 1993]. Because of this difference in strength between weapons of defense and offense combined with B, as well as because of the better opportunities for the defender than for an attacker to emplace B, the very use of B tends to magnify the difference in strength between the two types of combat (cf. *three-to-one rule*). It is therefore only logical that most *NOD* schemes envision extensive use of B.

BASIC PRINCIPLES AGREEMENT.   Agreement on Basic Principles of Relations Between the United States of America and the Soviet Socialist Republics, signed by the two superpowers in May 1972, during the same summit as the *SALT-1* and *ABM treaties.*

The BPA has been described as "the charter for *détente.*" It was valued by the *Soviet Union* as signaling U.S. acknowledgment of its status as an equal, and U.S. commitment to the principles of *peaceful coexistence* (a term that figured in the text). It committed each of the two superpowers to exercise restraint and to refrain from seeking unilateral advantages at the expense of the other, i.e., to respect the legitimate *security* interests of each other. The BPA might thus be viewed as a possible precursor to a commitment to *common security*. Furthermore, the two states acknowledged a special responsibility for the maintenance of *peace*, while claiming no special rights or advantages for themselves [in Allison & Williams 1990: 266–268; cf. Garthoff 1985: 290–298 (quote: 290); George 1988c: 588–590; 1988a: 668–670; Kolodziej 1991; Kremenyuk 1991: 48–50]. The principles of equal security, however, revealed themselves as so ambiguous (cf. *Balance*) that the USSR was able to use the BPA as a license to pursue more assertive policies in the Third World, which in turn undermined U.S. support for the BPA, and for détente in general. See also *Prevention of Nuclear War Agreement*.

BASTION.   Heavily defended and/or fortified area, ashore or at sea (see also *Territorial Defense*).

1. **Bastion-Type Land Defense.** The bastion-type land defense strategy is particularly relevant for countries with very low population density and/or very uneven distribution of the population (such as Russia with Siberia), where the population simply does not suffice for attaining a meaningful density of forces if distributed across the entire territory. Many areas in the Third World also suffer from the same problem, among which is the Middle East, with *Israel* as an exception [Cf. Conetta & al. 1991; 1991a]. Countries in such positions have to prioritize because they cannot possibly defend everything, hence the concentration on strongholds or Bs. A B-type defense may resemble territorial defense by its in-depth order of battle but leaves large tracts of land, perhaps even most of the country, undefended.

Such a selective defense, of course, raises numerous questions about what to defend and what not, the answer to which will depend on strategic and political considerations—the latter to at least the same extent as the former—which do not automatically point in the same direction. Politics would demand that major cities and, above all, the capital be defended, yet there are weighty damage-limitations reasons for not directly defending cities by military means but declaring them *open towns* in order to enjoy the protection of the *laws of war*. The best solution to this dilemma may be to establish a miniature *forward defense* around major cities and to have no military installations within their city walls.

Strategy would demand that the heavily defended strongholds should be distributed in such a way as to hamper the forward march of an invader as much as possible, and ideally make it impossible ever to consummate the victory. To achieve the latter, the strongholds should be contiguous, in the sense of exposing the entire territory between the strongholds to fire and small-scale attacks therefrom (cf. *checkerboard defense*). This would require some degree of mobility, but it should be as limited as possible in order to minimize offensive capabilities, and particularly it should not be tantamount to a genuine border-crossing mobility.

Such a defense posture might be likened to a mountain, jungle, or *archipelago defense*, regardless of whether or not such terrain happens to be available. What distinguishes these modes of defense is the advantage enjoyed by the defenders of relatively secure positions, from which they can direct their fire against an aggressor who is somehow handicapped: in mountainous terrain by having to move about on rather few, overcrowded roads below the defenders; in jungles by being unable to ascertain the positions of the defenders; and in archipelagos by having to operate in an entirely different environment, namely, at sea. The purpose of such a defense would be to impede the adversary's mobility by forcing him to maintain a defensive posture and to expend as much combat power on the protection of logistic chains.

2. **Maritime Bastions.** The term *bastion* has been used especially about the Soviet Union's form of protection of its nuclear submarines, i.e., the second-strike

capability. At least since the mid-1960s, and particularly since the 1970s, the *Soviet Union* had placed great emphasis on the maintenance of a secure second-strike potential, allowing it to ride out a nuclear attack yet still retaliate. Even though they may have assigned a higher priority to land-based nuclear forces, Soviet leaders shared the U.S. assessment of the value of sea-basing. Contrary to the U.S., however, the Soviets chose to concentrate their SSBNs (strategic submarines, ballistic, nuclear) in heavily defended Bs in adjacent seas rather than distributing them across the world's oceans [MccGwire 1987b: 98–102]. Two such SSBN bastions existed (and still exist): the Okhotsk Sea in the Far East and the Barents Sea, with the enormous Murmansk base complex. Even though the relative priority between these two was gradually shifting toward the former, the latter remained the most important. Thirty-seven SSBNs were operating out of the Barents Sea, including twenty-two Delta and four Typhoon SSBNs, equipped with long-range (6,500-8,300 km) SLBMs (submarine-launched ballistic missiles) that enabled them to remain within the confines of the B and still be within range of the continental U.S. The Typhoons, furthermore, were designed to operate under the ice cap so as to be able to exploit the excellent acoustic shelter provided by the marginal ice rim (where the underwater noise level is high and sonar detection of the SSBN hence difficult) [Skogan 1990: 32–33].

This basing mode was logical, considering the USSR's unfavorable geostrategic location. Moreover, the relative emphasis given to SSBNs compared with ICBMs and manned bombers in the Soviet strategic triad seemed sound. If anything, the West should therefore probably have preferred a further shift of emphasis away from ICBMs (intercontinental ballistic missiles) toward SLBMs, due to the latter's advantages in terms of *crisis stability*. Problems, however, arose from the ambivalent nature of the means of B defense, which required conventional naval forces, such as attack submarines, surface warships, maritime patrol aircraft, naval fighter-bombers, and even bombers, helicopters, antiship missiles, etc.

Even though their primary mission was undoubtedly the defense of the B, as well as of the Soviet homeland in general, against intruding U.S. submarines, *aircraft carriers*, and the like, they also had significant capabilities for more offensive utilization. Seen in a regional perspective, the surface ships might thus form invasion fleets for, say, attacks against *Norway* or Iceland, or other countries. These problems were compounded by the fact that the Soviet general-purpose forces were located at short distances from *NATO* territory, albeit as a matter of geographical coincidence.

Furthermore, from a global viewpoint, expendable parts of the Northern and Pacific Fleets might be brought to bear in, say, Third World contingencies, etc. If concentrated against weakly defended points, they might certainly have had an impact. Most important, seen from the angle of the NATO versus *Warsaw Pact* contest, the attack submarines (along with some of the long-range bombers and fighter-bombers) might be used for severing NATO's transatlantic *SLOC*s (sea lines of communication).

On the other hand, to have respected the Soviet bastions as sanctuaries would not have seriously jeopardized Western security, if only effective SLOC defense could have been provided by other means. Moreover, such Western respect would have allowed the USSR to build down its general-purpose naval forces, thus diminishing any threat to Western SLOCs. The U.S. Navy might have shown such respect by abandoning explicit planning for strategic *ASW* (antisubmarine warfare) under the auspices of the *Maritime Strategy* [Mearsheimer 1986; cf. for an attempted rebuttal: Brooks 1986]. Furthermore, various *disengagement* schemes could have been devised, according to which the Barents Sea might have been acknowledged as a "keep-out *zone*" for U.S. attack submarines, paralleled by a similar zone for

Soviet (Russian) submarines south of the GIUK (Greenland-Iceland-U.K.) line [Purver 1988; 1988a; 1990; Møller 1987].

BAUDISSIN, WOLF GRAF, (COUNT) VON.   See *Citizen in Uniform*.

BAYLIS, JOHN.   See *U.K.*

BEACH, HUGH.   See *U.K.*

BEAUFRE, ANDRÉ.   French general who had been a commanding officer in French Indochina (during the first stages of the *Vietnam War*), and the author of several works on military strategy that count as modern classics. His works include *Introduction to Strategy* [1963], *Deterrence and Strategy* [1965], *Revolutionary War: The New Forms of War* [1972], and *Strategy for Tomorrow* [1974].

AB advocated an indirect strategy very similar to *Liddell Hart*'s *indirect approach*, and he counted *guerilla strategy* as one of the most important instances thereof, to which his 1972 work was devoted. Revolutionary war (i.e., *guerilla warfare*) is just one among different forms of *limited wars*, in their turn made imperative by the specter of unbounded *escalation* [1972: 40–47; cf. 1965]. Not only in such wars but in all future wars, the goal could no longer be to win in any absolute sense (*vaincre*) but, rather, to persuade (*convaincre*). In this sense, howeverer, a war that is limited materially might well be unlimited in the psychological and moral dimensions. For this purpose, protracted war (or *manoeuvre par la lassitude*), e.g., of the guerilla type, would be the most powerful means. To be able to wage such a protracted war, the guerilla element is well-advised to economize with its forces to the extreme; it should force the opponent to disperse its forces by enlarging the combat area, and so on. Furthermore, whereas the decisive battle is the distinguishing feature of "classical war," in revolutionary war the military operations would rarely be decisive in a military sense. To the extent that they are offensive, then they tend to seek psychological rather than military goals. Guerilla warfare should thus be understood not a form of conquest but as "a tough form of negotiations" [1972: 56].

AB recommended a partial use of the same principles in the industrialized West in the form of a "territorial *militia*." Its mission should be twofold:

- To establish a *forward defense* along all threatened frontiers; and
- To establish an area-covering surface defense consisting of light forces responsible for the defense of a small area, equipped with light and inexpensive weapons and making extensive use of civilian vehicles for tactical mobility [1974: 75–87].

BEBERMEYER, HARTMUT.   Economist and member of the *SAS*, and author of several works on defense economics [1984; 1990; idem & Grass 1984] and *NOD* [idem & Unterseher 1986; 1989], as well as on *conversion* [idem & Unterseher 1992].

BELARUS.   Also known as Byelorussia. See *Soviet Union*, *CIS*.

BELGIUM.   The alternative defense debate in B has not been very intense and has focused predominantly on *NATO* strategy rather than on B itself. In the early 1980s, however, a Belgian general, *Robert Close*, presented a proposal for an *NOD*-like defense restructuring [Close 1981; 1981a; cf. 1976], inspired, inter alia, by his collaboration with German NOD groups and especially *Jochen Löser*.

The most important vehicle of the more recent debate has been GRIP (*Groupe de recherche et d'information sur la paix*: Group of Research and Information on Peace), founded in 1979, led by Bernard Adam, but with Eric Remacle as the *primus motor* as far as the debate on alternative defense, and conventional **arms control* in general, has been concerned. The main output of GRIP has been yearbooks on defense and security (*Memento De Defense—Desarmament*), in the tradition of the **SIPRI* yearbooks.

In his introduction to the 1990 edition, Adam (reviewing developments in the East) urged NATO to abandon **flexible response* and to take the initiative toward negotiation of a multilateral security doctrine based on defensive principles [Adam 1990: 17]. Remacle has also recommended NOD-inspired arms control measures, e.g., an inclusion of naval armaments and short-range nuclear forces in the agenda for the next round of the **CFE* negotiations, with a view to removing residual instabilities: "As long as the offensive capacity of one alliance will be superior to the defensive means of the others, we will have to face military instability" [Remacle 1990: 420; cf. 1990a; 1994].

Besides NOD proponents, B has consistently had its adherents of **civilian-based defense* (CBD) [e.g., Geeraerts 1977; 1978; 1986]. The most prominent of these has been the director of the Polemological Centre, Professor Johan Niezing. In 1987 he published a major work, *Social Defense as a Logical Alternative,* wherein he analyzed CBD in depth as a strategy and sought to assess its compatibility with NOD. He found that applied on a national scale, the two were logically (albeit not necessarily in practice) incompatible. The solution to this problem would presumably be to transcend the narrow boundaries of time and space, so as to create "an integrated, 'experimental' alternative defence system":

- Along the temporal dimension, NOD and CBD might be conceived of as the initial and a subsequent step in the same direction;
- Along the spatial dimension, NOD might be a recommendable posture for certain countries, whereas CBD would be preferable for others [1987; cf. 1982; 1989; 1990].

BERGFELDT, LENNART.   See **Sweden.*

BERLIN DECLARATION.   Statement, *On the Military Doctrine of the States Parties to the Warsaw Treaty,* issued by the **Warsaw Pact* from its meeting in Berlin 28–29 May 1987, i.e., after the Soviet shift (at least declaratorily) to a defensive military strategy [in J. Krause 1988: 74–76]. It was explicitly stated that the pact intended to "keep to the limits sufficient for defence and repelling any possible aggression." The creation of "**zones* of thinned-out arms concentration" was advocated as a first step toward "the returning of armed forces to their national territories." Finally, the pact challenged **NATO* to consultations with the task of analyzing and comparing the military doctrines of the two blocs (see also **Vienna Seminars*).

BERTELE, MANFRED.   Officer in the Bundeswehr who served on the German NATO liaison staff, but who was also temporarily affiliated with the Foundation of Science and Politics (*Stiftung Wissenschaft und Politik*), Ebenhausen. In the latter capacity, in 1982 he presented an analysis, *Structure and Deployment of Land Forces in Europe* [1982], the rationale for which was to create **conventional stability* through minimal changes of nuclear posture or military strategy.

MB proposed a division (to be agreed upon in the **MBFR* negotiations) of Europe into three deep **zones* on either side of the dividing line:

- A 250-km-wide forward zone, wherein each side might deploy a maximum of thirty-five divisions, predominantly new *infantry divisions without shock power (albeit including an armored regiment each), with an operational reserve of ten traditional armored brigades, capable of counterattack;
- A 750-km-wide zone without any armored forces, serving as a *buffer zone, preventing a surge deployment into the forward zone; and
- A rear area, for which there would be no limitations.

The new-type divisions in the forward zone should primarily be equipped with anti-tank missiles and guns, artillery, and *air defense systems, and preferably should fight from sheltered positions. The *tank regiments assigned to the infantry divisions should serve as fire brigades, and depend on the collaboration of the (stationary) infantry, and hence would have only a limited offensive capability: a combination pointing forward toward the *spider-and-web model of the *SAS.

BINTER, JOSEF.   See *Austria.

BIPOLARITY.   Structure of the international system under which all (or nearly all) states belong to either of two opposing *alliances, formally or otherwise. See also *Multipolarity.

BLACKABY, FRANK.   See *U.K.

BLITZKRIEG.   Literally, lightning war. Form of offensive maneuver warfare invented simultaneously in the 1920s and 1930s by British (*Fuller and *Liddell Hart), German, and Soviet (Triandifilov and Tuchachevsky) strategists [Perrett 1985; Bond 1977; idem & Alexander 1986; Hart 1951; cf. Mearsheimer 1988a; Geyer 1986; Vigor 1983; Simpkin 1983].

1. **Distinguishing Features.** B has been described as a form of *indirect approach because it is a way of winning a decisive victory without a decisive battle, thus a means for a materially inferior side to prevail in offensive warfare. B typically requires *surprise attack (either strategically or at least tactically), whereby the opponent is caught off guard, thereby allowing the attacker the initiative, which the latter should maintain until the end in order for the B to succeed. In reality, complete surprise is often unattainable, and the attack therefore has to start with a breakthrough of the defender's *forward defense, followed by a swift, and preferably deep, penetration into the defender's rear, whereby the latter could be thrown off balance and forced to respond to the attacker's moves, rather than the other way around.

2. **Soviet B Strategy.** The strategy of the *Soviet Union was, at least since the late 1960s (and especially in the early 1980s) a typical B strategy [Vigor 1983]. It emphasized swiftness of action, massed breakthrough assaults followed by deep penetrations, etc.—and it had abundant means of performing these missions: large mechanized armies with great shock power and integrated air cover, operational maneuver groups, etc. [Hines & Petersen 1984; Donnelly 1982; Shields 1985; Holcomb & Turbiville 1988; Petersen 1987]. The reasons for this Soviet disposition were obvious:

- Because of Soviet inferiority in terms of mobilizable potential, the only war the USSR stood any chance of winning was a B;
- Soviet allies were inherently unreliable and likely to become increasingly so the longer a war were to last; and
- The likelihood of *NATO's crossing the nuclear threshold (deliberately or

inadvertently) thereby negating any chance of victory, would also grow with time.

3. **Counter-B.** Taking this Soviet strategy as their point of departure, most Western *NOD advocates have assigned first priority to thwarting any attempted B. If only the attacker could be denied a swift and decisive victory (thus creating a *fait accompli*), the West would stand a good chance of prevailing in the ensuing *attrition contest* [cf. Afheldt 1976; 1983]. Furthermore, the demonstrable ability thus to transform a prospective B into a protracted war would presumably be a powerful means of war prevention through conventional deterrence (cf. *time factor*) because nearly all recorded cases of deterrence breakdowns were instances of beliefs in the option of waging a succesful B [Mearsheimer 1983].

From the imperative of thwarting attempted B, NOD proponents have derived a number of design criteria for the appropriate strategy and posture of the armed forces, including

- The *no-target principle*;
- The principle of an in-depth order of battle ensuring area coverage;
- The *nonbattle* principle of averting a decisive battle; and
- The *web* principle, whereby the isolation of individual units, and the resultant fragmentation of the defense forces, was to be prevented.

BLOCH, JEAN DE (IVAN).   Polish banker who in 1889 published a monumental six-volume work, *The Future War,* on the basis of the experience from the Crimean War (1854–1856) and the American Civil War (1861–1865). IB fairly precisely predicted the actual course of *WWI (1914–1918): that innovations in weapons technology (such as machine guns and barbed wire) would strengthen an entrenched defender to such an extent that offensives became close to suicidal, hence that a major European war would be a protracted and extremely costly war of *attrition [Bloch 1889]. Unfortunately, military professionals believed otherwise: that the offense was predestined to prevail, hence that war would be short and victorious, beliefs that were manifested in the fatal German *Schlieffen Plan and its counterpart, the French *Plan 17 [Wallach 1967; Howard 1985; Evera 1984; 1985; J. Snyder 1984; 1985].

BLOCKADE.   See *Sea Power, *SLOC.

BLUEPRINT DETERRENCE.   See *Jonathan Schell, *Existential Deterrence, *Minimum Deterrence, *Nuclear Deterrence.

BMD.   Ballistic Missile Defense. See *Strategic Defense; *GPALS.

BOA.   Bundesrepublik ohne Armee (The Federal Republic without an Army). Campaign launched by the *peace movements and *Die Grünen during the negotiations on German unification. The BOA opposed the maintenance of the Bundeswehr (regardless of any reductions and/or restructuring) and demanded a shift to *civilian-based defense [Frieden und Abrüstung, 1990(1): 52–53; cf. Buro 1990].

BOEKER, EGBERT.   Professor of the natural sciences, subsequently rector magnificus, at the Amsterdam University, and one of the first in the *Netherlands to publish (jointly with *Frank Barnaby) a concrete proposal for an *NOD-type defense [Barnaby & Boeker 1982; 1988; 1989]. He also wrote *European Defense: Alternatives to the Present Defense Policy* [1986].

EB has collaborated closely with both the *Pugwash* Study Group on Conventional Forces in Europe and the *SAS* [Boeker 1990a; idem & Unterseher 1986], as well as with the various study groups in his home country. He was, furthermore, the inventor of the term *spider-and-web*, used by the SAS to describe its NOD model [Boeker 1986: 62-65; cf. Grin & Unterseher 1988; Unterseher 1988b].

BÖGE, VOLKER. Leading member and ideologist for *Die Grünen* in the *FRG* and former researcher at the *IFSH*. In the latter capacity, he wrote (jointly with Peter Wilke) a comprehensive survey of alternative security political conceptions, including *NOD* [Böge & Wilke 1984]. In the former capacity, he advocated German demilitarization and neutralization, a dissolution of the military *alliances* [Böge 1990; 1991; 1991a]. Although acknowledging the advantages of NOD as compared with traditional military strategies, VB has been able to support it only as a lesser evil and has criticized various NOD proposals, in particular those of *Andreas von Bülow* [Böge & Schulert 1986].

BOK, SISSELA. Swedish-American professor of philosophy at Brandeis University. In 1989 SB published a major work, *A Strategy for Peace,* wherein she analyzed strategy from an ethical point of view, taking as her twin points of departure the writings of *Kant* and *Clausewitz*. SB urged her readers to learn from Kant about the dangers of war, and the power of a holistic view of the world, such as also implied by the conception of *common security* [Bok 1989: 76–78; cf. 1985; Kant 1788; 1795]. The most important lesson from Clausewitz was that of the supremacy of the defense over the offensive form of combat.

In continuity with her previous philosophical analyses of the phenomena of lying and secrets, SB emphasized the importance of transparency in the international system, hence recommended various forms of *CBMs*, transcending the narrow realm of military strategy to also include diplomatic, commercial, cultural, and other nonmilitary relations. As far as the narrow military field was concerned, however, she assessed *NOD* as a valuable contribution to furthering mutual trust [1989: 139].

BONIN, BOGISLAV VON (1908–1980). Colonel and staff officer in the "Amt Blank" (the predecessor of the West German Ministry of Defense).

In the 1950s BvB presented one of the first proposals for an alternative Bundeswehr strategy and posture, motivated by the fear that under the envisioned strategy, the *FRG* would be a defenseless glacis for *France*; that the use of battlefield nuclear weapons would destroy Germany; and that rearmament and *NATO* integration would prevent German reunification [cf. Wettig 1967; Schubert 1970]. BvB's alternative was a genuine *forward defense*, based on antitank units, deployed in a 50-km-deep *barrier* zone. The armaments should consist of unmistakably defensive antitank weapons: primarily jeep-mounted 105 mm recoilless rifles, mines, etc. Force requirements would amount to around 120,000–150,000 volunteers, as compared with the twelve conscripted divisions envisioned by the government and the "Amt Blank" [Bonin 1954].

When BvB's proposals were rejected by the authorities, he made his ideas public in the memorandum *Reunification and Rearmament: No Contradiction,* authored jointly with two colleagues, wherein they advocated "a change of guide-lines from an offensive force toward an outwardly clearly recognizable defensive instrument" [Bonin, Wietersheim, & Lüttwitz 1955: 24]. This led to the dismissal of BvB by the "Amt Blank," a step that became the subject of some controversy in the security committee of the Bundestag and elsewhere [Brill 1987: 150–165]. Having in the original "Bonin Plan" envisioned the employment of nuclear artillery, BvB soon shifted to advocacy of a *nuclear-weapons-free zone encompassing a neutral Germany along with *Poland and Czechoslovakia [Albrecht (ed.) 1986: 133–135; cf. Bonin 1955; 1955b]. One source of inspiration for the "Bonin Plan" was the Soviet strategy at the Battle of *Kursk; BvB in 1966 published an article, "The Battle of Kursk: A Model for the Defense of the Federal Republic," in Der Spiegel [1966].

BvB has served as an inspiration for many of the German *NOD proponents of the 1980s (albeit not quite to the extent that he would seem to deserve) thanks, inter alia, to the documentation provided by his biographer, Heinz Brill [Brill (ed.) 1976; 1987; (ed.) 1989; cf. Møller 1991: 52–54].

BOOTH, KEN.   Professor of international relations at the University College of Wales in Aberystwyth, a longtime supporter of the ideas of *common security and *NOD, and a collaborator with various groups in the *U.K. such as the *ADC and *Just Defence.

Against a solid background of classical strategy and traditional theory about international politics, KB has challenged both. First of all, he has on several occasions criticized the traditional paradigm of *Realism and proposed its replacement with a more holistic approach to *security that should emphasize international action over unilateral advantage, containment of enemies over rollback, international norms over national interest, empathy over ethnocentrism, and *defensive deterrence over offensive intervention [1991: xiv–xv].

Pointing out that "Realism is not necessarily realistic," KB has proposed "empirical realism" as a viable alternative to the prevailing "doctrinal Realism." Among his proposed "Ten Rules for Empirical Realists" was to "fear the man who fears you," i.e., to be aware of the impact of the *security dilemma and avoid causing fear in one's opponent [1987b; 1991a; Wheeler & Booth 1987]. Hence KB's consistent advocacy of *disarmament in general, and nuclear build-down in particular, as well as of common security and NOD [1985; 1987; 1990; 1991b; idem & Baylis 1988].

Among his concrete recommendations was one for Britain's unilateral abolition of its nuclear forces, combined with endeavors to bring about similar changes in the rest of *NATO, including, as a first step, a *no-first-use policy. However, to facilitate the latter, it would be important for the U.K. to remain "serious" about defense, e.g., by not weakening its contribution to the defense of NATO's Central Front (the BAOR: British Army of the Rhine) [Booth 1985: 43–47]. Another means of demonstrating seriousness might be for the U.K. to reintroduce *conscription, i.e., a national service (albeit with options for conscientious objectors to perform nonmilitary duties). In addition to allowing for a cost-effective increase in conventional defense capability, this would also point directly to a less offensive conventional posture, in casu a *territorial defense [1983; cf. idem & Baylis 1988: 109–112].

In 1988 KB elaborated on his conceptions of NOD in Britain, NATO and Nuclear Weapons: Alternative Defence versus Alliance Reform. It was written jointly with his colleague at Aberystwyth, John Baylis, who represented the traditional viewpoint, whereas KB stood for the alternatives [Booth & Baylis 1988]. KB emphasized the need for creating a worldwide political backing behind the call for strategic reform ("a critical non-nuclear mass") and found the prospects thereof encouraging, in view of emerging Soviet new thinking and the commitment to defensive principles on the part of social democratic parties across Europe. The U.K. should therefore shift to a nonnuclear, nonprovocative defense policy and work for NATO adoption of a defensive posture, particularly in Central Europe, in casu a layered system of three main *zones:

- A forward defense zone with a depth of around 60 km, saturated with sensors, mines, *barriers*, etc. (and preferably also modified through landscaping), in which only light infantry forces would operate;
- A strategic defense zone with a depth of 150 km, in which NATO should deploy mobile forces (including main battle *tanks* and other systems with shock power) in addition to missile squads capable of targeting invading forces that might penetrate the forward zone; and
- A strategic reserve zone, where army and air force reserves should be deployed, in as great dispersal as possible [Booth & Baylis 1988: 121–125].

KB has more recently been working on the application of NOD principles to the maritime environment, on the basis of his previous experience from the U.S. Naval War College [1977]. In addition to warning against interventionism by means of navies (the "globocop" approach) and advocating some defensive restructuring, he has also urged a further demilitarization, which might transform warships into "law ships" [1993; 1994].

BORG, MARLIES TER.   See *Netherlands*.

BORKENHAGEN, FRANZ H. U.   Major and *SPD* advisor on Bundeswehr issues in the Federal Diet. FHUB has been a consistent supporter of *NOD* within the party, based on his personal involvement with NOD research, first as a member of *Horst Afheldt*'s research team and subsequently as a member of the *SAS* [Borkenhagen 1984; 1984a; 1985a; 1987a].

In 1985 he elaborated on the modalities of NOD combined with a *collective security* system [1985]. The latter had to comprise supranational as well as national armed forces. With regard to national forces, role specialization should be promoted, so that no "balanced" armed forces would remain, thus ensuring a substantial reduction of offensive capabilities on the state level (cf. *multinationality*). Concerning the structure of national armed forces, however, FHUB recommended the *spider-and-web* model of the SAS [1987a; 1988a].

BOSERUP, ANDERS.   Danish physicist and peace researcher; research fellow at *SIPRI* in the late 1960s; subsequently leader of a small peace research institute in *Denmark* until 1974. After having been affiliated with the *Centre for Peace and Conflict Research* in Copenhagen from 1985 until around 1988, AB ended his career as director of the Danish branch of *EUCIS* from 1988 until his death in 1990. This was a reflection of his other main function in the international *NOD* debate: convenor of the *Pugwash* Study Group on Conventional Forces in Europe [cf. Boserup & Neild (eds.) 1990].

AB's first substantial contributions to the alternative defense debate were in the field of *civilian-based defense*; in 1974 he published (jointly with Andrew Mack) a modern classic, *War Without Weapons*. On the basis of a number of case studies of CBD, and on the strategic thinking of *Clausewitz*, the two authors concluded, first, that CBD was indeed a strategy, and second, that it contained valuable elements worthy of further study [Boserup & Mack 1974].

Although from the late 1970s on he shifted his attention from nonmilitary to military defense options, AB remained faithful to his Clausewitzian approach, and on more than one occasion sought to deduce the essence of NOD strategy from Clausewitz's posthumous *On War*. Particularly the notion of the suspension of fighting as a result of the supremacy of the defensive seemed to point toward a viable strategy for defense in what AB called a *subnuclear setting*: a situation where the inescapable risk of nuclear *escalation* made conceptions of classical strategy such as "decision" or "victory" meaningless [Boserup 1986; 1990; cf. 1981; 1985; 1986a; 1988a; 1989].

Besides this preoccupation with war prevention through the strength of the defensive, AB also devoted considerable attention to the *disarmament* prospects of NOD, e.g., through his critique of the prevailing notions of *balance*. On closer analysis, parity was irrelevant, whereas the key to stability lay in the simple formula: $D^A > O^B$ and $D^B > O^A$, signifying a situation in which A's defensive capability surpassed the offensive capability of B, and vice versa [1985].

In addition to such abstract and theoretical analyses of NOD, AB also made various concrete proposals for the restructuring of armed forces (land, sea, and air) [1987; 1990a; 1990c-d; idem & Graabaek 1990].

BOSTON STUDY GROUP.   See *U.S.*

BPA.   See *Basic Principles Agreement.*

BRADFORD SCHOOL OF PEACE STUDIES.   Subsequently Department of Peace Studies, Bradford University. See *Malcolm Chalmers*, *Michael Randle*, and *U.K.*

BRAUCH, HANS GÜNTER.   Dr. Phil., lecturer at the Goethe University of Frankfurt, former member of the *SAS* and chairman of the *AFES-PRESS*.

HGB has published extensively on such subjects as *chemical weapons* [1982a], the *INF* deployment issue and the post–INF Treaty modernization plans [1982; 1989; 1991b], and the *SDI* as well as its European counterpart, *Extended Air Defense* [1987b]. For a number of years he has also been actively involved in the *NOD* debate in the *FRG*, e.g., with elaborate surveys and analyses of the various NOD models [1988; 1988a; 1991a; idem (ed.) 1984; idem & Unterseher 1984]. HGB's direct contribution to the development of alternative defense and security thinking has, however, primarily been in the fields of *CBMs* (confidence-building measures) and *collective security.*

HGB's general approach to CBMs has been to advocate an enlargement of the concept from merely promoting transparency to actually removing destabilizing military options, i.e., what he called "Third Generation CBMs" or "CSBRRMs" (confidence- and security-building, and risk-reducing measures) pertaining to military doctrines, deployment patterns, and weapons development. Most important in these respects would be "overcoming the cult of the offensive in the tactical and operational deployment concepts of both military alliances" [1990b; cf. 1987; 1987a].

In light of the German unification process, HGB has devoted considerable attention to collective security and the future "architecture" of Europe. Writing prior to the 1990 Paris Summit (see *Paris Charter*), HGB proposed to create, on the basis of the *CSCE* and other existing institutions, a pan-European security system resting on three building blocks:

- Economic institutions, which should promote economic recovery in the East as well as East-West trade relations;
- Political consultative mechanisms; and
- The military *alliances*, i.e., the *Warsaw Pact* (already unmistakably in a process of dissolution), *NATO*, and the WEU, which might facilitate (and mitigate risks associated with) German unification, as well as coordinate the troops reductions following from the *CFE* Treaty and its successors.

From a longer-term perspective, the armed forces in Europe should be divided into two broad categories:

- Strictly defensive ("shield"), *territorial defense* forces; and
- "Common security" ("sword") forces under multinational command.

The initial ratio between national and multinational forces should, realistically, be 80:20 but could gradually approach 50:50, combined with a general reduction of forces to around 1.2 million in the whole of Europe by the year 2000 [1990; 1990a; 1991; 1991a; cf. Møller 1989a].

BRIDGE BUILDING.   Logistical item most often singled out for reduction or elimination in *NOD* proposals [e.g., E. Müller 1984]. The reason is that the possession of BB is a precondition for swift offensives over terrain intersected with rivers and canals, such as, e.g., Central Europe. Without BB, the (otherwise extremely offensive-capable) Soviet armed forces would be rendered virtually incapable of invading Western Europe.

BRIE, ANDRÉ.   See *GDR*.

BROSSOLET, GUY.   Former French military attaché in Beijing and author of *Essay on the Nonbattle* [1975 (quoted from the German translation)], which provided inspiration for the formulation of the seminal *NOD* proposal by *Horst Afheldt*. GB was obviously inspired by the Chinese military tradition from *Sun Tzu* to *Mao Zedong,* as well as apparently by the defensive school in the interwar strategy of *France* (see *Maginot Line*).

GB recommended a shift to an asymmetrical strategy: "to counter the enemy's rapidity with the depth of our deployment, his mass with our flexibility, his superiority with our activity" [1975: 158]. The defender should thus deny the aggressor the option of the large decisive battle, in favor of a multitude of small skirmishes, i.e., the *nonbattle* [1975: 173]. The sequential nonbattle would lead to a disproportionate cumulative *attrition* of the aggressor because the defender could select the time and place of each encounter, hence ensure that it would take place only on favorable terms.

The defense should therefore consist of modules, primarily fifteen-man squads, covering an area of approximately 20 sq km and constituting an "attrition web," capable of slowing down the attack. The web should cover the entire frontier area to a depth of about 120 km, so as to deny an aggressor any options of circumvention. The configuration as a network would facilitate communication, and the stationary deployment of the modules would allow them to capitalize fully on their familiarity with the terrain. The light modules should be equipped with man-portable ATGMs (antitank guided missiles), as well as light mortars, direct-fire systems, antitank mines, etc. Because the modules would be expected to take part in only one engagement before falling back, there would be no need for resupply and the logistical requirements would be minimal. The light stationary modules should be complemented with heavier, more mobile formations, including air-mobile modules mounted on helicopters. They should operate in a guerilla-like hit-and-run fashion, i.e., "like pirates" [1975: 168], with which recommendation GB showed his inspiration from nineteenth-century French naval strategy (see *Jeune Ecole*). Furthermore, the posture should include heavy armored modules, suitable for reinforcement and for delivering decisive blows. For this purpose, they should encompass *tanks* and upgraded *air defense* capabilities. On the other hand, neither ground-attack aircraft nor heavy artillery would be called for.

In total, it would presumably be possible to reduce in this way the conventional French forces by 50 percent to a mere 80,000 men [1975: 200, 217–223]. Although the proposal was, as a first step, intended specifically for France, GB did envision the possibility of, in due course, expanding the scope to encompass all of Western Europe.

BRUNDTLAND COMMISSION.   The World Commission on Environment and Development established in 1983 under the auspices of the *U.N.*, chaired by Norwegian Prime Minister Gro Harlem Brundtland and comprising ten members from the industrialized world (Europe, the U.S., Canada, and Japan) and twelve from the Third World.

In 1987 the BC published its report, *Our Common Future,* which spurred an international debate on broadening the notion of *security,* e.g., by highlighting imminent threats to the survival of humankind through pollution, resource depletion, environmental disaster, etc., for whose solution military power was quite irrelevant and that required a collaborative quest for global solutions. Furthermore, military power was not only irrelevant but also could be harmful in two respects: through the direct impact of military activities (most severely, war) on the environment, and through the opportunity costs of global armaments expenditures:

- An action plan to save the rain forests would cost the equivalent of half a day's global military expenditure; and
- For ten days' worth of military spending, clean water could be provided to millions of families in the Third World, thus improving health standards immensely, etc.

"Sustainable development" (the catchword of the BC) thus required concerted action on a global scale to curtail the *arms race* and to redirect resources to the promotion of global security, with which recommendations the BC continued along the path trod by its predecessors, the Brandt Commission and the *Palme Commission* [Brundtland Commission 1987: 290–307 & passim; cf. Brandt Commission 1980; Palme Commission 1982].

BRUNS, WILHELM.   See *Friedrich-Ebert-Stiftung.*

BRUSSELS DECLARATION.   Issued by *NATO* from its summit meeting in Brussels 2–3 March 1988, and accompanied by the statement *Conventional Arms Control: The Way Ahead* [*NATO Review,* 36(3) (April 1988): 30–33], outlining NATO's goals for the *CST* (Conventional Stability Talks), which appeared inspired by (or at least in line with) *NOD.*

The *Warsaw Pact* was criticized for its capabilities of *surprise attack* and large-scale offensive action, whereas allegedly NATO neither possessed nor aspired to such capabilities (§1). It was proclaimed that "military forces should only exist to prevent war and to ensure *self-defense,* not for the purpose of initiating aggression and not for the purposes of political or military intimidation" (§4). The CST should be guided by a coherent political vision that reflected these values (§9), and hence should seek "a secure and stable *balance* of conventional forces at lower levels; the elimination of disparities prejudicial to *stability* and security; and, as a matter of high priority, the elimination of the capability for launching surprise attack and for initiating large-scale offensive action" (§13). Accordingly, talks should focus on capabilities for the seizure of territory, i.e., on "the forward deployment of conventional forces capable of rapid mobility and high firepower" such as *tanks* and artillery (§14). The purpose should be through asymmetrical reductions to "establish a situation in Europe in which force postures as well as the numbers and deployments of weapon systems no longer make surprise attack and large-scale offensive action a feasible option" (§15): a set of demands very akin to those made by NOD proponents for several years.

BRZEZINSKI, ZBIGNIEW.   See *U.S.*

BUDAPEST APPEAL.   Issued by the *Warsaw Pact* Political Consultative Committee (11 June 1986) [in Krause, J. 1988: 67–71] as a follow-up to *Gorbachev*'s April 1986 proposals for large conventional force reductions. It was proposed, as a first step, to seek troop reductions in the range of 100,000–150,000 men, to be followed by a 25 percent cutback (about half a million men). The ground force reductions should be accompanied by reductions of tactical-strike aviation and nuclear weapons with ranges up to 1,000 km. Stringent verification would include on-site inspections. Negotiations would take place in the framework of a second round of the *CDE*, and the resultant reductions would improve military *stability*. However, this would also require changes in military doctrines: "In the interests of security in Europe and the whole world the military concepts and doctrines must be based on defensive principles." This was the first official Warsaw Pact endorsement of the NOD idea.

BUFFER ZONE.   *Zone* interposed between adversaries as a means of *disengagement* and thus a way of preventing hostilities, e.g., by extending available *warning* time for both sides. Before either could attack the other, it would have to traverse the BZ and thereby be slowed down. It would probably encounter various other problems, not least political ones. A BZ would in this way further *crisis stability* by removing pressures (motives as well as options) for *preemptive attack*.

BULGARIA.   See *Balkans*.

BÜLOW, ANDREAS VON.   Former state secretary for defense for the *SPD* during its government period, subsequently its Bundestag speaker on defense issues.

AvB was the first to go public with an advocacy of *NOD* within the SPD in 1984/1985, when he drafted an internal working paper that leaked to the press and became known as the Bülow Paper. AvB's ideas on NOD were close to those of *Jochen Löser* and *Albrecht von Müller*; he subsequently collaborated with the latter on several occasions. The ideas were also a reflection of his participation in the *Pugwash* Study Group on Conventional Forces in Europe.

The NOD scheme outlined in the B Paper was intended for negotiated implementation in a bloc-to-bloc format and envisioned a Soviet renunciation of its offensive strategy, in exchange for which NATO should abandon (at least) early first use of nuclear weapons and approach *minimum deterrence*. To facilitate this, AvB recommended a build-down of capabilities for *surprise attack* and deep offensive operations, while leaving open the option of *conditional offensives*. The *land forces* would emphasize antitank capabilities, to be concentrated in a 25–70 km-deep forward *zone* but deployed in relative *dispersion*. The equipment would include various forms of *ET* (emerging technologies), such as sophisticated mines, drones, etc. [1985; cf. 1984; 1986; 1986a].

In 1988, jointly with retired Col. *Helmut Funk*, AvB presented a concrete proposal for a *balance* of conventional forces in Europe (the *ATTU* area). It was not unlike what became the result of the *CFE* negotiations, but by virtue of its division of the area into zones, it also resembled various *disengagement* schemes [Bülow & Funk 1988; 1988a; cf. 1987; Funk 1988; 1988a]. AvB, furthermore, envisioned a greater role for the West German Territorial Army (as a presumed solution to the manpower shortage facing the Bundeswehr). The fact that he proposed deploying the *militia*-like reserve forces up front and with rather second-class equipment attracted criticism for using the reserves as cheap cannon fodder and for not really allowing for the promised reduction of offensive-capable forces [Gupta 1988; Thiman 1989a; idem & al. 1988].

In light of the unification process, AvB published a concrete proposal for the arms control process in Europe (albeit bearing the more ambitious title *European Security Year 2000*), authored jointly with *Egon Bahr* and *Karsten Voigt*, on behalf of the SPD, but nevertheless attracting serious criticism from within the party for being far too conservative [Bahr & al. 1989; cf. Wieczorek-Zeul & Scheer 1990; 1990a].

BUNDESWEHR. See *FRG*; *Citizen in Uniform*.

BUND FÜR SOZIALE VERTEIDIGUNG. Federation for Social Defense. Organization in the *FRG*, campaigning for "The FRG without an Army" (*BOA*). As an alternative to military defense, the BSV has advocated a shift to *civilian-based defense*, arguing that "even without weapons we are not defenceless."

BUNDY, MCGEORGE. See *U.S.*

BURO, ANDREAS. Professor of international relations in Frankfurt, a leading figure in the FRG's *peace movement*, and a cofounder of the *Committee for Basic Rights and Democracy*.

AB's general attitude has been distinctly disarmamentalist, and he has rejected all narrow military approaches to *security*. A shift to *NOD*, in his view, would therefore have to be combined with efforts to curb the influence of the *MIC* (military-industrial complex) through *democratization*. Moreover, a quest for "positive *peace*" should accompany any attempts at strengthening peace in the sense of an absence of war ("organised peacelessness") [Buro 1982: 85-86; 1983; 1983a; 1987; 1988].

These reservations notwithstanding, AB supported a shift to NOD, e.g., with a view to inducing the *Soviet Union* to adopt a more defensive posture vis-à-vis *NATO*, at least on the regional scale. A shift to NOD on the part of NATO should therefore be unconditional but embedded in a comprehensive process of transformation and confidence building, albeit not in the way envisioned by *gradualism* because this might allegedly contribute to cementing enemy images. Rather, unilateral action should be combined with multilateral consultations (up to the point of negotiations) [Buro 1987].

As a corollary of West European transarmament to NOD, AB envisioned and welcomed a withdrawal of U.S. troops from Europe, with a dissolution of NATO as a longer-term perspective. He declared that NATO was not a guarantor of peace and, e.g., denied all validity to the often-heard argument that the multinational posture of the alliance in any way detracted from its offensive capabilities [1987: 20-21].

BUSH INITIATIVE. U.S. President George Bush's statement 27 September 1991 (shortly after the signing of the *START* 1 Treaty) about nuclear weapons. The BI envisioned, inter alia, the following unilateral steps:

- Withdrawal of tactical nuclear weapons (including Tomahawk cruise missiles) from surface ships and attack submarines;
- Elimination of nuclear artillery munitions and warheads for short-range ballistic missiles; and
- A partial stand-down from alert.

Furthermore, the BI proposed negotiations with the *Soviet Union* on the elimination of MIRV (multiple, independently targeted reentry vehicles) for intercontinental missiles.

The BI was followed on 5 October by a Soviet announcement that responded positively to the negotiation proposals and promised reciprocal unilateral steps, including drastic reductions of both land-based and naval tactical nuclear weapons [*Survival,* 33(6): 567–570; cf. Thränert 1991; Arkin & al. 1991]. An additional follow-up on the BI was *NATO's decision (taken by the Nuclear Planning Group, 17–18 October 1991) to cut the arsenal of tactical nuclear weapons in Europe (excluding the French forces) by 80 percent, i.e., to below 800 warheads [*NATO's Sixteen Nations,* 36(6): 142].

BUZAN, BARRY.   Professor of international studies at the University of Warwick, and project director at the **Centre for Peace and Conflict Research* in Copenhagen.

BB has described himself as a neorealist, yet contrary to most analysts in the tradition of **Realism,* his focus has been on **security* rather than power. This was the topic of his 1983 work, *People, States and Fear* [2d ed.: Buzan 1991], constituting the most exhaustive analysis of the concept available. BB distinguished between security at three levels, between which conflict may well arise:

- Individual security, which might be both furthered and threatened through the quest for security at the second level, e.g., because of the **defense dilemma* [1991: 35–56, 270–293];
- National security, the quest for which might harm not merely individual security but also security at the highest level, inter alia because of the **security dilemma* [1991: 294–327]; and
- International security, i.e., the structure of the international system as such, which was basically anarchic. Nevertheless, a "mature anarchy" was conceivable, in which conflict might be minimized and norms shared, and where **regimes* might proliferate across issue areas, i.e., a situation resembling a "legitimate international order" or even a **concert* [1991: 175–181, 261–265; idem & al. 1990: 229–252; cf. Kissinger 1957; Jervis 1985; Mueller 1990; 1991; Kupchan & Kupchan 1991; Evera 1990].

BB also identified a fourth level, somewhere between the second and third levels: regional security, i.e., the security of what he called "security complexes." The European security complex with its traditional "bonds of amity and enmity" had until quite recently been "overlaid" by the bipolar rivalry. However, the overlay was being lifted as a result of developments in the East, leaving the region facing numerous challenges [Buzan 1991: 186–229; 1989; 1989a; idem & al. 1990; cf. Møller 1989f].

On several occasions, BB explicitly expressed himself in favor of *NOD as a suitable means for coming to grips with these post–Cold War challenges [Buzan 1987; 1987a: 276–288; cf. his foreword to Møller 1991].

BYELORUSSIA.   Also known as Belarus. See **Soviet Union,* **CIS.*

C

C³I.   (Pronounced C-cubed-I.) Command, Control, Communications, Intelligence.

The general technological level and the specific strengths and weaknesses of C³I are important determinants of the respective strengths of offense and defense, e.g., through the influence on mobility (tactical and strategic). An army's ability to attack another country is thus determined, inter alia, by its ability to command and control widely dispersed forces over long distances and to communicate with subordinate commanders and forces in the field, as well as by its level of intelligence, which determines both its ability to find worthwhile targets (at the right moment) and to protect itself from the defender's fire.

This opens up the possibility of fighting an enemy's C³I rather than its armed forces, thus incapacitating the latter by blinding and/or paralyzing them. *Liddell Hart* thus advocated anti-C³I tactics as an element in the *indirect approach*: "To destroy [an army's] lines of intercommunication, by which orders and reports pass, is to paralyse its sensory organization, the essential connection between brain and body. . . . To paralyse the enemy's military nerve-system is a more economical form of operation than to pound his flesh' [Hart 1967a: 183, 219]

C³I has experienced a radical development, particularly in wake of the industrial revolution (which brought, e.g., the telegraph and aviation) and as a result of the more recent development of microelectronics and the exploitation of space for military purposes [Creveld 1985]. In recent years a veritable "transparency revolution" has taken place, as a result of the introduction of powerful and multispectral sensors (facilitating data collection), combined with redundant means of communication (radio, satellites, telephone networks, fiber-optic cables, etc.) and capable means (computers) of data processing. Miniaturization has allowed for a wide-ranging *dispersion* of these means to small combat units as well as integration into the weapons themselves.

The effects of these developments relative to a future war are both disputed and inherently ambiguous, inter alia with regard to benefiting the defense or the offense. On the one side, the improved means of strategic *warning* tend to benefit the defender by allowing for a timely *mobilization*. On the other side, the same surveillance platforms may be used partly for long-range target acquisition, thus facilitating *surprise attack* or other offensive operations (see, e.g., under *FOFA* and *CDI*) [Herolf 1988; 1988a; Grin 1990: 135–196].

Tactically, *PGMs* (weapons with integral guidance mechanisms) may benefit the defender by, e.g., enabling small *infantry* units to engage enemy *tanks* and aircraft. Indeed, it was partly the impressive performance of PGMs in the 1973 Yom Kippur War that aroused interest in *NOD* [Digby 1975; P. Walker 1981; 1985; 1986; Barnaby 1986; 1990]. On the other hand, an attacker may also employ PGMs and conduct area surveillance in mop-up operations against the defender. The decisive factor thus becomes concealment, in its turn apparently inversely correlated with mobility.

NOD has been advocated, inter alia, as the appropriate way of capitalizing on the transparency revolution, but certain NOD schemes [e.g., Hannig 1984; 1986;

1986a; 1986b; H. Afheldt 1983; Acker 1984; A. v. Müller 1986; 1986b; 1987a; 1990b] seem to have been based on unfounded hopes in *E.T. Others have relied exclusively on well-tested technologies and emphasize structural changes in the defense posture (*decentralization, redundancy, etc.) as the decisive factors. Particularly the *SAS has emphasized the need for redundant (and by implication, rather cheap) means of intelligence and communication [Grin 1990; cf. SAS (eds.) 1984; 1989].

CAKE-SHARING.   Alternative approach to *arms control suggested by E. Singer [1963], *Stephen Salter [1986] and *Albrecht A.C. von Müller [1986: 212–214] but well known from everyday family life (whence the name). CS is presumably able to solve the problems of measurements and *balance, cheating, and mistrust.

As a preliminary, each party should disaggregate its forces (say, down to division or regiment level) and assign percentage values to each component. The two sides should then agree on a level of reductions, in terms of percentages, and each side should be allowed to select which of the other's force components should be dismantled. Percentage cuts would be more equitable than absolute figures because they would require the stronger side to reduce most. Furthermore, in the case of very uneven initial force ratios, the two sides might well agree on asymmetrical cuts (as they did as a preliminary to the *CFE negotiations), say, by cutting 10 and 15 percent, respectively. Most important, however, the incentive to cheat would be eliminated because the party trying to underrate high-value formations would run the risk of a disadvantageous bargain.

CS has never been attempted in actual negotiations but has merely been simulated (e.g., by the East-West participants in a *Pugwash workshop). In real life, it might well encounter obstacles, related, e.g., to the combined-arms and highly integrated nature of armed forces. It might prove difficult to find an appropriate level of disaggregation that would allow for meaningful percentage cuts, while not seriously weakening either side by completely eliminating military units (if, say, they were of a magnitude such as air defense units assigned to French forces in Germany). On the other hand, the focus on formations (rather than, say, weapons holdings or troop levels within formations) would tend to facilitate verification of an agreement.

CANADA.   Like, e.g., the *U.S. or *Australia, C has no defense problems in the sense of, say, the *FRG's because the only country with which C shares a border is the U.S. Because war between these two countries has been inconceivable for a very long time, they constitute a *security community, and neither feels a need to defend the common border.

1. **Issues.** Whereas some attention has been given to the defence of C, for the most part the defense and security debate in Canada has evolved around the following themes:

- The role (if any) of C, as a member of *NATO, in the defense of Europe;
- The repercussions of the naval superpower contest on Canadian security; and
- The impact of nuclear strategies on C.

Furthermore, C has always pursued a rather active policy in the field of *arms control. One of the more recent initiatives, in the aftermath of the *Gulf War, pertained to *arms trade regulation: C proposed that the industrialized countries stem the arms flow to the Third World (in particular the Middle East), with a special emphasis on *offensive weapons [Boulden & Cox 1991: 1–11, 35–39].

2. **Naval/Arctic Arms Control.** Even though C is party to the oldest agreement on *naval arms control* (the Rush-Bagot Treaty of 1817, limiting U.S. and Canadian naval operations on the Great Lakes), C has been reluctant to support proposals for naval arms control, lest this would alienate its NATO partners [Boulden & Cox 1991: 45–49].

However, the U.S. *Maritime Strategy*, with its emphasis on strategic *ASW*, has been a cause of some concern in C because the envisioned underwater battle might take place in or close to Canadian waters, and because inland installations would thereby become possible targets for Soviet attack. Hence the interest in naval arms control and *maritime C(S)BMs*, research on which has been carried out by the Canadian Institute for International Peace and Security (CIIPS) and in particular by Ronald Purver. With a view to averting ASW and thus to improving *crisis stability*, Purver proposed a combination of "keep-out" and "stand-off" *zones* for submarines: the U.S. Navy should keep out of (i.e., respect as sanctuaries) the Soviet *bastions* for their SSBNs (strategic submarines, ballistic, nuclear) in the Barents Sea and Okhotsk Sea, and reciprocally, the Soviet Navy should keep at a distance from U.S. shores [Purver 1988; 1988a; 1990; cf. Boulden & Cox 1991: 91–100; Møller 1987; 1989b].

3. **Alternative Defense.** Because of the absence of serious defense problems, the *NOD* debate in C has focused on conventional arms control in Europe [cf. Hamlin 1990], with two exceptions:

1. The New Democratic Party in 1988 published a party platform devoted to the themes of *common security* and NOD, wherein it outlined a strategy for a *territorial defense* of C [New Democrats 1988; cf. Thorburn 1986].
2. The independent researcher Peter Langille (collaborating with the *SAS*) in 1990 published *Changing the Guard: Canada's Defence in a World in Transition,* wherein he sought to apply the principles of NOD to C. This would require that C deliberately insert itself as a "security *buffer*" between the rival superpowers by assuming greater responsibility for patrolling and controlling its own territory. Canadian territory—land, airspace, and territorial waters—would thereby be made "militarily irrelevant" for each side. The Canadian navy would place greater emphasis on a coast-guard role, and the air force would focus on early *warning* and interception of aerial intruders. The enormous size of C would, for obvious reasons, rule out an area-covering stationary defense, such as sometimes is suggested for Germany; hence the need for enhancing mobility while scaling down the heavy equipment of the *land forces* [Langille 1990].

CANBY, STEVEN L.   Graduate of the U.S. Military Academy and Harvard University, private defense analyst (C & L Associates), and author of numerous articles on defense matters.

SLC has been a longtime critic of *NATO* and U.S. strategy in general and their nuclear and high-tech orientation in particular [1972; 1985; 1985a]. As early as 1972 he advised NATO to abandon its offensive posture in favor of "a defensive stance with a logistical system oriented for a short war but hedged against a less probable long war," which would also allow NATO to "trade *tanks* for anti-tank weapons" [1972: 46–47].

SLC's own alternative has been a greater use of *infantry*, especially "European-style" (i.e., *Jäger*) light infantry, configured as a "*checkerboard defense*," (i.e., a special variety of a defense *web*). Each division in the checkerboard (or close to the border, i.e., as a "cordon defense") would be responsible for the defense of around 50 km, and be equipped with various types of antitank weapons. The forces immediately at the border would have to be in a high state of readiness; those further back would largely consist of reserves. Behind these forces, furthermore, would be stationed operational reserves consisting of mechanized armored units capable of counterattack [1980; 1984; 1986a].

SLC has repeatedly warned against the temptations of high-technology gadgets, pointing out that "the quest for quality over quantity is self-defeating" and that "NATO's military inferiority is due purely to inadequate combat numbers" [1984: 130–131]. In consequence, he was also very critical toward NATO's *FOFA* doctrine and similar *ET*-oriented schemes for *deep strike* [1985; 1985a; 1987].

Even though SLC has not been a wholehearted supporter of *NOD* ideas [1986; 1994], his proposed defense scheme resembles closely the most sophisticated German NOD models, particularly that of the *SAS* (with whom there has been a regular exchange of views and partial collaboration).

CARTON, ALLAIN.   See *CIRPES*.

CARVER, MICHAEL.   (Field Marshall Lord.) See *U.K.*

CASCADING.   Neologism invented by *NATO* in connection with the *CFE* Treaty as a way of circumventing (quite legally) the need for scrapping modern excess TLIs (treaty-limited items) by handing them down to allies with less than state-of-the-art holdings.

Table 10 shows NATO's cascading plans as of 1991. The plans would seem to violate the spirit of the CFE, namely, to build down offensive capabilities, because it would simply imply transposing the problems from the former Central Front to the flanks.

Table 10.   Cascading

Releasing Nation	Equipment	Approximate Number	Receiving Nation
Germany	*Leopard-1*	400	TU, GR, DK
	M113	250	GR, TU
	LARS	150	PO, GR, TU
Netherlands	*Leopard-1*	170	GR
	M113	130	PO
U.S.	*M60A1/A3*	2000	SP, PO, GR
	M113	600	TU
	M110	180	SP, GR, TU, NO
			TU, GR, DA
Italy	*M60A1*	100	GR
	M113	100	TU

Source: Jane's Defence Weekly, 15 (22) (1 June 1991): 915, 918.
Note: TU = Turkey; GR = Greece; DK = Denmark; PO = Portugal; SP = Spain; NO = Norway.

CAT.   (Occasionally CATT.) Conventional Arms Transfer Talks conducted between the *U.S.* and the *Soviet Union*, 1977–1978. See *Arms Trade Regulations*.

CATT. See **CAT.*

CATEGORICAL IMPERATIVE. Principle of ethics formulated by **Immanuel Kant* in 1788 in his *Critique of Practical Reason:* "Act only on that maxim through which you can at the same time will that it should become a universal law" [Kant 1788: 53; cf. Piepmeier 1987; Donaldson 1993]. The CI has occasionally been seen as a precursor, albeit one with a wider application, of the principle of **common security* [Palme Commission 1982; cf. Møller 1992: 10–12, 28–32; H. Afheldt 1984b: 61].

CBD. See **Civilian-Based Defense.* Also the acronym used by the **SAS* for its **Confidence-Building Defense.*

CBM. Confidence-building measure. Generic term for various measures intended to build down mistrust between adversaries.

1. **Taxonomy.** One might, first of all, distinguish between a narrow and various broader senses of CBM [cf. Alford 1979; Holst 1983a; 1987; Baudissin 1982; Borawski 1986; Brauch 1987; IPRA-DSG 1980; Thee 1983a; Rittberger & al. 1990; Väyrynen 1985a; Birnbaum 1987b; Lutz 1982a; B. Meyer 1990a]:

- Narrowly defined, it refers to the actual CBMs codified in the 1975 **CSCE* Final Act, and additional CBMs negotiated, e.g., at the Stockholm **CDE* conference.
- In a broader sense, it may refer to all military measures intended to promote or actually contributing to the promotion of confidence: what might be called "functional **arms control*," distinguished from "structural arms control" (or merely "arms control") by dealing with activities, rather than postures.
- In the broadest sense, CBM might refer to all sorts of measures with the effect of building down mistrust: trade relations, **interdependence*, openness (*glasnost*), and **democracy* would be obvious cases in point.

Further distinctions (within the broader sense of CBM) might be made between the following (see Table 11):

- Formal and informal CBM: the former would refer to binding regulations, negotiated and codified in treaties and agreements; the latter might signify tacit understandings based on reciprocity, to the effect that each side intends to refrain from certain activities for as long as the other side does the same.
- Direct versus coincidental CBM: the former category signifying CBM negotiated as such; the latter referring to CBM-like effects of agreements entered into for entirely different purposes (e.g., verification clauses in arms control agreements).
- Communicative, transparency-furthering, and constraining CBM: the first type compromises attempts to improve communication between adversaries so as to remove possible misunderstandings; the second type consists of rules for preventing misunderstandings in the first place by clarifying the facts; and the third type consists of attempts to alter these facts, i.e., prohibit certain activities deemed destabilizing, a category also known as **CSBM* (confidence- and security-building measures).
- CBMs with universal versus CBMs with geographically and/or temporally circumscribed application.
- Bilateral versus multilateral CBMs.

- CBMs pertaining to nuclear and conventional forces, and to land, air, and naval forces, or any combination thereof.
- CBMs pertaining only to the activities of the armed forces, or to other military-related activities, such as the development of *military science*, military production, R&D (research and development), etc.

Table 11. CBM Categorization

STATUS	Formal		Informal
ORIGIN	Direct		Coincidental
NATURE	Communicative	Transparency-furthering	Constraining
APPLICATION	Universal		Circumscribed • Geographically • Temporally
SCOPE	Global	Multilateral	Bilateral
OBJECT	Nuclear		Conventional • Land forces • Air forces
	Land forces	Total posture	• Naval forces
	Military	Military-related	

2. **CSCE Process and CBM.** Even though de facto (coincidental) CBMs had been known before 1975, the CBM concept was introduced into the vocabulary of international law through the Helsinki Final Act, signed at the CSCE Conference in 1975 [in Freeman 1991: 147-150]. The declared purpose was "to contribute to reducing the dangers of armed conflict and of misunderstanding and miscalculation of military activities."

Concretely, the CBMs included in the 1975 agreement included provisions for the prior notification of, and for the exchange of observers to, major military maneuvers exceeding 25,000 participating troops, as well as advance notification of other military movements (such as relocation of military units). Controversial issues in this, and subsequent agreements, included

- Whether to expand the agenda to include declaratory CBM (such as nonaggression treaties and *no-first-use* declarations), as suggested by the *Warsaw Pact*, or to limit negotiations to "militarily significant measures," as maintained by *NATO*;
- Whether to continue with emphasizing the enhancement of transparency (NATO's position) or proceed to limit military activities, as proposed by the USSR;
- Whether to include air and naval forces, as proposed by the Soviet Union; and
- Whether to extend the area of application from the 250 km from a frontier with another CSCE member state to include all of the *Soviet Union* west of the Urals, and whether to include the *U.S.* and *Canada* [Berg 1986; Rotfeld 1986; 1987; 1989; 1989a; idem (ed.) 1986; Michejew 1987; Scherbak 1991].

The first breakthrough in these respects was the agreement reached in 1983 on the "Madrid Mandate" for the *CDE conference in Stockholm (1984–1986). It implied an extension of the area of application to the Urals, and vague phrases about the inclusion of naval activities insofar as these would "affect security in Europe as well as constitute a part of activities taking place within the whole of Europe."

The CDE was concluded in 1986 with a final document constituting a compromise between the opposing views. It contained the declaration "Refraining from the Threat or Use of Force," along with modest provisions for notification of amphibious operations exceeding a stipulated size (3,000 troops). With these partial exceptions, however, NATO's approach had generally prevailed [Dean 1987: 185–206; 1989c: 105–125; Borawski 1988; 1989; Ghebaldi 1989; Holst 1990; Lodgaard 1986; Freeman 1991].

The main concession to the Soviet point of view may have been the accompanying decision (taken by the Vienna follow-up meeting 1986–1989) to convene a conference on conventional armed forces in Europe, i.e., the *CFE talks, in parallel with the next round of CBM negotiations, both of which were to reach a decision before the subsequent follow-up meeting, scheduled to commence in March 1992 in Helsinki. The agenda for the new round of CBM negotiations was, furthermore, expanded to deal with CSBMs, i.e., constraining measures to complement the modest progress made in this field in Stockholm: to prohibit (with escape clauses) not duly notified activities [Buehring 1989; Ghebaldi 1990; Lachowski 1993].

The compliance record of the CBMs thus negotiated under the auspices of the CSCE has generally been good, and the CBMs have been almost universally appreciated as valuable, yet far too modest, contributions to improving *crisis stability* and mutual confidence.

3. **Other CBMs.** Besides the CSCE-related CBM, a number of other CBM-like measures are in existence [cf. S. Meyer 1984; H. Müller 1989]:

- Measures to improve communication, particularly between the U.S. and the USSR (subsequently Russia) with a special view to crisis situations, such as the "hotline agreement," which was signed in 1963 (in the aftermath of the Cuban missile crisis) and has subsequently been modernized [Ury 1985];
- Various rules-of-the-road agreements, such as the U.S.–Soviet Prevention of Incidents at Sea agreement, which has been followed by similar bilateral agreements [Lynn-Jones 1988; 1990];
- The *Vienna Seminars*, under whose auspices high-ranking military officials and defense ministers have discussed military doctrines [Hamm & Pohlman 1990; 1990a; Krohn 1991b; Lachowski 1992];
- The Open Skies Treaty, signed in 1992, permitting parties to overfly other parties' territory with reconnaissance aircraft [Kokoski 1993];
- The UN Conventional Arms Register, intended to keep track of arms transfers between countries, especially arms sales to the Third World [Anthony 1993; Chalmers & Greene 1993; Laurence & al. 1993; Wulf 1993] (see *Arms Trade Regulations*); and
- The verification clauses attached to most *arms control* and *disarmament* agreements [Altmann & Rotblat (eds.) 1989; Brauch (ed.) 1990b; Kokoski & Koulik (eds.) 1990]. The USSR used to be recalcitrant with regard to on-site inspection and similar forms of intrusive verification measures, but since 1987 it became at least as forthcoming as the West in these respects [Heckrotte 1986; Kokoyev & Androsov 1990]. The CFE Treaty thus constituted the most elaborate and intrusive scheme for mutual inspections ever

negotiated [Cleminson 1989; Graaf 1989; Grin 1990b; Krepon 1988; Altmann & Gonsior 1989; Kunzendorff 1989; 1989a; Lederman 1991; Smit 1990; Banner & al. 1990], a fact that will in itself promote confidence.

4. **New Types of CBMs and CSBMs.** In addition to the CBM already in existence or under negotiation, independent observers have made numerous proposals for supplementary CBM [Borawski 1988a; Brauch 1987a; 1990a-b; Gärtner 1989; Lodgaard 1987a; 1989; Streltsov 1989; Schütze 1987], such as

- CBMs relating to independent air force operations [Mason 1988];
- *Maritime C(S)BMs*, in the form of either a simple extension of the CBM rules to adjacent seas or of constraints on naval activities [Deyanov 1990; Grove 1990; P. Howard 1990; Haesken & al. 1988; Lodgaard 1989b; Lütken 1990; Miller 1990; Schmähling 1990f];
- Proposals for extending the European CBM regime to areas beyond Europe [Findlay 1990];
- Exercise restraints, e.g., with a view to preventing military maneuvers from being exploited for coercive purposes, i.e., for intimidating adversaries [cf. Blackwill & Legro 1989];
- Risk-reduction centers, intended to facilitate communication and fact-finding during crisis but also for prophylactic crisis prevention [Dean 1989c: 242–248; 1990f; cf. Twining 1986].
- Various proposals for military *disengagement*, such as maneuver-free *zones*, or the creation of restricted military areas, proposed by *Jonathan Dean* [Dean 1989c: 232–242].

5. **NOD and CBM.** The last category points directly to the contribution of *NOD* to achieving the same goals as sought through CBMs, only more directly. Whereas CBMs could be seen as attempts to avoid the misinterpretation of defensive military activities as attack preparations, NOD is intended to rule out attack altogether by eliminating offensive capabilities [cf. Müller 1982; Møller & Wiberg 1990].

CDE.   Conference on Disarmament in Europe, also called the Stockholm Conference [Brauch (ed.) 1987; Freeman 1991; Borawski 1988]. See *CBM*.

CDI.   Conventional Defence Improvement Initiative. An offspring of the so-called Weinberger Initiative, launched by U.S. Secretary of Defense Caspar Weinberger, in June 1982 (also known as the *ET* Initiative because of its emphasis on emerging technologies). After *NATO*'s political decision on the CDI in 1983, CNAD (Conference of National Armaments Directors) was assigned to develop a CMF (conceptual military framework), which was finished by 1985 [Flanagan 1986; 1987; Abshire 1987; cf. Tegnelia 1985; Canby 1985; 1985a; Berg & Herolf 1984; Herolf 1986a; 1988; OTA 1987]. It reflected to some extent divergent views between the Europeans and the *U.S.*: the Americans favored very deep and hi-tech options; the Europeans preferred shallower forms of *deep strikes*, using more conventional technologies. Nevertheless, by its trend toward deeper conventional strikes, the CDI could also be seen as an offspring of the *Rogers Plan*, which resulted in NATO's adoption of the *FOFA* doctrine in 1984, both of which were marketed as contributions to a *conventionalization* of NATO strategy and posture, i.e., as raising the nuclear threshold and *ipso facto* as steps in the direction of a *no-early-first-use* strategy [cf. Harris 1988; Møller 1987a; 1988a].

CDSA.   See *Christian Democrats in Favor of Steps Toward Disarmament*.

CDU.   Christlich-Demokratische Union (Christian Democratic Union). West German (and since 1989 all-German) conservative party, holding government power (except 1966–1982) since 1949 [cf. Møller 1991: 189–194, 217–258].

The CDU, along with its Bavarian sister party, the CSU (Christlich-Soziale Union), has all along belonged to the traditionalist camp and given first priority to integration with (and loyalty to) the Western system (*NATO*, WEU, EC, etc.), hence the opinion of the *FRG*'s NATO partners, and above all the *U.S.*, has been a decisive criterion for CDU-CSU security policy. Only since the late 1980s have the actual disagreements come out into the open, with the debate on short-range nuclear modernization [Griffith 1978: 31-106].

As might be expected, the CDU-CSU was therefore extremely critical of all alternative defense conceptions. Nevertheless, the party gradually came to accept some of the objectives advocated by the *NOD* community, including that of an abstention from offensive capabilities. In the 1988 party declaration *Our Responsibility in the World* it was stated that the aim should be to attain "a situation in which also the armed forces of the Warsaw Pact are capable only of defense" [*CDU-Dokumentation* 1988:19]. In 1990 Minister of Defense Gerhard Stoltenberg went a step further in the same direction by stating that "the political purpose of armed forces can and must in our time be nothing but defensive" [Stoltenberg 1990: 142].

CENTRAL AMERICA.   See *Latin America.*

CENTRE INTERDISCIPLINAIRE DE RECHERCHES SUR LA PAIX ET D'ÉTUDES STRATÉGIQUES.   Interdisciplinary Center for Peace Research and Strategic Studies (Paris). See *CIRPES.*

CENTRE FOR PEACE AND CONFLICT RESEARCH.   Previously Centre *of* Peace and Conflict Research at the University of Copenhagen. Founded in 1985 according to a directive of the Danish parliament in 1983. From 1985 to 1990 its research was concentrated on two projects: Nonoffensive Defense in Europe and Nonmilitary Aspects of European Security. Since 1991 there have been three research projects: Conversion, European Security, and Nordic Security.

The staff of the NOD project has included *Anders Boserup* (loose affiliation), *Bjørn Møller*, and *Håkan Wiberg* (director of the center). The CPCR, among other things, publishes the international research newsletter *NOD: Non-Offensive Defence;* since 1991, *NOD and Conversion.*

CFE.   East-West negotiations from 1989 until the present, the first round of which resulted in a treaty on conventional armed forces in Europe, the CFE-1 Treaty. It has been followed by the CFE-1A Treaty [J.M.O. Sharp 1990b; 1991; 1992; 1993].

1. **Background.** The negotiations came about as a result of converging proposals by the *Soviet Union* (echoed by the *Warsaw Pact*) and *NATO* (see *Berlin Declaration* and *Brussels Declaration*), wherein the general principles were outlined that became the guideline for the negotiations:

- That the talks should cover the whole of Europe from the Atlantic to the Urals (*ATTU*), rather than any artificial part thereof;
- That the two *alliances* by then in existence (and seemingly with a long lease of life) should, realistically, be the subjects of negotiations (even though French objections had to be taken into account);
- That the focus should be on major weapons systems (in the first stage)

rather than on manpower (as had been the case of the unsuccessful *MBFR* negotiations);

- That special attention should be given to such weapons systems as were of special importance for offensive operations and *surprise attack*;
- That asymmetrical cuts were conceivable; and
- That the goal should be major reductions [Blechman & al. 1990: 72-117].

2. **Mandate.** On the basis of these points of agreement, the concrete mandate was worked out in the course of the preparatory talks (by that time best known as the *CST*, Conventional Stability Talks). Among the controversial points were whether to restrict the talks to *land forces*. The USSR had initially proposed including both naval and air forces in the agenda, but it soon folded on the first matter (while reserving the right to raise it in another framework). NATO, in its turn, eventually compromised by agreeing to include aircraft in the mandate (while rejecting Soviet proposals for a distinction between offensive and defensive types of aircraft).

Another controversial issue was that of the *dramatis personae* and roles. Most of NATO favored the alliance-to-alliance format, but *France* insisted on a state-to-state format. Furthermore, the neutral states objected (at least formally) to being left out of the negotiations (albeit not to escaping the eventual limitations). The problem was solved by conducting the talks between the members of NATO and the Warsaw Pact (hence the name "23 Nation Talks") as states, and agreeing to consult the other *CSCE* member states regularly [R. Kaiser 1988; Mutz 1988; Dean 1988; 1989; 1989c; Blackwill & Larrabee (eds.) 1989; E.F. Jung 1988; Karpov 1989; J. Krause 1988; Nehrlich & Thomson (eds.) 1988; Freedman 1989; H-J. Schmidt 1988; Wittmann 1989; UNIDIR (ed.) 1989; Zellner 1989; Kamp 1989; Lapins 1989].

3. **CFE Treaty.** The treaty (which was also known as CFE-1) was finally signed in Paris on 19 November 1990, on the fringes of the CSCE summit meeting. The number of signatories was down to twenty-two as a result of the unification of the two German states [A. Arbatov, N. Kishilov & al. 1989; Blechman & al. 1990; Dean 1990d; 1990h; Ghebaldi 1990; 1990a; L. Hansen 1990; Rühl 1990; 1991; 1991a; H-J. Schmidt 1990; 1991; J.M.O. Sharp 1991; Zellner 1991; Barth 1990a; Böge & Pack 1990].

The treaty was intended to, and probably did, remove any capabilities for surprise attack on the part of the two alliances through a build-down to lower ceilings for each for five categories of major weapons systems: heavy *tanks* (20,000), armored vehicles (30,000), artillery pieces (20,000), combat aircraft (6,800), and combat helicopters (2,000). The CFE established ceilings on the holdings of major weapons systems (TLIs: treaty limited items) in the ATTU (Atlantic-to-the-Urals) region, and obliged the signatories to destroy their excess holdings. The distribution of reduction was extremely uneven (see Tables 12–16 and Figures 1–5).

4. **Ratification Controversies.** Several issues were raised by the process of ratification. First, the Soviet Union was accused (apparently correctly) of seeking to circumvent the treaty by transferring equipment from its land forces to its navy, a problem eventually solved by a compromise [*Atlantic News,* 19 June 1991, 1]. Second, and much more serious, the entire format for which the negotiations had been designed underwent a radical metamorphosis. Two of the original negotiating parties (the *FRG* and the *GDR*) merged into one; and one (in fact the main) negotiator, namely, the *Soviet Union*, dissolved nearly completely. Even though the successor states declared their intentions to abide by the treaty, it was far from clear what this meant. However, realizing the extreme fluidity of the situation, the other parties to the treaty moved to ratify with an unusual speed [Sharp 1992; 1993].

Table 12. CFE Treaty Tank Reductions

	November 1990	December 1992	Ceiling 1995	Reduction
		WTO		
Former USSR				
Armenia	258	77	220	38
Azerbaijan	391	278	220	171
Byelorussia	2,263	3,457	1,800	463
Georgia	850	75	220	630
Moldova	155	0	210	0
Russia	10,333	7,993	6,400	3,933
Ukraine	6,475	6,052	4,080	2,395
Total	20,725	17,932	13,150	7,630
NSWTO				
Bulgaria	2,145	2,209	1,475	670
Czech Republic	1,198	1,703	957	241
Slovak Republic	559	851	478	81
Hungary	1,345	1,331	835	510
Poland	2,850	2,807	1,730	1,120
Romania	2,851	2,960	1,375	1,476
Total	10,948	11,861	6,850	4,098
Total WTO	31,673	29,793	20,000	11,728
		NATO		
Europe				
Belgium	359	362	334	25
Canada	77	60	77	0
Denmark	419	499	353	66
France	1,343	1,335	1,306	37
Germany	7,000	6,733	4,166	2,834
Greece	1,879	2,276	1,735	144
Italy	1,246	1,276	1,348	0
Netherlands	913	813	743	170
Norway	205	205	170	35
Portugal	146	146	300	0
Spain	854	896	794	60
Turkey	2,823	3,234	2,795	28
U.K.	1,198	1,078	1,015	183
Total	18,462	18,913	15,136	3,582
U.S.	5,904	4,511	4,006	1,898
Total NATO	24,366	23,424	19,142	5,480
		ATTU		
Total	56,039	53,217	39,142	17,208

Source: Sharp 1993, 608–609.

Table 13. CFE Treaty Armored Combat Vehicle Reductions

	November 1990	December 1992	Ceiling 1995	Reduction
WTO				
Former USSR				
Armenia	641	189	220	421
Azerbaijan	1,285	338	220	1,065
Byelorussia	2,776	3,947	2,600	176
Georgia	1,054	49	220	834
Moldova	392	118	210	182
Russia	16,589	16,469	11,480	5,109
Ukraine	7,153	6,627	5,050	2,103
Total	29,890	27,737	20,000	9,890
NSWTO				
Bulgaria	2,204	2,232	2,000	204
Czech Republic	1,692	2,462	1,367	325
Slovak Republic	846	1,231	683	163
Hungary	1,720	1,731	1,700	20
Poland	3,377	2,396	2,416	961
Romania	3,102	3,143	2,100	1,002
Total	12,941	13,195	10,266	2,675
Total WTO	42,831	40,932	30,266	12,565
NATO				
Europe				
Belgium	1,381	1,267	1,099	282
Canada	277	72	277	0
Denmark	316	293	316	0
France	4,177	4,154	3,820	357
Germany	8,920	8,626	3,446	5,474
Greece	1,641	1,430	2,534	0
Italy	3,958	3,746	3,339	619
Netherlands	1,467	1,445	1,080	387
Norway	146	124	225	0
Portugal	244	280	430	0
Spain	1,256	1,057	1,588	0
Turkey	1,502	1,862	3,120	0
U.K.	3,193	3,003	3,176	17
Total	28,478	27,359	24,450	7,136
U.S.	5,747	4,800	5,372	375
Total NATO	34,225	32,159	29,822	7,511
ATTU				
Total	77,056	73,091	60,088	20,076

Source: Sharp 1993, 608–609.

Table 14. CFE Treaty Artillery Piece Reductions

	November 1990	December 1992	Ceiling 1995	Reduction
WTO				
Former USSR				
Armenia	357	160	285	72
Azerbaijan	463	294	285	178
Byelorussia	1,396	1,610	1,615	0
Georgia	363	24	285	78
Moldova	248	108	250	0
Russia	7,719	7,003	6,415	1,304
Ukraine	3,392	3,602	4,040	0
Total	13,938	12,801	13,175	1,632
NSWTO				
Bulgaria	2,116	2,085	1,750	366
Czech Republic	1,044	1,612	767	277
Slovak Republic	522	806	383	139
Hungary	1,047	1,037	840	207
Poland	2,300	2,309	1,610	690
Romania	3,789	3,928	1,475	2,314
Total	10,818	11,777	6,825	3,993
Total WTO	24,756	24,578	20,000	5,625
NATO				
Europe				
Belgium	376	378	320	56
Canada	38	32	38	0
Denmark	553	553	553	0
France	1,360	1,392	1,292	68
Germany	4,602	4,369	2,705	1,897
Greece	1,908	2,149	1,878	30
Italy	2,144	2,041	1,955	189
Netherlands	837	837	607	230
Norway	531	544	527	4
Portugal	343	354	450	0
Spain	1,373	1,219	1,310	63
Turkey	3,442	3,210	3,523	0
U.K.	636	502	636	0
Total	18,143	17,580	15,794	2,537
U.S.	2,601	1,773	2,492	109
Total NATO	20,744	19,353	18,286	2,646
ATTU				
Total	45,500	43,931	38,286	8,271

Source: Sharp 1993, 608–609.

Table 15. CFE Treaty Combat Aircraft Reductions

	November 1990	December 1992	Ceiling 1995	Reduction
WTO				
Former USSR				
Armenia	0	3	100	0
Azerbaijan	124	50	100	24
Byelorussia	243	389	260	0
Georgia	245	4	100	145
Moldova	0	29	50	0
Russia	4,161	4,387	3,450	711
Ukraine	1,431	1,650	1,090	341
Total	6,204	6,512	5,150	1,221
NSWTO				
Bulgaria	243	335	234	9
Czech Republic	232	231	230	2
Slovak Republic	116	116	115	1
Hungary	110	143	180	0
Poland	551	508	460	91
Romania	505	505	430	75
Total	1,757	1,838	1,649	178
Total WTO	7,961	8,350	6,799	1,399
NATO				
Europe				
Belgium	191	202	232	0
Canada	45	24	90	0
Denmark	106	106	106	0
France	699	688	800	0
Germany	1,018	946	900	118
Greece	469	458	650	0
Italy	577	542	650	0
Netherlands	196	175	230	0
Norway	90	88	100	0
Portugal	96	91	160	0
Spain	242	175	310	0
Turkey	511	355	750	0
U.K.	842	717	900	0
Total	5,082	4,567	5,878	118
U.S.	626	334	784	0
Total NATO	5,708	4,901	6,662	118
ATTU				
Total	13,669	13,251	13,461	1,517

Source: Sharp 1993, 608–609.

Table 16.   CFE Treaty Combat Helicopter Reductions

	November 1990	December 1992	Ceiling 1995	Reduction
		WTO		
Former USSR				
Armenia	7	13	50	0
Azerbaijan	24	6	50	0
Byelorussia	82	79	80	2
Georgia	48	3	50	0
Moldova	0	0	50	0
Russia	1,035	989	890	145
Ukraine	285	274	330	0
Total	1,481	1,364	1,500	147
NSWTO				
Bulgaria	44	44	67	0
Czech Republic	37	37	50	0
Slovak Republic	19	18	25	0
Hungary	39	39	108	0
Poland	29	30	130	0
Romania	13	15	120	0
Total	181	183	500	0
Total WTO	1,662	1,547	2,000	147
		NATO		
Europe				
Belgium	0	10	46	0
Canada	12	0	13	0
Denmark	3	12	12	0
France	418	376	352	66
Germany	258	250	306	0
Greece	0	1	18	0
Italy	168	177	142	26
Netherlands	90	90	69	21
Norway	0	0	0	0
Portugal	0	0	26	0
Spain	28	28	71	0
Turkey	5	11	43	0
U.K.	368	340	384	0
Total	1,350	1,295	1,482	113
U.S.	243	341	518	0
Total NATO	1,593	1,636	2,000	113
		ATTU		
Total	3,255	3,183	4,000	260

Source: Sharp 1993, 608–609.

Figure 1. CFE Treaty Tank Reductions

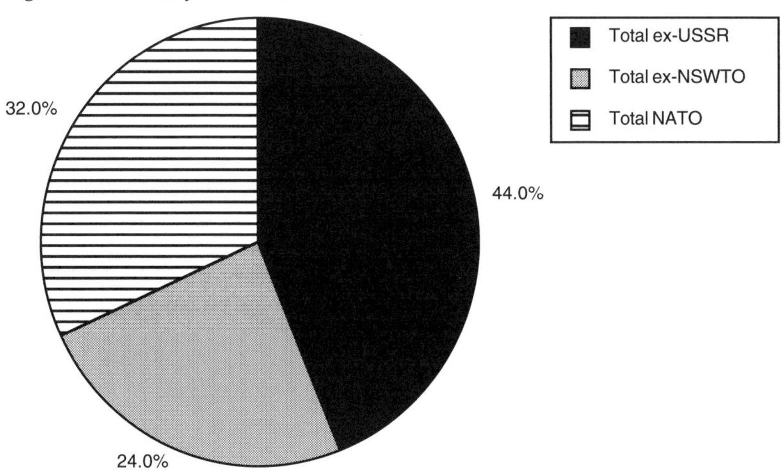

Figure 2. CFE Treaty Armored Combat Vehicle Reduct

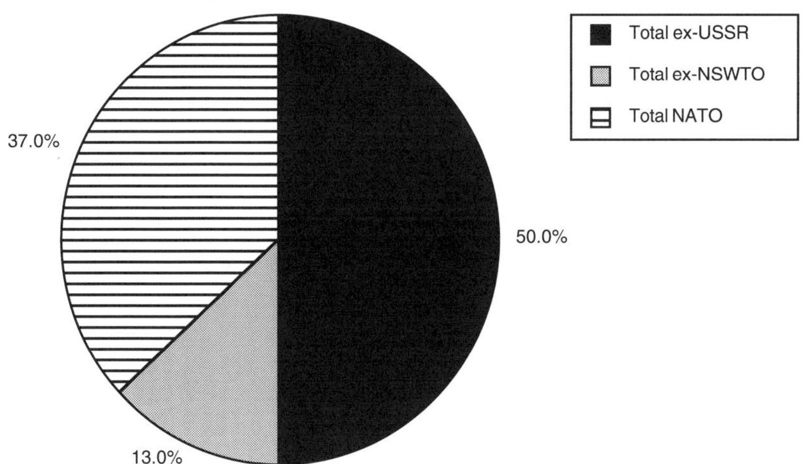

Figure 3. CFE Artillery Piece Reductions

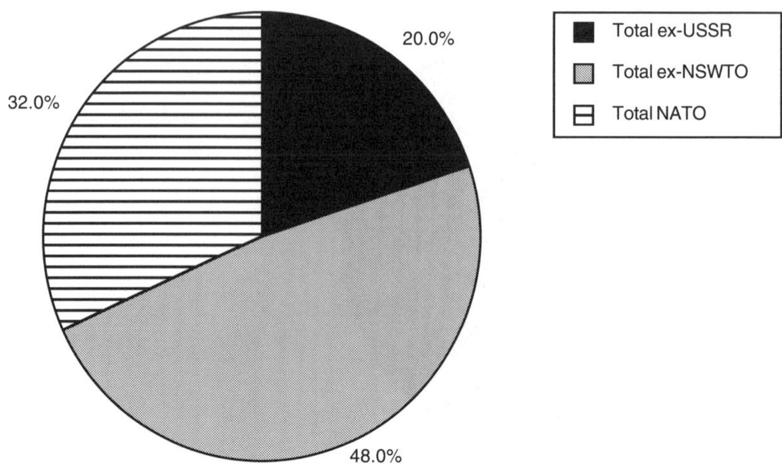

Figure 4. CFE Treaty Combat Aircraft Reductions

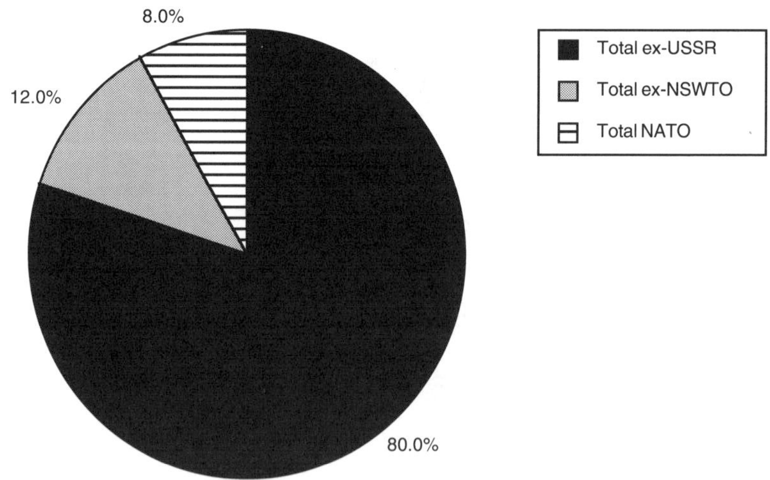

Figure 5. CFE Treaty Helicopter Reductions

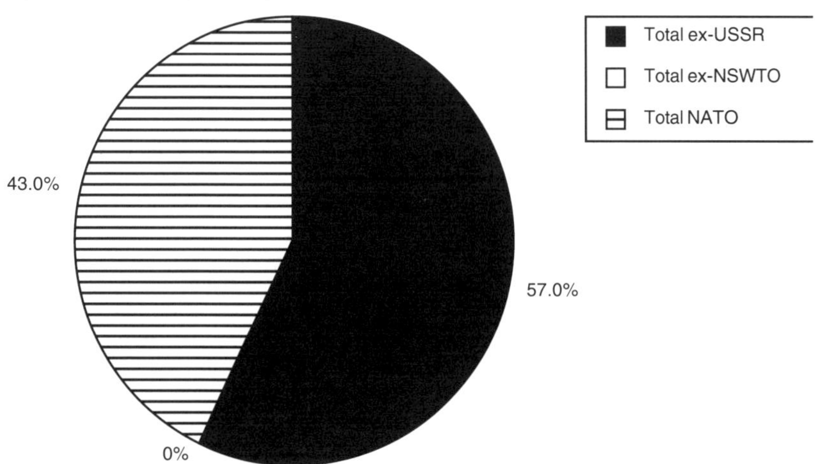

5. **Political Ramifications.** The CFE Treaty (CFE-1)was accompanied by a special declaration on Germany (see *Two-Plus-Four*) and declarations pertaining to land-based naval aviation and military personnel, as well as by a solemn joint declaration, wherein all twenty-two states proclaimed themselves to be adversaries no longer but to be willing to be friends. In this context they also committed themselves formally to what might best be described as an *NOD obligation: "They commit themselves only to maintain such military potentials as are necessary for war prevention and an effective defence. They intend to remain alert to the relationship between military potentials and doctrines" [*Bulletin,* 28 November 1990, 1425–1472; cf. ibid. 24 November 1990, 1422–1423].

6. **Follow-on Negotiations.** The CFE-1 Treaty was immediately followed by negotiations on the reduction of troops (CFE-1A), that were concluded in time for the CSCE follow-on meeting in Helsinki in 1992, albeit without achieving spectacular results (see Table 17) [Sharp 1993: 613–615]. Furthermore, under the auspices of the newly established Conflict Prevention Center in Vienna, a second seminar on military doctrines (*Vienna Seminar*) was conducted [Lachowski 1993].

7. **Evaluation.** It is possible to view the CFE-1 Treaty both as an astounding success and as a complete failure. On the one hand, a significant degree of actual *disarmament* was entailed by its provisions, which sets it apart from most previous *arms control* agreements, which have tended merely to establish rather generous ceilings. On the other hand, the CFE might well have simply codified what was bound to happen anyhow, namely, a drastic Soviet build-down of conventional forces that no longer made any sense after the sea change in security political conceptions [cf. MccGwire 1987; 1987a; 1988a; 1991].

CHALMERS, MALCOLM.   British economist, senior research fellow at the Department of Peace Studies, University of Bradford. MC was one of the first in the U.K. to advocate *NOD and has been a member of the *SAS for a number of years. His main objectives in this field have been to devise a defense posture that is nonnuclear, defensive, and affordable.

MC sought first to demonstrate that such a change was possible by analyzing official force comparisons critically; he found them to be based on flawed logic and questionable assumptions. On closer analysis (and taking qualitative differences into account), *NATO turned out not to be outspent or outmanned or outgunned by the *Warsaw Pact [1984: 219-221; idem & Unterseher 1987; 1988; 1989]. Second, MC proposed a scaling back of the *U.K.'s great-power ambitions (i.e., a lessened role for *out-of-area and none at all for the independent deterrent) in favor of a greater emphasis on the defense of the British Isles and the Central Front. Third, he proposed an alternative structure for the remaining forces, including the maritime components, which he suggested might consist largely of land-based aircraft, attack submarines, and mine barriers. Concerning the Central Front, he envisioned a layered structure, combining *forward defense with in-depth *territorial defense, and emphasizing antitank weapons, *barriers, and fortifications [1984: 224–228; 1985: 161–169; cf. idem & Stevenson 1989]. In a subsequent proposal (written in light of the new Europe) MC calculated the achievable savings to be in the range of 50–60 percent (from £21.2 to £9.4 billion in FY 2000/1), which would reduce the defense burden from the present 3.9 percent of GDP to 1.4 percent [Chalmers 1991].

MC also saw post–Cold War Europe as a challenge to innovative political thinking, pointing, inter alia, to the inadequacy of the traditional negotiated approach to *disarmament. As an alternative, he proposed "Reciprocal Unilateralism," i.e., a version of *gradualism [1989; cf. Osgood 1962; Etzioni 1962; George 1988b]. In 1990 he published a detailed proposal for a regional *collective security system: a European Security Organization. Even though it would not be primarily a military organization, MC nevertheless provided some detail about its military element, suggesting, e.g., the promotion of role specialization, both as a means of economizing and as "a means of ensuring that no single European country had the combined capabilities needed for successful unilateral action," i.e., as a contribution to reducing offensive capabilities (cf. *multinationality) [Chalmers 1990; 1990a; 1993; cf. Møller 1990; 1993a].

CHANGE THROUGH RAPPROCHEMENT.   (German: *Wandel durch Annäherung.*) Central term in the *SPD's conception of *Ostpolitik,* introduced by *Egon Bahr in 1963 [Bahr 1963].

The main idea was to abandon hopes (or perhaps rather proclaimed objectives) for bringing about a collapse of the communist systems in general, and that of the *GDR in particular, through isolation. Rather, by being woven into a web of collaborative arrangements, the communists would presumably be induced to reform the regime gradually, thus approaching Western-style *democracy. More immediately, improvements in human rights and living conditions in the GDR would be obtainable.

Since the demise of the SED regime and the *GDR (along with the other communist regimes in Eastern Europe), the SPD has been accused of having sought (obviously in vain) to stabilize communism as a corollary of CTR. SPD's attempted rebuttal has been that CTR allowed the East German opposition to operate more freely, thus paving the way for the 1989 revolution [cf. Oskar Lafontaine in *Der Spiegel,* no. 39 (1989): 21].

Table 17.   CFE-1A Reductions in Personnel

	Holding	Ceiling	Reduction
WTO			
Former USSR			
Armenia	7,101	not reported	n.a.
Azerbaijan	not reported	not reported	n.a.
Byelorussia	143,865	100,000	43,865
Georgia	not reported	40,000	n.a.
Moldova	not reported	not reported	n.a.
Russia	1,298,299	1,450,000	0
Ukraine	509,531	450,000	59,531
Total	1,958,796	2,040,000	103,396
NSWTO			
Bulgaria	99,404	104,000	0
Czech Republic	110,010	93,333	16,677
Slovak Republic	55,005	46,667	8,338
Hungary	76,226	100,000	0
Poland	273,050	234,000	39,050
Romania	244,807	230,000	14,807
Total	858,502	808,000	78,872
Total WTO	2,817,298	2,848,000	182,268
NATO			
Europe			
Belgium	76,088	70,000	6,088
Canada	4,077	10,600	0
Denmark	29,256	39,000	0
France	341,988	325,000	16,988
Germany	401,102	345,000	56,102
Greece	165,400	158,621	6,779
Italy	294,900	315,000	0
Netherlands	69,324	80,000	0
Norway	29,500	32,000	0
Portugal	39,700	75,000	0
Spain	177,078	300,000	0
Turkey	575,045	530,000	45,045
U.K.	288,626	260,000	28,626
Total	2,492,084	2,540,221	159,628
U.S.	175,070	250,000	0
Total NATO	2,667,154	2,790,221	159,628
ATTU			
Total	5,484,452	5,638,221	341,896

Source: Sharp 1983.

CHEESEMAN, GRAEME.   See *Australia.*

CHEMICAL WEAPONS.   Weapons that incapacitate or kill through their toxic effects, occasionally subdivided into choking, blistering, blood, and nerve agents [SIPRI 1973: 13; Bailey 1991b: 50-57]. For some analytical purposes, the category might be expanded to include riot control weapons such as tear gas or defoliants.

CW were first used extensively in trench warfare in *WWI* [Brauch 1982a: 62–75] but have not been used on a grand scale since. Even though both sides possessed ample stocks of CW, they were thus not used during *WWII* [Moon 1984; McFarland 1986], and they have been used only sporadically in other wars since 1945. Most allegations of use of CW [traced continuously in *SIPRI* Yearbooks] (e.g., against the *Soviet Union* for using them in *Afghanistan*) have remained unsubstantiated—the Iraqi use against Iran is an exception [Findlay (ed.) 1991: 139-143].

One inhibition against use may have been that of the *Laws of War*, several of which prohibit the use of CW: the 1899 and 1907 Hague regulations prohibited "poison and poisoned weapons" and "the diffusion of asphyxiating or deleterious gases"; the 1925 Geneva Protocol imposed a total ban on the use (but not the production or possession) of CW; and the 1972 Biological Weapons Convention forebade toxin weapons (which are actually CW) [De Lupis 1987: 211-226; SIPRI 1973]. However, the effect of these regulations was limited by their nonuniversal ratification (the *U.S.* ratified the 1925 protocol only in 1975); by the numerous reservations accompanying the signatures; by the lack of constraints on manufacturing and possession of CW; and by the obstacles to effective verification. Most of these problems have been removed by the 1993 *Chemical Weapons Convention.*

CW have almost universally been condemned as especially obnoxious, even though arguments have been made to the effect that they would tend to benefit the defender by virtue of their countermobility effects [Krause & Mallory 1992]. French General *Etienne Copel* envisioned the use of CW, even on long-range missiles, in the context of his otherwise *NOD*-like alternative defense scheme [1984: 158, 172]. With this exception, the entire alternative defense community has rejected any reliance on CWs, for war-fighting as well as for deterrence purposes.

CW possession has usually been justified because of its deterrent effect by *NATO* and the U.S., among others. In the early 1980s the latter launched plans for a new generation of CW, namely, binary weapons [Robinson 1982]. Consisting of two substances that were individually harmless and lethal only when mixed, binaries would presumably be less dangerous to civilians and others, hence more useful for war-fighting purposes. The U.S. Congress consented to the plan only on the condition of NATO endorsement, which required a withdrawal of existing stocks [Robinson 1986]. Besides NATO and the former Soviet Union, several other countries were in possession of CW. Moreover, by the late 1980s/early 1990s there seemed to be a trend toward a further proliferation that might well result in more widespread use, especially when seen in connection with ballistic missile proliferation [Findlay (ed.) 1991; Bailey 1991b: 50-95]. Hence the attempts at containing proliferation, inter alia through the "Australia Group," by intending to control the export of CW precursors [Robinson 1992]. The main accomplishment of these endeavors was the Chemical Weapons Convention of 1993 [Robinson & al. 1993].

CHEMICAL WEAPONS CONVENTIONS.   Convention on the Prohibition of the Development, Production, Stockpiling and Use of Chemical Weapons and Their Destruction [in SIPRI *Yearbook* 1993: 735–756], signed 13 January 1993 after protracted negotiations in the Conference on Disarmament, and scheduled to enter into force in 1995.

The CWC represented the culmination of consistent *arms control* endeavors over the past century to ban *chemical weapons* in recognition of their particularly obnoxious character. It thus supplemented the 1899 and 1907 Hague Conventions and the 1925 Geneva Protocol, which had prohibited the use of chemical weapons but left their possession unconstrained [De Lupis 1987: 211–226; SIPRI 1973]. The CWC outlaws chemical weapons altogether, and obliges present possessors to destroy their stocks by 2010, at the latest [Robinson & al. 1993].

CHEMICAL-WEAPONS-FREE ZONES.   *Zones*, usually along the border between two adversaries, in which no *chemical weapons* would be allowed [Trapp (ed.) 1987]. An example was the draft treaty to this effect that was signed by the West German *SPD* (in opposition) and the East German ruling party, the SED, in 1985; it was the first tangible result of the dialogue process between the two parties [SPD & SED 1985; cf. Voigt 1985; Lohs 1986].

CHECKERBOARD DEFENSE.   Term used by *Steven L. Canby* for his proposed weblike defense of Central Europe [1980].

CHICKEN.   Classical example of a *zero-sum game*, described, e.g., by Herman Kahn and formalized in *game theory*. Two youngsters are trying to demonstrate their courage by driving cars head-on toward each other in order to impress bystanders. Whoever turns first (i.e., "chickens out") loses (value –1); the one who continues wins (value +1). However, if neither turns, they collide and both probably die (value -10) (see Table 18).

Table 18. Chicken

		B	
A's Payoff, B's Payoff		Turn	Continue
A	Turn	–1, –1	–1, +1
	Continue	+1, –1	–10, –10

C has been regarded as a fair description of *nuclear deterrence* strategies in crisis situations, where backing down (i.e., turning) has been seen as weakening deterrence by revealing a lack of resolve. Because the threat to commit suicide (personally or as a nation) possesses limited credibility, a skillful strategy in playing C (according to Kahn) might be to relinquish control, the counterpart of which in deterrence strategy has been the notion of "threats that leave something to chance" (*Thomas Schelling*), in its turn part of the rationale for NATO's *flexible response* strategy [Rapoport 1989: 298–300, 317–318; Brams & Kilgour 1988: 38–59; cf. Kahn 1965: 11; Schelling 1960: 187–205].

CHINA, PEOPLE'S REPUBLIC OF.   Of all the world's countries, C probably has the longest history of *defensive defense*. In sharp contrast to the neighboring Mongol Empire under Ghengis Khan (1155–1227) and Tamarlane (1336–1405) [Bellamy 1990: 193–201], C has traditionally emphasized such defensive means of national defense as *infantry* and fortifications (e.g., the Great Wall).
     C has also fostered two of the most prominent precursors of modern *NOD* strategic thinking: *Sun Tzu* and *Mao Zedong*, who laid the foundations for the

indirect approach, protracted war, and the main principles of *guerilla strategy*. Actually, such guerilla warfare was waged long before Mao, e.g., during the *Taiping Rebellion* [Bellamy 1990: 218–228; Hsü 1975: 277–321].

Since the 1949 revolution, the defense strategy and posture of C has constituted a mixture of traditional and modern elements, in continuous conflict:

- On one side, there has been a strong tradition (e.g., during the Cultural Revolution, 1966–1969) of "people's war" [cf. Lin Piao 1965], emphasizing *militias* and depreciating modern weapons technologies [Dreyer 1982; Jencks 1984; Joffe & Segal 1985]: a largely defensive tradition.
- On the other side, there has been a desire for modernization, particularly strong after the death of Mao and the ascendency of Deng Xiaoping; this has resulted, e.g., in the attainment of nuclear status in 1964 [Segal 1991a], and in an emphasis on the acquisition of modern weapons technology, even at the expense of numbers of weapons [Segal 1985; Joffe 1987; Dellios 1989: 40–82; Segal & Tow (eds.) 1984], a considerably more offensive tradition.

These two trends in Chinese defense thinking seem likely to continue their competition in the years ahead, making it impossible to predict the future direction of Chinese strategy. Especially worrisome are the prospects of a continued Chinese military buildup that might, in due course, lead to *Japan's* abandonment of its self-imposed constraints, whence might ensue a full-blown regional arms race, perhaps also involving Vietnam, the two *Koreas*, and Russia, that would have implications for the entire Asia-Pacific region [cf. Wiseman 1994].

CHRISTIAN DEMOCRATS IN FAVOR OF STEPS TOWARD DISARMAMENT.   (CDSA: Christliche Demokraten für Schritte zur Abrüstung). Small group of (mostly local) *CDU* and CSU leaders who in 1986 published an anthology advocating notions usually associated with the *SPD*, including *common security* and *NOD* [Gohl & Niesporek (eds.) 1986]. The group included the television personality Franz Alt [1986; cf. 1983]; chairman of the party organization in North Rhine-Westphalia, Kurt Biedenkopf [1986]; and other, lesser-known, politicians.

Some of the group members explicitly advocated a shift to NOD, i.e., a defense that would serve only to dissuade aggression by exacting a high admission and staying price, and that would be entirely nonthreatening. As steps toward the political implementation of such an alternative defense, they proposed consultations between the two alliances on military doctrines, and/or a gradualist strategy of withdrawing particular offensive-capable weapons systems from the border areas, along with the creation of nuclear- and chemical-weapons-free *zones* [Deittert 1986; Niesporek 1986; Pfeiler 1986; Gohl 1986; Jaeschke 1986; Krüger 1986].

CHRISTIE, JOHAN K.   See *Norway*.

CIRPES.   Centre Interdisciplinaire de Recherches sur la Paix et d'Études Stratégiques: (Interdisciplinary Center for Peace Research and Strategic Studies, Paris), under the directorship of *Alain Joxe*.

C has been one of the very few research centers in *France* taking an interest in alternative defense. Besides the works of Joxe [1990; 1991], the studies by C have included analyses of the German *NOD* models [Carton 1984], and the defense structures and strategies of *Switzerland* [Cramer 1984] and *Yugoslavia* [Lukic 1984]. All the studies have been based on a common conception of "infranuclear dissuasion," which is almost synonymous with war prevention in what *Anders Boserup* called a *subnuclear setting*, or under conditions of what is known as *existential deterrence* [Joxe 1984; cf. Boserup 1981; 1986a; Bundy 1986]. Some inspiration from *André Beaufre* has also been discernible [Beaufre 1963; 1965; 1974].

CIS. Commonwealth of Independent States. Partial successor to the *Soviet Union*, founded at a summit in Minsk in December 1991 by Russia, Ukraine, and Belarus. By 1994, it had come to include all the former union republics with the exception of the *Baltic States*.

1. **Problems of Succession.** An orderly succession was hampered by the perceived need for making a clean break with the Soviet past, in particular for avoiding any semblance of supranational sovereignty. Nevertheless, the unprecedented dissolution of a superpower into its constituent parts raised several issues, some of which were of serious concern for the West [Bonesteel 1992; 1993]:

- Which successor state, if any, would assume the international rights and obligations of the USSR, such as the permanent membership in the *U.N.* Security Council, or the servicing of the national debt?
- To what extent were the successor states bound by treaties signed by the Soviet Union (such as the not-yet-ratified *START* and *CFE* treaties), and what were the concrete implications thereof, say in terms of numerical entitlements [Dunay 1993; Sharp 1993]?
- Which states were to assume control over Soviet armed forces in general, and nuclear weapons in particular?

Where no unanimous decisions could be reached at CIS summits, Russia in a number of cases assumed responsibility, almost by default, yet by so doing may have caused some concerns on the part of the other CIS states, including the strongest of them, Ukraine [Stenseth 1992; Zagorski 1993; Alexandrova 1992; 1994; Mihalisko 1993; Zlenko 1992]. The standing disputes between ideas about the proper scope and pace of cooperation were almost without exception resolved according to the principle of the lowest common denominator, making the CIS a very loose association, a far cry from a true confederation or federation.

2. **The Nuclear Issue.** Immediately after the founding of the CIS, it appeared that the vexing question of control over the nuclear weapons inherited from the USSR (deployed in Russia, Ukraine, Belarus, and Kazakhstan) would be resolved by placing them under CIS command. The three smaller states issued statements to the effect that they intended to dismantle their holdings of nonstrategic nuclear forces, which would imply transferring them to Russia for destruction, and declared their intentions to become *nuclear-weapons-free zones*. However, *security-dilemma* considerations seem to have played a role, in that all states seemingly preferred to maintain the weapons rather than relinquish control to any of the others. Tactical nuclear weapons were thus transferred to Russia without much ado; the transfer of strategic weapons proved much more problematic, especially in Ukraine, where a protracted dispute between Parliament and the president raged over whether—and if so on what terms—to transfer. Because the West, and especially the *U.S.* much preferred to have only one nuclear counterpart [Miller 1993a; for a dissenting view, see Mearsheimer 1993], they played a very active role in the negotiations, inter alia, by offering compensation to Ukraine, couched in terms of assistance for dismantlement [Batiouk 1992; Babst & Schaller 1993; "R." 1993; Umbach 1994].

3. **Conventional Forces.** All states embarked upon the task of raising indigenous conventional forces, in most cases by assuming control over former Soviet forces deployed on their territory. Russia was the last to decide (officially, at least) to raise its own army, apparently clinging to the hope for unified CIS forces (likely

to be dominated by Russia) [Holden 1993; Holoboff 1993; Konovalov 1993; Sorokin 1993]. Paradoxically, it was forced to do so by the need for a redistribution of CFE quotas—of which Russia was, of course, entitled to a large share. This presupposed national armed forces, however. By early 1994, all successor states had established indigenous armed forces, above all the Russian Federation and Ukraine, with target sizes of 1.5 million and 450,000 troops, respectively. The breakdown of Soviet *CFE* limits agreed upon in late May 1992 further served to fix the relative strength between these two states: of the 13,150 tanks permitted the Soviet Union, Russia was allowed 6,400 and Ukraine 4,080; ACVs, 11,480 and 5,050; artillery, 6,415 and 4,040; combat aircraft, 3,450 and 1,090; and helicopters, 890 and 330 [*Jane's Defence Weekly,* 6 June 1992, 966; Sharp 1993: 592–606].

A special problem arose with regard to the occupation forces in the former Baltic Republics, to a withdrawal of which the USSR and subsequently Russia had committed themselves, but which became effectively "an army without a state in a state without an army." The problem was exacerbated by difficulties with rehousing Russian soldiers, and by the occasional ill treatment of the troops as well as other Russian nationals in the Baltic states [*Jane's Defence Weekly,* 30 May 1992, 952; cf. Russian Defense Minister Grachev, quoted in *Atlantic News,* 20 May 1992, 4]. By early 1994, withdrawal from Lithuania had been completed, but that from Estonia and Latvia had been postponed, notwithstanding numerous demands from these states, supported by the West.

4. **Military Doctrine and Security Policy.** Immediately after the founding of the CIS, the constituent states also issued more or less wholehearted declarations to the effect that the strategy and configuration of their armed forces would be strictly defensive. The same held true for the joint armed forces of the CIS, whose commander in chief, Yevgeny Shaposhnikov, on more than one occasion commented to this effect [*Military News Bulletin,* 1 (January 1992); *Jane's Defence Weekly,* 15 February1992, 215; ibid., 11 April 1992, 597; *Atlantic News,* 18 March 1992, 3; ibid., 25 March 1992, 2–3; Fitzgerald 1992; Ries 1992; Romer 1992; Shevtsova 1992; Stepashin 1992; Zlenko 1992]. To the extent that they devoted their attention to security and defense policy at all, most CIS states had thus confirmed their support for the defensive doctrine adopted by the USSR in the late 1980s and expressed satisfaction with most elements of *New Thinking* (in fact, both the *IMEMO* and the *ISKAN* were taken over by Russia) [Danilovich 1992; Holoboff 1993]. All CIS members, and the Russian Federation in particular, were apparently eager to erase whatever might remain of the image of the Soviet threat, and they also embarked upon a confidence-building venture [Paznyak 1993] that included a Russian declaration of intent to apply for *NATO* membership; a retargeting of strategic nuclear weapons in Russian possession away from the U.S.; and numerous declarations of friendship with the West, which usually reciprocated, albeit not always in a manner as forthcoming as the CIS states might have hoped with regard to the accompanying requests for assistance. The *CSCE* member states, furthermore, in January 1992 invited the newly independent states to become members of the CSCE [*International Herald Tribune,* 1 February 1992], and NATO did likewise in the form of *NACC.* The question of the actual, as opposed to declared, military guidelines, however, remained unresolved until late 1993 [Dick 1992; Gareev 1992; Grachev 1992], when both Russia and Ukraine made public their new military doctrines, neither of which retained much of the former NOD-type commitments.

Russia. The new military doctrine was approved 2 November 1993 [for background, see Konovalov 1993; Raevsky 1993; Stepashin 1992; Trenin 1994; Ivlev 1994]. The document states, inter alia, that Russia "will not employ armed forces or other troops against any state other than for individual or collective self-defense, if an armed attack is made on the Russian Federation, its citizens, territory, armed forces, other troops or its allies." Further, a reduction of nuclear forces is envisioned "to the minimum level that would guarantee the prevention of large-scale war and the maintenance of strategic stability," holding out the prospect of an eventual abolition. An element of ambiguity was introduced by the definition of purposes for which military power might be employed: "sovereignty, territorial integrity and other vitally important interests." The last might be infringed in case of "the suppression of the rights, freedoms and legitimate interest of citizens of the Russian Federation in foreign states," which was regarded by many as tantamount to Russia's reserving for itself a *droit de regard* in the numerous other Soviet successor states hosting sizable Russian minorities. The completion through 1996 of the withdrawal of Russian forces stationed elsewhere was envisioned, but the option was left open of a swift redeployment. Hence the emphasis on mobility and of a growing share of professionals. The document underlined that forces should be trained for both defensive and offensive military operations, with a view to seizing the initiative and defeating the aggressor [*Summary of World Broadcasts, BBC Monitoring,* 29 November 1993, S/1-10].

The various unclear points were explained by Russian Defense Minister Pavel Grachev, who declared training for both defensive and offensive operations to be indispensable: "The Russian military doctrine avoids absolutizing any one category and method of waging combat operations." A partial justification was that the Russian armed forces had duties beyond that of deterring and repulsing aggression, such as **peacekeeping* operations, either under the auspices of the U.N. or of the CIS. Besides thus abandoning the commitment to NOD, Russia had also abrogated its no-first-use commitment. According to Grachev, Russian nuclear weapons might be used against a nonnuclear state in two cases: in the event of "an armed attack on the Russian Federation, its territory, armed forces or allies by such a state linked by alliance agreements with a nuclear state, and in the event of joint actions by such a state with a state that does possess nuclear weapons in carrying out or backing an incursion or military attack on the Russian Federation" [*Summary of World Broadcasts, BBC Monitoring,* 5 November 1993, 1–3]. One of the earliest Soviet proponents of NOD among the **instituchiki,* Andrey Kokoshin, now first deputy defense minister, attempted to downplay the change by stating (correctly) that the new guidelines for nuclear use were quite similar to those of the other nuclear powers [*Summary of World Broadcasts, BBC Monitoring,* 5 November 1993, 1–3].

Ukraine, having in 1992 announced its determination to opt for a strictly defensive posture and strategy, without any reliance on nuclear weapons [*Jane's Defence Weekly,* 15 February 1992, 215; cf. Honcharenko & Lisitsyn 1993; Izmalkov 1993; Pyskir 1993; Sauerwein 1993], in October 1993 finalized its new military doctrine. Its level of ambition was more modest than that of Russia, defining as the strategic task of the armed forces merely the defense of sovereignty, political independence, and territorial integrity. Moreover, Ukraine declared itself nonaligned, implying that it regarded no specific state as a potential enemy. Finally, it renouced any territorial claims against other states, while denying the validity of such claims against itself.

Militarily, Ukraine declared any use of nuclear weapons to be inadmissible and pledged its support for universal nuclear disarmament and for the establishment of "zones free of weapons of mass destruction." In the event of a war, its objectives were stipulated as "repelling armed aggression; inflicting damage on the aggressor; denying him an opportunity to continue war; stopping military operations on terms favorable to Ukraine." Hence the need for the armed forces to master "defensive, counteroffensive and offensive kinds of operations," allowing them to "seize the initiative from the enemy and keep it" [*Summary of World Broadcasts, BBC Monitoring,* 25 November 1993, 4–6].

5. **Unresolved Issues.** By early 1994 the situation seemed extremely unstable: large parts of the CIS were torn by internal conflicts and civil war; the economic situation was desperate; and the political situation was volatile, to say the least [Kennedy-Pipe 1993]. Even though the standoff between Parliament and president in Russia had been resolved in the latter's favor, the subsequent elections demonstrated around 25 percent electoral support for the imperialist policy of the Liberal Democratic Party of Vladimir Zhirinovsky, which espoused territorial claims against Russia's neighbors, above all *Poland. An especially serious element in this may be the factor that civil-military relations remain unresolved in most successor states, including Russia, and that the overwhelming majority of the officer corps supported Zhirinovsky [Arnett 1994; Heinemann-Grueder 1994; cf. for the background Erickson 1993; Tsypkin 1992; Nichols 1993].

CITIZEN IN UNIFORM.   Expression (*citoyen soldat*) dating back to the French Revolution (1789–1794) [Lynn 1984]. In the postwar period, CIU has primarily been a (West) German conception for the integration of the armed forces in society [Abenheim 1990; cf. Militärgeschichtliches Forschungsamt 1975: 52–57, 92–100, 104–127; Baudissin 1986; Bald & al. 1981].

Adoption of the CIU concept (fostered primarily by Count von Baudissin) by the renascent Bundeswehr in the 1950s was, inter alia, motivated by the desire to prevent a repetition of the situation before *WWI, when the status of the officer corps was close to that of a state within the state, and of the interwar years, when the *Reichswehr* also escaped control by the parliamentarians of the Weimar Republic, thus paving the way for Hitler [cf. J. Snyder 1984; 1985; Posen 1984b: 179–219; Møller 1991: 40–43].

The intentions behind CIU were to make the armed forces an integral part of society, enjoying the constitutional rights of free speech, etc. with as few limitations as possible. This facilitated the activities of numerous military "defectors" (see also *Military Profession), i.e., military officers who publicly criticized the defense policy of the *FRG. Examples include *Eckardt Afheldt, *Jochen Löser, *Johannes Gerber, *Franz Uhle-Wettler, *Günter Kiessling, and *Wilhelm Nolte. However, the experiences of *Bogislav von Bonin and *Elmar Schmähling demonstrated that there were limits to the freedom to disagree, even within the Bundeswehr [cf. Møller 1991: 135–155].

CIVIL DEFENSE.   The most direct form of *damage limitation, namely, a direct protection of the civilian population from the destruction ensuing from war, e.g., in the form of evacuation, shelters, protective clothing, and the like.

CD is, in itself, strictly defensive, yet it may nevertheless be a cause for concern on the part of adversarial states if combined with armed forces capable of offensive operations:

- The availability of adequate CD may free armed forces otherwise assigned to defensive tasks for offensive operations.
- A CD system that would be adequate (say, in combination with *strategic defenses*) for protection against nuclear attack would tend to negate the war-preventing effects of mutual *nuclear deterrence. Hence, during the Cold War, U.S. officials expressed concern about the presumed Soviet lead in CD [cf. R. Scheer 1982: 104–119 & passim; Holloway 1984: 52–53, 62–64; Cockburn 1984: 375–394; Freedman 1986: 196–197].

Combined with strictly defensive armed forces, i.e., with an *NOD posture, CD would be unambiguously defensive. Countries with largely defensive postures (e.g., *Austria and *Switzerland) have in fact tended to place great emphasis on CD, just as have most NOD proponents [D. Fischer 1982; cf. e.g., Löser 1981; Nolte & Nolte 1984; H. Afheldt 1988]. An exception is the trend (also within the NOD community) toward accepting at face value the *Totally Destructive Conventional War* (hypothesis) according to which no significant damage limitation is achievable, even in a conventional war [for a critique see Møller 1989c; 1989d].

CIVILIAN-BASED DEFENSE.   Form of national defense based on *nonviolence* as a functional substitute for military force [cf. Møller 1991a].

1. **Different Approaches.** There are at least five (by no means mutually exclusive) approaches to CBD [cf. Sternstein 1977; Muller 1984: 217–218; Geeraerts 1977; 1978; Rajewsky & Riesenberger (eds.) 1987]:

- A pacifist approach, which categorically denies the right to kill, even in *self-defense*. It is often, but not necessarily, based on religion, e.g., in the cases of Christian (cf. *Sermon on the Mount*, *Quakers*, *Tolstoy*), Hindu (cf. *Gandhi*), and Buddhist sects.
- A structural approach, which assumes that violence corrupts, either in the sense that armed forces corrupt society or that violence corrupts people as individuals, or both (cf. *nonviolence*) [Stinnes 1977; Lifton & Falk 1982; Krippendorf 1984; Kaldor 1987].
- A pragmatic approach, which recommends nonviolent forms of struggle for a variety of purposes, ranging from national defense to revolutionary struggle, from expedience, i.e., because they are deemed the most effective. *Gene Sharp* thus listed 198 nonviolent tactics for a variety of purposes [1973: 117-433; cf. for a critique: Wiberg 1977].
- A strategic approach, which takes as its point of departure scenarios based on assumptions about enemy objectives and means and devises nonviolent methods for countering these.
- A political approach, which according to CBD is merely a step in a more profound transformation of the international system. A state might, e.g., further global *disarmament* through unilateral disarmament and "*transarmament*" to CBD (e.g., by means of *gradualism*).

2. **History.** Interest in CBD and pacifism has tended to coincide with wars, i.e., to rise in the aftermath of major wars (according to the never-again logic) or during periods of intense preparations for war. Hence, both trends thrived in the interwar years as well as in the 1950s [Riesenberger 1987; Rajewsky 1987].

The interest in CBD (and particularly the structural approach) during the 1960s and early 1970s, however, was partly spurred by other factors. It was associated with the new social movements: the students' revolt, the women's liberation movement, the solidarity groups (especially with Vietnam), the ecological groups, and the *Green movement* and parties [Ebert 1978; 1978a; Grünen 1981; 1985; cf. Galtung 1986; Wasmuht 1987; Rosolowsky 1987]. Neo-Marxism also seemed to point toward CBD, e.g., as a way of overcoming the societal division of labor with its implications in terms of alienation. Hence the term "*social defense*," which should be "social" both in the sense of constituting a defense *of* society as a whole and of being a defense *by* the whole society [Galtung 1964: 333–334; Ebert 1970a; Jahn 1971]. CBD has also been depicted more instrumentally as the appropriate strategy for the proletarian revolution, or (more appropriately) for democratic revolutions, due to its intrinsically democratic character [Ebert 1968; 1972; cf. Muller 1984: 218–223; Mellon & al. 1985: 45–53].

3. **CBD as a Strategy.** CBM has occasionally been conceived of as a strategy of national defense [Boserup & Mack 1974: 148–182; Muller 1984: 212–215]. The

underlying assumption has been the thesis of *Clausewitz* that the end (*Zweck*) of war is to break the will of the adversary, thus, e.g., allowing the defender to define the center of gravity of the struggle, as well as to strike at the aggressor's will directly. Beneath this will would lie a complex political structure, consisting of rulers, opposition, public opinion, etc., parts of which might be induced to exert pressure on the immediate decisionmakers, with a view toward making them abandon their aggression or, in the last resort, to bringing about a political change in the opponent's political system [Sharp 1973: 665–678; Jahn 1984: 85].

The strategic case for CBD has also been argued with reference to the inevitable reliance of an occupying power on the collaboration of the occupied country's population and administrative apparatus. Hence the proposals for a differentiated approach to forces of occupation: all collaboration should be denied to the occupants *qua* soldiers, while fraternization with them *qua* human beings should be encouraged, which would presumably weaken the morale (and thereby the will) of the aggressor [Galtung 1971: 89; cf. 1965; King-Hall 1959: 27–30]. Concerning the defender's administrative personnel, "continued work without collaboration" has been recommended, also with a view to denying the occupant the spoils of conquest, thus, it is hoped, also contributing to *deterrence by denial* [Roberts 1967a: 243–245; Sharp 1973: 421–423; 1985: 90–100; Ebert 1968b: 80–82, 95–100; 1969: 103–105; 1970a: 148–152; 156–160].

CBD has been described by other proponents as an instance of the *indirect approach* and likened to political *jujitsu,* implying that enemy plans should be thwarted by asymmetrical means, and thus resemble *guerilla strategy* with its reliance on the powerful *nonbattle* [Sharp 1973: 109–113, 495–496; Ebert 1968: 16–17; Boserup & Mack 1974: 72–73; cf. Hart 1967a; 1967].

CBD seems to lack a satisfactory answer to the question how to thwart (or better: dissuade) an invasion intended merely to establish a military bridgehead. A few CBD advocates have acknowledged this as a problem [Galtung 1968: 384; Roberts 1967a: 226–227; Ebert 1980a: 188–189; Mellon & al. 1985: 34–35; cf. Boserup & Mack 1974: 41–42, 135], without, however, being able to provide a wholly satisfactory answer, apart from recommendations to blockade runways, etc. in a nonviolent fashion.

4. **Historical Experience.** CBD proponents have often referred to the (allegedly positive) historical experience with nonviolent struggles. In this connection, they have tended to focus on a handful of instances in three categories:

- Liberation struggles: *Gandhi*'s campaigns in India, *Martin Luther King*'s civil rights movement, the 1953 East German rebellion, and the Polish workers' struggle in 1956, 1970, 1976, and 1980–1981, as well as the struggle of Solidarnosc since 1981 [Sharp 1973: 82–86, 95–96; 1979; Hornung 1970; Grosse 1987; Ebert 1967c; Schmid 1985: 255–272; Rania 1978; Acheson 1981; Dross 1980; MacShane 1981].
- Resistance to internal coups: the Kapp putsch in Germany and the Algerian generals' revolt in 1961 [Boserup & Mack 1974: 120–124].
- Resistance to foreign occupation: the 1923 Ruhr campaign; the Danish, Norwegian, and Dutch resistance against German occupation (1940–1945); and the resistance to the invasion of Czechoslovakia in 1968 [Sternstein 1967; Sharp 1973: 81–82, 98–100; Boserup & Mack 1974: 92–116; Bennet 1967; Bergfeldt 1986; Poch 1970; Mellon & al. 1985: 96–121; Skodvin 1967; Ebert 1968b; 1969b; 1970; 1970a: 149–152; Klumper 1983; Roberts 1970; Schmid 1985: 335–366; Muller 1984: 283–294].

In all of these cases, CBD proponents have assessed the resistance as fairly successful. However, the selection of case studies has been extremely biased, and even in the selected cases the fact has been disregarded that CBD served merely as a supplement to military defense. Finally the meaning of "success" has to be stretched considerably in order for all the cases studied to be assessed as successes. Added to the problems of validity is one of relevance. Many of these cases concern internal challenges, and it is far from self-evident that the same means of struggle would be suitable for the defense against foreign occupants. Finally, many of the lessons (if so they are) may be outdated because they pertained to the prenuclear era.

5. **Transition to CBD.** Most CBD advocates have, realistically, anticipated a transitional period during which military forces would gradually be replaced with nonmilitary means of defense [Sharp 1985: 70]. The total defense posture would thus at any given moment consist of a mixture of military and CBD means or options, raising the question of how best to mix. Certain authors have warned against the dangers involved with mixing yet not explained how this might be avoided [Ebert 1970a: 183–187; 1974a: 84–87; cf. Niezing 1987: 24–36; Muller 1984: 296–297].

Some authors have advocated a temporal division of labor, recommending the use of CBD as a fallback defense option [Klumper 1983; cf. Schmid 1985: 375–382; ADC 1983: 228–230; Ebert 1969a: 41; Sharp 1985: 61–63; cf. Roberts 1967a]. Others have preferred a spatial separation of the two defense modes: armed forces defending the countryside and CBD being used in conurbations declared *open towns* (cf. *Wilhelm Nolte*) [Nolte & Nolte 1984]. Still others have simply underlined that CBD would be a useful "special option for special circumstances" [Roberts 1978; cf. Mellon & al. 1985: 184–187; Wiberg 1977: 117–119].

6. **CBD and NOD.** CBD proponents seem to have become gradually more favorably inclined toward combinations of CBD and *NOD* [ADC 1983: 227; Nolte & Nolte 1984; Fischer & al. 1987; Salmon 1988]. The two conceptions do indeed have much in common: both are inherently nonoffensive; both reject the use of nuclear weapons for military purposes; both are asymmetrical approaches to strategy; both attach major importance to the *time factor* and to the ability to wage a protracted struggle; and both seek to deprive the opponent of worthwhile targets.

There are also numerous differences and perhaps incompatibilities. NOD and CBD tend to take different scenarios as their point of departure and thus to emphasize different types of capabilities: CBD's paradigm case has been a 1939-type invasion, and CBD proponents have therefore emphasized deterrence by denial; NOD's paradigm case has been an inadvertent 1914-type war, and NOD advocates have thus emphasized *crisis stability*. Furthermore, CBD tends to be considered as a long-term perspective; NOD has normally been intended for almost immediate implementation. Finally, the techniques are, of course, entirely different: CBD rejects the use of weapons; NOD bases itself on military means.

CIVILIAN RESISTANCE.   See *Civilian-Based Defense, *Nonviolence*.

CIVIL WAR, U.S.   See *U.S.*

CLARKE, MICHAEL.   See *U.K.*

CLAUSEWITZ, CARL VON.   German classical military strategist (1780–1831) and author of the posthumously published (1832) *On War* [1980/1984; cf. Gat 1988; 1989: 156–250; Aron 1983].

CvC's theories of wars of annihilation and of boundless *escalation* have earned him a reputation as an apostle of absolute and offensive warfare [cf. Wallach 1967:13–51, 103–118; 1986]. He was, however, quite aware of the strength of the defensive [1980:360–371/1984: 357–366 (Book VI.1–3); cf. Aron 1983:144–171;

Gat 1988] and maintained that "the defensive form of warfare is intrinsically stronger than the offensive" [1980:361/1984:358 (Book VI.1.2)]. Because of the supremacy of the defensive, real-life wars (as opposed to the abstract absolute war) need not escalate beyond bounds because the defense might enforce a suspension of fighting. War being merely a means to political ends, in a protracted war the political end (which might be fairly modest) might thus take precedence over the military objective, always amounting to a subjugation of the enemy [1980:29–30, 34/1984:83–84, 87 (Book I.1.16–17 and 24)]. Although pure resistance would not suffice for achieving positive objectives, it might be adequate for a defense with merely negative objectives, such as maintenance or restoration of the status quo antebellum [1980:43, 193–198/1984:93, 216–219 (Book I.2, III.16)].

Among the reasons for the supremacy of the defense, CvC mentioned "the assistance of the theatre of war," i.e., the availability of fortifications and depots, the smaller operational space [1980:369/1984:365 (Book VI.3)], and the assistance of the people. The latter might sometimes be decisive, particularly in a "people's war," i.e., *guerilla warfare. In guerilla wars *concentration of force is of no avail, but forces have to be dispersed geographically and no decisive battle should be sought. Ideally "the principle of resistance exists everywhere, but is nowhere tangible." However, being able to envision only guerilla resistance in conjunction with (and in an auxiliary role to) regular warfare, CvC listed the following preconditions for success in a "people's war": that it take place in the heart of the country; that no single event is able to decide it; that the theater of war cover a large part of the country; that the national character favor this form of warfare; and that the country be of a broken and difficult nature [1980:521–528/1984:479–483 (Book VI.26); cf. Hahlweg 1986].

Even though the theses on *defense dominance and on the nature of guerilla warfare form central elements of the *NOD discourse, only a few modern NOD advocates have acknowledged their debt to CvC [Boserup 1986; 1990; Joxe 1984; 1990a; Møller 1992:131–159].

CLEMMESEN, MICHAEL H.   Lieutenant colonel in the Danish armed forces and member of SNU (*Det Sikkerheds- og nedrustningspolitiske Udvalg* [Commission on Security and Disarmament Policy]), subsequently military attache to the *Baltic States. Through his collaboration with *Anders Boserup and participation in the *Pugwash movement, MC gradually came around to supporting, with a number of qualifications, a stabilization of the conventional confrontation through *NOD [1986a; Boserup & Clemmesen 1984], particularly in a bilateral format, to be negotiated between the two sides [Clemmesen 1986a]. Subsequently, however (and probably reflecting the very heated public debate in *Denmark on NOD in 1986–1987), he swung around to a more critical attitude toward defensive restructuring, especially the most radical models of NOD, such as that of *Horst Afheldt and its derivative, the proposals of Boserup [1986; 1990].

CLOSE, ROBERT.   Belgian general (ret.), and author of two books: *Europe Without Defence* and *One Effort More and We Would Inevitably Have Lost the Third World War* [Close 1976; 1981; 1981a].

RC's point of departure was a scenario of a Soviet total surprise (bolt-out-of-the-blue) attack, to cope with which *NATO would be ill-adapted. He therefore proposed a defense along the lines of *Brossollet, the Swiss territorial forces, the *infantry forces of *Uhle-Wettler, and the area-covering defense of *Jochen Löser [Close 1976; 1981: 37–38, 168–173; 1981a; cf. Brossollet 1975; Löser 1981; Uhle-Wettler 1980], without, however, envisioning any abandonment of *nuclear deter-

rence. Indeed, he even supported a deployment of the (by then already abandoned) neutron (enhanced radiation) weapons [Close 1981: 147].

Even though RC's main concern was military efficiency (and certainly not to please the *Soviet Union*, of which he held a very somber view), he nevertheless emphasized the need for confidence building and argued in favor of his suggested *territorial defense* by referring to its strictly defensive nature [1981: 297]. To build down mutual mistrust, the armed forces of the West should be deployed in *zones* (thus accomplishing a certain *disengagement*): a screening force immediately at the border, behind which should be stationed autonomous territorial defense units possessing no offensive capabilities and thus serving as a *buffer zone* between the adversary and NATO's field armies, most of which would be stationed 300 km behind the East-West border. The latter should furthermore, as a *CBM*, be prohibited from being permanently supplied with ammunition. Munitions storage sites should be separated from the armored formations in peacetime [1981: 301–306, 311-312].

COASTAL DEFENSE.   See *Sea Power.*

COLLECTIVE SECURITY.   CS is a very old idea, the embryo of which is to be found, e.g., in late-medieval schemes [George of Bohemia 1464], that came to fruition for the first time in the *League of Nations.* The league's poor record in dealing with imperialist powers of the interwar years, such as Germany, *Italy*, *Japan*, and the *Soviet Union*, discredited CS for some time (particularly in the eyes of Realists [Carr 1946]. The *U.N.*, as a matter of deliberate choice, had more limited ambitions than its predecessor.

CS has all along had its supporters [Lutz (ed.) 1985], but it has been attracting growing interest (in the alternative defense community as well as elsewhere) since the disappearance of the East-West conflict, the incipient institutionalization of the *CSCE* process, and particularly in the wake of the *Gulf War* [Brauch 1990; 1991; Lutz 1990c; 1991b; 1993b; IFSH 1993; Senghaas 1990a–e; 1991a; Mueller 1990; 1991; Kupchan & Kupchan 1991; Schütze 1990; Chalmers 1990; 1993. For a critique see Betts 1992; Joffe 1990; 1992].

CS is the antithesis of a self-help system because it commits its members to assist one another in case of attack. Contrary to *alliances* (which are almost per definition directed against someone), CS is ideally all-encompassing (within a given system, be it regional or global). The fact that CS also includes potential aggressors makes it less confrontational, as well as perhaps less realistic.

CS may be a necessary complement to a shift to *NOD* on the national scale. The latter could go a long way toward averting aggression but would not eliminate size differentials. Very large states almost inevitably possess an offensive capability vis-à-vis very small ones, and the latter might thus need some additional insurance against attack. For CS to perform its role (and, by so doing, make individual states comfortable with strictly defensive postures), it needs military means at its disposal that would rightly be called offensive if fielded by individual states: forces with strategic mobility, with the ability to move forward under fire, to break through prepared defense lines, etc. Such capabilities have correctly been rejected as destabilizing on the national scale, but they would be much less threatening if fielded by a global organization [cf. Møller 1990; 1991: 270–273; 1992: 71–75;1993a]. See also *Multinationality.*

COMMAND AND CONTROL.   See *C³I.*

COMMITTE FOR BASIC RIGHTS AND DEMOCRACY.   (Komitee für Grundrechte und Demokratie.) West German group of peace researchers and others in the early 1980s, with close links to the *peace movement*, and comparable, e.g., to the British *ADC (Alternative Defence Commission). Members included *Ulrich Albrecht*, *Andreas Buro*, Ekkehart Krippendorf, Eva Senghaas-Knobloch, and *Egbert Jahn*.

In 1981 the CBRD published the report *Peace with Other Weapons: Five Proposals for an Alternative Security Policy* [Komitee 1981; abridged English version: Committee 1983; cf. ADC 1983; 1987]. The point of departure was a broad and citizen-oriented conception of *security*, including, e.g., a formulation of the main principles of *common security* (antedating the *Palme Commission*'s report): "Security as well as freedom necessarily include the security and freedom of the Other. . . . Security is not obtainable through a reduction of the security of others" [Komitee 1981: 18].

The deterrence strategy of *NATO* was criticized at length, even though the traditional view of the *Warsaw Pact* as the main threat to the *FRG* was accepted. The CBRD then compared five options for their contribution to meeting this threat without the dangers involved with NATO strategy. (1) *Disengagement*, (2) *nuclear-weapons-free zones*, and (3) *neutralism* were all assessed as realistic options, ripe for implementation. The FRG might, e.g., scale down its NATO commitments to the level of *Denmark* and *Norway*, i.e., remove nuclear weapons from its soil, along with support systems and offensive-capable conventional formations, thus creating a Central European disengagement zone [1981: 124]. (4) *Defensive defense* (i.e., *NOD*) was treated as primarily a political concept and assessed favorably for its contribution to a transformation of the deterrence system in the direction of *disarmament*. The CBRD, however, rejected the piecemeal approach to (5) *transarmament* and recommended a wholesale abandonment of deterrence and defense strategies relying on nuclear or other means of mass destruction. In their place, it proposed conventional armed forces resembling the *techno-commandoes* of *Horst Afheldt*, albeit only as a first step in the direction of a *civilian-based defense* [1981: 158–181].

As a collective body, the CBRD was unable to choose among the five options, all of which were nevertheless unanimously assessed as improvements on the present situation. Moreover, by virtue of different time perspectives and degrees of radicality, the options might be introduced sequentially: with disengagement and neutralism as the first steps on a path leading through defensive military defense and nonmilitary defense to an eventual abandonment of defense pure and simple [1981: 234–236].

COMMON EUROPEAN HOUSE.   Occasionally, Common European Home. A central conception in Soviet *New Thinking* during the *Gorbachev* period. The term could also be found occasionally in the security political discourse of the *SPD* and other proponents of *common security*, as well as in the writings of the SED regime and entourage in the former *GDR* [Gorbachev 1988: 194-195; Schmidt & Schwarz 1988; 1989c; Bahr 1987: 19; cf. Adomeit 1989; 1990a; Karaganov 1990; 1990a-b; Bezrukov & Davydov 1991; Lapins 1991].

CEH described a regional security arrangement in which cooperation would almost entirely displace competition, and where mutual *interdependence* would be deliberately promoted as a presumed contribution to both *peace* and prosperity. It has been regarded by its proponents as incompatible with deterrence, implying that military defenses—to the extent that they would remain indispensable—would have to be strictly defensive. CEH thus pointed directly toward *NOD*, even though the opposite is not necessarily true.

COMMON SECURITY.   The concept of CS became known through the *Palme Commission*'s 1982 report *Common Security: A Blueprint for Survival* [Palme Commission 1982].

1. **Status of CS.** CS is neither a full-fledged theory of international relations nor a political program, but on the other hand, it is more than a political catchword, even though it has occasionally been used as such. It might best be characterized as an approach to international politics from which a security political strategy might be derived [cf. Mutz 1986: 154–155; Møller 1992: 28–32].

The connotations and implications of CS remained disputed in the absence of any authority besides the aforementioned report, which reflected political compromises. Furthermore, the extensive use of the concept [Lutz 1986a; 1986f; idem & Theilmann 1986; Theilmann 1986], particularly by the Social Democratic parties (above all the *SPD* of West Germany), was not matched by any depth of theoretical analysis [Väyrynen (ed.) 1985; Bahr & Lutz (eds.) 1986; 1987; 1988; Nakarada & Øberg (eds.) 1989; Buzan 1987; Møller 1988; 1989a; 1990; 1992; Smoke 1991]. Also, some confusion arose from a proliferation of related terms, such as "collaborative security," "reciprocal security," and "mutual security," the distinction between which and CS (if any) is far from clear. It is, however, possible to distinguish between maximalist and minimalist conceptions of CS; the former have tended to dominate the political debate and to appeal to analysts of an idealist persuasion. Adherents of the latter have primarily been found in the academic community, and especially among analysts subscribing to *Realism* and neorealism [e.g., Møller 1992].

2. **Minimalist CS.** In its minimalist version, CS would presuppose no supranational authority over and beyond the individual states, i.e., it would be compatible with an anarchic international system. It would, however, entail a more long-term approach to the pursuit of national interest (i.e., a longer *shadow of the future*) than often presumed by Realists because it would take into account, and seek to circumvent, the *security dilemma*: "In the absence of a world authority . . . states have to protect themselves. Unless they show mutual restraint . . . the pursuit of *security* can cause . . . a reduction in security for all concerned" [Palme Commission 1982: 138; cf. E. Müller 1986: 164]. The implied restraint imperative was, inter alia, a result of the realization that "the security—even the existence—of the world is interdependent" [Palme Commission 1982: 5, 7]. Under these circumstances, for a state to pursue its national interest in disregard of those of its adversaries would only seemingly be rational because selfishness might well lead to undesirable reciprocation on the part of the adversaries (cf. *Prisoner's Dilemma*). The interests of all states would thus be served by reciprocal restraint; "restraint, out of respect for the right of others to security, but also in selfish recognition that security can be attained only by common action" [ibid.:9]. CS thus revealed its affinity to the concept of *security regimes*, which need not (but might of course) entail direct cooperation, but which might also rest exclusively on tacit coordination [Jervis 1982].

3. **Maximalist CS.** In the peace research community, as well as among political parties, however, CS has usually been conceived of as a more radical transformation of the international system. It has most often been assumed that CS meant that "cooperation will replace confrontation in resolving conflicts of interest," hence the near-synonym *security partnership* prominent in the SPD's political discourse. [Palme Commission 1982: 5–6; cf. H. Fischer 1987; Mutz 1986: 127]. Most maximalists have described CS as a replacement for deterrence, the maintenance of which would allegedly be incompatible with CS [Palme Commission 1982, 138-139; cf. Lutz 1986b: 51; Mutz 1986: 103]. On closer analysis, the view may be logi-

cally flawed. Were it not for mutual deterrence and the resultant shared vulnerability, a number of CS-related propositions would be obviously false: security might, e.g., be attainable by national means, and wars might well be winnable [cf. Lübkemeier 1988; Bertram 1987].

4. **CS and National Defense.** CS advocates have without exception been in favor of *disarmament*, but most have nevertheless accepted the need for military means of *self-defense*, at least for the time being [Palme Commission 1982: 7-8, 177]. Applying the CS criterion of restraint and concern for the security of one's adversary would, however, inevitably constrain a state's freedom of choice with regard to military posture. A state should, as far as the need for adequate defenses would permit, refrain from placing the security of its adversaries at risk, i.e., opt for a nonoffensive defense. Realizing that *NOD* was thus the logical military corollary of CS, a growing number of CS proponents advocated NOD as a suitable guideline for defense planning, just as most NOD proponents embedded their concrete proposals in the CS framework [Bahr & Lutz (eds.) 1988; Møller 1988, 1992; Mertes 1985; Lodgaard 1985].

5. **Nonmilitary Implications of CS.** CS advocates have tended to view security as something more than the absence of military threats to a state's security: not merely states but also societies and peoples might be insecure, and they might be endangered by other than military threats. Some attempts have therefore been made to apply the CS approach to fields other than traditional security policy [Bahr & Lutz (eds.) 1987].

In the field of psychology, CS has been seen both as requiring a dismantling of enemy images and as facilitating such an attitudinal transformation because CS would presuppose empathy with one's opponents. Furthermore, enemy images have been seen as reflections of the deterrence system that CS would presumably replace [E. Müller 1986: 170–171; R.J. Fischer 1990: 59–74; Richter 1987; Werbik 1985; 1987; Mutz 1986: 135–138]. The dismantlement of enemy images would not, however, automatically imply a denial of conflicting interests or even hostility but merely a commitment to a "peaceful culture of struggle" [cf. SPD & SED 1987; Eppler 1988; Birkenbach 1989].

With regard to economics, CS proponents have argued in favor of an expansion of East-West trade with a view to furthering *integration* between the blocs as well as common prosperity without risks to economic security [Palme Commission 1982: 71–99; Bolz 1987; Wilke 1987]. Most CS proponents have also taken environmental security very seriously and, e.g., viewed the 1987 *Brundtland Committee*'s report, *Our Common Future*, as an additional argument for cooperative ventures to salvage the common heritage of humankind [Brundtland Commission 1987; cf. Prins 1990b; Westing (ed.) 1989; Mische 1989; Soroos (ed.) 1990; cf. Thomas 1992; Prins 1992]. Finally, the CS approach has been applied to North-South relations, in which field the Palme Commission saw itself as a successor to the Brandt Commission. It acknowledged the links between the East-West and the North-South conflicts and proclaimed that "real progress will only be made nationally if it can be assured globally" [Palme Commission 1982: 12; Brandt Commission 1980; cf. Brandt 1985: 46-48, 99].

COMMON SECURITY PROGRAM.   See *Foundation for International Security*.

COMMON SECURITY REGIME.   Situation where the principles of *common security* have been adopted by all relevant states within a certain international system or subsystem, i.e., have become the dominating set of norms [Møller 1992: 28-43; cf. Lutz 1986; 1986f; 1987e. On regimes in general, see Krasner (ed.) 1982].

COMMONWEALTH INSTITUTE.   Small research institute in Cambridge, Massachusetts, with two codirectors and researchers, Carl Conetta and Charles Knight. The CI is affiliated with the *SAS and has collaborated quite closely with the *IDSS and the *IPIS. In 1991 the CI initiated the Project on Defense Alternatives, which resulted in a concrete proposal for a defensive restructuring in the Middle East. On the basis of recent military conflicts in the region (including North Africa), the authors suggested defense modernization guidelines that would be in line with the *NOD criteria yet take into account the special nature of the region. Contrary to widespread assumptions, the region (with the exception of heavily militarized *Israel and its immediate surroundings) is characterized by extremely low *force-to-space ratios. The implication is that neither *forward defense nor a genuine area-covering *territorial defense would be feasible, leaving a *bastion-type defense as the only alternative [Conetta, Knight, & Unterseher 1991; 1991a].

Another focus for the CI has been the concrete implications of *CFE for *NATO's defense needs in general, and for the U.S. forces in Europe in particular [Knight 1988; Conetta & Knight 1990; 1990a-b; 1991]. In these assessments, as well as in that of U.S. defense requirements in light of the *Gulf War, the CI team has been considerably more optimistic than establishment analysts with regard to possible savings, yet not quite so sanguine about the prospects for *disarmament and a *peace dividend as some European disarmament proponents [Conetta 1991; idem & Knight 1991a].

COMMONWEALTH OF INDEPENDENT STATES.   See *CIS.

CONCENTRATION.   One of the traditional *principles of war, according to which combat power must be concentrated at the decisive place and time, as formulated by *Jomini and others, and as formalized in the *Lanchester Laws, especially the Square Law. The traditional reading is that this requires large-scale (i.e., strategic or at least operational) mobility and large formations [cf. FM 1900-5 (1982)] and rules out *dispersion of forces, such as envisioned in most *NOD schemes [Afheldt 1983; SAS (eds.) 1989].

As pointed out by *Liddell Hart, however, concentration and dispersal are two sides of a dialectical whole: "The principles of war, not merely one principle, can be condensed into a single word: 'concentration'. But for truth this needs to be amplified as the 'concentration of strength against weakness'. And for any real value it needs to be explained that the concentration of strength against weakness depends on the dispersion of your opponent's strength, which in turn is produced by a distribution of your own that gives the appearance, and the partial effect of dispersion. Your dispersion, his dispersion, your concentration—such is the sequence, and each is a sequel. True concentration is the fruit of calculated dispersion" [Hart 1967a: 334]. Many NOD advocates have, furthermore, emphasized that what matters is the concentration of effort and firepower, not the physical concentration of weapons platforms, such as *tanks, because alternatively, firepower might be concentrated from dispersed launchers [Afheldt 1983; Acker 1984; Hannig 1984].

It is generally acknowledged that an attacker has to concentrate in order to attain a breakthrough of any linear defense. On the other hand, the increasing lethality of weapons has forced both sides to disperse for combat as the best way of limiting casualties [Bellamy 1990:45–50], a development that would tend to benefit the defender. Conversely, it is sometimes presumed that a limitation of lethality (such as would accrue from *no-first-use of nuclear weapons) would allow for a greater concentration, hence benefit the attacker. However, a dispersed NOD posture on the part of a defender would force the attacker likewise to disperse for mop-up operations, if not for the initial break-in.

CONCERT.   Loose form of *collective security*, based on a community of values (at least between the majority of states) within a system, yet without any far-reaching relinquishment of sovereignty or elaborate institutionalization. In the place of supranational authorities, a C of the great powers might make decisions with the (at least implicit) consent of the majority of states, as was the case during the C of Europe that lasted from 1815 until the Crimean War (1854–1855) [Jervis 1985; cf. Kissinger 1957; Albrecht-Carrié 1968; Holsti 1991: 114–174].

In the aftermath of the Cold War, there has been an abundance of proposals for the creation of a new C of Europe, institutionally formally founded on the *CSCE but with the real responsibility for managing the system resting with the great powers: the *U.S., the *U.K., the *Soviet Union (or Russia), *France, and the *FRG [Flynn & Scheffer 1990; Ullman 1990; Mueller 1990; 1991; Kupchan & Kupchan 1991; cf. Senghaas 1990e; 1991; 1991a–b].

CONDITIONAL OFFENSIVE SUPERIORITY.   Term employed by *Albrecht A. C. von Müller for the ability of each of two adversaries successfully to undertake *conventional retaliation (cf. *Samuel Huntington) after absorbing an attack from the other. It would thus presumably be a way of taking advantage of the strength of the defense that would severely attrite an invader, thus inverting the initial force ratio. Even though von Müller described COS as merely a manifestation of *mutual defensive superiority [Müller 1986b: 253–254; 1987a: 519; 1988c: 92–93; cf. Huntington 1983], most *NOD experts have dismissed the notion as quite incompatible with NOD principles, as well as with the political orientation of the West. It would presumably violate both *NATO's defensive policy and the *FRG's constitutional commitment to abstain from preparations for offensive war [cf. Lutz 1987d; 1989a; 1990; 1992; 1993; 1993a–b].

CONETTA, CARL.   See *Commonwealth Institute.

CONFERENCE ON DISARMAMENT.   See *CDE.

CONFERENCE ON SECURITY AND COOPERATION IN EUROPE.   See *CSCE.

CONFIDENCE- AND SECURITY-BUILDING MEASURE.   See *CSBM.

CONFIDENCE-BUILDING DEFENSE.   The term for *NOD used by the *SAS [1989]. See also *Spider and Web.

CONFIDENCE-BUILDING MEASURE.   See *CBM.

CONSCRIPTION.   Compulsory recruitment of personnel for national service, in most cases in the form of universal male C for the regular armed forces, with a length of around one year and with options of conscientious objection. C might be said to constitute the typical form of recruitment in modern Western-type societies (dating back to the *levée en masse* of the French Revolution), albeit with a number of exceptions:

- The U.S. and British armed forces are professional (all-volunteer), and the professional element in conscripted armed forces such as the Bundeswehr has been increasing, due to the growing sophistication of the equipment and the resultant need for technical skills.
- *Switzerland has no standing army but relies entirely on *militia forces

[Cramer 1984; Däniker 1987; Spillmann 1989], which also constitute a large part of the armed forces of *Austria and *Finland. Furthermore, most other countries have home guard forces to supplement the regular army [Fuhrer 1987; Danspeckgruber 1986; Kruzel 1989; Heegaard-Poulsen 1980].

C has several advantages over professionals, particularly for states with only defensive ambitions:

- Conscripted soldiers are cheaper than professionals (even though the opportunity costs of military service are difficult to calculate).
- Conscripted soldiers represent a fair sample of the (male) population with regard to level of education, etc. and in terms of attitudes; they are thus more difficult to abuse, say for *coup d'états.*
- C allows for the fielding of sizable armies in case of war, provided only that sufficient *warning time is available. The augmentation (inflation) potential is thus superior to that of professional armed forces.

Even though C is suitable for a defensive posture, and perfectly compatible with *NOD, only a few NOD advocates have recommended any major change in the traditional mode of recruitment. U.S. and British authors have tended to presuppose a maintenance of an all-volunteer army [Boston Study Group 1979: 274–276; ADC 1983: 138–140; Chalmers 1985: 158–173; cf. for an exception: Booth 1983; idem & Baylis 1988: 172–173], and German authors to assume that C would continue, proposing merely to add a militia element [Bülow 1986; Funk 1988; 1988a].

CONTADORA ACT.   See *Latin America.*

CONVENTIONAL ARMS REGISTER.   See *Arms Trade Regulations.*

CONVENTIONAL ARMS TRANSFER NEGOTIATIONS.   See *CAT.*

CONVENTIONAL DEFENSE (IMPROVEMENT) INITIATIVE.   See *CDI.*

CONVENTIONAL DETERRENCE.   Concept used to signify either any form of war prevention by conventional means or, more narrowly, a substitution of conventional for *nuclear deterrence,* i.e., a sort of *nuclear equivalence.*

1. **Offensive CD.** A substitution of conventional deterrence for nuclear deterrence was what many mainstream proponents of *conventionalization in the early 1980s were seeking with their proposals for *deep strikes [ESECS 1983; 1985; Golden & al. (eds.) 1984], *conventional retaliation [Huntington 1983], etc.: to make the costs of attacking appear so great as to outweigh (in the mind of the prospective aggressor) any conceivable gains, yet without recourse to nuclear weapons.

It has, however, been pointed out that conventional means of retaliation allow an attacker to calculate the risk, whereas only nuclear weapons provide the incalculable risk factor required for successful deterrence [Mearsheimer 1989]. Offensive CD schemes therefore often retain the option of nuclear *escalation, e.g., in the form of a *no-early-first-use strategy.

2. **Defensive CD.** A more genuine form of CD would combine *deterrence by denial based on improved defensive capabilities (or on a reassessment of the actual force-to-force ratios), with the war-preventing effects of posing no threat to the other side, thus not inviting *preemptive attack [cf. Betts 1985b; Epstein 1985;

1988; 1989; 1990; Garrett 1989; 1990; Hannig 1984; 1989; Mearsheimer 1979; 1982; 1983; 1986a–b; 1988; 1989; Posen 1984a; 1988; 1991]. In this sense, *NOD* would be a form of CD *par excellence,* even though most NOD proponents have preferred not to use the term "deterrence" because of its connotations.

CONVENTIONAL RETALIATION.   Proposal put forward by *Samuel P. Huntington* in 1983 for a retaliatory *NATO* offensive into *Eastern Europe* as a presumably more effective deterrence against Soviet attack than that provided by *flexible response.* CR has, however, been severely criticized for requiring NATO military capabilities that were neither available nor likely ever to become so, as well as for being too offensive to be compatible with NATO's defensive orientation and the *FRG*'s constitution [Huntington 1983; cf. Dunn & Staudenmaier 1985; Papp 1984; Mearsheimer 1981; cf. Lutz 1987d; 1989a; 1992; 1993; 1993a–b]. See also *Conventional Deterrence, *Conditional Offensive Superiority.*

CONVENTIONAL STABILITY.   Conception that became especially prominent in the years 1987–1989, i.e., while the *CST* (Conventional Stability Talks) were still in progress. A large number in the *NOD* community actually rephrased their conceptions by couching them in terms of CS [e.g., A. Müller 1989; 1989a–b; 1990; 1990a–c; E. Müller 1989], yet had to compete with alternative notions of CS [cf. Axelrod 1990].

   1. **Crisis and Arms-Race CS.** CS might be understood to mean a situation where neither side would have anything to gain by launching a war as compared with one side's waiting for its adversary to do so, without recourse to *nuclear deterrence*: a form of *crisis stability* based entirely on conventional forces. Or it might refer to conventional *arms-race stability*, where the propellants of the *armaments dynamics* were removed. Finally, it might also refer to a combination of the two, even though they are not automatically compatible.
   2. **Parity, Superiority, and CS.** *Balance* in the sense of numerical parity has traditionally been believed to further CS, even though it would satisfy the requirements of neither crisis nor arms-race stability. Nevertheless, its apparent fairness gave it such a political appeal that it became the predominant goal in the *CFE* negotiations.
   Certain analysts have (correctly) pointed out that even parity would not have ensured *NATO* against a *Warsaw Pact* attack (e.g., because of *mobilization* differentials between attacker and defender), and have therefore demanded very asymmetrical Eastern reductions as a precondition for conventional *disarmament* [Thomson & Ganz 1988; cf. Blackwill 1988]. Others have held more optimistic views on the balance of conventional forces, hence have regarded CS as relatively easy to achieve [cf. Epstein 1985; 1988; 1989; 1990; Mearsheimer 1982; 1988; 1989; Posen 1984a; 1988; 1991].
   3. **Defensive Superiority and CS.** Still others have pointed to the need for a structural, rather than quantitative, approach to CS, identifying as the relevant goal a situation of *mutual defensive superiority* [cf. Boserup 1985]. They have thus seen the restructuring of forces to (more) NOD-type postures as the only path to CS, yet in most cases have acknowledged the importance of numerical ratios as well as the (at least political) need for implementing restructuring through negotiated builddown of force levels [Afheldt 1988; Arbatov 1989d; Bülow 1989; Hofmann & Huber 1990; Huber 1986; 1988a; 1990a; idem (ed.) 1990; idem & Hofmann 1984; 1990; Kokoshin 1989; idem & Larionov 1990; Konovalov 1990; E. Müller 1989].

CONVENTIONAL STABILITY TALKS.   See *CST, *CFE.*

CONVENTIONAL WAR, TOTAL DESTRUCTION IN.   See *Totally Destructive Conventional War.*

CONVENTIONALIZATION.   Complete or partial shift from nuclear to conventional means of defense and/or deterrence.

The most radical form of C would be a total nuclear abolition, which would, however, be likely to come about only after a protracted transitionary phase during which the nuclear postures were configured according to the standards of *minimum deterrence* [cf., e.g., Feiveson, Ullman & Hippel 1985; Feiveson 1989; idem & Hippel 1990]. Even at this stage, and perhaps even subsequent to a complete abolition, some *existential deterrence* would remain and any future war would inevitably be fought in a *subnuclear setting* [Schell 1984: 119; Booth & Wheeler 1992; J. Miller 1988; Bundy 1986; Neild 1990: 46–73]. C would thus inevitably be a question of more or less, whereas total and irreversible C is inconceivable.

A substantial C would logically presuppose the adoption of a *no-first-use* policy with regard to nuclear weapons [Bundy & al. 1982; UCS 1983; Steinbruner & Sigal (eds.) 1983; Blackaby & al. (eds.) 1984; Lee 1988; Lübkemeier 1988]. Its principled rejection thereof notwithstanding, *NATO* since the early 1990s has seemed to be moving toward such a stand with its declarations that nuclear weapons should henceforth merely be means of the last resort, tantamount to a *no-early-first-use* (NEFU) policy [cf. Harris 1988; for a critique: Møller 1988]. Even though the *CDI* and *FOFA* programs were likewise marketed as contributions to C and NEFU, the subsequent policies seem more sincere in this respect. Furthermore, nuclear disarmament was, by the early 1990s, well under way, with the *INF* and *START* treaties, likely soon to be followed by SNF (short-range nuclear forces) negotiations and/or unilateral reductions in the SNF arsenals (see also *Bush Initiative*) [Brauch 1989; Bajusz & Shaw 1990].

These trends toward C might be contravened by an opposite trend toward nuclear proliferation, such as might appear through the dissolution of the *Soviet Union*, with the Ukraine, Belarus, and Kazakhstan suddenly appearing as independent nuclear states (notwithstanding their declared intentions to transform themselves into *nuclear-weapons-free zones* (cf. *CIS*). Furthermore, the Third World contains several undeclared nuclear states (*Israel*, as well as most likely Pakistan and *India*) and threshold states, such as North Korea, capable of becoming nuclear states on short notice [Barnaby 1993; Creveld 1993; Ham 1993; Spector 1990; D. Fischer 1992; Zagorski 1992a]. On the other hand, one state (besides Belarus and Kazakhstan) has willingly abandoned its (undeclared) nuclear status, namely, South Africa [*International Security Digest*, 1 (October 1993): 2].

In any case, with C apparently well under way in Europe, the question of how to configure the conventional forces inevitably acquires a greater topicality and importance, since they would now have to be conceived of as (almost) stand-alone forces. According to some analyses, C would make a quantitative expansion of the armed forces indispensable; other analyses have pointed to the opportunity for eliminating quantitative disparities in conventional forces through *arms control*. Still others have denied the very importance of numerical *balance* and recommended doctrinal and strategic changes [Neild 1986a; 1989a; 1990: 74–84, 114–116; Møller 1990f]. *NOD* may thus prove highly relevant in these respects, and numerous NOD proposals have actually been conceived of as contributions to C [e.g., Afheldt 1976; 1983; Löser 1981; 1982; SAS (ed.) 1984; 1989; Barnaby & Boeker 1982; 1988; 1989].

CONVERSION.   The term "conversion" signifies either a shift of productive resources from military to civilian production or, in a broader sense, a shift from military to civilian use of resources [e.g., Melman 1988; Dumas (ed.) 1988; idem & Thee (eds.) 1989; Köllner & Huck (eds.) 1990; Renner 1992; Lall & Marlin 1992; Southwood 1991; Markusen & Yudken 1992]. In an even broader sense, C may be taken to mean a shift of society as a whole from a wartime to a peacetime mode, i.e., the implications of the assumption that "war isn't coming" [Møller 1990a; 1991d].

1. **Rationale for C.** In a certain sense, C is the logical corollary of *disarmament* (because something has to be done with the henceforth redundant military resources), and as such a logical companion of *NOD*. As a longer-term perspective, it will undoubtedly become possible thus to transfer resources from wasteful military to productive civilian purposes, something that will prove beneficial for all economies and societies. *Peace movements* and others have tended to expect immediate benefits, i.e., to be able to cash in the so-called *peace dividend*. However, more pessimistic analysts have pointed to the extreme complexity of the task and, as a consequence, to the long transitionary period from the war to the peace system wherein the benign effects of disarmament might be felt. Particularly in the former *Soviet Union*, where the need for additional resources is the most acute, the obstacles to conversion are bound to be the most serious [Faramanzyan 1989; Izyumov 1991; Kireyev 1990; 1991; Ballentine 1991; Deger & Sen 1991; Schröder 1991].

2. **C from a War to a Peace System.** In its broadest sense, C is bound to be a long and multifaceted process because the war system has many manifestations that are not only directly military but also political, economic, societal, ideological, attitudinal, etc. (at home and abroad), each of which may be further subdivided.

The first step in C will often not even transcend the boundaries of the military sphere but remain a set of presumed military answers to what are still understood as military problems. Such intramilitary conversion may come in at least three varieties:

1. *Conventionalization*, i.e., a shift from partly nuclear to a strictly conventional defense.
2. *Transarmament*, i.e., a substitution of defensive for offensive military forces.
3. A partial replacement of standing with mobilizable armed forces. Such a shift has, incidentally, also been recommended as an element of the defensive transarmament, based on the presumption that a low state of readiness would *ceteris paribus* render otherwise offensive-capable forces less threatening [Löser 1981; Bülow 1988c]. It is, furthermore, envisioned under *NATO*'s new trifurcation of the armed forces into immediate-reaction forces, reinforcements, and *augmentation forces*.

The next step in C might be called "transmilitary C": a shift to nonmilitary solutions to traditional military problems, i.e., *civilian-based defense*, conceived of as a functional substitute for military defense.

A final step in the same direction might be to abandon defense as such, at least as something for which a state would need to be prepared.

3. **C of Productive Resources.** The need for C in the narrow sense depends, inter alia, on at which of the above steps the C process has arrived.

Step 1 (conventionalization) may require no C at all because nuclear arsenals are comparatively cheap: simply to do the same by conventional means as used to be done by means of nuclear weapons might easily require an arms buildup.

Step 2 (transarmament, most likely accompanied by at least some disarmament) would necessitate a certain C for several reasons. The overall numbers of troops would fall (albeit perhaps not proportionally to the budgetary savings because most NOD models are manpower intensive). To the extent that reserves substitute for standing forces, the need to provide employment for the thenceforth redundant personnel would further increase, particularly on the local scale. Leaving aside the (quite unrealistic) possibility of major investments in NOD-suitable equipment, weapons procurement would also tend to fall, hence the need for C schemes with regard to the present military plants [BRSF (ed.) 1991; Paukert & Richards (eds.) 1991; Wulf 1990; 1991].

Step 3 (reliance on mobilization) would require, first, a personnel pool from which to call up the reserves and sufficient stocks of military equipment to arm them. Second, it would require the ability to air or sealift forces to their destination, which in turn may place certain demands on naval strategies and postures: *SLOCs might need to be protected. Third, depending on the *time perspective* of a future war, there might be a need for industrial *mobilization schemes*. If war-prevention imperatives should dictate capabilities for waging a protracted war of defense (as asserted by most NOD advocates), then such capabilities would be particularly significant. The ability to start, continue, or expand military production during wartime might thus become important, and states might be ill-advised to opt for an irreversible C of industrial potential. Preferably plants should produce commodities for civilian use in peacetime (forever, it is hoped) but be able to shift to military production if needed. Paradoxically, the availability of such reconversion blueprints may be a precondition for converting in the first place [Møller 1990a; cf. OTA 1992. On the "follow-on system," see Kaldor 1981].

COOPERATIVE GAMES.   See *Game Theory*.

COOPERATIVE SECURITY.   Synonym for *common security* [Goodby 1988; Gottfried (ed.) & al. 1989; Grove 1988; Hopmann 1987; Huber 1990; Ragionieri 1991; Steinbruner 1988; Windass 1987; idem & Grove (eds.) 1987: 1-2].

COPEL, ETIENNE.   French general who retired voluntarily from his post as second in command of the air force to write *Defeat the War: It's Possible* [1984; cf. 1986].

EC proposed to model conventional forces on the example of the Swiss *territorial defense* forces in order to capitalize on the potentials of antitank and *air defense *PGMs* (precision-guided munitions): "Modern defensive weapons have such efficiency that they could literally make life impossible for whoever sought to invade a country with a prepared defence." That EC was nevertheless not entirely in line with the (rest of the) *NOD* community became evident from the fact that he cautioned against a Soviet use of chemical weapons and recommended that the West plan for retaliation "in the same coinage," i.e., with offensive *chemical weapons*, the delivery systems for which should preferably have a range allowing them to reach Soviet territory [1984: 137–157].

EC's use of the term "*defensive weapons*" was made even more questionable by his describing the neutron (enhanced radiation) weapon as "the defensive weapon *par excellence*." As a way of diminishing unintended offensive potentials, and thus of furthering confidence-building, however, he recommended what seemed like a paradoxical approximation to *no-first-use*: no-first-use beyond a state's own territory. Needless to say, it was not an idea that found much support in Germany, regardless of the fact that EC supported a German say on the use of nuclear forces stationed in Germany [1984: 111–137].

COUNTERATTACK.   See *Counteroffensive*.

COUNTEROFFENSIVE.   CO may be conceived of either as a generic term for offensive military actions intended merely to restore a previous position or as one species thereof. It is meaningful to distinguish at least three levels of CO [Reid 1991; Mackenzie 1991].:

- Counterattack, i.e., a tactical CO that merely drives the enemy back to the former front line.
- Counterstroke, i.e., an operational-scale CO that drives enemy forces back on a broad front.
- General, strategic-scale CO, which evicts enemy forces from the defended territory, perhaps even pursues them onto their home territory through a counterinvasion.

It has often (correctly) been pointed out that all defense presupposes CO, but it does not follow that all COs are defensive. Moreover, CO capabilities tend to be indistinguishable from offensive capabilities, hence the possibility for one state to misinterpret another state's peacetime preparations for a wartime CO as attack preparations, the kind of misperception that *NOD* seeks to avoid by reducing or eliminating offensive capabilities.

What distinguishes defensive strategies such as NOD is the scale of CO deemed permissible or advisable, a point that was also intensely debated in the *Soviet Union* after the political adoption of the new defensive doctrine [cf. Kokoshin & Larionov 1989; A. Arbatov 1988; 1990a]. With the exception of *Albrecht von Müller*, with his conception of *conditional counteroffensive*, all NOD proponents have agreed on the impermissibility of border-crossing CO: an invader should be driven back to its own territory but no further. Tactical CO, on the other hand, will always be regarded as both permissible and indispensable. Depending on the geographical setting, operational-scale COs would also be unproblematic, if only they remain distinguishable from strategic ones, something that may not always be the case. For a country the size of the USSR or Russia to evict an opponent from its territory, it might, e.g., need mobile forces with an effective range that would permit them, if deployed close to the border, to reach well into the territory of other states.

The *spider-and-web* principle of the *SAS* is intended to allow for a distinction between CO on a country's own territory (where its mobile forces, i.e., "spiders," remain within the "web") and beyond, i.e., on enemy territory [SAS 1984: 1989; Unterseher 1988b; 1989f; Boeker & Unterseher 1986; Grin & Unterseher 1988; 1990; 1990a].

COUNTERSTROKE.   See *Counteroffensive*.

CREDIBILITY.   See *Nuclear Deterrence*, *Self-Deterrence*.

CREVELD, MARTIN VAN.   Historian at the Hebrew University in Jerusalem. MVC has become known primarily for his analyses of the development of *logistics* [1977; 1991a], command and control [1985], and military technology (in the broadest sense of technologies affecting war) through the ages [1989].

In 1991 he published one of the most important works on war ever written: *The Transformation of War* [1991], wherein he philosophized about the phenomenon of war itself. Departing from the qualified optimistic belief that "large-scale, conventional war . . . may indeed be at its last gasp," he did not predict a peaceful world but, rather, a return to "WARRE in the elemental, Hobbesian sense of the word" [1991: 2, 22]. Even though the absence of large-scale conventional wars in Europe since 1945 might be partly due to *nuclear deterrence*, the influence thereof did not extend downward into the realm of *guerilla warfare* and other forms of unconventional war. The "trinitarian wars" (wars fought by the "Clausewitzian trinity" of people, army, and government) were the only really political ones. Moreover, they were exceptions rather than the rule because most armed conflicts were not fought by the state, for the state, or against other states.

MVC disagreed with *Clausewitz in many respects but agreed with him in regard to the inherent supremacy of the defensive. Furthermore, he supported the views of modern *NOD advocates on the importance of *dispersion as the best order of battle for a defender and shared the opinion that the *time factor tends to benefit the defender. Finally, he concurred with NOD advocates in the view of huge integrated weapons systems as dinosaurs that should be replaced by smaller and cheaper weapons [1991: 192-223].

CRISIS STABILITY.   In general terms, CS signifies the ability of a system to sustain stress and avoid turbulence. More particularly, it refers to the ability of an adversarial military system to remain under political control, even under conditions where decisionmakers take the possibility of war into account.

1. **War Prevention.** CS is a species of war prevention. One may distinguish between three main types of war: premeditated war, initiated by deliberate aggression, and two varieties of inadvertent war—*preventive war and *preemptive attack [Schelling & Halperin 1961: 9–17; Frei 1983; Møller 1992: 105–128]. The distinction between the second and third types is primarily one of timing and pace. A preventive war is based on the preference to fight now rather than later a war that is believed to be inevitable in the long run; nevertheless, the first blow represents a deliberate political act. A preemptive war, on the other hand, is motivated by a somewhat different logic. It is initiated by one party in the earnest, albeit often erroneous, belief that the other side has already started the operational preparations for attack, or even launched the actual attack. In such a situation, military logic may dictate that a state fire first in order to improve its chances.

Whereas it strained the imagination to envision states in the European orbit deliberately initiating wars against one another, during the Cold War, preventive or preemptive wars were somewhat less inconceivable (albeit far from likely). Such wars might, e.g., have sprung from chain reactions, the source of which might have been either a superpower crisis (e.g., of the Cuba 1962 type); a European crisis (similar, e.g., to the Berlin crises); a conflict in the Third World spilling over into Europe; or any combination thereof [Betts 1985; Nacht 1985: 55–117; Nincic 1985; Fukuyama 1985; Lodgaard 1987].

The problem of CS has commanded considerable attention, particularly in regard to the prevention of inadvertent nuclear war. *NOD proponents have, however, asserted that the basic principles of CS pertaining to the strategic level would also, *mutatis mutandis,* be applicable to the substrategic level, i.e., to conventional forces operating in a nuclear setting [Afheldt 1976; cf. Freedman 1989a: chaps. 9–16; Ball & al. 1987].

2. **Crisis Interaction.** CS or instability are features characterizing the interaction between two or several states under conditions where military means are employed for purposes other than actual fighting. Such interaction always consists of chains of actions, perceptions, and reactions: the actions of one side are perceived by the other, who responds with measures that are perceived by the first party, inducing it to respond, etc. [cf. Holsti & al. 1969]. Any such interaction is influenced by the *security dilemma [Herz 1950; Jervis 1978; Buzan 1991: 294–327] in that each side, by caring for its own security, tends to diminish that of its adversary, whereby the unilateral quest for *security may become counterproductive and both sides become trapped in a *circulus vituosis,* generating mounting

risks of inadvertent war. Furthermore, as compared with everyday interaction, interaction in periods of crisis tends to be more fraught with dangers, due to the short time span, the perceived urgency, and the high stakes involved [Cf. Lebow 1981: 7–13]. Moreover, a somewhat different logic is at work, in that military logic may take command in the midst of the crisis, generating self-sustaining and reinforcing incentives for preemption, as described by *Thomas Schelling*: "A modest temptation on each side to sneak in a first blow—a temptation too small by itself to motivate an attack—might become compounded through a process of interacting expectations, with additional motives for attack being produced by successive cycles of 'He thinks we think he thinks we think . . . he thinks we think he'll attack; so he thinks we shall; so he will; so we must'" [Schelling 1960: 207].

3. **Preconditions of CS.** To improve CS is tantamount to avoiding preemption by removing incentives and/or by eliminating options, i.e., by making it appear less urgent and/or less rewarding to preempt, preferably both. An opponent's preemptive options might be removed through an overwhelming military superiority; and its incentives could be eliminated through unilateral *disarmament*. That neither of these solutions is quite satisfactory is due to the inevitable uncertainty about the intentions of a state's adversary. The quest for superiority would be dysfunctional vis-à-vis a nonaggressive opponent (because it would provide incentives for preemption), whereas disarmament might tempt an expansionist adversary. Hence the need for solutions that provide an ample margin for errors of judgment, such as simultaneously minimizing options and motives for preemption on both sides. From these considerations one might deduce a number of criteria for crisis stability:

- No side should be capable of a disarming first strike against the other, but each side should possess secure second-strike capabilities, allowing it to absorb any attack and still retaliate.
- Both sides should avoid presenting important and vulnerable targets that might tempt the opponent to venture a disarming strike. The more dispersed and/or individually unimportant a state's military targets, the fewer the incentives for its adversary to attack.
- Both sides should hedge against *surprise attack* through effective surveillance because doing so would prolong the available time for political rather than military decisions. The effective *warning* time for both sides might also be prolonged through a deployment of forces requiring considerable time to prepare for attack.
- Because the perceptions of both sides are what matter in a crisis, transparency of military actions should be enhanced.
- *Self-deterrence* should be avoided, requiring a circumvention of what might be called the preparedness-versus-provocation dilemma. Both sides should have at their disposal defense options that would be adequate to protect against attack yet impossible to misconstrue as preparations for an attack that might provoke an otherwise avoidable war [Blaker 1987; Williams 1987].

4. **Elements of Crisis Instability.** A number of elements of the traditional military structures and strategies have violated the criteria of CS:

- Tactical nuclear weapons, first, because they have represented a considerable offensive capability, neutralization of which was bound to be a high-priority goal for the opponent; second, because both launchers and storage sites were extremely lucrative targets; and third, because the associated dual capability of a number of delivery systems introduced an element of ambiguity into the military equations [Afheldt 1978; cf. Kelleher 1987].

- *Blitzkrieg* strategies such as that to which the USSR apparently adhered until the mid-1980s, and that exposed *NATO* to the threat of a sudden surprise (bolt-out-of-the-blue) attack. Offensive conceptions of maneuver warfare such as the *AirLand Battle* doctrine pointed in the same direction [Vigor 1983; cf. Kipp 1987].
- *Deep strikes* such as favored by the USSR and as emphasized by NATO under the *FOFA* program [Rogers 1983; Farndale 1988; OTA 1987; Wijk 1986; Sutton & al. 1984; Burgess 1986; cf. Møller 1986a; 1988a].
- *OCA* (offensive counter air) conceptions, under which one offensive-capable air force was targeted against another, thus giving both an incentive to preempt [Brown 1986: 35–36; Cotter 1983: 220–225].
- The *SDI* (Strategic Defense Initiative), the achievement of whose proclaimed goals would have been tantamount to a genuine disarming first-strike capability. Furthermore, the extreme vulnerability of the associated *C^3I* systems would have provided the adversary with options of preemption [Cf. McNamara & al. 1985; Glaser 1986; Stockton 1986].
- The U.S. Navy's (Forward) *Maritime Strategy* in general, and in particular the elements of strategic *ASW* therein, threatened the survivability of the Soviet second-strike capability. Furthermore, the attack submarines and carrier groups assigned to these operations represented extremely lucrative targets for Soviet ASW and anti-carrier-group forces [Watkins 1986; cf. Brooks 1986; Friedman 1988; Stefanik 1987; for a critique: Mearsheimer 1986; Møller 1987].
- NATO's Rapid Reinforcement Concept of 1977, which envisioned crisis reinforcement of Europe, something that would have provided the USSR with incentives for preemption, and also because of the offensive capabilities of the assigned forces.

5. **Enhancing CS.** To abandon the above strategic conceptions would automatically have enhanced CS significantly. A further contribution might have been various *CBMs* (intended to improve predictability and transparency, and thus to avoid misunderstandings, particularly in crisis situations) and *CSBMs* (confidence- and security-building measures), e.g., in the form of exercise and deployment constraints, which would have materially hampered surprise attack [Brauch 1987a; E. Müller 1982; Borawski 1986; 1988a; Møller & Wiberg 1990].

6. **NOD and CS.** *NOD* would be an even more radical solution to the CS problem and was indeed conceived of by many authors primarily with this in mind. If a state were to relinquish offensive capabilities, its opponents would thereby be deprived of most preemptive motives. Further, most versions of NOD would also deny an adversary options of surprise attack, by virtue of presenting no lucrative targets (the *no-target principle*). Finally, a state with NOD-configured armed forces would be in a better position to respond adequately to any warnings received, by virtue of the elimination of self-deterrence: If defense preparations could not possibly be construed as threatening, they might be undertaken in a timely fashion without fear of provocative effects.

Many NOD proponents have advocated *disengagement*, e.g., through *nuclear-weapons-free zones*, accompanied by a similar withdrawal of offensive-capable conventional forces from the border areas [Lodgaard & Thee (eds.) 1983; Lodgaard & Berg 1983; 1985; Löser 1981; A. Müller 1986; Romberg 1986; Bahr 1987a; Lutz 1988a]. On the one hand, such *zones* would improve CS by making the required preparations for an attack more time-consuming, thereby prolonging the warning and mobilization time available to the other side. The ability to wait would thus be improved on both sides, and consequently the option of handling the crisis with political means would be subject to no military necessities. Furthermore, the withdrawal of offensive capabilities from the border areas need not seriously impair capabilities for *forward defense because strictly defensive means for this purpose are conceivable [cf. Hannig 1984; Barnaby & Boeker 1982; 1988; 1989]. Such defensive zones might serve as early warning devices because the introduction of troops or prohibited weapons systems into a zone might safely be taken as evidence of a pending attack. On the other hand, a mere peacetime withdrawal of heavy and offensive-capable formations would be problematic, to the extent that it would necessitate a forward movement of these forces in times of crisis, and this might lend itself to misunderstandings [Unterseher 1987a].

CROSSING-THE-T.   Geometrical law, according to which a defender (tactically) is able to bring a larger part of its firepower to bear on the approaching (i.e., attacking, at least tactically) forces than vice versa. The phenomenon manifested itself, e.g., in naval warfare, after the introduction of guns deployed in a broadside mode: the party waiting for the opponent to approach had the entire broadside battery available, whereas the attacker could only use a fraction of its guns. The CT thus benefited the tactical defender, and placed a premium on the defensive posture [Quester 1988: 85]. In a land battle the CT is a function of *force-to-space constraints: an attacker has to attack through corridors that are usually too narrow to accommodate its total force, hence the need for echeloning the attacking forces. However, doing so may allow the defender to bring superior force to bear against each successive echelon, even under conditions of overall inferiority, whereby it may stand a good chance of prevailing (i.e., fending off the attack) thanks to the square law (cf. *Lanchester's Laws) [Mearsheimer 1982: 144–145].

CSBM.   Confidence- and security-building measure. Technical term for constraining *CBMs, such as proposed by the neutral and nonaligned states (the *N+N group) during the *CDE conference. The term was also used in the Concluding Document from the Vienna *CSCE follow-on meeting as the agenda for a new set of talks, to be conducted in parallel with the *CFE negotiations from 1989 until the Helsinki CSCE follow-on meeting in 1992 [Freeman 1991: 203–204; Lachowski 1993].

CSCE.   Conference on Security and Cooperation in Europe. Originally a Soviet demand (intended to codify the status quo in Europe) to which the West acceded as a quid pro quo for Soviet acceptance of the *MBFR, intended (at least officially) to build down superior Soviet conventional forces.

1. **Helsinki Final Act.** In the final document signed in 1975 at the Helsinki Conference [in Freeman 1991: 147–150], all participating states committed themselves to the following principles with implications for the alternative defense debate:

- The inviolability of borders by forceful means, tantamount above all to a German acquiescence in the Oder-Neisse border;
- The illegality of the use for force in interstate relations;
- The objective, as a long-term goal, of general and complete *disarmament; and
- The need for stabilizing the military situation, e.g., by means of various *CBMs.

2. **CSCE Process.** The Helsinki conference was followed by regular follow-on conferences and accompanied by various sets of negotiations on special subjects

(including the *CDE* and the *CFE*) scheduled to coincide with the follow-on conferences: the so-called CSCE process.

The most important innovation occurred at the 1990 Paris Summit, where the member states signed the *Paris Charter* and inaugurated a certain institutionalization of the CSCE. At the 1992 Helsinki Summit, the member states proceeded along this path, inter alia by deciding to establish the *Forum for Security Cooperation*, located in Vienna [Möller-Gulland 1993; Ghebaldi 1993]. It was further officially declared that the CSCE regarded itself as a regional "branch" of the *U.N.* [Gaer 1993], and the possibility of CSCE *peacekeeping* activities was opened up, albeit with the qualification that they could not entail *peace enforcement* and presupposed the consent of the parties. It was envisioned to utilize the resources of the WEU and/or NATO for such operations, but even the *CIS* might be asked to act on behalf of the CSCE [*Atlantic News,* 14 July 1992, AT. DOC 80, 1–9].

3. **Perspectives.** The CSCE has remained one of the least controversial elements in the European security political landscape, albeit until recently with rather modest concrete accomplishments. The lack of success for preventing or solving the Yugoslav crisis has been regarded by some observers as discrediting it as a security political instrument; others have denied this, referring to the fact that the conflict may have been insoluble in any case [Remacle 1993]. Be that as it may, the almost unanimous satisfaction with what has been accomplished, as well as the fact that the CSCE was for a long time the only significant all-European forum, motivated numerous alternative defense and security proponents to devise schemes for an expanded role for the CSCE:

- CSCE has been depicted as a *security community,* if only *in nuce,* but with the potential of furthering *integration* and perhaps amalgamation [Schlotter 1982].
- CSCE has been described as a modern version of the *Concert* of Europe, in the context of which great-power collaboration with the consent of smaller powers should be promoted [Flynn & Scheffer 1990; Ullman 1990; Mueller 1990; 1991; Kupchan & Kupchan 1991; cf. Jervis 1985; Senghaas 1990e; 1991; 1991a–b].
- By virtue of its nonadversarial nature, the CSCE might be a viable replacement for the military *alliances* in Europe.
- Eventually, the CSCE might develop into a *collective security* system with responsibilities for upholding peace in Europe (and perhaps even beyond) [cf. Chalmers 1990; Brauch 1990; 1990a].

The CSCE experience has also prompted proposals for emulation in other parts of the world: "CSCM" for the Mediterranean [Armengol 1994; Ghebaldi 1993] and regional counterparts in Africa [Schimmelfennig 1993] or the Asia-Pacific [see Wiseman 1994].

CST. Conventional Stability Talks. Term used for the *CFE* negotiations during the preliminary talks (1987–1989) on the mandate.

D

DAMAGE LIMITATION. The quest for DL has been one of the main spurs for the development of alternative defense conceptions, many of which have been labeled "defense without suicide" and the like [E. Afheldt 1984; 1984a; cf. H. Afheldt 1971; 1971a; Weizsäcker (ed.) 1971; Löser 1981]. They have also been advocated as alternatives to what has been regarded as potential suicide (plus genocide, "ecocide," or even "omnicide") [Galtung 1984: 82]: an implementation of NATO's *flexible response* strategy with its predictable escalation to global thermonuclear war.

1. **Modes of DL.** In principle there are three main categories (or modes) of DL, which form a temporal continuum:

- DL through war prevention, i.e., prewar DL;
- DL through reducing the destructiveness of war, i.e., intrawar DL.
- DL through early *war termination*, i.e., postwar DL.

The three modes may not be entirely compatible. It might, e.g., well be the case that the more horrendous the prospects of war appear, the greater the aversion of the potential belligerents to war. Conversely, by ensuring effective damage limitation, states might unwittingly raise the likelihood of war by making it seem less of a disaster. Further, the more horrendous a war would actually turn out to be, the shorter one might expect its duration to be, whereas a relatively low-key war might conceivably go on almost indefinitely. The problem of assessing trade-offs and comparing alternative approaches is compounded by tremendous difficulties of measurement and the incommensurability of potential and actual damage.

On the other hand, the trade-offs are not so clear-cut as sometimes depicted. Although maximum war damage (such as sought through *nuclear deterrence*) might deter premeditated war (prior to which the stipulated prospective aggressor is presumed to compare the costs and benefits of war and nonwar), it might well be irrelevant or even dysfunctional for the avoidance of *preventive war* and *preemptive attack*, where the relevant comparison is between war now and war later. In addition, not merely the prospective attacker but also the defender (or deterrer) are bound to compare costs and benefits as well, and might find that surrender would be preferable to the costs of implementing the deterrent threat, hence succumb to *self-deterrence* [cf. Møller 1989c-d].

2. **Intrawar DL.** In principle, there are several ways of minimizing the damage of a war in progress:

- By terminating it as rapidly as possible.
- By waging it as far away as possible from one's own territory and population, hence the attraction of offensive (or at least *counteroffensive*) strategies, and even of preemption. The main problem with this approach is, of course, that by seemingly anticipating, a state could well bring about a war that might otherwise have been avoided; and that by preparations for large-scale counteroffensives, a state might give its adversary reason to fear an

impending attack, hence to attack preemptively, etc. An obvious compromise between a more destructive but less likely war on one's own ground, and a somewhat more likely yet less destructive war on the adversary's ground, would seem to be *forward defense*, such as envisioned by most NOD proponents.

- By preventing its escalation to nuclear war, i.e., by raising the nuclear *threshold*, something that has come to be nearly universally acknowledged as desirable [cf. Rogers 1982; Harris 1988]. A *no-first-use* policy would recommend itself as the obvious guideline but would have to entail more than a mere declaration, and be implemented in terms of strategy and exercise patterns, as well as weapons procurement and deployment [Bundy & al. 1982; UCS 1983; Steinbruner & Sigal (eds.) 1983; Blackaby & al. (eds.) 1984; Lee 1988; Lübkemeier 1988]. The choice of conventional postures would also have some impact on the nuclear threshold, and many NOD schemes have been devised with the explicit purpose of rendering the military use of nuclear weapons pointless (cf. the *no-target principle*) [Afheldt 1976; 1978; 1983; 1987]. To strive for the ability to limit nuclear war, on the other hand, would probably not contribute to DL because it would tend to lower the threshold—as well as be most likely infeasible [cf. Schelling 1960: 257-266; 1966: 109-116, 153-158; I. Clark 1982].
- By limiting deliberate conventional destruction.
- By limiting collateral damage.

3. **Conventional DL.** Even though there is a significant analytical distinction between the two last categories of intrawar DL, the means of minimizing deliberate and collateral damage are largely identical. Once again, there are several categories of measures:

- Strategic measures. Through a *dispersion* of its forces, the defender might dissuade massive, hence very destructive, conventional strikes, such as strategic bombardment. Furthermore, the defender might seek to avoid armed combat in urban areas, e.g., by declaring conurbations *open towns* and refraining from defending them militarily while perhaps using *civilian-based defense* methods [cf. Afheldt 1983: 133–143; Löser 1981: 133, 179; Nolte & Nolte 1984].
- Legal measures. The existing body of *laws of war* could be strengthened, e.g., by expanding the list of prohibited weaponry and modes of warfare [cf. De Lupis 1987; Best 1980; Walzer 1977]. Relevant new prohibitions might include bans on anti-personnel landmines, radiological, and EMP (electromagnetic-pulse) warfare; first-use of weapons of mass destruction; the use of incendiary weapons (such as fuel air explosives and napalm); the possession of *chemical weapons*; and the use of genetic manipulation for military purposes [Rotblat 1981; SIPRI 1980; Robinson 1982; 1986; Tsipis 1985: 98–101; Geissler 1984; 1985]. Additionally, the defender would have to respect scrupulously the *laws of war*, inter alia so as not to allow the aggressor to justify (politically, if not legally) violations as reprisals. Legal considerations would hence have to be taken into account in the planning for defense and might occasionally overrule military ones. The recommendation of a few NOD proponents [Galtung 1984: 180-184] for the use of paramilitary forces may thus be ill-advised because they would tend to blur the distinctions between combatants and civilians.
- *Civil defense* recommends itself as the ultimate means of limiting collateral damage to the civilian population. In addition to a "horizontal evacuation" from rural to urban areas a certain "vertical evacuation" into shelters

might be advisable. Although this would provide no meaningful protection against deliberate (and especially nuclear) attack, fallout shelters would afford some protection against small-scale nuclear weapons (i.e., inadvertent launches) or against (counterforce) chemical warfare, as well as against accidents in nuclear or chemical plants [cf. on the Swiss example: Cramer 1984: 23–28].

4. **Evaluation.** There are, then, various possibilities for DL that could ensure that defense could remain (or become once again) a rational option in case of an attack. Although there is considerable truth in the *Totally Destructive Conventional War* thesis, the predicted consequences of a conventional war are not inevitably tantamount to complete devastation beyond postwar recovery, especially if an *escalation* to nuclear warfare can be avoided.

DAMGAARD, KNUD.   Former speaker on defense issues for the Danish *Socialdemokratiet,* and the *primus motor* in its reorientation with regard to defense policy in the mid-1980s, when it came to embrace *NOD* in principle and began to apply it to short-to-medium-term defense planning [1984; cf. *Socialdemokratiet* 1986].

DÄNIKER, GUSTAV.   See *Switzerland.*

DANKBAAR, BEN.   See *Netherlands.*

DAVID-AND-GOLIATH PRINCIPLE.   Term used by *Norbert Hannig* and *Freeman Dyson* to describe the ability of a weaker defender to defeat a stronger attacker by means of an *indirect approach* and small precision-guided weapons, *PGMs* [Hannig 1984; 1989; Dyson 1984: 45–55; cf. original biblical story: 1 Samuel 17].

DDR.   Deutsche Demokratische Republik (German Democratic Republic), 1949–1990. Better known as East Germany. See *GDR.*

DEAK, PETER.   Colonel in the armed forces of *Hungary* and one of the first supporters of *NOD* within the ranks [1989; cf. Lemaitre 1989].

DEALIGNMENT.   A process of gradual loosening of bloc structures, based on the assumption that *neutrality* and *alliance* membership are poles on a continuum rather than discrete entities, i.e., that degrees of semialignment and/or semineutrality are thus conceivable [Ørvik 1986; cf. Møller 1987c].

D has been proclaimed as a political strategy by, e.g., the *ADC* and various (mostly German) neutralists, first among them, *Ulrich Albrecht* [ADC 1987: 45–48, 182–213; cf. Kaldor 1986]. D might take different shapes, such as conditional alliance membership (combined with threats to withdraw unless certain conditions are met); *disengagement;* partial or complete *neutralization;* *Finlandization,* etc. The logic behind such proposals was that the dealigned areas would serve as *buffers* between the gravest threats to peace, namely, the two superpowers. Before either might embark on aggression, it would have to reforge alliance links, reintroduce its forces into vacated areas, etc. Because the same logic pertains to various forms of *zones,* D and disengagement have often been combined with proposals for *nuclear-weapons-free zones,* tank-free zones, etc., as well as with *NOD* schemes (often incorporating zones).

Various schemes have been suggested, according to which the small *Warsaw

Pact countries would be "Finlandized," in reciprocation of which West Germany would be "Hollandized," and various other *NATO* states "Denmarkized" [Albrecht 1982; 1983; 1983a; idem & al. 1988; Lutz 1982; 1985; 1987e], i.e., would have assumed a status similar to that of *Denmark* and *Norway* with regard to the stationing of nuclear weapons and foreign troops [Møller 1985; 1987; 1989b; 1989f; Brundtland 1985; Flynn (ed.) 1985].

DEAN, JONATHAN.   Ambassador, (retired) former U.S. representative to the *MBFR* negotiations, subsequently arms control advisor to the Union of Concerned Scientists (see *UCS*).

JD was one of the first authors to present the German *NOD* debate to the Anglo-Saxon audience [1987a; 1989c: 64-86; 1989d; 1990e; 1990i]. Besides this, his main preoccupation has been with various aspects of conventional *arms control*, in which respect he has drawn on his personal experience from the MBFR and that of his colleagues from the *CDE* (Stockholm Conference) [1985a; 1987] with a special view to the perspectives of the *CFE* negotiations on which he has commented extensively [1988; 1989; 1989a; 1989e-f; 1990b; 1990h; 1990j].

JD, both personally and on behalf of the UCS, has presented many concrete proposals for implementation of NOD-type restructuring through the CFE and its successors [1994], including one for the creation of restricted military areas (RMAs, nearly synonymous with what he has on other occasions called "exclusionary zones"). The RMAs were conceived as a measure of *disengagement* and envisioned the pullback from a 150-km-wide zone (50 km on *NATO*'s side and 100 km on the *Warsaw Pact*'s) of units with surface-to-surface missiles, aircraft and helicopters, infantry fighting vehicles, airborne units, and self-propelled artillery, along with associated logistic installations and depots. RMAs would also entail reductions of manpower, but both sides would be free to deploy units with defensive equipment, such as antitank and *air defense* weapons, and to saturate the zone with various types of obstacles [1989c: 232–242; cf. 1988b]. The intention behind the RMA proposal was to hamper *surprise attack*, thereby improving *crisis stability*. With the same purpose in mind, JD also suggested various forms of *CBMs*, as well as the establishment of a joint NATO–Warsaw Pact risk reduction center, tasked above all with the handling of minor incidents with a view to averting unintended *escalation* [1989c: 242–248; 1990f].

His focus on conventional arms control notwithstanding, JD has also devoted considerable attention to nuclear build-down and to the raising of the nuclear *threshold*. In 1991, he published (jointly with Kurt Gottfried) an elaborate proposal, *Nuclear Security in a Transformed World*. The two authors asserted that the expansion of the nuclear arsenal no longer enhanced U.S. security, expressed the general opinion that *nuclear deterrence* was of declining relevance to the threats facing the *U.S.*, and proposed three key steps:

- Bilateral reductions of Soviet and U.S. nuclear arsenals to the level, by the year 2000, of reserve nuclear forces numbering no more than 1,000–2,000 strategic and tactical weapons;
- Restrictions on the arsenals of other nuclear powers, both declared and undeclared (with which expression the authors had *Israel*, *India*, and Pakistan in mind);
- A strengthening of the nuclear nonproliferation regime, i.e., the *NPT* [Dean & Gottfried 1991; cf. Dean 1988a].

DECENTRALIZATION.   Most *NOD* schemes have emphasized D of command structures as a means of enhancing survivability and endurance.

The *web-like configuration of most NOD models have tended to emphasize horizontal over vertical (i.e., hierarchical) chains of communication: units were supposed to alert one another and to provide one another with target coordinates directly rather than through some central *C^3I node [Afheldt 1983; Acker 1984; Freundl 1984], an arrangement that would hamper jamming and prevent a severance of communication. The required redundancy of channels could be ensured by means of an integration of military and civilian communication arrays, as well as by emplacement of fiber-optical cable networks in particularly exposed defense *zones [Grin 1990; 1990a].

Centralized command and control would, furthermore, be rendered less important by the reactive nature of most NOD defense models: units would be expected to fight only within a predefined square and no large-scale mobility would be envisioned. This would also allow for a more extensive use of directive control (*Auftragstechnik*) instead of more authoritarian forms of command, thereby enhancing flexibility [Simpkin 1986: 227–240; cf. Creveld 1985: 270–272]. Nevertheless, a few NOD proposals have envisioned a maintenance of centralization, e.g., for the coordination of long-range fire support [Hannig 1984]. However, this might violate the *no-target principle because the central command posts would be obvious high-priority targets for any attacker.

DEEP CUTS.   See *Minimum Deterrence, *Bush Initiative, *Year 2000 Plan, *START.

DEEP STRIKE.   Generic term for various forms of deep interdiction by means of missiles and/or aircraft, such as envisioned under *NATO's *FOFA doctrine. DS has also been envisioned, albeit on a smaller (operational) scale, as an element in the U.S. Army's *AirLand Battle doctrine. The latter, however, is primarily a doctrine for deep penetrations and maneuver warfare by means of ground forces, which would merely require DS in the form of extended fire support.

DS has been rejected by most *NOD advocates as being too offensive and as damaging *crisis stability: the adversary's territory would be held at constant risk, which would translate into a powerful incentive for *preemptive attack against the DS systems, most of which (missile sites and fixed airfields) would also be inherently vulnerable to preemptive strikes [Afheldt 1983: 81; 1987b: 220–229; Hannig 1986b; SALSS 1984a: 33–34; IFSH 1985; Schmähling 1988c: 70; Huber 1988: 25–28; Hagena 1990b: 92–97; Møller 1987a; 1988a; 1989b; 1990i]. Others have asserted that DS would constitute no real offensive capability (at least not when seen in isolation): without *land forces suitable for invasion and occupation, a state would possess no war-winning capability, regardless of any ability to destroy targets in the enemy's rear [Löser & Anderson 1984: 45, 55; Löser 1985: 525–26, 531; E. Müller 1984: 142–144].

The doubtless merits of DS notwithstanding, the line of argumentation seems to disregard the importance of mutual perceptions: if a state's adversary should feel threatened, even erroneously, the *security dilemma dynamics would be activated, with a low crisis as well as *arms race stability as the predictable result. Second, from the observation that without invasion capability whatsoever a state would automatically be nonoffensive, it does not follow that a state with, say, a small invasion capability plus DS systems would possess only insignificant offensive capabilities. The relationship may be nonlinear, and/or DS may serve as a multiplier rather than simply an additional capability [cf. Møller 1992: 182–187].

DEFENSE.   Generic term for the protection of values, such as life, property, freedom, national identity, and territorial integrity, against various threats from internal usurpers, external aggressors, or structural (systemic) factors (cf. *Security*). National D in the traditional sense is one species of this genus, dealing with the protection of territorial integrity and state sovereignty against attack from other states [cf. for a critique: Creveld 1991].

Depending on the nature of the threat and contextual factors, national D may assume a number of different forms. One might, e.g., subdivide D along the temporal dimension (when D is attempted prior to, during, or after an attack) into the following main categories:

- Preventive D, intended to dissuade, deter, or render impossible an expected attack before it occurs. A way of dissuading might be to opt for a strictly defensive military posture (i.e., *NOD*), thereby eliminating incentives for *preemptive attack* on the part of a prospective aggressor. To deter an attack might involve the threat of retaliation, either nuclear or conventional. To render attack impossible might be tantamount to *preventive war*, or at least to a preemptive strike against the adversary's means of attack.
- Direct D (the thwarting of an attack in progress) might take the form either of a *forward defense*, intended to stop the invasion at the border, or of a *territorial defense*, undertaken either in an area-covering or in a *bastion* mode.
- Restorative D would be tantamount to a recuperation of lost territory, either through *counteroffensive* operations or by political means, perhaps underpinned by the threat of using force.

It might also make sense to distinguish between active and passive modes of defense. With the partial exception of deterrence, which need not involve any active precautions (but might be no more than *existential deterrence*, based on the possession of means of retaliation), all of the above modes of defense are active. However, passive means such as fortifications (including the Chinese Wall and the Berlin Wall) may also be quite effective, albeit usually not without some backup of active forms of D (see *Maginot Line*).

Furthermore, it certainly makes sense to distinguish between offensive and defensive modes of D, with the aforementioned preventive war constituting the most extreme form of offensive D, and territorial D a clearly defensive one (cf. *offense/defense distinction*).

Finally, D need not be military at all but could rely on a strategy of *civilian-based defense*, or on diplomatic and other nonmilitary means, such as *arms control*, *alliance* membership, or usefulness and the cultivation of friendly relations with otherwise perhaps hostile neighboring states [see Galtung 1984].

DEFENSE DILEMMA. The dilemma ensuing from the prospects that even a successful defense may destroy the values it was intended to preserve and thus may not contribute to but, rather, endanger national *security* [Buzan 1991: 270–293].

The peacetime variety of the DD stems, above all, from the opportunity costs of defense, the expenditures for which become unavailable for other purposes of importance for national security (in all but the narrowest sense), such as environmental protection, economic security, etc. For a state to defend itself with one set of means against one type of threats (which are not necessarily the most serious ones) may detract from its ability for defense by other means against other (and possibly more serious) threats to its security. In addition to this cost perspective, the DD may also directly influence people's lives, e.g., by the immediate risks associated with hosting nuclear or chemical weapons [Møller 1991: 33–38; 1992: 22, 33]. Finally, the DD may manifest itself politically, if national defense is found to require (or is used to legitimize) suspension of democratic rights, implying that the defense of *democracy* might lead to a partial sacrifice of democracy [cf. Tatchell 1985; Kommitée 1981; Vogt 1989a–b; 1990a].

The wartime variety of the DD is related to the costs of war, which may well dwarf those of capitulation, particularly when the defense involves nuclear *escalation*, regardless of whether the defender or the attacker uses nuclear weapons [Weizsäcker (ed.) 1971]. Certain authors have asserted that even a conventional war in Europe would inevitably be suicidal, i.e., that there would be no escape from the wartime DD [M. Schmidt 1989; Benjowski & Schwarz 1988; 1989; Schwarz 1989; 1989a; S. Fischer 1990; Schmidt-Eenboom 1989; *S+F*, 7:4 (1989): passim]. This position is both questionable and logically incompatible with all forms of (even strictly defensive) military defense, such as *NOD* [Møller 1989c-d] (see *damage limitation*, *Totally Destructive Conventional War*). Finally, without adequate damage limitation, the DD might result in *self-deterrence* and thereby damage deterrence [Afheldt 1976: 324–325; 1977: 447; 1978: 642].

The severity of the DD (particularly in Germany) acted as a spur for a quest for alternative modes of defense. A number of NOD proposals have been advocated as providing "defense without self-destruction," i.e., as facilitating meaningful damage limitation and thereby opening an escape from the wartime DD [e.g., Spannocchi 1976; F. Afheldt 1984; 1984a]. NOD has also been advocated as mitigating the peacetime DD by being cheaper and requiring fewer personnel [Bebermeyer 1984; 1989; 1990; idem & Grass 1984; Grass 1984; 1990]. An alternative escape from the peacetime DD might presumably be dual use of the armed forces; to the extent that they might assist in natural disasters, in environmental protection, and the like, the opportunity costs of a military defense would be minimized [cf. Wellershoff (ed.) 1991: 222–260].

DEFENSE DOMINANCE.   A situation where the defense is stronger than the offense because of prevailing technology, terrain features, or for other reasons. According to *Clausewitz* [1980:361/1984:358 (Book VI.1.2)] and others, the defense is inherently stronger, but this rule holds true only for abstract war; the setting of any concrete war might well benefit an attacker. *WWI* was characterized by DD (thanks, inter alia, to the combination of railroads, barbed wire, and machine guns); *WWII* was, at least partly, characterized by the advantages of the offense (thanks to the combination of *tanks* and *air power*, utilized in a *blitzkrieg* strategy).

DD is a powerful inhibition against war because it prevents a would-be attacker from being able to count on a swift and cheap victory (presupposing, of course, that the attacker is aware of the DD). It thus decreases the likelihood of premeditated war while also diminishing incentives for *preemptive attack* because the party fearing an attack (but cognizant of the DD) would realize that its chances would be improved by remaining on the defensive. To the extent that the DD would render offensive capabilities superfluous for defensive purposes and hence allow for a shift to *NOD*, it would also eliminate an important element in the *security dilemma* [Jervis 1978; cf. Møller 1992: 24–28, 84–89]. A particular instance of the DD (with presumed universal application) is a situation of *mutual defensive superiority* between two contestants, which would promote stability in their reciprocal relations [Boserup 1985]. See also *Defense Efficiency Hypothesis*.

DD is, of course, not an absolute but a matter of degrees, degrees that may well vary over time as well as differ between environments. DD on land, then, does not rule out offensive dominance at sea and/or in the air, or vice versa; nor is it inevitably a fact in all theaters of war. A controversial numerical expression of the DD presumably pertaining to land forces has been the *three-to-one rule* [Mearsheimer 1988; 1989].

DEFENSE EFFICIENCY HYPOTHESIS.   Hypothesis of *Clausewitz and modern *NOD proponents, according to which the defense is inherently superior to the offense, implying that a defender needs fewer forces than a prospective attacker. The inherent superiority is, however, merely inherent and *ceteris paribus;* actual superiority depends, e.g., on technological factors and on the structure of the armed forces [Huber 1988a; idem & Hofmann 1990]. See also *Defense Dominance, *three-to-one rule.

DEFENSE RESEARCH TRUST.   See *Foundation for International Security.

DEFENSE WALL.   Term used by *Norbert Hannig for the forward-fire barrier constituting the main element in his *NOD scheme [Hannig 1981; 1984; 1989], as well as in the modified version developed jointly with *Johannes Gerber [Hannig 1986; 1986b; 1987; Gerber 1987; 1989; cf. Møller 1991:76–80]. See also: *Barriers, *Forward Defense.

DEFENSIVE DEFENSE.   Popular term for *NOD in, e.g., the Scandinavian countries (*defensivt forsvar*) and Germany (*defensive Verteidigung*). It has been used by, among others, *Horst Afheldt [1983] and, on certain occasions, *Anders Boserup and *Robert Neild [Boserup & Neild (eds.) 1990].

DEFENSIVE DETERRENCE.   Term used by the *ADC for a nonoffensive and nonnuclear defense strategy [ADC 1983: 9].

DEFENSIVE SECURITY.   Term used by the *U.N. in a resolution adopted by the General Assembly in 1990, yet with rather unclear connotations.

   Without defining the concept, the resolution stated that "security concepts and policies should be aimed at enhancing security and stability at progressively lower and balanced levels of armed forces and armaments" and that defense capabilities should reflect "true defensive requirements" [Resolution 45/58 O: "Defensive Security Concepts and Policies," passed by the 45th Session of the General Assembly 4 December 90]. To clarify the meaning of DS, an international group of experts was appointed to present a report to the 47th session. The work of the group has included consultations with some of the most prominent international proponents of *NOD, including *Alexei Arbatov, *Andrzej Karkoszka, *Bjørn Møller, *Robert Neild, *Jasjit Singh, and others.

DEFENSIVE SUPERIORITY.   See *Defense Dominance, *Mutual Defensive Superiority.

DEFENSIVE WEAPON.   Term used by a few *NOD advocates (most of them Americans), but above all by critics of NOD who have pointed out (correctly) that nearly all weapons can be used for both offensive and defensive purposes, depending on the circumstances. Attempts at defining a DW have so far been rather unsatisfactory and occasionally harmful, as was the case during the *World Disarmament Conference in 1932 [Noel-Baker 1979; Wiseman 1989:38–55; Borg 1992; Zubok & Kokoshin 1989].

   *Johan Galtung proposed a definition in terms of range and extent of destructive power, according to which a DW was a short-range weapon with a limited radius of destruction, whereas offensive weapons had long ranges and wide impact areas. Even Galtung had to acknowledge that "it would be naive to believe that any component of a weapon system is inherently defensive or offensive: it depends on the total system" [Galtung 1984:172–176]. Nevertheless, he was forced to reinterpret the strategic defenses envisioned in the *SDI program (which obviously relied on "DW" according to his own definition, but which was still "bad") as being merely a cover for "offensive weapons," capable of "igniting cities from space": an attempted rebuttal that revealed a surprising ignorance of the laws of physics [Galtung 1987b].

*George Quester has, alternatively, proposed to define a DW as one making a second-strike posture advantageous, whereas "offensive weapons" would reward striking first in individual engagements [Quester 1986:23]. Most NOD proponents have, however, prudently delegated the *offense/defense distinction to higher levels of analysis, such as military formations, total postures, or strategies. See also *Armaments Dynamics, *Robert Jervis.

DEMOCRACY.   D is sometimes asserted to be a sufficient inhibition against military aggression, and thus both a necessary and sufficient precondition of nonoffensiveness. *Kant thus called for constitutional governments in all states, along with a corresponding arrangement between states (federalism) as a hedge against aggressive war, on the assumption that the common man would shy away from the horrors and costs of war [Kant 1795; cf. Piepmeier 1987; Archibugi 1989; Donaldson 1993].

Modern research has confirmed the negative correlation between D and war involvement: although democratic states wage wars, they never do so against one another but only against dictatorships [Doyle 1993; Gleditsch 1992; Rummel 1979; 1985; 1991; Weede 1984; 1992; Chan 1984; Fukuyama 1991; Ember & al. 1992; Russett 1990; 1993; idem & Antholis 1992; Latham 1993; Mintz & Geva 1993; Starr 1992; Sørensen 1992]. Hence, the democratic revolutions in Europe in 1989 and 1991, according to certain analyses, have created powerful inhibitions against war in all European states and thereby laid solid foundations for *peace [Bächler 1990; 1990b; 1992a; Czempiel 1992; Eberwein 1992; Senghaas 1990e; cf. on the USSR: Gladkov 1991], and in consequence (according to a few analysts) made additional obstacles to war (such as *NOD) superfluous. Certain authors have pursued the ideas of Kant by demanding "global democracy," i.e., democratic forms of government both within and between states [Johansen 1991a-b; cf. Archibugi 1993; Barnaby (ed.) 1991].

Certain alternative defense proponents have explicitly advertised their defense schemes as more democratic than the present system. This has especially been the case for advocates of *civilian-based defense (CBD), which has been described as "a democratic defense for democracies" [Muller 1984: 218–223; Mellon & al. 1985: 45–53; cf. Ebert 1968; 1972]. The foundations for this claim have, first, been the affinity between, on the one side, the means of struggle employed by civil rights movements and in defense of democracy against coups, and, on the other side, the envisioned CBD means of national defense [Sharp 1973; Ebert 1967a–b; 1968; 1969; 1970; 1974; Roberts 1969; 1970; Sharp 1967; Sternstein 1967; 1977; Hornung 1970]. The logic behind such strategies has been based on the assumption that government is possible only by consent and that demonstrative withdrawal of consent would thus be a viable counter to usurpation [Sharp 1973: 8-31, 47]. Second, there may be a correlation between democracy and CBD because the methods of CBD could also be used in a democratic popular war and hence would be incompatible with widespread internal strife and oppression, i.e., would presuppose democracy [Ebert 1968b: 16-17; cf. Carter 1967; Jahn 1971, 1973; 1977; 1984; Wiberg 1977].

Certain advocates of military NOD have also declared there to be a junction between NOD and democracy, i.e., they either have seen NOD as a step toward demilitarization and thereby democratization [Tatchell 1985; cf. Lutz 1989: 27–28;

31–97] or have asserted that the special form of enrollment (allegedly) connected with NOD (a *militia*-like structure) would be conceivable only in democratic societies. Presumably, it would result not in "a militarization of society" but in a "socialization of armed forces" [Hanolka 1984; cf. Lippert, Schönborn & Wachtler 1984]. There does not, however, seem to be any strong correlation between political system and form of defense: European states with NOD-like postures have included both democracies like *Sweden*, *Austria*, and *Switzerland*, and less democratic (and even totalitarian) states like *Yugoslavia* and Romania. Furthermore, the adoption of NOD as (declaratory) military doctrine by the *Soviet Union* and the *Warsaw Pact* occurred prior to the democratic revolutions of 1989–1991.

DEMOCRATIZATION.   See *Democracy*, *Citizen in Uniform*.

DENMARK.   Small democratic capitalist country belonging to several international subsystems: Central Europe (by virtue of the links to Germany); Northern Europe (as part of the Nordic community, the *Nordic Balance* system, and the *Baltic Region*); and the North American/Arctic region (by virtue of its suzerainty over Greenland).

1. **Traditional Security Policy.** After *WWII* D abandoned its traditional neutrality, discredited by the failure to dissuade the German attack in 1940 and the subsequent occupation of 1940–1945. After a futile attempt to forge a Nordic defense community, D became a founding member of *NATO* in 1949. Even though this basic orientation has never been seriously questioned, the NATO commitments of D have all along been circumscribed by two main sets of reservations: to permit no permanent stationing of foreign troops on Danish territory (with the exception of Greenland, hosting U.S. bases [Claesson 1983; Petersen 1988]), and to permit no deployment of nuclear weapons in peacetime [Heurlin 1983; Boel 1988]. The reservations were explicitly motivated by the concern that a foreign military presence might be perceived by the *Soviet Union* as a provocation [cf. Møller 1990c]. Also, the Danish armed forces have observed a number of self-imposed constraints, likewise intended to minimize provocative effects, e.g., by allowing no allied military activities on the island of Bornholm and largely to refrain from a forward naval deployment in the Baltic Sea [Petersen 1988a; Møller 1990b; cf. Holbraad 1986].

2. **Breakdown of Consensus.** For the first time since 1945, the Social Democratic Party (*Socialdemokratiet*) from the early 1980s onward sided with the *peace movements* on all major issues. It rejected NATO's deployment of the *INF* missiles and the *SDI* program and supported *nuclear-weapons-free zones*, *no-first-use*, and *common security*. This resulted in a complete breakdown of the traditional consensus on security policy and the emergence of a strange constellation in the Danish parliament (*Folketinget*): the conservative-liberal coalition government thenceforth had a majority (Socialdemokratiet, the Socialist People's Party, the Left-Wing Socialists, and the Liberal Center Party) against it in matters of security policy. It was thus forced to insert dissenting footnotes into a number of NATO communiqués in matters pertaining to nuclear strategy, SDI, and the like [S.M. Christensen (ed.) 1984; Møller 1990c; idem (ed.) 1985; D.J. Adler 1984; Boel 1988; R. Pedersen 1982; O.K. Pedersen 1989].

This policy continued until 1988, when a motion was passed in Parliament requesting the government to notify visiting ships of the antinuclear policy of Denmark. The government called a general election on the matter, and covertly arranged with the U.S. and British governments to pretend to be alarmed over the new step. Defeated in the election, *Socialdemokratiet* gradually returned to its traditional compromise-seeking attitude, without, however, abandoning its principled positions of the early 1980s [Nordic Study Group 1990; cf. *Information,* 12–16 December 1991].

3. **NOD Debate.** The substance of defense policy used to be a matter of consensus between the major political parties, with only related financial matters being discussed. However, an intense debate on the structure and purpose of the Danish defense flared up in 1986–1987, spurred by a convergence of several (in reality nearly unrelated) trends:

- The publication of a set of proposals for a medium-term reform prepared by Socialdemokratiet (*Knud Damgaard,* with the assistance of a few independent officers, such as *Jens-Jørn Graabaek*) that contained a principled commitment to the principles of *NOD (defensivt forsvar,* i.e., *defensive defense)* [Socialdemokratiet 1986; cf. Damgaard 1984; Graabaek 1979; 1986; 1989].
- The first results of the research project on NOD were being disseminated by the newly established *Centre for Peace and Conflict Research,* particularly *Bjørn Møller* [1985; 1986; 1986a; 1987; 1987a–c; 1988; 1988a–b; 1989; 1989a–g; 1990; 1990a–i; 1991; 1991a–e; 1992; 1992a–b] and *Anders Boserup* [1981; 1982; 1985; 1986; 1986a; 1987; 1988; 1988a-b; 1990; 1990a-e; idem & Clemmesen 1984; idem & Graabaek 1990]. Even though the research did not deal with Danish defense planning (or Denmark at all) but with European matters, there was some convergence between its principled recommendations for more defensive naval postures [Møller 1987; 1989e; Boserup 1987; 1990a] and the proposal of Socialdemokratiet to mothball the largest Danish warships, the frigates.
- The conservative-liberal coalition government launched a vehement attack against the two aforementioned groups, led by Minister of Defense *Hans Engell* [1986; 1986a; cf. Konservative Folkeparti (ed.) 1986], whose lead was followed by most of the officer corps [Bergstein 1985; cf. Sørensen 1990; 1990a]. This was almost exclusively a political smear campaign, but a few officers (such as Lt. Col. *Michael Clemmesen*) [1986; 1986a; 1988; 1990; cf. Boserup & Clemmesen 1984] sought to combine intellectual rigor with a critical attitude toward NOD.

A partial compromise was reached when an independent study group was established under the auspices of the Defense Ministry, with the participation of one of the critics, and tasked with investigating the possibilities of a maritime defense based partly on land-based antiship missiles [FRAG 1986].

Although the alternative defense debate centered on the aforementioned actors, a number of other academics, politicians (particularly to the left of Socialdemokratiet), and representatives of the peace movements also took stands in the NOD debate [I.B. Andersen 1990; Bak-Nielsen 1990; Duus 1983; Godballe 1987; Hjarvad 1985; Thoft 1983; Thoft & Jacobsen (eds.) 1983; Høringsgruppen 1980; Jensen 1989; Kraft 1983; Pedersen 1981; Petersen 1981; 1982; B. Sørensen 1985; 1985a–b; 1986]. The peace movements, however, showed a more intense interest in *civilian-based defense* [F.S. Andersen (ed.) 1975; Boserup & Mack 1974; Callesen 1989; Hansen (ed.) 1990; Thoft (ed.) 1974], a precursor of which might be said to be the resistance movement during the German occupation 1940–1945 [Bennett 1967; Bennike 1946; Brandt & Christensen 1968; Haestrup 1954; 1959:1–2; Outze 1946; Poch 1970; Trommer 1971].

DEPLETION ZONE.   See *Disengagement, *Zones.*

DESERT SHIELD.   See *Gulf War.*

DESERT STORM, OPERATION.   See *Gulf War*.

DESERT WARFARE.   See *Lawrence of Arabia*, *Algeria*, *Gulf War*.

DÉTENTE.   D has been defined as "a continuation of conflict by means other than war, i.e., amity between enemies." In this minimalist version, D is perfectly compatible with *Realism*; D resembles what *Barry Buzan* called "a mature (or perhaps only half-matured) anarchy" [Goldmann 1988: 82; cf. Buzan 1991: 176–185]. It is thus something less than a *common security* regime, and certainly only *peace* in a narrow sense of the term. It allows states to continue in the pursuit of their national interests (even "defined in terms of power"), and even at the expense of one another, only not to the point of risking war. Accordingly, D may be conceived of as a continuum ranging from highly adversarial to more cooperative types of relations, whose most meaningful distinction is not either/or but more or less.

In the realm of politics there have been different definitions of D, indeed the different meanings attached to the term by the two superpowers played an important role in eroding the D regime between them through the late 1970s [Garthoff 1985: 24–68; cf. Gelman 1986]: The *Soviet Union* understood D as largely synonymous with *peaceful coexistence*, i.e., as a "continuation of the class struggle by other means," whereas the *U.S.* preferred to interpret is as a commitment to a status quo policy, hence saw any increase in Soviet influence around the world as a violation of its D obligations.

Furthermore, the USSR held a very broad view of internal affairs, including existing spheres of influence, a view that contrasted sharply with the U.S. conception, according to which *Eastern Europe* was certainly not "off-limits." What the U.S. regarded as ordinary foreign policy activities (such as the Carter administration's human rights campaign) were thus perceived by the USSR as intolerable "interference in its 'internal affairs'" [Kubalkova & Cruckshank 1990: 106–108, 145–154, 165–167; Jahn 1986: 448–451; Jegorow 1972; Nygren 1984].

A final limitation in the original conceptions (and practice) of D was that the political *rapprochement* was accompanied by no genuine *disarmament* but allowed the *arms race* to proceed nearly unabated, apart from certain *arms control* regulations. The absence of disarmament was logically compatible with the dominant conception of arms control (to create a stable military environment), but it led to a growing frustration in the public and the elites of both superpowers [cf. Blechman 1983].

The new D, which started around 1986/1987, rested on more solid foundations in several respects. Above all, *new thinking* in the Soviet Union during the "*Gorbachev revolution*" was far more compatible with Western conceptions of D:

- The previous taboo on the human rights issue was lifted;
- The dogmas of the inherent aggressiveness of imperialism was abandoned;
- The notion of *reasonable sufficiency* superseded the imperative of maintaining parity with the West, thus providing a wide margin for unilateral disarmament;
- The urgency of, and shared responsibility for, global problems was acknowledged; and
- *Glasnost* as well as the emergence of genuine *democracy* contributed to dismantling the enemy image held of the USSR by the West.

DETERRENCE.   See *Nuclear Deterrence*, *Deterrence by Denial*, *Existential Deterrence*, *Dissuasion*.

DETERRENCE BY DENIAL.   Also known as "deterrence by defense" as opposed to deterrence by (the threat) of punishment (or retaliation) [G. Snyder 1961: 14-16].

The underlying assumption is that a would-be aggressor undertakes cost-benefit calculations and will attack only if expected gains outweigh anticipated costs. Although deterrence by retaliation is tantamount to dissuading an adversarial state from aggression by increasing the costs thereof, say by counter-city nuclear attack, DBD seeks primarily to minimize the gains from aggression by an effective defense.

Even though the (primarily military) costs of aggression are automatically also increased through DBD, these costs tend to be in terms of battlefield casualties and material losses, whereas the (presumably innocent) civilian population of the aggressor is largely spared. Just as this is less questionable from an ethical point of view (cf. *Just War*) than to retaliate against civilians, it is also inherently more credible than a retaliation likely to be followed by counterretaliation (cf. *self-deterrence*, *nuclear deterrence*).

DEWA.   See *Defense Wall*.

DICK, CHARLES.   See *U.K.*

DIGBY, JAMES.   See *PGM*.

DISARMAMENT.   D signifies either a process of reductions of military inventories or one possible end point of such a process: a situation of general and complete disarmament.

1. **History of D.** D has been a controversial issue in the idealism versus *Realism* dispute. Idealism dominated the interwar period. The 1932 *World Disarmament Conference* constituted an ambitious quest for disarmament, above all in the categories of "offensive weapons," yet fell miserably short of its aspirations [Noel-Baker 1979; Neild 1990: 119–144; Borg 1992]. Idealism also exerted a considerable influence in the decade immediately after *WWII*, when all the founders of the *U.N.* shared (or at least proclaimed to share) the goal of general and complete D (GCD), as well as the more specific objective of nuclear abolition [Myrdal 1976: 66–84; Newhouse 1989; 63–64; Anon. (ed.) 1986: 106–117]. Since the 1960s, however, the goal of D has ceded its dominance to the less ambitious concept of *arms control*, which may, but need not, involve partial D [A. Carter 1989; cf. Blechman 1983; Schelling 1986; idem & Halperin 1961]. Only with the *INF*, *START*, and *CFE* treaties of the late 1980s and early 1990s did D again become an objective regarded as realistic.

2. **Pros and Cons of GCD.** The case for GCD is obvious and at first glance persuasive: if all states were to disarm completely, war would become inconceivable and the resources devoted to war preparation could be made available for civilian consumption (cf. *Conversion*). A closer analysis reveals several flaws in this argument:

- GCD is not tantamount to irreversible D. Indeed, if all states should disarm to zero (yet retain the capability for rearmament) a breakout from the D regime might be easy and quite rewarding, unless GCD were to be accompanied by effective *collective security* safeguards: implying, of course, that D would be neither general nor complete but merely that only supranational institutions would be allowed military means.
- The *security dilemma* militates strongly against GCD because each state

would tend to fear that its respective adversaries might have concealed weapons, or might clandestinely build some, a very persuasive argument for doing so oneself.

- Finally, armaments alone are not the cause of war but, rather, a manifestation of hostility and fear of war. Only if war came to be regarded as inconceivable, might GCD become politically realistic. This would, however, be exactly the circumstances under which it would be least useful.

3. **Partial D.** That GCD may be either unrealistic or irrelevant is not a valid argument against partial D, i.e., the reduction of particular types of weapons, either globally or within a certain region. The quest for such partial D has indeed been one of the most powerful motivating factors behind *NOD thinking, and a shift to NOD-type armed forces would presumably facilitate D by eliminating the action-reaction element in the *armament dynamics*.

DISENGAGEMENT.   D has two different, albeit related, meanings, distinguished by, inter alia, a difference in perspective. It can thus signify either

- D *of* somebody, i.e., the separation of two parties to a conflict (e.g., the superpowers or the *alliances*) by the interposition of a neutral *zone* between them; or
- D *from* something, i.e., the withdrawal of one of two sides from a conflict, e.g., a U.S. D from Europe.

1. **History.** D enjoyed a considerable popularity in the 1950s, when it was viewed as a means of lessening East-West tension in general, and in particular as a way of facilitating German (re)unification (see *Germany Plans*, *Rapacki Plan*, *Gomulka Plan*, *George Kennan*, *Eden Plans*, *Gaitskell Plan*, *Healy Plan*, *Bogislav von Bonin*) [Howard 1958; Albrecht (ed.) 1986; Møller 1991: 43–56]. The D ideas experienced a certain renaissance in the 1980s, once again with a view to decreasing tension, and thus as a defensive response to the *INF* deployment decision and the resultant deterioration of East-West relations. In addition to notions of *nuclear-weapons-free zones* and corridors (cf. *Palme Commission*), proposals abounded for D of conventional forces, as well as for political D [Møller 1991: 80-85].

2. **Political and Military D.** Notions of political D have been based on the assumption that the bloc system as such was a source of danger, which is why ways were sought of loosening alliance *integration* with a view to avoiding mutual entrapment [cf. G. Snyder 1984; J. Sharp 1987; 1987a]. Hence the numerous proposals for *dealignment* and/or *neutralization* [Albrecht 1982; 1983; 1983a; ADC 1987; Kaldor & Falk (eds.) 1987; cf. Ørvik 1986], some of which have also envisioned a certain military D.

Military D is characterized by the creation of a (in some sense) neutralized zone between the main forces of the two opposing sides: either a completely demilitarized zone or a zone depleted of certain types of armaments, such as *tanks* or offensive weapons in general, i.e., a "restricted military area." Because the required relocation of offensive-capable forces would necessitate modifications of residual arsenals, proposals for military D have often been accompanied by (or merely constituted components of) elaborate designs for force restructuring, usually in the direction of *NOD*-type forces [e.g., Löser 1981; E. Afheldt 1984; Lodgaard & Berg 1985; Brzezinsky 1988; Bahr 1987a; idem & al. 1988; Dean 1989c: 232–242; 1988b].

3. **D Logic.** The logic behind military D is that otherwise offensive-capable and mutually threatening forces would be less so, the greater the distance between them

and the dividing line, e.g., because spatial separation would impede *surprise attack*. The resultant need for *mobilization* and/or forward movement of attack formations would provide the defender with valuable additional *warning* time. Because this would allow the defender (in which role both sides might well see themselves) to build down its state of readiness, each side would come to constitute a diminished threat to the other, and an improvement of *crisis stability* would ensue. This would, however, presuppose that effective yet strictly defensive means of *forward defense* would be installed in the place of the withdrawn units.

DISPERSION.   Distribution of the armed forces in small units over an extended surface, i.e., the opposite of *concentration.

As a principle of deployment in most *NOD models, D has been motivated by the intention to present as few lucrative targets to enemy strikes as possible (by substituting many, individually insignificant targets for a few concentrated targets: the *no-target principle), thereby enhancing survivability and making *preemptive attack seem less rewarding. Furthermore, the D of the armed forces also provides the adversary with longer *warning because forces would have to be concentrated before an attack could be launched. D would thus be a contribution to *crisis stability (see also *Guerrilla Warfare, *Horst Afheldt).

DISSUASION.   Generic term for various forms of war prevention through the manipulation of a prospective aggressor's motivation. Even though D is simply the French term for deterrence, it and its German synonym *Abhaltung* are often preferred (especially by alternative defense proponents) to "deterrence" because it does not carry the same connotation of terrifying to an opponent by a threatening posture [Lutz 1988d; 1989; Schmückle & Müller 1988; Mellon & al. 1985; cf. Däniker 1987; 1990a].

DREGGER, ALFRED.   Parliamentary leader of the right-wing "Steel Helmet Faction" (*Stahlhelm Fraktion*) within the *CDU.

A generally hawkish attitude toward defense and security matters notwithstanding, AD was one of the first to include at least parts of the *NOD vocabulary in the party's security political discourse. In a 1988 article, he set out his recommendations for a strategy for *NATO, wherein he included the objective of "bringing about a mutual non-aggression capability" [Dregger 1988: 411; cf. 1989].

During the unification debate in 1990, AD furthermore came quite close to an advocacy of NOD as a means to facilitate unification. He emphasized, e.g., that "we Germans are in any case prepared to make military concessions to our eastern neighbors and the *Soviet Union, as far as possible," and stressed the importance of "*disarmament to a minimum of conventional and nuclear weapons, as well as a transformation of the structures of armaments, implying a strengthening of the defender and a weakening of the potential aggressor, depriving either side of any invasionary potential vis-à-vis the respective other side" [*Die Welt,* 31 January 1990].

AD also suggested the replacement of *forward defense with an alternative defense posture: the forward area (i.e., the territory of the former *GDR) would be defended by "territorial forces, not under NATO command," whereas allied forces were to be organized in multinational mobile formations, stationed in the rear, "not merely on German territory" [*Frankfurter Allgemeine Zeitung,* 22 March 1990]. The proposal was reminiscent not only of the *Genscher Plan but also of various NOD models, particularly those of *Löser, *Bülow, and other *disengagement advocates.

DRONES.   See *RPVs*.

DUMAS, LLOYD J.   Professor of economics and political economy at the University of Texas. LJD is one of the most prominent experts on *conversion* in the U.S. [1988; 1989; idem (ed.) 1988; idem & Thee (eds.) 1989] and has also devoted some attention to the peace-promoting effects of *interdependence* and to the need to face up to the global challenges [1990].

DUNAY, PÁL.   See *Hungary*.

DÜNNE, LUDGER.   Captain in the Bundeswehr, temporarily affiliated with the Hamburg peace research institute *IFSH*. In 1988, under the auspices of the IFSH, he presented an elaborate proposal for an alternative defense [1988; 1988a].

LD's point of departure was a permissive definition of *NOD* as the incapability of border-crossing operations with political or military significance, implying that neither minor invasions nor even large-scale air or missile strikes would be impermissible. Skeptical toward *forward defense* (with the fallback option of nuclear *escalation*), LD also rejected stationary defense schemes (such as that of *Horst Afheldt*) because of their inability to reconquer lost terrain and because of (allegedly) prohibitive costs [1988a: 13–14, 22–26, 32–40].

LD's alternative was a defense adapted to the terrain and employing only well-tested military technologies. The main weapons systems would include (besides ordinary *infantry* equipment) armored *air defense* vehicles mounted with both guns and missiles; helicopters for combat (especially antitank); multiple-launch rocket systems (MLRS); infantry fighting vehicles; and small *tanks* in light, airmobile versions. Countermobility measures would be very important, for which purpose he envisioned the MLRS being used for minelaying, in addition to (as a long-term perspective) landscaping, i.e., the transformation of tank-friendly into tank-hostile terrain by the planting of trees, the digging of canals, etc. [1988a: 123–124]. The posture would be a genuine combined-arms posture, wherein helicopters would play a very important role.

That this would presumably entail no offensive capabilities was due to the lack of ordinary tanks, without which major border-crossing operations were inconceivable. The airmobile armored vehicles would be inferior in term of firepower to the enemy's tanks on his territory as well as only tactically mobile [1988a: 128–129]. LD's proposal thus resembled both the model of *Norbert Hannig* (in the permissive attitude toward aerial means of combat) and that of the *SAS, the "spiders" of which were deliberately handicapped on enemy ground, just as were LD's heliborne tanks.

DÜRR, HANS-PETER.   West German scientist at the Max Planck Institute in Starnberg, who has, inter alia, published many appeals for a redirection of resources from the *arms race* to the solution of global problems [1987; 1988]. In the mid-1980s, HPD was directly involved in the German *NOD* debate in his capacity as codirector of the research project Stability-Oriented Security Policy at the Max Planck Institute, the administrative director of which was *Albrecht A. C. von Müller* [Dürr & al. 1984].

DYSON, FREEMAN.   Professor of physics at Princeton University for many years, and consultant to the Defense Department and the Arms Control and Disarmament Agency of the *U.S.* In *Weapons and Hope* [1984] he advocated radical nuclear *disarmament*, combined with a *NOD*-like conventional posture, making nuclear deterrence dispensable. This was, according to him, a *David-and-*

Goliath approach whereby the weak could defeat (or at least stand up to) the more powerful. Although it might also be an argument in favor of a shift to *civilian-based defense*, he regarded it only as a realistic objective for the long term because, e.g., it required for its success "an improvement in the moral standards of mankind" [1984: 45–55; 258–285].

E

EASTERN EUROPE. Generic (and geographically misleading) term for the countries belonging to the former *Warsaw Pact* and the Comecon (*Poland*, the former *GDR*, former Czechoslovakia, *Hungary*, Bulgaria, and Romania), along with the nonaligned but communist states, the former *Yugoslavia* and Albania. Occasionally, the former *Soviet Union*, or parts thereof, was counted as belonging to EE.

EBERT, THEODOR. Professor at the Free University in Berlin and the leading German advocate of *civilian-based defense* (CBD) and *nonviolence* (NV) since the late 1960s.

TE's writings on CBD are numerous and cover most angles of the topic. He has, e.g., analyzed at length the methods of NV for purposes of internal political struggle in *Non-Violent Rebellion: An Alternative to Civil War* and various other writings on the struggle against capitalism [1968; cf. 1972]. However, to at least the same extent, he has followed the experience of NV resistance against communist dictatorships, with the 1953 East German rebellion and the Czechoslovak resistance against the 1968 invasion as the outstanding examples [1967a–b; 1969; 1970]. In continuity therewith, he has surveyed the use of NV methods by the new social movements of the 1970s, such as the women's movement and the environmental groups [1978; 1978b].

TE saw a direct link between the internal use of NV and a national defense by means of CBD: "the best training in social defense is the 'non-violent rebellion' against exploitation, authoritarian rule and discrimination in one's own country. Whoever has rehearsed independent action of the 'citizen against establishment type' will be best capable of finding appropriate forms of resistance against putschists and aggressors in a crisis situation" [1967: 94]. TE also regarded CBD as the appropriate strategy for national defense, not least because of its *damage limitation* potentials, and because it constituted a viable replacement for *nuclear deterrence*. CBD would, furthermore, be more relevant than military defense because it would defend the most important values, namely, human beings and social structures rather than borders and regimes, hence the term *social defense*, signifying a defense *of* society as a whole *by* the entire society [1970a; 1972; 1974a; 1978a; 1980a; 1984a; 1987]. In this respect, CBD resembled *guerilla strategy*, indeed constituted a form of "democratic people's war" suitable for industrial societies: "You maximize the nonviolent, political and psychological, civilian and social elements of *guerilla warfare*, and minimize the violent, military and territorial elements," whereby "the quantitative change in guerilla warfare at a certain stage is transformed into a new quality of social defense" [1968: 16–17].

The seeming affinity between CBD and *NOD* notwithstanding, TE was far from enthusiastic about the new defense concepts of *Horst Afheldt* and others, albeit on balance favorably inclined. He described it as "by far the lesser evil, compared with *forward defense*," and appreciated the better opportunities NOD provided for a combination of military with CBD methods, and thus for gradually replacing the former with the latter [1974a; 1977a; 1978a; 1983; 1984b; 1987b]. In light of the emerging prospects of German unification, TE joined forces with *Die Grünen* (the Green Party) and the *peace movements* in demanding a demilitarization of Germany [1990], inter alia under the auspices of the *BOA* campaign.

ECONOMIC SECURITY.   See *Security*, *Total Defense*.

EDEN PLAN(S).   The first EP (named after British Foreign Secretary Anthony Eden) launched the WEU; the second (1955) was intended to ensure "that the reunification of Germany, and its right to join with other countries at its own choice, does not entail dangers for anybody else." It envisioned negotiations about the level of armed forces in Germany and in the neighboring countries, and even opened up the possibility of creating a demilitarized *zone* between East and West [Albrecht (ed.) 1986:68; Schramm & al. (eds.) 1972:5–6].

EDI.   European Defense Initiative. See *Extended Air Defense*.

EGRIT.   See *Extended GRIT*.

EMERGING TECHNOLOGIES.   See *E.T.*

END.   European Nuclear Disarmament. See *Peace Movements*, *Mary Kaldor*.

ENDURANCE.   See *Time Perspective*.

ENEMY IMAGES.   See *Common Security*.

ENGELL, HANS.   Former minister of defense (for the Conservative People's Party) in *Denmark*, subsequently minister of justice. Since 1993 leader of the party, now in opposition.
    In 1986, HE launched a campaign against *NOD* ("*defensive defense*"), motivated, e.g., by the publication of a new social democratic set of proposals for medium-term defense planning that contained a principled commitment to NOD, albeit only as a long-term goal [Socialdemokratiet 1986]. HE published the pamphlet *Defensive Defense: A Defense Without Credibility* and a book in which he attacked the whole range of alternative defense and security proposals: NOD, the Nordic *nuclear-weapons-free zone*, etc. [Engell 1986; 1986a]. Thanks to his position, HE managed to make NOD a "hot" issue, on which the majority of the Danish officer corps expressed opinions in the press, almost without exception in agreement that NOD was a dangerous conception whose adoption would inevitably result in Denmark's being expelled from *NATO* [cf. Sørensen 1990a].

ENGLAND.   See *U.K.*

EPPELMANN, RAINER.   Protestant priest in the *GDR*, pacifist and former conscientious objector, for which "crime" he was imprisoned for eight months. After the 1989 revolution, RE was appointed minister of defense of the coalition government, a post he accepted only on the condition that the ministry be renamed Ministry of Disarmament and Defense (as a matter of principle, in that order) [*Der Spiegel*, no. 17 (1990): 22–24; no. 30 (1990): 33–37]. Upon unification, RE's ministry was closed.
    During RE's tenure, the ministry began issuing the journal *Militärreform in der DDR* (Military Reform in the GDR), featuring an intense debate on *NOD* and simi-

lar matters. In a programmatical speech on 2 May 1990, RE described the new government's goals as *disarmament* and *conversion*, the latter in the dual sense of "conversion from offensive to defensive structures" and "conversion from military to civilian purposes." Nevertheless, he proclaimed the intention of maintaining an independent army "of an appropriate size and strictly defensively structured." The rationale was that "on the territory of the present GDR, there must be no military vacuum, wherefore demilitarization was unacceptable to us" [*Militärreform in der DDR,* 1990:17]. Paradoxically, the pacifist RE now argued in favor of *conscription*, whereas the leader of the former SED, now rebaptized PDS (the party that had imprisoned RE for conscientious objection) advocated its abolition [*Der Spiegel,* 1990:24, 29–31; *Neues Deutschland,* 23 April 1990. See also Møller 1991: 196–259].

EPPLER, ERHARD. Member of *SPD* presidium, and one of the very first in the party to take *NOD seriously.

In *Ways out of Peril* (1981), EE recommended, as a presumed escape from the *arms race,* adoption by the *FRG of *defensive defense* (conceived along the lines of *Horst Afheldt's* proposals). He further argued in favor of a West German endeavor toward *Europeanization,* meaning that "the European shirt must be closer to her body than the Atlantic coat" [Eppler 1981: 82–90, 211, 213–216].

EPSTEIN, JOSHUA M. Senior fellow of the Brookings Institution (Washington) and author of several works on the methodology and practice of force planning as well as on *arms control* [1985; 1987; 1988; 1989; 1990].

JME's application of so-called dynamic analysis (analysis of the development of force ratios over time, determined, inter alia, by the speed of *mobilization,* deployment schedules, etc.) to the measurement of the conventional force balance in Europe rendered rather optimistic results, showing that *NATO had quite a fair chance of thwarting a Soviet attack without recourse to nuclear *escalation* [1988; cf. 1985]. At the same time, he criticized "simplistic" models of *defense dominance,* such as the *three-to-one rule* [1989].

In 1990 JME published a detailed analysis of the prospects of the *CFE negotiations: *Conventional Force Reductions: A Dynamic Assessment.* His conclusion was that the security of the West could be considerably improved, and at substantially reduced costs, as a consequence of the unilateral *Warsaw Pact* reductions, the CFE limits, and the additional 50 percent reductions that JME recommended as the appropriate goal for a "CFE-II" agreement [1990: 2–3]. One premise of this optimistic conclusion was that the existing (1989) balance was favorable, and that the decisive factor was "dexterity," i.e., NATO's ability to detect, and determine the size and direction of, an attack in progress, decide on the appropriate response, and actually move its forces into position [1990: 14, 24].

Whereas dexterity would automatically improve as a result of Warsaw Pact cuts and CFE ceilings (effectively depriving the East of its *surprise attack* option), further improvements might be achievable through modifications of NATO's posture as well as (through arms control) that of the East. JME especially recommended a more extensive use of various tank *barriers,* and a defensive restructuring of armored divisions, *in casu* a substitution of antitank and *air defense* weapons for *tanks.* Furthermore, a withdrawal of ammunition stocks and reductions in *bridge-building* and mine-clearing equipment would severely reduce offensive capabilities [1990: 64–72].

JME hence recommended *NOD-like transformation of force postures in Europe as an essential building block for the emerging new Europe: "Obstacle-bolstered defenses distributed in depth might well prove to be the sinews of *collective secu-

rity in Europe. If all major potential adversaries were to adopt such defenses, the likelihood of crises between any two such actors escalating to war would almost surely be reduced" [1990: 79].

ESCALATION.   Generic term for the expansion and/or intensification of a conflict. In principle, there are two main categories of E, even though the term is often used exclusively about vertical E:

- Vertical E. The intensification of conflict, e.g., from conventional through tactical to strategic nuclear war, such as implied by the *flexible-response* strategy of *NATO* [cf. Stromseth 1988; Daalder 1991].
- Horizontal E. The geographical expansion of the battlefield, such as that envisioned by the *U.S.* in the late 1970s and early 1980s: through a combination of the Rapid Deployment Force (subsequently renamed CentCom) and the U.S. Navy, it was planned to respond to Soviet aggression both at the scene and by "taking the war to the enemy" by posing a threat to Soviet interests elsewhere, e.g., by entering submarine *bastions*, such as envisioned under the *Maritime Strategy* [Epstein 1983; 1981; 1987; cf. Klare 1981; Mearsheimer 1986: 74–76].

Whereas both vertical and horizontal E are highly offensive and destabilizing, alternative defense advocates have also envisioned (more benign forms of) E, sometimes without realizing it, and certainly without using the term *escalation*:

- The *web*-like configuration of most *NOD* schemes would actually be a means of horizontal E, albeit only on the local ("tactical") scale. While deliberately refraining from strategic *counteroffensives*, the web would ensure the possibility of tactical counterattacks, thereby making it impossible for an invader to "close the front" because each defending unit would be in a position to reinforce its neighbors.
- Furthermore, by taking advantage of *existential deterrence* (albeit perhaps only in the form of weaponless "blueprint deterrence"), i.e., by designing defenses to be effective in a *subnuclear setting*, NOD advocates capitalized on the possibility of E to the level of nuclear war, even though to avert this development remained their primary objective.

ESECS.   European Security Study Group. High-level independent study group, the steering members of which included, among others, Hugh Beach, McGeorge Bundy, Lord Carver, Lawrence Freedman, the late Johan Jørgen Holst, Michael Howard, William Kaufmann, and Richard Ullman. In 1983 the group published its influential *Strengthening Conventional Deterrence in Europe: Proposals for the 1980s,* which was followed in 1985 by a report of a special panel [ESECS 1983; 1985].

Even though ESECS accepted *flexible response*, it supported a raising of the nuclear *threshold* by strengthening conventional forces, particularly through deep interdiction of *Warsaw Pact* follow-on forces [1983: 9, 19]. It explicitly argued that "NATO needs response measures which are . . . non-provocative," yet lent its support to *deep strike* operations, including *OCA* (offensive counter-air), i.e., long-range missile attacks against the Warsaw Pact's main airfields, albeit only after the first (offensive) sortie of the latter. By destroying the return airfields, *NATO* would force the enemy aircraft to use secondary airfields, where they could be destroyed by means of aircraft [1983: 20–22]. The *FOFA*-like elements of the recommendations of ESECS were, furthermore, argued to be compatible with a defensive posture by virtue of "blocking Soviet offensive options without itself posing an offensive threat" [1983: 205]. The recommendations were primarily for high-tech or even *E.T.* weapons programs, and both the envisioned schedules and cost estimates have been criticized for unfounded optimism (they were modified in the 1985 report) [1983: 24–31, 197–208; Cotter 1983; cf. ESECS 1985].

ESECS could be viewed as an example of the mainstream strategic community's approaching some of the notions prevalent in the alternative defense community, in this case the *no-first-use* proposals.

ESTONIA.   See *Baltic Region.*

E.T.   Emerging technologies. Generic term for various technologies under development that have military implications, particularly in the field of microelectronics and directed energy.

ET have obvious potentials for both offensive and defensive military purposes. Their offensive utility would primarily be a function of the improved destruction technologies (fuel air explosives, submunition dispensers, etc.) that allow for long-range destruction and *nuclear equivalence,* which makes ET suitable, e.g., for *deep strikes,* such as planned for under the *FOFA* doctrine (or perhaps rather program) of *NATO* [De Lauer 1986; Tegnelia 1985; Boyer 1985; Brown & Leney 1984; Gordon 1984; Burgess 1986; ESECS 1983 (particularly Cotter 1983); OTA 1987; Berg & Herolf 1984; Herolf 1986a; 1988; Borg & Grin 1986; Schulte 1990]. It has, however, been questioned whether these potentials have been exaggerated and/or the opportunity costs underestimated with regard to conventional improvements in NATO's conventional defenses, in terms of, say, greater amounts of ammunition [Canby 1985; 1985a; 1987].

The defensive utility of ET is primarily a function of miniaturization, which would allow for a subdivision of firepower. Even small *infantry* teams might, for instance, acquire the ability to engage enemy *tanks,* aircraft, and helicopters. This would allow for a largely stationary, *web*-like defense posture with no significant offensive capabilities. *PGMs* (precision-guided munitions) for antitank, antiship, and *air defense* purposes have thus played an important role in most *NOD* schemes [Barnaby 1986; 1990; Dankbaar 1982; 1986; Hannig 1981; 1984; 1986; 1989; Hofmann & Huber 1990; Afheldt 1976; 1983; Acker 1984; Löser 1986a; A. v. Müller 1985; 1986; 1987a; Neild 1986; Saperstein 1985; Walker 1981; 1985; 1986; Boskma & Meer 1986; Brünner 1988; Bölkow 1990; Herolf 1983; 1988; 1988a]. Indeed, it may have been the demonstrated potentials of PGMs, especially during the 1973 Yom Kippur War, that spurred the initial interest in NOD-like defense schemes [Weller 1974; Morse 1975; 1976; Kennedy 1979; Miksche 1977; Mearsheimer 1979; Digby 1975; cf. Klippenberg 1977; Cameron 1983].

Gradually, other applications of ET have met with growing interest in the NOD community, including *RPVs* (remotely piloted vehicles, i.e., drones) and sophisticated mines. At the same time, parts of the NOD community have abandoned their high hopes for ET and focused more on strategic, tactical, and structural features, coming to the realization that ET may develop much slower than anticipated and often at higher costs, and that ET are susceptible to countermeasures [Unterseher 1987c; Møller 1990i; Grin 1990a; Orbons 1986; cf. Uhle-Wettler 1980; Gummett 1988; Dick 1986; Karber 1986a; Davis 1990]. A final field of ET application (between the offensive and defensive) is that of strategic defenses, primarily BMD (ballistic missile defense). See also *SDI.*

ETHICS.   See *Just War.*

ETZIONI, AMITAI.   American sociologist and one of the founders of *gradual-*

ism, with *The Hard Way to Peace* [1962] and with his case study of the prehistory of the first Test Ban Treaty (the so-called Kennedy Experiment) as an instance of successful gradualism.

EUCIS.   European Centre for International Security. Located in Neubiberg, Germany, with *Albrecht A.C. von Müller* as director (and effectively only employee). Until the death of *Anders Boserup*, EUCIS had a branch in Copenhagen (with AB as director and sole staff member).

EUROPEANIZATION.   E can refer either to a redistribution of burdens and/or influence within *NATO* from the *U.S.* to the European members or to the development of a European "identity," e.g., in defense and security matters.

   1. **Burdens and Influence.** Ever since its creation, NATO has been characterized by hegemony, i.e., by a U.S. dominance with which the Europeans have become increasingly weary [Kaldor 1983; 1987]. As long as the U.S. was clearly stronger (economically, politically, and militarily) than the Europeans, and as long as nuclear weapons were regarded as indispensable, the Europeans expressed their resentments sotto voce. However, as both these pillars of hegemony became shaky—with the EC now surpassing the U.S. along most dimensions, and with the Soviet threat receding (accompanied by a lesser perceived need for nuclear deterrence)—previously latent disagreements have come out into the open [cf. Kennedy 1988: 514–535; Kaldor 1978: 14–28; Senghaas 1986: 23–50; 1988: 94–110; Ramsbotham 1989; Lucas 1988; 1990; Binnendijk 1989; Brauch 1989; J. George 1990].

   Another factor pushing for E has been the ongoing U.S. dissatisfaction with the European contribution to the alliance, hence a persistence of isolationist trends among the public and among the elite [Kelleher 1983; cf. Krauss 1986: 17–28]. A partial explanation of the U.S. impression of the Europeans as free riders has been the disagreement on the *out-of-area* question. Most European states have abandoned previous aspirations for global power, and hence have felt no obligation to reward or reimburse the U.S. for expenses involved with the maintenance of U.S. global power. Further, they have tended to doubt the value of interventions [Bills 1979; Bentinck 1986; Luard 1987; Sherwood 1990: 149–183].

   2. **E Within NATO.** Numerous proposals have been made by Europeans and Americans for E, ranging from modest reforms such as the appointment of a European instead of an American SACEUR (Supreme Allied Commander Europe) to radical suggestions that the U.S. should leave NATO for good, thereby transforming it into an exclusively West European alliance [Kissinger 1984; H. Schmidt 1986: 176–177; 1987; Eberle 1988; cf. Calleo 1987: 162–163; Taylor 1988]. The 1984 resurrection of the WEU (Western European Union) from oblivion, as well as the Franco-German defense cooperation reflect the same trend toward E [Cahen 1986; 1986a; Hintermann 1988; Kaiser & Lellouche (eds.) 1986; Lucas 1988: 209–216].

   3. **Pan-E.** The prevailing attitudes in the alternative defense community toward the aforementioned forms of E have ranged from skeptical to negative. The WEU has been criticized for being too militaristic and obsessed with *nuclear deterrence* [Grünen (eds.) 1985], and Franco-German collaboration has been seen as either infeasible or undesirable: Infeasible because of the unbridgeable gap between the French "nuclear-infected" strategy and German conventional strategy; undesirable, if France should succeed in persuading the Germans to place greater emphasis on nuclear weapons. Furthermore, any exclusively Western form of E has been criticized for blocking the route to broader forms of E, and thus for merely cementing

the division of Europe into two adversarial blocs [Böge 1991a; Borkenhagen 1988; Brock & al. 1988].

As an alternative, most alternative defense and security proponents have pointed to pan-European solutions (especially after the 1989 and 1991 revolutions), which should involve the (former) *Warsaw Pact* countries as well as the neutrals and nonaligned. Additionally, they have underlined the need for deemphasizing the military dimension, thereby ruling out the WEU as a suitable vehicle for E. Much preferable would be an all-European system, based on the principle of *collective security* (and *NOD* insofar as national armed forces are concerned) and resting on the *CSCE* as its main institutional foundation, albeit perhaps with the EC (now EU) playing a significant role [H. Afheldt 1989a; Ammon 1990; E. Bahr 1988a; H-E. Bahr 1990; Barth 1990; Birkenbach & al. 1986; Böge 1991; Booth 1990; 1991d; Borkenhagen 1991; Boudreau 1987; 1987a; 1990; Brauch 1985; 1990; Buro 1990; Chalmers 1990; 1990a; Ehrhart 1991a; Evera 1990; Faber 1990; Galtung 1989b; Glotz 1990; Hopman 1991; Hyde-Price 1991; IFSH 1990; Kaldor 1990a; Kiessling 1989a; 1990a; Kirby 1991; Krause 1990b; Kupchan & Kupchan 1991; Lübkemeier 1990b–c; Lutz 1990b; Mueller 1990; A. Müller 1990e; Møller 1990; 1991b; Remacle 1990; B. Richter 1991; Risse-Kappen 1990; 1991; Rotfeld 1991; Schütze 1990; Senghaas 1990a–e; Sivonen 1990; P. Smith 1990; J. Snyder 1990; 1990a; Ullman 1991; Voigt 1989; 1990a–f].

EUROPEAN CENTRE FOR INTERNATIONAL SECURITY.   See *EUCIS*.

EUROPEAN NUCLEAR DISARMAMENT.   END. All-European *peace movement*. See also *Mary Kaldor*.

EUROPEAN SECURITY STUDY GROUP.   See *ESECS*.

EVENSEN, JENS.   Former Norwegian minister of the Law of the Sea (for the Social Democrats) and author of a draft treaty for a *nuclear-weapons-free zone* in the Nordic region [1983].

EVERA, STEPHEN VAN.   Assistant professor at MIT, and former managing editor of *International Security* (1984–1987).

SVE's point of departure was a historical and theoretical analysis of the implications of offensive doctrines, with the pre–*WWI* "cult of the offensive" serving as the classical example of the dangers inherent in this approach (cf. *Schlieffen Plan*, *Plan 17*) [1985; cf. 1984; 1985a]. One of the lessons that SVE drew from this example was that counterforce (the nuclear counterpart of offensive conventional strategies) was dangerous, whereas pure deterrence (MAD: mutual assured destruction) was far more stable.

He has subsequently applied the same principles to conventional force planning, e.g., in a detailed critique of defense policy that charged the Reagan administration with abandoning the (basically defensive) objective of containing the Soviet threat in favor of a more offensive political strategy. Furthermore, "the Administration's emphasis on conventional offense, counterforce, and nuclear warfighting raises the risk that a conventional conflict will escalate to a general thermonuclear war" [Posen & Evera 1983: 59]. As pointed out by SVE, even if certain countries (such as *Israel*) might in fact need offensive capabilities, the *U.S.* was certainly not one of them. On the contrary, it would be "uniquely able to rely on a defensive strategy," just as would the *Soviet Union* [Evera 1989: 102].

Whereas most *NOD* proponents have asserted that defensive strategies are always preferable to offensive ones, SVE's advocacy of the defensive has been somewhat more qualified. Occasionally, the offense has the advantage (say, for technological or geopolitical reasons), in which case states would be penalized for being excessively defensive. However, Europe did not belong to this category, according to SVE, who in the wake of the East European 1989 revolutions assessed the prospects for *peace on the Continent as favorable, nothwithstanding the dangers flowing from nationalism, *multipolarity, etc. A precondition would be, however, that the U.S. would remain in Europe and that *NATO would be retained, albeit modified to become a *collective security mechanism [1990; cf. 1990a].

EXISTENTIAL DETERRENCE.   Effects of the very existence of nuclear weapons, regardless of their numbers and deployment modes, and the doctrines for their use [Bundy 1986; cf. Jervis 1988; or for a (partly sympathetic) critique: Freedman 1988; 1991; or J.N. Miller 1988]. According to the theory of ED, the possibility of *escalation to the use of nuclear weapons would inevitably influence the calculus of military planners and prevent unlimited warfare (such as large-scale invasion and/or terror bombing) from appearing worthwhile because nuclear escalation would remain possible and become increasingly likely the closer the aggressor would come to victory. By implication, conventional wars (at least in Europe) would always be fought in the "nuclear shadow," i.e., in a *subnuclear setting [Afheldt 1988:402; Boserup 1989; 1990b; Neild 1990:46–73; cf. Beaufre 1963:73–75; 1965: 107–123], where the rules of classical (Clausewitzian) warfare would no longer fully apply. Belief in the ED thesis might thus facilitate a shift to a *no-first-use strategy and/or *minimum deterrence, or even to a "blueprint deterrence" [Schell 1984:112–123]. It remains disputed, however, whether a minimum of actually deployed and usable nuclear weapons might be required for ED, as well as whether *nuclear deterrence, in whatever form, might be required at all, in the face of numerous nonnuclear inhibitions against large-scale war [Mueller 1988; 1989]. See also *Totally Destructive Conventional War.

EXTENDED AIR DEFENSE.   *NATO plans from the mid-1980s for creating a defense against tactical and short-range ballistic missiles and, in some versions, also against cruise missiles: an ATM (antitactical missile) or ATBM (antitactical ballistic missile) defense.

By means of its SS-21, -22, and -23 missiles (partly deployed in response to NATO's deployment of *INF missiles since 1983) the *Soviet Union would presumably be in a position to launch a disarming strike against NATO's airfields, command posts, nuclear storage sites, etc. without crossing the nuclear *threshold. Against this new type of threat *nuclear deterrence was believed to be ineffective because the onus for initiating nuclear war would be placed on NATO. The stipulated Soviet plans, furthermore, seemed to fit the picture of a Soviet increased emphasis on *deep strikes and *blitzkrieg conceptions [Wörner 1986; Enders 1986; Daalder 1987; Hafner & Roper (eds.) 1988].

EAD, however, also fit in closely with the *SDI (Strategic Defense Initiative) program of the Reagan administration and was even occasionally couched in terms of a "European Defense Initiative" (EDI) [High Frontier Europe 1986]. EDI might thus be interpreted as a U.S. scheme for developing the first stages of a full-fledged SDI (for defense against Soviet strategic missiles) in circumvention of the *ABM Treaty and without congressional approval [Brauch 1987b; Borg & Smit (eds.) 1987; Altmann & al. 1987].

The original rationale for EAD vanished along with the Soviet missiles it was designed to defend against, but something like EAD may be nearly the only legacy of the original SDI plan: a limited (and far-from-leak-proof) defense against minor nuclear powers and other states in possession of ballistic missiles, i.e., Third World powers such as Iraq, renamed *GPALS*, Global Protection Against Limited Strikes. The Patriot *SAM* (envisioned as an important component of EAD) did indeed demonstrate similar potential against the Iraqi SCUD missiles during the *Gulf War* [Friedman 1991: 349-342; cf. Postol 1991], and many states might want to procure such limited defenses. However, there are other ways of meeting the problem of ballistic missile proliferation [Nolan 1991], e.g., through *arms control* in general, and a strengthening of the Missile Technology Control Regime (*MTCR*) in particular [Forsberg 1987b; Skjold 1989; Anthony 1991a; Bailey 1991a; Carus 1990; A. Karp 1991a; 1992; Kniest 1992; Mack 1991b; Nolan 1992; Shuey 1992; Zimmerman 1992; Findlay (ed.) 1991; Neuneck & Ischebeck (eds.) 1992].

EXTENDED DETERRENCE.   Dissuasive effect of one state's *nuclear deterrence* on contingencies other than a direct nuclear attack against itself [cf. Huth 1988; Cimbala 1989].

1. **Categorization.** In principle, there are two main categories of ED, as well as various combinations thereof:

- Vertical ED, the deterrence of nonnuclear aggression; and
- Horizontal ED, the deterrence of attacks against other states by virtue of the implicit threat of nuclear retaliation.

Usually, though, the term signifies the second category and has been applied almost exclusively to the extension of U.S. ED over its allies, above all *NATO*-Europe and South Korea. This ED has constituted almost the *raison d'être* of NATO ever since its foundation in 1949.

2. **Credibility Problem.** ED suffers from a serious credibility problem because the stakes for which the provider of ED should presumably be willing to risk war (and thereby perhaps annihilation) would by definition be less than vital interests, implying the need "to sacrifice New York for the sake of Munich" [Kissinger 1979; cf. Sigal 1983b: 7–23]. Hence the perennial disagreements between the *U.S.*, eager to raise the nuclear *threshold* (thereby postponing the moment of truth when the credibility of its ED might be put to the test), and the Europeans, who have been unwilling to strengthen their conventional defenses sufficiently to render ED superfluous. This reluctance has been due not only to European desire for a "defense on the cheap" but also to the belief that sufficient conventional forces would further weaken the credibility of the nuclear *escalation* option, thereby eroding deterrence.

A fragile compromise between these conflicting interests was found in 1967 in the form of the *flexible response* strategy, on which the U.S. had unilaterally based its military planning since around 1962 [Stromseth 1988; Daalder 1991; cf. Freedman 1989a: 227–244]. The implied flexibility logically presupposed a belief in the possibility of limiting escalation, i.e., of waging a limited nuclear war, in order to possess options of calibrated responses to various levels of provocation, and of making a conventional war in progress appear increasingly unattractive to the aggressor, thereby, it was hoped, facilitating *war termination* on acceptable terms.

Flexible response was also believed to enhance the credibility of ED by constituting "a threat that leaves something to chance" [Schelling 1960: 187–203], i.e., by causing uncertainty in the mind of the prospective aggressor of the potential response to aggression, thus rendering it impossible to calculate the costs thereof. Furthermore, ED was believed to require a whole panoply of nuclear forces of different ranges, with multiple levels of destructive power, and deployed at many different locations (preferably *in situ*). Such deployment was inherently unstable because tactical nuclear launch authority would have to be delegated to field commanders at a certain point, something that would, however, merely strengthen deterrence by increasing uncertainty [cf. Kelleher 1987; 1988; Bracken 1987a] while, on the other hand, damaging *crisis stability*.

3. **ED, No-First-Use, and NOD.** The requirements of ED have been used as arguments against nuclear *disarmament*, *no-first-use*, *nuclear-weapons-free zones*, etc. According to skeptics, however, to the extent that ED has ever been a reality, it has been profoundly political, i.e., a matter of an adversary's perceptions of the likely political deliberations on the part of the holder of ED [cf. Ravenal 1988; Cimbala 1990: 72–95]. There would thus be no particular reason to expect ED to suffer from, say, the adoption of a no-first-use *policy* because the inalienable first-use *option* would remain.

EXTENDED GRIT.   Special variety of *GRIT*, described by *Joshua Goldstein* and *John Freeman* [1990: 15–16, 31–32, 130–136], that differed from the original proposal by being sustainable over a long period. It would consist of rather sporadic, but repeated, unilateral cooperative initiatives, which would tend to transform the image held of a state by its adversaries, i.e., to dismantle an unjustified enemy image. See also *Progressive GRIT*, *Super-GRIT*.

F

FABIAN STRATEGY. (Also F tactics). Strategy employed by Quintus Fabius (known as "*Cunctator*," i.e., "The Hesitant") during the Second Punic War of Rome against Hannibal's Carthage. The FS was a way of exploiting the strategic advantages of the defense against a (at least tactically) superior invader by refusing any decisive battle and striking against the enemy's weak points, including its *logistics*, while enjoying the advantages of the home ground, not least the availability of fortified cities and a loyal locally conscripted militia: "the astute Fabius, having determined not to expose himself to any risk or to venture on a battle, but to make the safety of the army under his command his first and chief aim, adhered steadfastly to his purpose" [Polybius: *Histories* III: 86.9–10, 87.1–2; A. Jones 1988: 65–70; Strauss & Ober 1990: 146–161]. These qualities made the FS a forerunner, in certain respects, of modern *guerilla warfare* (cf. *Mao Zedong*) and other manifestations of the *indirect approach*, as well as of the *nonbattle strategy* of modern *NOD* advocates such as *Brossollet* [Hart 1967: 26-28; Brossollet 1975].

FALK, RICHARD A. See *U.S.*

FDP. Freie Demokratische Partei (Free Democratic Party) of the *FRG*, the leader of which has been the former foreign minister *Hans Dietrich Genscher* (HDG). Although aligned with the *CDU*, the FDP had a more favorable assessment of *NOD* and other security political alternatives. This has probably reflected the FDP's experience with *Ostpolitik* and HDG's profound understanding of the requirements of *arms control* [cf. Møller 1991: 194–195, 232–234].

As early as 1986, HDG proclaimed NOD as the relevant goal for Europe: "It is necessary to create cooperative security structures in Europe. The existing *alliances* should continue, but with regard to the weapons systems, the *logistics*, the force structure, the geographical deployment, and the respective doctrines, our armed forces should be shaped in such a way that both sides have the capabilities only for defense, but not for offense and invasion" [quoted in *NOD*, 6 (1986): 4].

Since 1988–1990, military doctrines and NOD have been part and parcel of the arms control agenda (indirectly in the *CST* and *CFE* negotiations and directly in the *Vienna Seminars*), in which connection HDG argued in favor of deeper cuts than those favored by the CDU. He also proposed a second round of the CFE that would focus directly on the defensive restructuring of armed forces [*Der Spiegel,* 30 November 1987; cf. Genscher: speech to the CSCE follow-up conference 5 July 1988, *Bulletin,* 5 July 1988, 858]. In his elaborations on the specifics of conventional arms control objectives, HDG revealed a clear inspiration from the NOD debate, the greatest resemblance being to the *SAS*. He pointed, e.g., to the need for modifications of *C^3I* structures, logistics, engineering support, and the like so as to tie the mobile forces to their home territory [*Bulletin,* 14 June 1988, 785; cf. *Frankfurter Rundschau,* 7 April 1988].

Generally, the FDP was quite willing to allay Soviet concerns, e.g., with regard to German unification, hence the *Genscher Plan*, which envisioned a special status for the territory of the *GDR, as well as quite deep cuts in Bundeswehr strength [*Atlantic News,* 28 March 1990; *Der Spiegel,* no. 8 (1990): 17-18; no. 17 (1990): 20].

With regard to political conceptions the FDP has been more favorable than the CDU toward cooperative and alternative conceptions. While deeming alliances useful for the transition period, the FDP has never conceived of them as perennial but has envisioned their eventual replacement with a *collective security* system. Toward that goal, the West should collaborate with the East in creating "cooperative security structures," in which connection HDG also spoke quite approvingly of *Gorbachev*'s notion of the "*Common European House*" [Genscher 1987: 443; *Atlantic News,* 24 March 1990; *Der Spiegel,* no. 10 (1990): 26–31; no. 20 (1990): 29]. In the same vein, the FDP also resurrected from near-oblivion its traditional notion of European federalism, finding this to be a suitable framework for facilitating German unification by embedding it in a broader context: a federation of all Germans within a grand European federation [*Der Spiegel,* no. 39 (1989): 26].

FEDERAL REPUBLIC OF GERMANY.   See *FRG.*

FIGHTER AIRCRAFT.   See *Air Defense.*

FINLAND.   Neutral, democratic, capitalist state that has alternated through the centuries between Swedish and Russian domination. F acquired independence in connection with the 1917 Russian Revolution and since then has been a member of the Nordic community (and the *Nordic Balance*) alongside *Sweden,* *Denmark,* and *Norway.*

1. **Geography.** F's geographical and geopolitical situation is the most constant factor that policymakers have had to take into account for centuries. First, its proximity to Russia (the *Soviet Union*) in general and St. Petersburg (Leningrad) in particular, hence its strategic importance to the USSR and the periodic territorial disputes (Karelia, Porkkala). Second, its harsh climate and terrain (lakes, marshes, forests) have made the country difficult for an attacker to occupy. Third, the uneven distribution of the population: the North has always been very sparsely settled, the South host to the great majority.

2. **The Winter War.** In 1939 the Soviet Union attacked F over territorial claims. Even though they eventually lost, the Finnish defense forces did remarkably well, exploiting their familiarity with the terrain and the climate (ski patrols, etc.). After its defeat in 1940 and the German attack on the Soviet Union in 1941, F aligned itself with Germany by counterattacking the USSR [Vuorenmaa 1985; 1985a; Juutilainen 1985; Ahto 1985; Ries 1989: 1–171].

3. **Paasikivi-Kekkonnen Line.** As a former German ally, F had certain constraints placed on its defense and security policy in the 1947 Paris Peace Treaty (as was the case for other such allies) [Visuri 1991]. Furthermore, in 1948 F was forced to sign the FCMA (Friendship, Cooperation, and Mutual Assistance) Treaty with the Soviet Union whereby it was committed to prevent the use of its territory for attacks against the USSR. However, successive Finnish governments managed to avoid any actual application of both the consultation and the mutual-assistance clauses, which might have entailed the stationing of Soviet forces, and perhaps even nuclear weapons, on Finnish territory [Möttölä 1984; R. Allison 1985: 174–175; Joenniemi 1989; 1989a].

The Finnish policy of equidistance between the East and West (in security political terms) has far exceeded the implications of *neutrality* in international law [Hakovirta 1988; Joenniemi 1986; 1988; 1989; 1989a; idem & Vesa (eds.) 1988; Mouritzen 1988; Möttölä 1984; R. Väyrynen 1983; 1987; 1987a; P. Väyrynen

1989]. This has been not so much a function of enforced concessions as a policy of deliberate adaptation to the unchangeable circumstances, and of conflict prevention (rather than solution): the so-called Paasikivi-Kekkonen line, which has constituted a blend of *dissuasion* and nonprovocation vis-à-vis the Soviet Union.

4. **Defense Doctrine.** Finnish defense policy has always emphasized the dual tasks of efficiency and nonprovocation, above all in order to relieve the USSR of any fear of (or basis for legitimation based on) attack by or through Finland. Paradoxical though it may seem, Finland has even to some extent been able to play out one constraint against the other by using the FCMA obligations (requiring, inter alia, an effective defense) to mitigate the Paris Treaty's constraints on weaponry, which might have resulted in inefficiency. It developed a defense structure that was strictly nonoffensive, yet (according to most assessments) quite effective in spite of its modest consumption of financial resources [Kanninen & Tervasmäki 1985; Visuri 1985; 1990; 1991; Väyrynen 1982; 1989; Ries 1989; 1989a: 175–375; Kemp (ed.) 1990].

The defense policy has also mustered a considerable consensus. The alternative defense debate (including that on *NOD*) has focused on alternatives for *NATO*, as well as on general principles, in which connection Pekka Visuri and Artto Nokkala have stood out as the main proponents [Nokkala 1990; 1991; Visuri 1989; 1990a]. In addition, there have been isolated voices proposing a unilateral *disarmament* [Haukijärvi 1989].

5. **Post–Cold War Developments.** Probably in an attempt to reap the fruits of the new *détente*, the Finnish government on 21 September 1990 unilaterally declared the Paris treaty "obsolete" with reference to the developments in Germany. Simultaneously, President Mauno Koivisto also termed the FCMA treaty obsolete, while acknowledging the continued validity of its objectives. Nobody, not even the Soviet Union, seemed to take notice of this peculiar instance of small-power *unilateralism* [*Europa-Archiv*, no. 20 (1990): Z-204].

FINLANDIZATION.   Pejorative term for a partial *neutralization* of a *NATO* member state, conceived of as an enforced concession to the *Soviet Union*. The term has thus been based upon a profound misunderstanding of the actual security policy of *Finland* [Mouritzen 1988; cf. Quester 1990].

The term has been used about states such as *Denmark* and *Norway* to describe the (allegedly) logical outcome of the trends toward strengthening the nuclear reservations and minimizing offensive capabilities. It was also used about the *FRG* as a caricature of what might have happened had the *SPD* and the *peace movements* succeeded in preventing the *INF* deployment in 1983 [cf. e.g. Holmes 1984; Pfaltzgraff & al. 1983; Kaltefleiter & al. 1989; Herf 1991].

The term has also been used, with positive connotations, about a possible (and desirable) security political guideline for former *Warsaw Pact* member states (minus, of course, the former East Germany). They might become militarily neutral and nonaligned, underpinned, e.g., by bilateral nonaggression treaties with the USSR, and with close economic and political, but no military, ties to the West [Albrecht 1982; 1983; 1983a; 1985; Lutz 1988a; Møller 1991b]. Furthermore, through a shift to *territorial defense* and *NOD* strategies and postures, resembling those of Finland or *Austria* [cf. Roberts 1976; Danspeckgruber 1986; Fuhrer 1987; Kruzel 1989; Bundeskanzleramt 1985; Rotter 1984; Vetschera 1985], the states might remove any conceivable justification for preventive Soviet attack. Furthermore, were their defense efforts genuinely *tous azimuts,* they would even shield the USSR from an attack from the West, a shield that the West might well know that the USSR did not need, but that Soviet conservatives might still find comforting (and that would strengthen the position of Soviet reformers).

The rethinking of military doctrine under way in the Warsaw Pact by the time of the 1989 upheaval might thus simply be continued by the new noncommunist governments, which might even be able to build upon certain national traditions that were noticeable in the 1950s and 1960s, such as the notions of the "Polish Front" and the conception of territorial defense contained in the 1968 *Gottwald Memorandum* [C. Jones 1981; 1981a; 1984; cf. Holden 1989]. The former *Eastern Europe* would thus come to constitute a wide *buffer zone*, wherein neither the USSR nor NATO (nor the WEU) would have a military foothold. Moreover, were these countries to maintain some economic and political links with the Soviet Union, they might add the function of bridge to that of a buffer, i.e., not merely separate the potential adversaries militarily but also link them together and help forge cooperative relations.

FIRE BARRIER.   Potential "killing zone" to be established at the onset of hostilities in a predefined area (most often along the border) so as to prevent an invasion. The means of doing so might be barrage fire (from either tube or, more often, rocket artillery), precision-guided indirect fire by means of sophisticated missiles, and/or minefields (including perhaps artillery-emplaced mines). The FB conception has played a central role particularly in the *NOD* schemes of *Norbert Hannig* [1984; 1986; 1986b] and *Albrecht A.C. von Müller* [1986; 1986b]. There has, however, been near-unanimity that even the best FB would be insufficient on its own because of its inevitable vulnerability to countermeasures, e.g., directed against the defender's *C^3I nodes.

FIRST WORLD WAR.   See *WWI*.

FISCHER, DIETRICH.   Professor of economics in the U.S., but of Swiss origin, and with a continuing interest in the security and defense policy of *Switzerland* [1982].
    DF's main work about *NOD* was *Preventing War in the Nuclear Age* [1984], which was inspired by *Johan Galtung*. It provided an elaborate criticism of traditional notions of *balance* as a guarantee of peace and demonstrated how the quest for balance acted as a spur to the *arms race*. This presupposed, however, that the emphasis was placed on what he (unfortunately) called "offensive weapons," such as *tanks*, bombers, and amphibious craft, which were supposedly characterized by long ranges and rewards for the initiator of combat. A shift of emphasis toward *defensive weapons* (e.g., antitank weapons, mines, *air defense* weapons, coastal artillery, and small patrol craft) combined with abundant *civil defense* measures would, on the other hand, allow one side to improve its security without endangering that of its opponent, thereby facilitating *disarmament* [1984: 47–71].
    On the basis of this analysis, DF recommended a *transarmament* to a "general defense" posture consisting of *territorial defense* forces (*militias* and guerillas) collaborating with the civilian population, in its turn engaged in *civilian-based defense* [1984: 108–132, 154–170]. In 1987 DF made largely the same recommendations in the context of a work authored jointly with *Wilhelm Nolte* and *Jan Øberg* [Fischer & al. 1987].
    In 1993 DF published a major work on nonmilitary aspects of *security*, in which he maintained his advocacy of NOD yet urged a refocusing of attention away from military matters toward other types of security problems, related, inter alia, to the environment [1993].

FISCHER, SIEGFRIED.   East German naval captain and lecturer at the Military Academy until German unification. After the 1989 revolution, he joined the antimilitarist study group *SES. Subsequent to its disbandment and the dissolution of the East German navy, DS was affiliated with the private research foundation BITS (Berliner Informationszentrum für Transatlantische Sicherheit [Berlin Information Center for Transatlantic Security]) until his emigration to the Canary Islands in 1993. He has been one of the most prominent proponents of the *Totally Destructive Conventional War hypothesis* [1990; 1994].

FLANAGAN, STEPHEN J.   See *U.S.

FLEXIBLE RESPONSE.   One of the two main pillars (alongside *forward defense*) of the strategy of *NATO, according to which the defense of the alliance against an attack would proceed in three stages:

- direct conventional defense;
- *escalation* to the use of tactical and theater nuclear weapons; and
- full-scale escalation, entailing the use of strategic nuclear weapons.

1. **History.** In 1967 FR superseded "massive retaliation" (which had guided U.S. strategic planning since the early 1960s) because of the latter's inherent credibility problems. It presupposed a U.S. willingness to risk a war against the *Soviet Union* (by then also a nuclear power), which might well be tantamount to national suicide for less than vital interests. This was a general problem with *nuclear deterrence*, but it was particularly severe in the context of an *alliance*. FR was contrived as a compromise between the U.S. interest in raising the nuclear *threshold* through a strengthening of conventional defense and the European interest in preserving, and, it was hoped, strengthening *extended deterrence* by coupling the U.S. as inseverably to the security of Europe as possible [Stromseth 1988; Freedman 1989a: 285–329; Cimbala 1990: 72–95; Ravenal 1988; Daalder 1991].

Although NATO has not officially abandoned FR, FR has been significantly modified as a result of the collapse of the Eastern bloc. NATO has thus acknowledged that nuclear escalation will henceforth merely be a last-resort option (cf. the *London Declaration*). Even though it might be argued that this has all along been the case (the question being when to resort to the last resort), the acknowledgment undoubtedly signified the intention to raise the nuclear threshold.

2. **Problems and Inconsistencies.** Understandably for a political compromise, FR has been fraught with logical inconsistencies:

- Even though European NATO members have all along rejected the conceivability of limited nuclear war, FR has presupposed precisely this, namely, that the second rung on the escalation ladder would be discernible from the third. If not, it would make no sense not to launch full-scale retaliation in response to an attack, as envisioned under massive retaliation.
- Even though maintenance of strict political control has been envisioned through all three stages, credibility would require that the threats "leave something to chance," implying, e.g., a subdelegation of tactical launch authority to field commanders past a certain point [cf. Schelling 1960: 187–203].
- Even though what would presumably deter a prospective aggressor was the possibility of strategic nuclear war (by means of strategic nuclear weapons), FR was believed to necessitate the forward deployment of nuclear forces, some of which (battlefield nuclear weapons) were conceived of as means of war fighting as well as coupling, whereas others (the *INF weapons) were primarily substrategic means of coupling.
- Although escalation should be postponed, and nuclear forces thus withheld

during the initial fighting, it should not be possible to withhold them indefinitely. That field commanders were bound to face "use 'em or lose 'em" dilemmas has been regarded as strengthening deterrence, but it would also tend to increase the risk of crossing the nuclear threshold inadvertently [Posen 1991].

* Although various safeguards were built into these forces (separation of warheads from launchers, special custodian teams in charge of the warheads, installation of permissive action links, etc.) with a view to preventing unauthorized launches, all such safeguards would gradually have to be relaxed as the war progressed (or perhaps even under severe crisis conditions) in order to preserve the deterrent effect of tactical nuclear systems [Sigal 1983; Kelleher 1987; 1988; Bracken 1987].

3. **FR, No-First Use, and NOD.** FR logically presupposed that the option of nuclear first-use was kept open, and was thus incompatible with a *no-first-use* policy [Kennan 1982:3–10; cf. Bundy & al. 1982; Steinbruner & Sigal (eds.) 1983; Blackaby & al. (eds.) 1984]. It was, however, perfectly compatible with raising the nuclear threshold considerably, i.e., with a *no-early-first-use* policy [cf. Harris 1988] such as that adopted by NATO in 1990.

Nearly all alternative defense (and especially *NOD*) schemes have presupposed an abandonment of FR, by envisioning the relegation of nuclear weapons to the sole function of deterring nuclear attack (through a *minimum deterrence* strategy) or even abolishing them entirely, leaving behind "merely" a degree of *existential deterrence* based on the possibility of their being rebuilt, i.e., a "blueprint deterrence" [Bundy 1986, Schell 1984].

FM 100-5.   U.S. Army's *Field Manual 100-5: Operations.* The 1982 edition of the manual constituted the most authoritative exposé of the *AirLand Battle* doctrine. In 1986 it was superseded by a new edition containing marginal revisions in, but retaining the particular (extremely offensive) orientation of, AirLand Battle [FM 100-5 1982; 1986].

FOFA.   Follow-on forces attack. New doctrine adopted by *NATO* in 1984 as one element of the *Rogers Plan* [Rogers 1983; 1985; Berg & Herolf 1984; OTA 1987; Cardwell 1986; Farndale 1988; Wijk 1986; 1986a; 1989; Flanagan 1986; 1987; Abshire 1987; cf. Tegnelia 1985; Canby 1985; 1985a; Herolf 1986a; 1988; Burgess 1986; Dugan 1990; Sharfman 1991; Skingsley 1991], and "marketed" as a contribution to raising the nuclear *threshold.*

FOFA has envisioned *deep strikes*, by means of missiles and aircraft, against advancing *Warsaw Pact* forces, supposedly organized in echelons on all levels: tactical, operational, and strategic (inter alia because of *force-to-space* constraints). Through such deep interdiction, NATO could presumably prevent the enemy follow-on forces (particularly the second operational echelon and reserve forces) from ever reaching the FEBA (forward edge of battle area), thus facilitating *forward defense* by improving the effective force-to-force ratio (see *Force Comparison*). According to the least ambitious versions of FOFA, these strikes would be directed against choke points (bridges, railway junctures, and the like); more ambitious versions envisioned striking enemy forces directly, while they were on the move.

Notwithstanding the NATO consensus on FOFA in principle, there remained a considerable divergence of opinion between the *U.S.* and Euro-NATO with regard to specifics. The U.S. tended to prefer very deep interdiction, primarily with hi-tech means (preferably missiles and *E.T.* systems); the Europeans tended to prefer aircraft and other proven technologies. Furthermore, they were concerned about the opportunity costs of striking deep in terms of thinning out forward defense at the FEBA.

FOFA has been criticized by technology skeptics [Canby 1985; 1985a; 1989] and the *SPD [Bahr 1984: 281, 283; Voigt 1984; 1987a; Glotz 1987], as well as by the entire alternative defense community [Afheldt 1983: 81; 1987b: 220–229; Hannig 1986b; SALSS 1984a: 33–34; IFSH 1985; Schmähling 1988c: 70; Huber 1988: 25–28; Hagena 1990b: 92–97; Møller 1987a; 1988a; 1989b; 1990i], with a few exceptions [Löser & Anderson 1984: 45, 55; Löser 1985: 525–26, 531; E. Müller 1984: 142–144]. *NOD proponents have charged FOFA with being too offensive and with jeopardizing *crisis stability because the capabilities for deep strikes would simultaneously create (1) incentives for *preemptive attacks (in order to prevent disarming strikes) and (2) options of doing so, say, by striking at the missile sites and airfields.

Whatever rationale FOFA may have had seems to have rapidly disappeared with the Soviet conventional force reductions (see *CFE) and the dissolution of the Warsaw Pact. Furthermore, planning for strikes against (Soviet formations in) *Poland and what used to be the *GDR (such as was originally the intention) was also rendered absurd by the withdrawal of Soviet forces from, and *democratization of, these countries. Nevertheless, by early 1994 NATO had not yet abandoned FOFA but merely modified its official rationale. According to General Eberhard Eimler, deputy SACEUR, FOFA would be even more sensible than ever because of the increased need for rapid *concentration under conditions of lower force density [*Jane's Defence Weekly*, 19 May 1990, 984].

FOLLOW-ON FORCES ATTACK.   See *FOFA.

FORCE COMPARISON.   Assessment of numerical force-to-force ratios between two or several states. FC has achieved political saliency as a corollary of (the traditional approach to) *arms control and the associated quest for *balance. On closer analysis, FC tells very little about the actual *security of states, or about one side's chances of prevailing in an actual contest of strength.

1. **Units of Measurement.** FC has frequently been the object of manipulation by both *NATO and the *Warsaw Pact, not merely for propagandistic purposes (cf. the critique of NATO's and the Pentagon's FCs) [Gervasi 1987; Cockburn 1984; Bülow 1984a] but also because of genuine difficulties involved with FC [cf. Epstein 1985; 1987; 1990: 6–29; Biddle 1988; Mearsheimer 1982; 1988; idem & al. 1989; Posen 1984a; 1988; Blackwill 1988; Clemmesen 1988; Dupuy 1979]:

- The almost inevitable asymmetry of postures (in its turn a reflection of geostrategic asymmetries) implies a need for comparing incommensurables: firepower with mobility, long-range with short-range mobility, manpower with weapons systems, conventional with nuclear weapons, aircraft with ground-forces equipment, warships with artillery, etc.
- Qualitative factors are obviously important, yet are very difficult to measure objectively, as well as to weigh against numbers: whether one high-quality *tank equals, say, two inferior ones or not depends on concrete circumstances [Chalmers & Unterseher 1987; 1988; 1989].
- The strength of a given unit or weapon depends on the use to which it is put, e.g., on whether it is part of a stipulated offensive or a defensive force posture (cf. *offense/defense distinction).

Attempts at assessing weighted unit values for different types of military formations have been made, the best known of which are the U.S. Army's "Armored Division Equivalents" (ADE). However, synergies between weapons systems and formations are difficult to take into due account, and to compare ADEs with, say, naval power makes little sense.

2. **Static versus Dynamics FC.** Peacetime force ratios are inevitably only very crude approximations of the actual ratio between the forces engaging each other in battle. An aggressor would have mobilized prior to an attack (yet probably not to maximum strength, for fear of thereby alerting the defender); the defender would usually be at least a few days behind. The *mobilization* and force-deployment schedules would thus be determinants of combat force-to-force ratios of at least the same importance as peacetime strengths [Epstein 1985; 1987; 1990]. However, because such comparisons presuppose the stipulation of one side as the aggressor and the other as the defender, the estimates by both sides will automatically differ considerably, and it becomes almost impossible to identify mutually acceptable force ratios [see Møller 1992: 79–84].

FORCE-TO-FORCE RATIO.   See *Force Comparison.*

FORCE-TO-SPACE.   Density of the deployment of military forces, measured as the number of troops or formations per square kilometer, or kilometer of front.

1. **Constant and Variable Factors.** States are not entirely free to choose the appropriate FTS ratio, inter alia, because of differences in population density: the *Soviet Union* and *Canada* (to say nothing of Greenland), e.g., have extremely low ratios; "crowded" countries such as the *Netherlands* or *Israel* are in a position to opt for high ratios. Furthermore, not merely the aggregate FTS ratio for each country matters. Equally important are the ratios along a specific front (in the case of a *forward defense*) or within a specific area (in the case of a *territorial defense*). Unfortunately, the actual distribution of the population rarely fits the ideal distribution of the armed forces because population densities are sometimes low along the most likely avenues of attack.

2. **FTS Minima and/or Maxima?** In the course of the *CFE* negotiations (especially in the preparatory phase) arguments abounded to the effect that *NATO* had no margin for reductions because of FTS constraints: Western forces were allegedly already spread too thin along the Central Front, i.e., were close to an FTS minimum, usually set at one division per 25 km of front [Epstein 1990: 57]. At lower densities than this, holes would appear in the front through which an aggressor might penetrate. Hence the demand for extremely asymmetrical reduction rates, almost tantamount to a unilateral Eastern build-down [Thomson & Ganz 1988; Galvin 1989]. One might, however, also assert the existence of absolute FTS maxima because there are limits (depending, among other things, on terrain features) to the number of forces an aggressor would be able actually to bring to bear against the other side (cf. *Crossing-the-T*).

3. **FTS, Strategy, and Posture.** Quite a convincing case has been made against the assertion of absolute FTS minima, e.g., by *Joshua Epstein* [1990: 51–67]. First, defensive FTS minima are not independent of force-to-force ratios, inter alia because of offensive FTS minima. Even if a local penetration of the forward defense line (through a hole) might be possible, no meaningful exploitation would be possible unless the penetrated forces were of a certain size. Furthermore, the defender would be in a position unilaterally to modify the FTS minima, e.g., by emplacement of obstacles, such as tank *barriers*, or through *terrain modification*; or he might seek to do so by means of negotiations with a view to depriving opposing forces of their shock power. Finally, by shifting to an area-covering territorial

defense [e.g., H. Afheldt 1983], the defending side could make penetrations by enemy forces less of a problem. (See also *Concentration, *Raj Gupta).

FORSBERG, RANDALL WATSON. See *IDDS.

FORTIFICATIONS. See *Barriers.

FORUM FOR SECURITY COOPERATION. All-European institution with headquarters in Vienna, established in September 1992 by the *CSCE, according to the decision at the Helsinki Summit in July [Kuglitsch 1992; Ghebaldi 1993; Möller-Gulland 1993].

The purpose of the FSC is to provide an integrated forum for negotiations on *arms control, *disarmament, and *CBSMs (confidence- and security-building measures), as well as to facilitate "a goal-oriented dialogue on security enhancement and cooperation." In the Program for Immediate Action, it was envisioned to give early attention to cooperation in defense *conversion [Focus on Vienna, 28 (November 1992): 1–5; Atlantic News, 25 September 1992, 2]. Other items under discussion have included nonproliferation of weapons and *arms trade regulations, and the development of a code of conduct in military-matters security. In the latter connection, several conceptions stemming from the *common security and *NOD debate were proposed [Focus on Vienna, 29 (April 1993): 2–4].

FORUM ON THE PROBLEMS OF PEACE AND WAR. Forum per i problemi della pace e della guerra. Group of researchers, based in Florence, Italy, who have dealt with *NOD and *common security with a special emphasis on *Italy and the Mediterranean region [Ragionieri (ed.) 1989; Cerutti & Ragionieri (eds.) 1990]. Its members include *Rodolfo Ragionieri and Furio Cerutti.

FORWARD DEFENSE. FD may signify either any form of defense based on forward deployment or a special form of national defense that seeks to deny an aggressor access to the country by means of a strong frontier defense.

1. **Forward Deployment.** The rationale for a forward deployment, such as that envisioned under the U.S. Navy's *Maritime Strategy, may be to tie up opposing forces in a particular location, thereby maintaining the initiative. Furthermore, it may be motivated by a quest for *damage limitation because it would ensure that whatever combat were to take place would cause a minimum of damage in the territory and society to be defended. Finally, forward deployment might reflect *alliance commitments in general and might presumably support *extended deterrence. Accordingly, the stationing of U.S. forces and nuclear weapons in Europe was valued by the *U.S. for extending the U.S. military reach as well as for its contribution to damage limitation; the Europeans tended to emphasize the coupling effects of forward deployment, which would presumably make it impossible for the U.S. to stay aloof from any major war in Europe.

Forward deployment has drawbacks as well. First, it may be militarily disadvantageous because it hampers logistical and other support for the defending forces, operating as it does far from base and with long lines of lateral communication and reinforcement [Gupta 1993]. Second, it may appear provocative because a forward defensive posture may be indistinguishable from an offensive one. Just as the West was concerned about the forward stationing of Soviet forces, especially in East Germany [cf. Lippert 1987; MacGregor 1989], the *Soviet Union may have been similarly uncomfortable with being surrounded by U.S. forward bases [Harkavy 1989; Duke 1989].

2. **Defensive FD.** FD may, however, also be entirely defensive if it is tantamount to a FD on a state's own territory. To commence defense immediately at the border would have several advantages over an in-depth *territorial defense*:

- If successful, FD would spare most of the defender's own territory, thus limiting war damage.
- It would slow down an invasion, allowing the defender to mobilize the rest of its forces, assume combat positions, etc. FD might also allow the defender to maintain a lower state of overall readiness (i.e., to cadre most of its forces), thereby minimizing its own capability for short-warning or complete *surprise attack*.
- FD, finally, could signal resolve and might be appropriate for crisis management because of its suitability for parrying and containing probing attacks [cf. Krause 1984; 1987a; 1988].

For all its merits, FD also has its drawbacks. First, even though it is a rather simple standard when applied to national defense, its application to alliances is bound to introduce ambiguities because "forward" may mean different things to different member states. For the *FRG* it had to mean a defense along the inner-German border; but it might make perfect sense for *France* to establish a defense perimeter on the Rhine, for the *U.K.* to do so at the Channel coast, and for the *U.S.* and *Canada* to do so somewhere in the middle of the Atlantic. Hence the German insistence on the principle of FD along the inner-German border as a precondition for its *NATO* membership [Møller 1991: 43–56; cf. Schubert 1970; Wettig 1967].

Second, unless allies actually station all forces assigned to alliance FD on the envisioned perimeter, they will require a substantial long-range mobility that might be nearly indistinguishable from that required for offensive purposes. U.S.–based divisions assigned to the defense of the FRG might thus easily be suitable (albeit, of course, not sufficient) for an invasion of Poland or the USSR.

Third, unless specifically designed to be strictly defensive, mobile forward-deployed forces tend to possess some offensive, border-crossing capabilities and hence might lend themselves to misunderstandings. This feature may be especially problematic if the forward defenses need to be reinforced in the face of an expected attack, such as actually envisioned by a number of *NOD* proponents. The low state of readiness would reduce capabilities for surprise attack, but *crisis stability* might be seriously damaged.

Finally, a genuine FD is by its very nature linear, thus almost inviting penetration attempts, which might seem all the more meaningful the shallower the depth and the less defended the rear. There are thus inescapable trade-offs between FD and in-depth defense.

3. **NOD and FD.** The proponents of *civilian-based defense* have advocated abandonment of FD, but only a few NOD proponents have recommended this [Afheldt 1976; 1983]. Most have merely emphasized the need for, and devised the means of, a strictly defensive FD. They have, e.g., pointed to *barrier*-type defenses, shallow fire belts (i.e., "killing zones"), *infantry* *webs*, etc. [Barnaby & Boeker 1982; 1988; 1989; Hannig 1984; 1986; Gerber 1984; 1987; A. Müller 1986; Bülow 1986; 1986a; SAS (ed.) 1984; 1989].

FOUNDATION FOR INTERNATIONAL SECURITY. British alternative defense and security research group based in Adderbury, Oxfordshire; founded by the former member of the British *Just Defence*, Stan Windass, under the auspices of the Defence Research Trust (DRT). In 1985, the DRT published *Avoiding Nuclear War: Common Security as a Strategy for the Defence of the West* [Windass (ed.) 1985] as a preliminary to its *Common Security* Programme, inaugurated in

1986. From the outset, Windass placed the focus on improvements at *NATO's conventional defenses through what he called *"*defensive deterrence*," which might presumably pave the way for a *no-first-use strategy and allow for elimination of tactical nuclear weapons [Windass 1985; 1985a–b].

An elaborate institutional network was subsequently set up, including a group of advisors, among whom were the Norwegian defense minister, the late Johan Jørgen Holst; former U.S. secretary of defense, James Schlesinger; later presidential security advisor Brent Scowcroft, and several generals [Defence Research Trust 1986]. The FIS's next major output was a series of "consultancy papers," written by some of the world's leading experts on *NOD [Windass & Grove (eds.) 1987:1–2; 1988]. Besides elaborating on the concepts of common security and cooperative security, the FIS suggested a somewhat unusual approach to improving *crisis stability, applied to both land and naval forces: "routinization," whereby it meant to "progressively extend normal times by anticipating and ritualizing crisis and war" [Windass 1987; cf. Grove 1987; 1988].

The most innovative part of the FIS's work has been its analyses of naval strategy, directed by one of the U.K.'s leading naval experts, Eric Grove [1987; 1988; 1990; 1990a–b]. Initially, EG advocated "routinization" for the U.S. Navy's *Maritime Strategy (calling for regular U.S. naval operations in the Norwegian Sea) [Grove 1987; 1988; 1989; 1990a]. These ideas were aired in the context of a series of meetings on naval strategy, held in Adderbury in 1988–1989 under the auspices of the FIS, and involving both U.S. and Soviet naval specialists [in Grove 1990a: 91–160].

In 1990 Grove published what was undoubtedly one of the most important works on naval strategy for many years, *The Future of Sea Power,* based on his study of Julian Corbett's works [1990b; cf. 1988a]. Besides detailed analyses of the potentials of various types of naval armaments, it contained elaborate strategic and grand strategic analyses of the role and special nature of *sea power. This amounted to a dismissal of the Mahanian school's emphasis on the offensive in general, and the decisive naval battle in particular, in favor of Corbett. The latter acknowledged the defensive as the stronger form of war, and therefore realized that "it is *prima facie* better strategy to make the enemy come to you than to go to him and seek decision on his own ground" [Grove 1990b: 14].

Of particular relevance from an NOD point of view was Grove's well-argued case for the employment of a convoy-with-escorts system for *SLOC defense. This was politically and strategically defensive, and *ipso facto* appropriate for NATO; it allowed the defender to be on the tactical offensive and, furthermore, provided better opportunities for inflicting decisive defeat on a strategically offensive adversary than going into defended *bastions in a close blockade mode (such as envisioned under the Maritime Strategy).

FRAG.   Forsvarsministerens Rådgivnings- og Analysegruppe (Advisory and Analytical Group of the Minister of Defense). See *Denmark.

FRANCE.   The alternative defense debate in F has been most conspicuous by its almost complete absence [cf. Joxe 1990a]. Nevertheless, a few instances deserve mention.

1. **Tradition.** Alternating with periods dominated by a veritable "cult of the offensive" (see *Plan 17*), F has for long periods adhered to defensive strategies. The *Jeune École* with its emphasis on sea denial (*Guerre de Course*) rather than sea command (cf. *sea power*) was an instance of this in the realm of naval strategy [Ropp 1941; Ceillier 1990], and the *Maginot Line* was a manifestation of a defen-

sive strategy for the land battle [Gibson 1941; Posen 1984a: 105–140; Bond & Alexander 1986; Bellamy 1990: 44–45; Pedroncini 1991; cf. Porch 1991].

Furthermore, ever since the 1789 revolution, there has been a strong democratic tradition in French strategy, with the *citoyen-soldat* conception of the 1789 revolution (see *Citizen in Uniform*) as an early manifestation [Lynn 1984], and the ideas of the Socialist Party (especially *Jean Jaurés* [1915]) of a "New Army" as a more recent one.

2. **Defense Policy.** Both the *Parti Socialiste* and the PCF (*Parti Communiste Francaise*) previously were in opposition on defense policy [Hanley 1984; Howorth 1984; 1984a], but for several years there has been an almost total consensus on the centerpiece of French security policy: the maintenance of the independent nuclear deterrent, the *Force de Frappe*. In the late 1980s and early 1990s, this was, furthermore, undergoing a comprehensive modernization-*cum*-expansion, which (in the context of U.S. and Soviet/Russian reductions) may land F in almost the same league as the superpower(s).

Even though F has traditionally adhered to the doctrine of "the weak's *dissuasion* of the strong" (*Dissuasion du faible au fort*), the expanded arsenal would provide F with an increased targeting flexibility and a number of options besides that of deterring an attack on F itself [Andrets 1986; Kolodziej 1987; 1989; Nehrlich 1986; Yost 1985: 1–2]. F apparently sought to exploit this by offering the *FRG* a modicum of *extended deterrence*; it did not achieve the intended effects but, rather, alienated the Germans. They would much have preferred F's pledge of assistance for the *forward defense* of the FRG to being included in yet another power's target list [Andrets 1986; Brigot 1989; Ehrhart 1987; 1989; Gloannec 1986].

3. **Alternative Defense.** France has traditionally neglected conventional defense in favor of *nuclear deterrence*, with only a few dissenting voices being heard [Lellouche 1985]. Indeed, some of the alternative defense proponents (e.g., *Alain Joxe*) have reached the disillusioned conclusion that they have to accept the Force de Frappe and can hope only to "ritualize" it.

Some of the foundations of the *NOD* debate elsewhere were nevertheless laid in F by authors such as *André Beaufre, *André Glucksmann*, and the precursor of *Horst Afheldt, *Guy Brosollet* [Beaufre 1963; 1965; 1972; 1974; Brosollet 1975; Glucksman 1967; cf. 1983]. Yet apart from a certain interest in the general topics of *conventional stability* and *arms control* [J. Klein 1988; 1989; 1990; 1990a] and some interest in the German debate (particularly in light of the evolving Franco-German security cooperation [Faivre 1987; Mannfrass-Sirjaques 1991; Schütze 1987; 1988; 1989; 1990], only a very few have taken any interest in the quest for defense alternatives.

- There was, of course, a *peace movement* (especially CODENE [*Comité pour le Désarmement nucléaire en Europe,* i.e., Committee for Nuclear Disarmament in Europe]), but it was considerably smaller and weaker than those of the other European countries and of the *U.S.* Furthermore, it showed little, if any, interest in conventional defenses [Bourdet 1984; Fisera 1984; Howorth 1984: 48-83; Mellon 1984].
- A few proponents of *civilian-based defense* have been close to the peace movement, with the author of *Did You Say Pacifism?*, Jean-Marie Muller, as the most innovative thinker. In 1985 he published a major work, *Civilian Dissuasion,* jointly with Jacques Semelin and Christian Mellon [Mellon & al. 1985; Muller 1984].
- There have been a few dissenting officers (albeit fewer than in the FRG), such as *Etienne Copel* [1984; 1986] and Admiral Antoine Sanguinetti, a member of the international organization *Generals for Peace* and an out-

spoken critic of French defense policy [1984; 1988; interviewed in Kade 1981: 281–321]. He criticized not only the nuclear strategies of France and *NATO* but also offensive conventional strategies such as the *AirLand Battle* doctrine. His positive recommendations included one for a dissolution of both military *alliances*, and for a closer Franco-German collaboration in creating a defense based on the model of Brosollet and Horst Afheldt [1988: 149]. In 1976 he was prematurely retired because of his dissent, yet subsequently was reinstated with full honors.

The most important contribution to the international NOD debate has, however, been that of the interdisciplinary research group *CIRPES* and its leader *Alain Joxe* [Carton 1984; Cramer 1984; Lukic 1984; Joxe 1984; 1990; 1990a; 1991].

FREEDMAN, LAWRENCE.   See *U.K.*

FREEMAN, JOHN R.   Professor of political science at the University of Minnesota; coauthor (with *Joshua Goldstein*) of *Three-Way Street: Strategic Reciprocity in World Politics* [Goldstein & Freeman 1990].

FREEZE.   U.S. *peace movement* founded in 1980s as a reaction to the new round in the (especially nuclear) *arms race* heralded by the election of Ronald Reagan [Solo 1988].

The strategy of the F was to enlist as many people as possible (manifested in the great rallies of 1982) in its call for rather modest policy objectives (the "lowest common denominator"): neither *disarmament* nor even significant reductions of the holdings, but merely a halt to a further expansion of the arsenal (albeit usually conceived as just a first step). The Freeze proposal had first been put forward by Gerard Smith in the early 1970s and subsequently been promoted by Richard Barnett. The author of the pamphlet *A Call to Halt the Arms Race,* which launched the F, was Randall Forsberg (cf. *IDDS*), who also played a leading role in organizing the movement [cf. Solo 1988: 44–46].

Gradually, most leading members of the F acknowledged the need for an expanded agenda, in which connection several (among them Mrs. Forsberg) pointed to the need for curbing the conventional arms race as a precondition for attaining substantial nuclear reductions [Forsberg 1984; 1985; 1986; Ford & al. 1982], in which connection they occasionally pointed to restructuring (i.e., *NOD*). Above all, Mrs. Forsberg highlighted the need for a delegitimation of the use of armed force (e.g., through principled *nonintervention*) as a precondition for build-down [Forsberg 1986; 1987; 1987a–b; 1988; 1988a].

FRG.   Federal Republic of Germany. Beyond comparison the country with the most intense alternative defense debate, ever since the 1950s, as well as the focus of most non-German *NOD* proposals [cf. for a comprehensive analytical survey: Møller 1991].

1. **Background.** There are many reasons for this centrality of Germany, which is acknowledged by Germans and non-Germans alike:

- The division of Germany symbolized the division of Europe, hence the answer to the "German question" would presumably provide the clue to a unification of Europe.
- *NATO*'s inherent *multinationality* was particularly visible in the FRG, where force contingents from several allies were deployed side by side. Any reform of the FRG's military strategy would influence that of the others, and vice versa.

- Neither friends and allies nor adversaries were ever entirely comfortable with Germany, not merely because of her (less-than-confidence-inspiring) past but also because of her size and central location. Any German escape route from the *security dilemma* might therefore presumably untie many other "knots" in the European security system.

It is even less surprising that the Germans have always taken defense and security more seriously than most other nations, even though escapes from their dilemmas have been hard to find:

- Ever since 1949 the FRG has found herself caught in the *security dilemma*, hence her apparent oscillation between a Cold War attitude and a quest for *détente*, most visibly in the relations (or nonrelations) with the *GDR*. On the one hand, the FRG sought détente as a way of improving the lot of the East Germans, as well as of making the *Soviet Union* more forthcoming, inter alia, with a view to allowing unification. On the other hand, *détente* risked cementing the GDR regime to the point of excluding unification.
- Concretely, the security dilemma manifested itself in the perennial problem of the right size of German military power. On the one hand, it should be strong enough to deter (albeit with allied assistance) the Soviet Union; on the other hand, it should be so weak as to pose no threat to small neighbors such as *Denmark*, *Austria*, the *Netherlands*, or *Belgium*.
- The *defense dilemma* was particularly severe and complicated in the FRG because almost any conceivable war (also one of defense) could easily prove suicidal, as a result of the vast stocks of conventional, chemical, and nuclear weapons stored on German soil [Mechtersheimer & Barth 1986; Arkin & Fieldhouse 1985; Duke 1989: 56–148], and because the FRG was predestined to become the main battlefield in any East-West war. The USSR stationed more than nineteen divisions in the GDR (the GSFG), poised for a fast (and probably nuclear-assisted) invasion of the FRG as the first move in a future war [Lippert 1987; Rühl 1991c–d].
- Finally, the security and defense policy of the FRG had to strike a balance between three sets of policies that were far from always compatible: *Westpolitik, Ostpolitik,* and *Deutschlandpolitik.*

2. **The 1950s.** The 1950s saw an abundance of NOD-like proposals for German security policy and defense posture in the making, intended both to avert the extensive deployment of tactical nuclear weapons and to avoid blocking the way for (re)unification by the envisioned unconditional NATO membership of the FRG:

- All the FRG's major political parties, as well as the Soviet Union, *Poland,* and a host of independent foreign observers published *Germany Plans, *disengagement* schemes, and proposals for *nuclear-weapons-free zones* (cf. *Gaitskill Plan, *Eden Plan, *Healey Plan, *George Kennan, *Rapacki Plan, *Gomulka Plan, *Pfleiderer Plan).
- The nascent Bundeswehr saw its first dissenting officer, who was sacked without delay: *Bogislav von Bonin.*

Nevertheless, the federal government (under Konrad Adenauer, *CDU) and its allies proceeded with rearmament plans, as well as with *integration* with the West [Fochepoth (ed.) 1988; Schubert 1970; Wettig 1967; Brill 1987]. The rearmament preparations resulted, e.g., in the Himmeroder Memorandum, which envisioned mobile and quite offensive maneuver warfare, relying on mechanized, armored forces: "The defense must, wherever possible, be conducted offensively. This means that attacks must be launched everywhere, and from the very beginning, wherever it is feasible. . . . It must be attempted by all means to carry the fighting onto East German territory" [Rautenberg & Wiggershaus 1977: 40; Fischer (ed.) 1975: 32–34; Schubert 1970: 65–66; Brill 1987: 90–93].

Such plans met with a wave of protest, e.g., from the *SPD, which was most disturbed by the fact that a defense perimeter on the Rhine was planned for, which would provide no protection for most of the FRG. Its concerns temporarily made the SPD recommend an even more offensive strategy, through which war would be carried into the aggressor's home territory [Schubert 1970: 42–48; Brill 1987: 82, 190–192].

3. **Ostpolitik.** After a decade of Cold War, the FRG under the chancellorship of Willy Brandt in the late 1960s and early 1970s embarked on a quest for normalized relations with the East, under the catchword (first used by *Egon Bahr* in 1963) *change through rapprochement* (*Wandel durch Annäherung*) [Bahr 1963].

After a series of cautious overtures to the USSR, Poland, and the GDR, a set of treaties (*Ostverträge*) were signed, in which the FRG acknowledged (at least de facto) the GDR and the Oder-Neisse border. Only then did the FRG renounce its revisionist ambitions and become a status-quo power, thereby also paving the way politically for a more defensive military strategy [Longerich (ed.) 1990: 230–233; Griffith 1978: 115–120, 131–225; Rosolowsky 1987: 43–44; Schramm & al. (eds.) 1972: 150–154, 166–168, 217–227; Joffe 1984: 175–177]. The *Ostverträge,* furthermore, transformed the FRG from an opponent into one of the most wholehearted supporters of détente and *arms control* in the Western camp. Thus were also laid the foundations for a certain tension in relations with the *U.S., as was manifested in the debate in the mid-1970s about the possible deployment of the neutron bomb [Newhouse 1989: 309–311; Szabo 1990: 22–27], and the heated *INF* controversy from 1979 to 1983 [Brauch 1983; Joffe 1988: 60–90; Szabo 1990: 27–30]. The long-term effects of these was an irreparable breakdown of the previous near-consensus on security politics.

4. **NOD Debate in the 1980s.** Such was the background against which the debate on security and defense policy was conducted throughout the 1980s, in which the themes of nuclear *disarmament*, disengagement, and *NOD* (usually under the German term *Strukturelle Nichtangriffsfähigkeit* [structural nonattack capability]) were in focus. The central actors in this debate were:

- Independent NOD proponents, most of them from academia: *Horst Afheldt*, *Lutz Unterseher*, and other members of the *SAS*; researchers at the *IFSH* and the *Friedrich-Ebert-Stiftung*; *Albrecht A. C. von Müller*; *Alfred Mechtersheimer*; *Dieter Senghaas*; and others. The group also included a number of (active or, more often, retired) Bundeswehr officers: *Eckardt Afheldt*, *Johannes Gerber*, *Norbert Hannig*, *Günter Kiessling*, *Jochen Löser*, *Elmar Schmähling*, *Franz Uhle-Wettler*, and others [cf. Møller 1991: 63–155].
- Independent defenders of the status quo, within the academic establishment: Karl Kaiser [1989; idem & al. 1982], Josef Joffe [1988], the *Konrad Adenauer Foundation*, and others.
- The general public and the *peace movements*.
- The political parties, for whose favor the aforementioned actors, in a certain sense, all vied.

Of the political parties in the FRG, only the SPD came out wholeheartedly for NOD; *Die Grünen* (the Green Party) was ideologically committed to *civilian-based defense* and *neutrality*. Even though the party of Federal Chancellor Kohl, the CDU, remained critical toward NOD, a significant number of party members came out in support of the conception (cf. *Christian Democrats in Favor of Disarmament*). Finally, Foreign Minister *Genscher*'s *FDP* remained uncommitted but on several occasions adopted political standpoints clearly inspired by NOD and *common security* ideas [Møller 1991: 156–196].

5. **Unification Debate.** The NOD debate, fought, e.g., over Bundeswehr long-term planning, remained undecided, inter alia, because it was cut short by the interference of the debate, from the summer of 1989 until October 1990, over German unification.

Even though the focus was on the economic and political terms of unification, security and defense political aspects also entered into the debate, both in the two Germanys and abroad, inter alia, in the context of the *Two-Plus-Four* negotiations [Møller 1991: 212–259]. Furthermore, the actual outcome (modeled on the *Genscher Plan*) contained quite a few elements inspired by NOD literature.

In the course of the unification debate, a number of the most prominent NOD proponents modified their models with a view to accommodating the GDR [Lutz 1990; 1990f–g]; the Greens and parts of the peace movement campaigned in favor of *neutralization* and/or complete demilitarization of the united Germany in the making (cf. *BOA*, *Andreas Buro*).

6. **Postunification Debates.** After unification, the alternative defense debate, indeed the entire debate on security and defense policy, faded out, overshadowed by the economic, social, and political problems created by the incorporation of the GDR [see Møller 1994; Heydrich & al. (eds.) 1992]. What little debate there has been has mainly concerned the following issues:

- The size of the defense budget. The debate has been about how much and how fast to cut, yet not whether to cut at all, about the necessity of which there has been consensus.
- Whether, and if so when, to abandon conscription in favor of professional armed forces [Klein (ed.) 1991; Kaldrack & Klein (eds.) 1992], a question that is closely related to the following issues.
- The appropriate size of the Bundeswehr. There is little doubt that the ceiling stipulated in the Two-Plus-Four agreement and the CFE of 370,000 men under arms will not be met, yet how far below this the number will be remains to be seen.
- How to reap the hoped-for *peace dividend* through *conversion* plans [Giessmann (ed.) 1992; Köllner & Huck (eds.) 1990; Bebermeyer & Unterseher 1992].
- Whether, to what extent, and how to integrate the personnel of the *NVA* (*Nationale Volksarmee*, National People's Army). Above all by simply sacking employees, but to a limited extent also through integration in the Bundeswehr, something that requires extensive retraining [Scheven 1992; Schönbohm 1992; Giessman 1992; Klein & Zimmerman (eds.) 1993].
- The implications of the progressive multinationalization, materialized, e.g., in the Eurocorps. For a state so accustomed to integration with others in security political matters as the FRG, this may not be a major problem, at least so long as the defense missions are not expanded beyond the NATO area.

7. **Out-of-Area Debate.** The last issue has probably been the most controversial: whether to participate in collective security-type operations beyond the former boundaries, either under the auspices of the *U.N.*, i.e., as "blue helmets," or under those of the CSCE, in either case perhaps in the framework of NATO and/or WEU joint task forces [Lutz (ed.) 1993]. In this *out-of-area* debate, there have for some

time been numerous and strong calls for a more assertive German military role [Souchon 1990; 1992]. Proponents have occasionally argued in terms of a normalization, i.e., of shedding the last remnants of singularization in the form of military constraints applying merely to Germany. Critics, on the other hand, have couched their position in legal and political terms. Legally, the FRG constitution prohibits German military involvement beyond the boundaries of NATO, at least according to some interpretations [Lutz 1990a; 1992; 1993; 1993a–b]. Politically, critics regard the constraints in force to be indispensable, or at least valuable, safeguards against a resurgent German bellicosity [Böge 1991b; Bastian 1993; Wette 1993], signs of which critics have found in Bundeswehr planning [Nassauer 1992]. An additional, complicating aspect of the debate has been whether Germany should demand a higher international status, for instance, through a permanent seat in the U.N. Security Council, as a precondition for, or at least a consequence of, a more active involvement in world affairs [Kaiser 1993; Wagner 1993].

What has not been controversial is that the united Germany has a special responsibility for ensuring peace in Europe, inter alia, through a reintegration of the former East. Hence the rather intense debate on the appropriate security architecture, in the course of which there has been nearly complete consensus on the need to avoid renationalization, above all to embed Germany in larger European contexts [Genscher 1991; Albrecht & al. 1992; Böge 1991; IFSH 1990; 1993; Lutz 1991; 1991b; Voigt 1990; 1990a; Waever 1990; Brauch 1990; 1991a; 1992; 1992a; Grosser (ed.) 1992; Heisenberg (ed.) 1991; Linnenkamp 1992; Lübkemeier 1993; Merkl 1993; Rotfeld & Stützle (eds.) 1991; Stares (ed.) 1992; Szabo 1993].

FRIEDRICH-EBERT-STIFTUNG.   Foundation located in Bonn and affiliated with the *SPD*, including a research unit serving as a think tank for the party [Enders & al. 1990:141–143]. Among its researchers and central in the alternative defense debate, one could mention the late Wilhelm Bruns [1989; 1990; 1990a–b; idem & al. 1984], General (ret.) *Christian Krause*, and Eckhard Lübkemeier.

FULLER, J. F. C.   Major general in the British Army and one of the most famous strategists of the twentieth century. In 1924 he codified the *principles of war* in approximately their present formulation [Reid 1987: 89–100].

JFCF has been credited with inventing the *blitzkrieg*. At the very least he was one of the first to see the prospects of the *tank* for future offensive maneuver warfare [Bond & Alexander 1986: 601–602]. While denying that a distinction between offensive and *defensive weapons* made any sense [Fuller 1932], he nevertheless took a certain interest in the defensive, e.g., in antitank defenses. He likened this to an *archipelago defense* [1944] and described how antitank forces should "entangle the enemy's forces in a tactical net of fine mesh, from which, should he break through it, he will find himself caught in a strategical net of wider mesh" [quoted in Reid 1987: 192], thus pointing forward to modern *NOD* conceptions such as that of *Horst Afheldt*. See also *Web*.

FUNK, HELMUT.   Colonel in the Bundeswehr (ret.) and collaborator of *Andreas von Bülow*, the concrete implications of whose defense proposals he calculated [Funk 1988; 1988a; cf. Bülow & Funk 1987; 1988; 1988a] (albeit not correctly, according to several critics) [Gupta 1988; Thimann 1989a; idem & al. 1988].

G

GAITSKELL, PLAN.   Proposal launched by the chairman of the Labour Party of the *U.K.*, Hugh Gaitskell, in 1957 in response to the *Rapacki Plan*. It envisioned a piecemeal withdrawal of foreign troops and nuclear weapons from the two Germanys (envisioned as united), *Poland*, *Hungary*, and Czechoslovakia, as well as a patchwork of nonaggression treaties between the involved states [Schramm & al. (eds.) 1972:13–17; cf. Albrecht (ed.) 1986:71–76].

GALTUNG, JOHAN.   Norwegian sociologist and peace researcher, founder of PRIO and subsequently professor at various universities around the world, most recently the University of Hawaii.

Besides his theories on structural versus direct violence, positive and negative *peace* [1965, 1969; 1987a], imperialism, *integration theory*, and structural-sociological analyses of, e.g., the *peace movements* and the Green movement [1984: 19–32; 1986; 1988b], JG's most direct contribution to the alternative defense debate has been his writings on *nonmilitary defense* (i.e., *civilian-based defense*) in the 1960s [1959; 1964; 1968; 1971; cf. 1989a; 1990b] and on *transarmament to *NOD in the 1980s [1981; 1983; 1984; 1984a; 1988; 1989; 1989c].

JG's main work in the latter field is *There Are Alternatives: Four Roads to Peace and Security* [1984]. His point of departure here was a rank-ordering of European countries according to degree of security.

* Most secure: *Switzerland*, *Yugoslavia*, Albania.
* Second: *Finland*, *Austria*, *Sweden*, Malta.
* Third: *France*, Greece, Romania, Iceland, Ireland, *Spain*, Cyprus.
* Least: *NATO* and *Warsaw Pact* member states [1984: 13].

This (perhaps somewhat surprising) assessment was based on a weighted index of the country's defensive capability, its degree of decoupling from the superpower conflict, *invulnerability*, and usefulness. Romania, e.g., scored high, leading JG to the assertion that "Romania under the leadership of Nicolae Ceausescu is an important part of the general peace movement in Europe" [1984: 27].

Concerning defense structure, JG recommended a combination of three elements: (1) conventional military defense; (2) paramilitary defense; and (3) *nonmilitary defense*.

The first category should be "short-range, mobile, small, local, quick, dispersed and autonomous," and armed exclusively with *defensive weapons*. The latter (unfortunate) term, JG suggested, should be defined by range and radius of impact area. On the other hand, he acknowledged that only the total system, not its components, could be categorized as offensive or defensive [1984: 172-176] (cf. *offense/defense distinction*) (see Table 20).To make the facts fit this definition, JG subsequently sought to reinterpret "bad" armament programs, such as the *SDI*, as being based on "offensive weapons" [1987b]. However, his explanations of how orbiting lasers could be used to "ignite cities from space" revealed a profound ignorance of the most elementary laws of physics.

Table 20. Offensive and Defensive Systems

	Range		
Impact Area	Immobile	Short	Long
Extensive	Offensive	Offensive	Offensive
Limited	Defensive	Defensive	Offensive
Local	Defensive	Defensive	Offensive

The paramilitary forces should be even more local and guerilla-like than the regular forces (albeit wearing some sort of uniform to set them apart from civilians). Finally, the nonmilitary component should wage a defensive struggle against aggression by means of civilian-based defense methods. Other than this very sketchy outline of an alternative defense posture, JG has contributed nothing to NOD theory.

GAME THEORY.   In actual fact not so much a theory as a terminology suitable for describing and analyzing strategic moves, developed by John Neumann and Oskar Morgenstern in *Theory of Games and Economic Behaviour* (Princeton University Press, 1944). GT presupposes individual rationality, and its level of analysis is not actual human decisionmaking but the behavior of an abstract *homo strategicus,* similar to the *homo economicus* of economic theory. GT parallels *Realism*, with its assumption of states as unitary and rational actors pursuing their national interests in an anarchic setting. The matrix of a typical two-person game appears in Table 21.

Table 21.   Game Matrix

		B	
		Alternative 1	Alternative 2
	Alternative 1	A's payoff, B's payoff	A's payoff, B's payoff
A	Alternative 2	A's payoff B's payoff	A's payoff B's payoff

Common distinctions have been made between "games against nature" (where one player is passive and unresponsive) and "games of strategy" (where the two players interact), and between *zero-sum games*, *mixed-motive games*, and *cooperative games.*

During the Cold War, attention was focused on two-person, zero-sum games, which appeared to describe the bipolar and highly antagonistic set of international relations, in which military factors (e.g., nuclear strategies) played a decisive role. GT was thus used to legitimize nuclear strategies by, it was alleged, demonstrating their rationality (cf. *Chicken*). This "nuclear strategy-with-GT complex" was accused of cynicism, e.g., by one of the most prominent game theorists, *Anatol*

Rapoport: "We have translated the game of strategy (where men may engage in ruthlessness and cunning to their heart's content because it is only a game) into a plan of genocidal orgies, and we call the resulting nightmare 'realism'" [1970a: 224; cf. idem 1989: 247–346].

GT has, however, also been used to analyze less confrontational games, i.e., cooperative and mixed-motive games (cf. *Prisoner's Dilemma*, *Tit-for-Tat*), thus providing legitimation for *arms control* and cooperative security arrangements, etc. [Schelling 1960; Axelrod 1984; Brams & Kilgour 1988; Rapoport 1989: 304–324; Jervis 1988; Goldstein & Freeman 1990; Stein 1990].

GANDHI, MOHANDAS K.   (1869–1948). Leader of the Indian movement against British rule.

MKG both taught *nonviolence* through his writings and practiced it in his struggle against British rule of *India* [Galtung & Naess 1955; Sternstein 1970; Sharp 1979; Joccheim 1987]. His teachings on *satyagraha* were based on Hinduism (blended with some aspects of the thinking of others, e.g., *Tolstoy* and *Thoreau*), and hence on an assumption of a fundamentally harmonious world. Because conflict was merely a surface phenomenon that obscured the harmony of the world, MKG's strategy was intended to promote conversion of opponents by opening their eyes to this truth rather than to coerce their compliance with one's will.

MKG's teachings might also be interpreted as a political strategy, with power constituting a basic variable and with nonviolent insubordination representing merely an expedient form of power [Sharp 1979]. MKG seems to have been the first to formulate the thesis that government presupposed consent, and that disobedience would thus be a strong counter to usurpers of power. As early as 1909, he recommended such a strategy against the British rulers, whom he addressed thus: "You have great military resources. . . . If we wanted to fight with you on your own ground, we should be unable to do so, but if the above submissions be not acceptable to you, we cease to play the part of the ruled. . . . If you act contrary to our will, we shall not help you; and without our help, we know that you cannot move one step forward" [quoted in Sharp 1973: 84].

GARRETT, JAMES M.   Professor of international relations at Millersville University (U.S.) and fellow of the Center for Strategic and International Studies, author of two books on *NATO*'s defense strategy with an emphasis on Central Europe: *The Tenuous Balance* [1989] and *Central Europe's Fragile Deterrent Structure* [1990].

In *The Tenuous Balance*, JMG analyzed NATO's conventional defense options, taking into account the special terrain features of the most likely avenues of a *Warsaw Pact* attack. Contrary to the traditional view, he asserted they would be quite defensible if the allies tailored their force postures and strategy to the circumstances. He assessed both *AirLand Battle* and *FOFA* as ill-advised, not so much because of their negative security political implications as because of their poor efficiency [1989: 45-47]. His view of the German *NOD* proposals was far more favorable, even though he emphasized that they could serve as only a supplement to, not a substitute for, existing forces. They did, however, have the political advantage of "pulling the teeth of the European anti-defense left" [1989: 50], and he recommended the following solution to NATO's defense problem: "an area defense backed by substantial operational reserves, sufficient counter-battery firepower to prevent Pact artillery from neutralizing dispersed NATO anti-tank defenses, and sufficient tactical *air power* to gain and hold superiority in critical sectors" [1989: xi].

To this conventional posture, JMG in his subsequent work proposed to add a

European nuclear deterrent, based on British and French nuclear weapons, as well as a limited offensive chemical warfare capability, albeit with a view only to deterring the Warsaw Pact from using *chemical weapons [1990: 63–72, 121–130].

GATES, DAVID.   Lecturer for the Ministry of Defence at the University of Aberdeen, and author of works on the *Peninsular War [1986] and the potentials of light *infantry [1989; cf. 1990], as well as of critical analyses of *NOD [1987; 1987a; 1991].

His *Non-Offensive Defence: An Alternative Strategy for NATO?* was the most comprehensive critique to date of NOD. It was based on a rather extensive (but unfortunately not quite adequate) reading of especially German NOD literature. His main critique was that "'Defensive Defence' concepts . . . assume a degree of mutual trust which, if it existed, would make defence capabilities beyond those necessary for internal security quite irrelevant" [1991: 110]. Further, the fact that NOD implied an acceptance of fighting on one's own territory would, according to DG, lead to "immense structural damage and loss of life," wherefore large-scale *counteroffensives would have to be planned for, if only as a means of *damage limitation. Finally, DG criticized NOD for many of the deficiencies inherent in certain, but not all, concrete models, by emphasizing the wide range of tactical, operational, and strategic countermeasures that would be available to an aggressor.

While acknowledging the validity of certain elements in some NOD schemes, DG concluded by sarcastically predicting NOD's falling into oblivion: "Far from being a strategy for tomorrow, it would appear to be a concept founded on the permafrost of the Cold War; and as the one melts away, so will the other" [1991: 186].

GDC.   General and complete disarmament. See *Disarmament.

GDR.   The German Democratic Republic (*Deutsche Demokratische Republik*) was established in the Soviet occupation zone of Germany in 1949, and until 1989 was ruled by the Socialist Unity Party, SED (*Sozialistische Einheitspartei Deutschlands*). The peaceful revolution in 1989 set in motion an accelerating process toward unification with the FRG, which was accomplished in October 1990, when the GDR ceased to exist.

Alternative thinking on defense and security matters had been banned for most of the GDR's existence, but a few glimpses thereof had been visible in the last years of the SED regime. Alternative thinking flourished during the short interregnum, 1989–1990 [cf. Møller 1991: 197–211].

1. **Until 1989.** In the late 1980s the SED had come to support the themes from the West German security debate that amounted to a stabilization of the East-West conflict (including the survival of the communist regimes and the continuing division of Germany), such as denuclearization (e.g., in the shape of a Central European *nuclear-weapons-free zone). Accordingly, the SED had negotiated a draft "treaty" with the West German *SPD, just as it had with regard to *chemical weapons [SPD & SED 1985; 1987; 1987a; Voigt 1985; 1987; Hahn 1987; T. Meyer 1987; Birkenbach 1989; Seidelmann 1987; Eppler 1988].

After a period of ignorance and deliberate disregard, the *NOD theme was also adopted by SED, albeit only after the Soviet endorsement of the notion. This led to the publication of numerous analyses on offensive elements in Western military strategy and posture; self-criticism was conspicuous by its complete absence (cf. *IPW, *Max Schmidt). Apart from these propagandistic and diplomatic maneuvers, the GDR also embarked on genuine arms reductions and a restructuring of the NVA. In early 1989 the GDR promised a 6 percent reduction of the NVA and a 10 percent cut in the defense budget, entailing, inter alia, a dismantling of six tank regiments and a certain defensive reconfiguration [*Jane's Defence Weekly*, 2 September 1989; 23 September 1989]. This might indicate that the SED was, at least partly, serious.

Inspired by Soviet *new thinking*, the ideology of the SED underwent a certain revision. Emerging themes in the official and semiofficial discourse included a reassessment of the capitalist propensity for war (hence a potentially longer "*shadow of the future*"); a favorable attitude toward *interdependence* as a contribution to *peace* as well as development; the notion of the "*Common European House*"; acknowledgment of global problems independent of the "class struggle"; the theory of the *totally destructive conventional war* (according to which industrial and postindustrial societies were unable to wage war); the concept of *common security*; and, finally, the concept of NOD. All of them were well-known themes from the Western alternative debate that now made their first appearance in the East [D. Klein 1988; 1988a; Heininger 1989; Schmidt 1989; idem & Schwarz 1988; 1989a-d; Benjowski & Schwarz 1988; 1989; Türpe 1989; Manfred Müller 1988; 1988a; 1989; Peter 1989]. The focus was on NATO's offensive conceptions (such as *FOFA*), and the *Warsaw Pact* military posture was taken into account only as far as *arms control* was concerned (*CSBMs* and/or *disengagement* schemes); *unilateralism* remained a nonissue [Schirmeister 1989; Brie 1989; Schmidt & Schwarz 1989; 1989b; 1990; 1990a–b; Schwarz 1988; 1989; 1989a; Scheler & al. 1989: 196–201, 218–223].

Only under the auspices of the Evangelical Church did some independent research take place, with the mathematician *Walter Romberg* authoring the first full-fledged NOD proposal in the entire Eastern bloc [1986].

2. **1989.** After the replacement at the pinnacle of power of Erich Honecker with Egon Krenz and subsequently Hans Modrow, the initiated defense reform continued. The new defense minister, Theodor Hoffmann, inter alia declared that the GDR would adopt a "defensively oriented" military doctrine [*Jane's Defence Weekly*, 9 December 1989, 1275]. Of even greater importance was the emergence of a free and open defense debate with participants from the (previously clandestine) opposition.

3. **1990.** After the March 1990 elections (which led to the creation of a coalition government led by the GDR's *CDU*), the new leader of government, Lothar De Maiziere (CDU), stated that "the road (to German unification) must pass through European *disarmament*" rather than through *neutrality*. Furthermore, he emphasized that "it is the task of the government of the GDR to pursue a policy of promoting the replacement of the military blocs with alliance-transcending structures, in their turn constituting the beginning of an all-European security system." A subsequent government declaration even included a commitment to NOD: "For a transitionary period, there will, in addition to the Soviet armed forces, be a radically reduced and strictly defensive NVA, with the task of defending this territory" [*Der Spiegel*, no. 6 (1990): 20; *Europa-Archiv*, no. 10 (1990): D249, D258–D259].

The defense debate until unification in October 1990 was characterized by the participation of the new pacifist defense minister, *Rainer Eppelmann*; the aforementioned Romberg (now as finance minister); high-ranking NVA officers suddenly posing as pacifists; and leading SED officials denouncing previous policies. The debate was influenced by frequent interventions by the West German academic and political elite.

GEERAERTS, GUSTAV.   See *Belgium*.

GENERAL AND COMPLETE DISARMAMENT.   See *Disarmament*.

GENERALS FOR PEACE AND DISARMAMENT.   International group of high-ranking officers (generals and admirals) from East and West, formed in 1981. The GPD has held both multilateral and bilateral (U.S.–Soviet) conferences and seminars, in which the following generals and admirals, among others, have participated: Wolf von Baudissin, G. Wollmer, and Gert Bastian (FRG); Francisco da Costa Gomes (Portugal); Michael Harbottle (U.K.); Georgios Houmanakos, A. Papspyrou, Papathanassiou, and M. Tombopoulos (Greece); John Marshall Lee (U.S.); M. H. von Meyenfeldt (the Netherlands); Nino Pasti (Italy); Antoine Sanguinetti (France); Johan Christie (Norway); and Vadim Makarevsky, Mikhail Milshstein, Rair Simonioan, and Alexander Shevchenko (Soviet Union).

In joint statements the GPD has, e.g., recommended various nuclear *arms control* measures such as *no-first-use*, *nuclear-weapons-free zones*, and a complete elimination of nonstrategic nuclear weapons [Kade (ed.) 1981; Harbottle & al. 1984; Fedorenko (ed.) 1988]. Furthermore, it has also expressed itself in favor of *NOD*. In the final document from the Second Conference of Soviet and U.S. Admirals and Generals in 1988, it was stated that "conventional forces must be reduced . . . in order to make it impossible for either side to launch a *surprise attack* or to conduct offensive military operations successfully against the defense forces of the other side."

GENEVA CONVENTION.   See *Laws of War*.

GENSCHER, HANS DIETRICH.   See *FDP*.

GENSCHER PLAN.   Plan devised by FRG Foreign Minister Hans Dietrich Genscher (of the *FDP*) in the spring of 1990. It was intended as a compromise between Soviet demands that the *GDR* should not become part of *NATO* upon unification with the *FRG* in 1990, and Western (including West German) demands that the FRG should remain a member of NATO. The GP envisioned a special status for the territory of the GDR, implying that forces should not be integrated into NATO's command structure until after 1994, and that neither allied forces nor nuclear or *chemical weapons* should be allowed thereon [*Der Spiegel*, no. 8 (1990): 17–18; no. 17 (1990): 20]. It implied a certain form of *disengagement*. Most of its provisions were included in Art. 5 of the final treaty (signed 12 September 1990) as a conclusion of the *Two-Plus-Four* negotiations [in Rotfeld & Stützle (eds.) 1991: 183–187].

GERBER, JOHANNES.   Expert in defense economics and general (ret.) in the Bundeswehr. In the mid-1980s he collaborated with *Norbert Hannig* on a *NOD* model [Gerber 1987; cf. Hannig 1986; 1986b], and since 1990 he has been involved in research on *conversion* under the auspices of the Society for Military Economics (Gesellschaft für Militärökonomie) in Berlin.

As early as 1973 JG examined the future cost-effectiveness of the traditional form of mobile warfare in general and of *tanks* and surface warships in particular [1973: 62], albeit without elaborating on this theme until the early 1980s. Contrary to most other NOD proponents, JG questioned neither *flexible response* nor *forward defense* as strategic guidelines but merely criticized current defense planning as ineffective and not taking into due account the impending budgetary and demographic squeezes facing the Bundeswehr. The way to do so, according to JG, would be to switch from a manpower-intensive to a more capital-intensive defense; to capitalize on the most cost-effective form of military technology, namely, rockets; and to abandon the obsolete division of the armed forces for separate services. There would then be no justification for expensive procurement of new frigates, tanks, or fighter-bombers, nor for the maintenance of the navy, army, and air force responsible for mismanagement of funds [1982; 1982a-b; 1984; 1988].

JG's proposed alternative was an integrated *fire belt* emphasizing antitank and *air defense* (similar to that of Hannig). It was to be deployed along the border by the defenders with the help of sensor arrays that would provide target coordinates for a system of rocket launchers with different ranges and stationed at variable distances from the fire belt. He envisioned, however, maintenance of a number of armored regiments as well as antitank helicopters, at least for an interim period, to serve as hammers against any enemy columns that might succeed in penetrating the fire barrier, in its turn conceived of as an anvil [1982; 1982a: 83–89; 1982b: 94–95; 1984: 107–114; 1987; 1988; 1989a: 196–197; 1989b].

JG further envisioned a similar firepower-intensive maritime defense (particularly of the North and Baltic Seas), by means of land-based antiship missiles as a substitute for costly and vulnerable surface warships; a proposal since elaborated upon by Hannig and Danish authors [1982a: 78–83; cf. Boserup 1987; 1990a; Møller 1987; 1989e; 1990b; 1990i; FRAG 1986].

GEORGIA.   See *CIS*.

GEORGE, ALEXANDER L.   See *U.S.*

GEORGE OF BOHEMIA, KING.   Author of one of the world's oldest recorded proposals for a *collective security* system (1464), *Tractatus Pacis Toti Christianitati Fiendae* (Treaty on the Establishment of Peace Throughout Christendom), for which he strove unsuccessfully from 1462 to 1464 [George of Bohemia 1464; cf. Vanecek 1964].

The purpose of the proposed treaty was "the establishment among Christians of true, pure and lasting *peace*, unity and love." It was primarily directed against "the most vicious Turk" but also expressly forbade the signatories to "resort to arms or allow any man to resort to them in our name" against one another and obliged the parties to punish contravenors. Furthermore, it included a mutual assistance obligation by stipulating that if a third party should attack one of the signatories, the others would "dispatch . . . envoys to settle the dispute." Should this prove fruitless, "all of us shall help our attacked or self-defending companion." An international court would be set up to settle disputes between the signatories ("the cult of peace is unthinkable without justice"), and to facilitate implementation of the covenant, an assembly would be established and rotate among the signatories at five-year intervals. Even though the proposed treaty was motivated by the instability prevailing at the time ("How quickly empires change, how quickly kingdoms succeed each other, how quickly governments deteriorate"), it was intended to be of indefinite duration: "If any of us should be called to the heavenly abode," the successor should be obliged to pledge allegiance to the treaty before his coronation.

GERMAN DEMOCRATIC REPUBLIC.   See *GDR*.

GERMANY.   See *GDR*, *FRG*, *Germany Plan(s)*.

GERMANY, FEDERAL REPUBLIC OF.   See *FRG*.

GERMANY PLAN(S).   In the mid-1950s a number of proposals were formulated with a view to preventing German rearmament from jeopardizing the chance of reunification [Fochepoth 1988; Griffiths 1978; Møller 1991: 43-55]. Most of them coincided with the Polish *Rapacki Plan(s)*, as a response to which they were often conceived.

1. **Soviet Union.** In notes of 10 March and 9 April 1952 addressed to the three Western occupation powers, the *Soviet Union* proposed negotiations on a peace treaty with Germany along the following lines: Germany should be reunited and accorded full sovereignty, and free elections should be held prior to unification, supervised by the four powers. It would be permitted to field armed forces (land, air, and maritime), as well as to create an indigenous arms industry, albeit subject to certain constraints: only forces required for national defense (to be stipulated in the peace treaty), and military production only for its own needs. The proposal was turned down by the FRG and the Western powers, but not without some public debate and dissent within their own ranks, as well as from the *SPD* [Griffith 1978:55–56; Fochepoth 1988:45–50; Schubert 1970:165–175].

In 1957 the Soviets proposed a "*zone of limited armaments" combined with a prohibition of nuclear weapons in Europe, as a corollary of which a series of nonaggression treaties were to be signed (it was tabled in the *U.N.*). In 1959 the USSR once again proposed a peace treaty with Germany, envisioning a confederate structure as a step toward reunification. The move was combined with a demand for German *neutralization*; constraints on the armed forces (a prohibition of weapons of mass destruction, rockets, guided missiles and launchers, bombers, and submarines); and a withdrawal of foreign troops [Schramm & al. (eds.) 1972: 388, 394–401, 402–407].

2. **SPD.** The SPD presented its GP in March 1959, calling for creation of a *détente* zone comprising the two Germanys, *Poland*, Czechoslovakia, and *Hungary*. Foreign troops (whose inventory of nuclear weapons would be frozen) would be gradually withdrawn from the zone. Within it, indigenous troops would be subjected to various constraints, above all a prohibition of the acquisition of nuclear weapons. A *nuclear-weapons-free zone* would thus be created and would be underwritten by a *collective security* pact, which presumably would facilitate German reunification [Albrecht (ed.) 1986: 17–18].

During the debate on the proposal in the *Bundestag* (5 November 1959), SPD spokesman Helmut Schmidt stressed the continuity with NATO doctrine as well as with traditional West German policy, asserting that no "singularization" of Germany was contemplated. Further, that the zone could be adequately monitored, inter alia, by means of a system of reciprocal aerial surveillance; that the build-down of foreign troops would be very gradual; and that many nuclear weapons would remain. The GP would thus be compatible with NATO's "*sword and shield*' strategy, only the shield would be somewhat strengthened because either side would have to transit through the zone before embarking on actual attack, thus postponing the political decision to use nuclear weapons [Schramm & al. (eds.) 1972: 22–32].

3. **FDP.** The *FDP* presented its GP on 20 March 1959, likewise inspired by the Polish Rapacki Plan. It envisioned a nuclear-weapons-free zone, combined with a withdrawal of foreign troops. Furthermore, the reunited Germany would commit itself to *neutrality*, while being prepared to participate (on a nondiscriminatory basis) in a future all-European security system that would include both superpowers [Schramm & al. (eds.) 1972:18–19].

4. **Government.** Defense minister and leader of the CSU Franz-Joseph Strauss felt compelled by March 1958 to respond to the Rapacki Plan(s). While not dismissing the concept of nuclear-weapons-free zones as such, he demanded the inclusion of the Soviet satellite states. Also, the conventional forces within the zone should gradually be built down to the same level as the stationed forces in the FRG [Schramm & al. (eds.) 1972:11–13, 19–20].

GIAP, VO NGUYEN.   General, leader of the North Vietnamese armed forces during the *Vietnam War*, and author of several theoretical studies of *guerilla strategy* and practical guidelines for *guerilla warfare* [collected in Giap 1970].

In 1961 VNG retrospectively described the grand strategy of the Vietminh (the national liberation forces) for the war against French rule: "We . . . advocated the wide development of guerilla warfare, transforming the former's rear into our front line. Our units operated in small pockets, with independent companies penetrating deeply into the enemy-controlled zone to launch guerilla warfare, establish bases, and protect local people's power. . . . The enemy mopped up; we fought against mopping-up. . . . We gradually formed a network of guerilla bases. . . . There was no clearly defined front in this war. It was wherever the enemy was. The front was nowhere, it was everywhere. . . . The enemy wanted to concentrate their forces. We compelled them to disperse" [1970: 87–88, 91].

The above was a classical description of guerilla warfare, but the context gradually changed with the establishment of the Democratic Republic of Vietnam (North Vietnam), and the continuing struggle for unification (in the form of support for the Vietcong in the South). VNG was therefore in full support of the gradual admixture of what might be called large-scale guerilla warfare (mobile warfare) and regular warfare [1970: 94–95, 106, 109, 141]. This included major offensives against enemy strong points (Dien Bien Phu, Khe Sanh, and others) and the 1968 Tet Offensive. By virtue of his presiding over this (extremely effective) combination, VNG has often been counted among the "great captains" of the twentieth century [cf. Bellamy 1990: 230–235].

GILGES, KONRAD.   Member of the *SPD*, belonging to the left-wing "Frankfurter Circle" and leader of its "Working Group for Peace" in the mid-1980s.

In 1985 KG elaborated on the security policy of the working group in *Peace Without NATO*. From his vantage point of a Marxist ideology, he supported neither the EC nor *NATO*, as capitalist organizations [Gilges 1985: 22], which put him at odds with the SPD majority (the decisions of which he criticized at length) over its call for *security partnership* within the framework of the existing military blocs. He further dissented from its demand for *minimum deterrence*, which he found far too modest; from the long transition period envisioned for adoption of *NOD* (more precisely: *structural inability to attack*, i.e., *StruNA*); and from the narrow military sense in which the concept tended to be employed. According to KG, *StruNA* should, rather, be conceived of as a comprehensive strategy for *peace* and *détente*, calling for changes in national as well as international policies. He came close to *Dieter Lutz*'s conception of "structural inability to attack in the broader sense," but he also supported StruNA in its narrow military sense (i.e., NOD), while emphasizing the need for unilateral implementation. StruNA might thus be accompanied by a radical military build-down, reducing the standing forces of the *FRG* to 100,000–200,000 troops, combined with a *militia*-like structure [1985, 37–49, 123–126].

GLACIS.   A forward zone beyond the actual defense perimeter of a state that an aggressor would have to traverse prior to reaching the state. E.g., it has been speculated in the *FRG* since the 1950s that *France* might regard the FRG simply as G because she was not prepared to commit herself to a *forward defense* along the inner-German border [cf. Schulze 1986].

GLOTZ, PETER.   Leading member of the *SPD*, and editor of the party's journal *Die Neue Gesellschaft* (The New Society). PG seems to have been the first in the SPD to use the unfortunate term for *NOD* "structural non attack capability" (*Strukturelle Nichtangriffsfähigkeit*), devised by *Albrecht A. C. von Müller*. PG supported the notion, both as a viable alternative to *deep strike* (by that time appearing as the dominant trend in conventional strategy) and as a valuable contribution to *détente*. In the latter role, NOD would recommend itself as the military component of a "second *Ostpolitik*," which in its turn would pave the way for a system of *collective security* [Glotz 1985: 113–114; 1987; idem quoted in Lutz 1989: 19–20].

Under the impression of an improving international climate, PG in 1990 pushed the latter conception, e.g., as a solution to the so-called *stability-instability paradox*. For it to have the desired effect, however, collective security should be embedded in the framework of a pan-European federalism [Glotz 1990].

GLUCKSMAN, ANDRÉ.   French strategic philosopher; author of, inter alia, *The Discourse of War* (1967), wherein he analyzed the concept of war on the basis of the writings of, e.g., *Clausewitz, Hegel*, and *Mao Zedong*.

In his reflections concerning Clausewitz, AG emphasized the stabilizing impact of the defensive, thanks to which stability (and ultimately *peace*) was achievable: "The equilibrium of forces stems from the difference between the two 'forms' of warfare and their inequality. By virtue of these, the power of the strongest is not directly proportional to the weakness of the weakest: Two combatants may balance each other by both being the strongest, one in the offensive, the other in the defensive . . . the asymmetry between offensive and defensive allows the defense to equalize an offensive power greater than itself" [Glucksman 1967:42–43]. This was apparently the first formulation of the principle of *mutual defensive superiority*, usually attributed to *Anders Boserup*. The strength of the defensive, according to AG, stemmed basically from the defender's ability to define the point of gravity of any struggle and opt for protracted war rather than to risk everything in a decisive battle, such as exemplified by the Chinese guerilla war [1967:32, 50–51, 62, 239, 291, 305; cf. Brosollet 1975; Mao 1938].

Having laid the theoretical foundations for a defensive strategy, AG subsequently shifted to an advocacy of *nuclear deterrence* and to fierce attacks against *pacifism* [Glucksman 1983].

GOLDSTEIN, JOSHUA S.   Professor of international relations at the University of Southern California, author (with *John Freeman*) of *Three-Way Street: Strategic Reciprocity in World Politics* [Goldstein & Freeman 1990].

JSG and Freeman analyzed the potentials of various cooperative strategies, such as *Tit-for-Tat* and *GRIT*, both theoretically and compared with the actual policies pursued by the three global powers: the *U.S.*, the *Soviet Union*, and *China* (a broader framework than the traditional two-person games). They found evidence that all three had tended to reciprocate one another's behavior, but that all had occasionally played according to these cooperative strategies. They had tended to benefit from this, because of "an overall superiority of soft-line strategies for eliciting cooperation" [1990: 142]. However, neither of the classical strategies was ideal, according to the two authors: tit-for-tat was too much of a "one-shot strategy" and might easily revert to confrontation; GRIT tended to "wear off" too easily and to produce no lasting improvement in relations. JSG and Freeman therefore proposed a "*Super-Grit*" as the best way to exploit the unique opportunities provided by the Soviet "peace offensive" as well as by the historical coincidence that the three powers were simultaneously inclined toward cooperation [1990: 144–158].

GOMULKA PLAN.   Proposal made by Polish First Secretary Vladislav Gomulka, 28 December 1963, for a freeze on nuclear armaments in Central Europe [Schramm & al. (eds.) 1972:413–415], which should be seen in connection with the *Rapacki Plan*.

GOOD SOLDIER SCHWEYK.   Fictitious character in the play *Schweyk in zweiten Weltkrieg* (Schweyk in the Second World War), written by Bertoldt Brecht (1898–1956) in 1944 and first performed in 1956. S conducted a primitive (?) form of *nonviolent resistance* against the German occupants of his native Czechoslovakia [cf. Hart 1967:207]. He faked stupidity and thereby denied the Germans his assistance, and he "attacked" their logistical chains, e.g., by confusing the wagon count of the German soldiers (Scene 6).

GORBACHEV, MIKHAIL.   Secretary general of the Soviet Communist Party (CPSU) since 1985, subsequently also president of the *Soviet Union*. The abortive coup d'état against him in August 1991 was followed by his reinstatement as president, but the event had set in motion a revolutionary momentum that rapidly led to the banning of the CPSU and the dissolution of the Soviet Union. After his forced retirement, MG set up the Gorbachev Foundation and engaged in collaborative research ventures with Western counterparts. He is the author of the international bestseller *Perestroika: New Thinking for Our Country and the World* [1988].

MG's advocacy of *common security* and *NOD* developed gradually. It was paralleled by a similar development in the policy of the USSR and the *Warsaw Pact* and accompanied by initiatives in the realm of nuclear *disarmament*, with the *Year 2000 Plan* and the *INF* compromise as the most prominent examples. In addition to these global initiatives, MG also made many openings concerning regional *security*, including the *Murmansk Initiative* and the *Vladivostok Initiative* [Holden 1989:97–146; 1991; Bluth 1990].

In his report to the CPSU Congress in February 1986, MG made his first cautious opening with the promise "In the military sphere we intend to act in such a way as to give nobody ground for fears, even imagined, about their security" [Gorbachev 1986:85]. In his speech in April 1986 to the SED Congress, he made some further openings with his proposals for deep cuts in the conventional forces [in Gorbachev 1987:81–87]. These proposals were followed up by the Pact with the *Budapest Appeal* (11 June 1986).

In a statement to the international forum convened in Moscow in February 1987, MG became more explicit in endorsement of NOD-like conceptions: "The most dangerous offensive arms must be removed from the *zone* of contact" [*TASS,* 17 February 1987]. The WTO quickly followed suit with its *Berlin Declaration* (May 1987). In an article published jointly by *Pravda* and *Izvestija,* 17 September 1987, MG clarified his views by advocating *reasonable sufficiency* and defining this as tantamount to "a structure in the military forces of countries that suffices for the prevention of possible aggression but is insufficient for attack." In addition, he proposed negotiations on "a controlled withdrawal from the border areas of nuclear weapons and other weapons of attack with the succeeding establishment of a belt along the frontiers with restricted military deployments and demilitarized zones between the potential adversaries," pointing toward "a disintegration of the military blocs and abandonment of bases on foreign territories and withdrawal of all troops from abroad."

Having thus fully endorsed the concept of *NOD*, MG in an address to the 43d Session of the U.N. General Assembly in December 1988 announced a Soviet decision to reduce its armed forces by 500,000 men over two years, accompanied by a "substantial reduction" in conventional armaments, unilaterally and without relation to the impending *CFE* negotiations. Agreement had been reached with the Soviet allies to withdraw and disband by 1991 six tank divisions from the *GDR*, Czechoslovakia, and *Hungary*, along with "assault landing troops" and "assault crossing units." The number of Soviet forces stationed in those countries was to be reduced by 50,000 men and armaments by 5,000 *tanks*. Further, the remaining

Soviet divisions would be reorganized: "After a major cutback of their tanks [they] will become clearly defensive." This amounted to a reduction in the European USSR and the Pact of half a million men, 10,000 tanks, 8,500 artillery systems, and 800 combat aircraft [in *Survival,* 31:2, 171–176]. The declaration was nearly universally accepted as conclusive proof of the sincerity of the Soviet commitment to NOD.

GOTTWALD MEMORANDUM.   Planning document from the Czechoslovak Ministry of Defense written during the "Prague Spring" in 1968. The GM envisioned a partial shift to a *territorial defense* doctrine, a reform that would have weakened ties to the *Warsaw Pact* and might have contributed to a certain *neutralization* [Jones 1981: 95–100; Holden 1989: 85–94; cf. Harman 1974: 196–205]. These ideas may have been a contributory cause of the Soviet postinvasion decision to purge the armed forces of Czechoslovakia thoroughly [Johnson & al. 1982: 114–121, 126–137].

GPALS.   Global Protection Against Limited Strikes. Successor to the *SDI* adopted as the guideline for R&D and deployment (and funding) by the Bush administration in 1991 [Wright 1992]. GPALS has three components: ground-launched missiles for the interception of long-range missiles; space-based "brilliant pebbles" interceptors; and theater-defense missile systems, i.e., a form of an *extended air defense*. It should provide a certain (but not leakproof) defense of the U.S. and selected allies against ballistic missile attacks. The program has primarily been argued with reference to the proliferation of ballistic missiles in the Third World, hence either as an alternative or a supplement to the *MTCR* [Allgeier 1992; cf. for a critique: Wright 1992; Pike & Stambler 1992; Pike & al. 1992].

The implications for the *ABM Treaty* remain unclear, especially regarding the space-based components but also because deployment outside the permitted areas of weapons suitable for a defense against strategic ballistic missiles [Brauch 1987b]. However, judging by Russian statements in 1992–1993, the other party to the treaty might be more interested in partaking of the GPALS than in enforcing a rigid compliance of the treaty ["Documentation," *Comparative Strategy,* 12:1 (January–March 1993): 109–113; cf. Savelyev 1993].

GRAABAEK, JENS JØRN.   Major in the Danish armed forces, and one of the first in *Denmark* to advocate defense reform. JJG's main point of departure was the emerging trends in weapons technologies that tended to make traditional weapons platforms obsolete (they would presumably be incapable of competing with antiship and *air defense* missile technology) [1979; 1986; 1989]. While also coming out in support of the proposals for a Nordic *nuclear-weapons-free zone*, JJG came to advocate the notion of *NOD*. He also collaborated with the *Centre for Peace and Conflict Research* and particularly with *Anders Boserup* [cf. Boserup & Graabaek 1990].

Because his thinking amounted to an implicit (as well as occasionally explicit) critique of Danish defense planning, JJG became the victim of a campaign waged against him by his fellow officers and the right-wing political parties. He was, however, supported by the Social Democrats (cf. *Socialdemokratiet*). As an element in a defense agreement, the latter managed to institutionalize an independent advisory group under the auspices of the Ministry of Defense (*FRAG*), and to secure a position for JJG in it. In this capacity, he was instrumental in the creation (on an experimental basis) of land-based antiship missiles for coastal defense [cf. FRAG 1986], intended as a partial substitution for the mothballed Danish frigates that had thitherto served primarily as missile platforms.

GRAAF, HENNY VAN DER.   See the *Netherlands*.

GRADUALISM.   Generic term for various proposals made by *Charles Osgood*, *Amitai Etzioni*, and others for breaking what was seen as a vicious circle of actions and reactions in the East-West relationship. Presumably one side would be able to break the circle through calibrated *unilateralism* that would presumably be reciprocated by the opposing side [Osgood 1959; 1962; Etzioni 1962; cf. Böge & Wilke 1984: 26–55; George 1988b; Goldstein & Freeman 1990]. See also *GRIT*, *Extended GRIT*, *Progressive GRIT*, *Security Regime*, *Tit-for-Tat*.

GRAND STRATEGY.   See *Military Science*.

GRAY, COLIN S.   See *U.S.*

GREAT BRITAIN.   See *U.K.*

GREECE.   See *Balkans*.

GREENE, OWEN.   See *U.K.*

GREEN PARTIES.   The GP in most West European countries (and partly in the former "East") are offsprings of the environmental and *peace movements*. As such, they are not political parties in the traditional sense but, rather, hybrids of parties and mass movements [Galtung 1986; Holmes 1984; Rosolowsky 1987: 77–97; Wasmuht 1987: 139–141; Rochon 1989: 83–86; Møller 1991: 185–187].

On the one hand, the GP have seen themselves as parts of the peace movements and the green movement, sharing their antiauthoritarian and almost antipolitical sentiments, as well as their basic democratic structure. Compared with the grass-roots movements, however, the GP have sought to integrate a panoply of causes (ecology, peace, international solidarity, etc.) into a holistic alternative, presumably transcending the left/right and socialist/liberal dichotomies. The GP have also had to compete for power positions in local constituencies, national parliaments, and even the European Parliament (the "Rainbow Group"), which has forced them to play according to the rules of the game, particularly when opportunities have opened up for wielding power in coalition with other parties. Such challenges have frequently led to controversies between fundamentalists and pragmatists.

In security political and defense matters, the GP have tended toward *pacifism* in general and nuclear abolitionism in particular, often advocating *civilian-based defense* as an alternative. They have occasionally also lent their support to *NOD* as presumably a lesser evil than the prevailing offensive strategies, and thus a step toward *demilitarization*. The GP have, furthermore, tended to support *neutrality* (yet also to deny the need for underpinning neutrality with a defensive capability), and they have occasionally supported *neutralism* and/or *dealignment* as steps in this direction. Finally, the GP have opposed any form of military intervention abroad and have occasionally taken a very activist stand on these matters, to the point of advocating mutinies. See also: *Die Grünen*.

GRIN, JOHN.   Physicist, Amsterdam University, affiliated with *INSTEAD*, the *AFES-PRESS*, and the *SAS*, on whose behalf he has on several occasions published presentations of the *spider-and-web* model [Grin & Unterseher 1988; 1989; 1990; 1990a].

JG has specialized in *C^3I* (command, control, communications, and intelligence) and in 1990 published by far the most comprehensive analysis of the C^3I requirements of an *NOD* posture. His example was the SAS model of the land forces, which he compared with *NATO*'s C^3I plans associated with the *AirLand Battle* and *FOFA* doctrines [Grin 1990a; 1990; cf. Borg & Grin 1986; SAS 1984; 1987; idem (ed.) 1989; Unterseher 1988b; 1989b–d]. The latter were criticized for their unjustified trust in the potentials of *E.T.* (emerging technologies) and complex high-technology systems, as well as for their negative security political repercussions. C^3I in an NOD-type defense, according to JG, would be inherently less demanding because it could discard target acquisition beyond the border along with cross-border mobility of systems and communication arrays. Also it could assume a network structure, similar to that of the defense *web* itself. Dug-in fiberglass cables would be comparatively cheap, jamming-resistant, unseverable, and quite sufficient to insure cohesion and coordination within the web. In addition, such means of communication would by their very nature be stationary, hence confidence-building in the sense that otherwise mobile forces would have access to the C^3I array only when operating defensively on their own territory [1990a: 253–319].

Under the auspices of INSTEAD, JG has for some time also been involved in research on verification principles and procedures pertaining to conventional forces. In this connection he has focused on possible ways of ascertaining the offensive or defensive nature of military formations, inter alia, through monitoring their *logistics*: ammunition supplies, airfield equipment, transport vehicles, etc. [1990b; cf. idem & Graaf (eds.) 1990].

GRIP.   *Groupe de recherche et d'information sur la paix* (Group of Research and Information on Peace). See *Belgium*.

GRIT.   Gradual Reciprocated Initiates in Tension Reduction. Strategy developed by *Charles Osgood* [1962; cf. 1959], and the best-known variety of *gradualism* [cf. Böge & Wilke 1984: 26–55; George 1988b; Goldstein & Freeman 1990; R. Fischer 1990: 196–202]. See also *Progressive GRIT*, *Extended GRIT*, *Super-GRIT*.

GROUND FORCES.   See *Land Forces*.

GROVE, ERIC.   See *Foundation for International Security*.

GRÜNEN, DIE.   The (West and subsequently all-) German *Green Party*, which is represented in the Federal Diet (Bundestag) as well as in the European Parliament. On a few occasions, DG have even formed coalitions with the *SPD* in local and *Länder* elections and thereby gained positions of political power [Holmes 1984; Rosolowsky 1987: 77–97; Wasmuht 1987: 139–141; Rochon 1989: 83–86; Møller 1991: 185–187].

In security political matters, DG may have served as a catalyst, inducing the SPD to take a stand on various otherwise neglected issues. DG have thus shared most of the SPD's critical positions with regard to *NATO* strategy in general, and nuclear strategy as well as *deep strike* in particular [Grünen im Bundestag 1984; Grünen 1985; Rosolowsky 1987: 77–79; B. Meyer 1987: 47–52]. On the other hand, contrary to the SPD, DG have supported *neutrality* and hence opposed not merely NATO membership but also the various initiatives (e.g., on the part of the SPD) toward *Europeanization*. Furthermore, being ideologically committed to *disarmament*, DG have been unwilling to support the SPD's advocacy of *NOD*, not even as a lesser evil. While following the NOD debate, DG have continued to lend their support to *social defense* (i.e., *civilian-based defense*) [Grünen im Bundestag 1988; Schulze-Marmeling 1987: 214–236; Böge & Schulert 1986: 163; Schierholz 1986; Statz 1988; Grünen 1981; cf.Bruckmann 1984].

In connection with the German unification debate, DG advocated a complete demilitarization and *neutralization* of the future Germany and were involved in the *BOA (*Bundesrepublik ohne Armee* [Federal Republic without an army]) campaign [*Frieden und Abrüstung,* no. 1 (1990): 52–53; cf. Buro 1990]. DG have likewise been vehemently opposed to any form of intervention and hence to any *out-of-area* role for the Bundeswehr, even under *U.N.* auspices. During the U.N. blockade of Iraq and the ensuing *Gulf War,* DG urged conscripted soldiers (of both the Bundeswehr and the East German NVA) to desert if they were to be sent to the Persian Gulf [reprinted in *Blätter für deutsche und internationale Politik,* December 1990: 1523–1526; cf. Pax Christi, "Für eine neue Friedenspolitik," *Publik-Forum,* (5 October 1990)].

GRUPPE FÜR EINE SCHWEITZ OHNE ARMEE.   Group in Favor of Switzerland without an Army. See *Switzerland.*

GSOA.   *Gruppe für eine Schweitz ohne Armee* (Group in Favor of Switzerland without an Army). See **Switzerland.**

GUERRILLA STRATEGY.   Particular *strategy* for waging *guerrilla warfare.*

1. **Prehistory.** The first theoretical treatments of GS appeared in the wake of the *Peninsular War,* i.e., Napoleon's war in Spain, in the course of which he was opposed by guerrillas. Among the founding fathers of GS were Von Valentini and De Corvay, but *Jomini* and *Clausewitz* also made valuable contributions to the theory of GS.

Denis Davydov contributed to the theoretical development of GS on the basis of the experience with anti-Napoleonic guerrilla warfare in Russia, spurred by the French invasion. He emphasized the advisability of exploiting the vulnerability of the long supply trains of the invader and urged the guerrillas to ignore both the supply field and the battlefield because both would be heavily defended. Rather, they should to seek to sever the links between the two, which, because of their length, could not possibly be adequately defended [excerpted in Laqueur (ed.) 1978: 49–57]. Guerrilla methods were employed on a large scale in the nationalist revolutions of the nineteenth century, e.g., in Italy and Poland, where they were preached by Carlo Bianco, Guiseppe Mazzini, Wojcheic Chrzanovski, and Karol Stolzman, among others [Laqueur (ed.) 1978: 66–89].

Although these revolutionary events presaged the upsurge of revolutionary guerrilla warfare in the twentieth century, the nineteenth-century advocates of proletarian revolution in actual fact had no great faith in guerrilla methods. Karl Marx and Friedrich Engels were impressed by the Spanish guerrillas and the so-called 'Franc-Tireurs' (guerrilla-like snipers operating during the 1870 Franco-German War), but neither advocated guerrilla warfare for the ultimate overthrow of capitalism. Lenin (in line with the French socialist Auguste Blanqui) paid some attention to urban guerrilla warfare between the 1905 and 1917 revolutions, yet he did not place his faith in this form of armed struggle for the revolution. Indeed, on the verge of the 1917 revolution he explicitly warned against its employment. Moreover, although guerrilla methods were occasionally used by the communists in the course of the civil war, they were seen as a regrettable necessity, and Lenin (as well as Trotsky) thenceforth wholeheartedly supported the creation of regular armed forces.

2. **Twentieth-century GS.** In connection with the prevalence of guerrilla warfare (inter alia, as an element in anticolonial struggles), the theoretical literature on GS has burgeoned. Among the main authors one could mention *Lawrence of Arabia,* *Mao Zedong,* *Josip Broz Tito,* *Vo Nguyen Giap,* and *Ernesto Che*

Guevarra. Even distinctly Western strategists have taken an interest in GS; these fall into three categories.

- One group has been impressed by the accomplishments of Third World guerrillas and has been discontented with *NATO* strategy; hence it has proposed a partial assimilation of the latter to the former. In addition to "genuine" proponents of alternative defense (cf. below), this group has included some of the most prominent strategic thinkers: the french general *André Beaufre* [1972], *Liddell Hart* [1967a], and *Richard Simpkin* [1986: 311–321].
- Another group has sought to develop viable strategies for counterinsurgency warfare, the means to defeat guerrilla movements [Zawodny 1962; Lindsay 1962; cf. Gibson 1988: 76–81].
- Finally, quite some thought was apparently given (albeit nearly always clandestinely), particularly by the two superpowers, to the utilization of guerrilla strategy for more offensive purposes, namely, as an instrument in their contest for influence, above all in the Third World. One superpower's support for insurgents with guerrilla characteristics in countries controlled by the other superpower has accordingly been regarded as a viable substitute for direct confrontation.

3. **GS and NOD.** Many *NOD* proponents of the 1980s have been inspired by guerrilla warfare and strategy [e.g., Spannocchi 1976; Löser 1981]. There are, indeed, many similarities between the two. Both are asymmetrical strategies (i.e., instances of the *indirect approach*) that anticipate countering the opponent's military forces by different means: *tanks* with antitank weapons, aircraft with *air defense* guns and missiles, as well as with *dispersion*, etc. Both seek to avoid decisive battles, and seek a decision by way of *nonbattle*, i.e., the cumulative *attrition* of the enemy in the course of a protracted war. Both emphasize the importance of adapting to the home terrain, at the inevitable expense of general purpose capabilities. Both lack strategic mobility, while maintaining or even increasing tactical mobility and agility.

There are also a number of important differences. Guerrilla warfare may be used offensively, as a means of overthrowing a regime; NOD is not only militarily but also politically defensive. GS deliberately blurs the distinction between combatants and noncombatants; (most versions of) NOD have emphasized the importance of rigid distinctions, for the sake of *damage limitation*. And GS is the direct antithesis of forward (i.e., frontier) defense; most versions of NOD have placed great (although not exclusive) emphasis on *forward defense* [Møller 1991c].

GUERRILLA WARFARE.   Particular form of warfare usually associated with the postwar Third World but actually both considerably older and of much broader scope.

1. **Early History.** The history of GW (or "small war," as it used to be called) goes back for millennia. King Saul's war against the Philistines, e.g., resembled modern GW in its reliance on ambushes and small forces, just as did the Gallic chief Vercingetorix's strategy of resistance against Caesar [Beaufre 1972: 86–87]. Even the Romans occasionally resorted to strategic or tactical conceptions similar to those of GW (cf. *Fabian Strategy*) [Hart 1967a: 26–27]. Most wars during the Middle Ages also resembled GW in their propensity for battle avoidance.

During the American Revolution, guerrilla raids were employed on a grand scale, and they played an important auxiliary role for the armies of revolutionary France after 1791, as well as for the Vendée revolt of the French aristocracy

[Beaufre 1972: 87–100]. Finally, GW came of age during Napoleon's war in Spain (the *Peninsular War*), in which the guerrillas accomplished amazing successes, thus giving birth to the first theoretical analyses of *guerrilla strategy* [Gates 1986; Hart 1967a: 110–119]. In Russia as well, guerrillas played a notable part in defeating Napoleon's armies. Guerrilla struggles also occasionally flared up throughout the nineteenth century, e.g., in connection with the national independence struggles and as "sideshows" in, e.g., the Franco-German War. In the first half of the twentieth century, GW was waged on a large scale, e.g., in the Middle East (see *Lawrence of Arabia*) and in China (see *Mao Zedong*).

2. **GW During World War II.** During *WWII*, GW ("partisan warfare") was employed on a small scale (in the form of sabotage and the like) in most German-occupied countries, as well as on a large scale, e.g., in the *Soviet Union* and *Yugoslavia*. In all cases it was supported materially as well as morally by the Allies, who regarded GW as a valuable contribution to the overall war effort.

The USSR and the European communist parties had been rather negative toward GW, although partisan units had played an important part in the civil war in Soviet Russia until 1921 [Armstrong 1964: 10–11]. In the Spanish civil war (1936–1939) the communists and the International Brigades had supported a strategy of positional warfare, and it was only in the very last stages, and on a small scale, that guerrilla warfare was tried, e.g., in Asturia [Brouée & Temine 1968: 516–524; Jackson 1965: 265–266]. The negative attitude should probably be attributed to the seeming incompatibility between effective centralized control and the anarchic character of irregular warfare, as well as to an aversion against an independent role for the armed forces.

Nevertheless, from 1941 to 1945, the USSR had to resort to GW on a grand scale, especially in the western and southern parts of the country that had been swiftly overrun by the German armies. In this endeavor, the military leadership (decimated by the Stalinist purges a few years earlier) was able to utilize some of the experience and personnel from the Spanish civil war, as well as, to a certain extent, the Chinese experience. However, when trying to apply these lessons, the Soviet planners faced a number of problems:

- The Stalinist atrocities had turned the overwhelming majority of the peasantry against the regime, hence its reluctance to support Soviet rule against the Germans. Gradually, however, the atrocities of the invaders persuaded the peasantry to support the regime as the lesser evil.
- The lack of spontaneous support meant that the resistance had to be organized from above by means of flown-in officers and party staff, hence its fragile basis. The situation was exacerbated by the fact that the imported personnel consisted to a large extent of deserters from the regular forces, former counterrevolutionaries, and the like, who hoped to regain the favor of the regime.
- The flat Soviet terrain with its lack of cover was unsuited to guerrilla operations, and only the unavailability to the Germans of sufficient *air power* saved the partisans from being mopped up in many cases.
- The harsh climate and barren soil made it extremely difficult to live off the land without very severe strains being put on relations with the local population.
- The above factors and reliance on air delivery of munitions and other supplies precluded *dispersion* and thus elusiveness because the guerrilla units had to remain concentrated and close to airfields.

On the other hand, the partisan struggle was facilitated by the overall superiority of Soviet forces, which implied that the guerrillas were inferior only locally.

Moreover, the German forces stationed in the areas bypassed by the main forces (i.e., the regions where partisan warfare was most intense) were distinctly second-rate as well as poorly equipped. Hence the partisans in most cases enjoyed air superiority, quite an exceptional situation for a guerrilla struggle. Although the partisans generally shunned direct encounters with the invaders, they did attack the German logistical chains into the depths of the country. The disputable overall cost-effectiveness notwithstanding, the partisans were thus able to inflict great losses on the Wehrmacht [Armstrong 1964; idem (ed.) 1964: 14–15, 31–34; idem & DeWitt 1964; Ziemke 1964; Weinberg 1964].

Partisans also operated in a number of other countries during WWII, in most cases in rather close coordination with the Allies and receiving Allied weapons. The spectrum ranged from full-fledged guerrilla forces (mostly in southern and eastern Europe) to minuscule groups of saboteurs in occupied countries such as *Denmark*. The latter almost never sought direct engagements with the German forces but limited themselves to disrupting communication lines, destroying war-making potentials, etc. To the extent that such saboteurs should be counted as armed forces at all, they constituted guerrillas par excellence because they emphasized elusiveness and hit-and-run tactics to an even greater extent than full-strength guerrilla formations did [Kühnrich 1968. On Denmark, see Bennike 1946; Outze 1946; Haestrup 1954; 1959; Brandt & Christensen 1968; Trommer 1971].

During most of WWII, factional disputes tended to be put aside in favor of national defense. However, as the focus gradually shifted from defeating Germany to a satisfactory postwar settlement, the same splits that divided the Allies infected the partisan movements as well. In some cases, competing provisional governments were set up and internal strife erupted, albeit on a small scale. The majority of these conflicts were effectively settled at the Yalta and Potsdam Conferences, but in Greece a civil war erupted, in which GW played a predominant role. Still, GW achieved a truly decisive victory in only one country: Yugoslavia.

3. **Postwar GW.** In the postwar period GW has, on the one hand, spread across the globe, albeit distinctively in the Third World. A certain bifurcation seems to have occurred. On the one side, GW methods have been used extensively for national liberation purposes (i.e., for what might be called defensive ends), e.g., in *Vietnam* or the Portuguese colonies. On the other side, similar methods have also been used in attempts to overthrow existing regimes—be they just and legitimate or not—something that is hardly politically defensive (no matter how just the cause may be); hence the synonym "revolutionary war." Cuba, Angola, and Mozambique are just a few examples of a very general trend in postwar international history. A particularly bloody example has been the Peruvian guerrilla movement Shining Path, the methods of which have barely, if at all, been distinguishable from terrorism [Chaliand 1977; 1982a; 1991; Bejar 1982; Møller 1991c; Beaufre 1972; Shy & Collier 1986; Baylis 1987].

4. **Distinguishing Features.** GW has most often been resorted to by the weaker side, usually by means of irregular forces and nearly always on home territory. There are exceptions. It does happen, e.g., that the stronger side resorts to offensive use of guerrilla proxies (such as the Nicaraguan Contras being utilized by the *U.S.*); regular forces sometimes resort to guerrilla tactics, such as is the case in various *territorial defense* schemes (inter alia, those of *Sweden* and *Finland*); and GW may well be used on enemy territory, as the Nicaraguan example shows.

The distinction between regular and guerrilla warfare is thus not so much one of political context but of mode of combat, as well as configuration of the armed forces. Among the particularities of GW are the following:

- GW forces tend to be small, i.e., subdivided to a greater extent than regular forces: combat units numbering tens or at most hundreds are the rule. An implication thereof is that guerrillas almost never attack the enemy's main force but attack isolated detachments or supply chains.

- Regular forces defend or attack a specific front, but guerrillas tend to direct their attention to the enemy's armed forces, hitting them whenever and wherever possible.
- GW tends to be fluid, there are neither fronts nor rear areas but one extended combat area with sporadic fighting. An implication is that the distinction between combatants and civilians is difficult to maintain (and occasionally is deliberately blurred); hence the tendency of antiguerrilla warfare to become "dirty," i.e., to violate the *Laws of War* (as well as human dignity) and to exact a heavy toll of civilian casualties.
- Strategic mobility tends to be very low (with the Chinese "Long March" as almost the only exception), whereas tactical mobility is often higher than that of regular forces. This tends to make GW a more appropriate means of national defense than of expansionist policies.

GUERRE DE COURSE.   Cruising warfare, i.e., a naval strategy directed against an opponent's commercial shipping, and reminiscent of *guerrilla warfare* in its avoidance of the decisive battle. It has been advocated by the *sea power* partisans such as Alfred T. Mahan. The GC's alternative is a *nonbattle*, small hit-and-run encounters generating a cumulative *attrition*.

Such a naval strategy (developed by the so-called *Jeune Ecole*) predominated in France in the nineteenth-century, when France was clearly inferior to the *U.K.* in terms of maritime power [Ropp 1941; Ceillier 1990]. More generally, GC has traditionally been regarded as an appropriate strategy for the weaker of two opponents. Accordingly, it has been assumed that the *Soviet Union* had contingency plans for cruising warfare against *NATO*'s transatlantic *SLOC*s (sea lines of communication) in the event of a major war in Europe. There was no solid evidence of such plans but many indications that anti-SLOC operations were always far down on the Soviet list of wartime naval priorities [MccGwire 1979; 1981].

GUERTNER, GARY L.   See *U.S.*

GUEVARA, ERNESTO CHE.   Cuban guerrilla leader (1928–1967) and a central participant in the 1959 Cuban revolution. He was executed in Bolivia for trying to start a revolution. Besides an autobiographical work, *Reminiscences of the Cuban Revolutionary War* [1968], he wrote several theoretical treatises on *guerrilla strategy* [1976].

Both the Cuban revolution and the subsequent unsuccessful would-be revolutions that ECG sought to instigate (under the catchword "One, Two, Many Vietnams") were much more adventurist endeavors than most other guerrilla wars. The Cuban revolution was launched with the expedition of a small vessel, the *Granma*; in December 1956 it landed a group of prospective guerrillas in Cuba that was rapidly decimated by the alerted Batista troops. During the following years, however, the guerrilla movement gradually spread, so that in 1959 it was finally able to overthrow the dictatorship. Although the guerrilla struggle took place in the countryside, it collaborated closely (albeit not without considerable friction) with the workers' and students' movement in the cities, so that it was a joint effort that accomplished the revolution [Guevara 1968: 36–41, 196–211].

When formulating his own strategy (or grand strategy), as well as that of the revolutionary Movement of the 26 July, ECG consistently emphasized the political nature of *guerrilla warfare*: the role of the *guerillero* was that of a social reformer, at least to the same extent as that of an ordinary soldier. Nevertheless, the armed struggle would, in the final analysis, prove decisive. In due course, the guerrilla forces should therefore transform themselves into a regular army, able to strike the final blow against the ruling classes [1960: 28–30, 32, 52–54, 84–94]. During the guerrilla phase, revolutionaries should initially seek, above all, to preserve a nucleus of fighting men and to establish a secure operating base. Further, it was important to emphasize the light, elusive, and independent character of the armed forces. The fighters should remain dispersed, with the ideal unit about ten men, so as to make the enemy's air supremacy irrelevant because of an absence of targets. Weaponry should be light and include rifles, bazookas, Molotov cocktails, antitank mines, and mortars. The main source of weapons supplies should be the enemy, in order for the guerrilla to avoid the burden of cumbersome *logistics* and/or dependency on foreign suppliers [1960: 33–36, 39–40, 44, 47, 64–65, 74–75].

Contrary to other Latin American guerrilla strategists and movements (such as the Tupamaros in Uruguay [cf. Guillen 1973; Marighella 1978]), ECG remained skeptical of urban guerrilla warfare, even though it might seem to recommend itself because of the rapid proliferation of cities. He regarded it as, at most, an auxiliary form of struggle, while realizing that "the importance of the struggle in the urban zone . . . is very high. A good work . . . could practically paralyze the commercial and industrial life of the sector entirely and . . . make [the population] desire violent events. . . ." On the other hand, he recommended extreme caution regarding employment of terrorist assassination and sabotage (without, however, renouncing such means entirely) because they invited retaliation against civilians [Guevara 1960: 51, 98-100].

After the Cuban revolution, ECG supported creation of a strong army capable of defending the state against foreign intervention. The armed forces should be tantamount to "the people in arms" and maintain guerrilla struggle as a fallback option should regular military operations fail. In that case, it should seek to maintain control over the most inaccessible regions, whence the armed resistance could gradually spread [1960: 124–129].

GULF WAR.   War between Iraq and the *U.N.*–sponsored coalition in January and February 1991. The term is also sometimes used in reference to the Iran-Iraq war (1980–1988), which preceded it and was a contributory cause of Iraq's (i.e., Saddam Hussein's) decision to attack Kuwait on 2 August 1990 [Cooley 1991; cf. Karsh 1987; Freedman & Karsh 1991; Friedman 1991: 10-42; Gow (ed.) 1993].

1. **Crisis.** The international community reacted promptly to the invasion, and the recent end of the Cold War made the U.N. a more effective instrument than it ever had been. A series of Security Council resolutions that called upon Iraq to withdraw immediately and unconditionally were passed with near-unanimity [in *Survival* 32(6): 558–559; 33(1): 78–79; Weller (ed.) 1993]. The U.N. also imposed a number of economic sanctions on Iraq, implemented through a naval blockade, in which the *U.S.* and the WEU played central roles and naval forces from small states such as *Denmark* were also involved [Eekelen 1990; Burgelin 1991; Friedman 1991: 310-323].

Simultaneously with the embargo intense diplomatic efforts to achieve a peaceful solution were made in parallel with military preparations for a war against Iraq, should diplomacy fail. And Saudi Arabia's defenses against an Iraqi attack were bolstered, the so-called Desert Shield [Friedman 1991: 74–107]. In its diplomacy, the U.S. seemed to place greater emphasis than had been the case in previous crises on acting in concert with regional states as well as on involving both allies and former adversaries (above all the *Soviet Union*), with considerable overall success [Fuller 1991]. Nevertheless, in the *peace movements* and parts of the peace research community, many critical points were raised, some of which were based on a somewhat twisted logic:

- The U.N. was charged with having become "a virtual tool of US foreign policy" [Richard Falk in *The Guardian,* 17 January 1991].
- The U.S. and the rest of the West were charged with applying a double standard because others (including the U.S.) had previously gotten away with similar violations of international law.
- It was (undoubtedly correctly) pointed out that ulterior motives were involved, first among which was the concern about an unbroken supply of oil [Lutz 1990a; B. Müller (ed.) 1991].

2. **War.** U.N. Security Council Resolution 678 (29 November 1990) had authorized "member states cooperating with Kuwait" to "use all necessary means . . . to restore international peace and security in the area" unless Iraq withdrew by 15 January 1991, at the latest [*Survival* 33(1): 78]. This was interpreted by the U.S.–led coalition as sufficient justification for initiating combat.

The war ("Desert Storm") began with a gigantic nocturnal *OCA* air attack (16–17 January 1991) that effectively wiped out Iraq's *air defenses* and was followed by a longer air campaign against military targets in Kuwait and Iraq. Throughout the air war, the allies sought to persuade Iraq to surrender, but it took a large-scale ground offensive (starting 24 February) to force Saddam Hussein's capitulation [Friedman 1991: 147–235; Freedman & Karsh 1991; *International Defense Review,* no. 9 (1991): 979–981].

3. **Military Lessons.** The GW was a clear-cut instance of the *offensive-turned-defensive* in that Iraq had fortified occupied Kuwait heavily and employed many of the traditional means and methods of the defender, including a perverted form of *scorched-earth strategy* [Friedman 1991: 108–119; *International Defense Review,* 1991(9): 989–994]. Still, the allied victory was surprisingly swift and bloodless: even though the number of direct war casualties on the Iraqi side remain unknown or disputed, their number seems to have been lower than the number of victims (e.g., among the Kurds) of Iraqi repression.

Many, and partly contradictory, lessons have been drawn from the experience of the GW [Aspin & Dickinson 1992; Bundy 1991; Perry 1991; Vuono 1991; Freedman & Karsh 1991; Friedman 1991: 236-260; Sinn 1991; Poirier 1991; Géré 1991; Colson 1991; Dumoulin 1991; Herrmann 1991; McCausland 1993]:

- Diplomatic means of conflict prevention should be used more effectively [Cooley 1991; Record 1993; Kemp 1990].
- Nuclear deterrence is ineffective, hence conventional forces will be what matters in future wars [Bundy 1991].
- *Air power* in general, and offensive counter-air (*OCA*) in particular, could decide a war, but it would take *land forces* to win it.
- The threat posed by ballistic missiles (the Iraqi *SCUDs*) might have been exaggerated [cf. Postol 1991].
- High-tech weapons made a difference.
- Morale was of utmost importance and tended to benefit democracies over authoritarian regimes.
- Collateral damage to civilians could be kept limited, provided care was taken in the selection of targets and that precision-guided weapons were available.

Although some of these lessons seem to constitute arguments against *NOD* (or at least call for modification of existing NOD models), the poor defense performance of the Iraqi armed forces was to a large extent due to particular circumstances rather than to a general weakness of the defensive form of combat:

- The air defense posture of Iraq had been neglected, inter alia, because of Saddam Hussein's need to be able to suppress insubordinate air force units himself [Friedman 1991: 158–159].
- The officer corps was selected politically rather than on criteria of merit.
- The terrain was very unfavorable to defensive warfare, especially to the defense of fixed positions.
- Above all, the Iraqi armed forces were not defenders but aggressors who had hastily assumed defensive positions.

4. **Security Political Lessons.** The GW has also led to a political quest for stabilization of the region, based primarily on local security arrangements but with a certain foreign (above all, U.S.) contribution. Furthermore, to the realization that the arms flow to the region had to be contained (cf. *arms trade regulations*) [Kramer 1992], the armistice agreements sought to prevent a future war instigated by Iraq by subjecting it to armament constraints reminiscent of those in the Paris treaties pertaining to Germany's allies in *WWII* [Molander 1992].

Finally, the crisis led to a better understanding of the need for improving the U.N. apparatus. Even though the U.N.'s performance throughout the crisis and war was adequate, it was nevertheless *ad hoc* and amounted to little more than a subdelegation to a hastily assembled group of nations of the implementation of its resolutions; hence the quest for improved *collective security* arrangements [Steinbruner 1991; Chubin 1991; Rubinstein 1991; Urquhart 1991].

GUPTA, RAJ.   Resident scholar at the Brookings Institution, *U.S.* In 1993 he published *Defense Positioning and Geometry: Rules for a World with Low Force Levels* [1993], in which he challenged the argument often encountered in connection with the *CFE* negotiations: *NATO*'s defenses were already stretched so thin that there was very little room for force reductions. According to Gupta's calculations, several means are available to a party with strictly defensive intentions that might allow it to cope with reduced force densities: passive defenses, including a system of bunkers with automatic antitank firing systems (somewhat reminiscent of the *archipelago defense* of *J. F. C. Fuller*). Combined with a less forward deployment of forces, the alleged *force-to-space* constraints would lose their relevance, provided only that force-to-force disparities were reduced. The task might be further facilitated by a number of collaborative *arms control* agreements, including a subdivision of the total frontage separating two adversaries, not merely (as envisioned by most *NOD* proponents) vertically, i.e., into *zones* at various distances behind the border. Of even greater effect would be a horizontal subdivision. Such schemes would have the advantage that they would make the most effective defense posture available to each party identical to the most defensive one, thereby putting a premium on defensive restructuring that would benefit *stability*.

GYSI, GREGOR.   Chairman of the successor to the ruling party of the *GDR*, the SED: the PDS (*Partei des Demokratischen Sozialism* [Party of Democratic Socialism]). In January 1990 GG published *German-German Security Model 2000*, a model that he set forth as an element in the "contractual community" between the two German states by then under consideration. The model resembled *SPD* policy as well as the notions of West German *NOD* proponents quite closely. GG proposed, e.g., to cut the armed forces of both German states in half; to terminate all modernization programs; to withdraw all military formations above company level from a strip of 50–80 km on both sides of the border; to ban German submarines and amphibious vessels from the Baltic Sea; to demand a complete withdrawal of *NATO as well as *Warsaw Pact forces from Germany by 1999, starting with the border areas; and to call for a complete withdrawal of nuclear and *chemical weapons before 1991 [*Blätter für deutsche und internationale Politik,* no. 2 (1990): 237–238].

H

HAGENA, HERMANN. Brigadier-general in the Bundeswehr (ret.) and author of one of the few comprehensive analyses of *air power* from an *NOD* point of view: *Low-Altitude Flying in Central Europe* [1990b; see also 1994].

HH's vantage point was a critical analysis of low-altitude flying, which constituted a serious inconvenience for the German civilian population [Cf. Mechtersheimer & Barth 1986; 167–180; Szabo 1990: 56]. The emphasis on this type of operations was warranted only with a view to offensive missions (such as offensive counter-air, *OCA*), and even for this purpose its efficiency was questionable, especially if the aircraft were up against state-of-the-art *SAMs* [Hagena 1990b; 32–53]. In addition to the technical obstacles, there were serious security political arguments against OCA. Its placing a premium on surprise put *NATO* (as the stipulated defender) at a severe disadvantage, while rewarding the prospective attacker [1990b: 82–83]. The defender would do better to rely on defensive counter-air (i.e., *air defense*), for which purpose low-altitude flying was irrelevant. Contrary to most air power proponents, HH did not change his opinion in light of the *Gulf War* [1994].

In addition to recommending a change of strategy, HH also urged NATO to accept negotiations on conventional air power, which should focus on a build-down of offensive aerial potentials. Their purpose would be to bring about a shift of emphasis from OCA and *deep strike* to air defense, battlefield interdiction, and close air support [1990: 92–97; 1994; cf. for a very similar set of proposals: Møller 1989].

HAGUE CONVENTIONS. See *Laws of War*.

HANNIG, NORBERT. Lieutenant-colonel and staff officer of the Bundeswehr (ret.), author of *Deterrence with Conventional Weapons: The David-and-Goliath Principle* [1984; cf. for the background: 1979; 1981. English condensation: 1989], in which he launched an alternative *NOD* paradigm alongside that of *Horst Afheldt*'s *territorial defense*: a linear border defense based on *fire barriers*. The essence of the original proposal has been maintained since then, but the details slightly modified in the context of a collaborative project with *Johannes Gerber*, resulting in the *DEWA* (Defense Wall) model [Hannig 1986; 1990].

By accepting the *forward defense* imperative, NH was more in line with the *FRG's* political orientation. Nevertheless, his proposals pointed in the direction of a radical break with the Bundeswehr tradition by envisioning a complete abandonment of mechanized, armored formations. In contrast to historical precedents (such as the *Maginot Line*), NH's linear border defense would be a potential fire barrier, i.e., a form of killing *zone*. The fire forming the barrier would above all consist of antitank mines, with which selected parts of the frontier zone would be saturated (to a density of 0.4–0.8 mines per sq. m), and which were to be emplaced by means of indirect-fire systems.

In addition to serving as submunition dispensers with a variety of antitank, air defense, and antipersonnel munitions, the rockets and missiles (and especially mul-tiple-launch rocket systems) would also direct precision strikes against point targets. The missile systems would have a broad panoply of ranges and be deployed accordingly: those with the longest ranges would be farthest from the designated fire zone, which would minimize the threat to enemy territory. Typical ranges would be from 20 to 200 km, but even longer ranges were deemed indispensable, e.g., for *OCA (offensive counter-air) missions. Its offensive element notwithstanding, the scheme was still declared to be nonoffensive by virtue of the absence of invasion capabilities, i.e., armored, mechanized *land forces.

NH envisioned using basically the same systems for *air defense and coastal defense purposes, which would effectively render the air force and navy superfluous for other than surveillance purposes. Hence the division of the armed forces into separate service arms might be abandoned in favor of an integrated hierarchical system of command.

HARBOTTLE, MICHAEL.   See *U.K.

HARMEL REPORT.   See *NATO.

HART, BASIL LIDDELL.   See *Liddell Hart, Basil.

HEALEY PLAN.   On behalf of the Labour Party of the *U.K., Denis Healey in January 1958 issued a plan for military *disengagement. Called *A Neutral Belt in Europe,* it envisioned a *neutralization of the two Germanys, along with *Poland, Czechoslovakia, and *Hungary, with a possible extension to Romania and *Denmark. In the defined area no nuclear weapons would be deployed, and conventional forces would be built down to a level just adequate for frontier defense. The only viable deterrence of full-scale attack would presumably consist in security guarantees on the part of major powers, underwritten by capabilities for limited nuclear war [Albrecht (ed.) 1986:81–88; cf. Howard 1958]. See also *Gaitskell Plan, *Germany Plans, *Rapacki Plan.

HELSINKI FINAL ACT.   See *CSCE.

HERZ, JOHN.   American political scientist, usually categorized as belonging to the *Realism school; one of the first to write at length about the *security dilemma [1950; 1951].

HO CHI MINH.   See *Minh, Ho Chi.

HOLDEN, GERARD.   Former research fellow at the University of Sussex, subsequently at the Peace Research Institute, Frankfurt (PRIF). GH was the author of two of the main works published in the West on Soviet *New Thinking, *reasonable sufficiency, *NOD, and Soviet *alliance policies: *The Warsaw Pact: Soviet Security and Bloc Politics* [1989; cf. 1990; 1990a; 1991a] and *Soviet Military Reform* [1991; cf. 1990b]. In both works, he sought to open the "black box" of Soviet security political decisionmaking, and this uncovered quite profound differences of opinion between the Communist Party leadership, the *institutchiki, and the armed forces over such issues as the appropriate scale of *counteroffensives, and the relative emphases on offensive and defensive operations.

HOLLAND.   See the *Netherlands.

HOLLINS, HARRY B.   See **U.S.*

HOLST, JOHAN JØRGEN.   See **Norway.*

HOOVER, HERBERT.   See **U.S.*

HOWARD, MICHAEL.   See **U.K.*

HUBER, REINER K.   Professor at the **IASFOR* (Institute for Applied Systems and Operations Research) at the University of the Bundeswehr in Munich. His main contribution to the **NOD* debate has consisted of technical and formal analysis of **conventional stability*, including the potential trade-offs between the arms race and **crisis stability* [1986; 1986a; 1988; 1988a; cf. Møller 1991: 151–154].

RKH found stability to be unattainable between antagonists equipped with general-purpose (i.e., also offensive-capable) forces, such as was already the case for both **NATO* and the **Warsaw Pact* in Europe [Huber 1986:11; 1988a:25–28]. Furthermore, crisis instability would be exacerbated by **FOFA* and other programs that would increase NATO's number of vulnerable and high-priority targets while weakening **forward defense*. Numerical reductions through **arms control* would not really improve stability, regardless of how asymmetrical but might even worsen the situation. RKH criticized his former collaborator **Albrecht von Müller* for neglecting the qualitative side of arms control and particularly for disregarding the importance of the **defense efficiency hypothesis*, which pointed to the possibility of exploiting the inherent superiority of the defense [Huber & Hofmann in *Süddeutsche Zeitung,* 20 February 1988]; it alone would allow for a risk-free transition to an **NOD*-type defense [Huber 1988a; idem & Hofmann 1990].

Since the end of the Cold War, RKH has devoted considerable attention to the requirements of stability in a setting of **multipolarity*, finding this to call for defensive restructuring accompanied by elements of a **collective security* arrangement [1990a; 1991; 1993; 1993a; idem & Avenhaus 1993; idem & Linnenkamp 1994; idem & Schindler 1993].

HUNGARY.   Small country in East-Central Europe; former member of the **Warsaw Pact.*

1. **Background.** The position of H within the Soviet **alliance* system was ambiguous for decades, and particularly so after the uprising against Stalinism and Soviet domination in 1956, in the course of which H proclaimed its withdrawal from the recently (1955) established Warsaw Pact [Harman 1974: 124–187; Jones 1981a: 72–79; Larrabee 1984]. The Soviet reluctance to have to repeat the bloody suppression of the revolt, reinforced by Khrushchev's quest for a **détente*, which was obviously incompatible therewith, made the USSR acquiesce in a considerable Hungarian diversion from the standards of the pact. H apparently sought to increase its leeway by stressing that it belonged to southern rather than central Europe because such a less important strategic position would tend to increase its freedom of action over the (very circumscribed) one enjoyed by the "Northern Tier" [Dunay 1990: 65–66; cf. Johnson & al. 1982].

H was permitted a more market-oriented economic system, as well as a certain independence with regard to security and defense policy: shorter military service than in most other pact states, high priority for **air defense* over offensive aerial missions, and a somewhat less offensive structure of the army. Furthermore, H was a firm supporter of **disarmament* and **détente*, objectives that it pursued through cultivating economic, political, and cultural ties with the West, not least its former companion in the Hapsburg monarchy, *Austria* [cf. Lemaitre 1989; 1990].

2. **Alternative Defense Debate.** Thanks to the relative freedom of speech, a genuine *NOD* debate developed in H well before the USSR began to take an interest in the matter. There were a few officers in the armed forces advocating defensive doctrines [Deak 1989], but the center of the debate was the Eötvös University of Budapest, which hosted two of most prominent (and first) NOD advocates of the entire Eastern bloc: Laszlo Valki and Pál Dunay.

In addition to criticizing offensive Western doctrines, such as *flexible response*, *FOFA*, and *AirLand Battle* [Valki 1988; cf. Vajda 1987], Valki and Dunay were critical of Warsaw Pact doctrine as well as interested in the topic of NOD as such. Their main critique of the West German proposals was that they might not be realistic in view of U.S. offensive orientation, and that they might require unaffordable investments in new hardware. The two authors hence consistently advocated the combining of *transarmament* with *disarmament* and *arms control* [Valki 1985; 1986; 1988; 1990; 1990a; Dunay 1988; 1991; idem & Valki 1990; cf. G. Horn 1989; 1989a].

3. **Post–Warsaw Pact Developments.** After the 1989 revolutions, H pushed hard for dissolution of the Warsaw Pact, threatening to withdraw unilaterally [*Atlantic News,* 27 July 1990]. Since its dissolution, H has sought to forge new security political ties in two directions: with its former allies in Eastern Europe, in particular *Poland* and the Czech and Slovak Federal Republic (subsequently, its two successors), and with *NATO* and the WEU [*Atlantic News,* 20 February 1991; 9 October 1991; Valki 1992; 1993; Dunay 1993b].

Further, H has embarked upon a reconfiguration of its armed forces [Köszegevári 1992; Schulz (ed.) 1993], accompanied by a complete reevaluation of its military doctrine. The guidelines for the latter were described by the deputy head of the Hungarian *CSCE* delegation in terms very similar to those used by Western NOD proponents:

> Hungary must prepare for defense in depth, knowing this means the loss of some territory and resources. Nevertheless, the deterrent value of a defensive strategy depends very much on a readiness to fight. . . . The Hungarian military should consist of two parts: high-readiness forces and *territorial defense* forces. The high-readiness forces should be deployed in accordance with the requirements of "concentric defence." . . . The territorial forces, based on local battalions, should train all the remaining conscripts. These forces, with mostly light armament, would defend their home villages using the defensive possibilities of densely populated areas. In case of an attack, the high-readiness forces would engage the enemy first. As a result, the aggressor would either decide to withdraw, or would press on and face the *guerilla warfare* tactics of the territorial defense forces [*Defense & Disarmament Alternatives,* April 1990; cf. Defense Minister Lajos Für interviewed in *Jane's Defence Weekly* (30 November 1991), 1076].

HUNTINGTON, SAMUEL P.   Professor at Harvard University and prominent authority on civil-military relations and interservice rivalry phenomena.

In 1983 SPH proposed that *NATO* should adopt a strategy of *conventional retaliation* as a partial substitute for nuclear retaliation. On the one hand, nuclear deterrence by retaliation was both politically problematic because of popular opposition and of questionable credibility. On the other hand, conventional *deterrence by denial* would allow a prospective aggressor to calculate the costs, and its deterrence value would thus be dubious. Deterrence by retaliation, however, presumably did not require nuclear weapons but could be accomplished through deep offensive thrusts into *Eastern Europe by means of conventional *land forces. The strategy would possess "unimpeachable credibility" and would "capitalize on the uncertainties and fears that the Soviets have concerning the reliability of their Eastern European allies" [Huntington 1983: 260–261].

SPH frankly admitted that conventional retaliation was an offensive military strategy (and hence a perfect match for *FOFA and *AirLand Battle), while stating that "there is no reason why a politically defensive alliance cannot have a militarily offensive strategy" [1983: 264]. Others have been less sanguine in this respect and have criticized SPH for recommending something incompatible with NATO's defensive orientation (as well as for envisioning operations well beyond NATO's means). The proposal that West German forces would be responsible for a thrust into East Germany seemed especially to violate the *FRG's constitution, which prohibited preparations for aggressive warfare [Dunn & Staudenmaier 1985; Papp 1984; Mearsheimer 1981; cf. Lutz 1987d; 1989a; 1990a; 1992; 1993; 1993a–b]. SPH's ideas were condemned by a nearly unanimous *NOD community, yet *Albrecht A. C. von Müller (who is alleged to belong to the community) advanced very similar ideas, a "*conditional offensive superiority."

HYDE-PRICE, ADRIAN.   Lecturer at the University of Southhampton and author of one of the most elaborate analyses of the future European "architecture"; it took into account German unification and *Warsaw Pact dissolution: European Security Beyond the Cold War: Four Scenarios for the Year 2010 [1991; extracted in 1991a; cf. for a comparable work: Buzan & al. 1990].

AHP's premise was that everything was in a state of flux, and that there might be a need for a paradigm shift because *Realism might be an inadequate tool for understanding the complexities and predicting the future [Hyde-Price 1991: 13–16; cf. Booth 1991c]. To avert the dangers of anarchy, AHP recommended a combination of military and nonmilitary steps: creation of "a constructive synergy" between the process of West European *integration on the one hand, and overcoming the legacy of the East-West divide on the other [1991: 240]. This did not call for any grand design or architectural blueprint, but might presumably best be accomplished through the application of ten principles, including those of *common security, institutionalized *interdependence, regional differentiation and functional diversity, and integration. Concerning military postures, AHP urged a restructuring with *mutual defensive superiority as the goal but with an underpinning of nuclear weapons configured according to *minimum deterrence standards [1991: 247–258].

I

IASFOR. Institute for Applied Systems and Operations Research of the University (formerly, Academy) of the Bundeswehr in Munich (FRG) (see also: *Reiner K. Huber*). The IASFOR has on several occasions collaborated with the research team(s) around *Horst Afheldt* and *Albrecht A. C. von Müller*. Furthermore, it has been responsible for the first elaborate computer-aided simulation of various *NOD models, inter alia, with a view to determine their compatibility with the *defense efficiency hypothesis* [Huber & Hofmann 1984a; Hofmann, Huber, & Steiger 1986; cf. Møller 1991: 151–154].

For all its merits, the simulation study had a limited scope. All scenarios dealt with the tactical (i.e., regiment and battalion) level, and neither the operational nor the strategic levels were analyzed. Only components of the various models were tested (thereby disregarding potential synergies); the components were evaluated as supplements to, rather than as substitutes for, traditional armored forces, and were assessed for their *forward defense* potentials (even though some models envisioned abandoning forward defense). No account was taken of conceivable benign security political effects of NOD, and intangibles (such as the surprise factor and troop morale) were disregarded completely. The formations compared were, on the one side, four types of traditional armored *infantry* and *tank* battalions, and, on the NOD side, the following: *Horst Afheldt*'s *techno-commandoes*; the infantry forces of the Swiss and Austrian *territorial defense*; the infantry and cavalry battalions of the *SAS* model; the "shield forces" of *Löser*; the antitank groups of Füreder's "cellular defense"; the "*fire barrier* forces" of *Gerber* and *Hannig*; and a light infantry battalion forming part of a barrier brigade intended to create a selective fire barrier (a proposal of the IASFOR team itself). The simulation runs showed that modernized armored forces were reasonably effective; that stationary forces were incapable of concentration; and that fire barriers were incapable of preventing breakthroughs. However, the partly mobile forces (the SAS formations and the Swiss infantry) were rather effective, surpassed only by the mixed formations proposed by the IASFOR itself.

Since the end of the Cold War, researchers at the IASFOR have analyzed the implications of *multipolarity* for *stability* [Huber 1990a; 1991; 1991a; 1993; 1993a; idem & Avenhaus (eds.) 1993], inter alia, in the context of an international subsystem comprising Russia, Ukraine, Belarus, and Poland [Huber & Schindler 1993]. The implications seemed to be that stability might be achievable through a combination of defensive restructuring with a modicum of *collective security* mechanisms.

IDDS. Institute for Defense and Disarmament Studies in Brookline (subsequently Cambridge), Massachusetts (*U.S.*), founded in 1979 by Randall Watson Forsberg, who has since served as its director.

The IDDS has always taken *arms control* very seriously, manifested in its *Arms Control Reporter,* edited by Chalmers Hardenberg, which keeps track of all negotiations in progress, as well as political statements pertaining thereto. It has routinely reported the evolution of, inter alia, the Soviet attitude toward *NOD. The *CFE negotiations have enjoyed the special attention of the IDDS, which has followed them closely in the newsletter *ViennaFax*, in *A Chronological History of the CFE Negotiations* [IDDS (ed.) 1991], and in detailed analyses of various aspects [Conetta & Knight 1990; 1990a–b; Lilly-Webber & Remme 1990; Kahl 1990].

Reflecting its close connections with the peace movement (in particular *Freeze*, founded by Forsberg), the IDDS has also published a wide range of material for the general public, including *Peace Resource Book* [Conetta (ed.) 1988] and the newsletter *Defense and Disarmament Alternatives*.

The driving force in the IDDS all along has been Forsberg, who from 1968 to 1974 had worked at *SIPRI and later participated in the Boston Study Group, which in 1979 presented a detailed critique of U.S. defense policy, accompanied by recommendations for a more defensive posture [Boston Study Group 1979]. Subsequently, she focused primarily on nuclear disarmament, designing the (deliberately moderate) "freeze" proposal [Forsberg 1984; 1985; cf. Solo 1988], which was envisioned as a first step toward nuclear build-down. She gradually became convinced, however, that a disarmament strategy needed three pillars to be realistic:

- Nuclear disarmament via *minimum deterrence* and *no-first-use* to complete abolition.
- *NOD, to reduce the perceived need for nuclear deterrence.
- A *nonintervention regime* pertaining to the Third World, intended to delegitimize military force as a tool of politics, thereby paving the way for further disarmament [Forsberg 1984; 1985; 1986; 1987; 1987a-b; 1988; 1988a].

IFSH. *Institut für Friedensforschung und Sicherheitspolitik an der Universität Hamburg* (Institute for Peace Research and Security Policy at the University of Hamburg, *FRG). Independent entity, albeit with close links to the *SPD. Its director for a number of years has been *Egon Bahr*, one of the party's leading ideologists, and his deputy has been *Dieter S. Lutz*, likewise a member of various party commissions. In 1994 the latter took over the post as director.

Since the mid-1980s, the focus of the IFSH's research has been *common security* and *NOD. The IFSH has contributed to the theoretical development of both concepts, above all by hosting a number of symposia, the proceedings of which have subsequently been published [Bahr & Lutz (eds.) 1986; 1987; 1988], but also through its own research. The most substantial contribution to common security thinking has been the analyses of *Reinhard Mutz*; the only member of the the institute staff to contribute substantially to NOD theory has been *Erwin Müller*.

The IFSH on several occasions has also issued policy recommendations, e.g., on the *Rogers Plan* and *deep strike* [IFSH 1985], both of which were strongly criticized. It also took a unified stand on the modalities of German unification with a set of recommendations written in light of the *Two-Plus-Four* negotiations, then in progress. The IFSH suggested embedding unification in a regional *collective security* system, covering an enlarged central Europe. Within this system, "the bulk of the national armies would be defensively structured and equipped and tasked with the traditional defense. A smaller contingent of more mobile forces could perform the function of an 'intervention force' or a 'Euro-police force,' and would be available for assistance in case of attack." This restructuring should be accompanied by deep cuts in the size of the armed forces, those of the united Germany to 105,000 soldiers, within a Europe where national armed forces would not exceed 500,000 troops [IFSH 1990; cf. "European Security Concept," in *Tageszeitung,* 28 April 1990; Kaestner 1990. For a critique, see Martin Müller 1990; and for a rebuttal,

Mutz 1990]. These proposals were followed up in 1992 with an elaborate plan for a "European Security Community" that likewise combined NOD and collective security [IFSH 1993].

IMEMO.   Institute of World Economy and International Relations at the USSR Academy of Sciences (since 1991, at the Academy of Sciences of the Russian Federation).

The huge research establishment of IMEMO (with a staff of around one thousand) has a Department for Disarmament Studies, headed by *Alexei Arbatov*. Other researchers at the IMEMO who have contributed significantly to the alternative defense debate include Nadezhda Arbatova, Oleg Amirov, Vladimir Baranovsky, Andrei Chervyakov, Nikolai Kishilov, Gennadi Kolosov, Alexander Kokoyev, Vadim Makarevsky, Nikolai Shaskolsky, Yuri Streltsov, and Yuri Usachev [A. Arbatov 1988; 1989; 1989a–d; 1990; 1990a; idem, Amirov & al. 1989; idem, Kishilov & al. 1989; Arbatova & al. 1989; Baranovsky 1990; idem & Streltsov 1988; Kokoyev & Androsov 1990; Makarevsky 1988; 1990; Shaskolsky 1989; Streltsov 1989]. Since 1986 IMEMO has been publishing a yearbook (in the *SIPRI* tradition), *Disarmament and Security* [IMEMO 1987; 1989]; it is the best available source in English on Soviet thinking about *reasonable sufficiency* and *NOD*. IMEMO also often provided concrete policy guidance for the entourage of *Mikhail Gorbachev* and was thus instrumental in the development of *New Thinking* [Bialer 1988; Oudnaren 1991; Meyer 1988; Shenfield 1987; Bluth 1990: 22–35; Holden 1990b; 1991: 8–17, 52–58; Holloway 1989; 1989a; Krakau & Diehl 1989a; Nichols & Krasik 1990; Conner & Poirier 1991; Bellamy 1992; Diehl 1993].

INDEPENDENT COMMISSION ON DISARMAMENT AND SECURITY ISSUES.   See *Palme Commission*.

IN-DEPTH DEFENSE.   See *Territorial Defense*.

INDIA.   The world's second-largest state in terms of population, yet only a medium-range military power that has only very rarely constituted a threat to other states.

1. **Defensive Tradition.** Throughout most of its recorded history (nearly five thousand years), the armed forces and military activities of I have been predominantly defensive: based on fortifications and stationary forces, configured in a *bastion* mode rather than as frontier defenses [Singh 1989a; cf. Bellamy 1990: 209–211]. The defensive traditions were strengthened through the influence of *Mohandas Gandhi* and his followers, as well as by the principle of nonalignment (cf. *neutrality*) championed by Jawaharlal Nehru and his successors. Both through the nonaligned movement and on its own, I has also consistently worked for nuclear abolition, peace *zones*, and *disarmament*.

The present military posture of I is largely defensive, even though its two main neighbors, *China* and Pakistan, have attacked it in the postwar period. The establishment of a *forward defense* around the entire perimeter was obviously infeasible because the border extended 16,500 km (including 7,000 km facing states with whom I has territorial disputes). I has therefore chosen a bastion-type defense strategy. Furthermore, offensive-capable formations constitute only an insignificant share of total military power: two *tank* divisions (out of an army of 1.2 million troops) deployed at great distance from the border; an *air defense* system relying primarily on ground-based systems; and a navy consisting mostly of small (frigate-size and below) vessels and only four destroyers [Singh 1989a: 230-233].

I's nuclear capability is ambiguous, which, inter alia, was demonstrated in 1974 by the testing of a nuclear "device" (as a matter of principle, not a bomb). The intention was, above all, to dissuade (apparently unsuccessfully) Pakistan from acquiring nuclear weapons, but the explosion was also meant as a protest against the nuclear nonproliferation treaty (*NPT), which was regarded by I as discriminatory. To date I has not officially crossed the nuclear threshold, even though several observers regard it likely that I has nuclear weapons stockpiled, just as Pakistan does [Singh 1989b: 232–233; 1993; 1993a; 1994; Mohan 1989; cf. Myrdal 1976: 159–207; Spector 1984: 19–110; 1990: 63–117; Barnaby 1993: 68–74; Bidwai & Vanaik 1991].

2. **NOD Debate.** Indian academics and military officers have taken some interest in the Western *NOD debate, particularly at the Institute for Defence Studies and Analysis in New Delhi under the directorship of *Jasjit Singh. In addition to following the European NOD debate and critically analyzing offensive conceptions such as *deep strike [Gupta 1989; Mohan 1989; Singh 1989; 1989b; Subrahmanyam 1989], Indian researchers have suggested broadening the debate, e.g., by taking armed intervention and coercive diplomacy into account, as well as by applying the conceptions to non-European areas of conflict [Singh 1989b; 1990; 1994; Banerjee 1989; Mohan 1989: 52–53].

INDIRECT APPROACH.   The basic idea of asymmetrical response goes back to *Sun Tzu [500 B.C.], but in the twentieth century the concept has been associated with, above all, *Liddell Hart, who coined the term "IA" [1967a]. Alternative terms for basically the same concept have included "indirect strategy" (*André Beaufre) [1963; 1972; 1974], and "paradoxical approach" (*Edward Luttwak) [1987].

Asymmetry may mean many things, such as responding "in a different coinage," e.g., by opposing *tanks with antitank weapons rather than tanks, warships with coastal artillery, aircraft, etc. Or it may refer to (special forms of) maneuver warfare with an emphasis on outflanking and enveloping an enemy rather than attacking him head-on. Because the ID tends to go against the usual military preference for matching an opponent in terms of hardware and numbers, it has appealed above all to irregular military forces. Historically, the asymmetrical approach has primarily been associated with *guerrilla warfare, with *Mao Zedong the most explicit advocate of ID.

INDIRECT STRATEGY.   See *Indirect Approach.

INF.   Intermediate-Range Nuclear Forces, also known as Long-Range Theater Nuclear Forces (LRTNF), i.e., nuclear weapons with ranges between 1,000 and 5,500 km. Although many aircraft also belong to this category, the focus came to be placed on three systems: *NATO's Pershing-2 ballistic missile and the Tomahawk ground-launched cruise missile, and the *Soviet Union's SS-20 multiple-warhead ballistic missile [Lodgaard & Berg 1982; Holm & Petersen (eds.) 1983].

1. **Early Negotiations.** INF first became an issue when in 1977 Federal Chancellor Helmut Schmidt pointed to the danger of decoupling the *U.S. from Europe as a result of strategic parity (codified in the SALT 1 and 2 Treaties), combined with substrategic Soviet nuclear superiority resulting from the large-scale deployment of SS-20s. NATO's solution to this problem was to plan for deployment (from 1983 onward) of 108 Pershing-2s and 464 Tomahawks, but before that to negotiate with the Soviet Union on reductions or an elimination of INF missiles. For a variety of reasons, the negotiations accomplished nothing, and NATO started deployment in 1983, which led to an almost total breakdown of the previous

détente when the Soviets walked out of virtually all East-West negotiations [Talbott 1984: 21–208; Garthoff 1985; 849–886; Brauch 1983; Dean 1987: 125–152; A. Carter 1989: 172–204].

Another casualty was the previous near-consensus on the strategy and defense policy of NATO, which broke down almost everywhere in the course of the INF debate [cf. Risse-Kappen 1988]. One of the reflections of this was the rapidly declining support for *nuclear deterrence* as such. The debate in the early 1980s about *freeze*, *no-first-use*, and *nuclear-weapons-free zones* made manifest that nuclear weapons were increasingly becoming anathema to the general public and large sections of the elite. An indirect effect was the debate on *NOD, spurred by the quest for benign substitutes for nuclear deterrence.

2. **INF Treaty.** Negotiations were resumed in 1985 after a Soviet reassessment of the matter, this time in a much more constructive spirit [Dean 1988c; A. Carter 1989: 205-229]. The Soviet Union accepted the "zero solution" originally proposed by President Reagan and even added an offer to include short-range INFs (SRINF), i.e., missiles with ranges between 500 and 1000 km. In 1987 agreement was reached on the INF Treaty, tantamount to a "double-zero," and constituting a break-through in verification [SIPRI 1988: 395-485].

3. **Post-INF Strategic Debate.** The elimination of INF missiles once again raised the problem of coupling and *extended deterrence*, reflecting the fact that NATO's deployment decision had never been closely related to the Soviet SS-20s. Hence the quest for other means of coupling, and of preserving the "seamless web" of deterrence, by means of nuclear weapons deployed in Europe, but capable of posing a threat to the USSR [cf. J. L. George 1990: 112–119; Garrett 1990: 49–72; Blackwill 1990].

Even though NATO made plans for INF replacements such as FOTL (follow-on to Lance) and TASM (tactical air-to-surface missile), the supporters of deterrence increasingly had to fight a rearguard struggle against proponents of a "third zero," pertaining to short-range nuclear forces (SNF) [Enders 1988; Lodgaard 1989d; Risse-Kappen 1991a; Lucas 1988; 1990; Ramsbotham 1989; Binnendijk 1989; Brauch 1989].

Political developments, finally, settled the issue:

- German unification implied that nuclear forces with ranges that would allow for reaching only one part of the united Germany from the other were rendered unacceptable.
- The democratic revolutions in Eastern Europe in 1989 rendered it absurd to target these former Soviet but present (de facto) Western allies.
- The paritetical ceilings for major weapon systems of the *CFE* Treaty (which translated into Western conventional superiority because of the *Warsaw Pact*'s collapse); the unilateral Soviet *disarmament* initiatives; and, perhaps most decisive, the *democratization* and subsequent dissolution of the Soviet Union reduced the perceived need for nuclear deterrence to such an all-time low as to deprive modernization of all justification.

INFANTRY.   Traditionally the main branch of all armies, consisting of foot soldiers, usually armed with small arms (rifles, bayonets, etc.).

1. **History.** Technology has radically transformed the nature of the I over the ages, as may be gathered from a comparison, say, of ancient Greek infantry (the *hoplites*) and modern infantry in the motorized rifle, mechanized infantry, or *Panzergrenadier* formations of the *Soviet Union*, the *U.S.*, and the *FRG*, respectively:

- The complexity, range, and lethality of the weapons of the I have increased by several orders of magnitude: from the javelins, swords, and shields of ancient and medieval times to the modern I's wide panoply of weapons for both close and longer-range combat; handguns, machine guns, and bayonets for "in-fighting" and self-defense; mortars, ATGMs, and portable *SAMs to be used for "stand-off combat"; mine-laying and mine-clearing equipment, etc. [Jones 1988: 1–81; Dupuy 1980: 10–15, 288–289; Dunnigan 1983: 18–45].
- The synergies inherent in combined-arms contexts have become increasingly important. The I used to be more or less on its own (except for the occasional assistance of archers and cavalry), but modern infantry divisions include integrated *tank battalions (i.e., modern "heavy cavalry"), artillery, and occasionally helicopters (airborne "light cavalry"), all integrated into the "infantry" formations, which thus almost constitute miniature armies with several service arms.
- Mobility has increased immeasurably: the I in ancient Greece marched on foot; the modern I is almost entirely mechanized and tends to fight from mounted positions.
- Finally, technology has enabled field commanders to control large and scattered forces to a much greater extent than before. The Greeks relied on visible and audible signs for tactical *C^3I, and on messengers for long-range communication, hence their strategic emphasis on short distances, small formations, and a steady pace of combat. Today, field commanders can communicate instantaneously through radio and space-based communications systems and may have an almost real-time view of an extended battlefield [cf. Creveld 1985: 17–57; Grin 1990], hence the emphasis on long distances, large formations, and high-speed combat.

2. **Light and Heavy I.** Some of the above features pertain only to the regular (i.e., heavy) I, whereas light I (such as the German *Jägers*) tends to be less mechanized and more lightly armed and to be trained for fighting without aerial or armored support [cf. Canby 1982; 1986a]. These features have received growing attention in recent years, not least because of the suitability of light I for low-intensity combat. The U.S. Army has for a number of years experimented with "light divisions" consisting largely of I and intended, inter alia, for purposes of intervention. Their equipment is generally lighter, thus more air-mobile, and they have only limited armor protection and/or tactical air support [W. E. Dupuy 1985; Kafkalas 1986; Mazarr 1990; W. J. Olson 1985]. Such light forces have primarily been designed for Third World contingencies, but some thought has also been given to using them in Europe, e.g., for fighting in difficult terrain, including cities [Guertner 1982; Petraeus 1984; Wickham 1985; Downing 1986; Gates 1989].

3. **NOD and I.** Even though I is thus not automatically defensive, *NOD advocates have always emphasized I over armored forces, above all because of its lack of shock power. I deprived of armor, artillery, and tactical air support simply cannot break through well-prepared defenses or defensive artillery barrages. On the other hand, I equipped with armor-breaking *PGM, mine-laying equipment, and *air defense weapons, fighting from prepared combat positions, and with some artillery (usually rocket artillery) support, may well be able to halt an armored assault, or at the very least slow it down considerably.

Many NOD schemes have therefore envisioned a substitution of tanks with I [Uhle-Wettler 1966; 1980; 1990; Canby 1980; 1983; 1984; 1986a; 1990; H. Afheldt 1983; E. Afheldt 1984; 1984a], often in the form of defending a forward *zone by means of I while holding back armored forces to serve as a strategic reserve [Löser

1981; 1986a; Bülow 1986; 1986a; A. Müller 1986] or as a source of internal reinforcement [SAS 1984; 1987; Unterseher 1987; 1987c; 1989b–d; 1989h].

INSTEAD. Interuniversity Network for Studies on Technology Assessment in
Defense, involving institutes, groups, and individual researchers at five universities
in the *Netherlands* [Borg & Tulp 1987]. INSTEAD has collaborated quite closely
with alternative defense study groups in the *FRG*, such as *AFES-PRESS* and
SAS. Furthermore, it has been a central node in the Dutch, and to some extent
international, scholarly debate on *NOD* by arranging a number of seminars and
publishing anthologies as well as monographs on NOD and related subjects
[Barnaby & Borg (eds.) 1986; Borg & Smit (eds.) 1987; 1989; Grin 1990; Grin &
Graaf (eds.) 1990]. Besides Chairman Wim Smit of the University of Twente [1986;
1989; 1989a; 1990; idem & al. 1987], INSTEAD has included, at various points,
other preeminent experts on NOD in the Netherlands, such as Sjef Orbons [1986;
1989], Marlies ter Borg [1989; 1992], *John Grin* [1990; 1990a–b; idem &
Unterseher 1988; 1989; 1990; 1990a], and the retired brigadier general Henny van
der Graaf [1989]. Borg has since become director of the Foundation for Building
Peace (a nonprofit organization involved with housing assistance for decommis-
sioned Soviet officers, i.e., a special instance of *conversion*), and Grin and Graaf
have been engaged with the Center for Verification Technology set up under the
auspices of INSTEAD.

INSTITUTCHIKI. Nickname for the civilian analysts of military affairs in the
Soviet Union (e.g., at *IMEMO* and *ISKAN*) who were instrumental in developing
Soviet *New Thinking*, including the endorsement of *NOD* [Bellamy 1992; Diehl
1993].

INSTITUTE FOR DEFENCE STUDIES AND ANALYSIS. See *India*,
Jasjit Singh.

INSTITUTE FOR DEFENSE AND DISARMAMENT STUDIES. See
IDDS.

*INSTITUT FÜR FRIEDENSFORSCHUNG UND SICHERHEITSPOLITIK
AN DER UNIVERSITÄT HAMBURG.* (Institute for Peace Research and
Security Policy at the University of Hamburg). See *IFSH*.

INSTITUT FÜR INTERNATIONALE POLITIK UND WIRTSCHAFT.
(Institute for International Politics and Economy). See *IPW*.

INSTITUTE FOR INTERNATIONAL POLITICS AND ECONOMY.
(Institut für Internationale Politik und Wirtschaft). See *IPW*.

INSTITUTE FOR PEACE AND INTERNATIONAL SECURITY. See
IPIS.

**INSTITUTE FOR PEACE RESEARCH AND SECURITY POLICY AT THE
UNIVERSITY OF HAMBURG.** See *IFSH*.

INSTITUTE FOR U.S. AND CANADIAN STUDIES. See *ISKAN*.

**INSTITUTE OF WORLD ECONOMY AND INTERNATIONAL
RELATIONS.** See *IMEMO*.

INTEGRATION.   Generic term signifying the gradual merging of several disparate entities into a unified one. The term has usually been applied to the I of states into larger international or supranational organizations. I has been conceived of as a means to *peace* as well as to economic prosperity, by virtue of its erosion of boundaries between states and creation of a community of values. The notion of federalism (stemming from *Immanuel Kant* [1795]) played an important role in the development of the European Communities, especially in the first decade of the EEC [Burgess 1989].

Other, more empirically oriented I theorists (many belonging to the tradition of *Realism*) have stressed the economic benefits of I and envisioned emergence of a *security community* as an intermediate stage between a system based on sovereign nation-states and one characterized by supranationality [Deutsch & al. 1957]. According to "functionalists" and "neofunctionalists," I was to be pursued gradually and with an emphasis on "low politics." Gradually, the integrative momentum—and growing *interdependence*—would presumably generate expectations among elites, who would strive for further integrative steps, some of which might belong to the sphere of "high politics." The first accomplishment might be a pluralistic security community, in which war between members would cease to be a genuine option, but no relinquishment of national sovereignty would occur. Eventually, the way would be paved for proceeding with establishment of an amalgamated security community, i.e., a confederation of states, or even a federation, wherein people would (at least partly) transfer their loyalties and sense of identity to the larger unit, thereby making the nation-state obsolete and laying the foundations for peace [Deutsch & al. 1957; Haas 1966; 1970; 1987; R. Hansen 1972; Schlotter 1982; Tranholm-Mikkelsen 1991].

Other analysts of I have been less sanguine about its peace-promoting effects, pointing out that peace between the participants may well go hand in hand with an aggressive attitude toward nonmembers. The EC might thus be "a superpower in the making," as prophesied by *Johan Galtung* [1973]. Yet another group of analysts have found complete I of Europe to be unrealistic, while nevertheless crediting EC integration with a largely peace-promoting role. It could, e.g., be conceived of as the epicenter of a set of "concentric circles," as a center of gravity in Europe [Buzan & al. 1990: 202–228].

INTERCEPTOR.   See *Air Defense.*

INTERDEPENDENCE.   Mutual dependence of states and societies in various fields.

According to I theory, nations are becoming increasingly interdependent and "woven into" one another to such an extent as to partly invalidate the paradigm of *Realism*: an anarchic international system consisting of sovereign nation-states. "Beneath" the state level, transnational corporations and INGOs (international non-governmental organizations) play an increasingly influential role; "above" the state level, IGOs (international governmental organizations) are gaining in importance as well. This growing I might open up prospects for a more enduring and robust *peace* structure between states [Keohane & Nye 1977; cf. Tromp 1988; Wilde 1991. For a critique, see Haas 1987]. In this sense, the Soviet reassessment of I as an element in *New Thinking* seemed to bode well for *security* in Europe [Bezrukov & Kortunov 1991].

At the same time, I also creates a wide range of security problems for the handling of which military force, the traditional instrument of state power, is ill-suited. These include structural threats, which are integral to the system as such rather than stemming from the concrete behavior of individual states: environmental decay; depletion of resources; uncontrolled migration; etc. [Brundtland Commission 1987; Mische 1989; Westing (ed.) 1989; Soroos (ed.) 1990; Prins 1990b]. The strategy of *common security* constituted an attempt to manage the challenge of complex interdependence in a cooperative mode.

INTERDISCIPLINARY CENTER FOR PEACE RESEARCH AND STRATEGIC STUDIES.   (Centre Interdisciplinaire de Recherches sur la Paix et d'Études Stratégiques). See *CIRPES*.

INTERIOR LINES.   Theoretical conception usually attributed to *Antoine de Jomini* but in actual fact discovered by strategic thinkers of the Enlightenment, such as Henry Lloyd (1718–1783), Adam von Bülow (1757–1807), and the Archduke Charles (1771–1847), and practiced at least as far back as the Roman Republic (207 B.C., Consul Nero vs. Hasdrubal and Hannibal) [Gat 1988: 67–135; Jones 1988: 70–72 & passim].

IL of operations are presumably inherently superior to exterior lines because they are shorter and more protected, with regard both to the movement of the armed forces themselves and to *logistics*. The almost automatic availability to a defender of IL is part of the explanation for the phenomenon of *defense dominance*, i.e., for the fact that "the defensive form of war is intrinsically stronger than the offensive," as asserted by *Clausewitz*. He distinguished between the "convergence of attack and divergence of defense" and acknowledged that "once the defense has embraced the principle of movement . . . the benefit of greater concentration and interior lines becomes a decisive one" [1980: 361, 373/1984: 358, 368 (Book VI.1.2, VI.1.4)].

The principle of the superiority of IL is presumably of universal validity, but it became particularly manifest in the age of the railway: the defender was able to resupply large forces at the front swiftly by rail via IL; the attacker experienced enormous difficulties on exterior lines, where rail transport was often not available. This was a phenomenon of considerable importance in both the American Civil War and *WWI* [cf. McNeill 1982: 242–244; Creveld 1989: 168–170; Thompson 1991: 29–33, 36–38, 52–55].

All *NOD* models have envisioned extensive use of IL by planning for military operations exclusively on friendly territory and, further, exploitation of the invader's need to rely on exterior lines by waging "counterlogistical warfare," i.e., by seeking to sever the logistical chains of the aggressor [e.g., H. Afheldt 1983; Unterseher 1989b].

INTERMEDIATE-RANGE NUCLEAR FORCES.   See *INF*.

INTERVENTION.   See *Nonintervention Regime*.

INTIFADA.   See *Israel*.

INVULNERABILITY.   According to several alternative defense proponents, a country's I is one element of its *security* and could complement its (active) defensive capability, perhaps even to the point of rendering the latter superfluous, thus constituting an entirely defensive alternative to military power [Galtung 1984: 192–198; D. Fischer 1982; 1984: 142–153; Øberg 1983: 125–142]. I, however, is not an either/or feature but a feature that may characterize countries to a greater or lesser extent. Furthermore, there are various forms of I, some (but not all) of which lend themselves to modification [cf. Agrell 1984b]:

- Insularity, which for obvious reasons is a constant;
- Other forms of geographical inaccessibility, which may be modified (slightly), e.g., through *terrain modification*;
- Economic self-sufficiency, which can be enhanced through stockpiling strategic goods and/or promotion of autarky;
- Political I, signifying a country's ability to stand up to psychological warfare, to which end *democratization* might be a central means; and
- Structural I, i.e., a society's cohesiveness in the face of attempts at disruption, to which end *decentralization* might be a contribution.

IPIS. Institute for Peace and International Security. Small research institute in Cambridge, Massachusetts (*U.S.*) under the joint directorship of Pam Solo and Paul Walker. The IPIS has devoted a large part of its work to research on the *peace movements* [Solo 1988], *common security*, and *NOD* [P. Walker 1981; 1985; 1986].

Like so many other U.S. institutions, the IPIS has also devoted substantial effort to enlisting political and academic support for its ideas, thereby contributing to disseminating information about alternative defense. In 1988, e.g., it issued the report *New Directions for NATO: Adapting the Atlantic Alliance to the Needs of the 1990's* [Thompson & Solo (eds.) 1988], which included a chapter on NOD.

In the same vein, the IPIS formed the Committee on Common Security, which issued *Call to Action: Common Security and Our Common Future*. The committee's purpose and goals were "to give voice to a new consensus among the American people," based on the realization that "disarmament, ecology and economic justice are indivisible." It recommended radical reductions in the levels of nuclear and conventional arms through nuclear-free zones, the restriction of all theater and tactical nuclear weapons to home territories, publicly declared policies of *no-first-use*, and harsh sanctions against nuclear proliferation. Regarding conventional forces, the committee urged a build-down "to levels reasonably sufficient to deter attack, assure security, and encourage verifiable arms control and disarmament, but not so large and so threatening as to instill mutual anxiety, suspicion, and instability." Additionally, "military doctrines and remaining conventional force structures must be made nonprovocative and undeniably defensive."

This explicit "NOD manifesto" received the endorsement of a number of prominent academic figures, including John Kenneth Galbraith, Richard J. Barnet, Elise and Kenneth Boulding, Thomas B. Cochran, William Colby, Daniel Ellsberg, *Dietrich Fischer*, George Rathjens, Bruce Russett, Arthur Schlesinger, and Victor Weisskopf.

IPW. *Institut für Internationale Politik und Wirtschaft* (Institute for International Politics and Economy) in the former *GDR*, which until the revolution was very closely affiliated with the ruling party, the SED. IPW staff regularly served as speech writers for Erich Honecker and his entourage.

During the negotiations on the envisioned "contractual community" in the early stages of the unification process, the IPW was tasked with elaborating on the plan of Prime Minister Modrow, and in December 1989 it issued a set of recommendations that betrayed an inspiration from the West German *NOD* discourse. E.g., the need for substantial contributions by the GDR and the FRG to the piecemeal creation of "mutual incapability of attack" was emphasized [*Blätter für deutsche und Internationale Politik*, no. 2 (1900): 243].

After unification the IPW was closed down, and most of its staff were blacklisted, something that affected the younger members most severely (see, e.g., *Wolfgang Schwarz*); senior members could simply go into early retirement, even though they undoubtedly had been most to blame for the mistakes of the IPW (see, e.g., *Max Schmidt*).

ISKAN.   Institute for U.S. and Canadian Studies in Moscow. Soviet (subsequently Russian) research institute founded in 1968, with a research staff of around three hundred [Smoke & Kortunov (eds.) 1991: 393–394; Holden 1991: 52–58]. In addition to the director, *Georgy Arbatov*, the research staff has included, inter alia, Alexander Konovalov, Andrei Kortunov, Valentin Larionov, Valeri Mazing, and Andrej Kokoshin [G. Arbatov 1990; 1991; idem & Oltsman 1983; Kokoshin 1988; 1988a; 1989; 1989a–b; 1990; 1991; Konovalov 1989; 1990; 1990a; 1991; idem & Mazing 1989; Bezrukov & Kortunov 1991; Mazing 1989]. After the accession of Boris Yeltsin, Kokoshin was appointed deputy minister of defense.

Among the most interesting studies from ISKAN was a comparison by Kokoshin and the retired general Larionov of four strategies and postures, ranging from a predominantly offensive to a defensive orientation [Kokoshin & Larionov 1989; 1990]:

- Option 1. Immediate large-scale *counteroffensive*, seeking to transfer combat to enemy territory; the traditional Soviet strategy.
- Option 2. Initial defensive battle, followed by a decisive counteroffensive, modeled on the Battle of *Kursk* (which the two authors had analyzed in depth in a previous article [Kokoshin & Larionov 1987]).
- Option 3. Emphasis on defensive combat, including counterattacks on the defended territory, but not beyond, as envisioned by most German *NOD* proponents.
- Option 4. A strictly *defensive defense*, capable of only tactical counterattack, resembling the most radical Western NOD proposals, such as that of *Horst Afheldt*.

Warning against the dangers flowing from option 1, the two authors recommended option 2, or a combination thereof with option 3; they regarded option 4 as militarily inadequate.

ISRAEL.   State established in 1948 on the basis of a *U.N.* resolution that terminated the previous British Mandate.

1. **Wars.** I has been one of the countries with the highest frequency of wars with its neighbors, in most cases initiated by I, albeit usually in anticipation of an Arab attack [Carver 1981: 234–272; Horowitz 1993]:

- The 1948–1949 "War of Independence" between I and the League of Arab States;
- The 1956 Suez Crisis, involving an Israeli invasion of Suez;
- The 1967 June War against Egypt, Syria, Iraq, and Jordan, initiated preemptively by I and won in less than a week, hence also known as the Six-Day War [O'Ballance 1972];
- The 1973 Yom Kippur War, initiated by the Arab states [Cordesman & Wagner 1990(1): 14-116; O'Ballance 1978].
- The 1982 invasion of Lebanon [Cordesman & Wagner 1990(1): 117–228].

I won all of these wars militarily (less convincingly in 1973 than in the other cases) as a result of several factors: political and economic support from the West, above all the U.S.; technological superiority; a more intensive exploitation of its (inferior) military personnel resources; what many regard as the most competent officer corps in the world; and, not least, a very skillful application of strategy and operational art.

2. **Military Doctrine.** The military doctrine of I has above all been determined by its geostrategic position, which (at least until 1967) was characterized by a very shallow depth, implying that strategic targets were located very close to I's enemies. The armed forces (IDF: Israeli Defense Forces) have traditionally seen the only solution to this problem in *preemptive attack* (occasionally even *preventive war*) and in "taking the war to the enemy" as quickly as possible. Hence the initiation of the June War with a devastating preemptive *OCA* (offensive counter-air) strike and the general emphasis on large-scale maneuver warfare. Furthermore, I has shown a general lack of respect for international law and the *laws of war*, examples of which have been the frequent punitive raids against neighbors and the 1982 invasion of neutral Lebanon.

In conformity with this offensive orientation, the Israeli posture features, inter alia, ballistic missiles (the indigenous Jericho missiles) and has a very high density of main battle *tanks*, heavy artillery, offensive-capable aircraft, etc. [Cordesman & Wagner 1990(1): 230–274]. I has further relied on *nuclear deterrence*, based on an undeclared yet well-known nuclear capability [Evron 1991; Cohen 1993], accompanied by research, development, and possible production of (biological and) *chemical weapons* [Cordesman & Wagner 1990(1): 359].

The main emphasis has been on the offensive, but defensive measures have not been entirely disregarded. Prior to the 1973 war, I constructed one of the most elaborate systems of *barriers* in modern times on the occupied Sinai Peninsula: the Bar Lev Line. Likewise, the Golan Heights were saturated with tank obstacles [Cordesman & Wagner 1990(1): 39–44]. *Air defense* has also received considerable attention, including (since the *Gulf War*) *extended air defense* against ballistic missiles.

3. **Alternative Defense Debate.** There has been little debate on the military policy of I, even though I is undoubtedly the home of some of the world's most able military practitioners, as well as one of the most prominent military historians, *Martin Van Creveld*.

One of the few exceptions has been Ariel Levite of the Jaffee Center for Strategic Studies at Tel Aviv University, who in 1990 published *Offense and Defense in Israeli Military Doctrine* [1990]. Levite charged that the doctrine and military conceptions of the IDF were excessively offensive. The advantages for deterrence purposes of being able to "take the war to the enemy" notwithstanding, actual offensive operations would not be advantageous under all circumstances. Second, the emphasis on preemption was misplaced and ignored the political imperative of allowing a would-be aggressor the first blow and had negative effects on *crisis stability*. He therefore recommended a shift to more defensive capabilities and operational plans, in recognition of the improvement of the geostrategic position of I resulting from the acquisition of some strategic depth (e.g., in Sinai and the Golan Heights). To plan for an orderly initial retreat had thus become conceivable and indeed inevitable because the linear *forward defense* had always been a fiction. Second, military technology (scatterable mines, *SAMs*, *PGMs*, and the like) had magnified the relative advantages of the defender; this had rendered offensive operations, such as those thitherto planned for, increasingly demanding, while at the same time making an alternative, defensive strategy more practicable. Accordingly, I would presumably be well-advised to plan for absorbing an initial attack and then regrouping its forces and launching a *counteroffensive*.

A reassessment of geostrategic realities has apparently been undertaken by the Israeli government as well, especially in the aftermath of the Gulf War. Defense Minister Moshe Arens in 1991 openly acknowledged that "the Iraqi potential contribution has been decreased very significantly as a result of the Gulf war. The sum total of armed might that could be brought to bear in aggression against Israel has gone down not insignificantly" [*Jane's Defence Weekly,* 8 June 1991, 992; cf. ibid., 15 February 1992, 256].

4. *Intifada.* Besides its external security problem, I throughout its existence has had an internal one: the Palestinian problem, which has been particularly severe since the occupation of the West Bank and the Gaza Strip after the 1967 war. It has been further exacerbated by the policy of Jewish settlement in the occupied territories pursued by the Likud government [Harkabi 1988: 43–52; 116–128; McDowall 1989; Tessler 1989; 1989a; idem & Lesch 1989], and motivated to some extent by Judaic fundamentalism [Harkabi 1988: 141-200]. All along there had been some Palestinian resistance, but it erupted into a full-scale uprising in 1987. The *Intifada* [Lockman & Beinin (eds.) 1989; Hunter 1991] was predominantly nonviolent and employed many of the methods connected with *civilian-based defense* [cf. Awad 1985; Galtung 1989a]. It had tremendous political advantages for the Palestinian movement compared with the movement's previous use of violent (and partly terrorist) methods because it placed the onus of initiating violence on the Israeli authorities, thereby influencing world opinion favorably.

5. **Peace Process.** The breakthrough achieved in the peace process in September 1993 most likely was a consequence of the strains of the *Intifada* on Israel's international standing and the improved military situation as a result of the Gulf War. The mutual recognition between Israel and the PLO, brought about through the mediation of the late Johan Jørgen Holst, was accompanied by a plan for partial self-government of the Gaza Strip and the Jericho area ["Documentation," *Survival,* 35:4: 163-168]. The defusing of the Arab-Israeli conflict ideologically and religiously may conceivably open the doors for more far-reaching peace agreements, for instance between Israel and Syria.

ITALY.   The alternative defense debate in I has primarily been the making of *Forum on the Problems of Peace and War* [Ragionieri (ed.) 1989; Cerutti & Ragionieri (eds.) 1990]. In addition to working with the conceptualization of *NOD* and *common security,* as well as with the general problem of nuclear *disarmament* (with a special emphasis on the U.S. cruise missiles deployed in I) [Ragionieri 1989; Minni & Salvato 1989], the forum has primarily devoted its attention to naval and regional matters. The obvious reason for this emphasis has been that the only serious threats to Italian security have been the risks of a spillover from war in Northern Africa and/or of a superpower conflict in the Middle East involving naval forces in the Mediterranean. Neither the *Warsaw Pact* nor *Yugoslavia* would ever have been able to launch an invasion of I [Ragionieri 1989b; 1990; 1990a; 1991; cf. Unterseher 1988a]. Nevertheless, the forum did undertake some exploratory work on the possible employment of *territorial defense* and *civilian-based defense* for meeting the invasion threat, finding a combination of the two to be quite adequate for satisfying Italian security concerns in an entirely defensive manner [Farinella 1989; Spreafico & Farinella 1989]. The forum has further proposed a general shift to less offensive naval postures in the Mediterranean region, particularly with regard to the U.S. Navy, combined with an emphasis on coastal defense regarding the littoral states (including I). Further, it has recommended creation of a Mediterranean *nuclear-weapons-free zone* and a range of *CBMs* (confidence-building measures) for the region, which might thereby be stabilized and the asserted need for foreign intervention diminished [Ragionieri 1990; 1990a; 1991; Minni & Salvato 1989; Nardulli 1989; 1990].

J

JAHN, EGBERT. German peace researcher, professor of Soviet studies at the University of Frankfurt, temporarily affiliated with the Peace Research Institute Frankfurt (FRIF), and for a short period project leader at the *Centre for Peace and Conflict Research* in Copenhagen. EJ has primarily dealt with *civilian-based defense*, albeit without adding anything noteworthy to the strategy except for admonitions that it is unlikely to be implemented unless accompanied by a rather profound transformation of society [1971; 1973; 1977; 1984].

JAPAN. What makes J interesting from an alternative defense point of view is not so much the (apparently almost nonexistent [cf. for an exception: Hiroshi Doi, in Barnett 1984: 100–103]) debate on the subject as the actual defense and security policy of J.

1. **Demilitarization and Rearmament.** Because of its role as aggressor in *WWII*, J was completely demilitarized in 1945, and far-reaching measures of *democratization* were taken to prevent the reemergence of a military-industrial complex (*MIC*) [Domange 1985; McIntosh 1986; Coulmy 1991: 39–60; Chinworth 1992; Ueta 1992]. Demilitarization was even inscribed in the 1946 constitution as an unconditional obligation of infinite duration [§ 9, in Coulmy 1991: 53]. However, in light of the Korean War (1950–1953) and the apparently mounting Chinese threat, the *U.S.* became interested in a Japanese military contribution to the defense of the West and in 1954 signed a mutual defense agreement with J that envisioned Japanese rearmament.

The Japanese armed forces were created for only one mission, namely, *self-defense (jieitai)*. Moreover, the principle of nonoffensiveness (cf. *NOD*) has been regarded as almost sacrosanct and of perennial validity, albeit open to different interpretations. Finally, the nuclear-free status of J has been a matter of principle for all significant political forces because nuclear weapons are even more anathema to J than to any other state because of Hiroshima and Nagasaki [Berger 1993]. J constitutes a de facto *nuclear-weapons-free zone*.

2. **Present Posture and Strategy.** Even though J's defense expenditures have remained below 1 percent of GDP since 1945 (with a few insignificant exceptions), the rapidly rising GDP has nevertheless made large resources available for the military forces, making Japan the world's third-largest military power.

Compared with that of other major powers, however, the structure of the forces is clearly more defensive (see Table 20): only a few armored divisions (with only twelve hundred tanks) and long-range ground-attack aircraft, no *aircraft carriers*, etc. [Coulmy 1991: 89–106, 119–160; Defence Agency 1990]:

3. **Future Prospects.** J possesses a nearly unique defensive strength (in the sense of *invulnerability*) by virtue of its insularity and the fact that 70 percent of its surface consists of mountains. Further, its defensive posture appears to afford it satisfactory protection because the military threats in the area are quite manageable.

Table 20.   Selected Weapons Holdings of Japan

	Types of Armament	Quantity
Land Forces	Battle tanks	1,200
Offensive	Other armored vehicles	700
	Artillery	540
Defensive	Mortars	1,900
	Antitank missiles	2,900
	Air defense guns	120
	Antitank helicopters	48
Navy	Destroyers, frigates	58
	Minelayers, sweepers	41
	Patrol boats	14
	Submarines	14
Air Force	Fighers/interceptors	256
	Ground support	74

The main challenge to the defense-only principle may, ironically, be the U.S.'s desire to share the burdens of upholding peace in the area with its allies. The U.S. has been exerting some pressure on J to assume a more active role, especially in naval affairs. There are also forces in J in support of the view that as a world power, J must live up to its international obligations by taking part in *out-of-area* activities [Drifte 1990; Tamamoto 1990; 1990a; Funabashi 1991; Katzenstein & Okawara 1993; Okawara 1993; Ungar & Blackburn (eds.) 1993]. In 1992 a small breakthrough was achieved in this direction: the Japanese Senate approved a bill authorizing the dispatch of two thousand troops for *peacekeeping* missions under *U.N.* auspices. Even this type of out-of-area mission remains controversial [*Atlantic News*, 10 June 1992, 2–3]. Japan is also aware of the need for reassuring its neighbors; hence, for instance, the declaration by Prime Minister Miyazawa, on 9 June 1992, that "it is hereby reaffirmed that Japan will continue to strictly abide by its basic policy not to become a military power which threatens other countries, by maintaining a strictly defensive posture (a rather clear NOD commitment, BM)" [*Pacific Research*, 5:3 (1992): 19].

JARUSZELSKI PLAN.   Polish plan presented by General Jaruszelski and formally presented to the other *CSCE* states in 1987. It was tantamount to an updated version of the 1957 *Rapacki Plan*, but *disengagement* was now envisioned to include conventional as well as nuclear weapons, and the focus was to be placed on "those of the strongest power and precision on destruction, allowing a *surprise attack*" [in J. Krause 1988:77–79; cf. Karkoszka 1988b; 1989c; Dobrosielski 1990]. Its purpose was a reduction of offensive capabilities.

JAURÉS, JEAN (1859–1914).   French socialist leader in the late nineteenth century and early twentieth. JJ opposed the growing bellicosity in the French officer corps prior to *WWI* (the *offensive a l'outrance* school behind the notorious *Plan 17*) and in *The New Army* proposed a shift to a *militia*-like structure of the armed forces. He was assassinated in 1914, most likely for his antimilitarist stand [Jaurés 1915; cf. Wild 1987].

JERVIS, ROBERT.   Professor of political science at Columbia University.

Without ever dealing directly (at any length) with alternative defense, RJ has contributed to the theoretical underpinning of *NOD* and *common security* in several ways:.

He has, first, challenged the image of the state as a rational, unitary actor by analyzing how "the logic of images" as well as misperceptions influence foreign- and defense-policy decisionmaking. The objective *security dilemma* thus becomes "overlaid by reinforcing misunderstandings." Because states are unable to ascertain likely opponents' intentions directly, they infer the intentions from actual conduct and military capabilities. Unable to empathize with the fears of likely opponents, states tend to see their military potentials as signs of malign intentions requiring reciprocal action. Hence the proclivity of security policies for spiraling *armaments dynamics*, competitive *alliance* building, etc. [1976: 58–93; cf. 1970; 1978]: "When states seek the ability to defend themselves, they get too much and too little: too much, because they gain the ability to carry out aggression; too little because others, being menaced, will increase their own arms and so reduce the first state's security" [1976: 64].

RJ pointed to a possible escape from the dilemma, namely, a distinction between offensive and defensive postures, as well as what has subsequently become known as *mutual defensive superiority*: "When defensive weapons differ from offensive ones, it is possible for a state to make itself more secure without making others less secure. And when the defense has the advantage over the offense, a large increase in one state's security only slightly decreases the security of others" [1978: 197].

RJ has, furthermore, challenged the assumptions of rationality inherent in the deterrence logic, e.g., in *The Meaning of the Nuclear Revolution: Statecraft and the Prospects of Armageddon,* in which he came out in favor of a maintenance of *nuclear deterrence* as a caution-instilling factor yet warned against the prevalent notions of counterforce, calibrated response, and the like, which merely threaten to jeopardize *crisis stability* and/or *arms race stability* [1989; cf. 1984; 1987; 1988, 1989a; 1990].

The possibilities of cooperation between adversaries have long received RJ's attention [1978; 1985; 1991; 1991b; idem & Snyder (eds.) 1991]. He has consistently pointed to the strength of the reciprocity principle: states tend to do to others what they expect others to do to them; hence the aforementioned security dilemma, but also opportunities for more stable relations, including *regimes*. RJ saw a *security regime* as a realistic possibility, defined as a set of "principles, rules and norms that permit nations to be restrained in the belief that others will reciprocate" [1982]. This has precisely been the logic underlying the more "Realist" versions of *common security regimes*, "mature anarchy," etc. [cf. Buzan 1989; 1991a; Booth 1990; 1991b; Møller 1992], notions to which RJ also devoted some attention after the end of the Cold War, when he came out in (qualified) favor of steps toward *collective security*, e.g., in the form of *concert*-like arrangements [Jervis 1991d].

JEUNE ÉCOLE.   See *Guerre de Course, *Sea Power, *France.*

JOHANSEN, ROBERT C.   See *U.S.*

JOMINI, ANTOINE-HENRI (BARON DE) (1779–1869).   Classical French military strategist, author of the monumental *Treatise on Grand Military Operations* [1811].

AHJ is known above all as the discoverer of the *principles of war* that have a presumably perennial validity—even though they were first formulated in approximately their present form by *Fuller*. The central principle was that of force *concentration* (with which modern *NOD* proponents have tended to be at odds). With his emphasis on the superiority of *interior lines*, however, AHJ laid part of the foundations upon which modern NOD strategists have rested their claims to achieve a stronger defensive capability by renouncing capabilities for long-range offensives: by specializing on the defense of the home country, they recommend to exploit the advantages of interior lines [1811(4): 275–286; cf. Reichel 1988; Shy 1986; Gat 1988: 106–135].

Furthermore, his advocacy of concentration notwithstanding, AHJ was well aware of the strength of dispersed forces and guerrillas. He had participated in the *Peninsular War* and been enthralled by the spectacle of a "spontaneous uprising of a nation." For an invader possessing "only" an army, this amounted to an unequal struggle:

> His adversaries have an army, and a people wholly or almost wholly in arms, and making means of resistance out of everything, each individual of whom conspires against the common enemy. . . . No army, however disciplined, can contend successfully against such a system applied to a great nation, unless it be strong enough to hold all the essential points of the country, cover its communications, and at the same time furnish an active force sufficient to beat the enemy wherever he may present himself.

For all its strength, *guerrilla warfare* seemed terrifying to AHJ because of its lack of organization. As an alternative, he recommended the creation of a *militia* (*Landwehr*) that would incorporate the popular element in an organized framework [Jomini: *Précis de l'Art de Guerre*, excerpted in Laqueur (ed.) 1978: 42–44].

JONES, CHRISTOPHER D.   Professor at the University of Washington and one of the most prominent analysts of the *Warsaw Pact* [1981; 1981a; 1984].

CDJ has been the main proponent of the interpretation of the Eastern military *alliance* (and by implication, Soviet strategy) according to which its main rationale was to maintain the cohesion of the bloc and Soviet hegemony therein. This might require support for communist regimes against popular uprisings (as in the *GDR* in 1953) or ultimately armed intervention (as happened in *Hungary* in 1956 and in Czechoslovakia in 1968). Soviet armed forces were thus not primarily intended for use against the West but for bloc-internal oppression, hence the need for doctrinal uniformity and tight integration of non-Soviet pact forces under Soviet command. Hence also the need for preventing pact countries from adopting defense strategies and postures that might enable them to resist Soviet intervention: *in casu* strategies of *territorial defense* like that of *Yugoslavia* (and to some extent Romania) and like those apparently contemplated in *Poland* in the mid-1950s and by the Dubcek leadership in Czechoslovakia in 1968 (cf. the *Gottwald Memorandum*).

In his most recent (retrospective) analysis of the (by then dissolved) Warsaw Pact, CDJ explained the circular logic of Soviet hegemony: "WTO doctrine was thus part of a self-reinforcing circle of policies. Soviet offensive capabilities guaranteed the survival of the East European communist regimes against internal and external threats. . . . The common doctrine deprived national defense ministries of the capacity to resist Soviet interventions" [1992: 116].

JOXE, ALAIN.   Professor of social sciences at the Ecole des Hautes Études in Paris and one of the most prominent peace researchers in *France*, inter alia, in his role as founder and director of *CIRPES*, apparently the only research institution in France taking an interest in *NOD* [1990a].

AJ's main contribution to NOD theory has been elaboration of the strategic principles of "infranuclear dissuasion," i.e., what others have called war prevention in a *subnuclear setting* [1984]. The principles are sui generis and differ from those of both classical strategy [1991] and of nuclear strategy [1990].

JUS AD BELLUM.   Just cause of war. See *Just War*.

JUS IN BELLO.   Justice in war. See *Just War*.

JUST DEFENCE.   British study group on *NOD* under the chairmanship of *Frank Barnaby* [1986; 1990; idem & Boeker 1982; 1988; 1989; idem & Borg 1986], and with Peter Smith serving as secretary until his death in 1991. JD has all along been close to, and has emphasized the lobbying of, the Labour Party [cf. various issues of the *Just Defence Newsletter;* P. Smith 1990; 1991; idem & Gapes 1984].

JUST WAR.   Originally a Roman concept (*bellum justum*) that subsequently formed the core of a tradition in Christian thinking about war and peace, traceable back to Clement of Alexandria but usually attributed to Saint Augustine [J. T. Johnson 1987: 50–66; cf. 1981]. However, counterparts of the Christian JW tradition are also to be found in other religions, including Judaism, Islam, Hinduism, and Buddhism [Smock (ed.) 1992].

1. **Status.** In addition to constituting, until the present day, an element of the dogma of the Roman Catholic Church, the basic principles of JW have become part and parcel of the modern attitude toward war [cf. Walzer 1977; Best 1980; Paskins (ed.) 1986]. It is, inter alia, enshrined in international law and the *laws of war* and reflected in the military code of ethics that is taught in military academies worldwide [Matthews & Brown (eds.) 1989].

The point of departure for the JW tradition is the (apparent or genuine) contradiction in the Christian scriptures between, on the one side, the prohibition against killing (cf. *Sermon on the Mount*) and, on the other side, the admonition to servitude vis-à-vis secular authorities and the numerous bellicose passages in the Old Testament. As a general rule, killing is forbidden, but there are exceptions to the rule, specified by the stipulation that Christians are allowed to wage only "just" wars.

2. *Jus ad Bellum* signifies the requirement for a just cause of war (*casus belli*), which is one part of the JW criteria. *Jus ad Bellum* presupposes:

- Legitimate authority. Only a legitimate ruler (e.g., divinely instated, representing "the vanguard of the proletariat," or democratically elected) is entitled to wage a JW.
- Just motivation. JW can be fought only for the purpose of righting a wrong, such as to punish the infidels or to liberate Kuwait; ulterior motives are, in principle, incompatible with JW.
- Just consequences. This requirement is a hedge against stupidity and/or deception; it implies that governments are not absolved by intentions alone but should also have grounds for believing that war will actually further the end fought for.
- Last resort. Even a legitimate government, motivated by the best of intentions and reasonably certain of being able to achieve the end sought, is allowed to resort to war only when there are no other ways of achieving the same end.

3. *Jus in Bello.* Justice in the waging of war is an indispensable second part of JW because even the most just cause does not give an unqualified "licence to kill."

- The requirement of legitimate authority also pertains to the waging of war: combatants should be subjected to command and thus accountable to superiors.
- Noncombatant immunity implies that war may be waged only against the other side's soldiers, not against its civilian population or soldiers-turned-civilians through surrender. Also known as the principle of discrimination.
- The call for proportionality is, on the one hand, an acknowledgment that noncombatants do suffer from war, at least in the form of collateral damage, which therefore should be proportional to the gains and kept as low as possible. On the other hand, *Jus in Bello* is a requirement for proportionality between crime and punishment.

4. **Nuclear Deterrence and JW.** Ever since the emergence of weapons of mass destruction, it has been questioned whether nuclear weapons are compatible with JW criteria [cf. Hardin & al. (eds.) 1985; Shue (ed.) 1989]: Nuclear weapons are obviously indiscriminate in their effects, thereby presumably violating the principle of noncombatant immunity, as well as that of proportionality. However, attempts have been made to justify reliance on *nuclear deterrence* with reference to its war-prevention effects, by distinguishing between threat and actual use, and through more discriminate target selection [McCall & Ramsbotham (eds.) 1990].

5. **NOD, CBD, and JW.** The attitude of alternative defense proponents toward JW has been far from uniform:

- Advocates of *civilian-based defense* (as well as all principled pacifists) have tended to deny the relevance and validity of JW, either by rejecting the logic altogether or through a *reductio ad absurdum:* if *limited war* is inconceivable, the damage resulting from war will always surpass whatever good might come of it, wherefore the criterion of proportionality is never met.
- *NOD* advocates have tended to accept the JW principles and to criticize existing strategies and postures for violating them, in particular insofar as weapons of mass destruction are involved. Presumably, NOD would meet the JW criteria because only national defense would be envisioned (which is per se sufficient *jus ad bellum*). Furthermore, the fact that no *counteroffensives* onto enemy territory would be undertaken would automatically spare the civilian population of the aggressor, thereby satisfying the criteria of discrimination and proportionality. Finally, the use of predominantly short-range and precision-guided weapons would improve possibilities of discrimination on the battlefield.

K

KALDOR, MARY. British peace researcher and political scientist at the University of Sussex. She has played a leading role in the END (European Nuclear Disarmament) movement since its genesis, as well as in the Alternative Defence Commission (*ADC). MK has primarily contributed to alternative defense thinking in two fields:

• In *The Baroque Arsenal,* she analyzed the *armament dynamics with a special emphasis on the internal dynamics, demonstrating how weapons acquisition was governed more by follow-on imperatives than by sober assessment of threats and requirements. As a way of breaking this follow-on logic, MK recommended a shift of emphasis in the direction of smaller, cheaper, and more defensive weapons systems, such as *PGMs [1981: 226–230].

• Ever since publication of *The Disintegrating West* [1978], MK has advocated a reformation of the *alliances, based on an interpretation of both *NATO and the *Warsaw Pact as instruments of hegemony. As elaborated upon in *The Imaginary War* [1990; cf. 1987], each alliance had been a means of superpower control over minor allies rather than directed against the other, albeit justified with reference to external threats. Hence the need for breaking the self-perpetuating "Cold War logic," e.g., through "*détente from below" (the catchword of the END movement) [cf. Kaldor, Holden & Falk (eds.) 1989] and/or through *dealignment [1986; cf. 1983; 1985; idem & Falk (eds.) 1987].

KANT, IMMANUEL (1724–1804). German philosopher of the late Enlightenment who, in his main work on ethics, *Critique of Practical Reason,* formulated the categorical imperative "Act only on that maxim through which you can at the same time will that it should become a universal law" [Kant 1788:53]. This has often been interpreted as a generalized version of the principle of *common security [Palme Commission 1982; cf. Møller 1992: 10–12, 28–32; H. Afheldt 1984b: 61].

IK furthermore described standing armies as an important cause of war, and in his pamphlet *On Perpetual Peace* urged their disbandment in favor of something reminiscent of a *militia. He also recommended constitutional, representative government (*democracy) as a valuable inhibition against aggressive war, particularly in a broader context of federalism [Kant 1795:17–18; cf. Piepmeier 1987; Donaldson 1993; Roy 1993; Sørensen 1992a; 1993]. IK might thus be seen as the father of the contemporary school of international relations, according to which democracy is the most powerful of all inhibitions against war because democracies do not fight one another [Archibugi 1989; Bächler 1990; 1990b; 1992a; Czempiel 1992; Eberwein 1992; Ember et al. 1992; Fukuyama 1991; Gleditsch 1992; Johansen 1991a; Latham 1993; Mintz & Geva 1993; Russett 1990; 1993; idem & Antholis 1992; Senghaas 1990e; Starr 1992; Weede 1984; 1992].

KARKOSZKA, ANDRZEJ. Polish peace researcher, formerly at *SIPRI, subsequently at the Institute for East-West Security Studies in New York. AK has primarily dealt with *disengagement, in the form of *nuclear-weapons-free zones and as a means of defensive restructuring, in both cases with a special emphasis on the national interests of *Poland [1988b; 1989b-c]. In this field, e.g., he collaborated in the development of (probably drafted) the *Jaruszelski Plan.

KAUFMANN, WILLIAM W.   See *U.S.

KAZAKHSTAN.   See *Soviet Union, *CIS.

KEEP-OUT ZONE.   See *Maritime C(S)BM.

KEKKONEN PLAN.   Proposal made by Finnish President Kekkonen in 1963 (reiterated in 1978) for establishment of a *nuclear-weapons-free zone in the Nordic region [Väyrynen 1983; Möttölä 1984; cf. Møller 1985].

KENNAN, GEORGE FROST.   U.S. ambassador and father of containment strategy. In the early 1950s K proposed a *no-first-use policy for nuclear weapons, which he followed up in 1982 with an elaborate proposal to that effect, issued jointly with three other prominent retired government officials: the so-called Gang of Four comprising, besides GK, McGeorge Bundy, Robert McNamara, and Gerard Smith [Kennan 1982:3–10; cf. McNamara, & al. 1985].

In 1957 on the BBC K proposed a *disengagement of forces in Europe, through a combined withdrawal of superpower forces and the restructuring of indigenous forces into "para-military ones, of a territorial-*militia type, somewhat on the Swiss example." Their main mission would not be *forward defense but, rather, deterrence by denial by means of *territorial defense: "the defense at every village crossroad" [Kennan 1958:65–66; cf. Brinckley 1987].

KENNEDY EXPERIMENT.   See *Amitai Etzioni.

KIESSLING, GÜNTER.   General in the Bundeswehr (ret.) and former deputy SACEUR (Supreme Allied Commander Europe).

In 1989 (before the East German revolution), GK published *Neutrality Is Not Treason: Sketch of a European Peace Order,* wherein he advocated *neutrality as a way of furthering German unification. It was followed in 1990 by a sequel with the mystifying title *NATO or Elbe? Model of a New European Security System.* Unification presumably would presuppose the acquiescence of both East and West, and GK regarded it as inconceivable that the *Soviet Union would accept a unilateral withdrawal from "its Germany" [1989: 384–389; 1989a: 90–99, 198–200]. The four powers (the *U.S., the *U.K., *France, and the USSR) would have to guarantee the status of a united Germany, which should be permitted to enter no *alliance directed against any one of the signatories.

GK further assumed that the four powers would prefer an armed united Germany to a military vacuum, wherefore it should, as a minimum, possess the wherewithal of an effective border defense. Preferably, however, Germany's defense should have sufficient endurance to obviate the need for early employment of nuclear weapons [1989: 390; 1989a: 160, 228, 264–265]. It was, however, just as imperative to avoid that united Germany be perceived as a military threat by its neighbors, hence the need for refraining from the acquisition of, or access to, nuclear weapons. The size of the conventional forces should not exceed 300,000 troops, with the main emphasis on the *land forces; the navy should be little more than a coast guard; and the air force little more than an *air defense system. Offensive capabilities should be minimized, e.g., through a prohibition of long-range rockets and a ceiling of two thousand on the number of *tanks [1989a: 213, 264–276].

The 1990 work was written while the *Two-Plus-Four negotiations were in progress, for which GK made a number of recommendations; few of which were taken into consideration. Under the new auspices, GK envisioned a German *NATO membership, albeit accompanied by a de facto military *neutralization through a radical build-down of military forces on German territory. Stationed troops should not exceed 100,000, and German forces not 300,000. The whole should be embedded in a pan-European *collective security structure, for which the author coined the acronym GESSYM (*Gesamteuropäisches Sicherheitssystem* [All-European Security System]). An element would be the so-called MITSYM (*Mitteleuropäisches Sicherheitssystem* [Central European Security System]) comprising Germany, *Poland, *Hungary, Czechoslovakia, *Austria, the Benelux countries, and (preferably) *Denmark.

KING, MARTIN LUTHER.   African-American civil rights leader, pacifist, and one of the main modern proponents of *nonviolence. For his outstanding achievements, MLK was awarded the Nobel Peace Price in 1964. He was assassinated in 1968 [Grosse 1987].

Besides Christianity, MLK's sources of inspiration included *Thoreau, *Gandhi, and *Tolstoy. The nonviolent methods suggested by these authors were mostly used by MLK for the civil rights struggle, but he also became actively engaged in campaigning against U.S. involvement in the *Vietnam War. Contrary to many other proponents and practitioners of nonviolence (who have emphasized the power element), MLK and his movement focused their attention on the symbolic or communicative impact of their actions. Further, they emphasized the importance of the moral high ground, e.g., by proscribing all violent forms of struggle from their activities but attempting nevertheless to show strength and self-confidence. Doing so was deemed to place the onus of initiating violence on the opponent, with negative repercussions in terms of image-building and public relations.

KING-HALL, STEVEN.   Commander in the British armed forces, inter alia, during *WWII. The nuclear bombs over Hiroshima and Nagasaki in 1945 convinced him that "Total War has abolished itself" [*King-Hall Newsletter,* 16 August 1945, in King-Hall 1959: 11]; other than military means of defense had to be found. SK-H thus became one of the first British proponents of *civilian-based defense, conceived of as a functional substitute for nuclear and conventional armaments: if the end of war was to "change the enemy's mind" (a broadened version of "imposing one's will on the opponent"), then the most economical means of doing so was not physical force. Rather, it was *nonviolence, whereby the enemy's psychological warfare could be thwarted and world opinion influenced most effectively [1959: 23, 30, 190–205].

SK-H envisioned a *collective security system called the "European Treaty Organization" (ETO) as the most appropriate framework for such a shift to nonviolence. However, during a transition period the ETO should have at its disposal conventional armed forces sufficient for dealing with internal threats as well as for "putting up a token resistance" against a conventional Soviet attack [1959: 145–152].

KLEIN, JEAN.   See *France.

KLEINKRIEG.   Small war. Classical term for *guerrilla warfare.

KLUMPER, ANTONIUS ALBERTUS.   See the *Netherlands.

KNIGHT, CHARLES.   See *Commonwealth Institute.

KOKOSHIN, ANDREI.   See *ISKAN, *CIS.

KOMITEE FÜR GRUNDRECHTE UND DEMOKRATIE.   See *Committee for Basic Rights and Democracy.

KONOVALOV, ALEXANDER.   See *ISKAN.

KONRAD ADENAUER FOUNDATION.   German research institution with close links to the *CDU, serving as a think tank for the party [Enders & al. 1990: 141–143]. The staff of the KAF have primarily been engaged in redesigning nuclear arsenals in the aftermath of the *START and *INF agreements. The arsenals should preferably maintain deterrence and the (allegedly benign) effects of nuclear weapons, while meeting German concerns. The latter, it had to be acknowledged, required a build-down of short-range and battlefield nuclear forces [Enders 1988; 1990; idem & Siebenmorgen 1988]. In the quest for salvaging nuclear deterrence, some of these CDU-affiliated researchers even came to endorse the notion of *NOD, albeit only conditionally [Enders 1990a].

KOREA.   Since the Korean War (1950–1953) K has been in one of the closest non-European counterparts to the situation of Germany throughout the postwar period: a country divided along ideological lines into communist and capitalist (approximate) halves, each aligned with its superpower, albeit less closely as far as North K was concerned than was the case for the *GDR. Recent years have seen numerous initiatives for ameliorating the situation, not least from the North, apparently anxious to avert internal crisis through improved relations with the prosperous South.

- In 1990–1991, North K lobbied for a ten-point nonaggression declaration, including several *CSBMs (such as a "hotline" agreement) as well as measures of *disengagement. The 1953 Military Demarcation Line should be recognized as a "nonaggression demarcation line" (i.e., frontier), and the Demilitarized Zone should be expanded into a "peace *zone" [*Pacific Research*, 4:1 (1991): 20].
- North K has also proposed establishment of a *nuclear-weapons-free zone covering the whole peninsula, to be established by means of bilateral talks between the two states, and subsequently guaranteed by the *U.S., the *Soviet Union, and *China [*Jane's Defence Weekly*, 14 September 1991, 492; cf. Shim 1991].

This stood in sharp contrast to the vacillations of North K in 1993–1994 about its possible withdrawal from the *NPT. In actual fact, North K may be quite close to "going nuclear" or may even have crossed the line: a development the rest of the world has been very anxious to prevent. Hence, inter alia, the negotiations between North K and the U.S. about possible compensation for not exercising the nuclear option and for accepting inspection teams in nuclear facilities [cf. Barnaby 1993: 94–99; Bracken 1993; Cheon 1993]. Prior to this, the U.S. seems even to have contemplated a *preemptive attack* againt these facilities yet to have decided against it [*Jane's Defence Weekly*, 18 December 1993, 5]. As an alternative, the U.S. was also considering providing South K with Patriot missiles for an *extended air defense against ballistic missiles launched from the North [*International Herald Tribune*, 26 January 1994, 1].

Western observers have also proposed a wide range of *arms control* and confidence-building measures for K [Mack 1991; 1991a; C. M. Lee 1991; B. N. Garrett 1991; Segal 1991; Wendt 1992], among which were recommendations for shifting to defensive postures, i.e., for creating a situation of *mutual defensive superiority* by means of *NOD* postures on both sides. This would presumably not merely halt the *arms race* in progress but also might pave the way for unification [Wiberg 1989c; 1990c; Zagoria 1991: 179–180; Jung 1992; Kwak 1993; Suh 1992].

KRAUSE, CHRISTIAN.   Retired general of the Bundeswehr, member of the *SPD*, and on the staff of the party's research institution, *Friedrich Ebert Stiftung*.

CK's role in the SPD has been that of a "resident disbeliever" with regard to *NOD*. While acknowledging, in principle, the utility of defensive military strategies, CK nevertheless subjected the entire NOD discourse to a comprehensive and very competent critique [1987; 1987a; 1988; 1988a; cf. 1984: 126]. First, he disagreed with NOD proponents that NATO had a problem because both alliances were already actually incapable of successful attack. Because *conventional stability* was a fact, *NATO* might simply abandon nuclear first-use strategy without further ado [1986; 1987: 49]. Second, CK criticized the NOD models (and particularly that of *Horst Afheldt*) for excessive reliance on *E.T.* (emerging technologies), as well as for an unwarranted optimism with regard to the efficiency of their models. He was even more skeptical about the *von Bülow* and *von Müller* proposals because of their excessive offensive potentials [1987; 1988a: 189]. Finally, no NOD advocate had paid sufficient attention to what was, in CK's opinion, the most serious scenario, namely, that of limited attacks with limited goals. Such contingencies called for crisis management capabilities, duly accompanied by *CBMs* and, at most, some redeployment of forces, but certainly for no fundamental restructuring of the armed forces [1987: 48; 1987a].

In the course of the debate on German unification in 1990, CK's attitude toward NOD became somewhat more favorable. When trying to devise a stable security architecture for a truly pluralistic Europe, he had to acknowledge the need for *tous azimuts* defense postures that should be as defensive as possible. While thus conceding that NOD would be useful, he also pointed to an inherent contradiction in the notion of bilateral *transarmament* to defensive structures: "When there are no longer any armed forces capable of attack, there will also no longer be any need for defense. In the final analysis the implications of this conception were to make the armed forces redundant" [1990; 1990a; 1990b].

KROHN, AXEL.   West German peace researcher, for a time a researcher at *SIPRI*. AK's main contributions to the alternative defense debate have been analyses of the Nordic *nuclear-weapons-free zone* and future developments in the Baltic Sea region [1989; 1991; 1991c; 1992; 1993; 1993a], as well as problems of restructuring armed forces as a corollary of the *CFE* [1991a-b; Wachter & Krohn 1989].

KRÖNIG, VOLKER.   Senator from Bremen for the *SPD*, member of the *SAS* [1989; 1989a].

KURSK, BATTLE OF.   Initiated in July 1943, when the German *Wehrmacht* embarked upon a gigantic offensive against the Soviet fortifications on the Kursk Salient. The fortifications were constructed in great depth and consisted primarily of tank *barriers* and antitank forces backed up by strong operational reserves.

After the (largely unsuccessful and extremely costly) German assault against the fortifications, the Red Army went over to a large-scale (operational-scale) *counteroffensive.*

In 1966 *Bogislaw von Bonin* acknowledged K as a source of inspiration for his NOD-like proposals [Bonin 1966]. Renewed attention was drawn to K by Soviet proponents of *NOD* in the late 1980s. By pointing to the Kursk experience, they questioned the dominant image of *WWII* as one of unqualified offense dominance and sought to refocus military history on the defensive lessons that might be learned from Soviet history [Kokoshin & Larionov 1987].

L

LABOUR PARTY. See *U.K.*

LAFONTAINE, OSKAR. Prime minister of Saarland (a constituent entity in the *FRG), deputy chairman of the *SPD, and for a while its candidate for the chancellorship [cf. Møller 1991: 171–172, 238–239]. Subsequent to the SPD's electoral defeat in 1990, OL resigned and withdrew to his post as premier in Saarland [*Der Spiegel*, 1990:50, 18–25].

In his 1983 bestseller, *Fear of One's Friends,* OL sharply criticized *NATO and the *U.S.*, above all for their nuclear strategy [Lafontaine 1983]. In a work published five years later, OL devoted himself to elaborating on the alternatives, identifying as one of the main tasks for the coming decades to achieve a more independent European stand. This presupposed defense cooperation in the EC, the short-term core of which might be Franco-German collaboration. In the longer term, only a pan-European perspective would be appropriate. To pave the way for such a development, the West should embark on a defensive restructuring of its armed forces as well as establish a *nuclear-weapons-free zone* ("corridor") in central Europe [1988: 126–133, 142–147, 151].

Having been elected as SPD candidate for chancellor in the decisive year 1990, OL went out of his way to change his image as a semineutralist, e.g., by proclaiming himself expressly against *neutrality* for the united Germany; neutrality would merely make Germany a sort of *cordon sanitaire* between NATO and the *Warsaw Pact.* Indeed, neutrality was an obsolete notion; the need was for pursuing *integration* to its logical conclusion. In the fullness of time, integration might transcend the East-West division and lead to a pan-European security system including the *Soviet Union*, the U.S., and *Canada.* OL foresaw East European states joining NATO and could even conceive of collective *Warsaw Pact* membership in NATO [*Atlantic News,* 30 March 1990]. On other occasions, OL seemed to be arguing against German NATO membership and advocating, as his preferred alternative, a "European Defense Community," i.e., a pan-European *collective security* system [*Frankfurter Allgemeine Zeitung,* 5 March 1990; *Atlantic News,* 21 March 1990; cf. *Der Spiegel,* 1989:52, 67].

LANCHESTER'S LAWS. Mathematical formulation of central *principles of war* [Lanchester 1916; cf. Lepingwell 1987; Dupuy 1990:216–218], which presumably makes them amenable to modeling, thus allowing for battle *simulation* [Blechman & al. 1990:142–144; Epstein 1990; Weiner 1986; Hofmann & al. 1986].

Lanchester's "Linear Equation" refers to direct fire, i.e., battle circumstances under which the two sides can see each other. The formula is dR/dt = K(RB) or dB/dt = K'(BR) (R and B referring to the strength of Red and Blue forces, respectively; t, to the time interval applied to measuring changes therein; and K and K', to two constant coefficients). Accordingly, the equation describes one side's *attrition* as a function of the other side's strength.

The so-called Square Law applies to indirect fire, under which conditions the formula is dR/dt = CB and dB/dt = C'R (C and C' referring to constant coefficients). This means that the attrition of one side is proportional to the square of the opponent's numerical superiority, implying that an initial force imbalance will be multiplied through successive engagement rounds. The Square Law has been used as an argument against the *dispersion* envisioned under *NOD*, and in favor of a continued emphasis on the ability to concentrate forces for battle.

NOD proponents have questioned the absolute validity of LL, pointing out that the laws presuppose a low accuracy of indirect fire. With the increasing accuracies (and hence kill probabilities) associated with, e.g., *PGMs*, the Square Law presumably no longer applies and the defender can safely disperse without suffering disproportionate attrition. The inevitable concentration on the part of the attacker would, on the contrary, only make its forces all the more rewarding as targets, thus accelerating its attrition [Neild 1986; 1990:50–51; cf. Davis 1990; Payne 1990; Boserup 1990c]. Furthermore, it has been pointed out that attrition (even in individual battles) is only one determinant of victory or defeat [Dupuy 1990: 207–218 & passim]. See also *three-to-one rule*, *Concentration*, *Force-to-Force Ratio*.

LAND FORCES.   The configuration of the LF is at the very heart of the alternative defense debate. The possession of offensive-capable LF that can conquer and hold ground is a sine qua non of genuine offensive capability, i.e., of winning wars of aggression. It is therefore in the realm of LF that *NOD* proponents have envisioned the most radical changes.

1. **Taxonomy.** LF may be categorized according to various criteria, reflecting alternative strategies and tactics for the land battle. Two dichotomies stand out in most taxonomies: those of positional versus maneuver warfare, and those of firepower versus mobility. Table 23 indicates a few interconnections and incompatibilities between the four terms.

Table 23.   Taxonomy of Land Warfare

	Firepower	Mobility
Positional	yes	(no)
Maneuver	(yes)	yes

- Positional warfare (through which the defense of certain positions is sought) cannot emphasize mobility, even though forces engaged in positional warfare are not entirely stationary. They therefore tend to give precedence to firepower over mobility because firepower alone can accomplish the *attrition* of enemy forces, which is the purpose of positional warfare.
- Maneuver warfare, on the other hand, does not seek to prevail through attrition but through outmaneuvering (and perhaps outsmarting) the opponent, ideally without fighting at all. Mobility is therefore of the essence, but maneuver without firepower is, of course, meaningless.

2. **Traditional LF.** Just as is the case for maritime and air forces, states tend to seek a maximum of flexibility for land forces, hence the traditional emphasis on mobility and the wherewithal for its accomplishment. Because what matters most is mobility under enemy fire, not only mechanization plays a role but also protection, i.e., armor and/or protective artillery fire. All modern armies hence field large numbers of *tanks*, APCs, and self-propelled artillery, assisted by aircraft and heli-

copters, and they are provisioned logistically by means of truck convoys, aircraft, pipelines, etc. [Bellamy 1987]. The quest for mobility is usually strictly defensively motivated, e.g., by the need to be able to evict an invader by force. Furthermore, mobility provides protection because mobile forces and/or weapons systems are more difficult for an enemy to destroy. Finally, in many countries, fairly long-range mobility is a sine qua non of area coverage, either because of sheer size or because the distribution of the population fits poorly with the strategic importance of different areas.

On the other hand, excessive emphasis on long-range mobility under enemy fire may be tantamount to offensive capabilities, i.e., to the ability to invade other states. The strategy, operational art, and posture of the *Soviet Union (particularly in the 1970s and early 1980s) were extreme examples of LF streamlined for large-scale, fast-speed (i.e., nearly *blitzkrieg-like) operations [Sidorenko 1970; cf. Vigor 1983; Lebow 1985; Leites 1981; Hines & Petersen 1984; Petersen 1987; Bellamy 1987: 105–124]: an all-mechanized army suitable for penetrating positional defenses [Simpkin 1983]; successive echelons intended to maintain the momentum of the initial assault [Vigor 1982; Hines 1982]; OMGs (operational maneuver groups) designed for the swift exploitation of any penetration [Donnelly 1982; Shields 1985]; and air-mobile and similar formations for *desant* operations [Holcomb & Turbiville 1988; cf. Suvorov 1983]. All this pointed to an emphasis on attaining swift and operationally decisive results.

The strategy and posture of *NATO have been predominantly positional (cf. *Forward Defense*, *Active Defense*) and firepower-oriented, but the early 1980s saw a growing interest in maneuver-type warfare, with the *AirLand Battle* doctrine of the U.S. Army standing out as the most extreme example [FM 100-5 1982; cf. Czege 1984: 105–108; Lindner 1984; Mallin 1990: 234–239; Garrett 1989: 45–47; Bellamy 1990: 131–138; Sutton & al. 1984; Strachan 1984: 34–38; Richardson 1986].

3. **NOD and Mobility.** Although all NOD advocates have agreed on the need for minimizing the offensive capabilities of LF, they have disagreed on both the appropriate standards and the means [cf. Møller 1991: 63–155]. Some have viewed only large-scale invasion and occupation capabilities as impermissible [e.g., Bülow 1986; 1986a; 1988c; 1989a; Hannig 1984; SAS (eds.) 1984; 1989]; others have maintained the need for a complete incapability of crossing the border, be it with LF or "stand-off" means such as missiles or aircraft [e.g., Afheldt 1983; Lutz 1988b; Buro 1983a; 1987]. The latter NOD schemes have tended to be almost completely stationary, relying exclusively on firepower; the former have envisioned the maintenance of mobile systems (both for internal reinforcement and for *counteroffensives*) and fairly-long-range (say, up to 100 km) "stand-off" systems and therefore have had to ensure the absence of offensive capabilities through various precautions:

- Withdrawal of particularly offensive-capable forces from forward positions [Löser 1981; Senghaas 1987; Bahr & al. 1988].
- Creation of a *missing link*, deprived of which the LF would be rendered unable to operate on enemy territory [E. Müller 1984].
- A *spider-and-web* configuration, implying that the mobile forces would be severely handicapped beyond a stationary *web*, hence not effectively offensive [SAS (eds.) 1984; 1989; Boeker & Unterseher 1986; Grin & Unterseher 1988; Unterseher 1988b; 1989b].

4. **NOD-Type LF.** NOD is not a function of *defensive weapons* but, rather, a complex synergetic unity of *strategy*, operational art, tactics, and posture. Nevertheless, the weapons profile of strictly defensive LF would be significantly

different from that of a traditional army. Some forms of equipment would be entirely absent, and the relative emphases on other types would be different (see Table 24).

Table 24.   NOD Compared with Traditional Ground Forces

	None	Fewer	More
Weapons platforms		MBTs, IFVs, APCs, combat helicopters	Trucks, motorcycles
Weapons	Long-range missiles	Heavy artillery	Antitank and air defense weapons
Munitions	Nuclear, chemical, FAE		Mines, armor-piercing warheads
Logistics, C^3I	Tank transporters	Bridge-building equipment	Dispersed depots

Note: MBT = main battle tank; APC = armored personnel carrier; IFV = infantry fighting vehicle; FAE = fuel-air explosives.

LANDSCAPING.   See *Terrain Modification*.

LAND WARFARE.   See *Land Forces*.

LANGILLE, HOWARD PETER.   See *Canada*.

LARIONOV, VALERI.   See *ISKAN*.

LA ROCQUE, GENE R.   See *U.S.*

LATIN AMERICA.   Generic term for Central and South America, one of the world's least militarized regions, albeit also one where the armed forces have played a very decisive political role.

1. **Background.** The cultural homogeneity and the absence of major territorial disputes, as well as the low saliency of the East-West conflict, have all contributed to making LA a relatively peaceful region. However, in the few instances when the U.S. preponderance in the region has been challenged (Cuba, the Dominican Republic, Nicaragua, Grenada), the result has been intense conflict, and direct or covert U.S. intervention. Such instances have largely been confined to Central America and the Carribean, which have generally been more volatile than South America.

2. **Central America.** In an attempt to create a regional security order, the Contadora Act for Peace and Cooperation in Central America was initiated (but has not yet been ratified) by Costa Rica, El Salvador, Guatemala, Honduras, and Nicaragua [Child 1992]. The signatories committed themselves "to refrain from introducing new arms systems which qualitatively and quantitatively modify present inventories of war material" and to maintain the region free of nuclear, chemical, and biological weapons [Anthony 1991: 11–12]

Furthermore, there have been a few unofficial proposals for applying *NOD* criteria to the military postures in the region [Herrera-Lasso 1990].

3. **South America.** As a reflection of the absence of major conflicts, as well as of the dangers of nuclear proliferation, the South American countries in 1967 signed the Tlatelolco Treaty, thereby establishing the world's first *nuclear-weapons-free zone* [*SIPRI Yearbook* 1969/70: 237–256; Robles 1969; Calderon 1982]. Even though not all Latin American states are parties to it, and notwithstanding the fact that the nuclear powers have attached a number of reservations to the Additional Protocol II (defining their obligations) [*SIPRI Yearbook* 1991: 688–690], it has been effective ever since and may have played an important role in curtailing the nuclear ambitions of Brazil and/or Argentina [Spector 1984: 195–273; idem & Smith 1990: 221–263; Gamba-Stonehouse 1991; on Argentina's ratification in 1993, see *Financial Times,* 12 November 1993].

In addition to the Tlatelolco Treaty, Argentina, Brazil, and Chile in 1991 signed a declaration banning the production or use of chemical and biological weapons [*Jane's Defence Weekly,* 14 September 1991, 451]. Peru has proposed similar prohibitions with a continentwide scope. The Peruvian proposal (1991) furthermore envisioned prohibition of various forms of conventional military equipment, including: short- and medium-range ballistic missiles; fuel-air explosives; laser-guided bombs; long-range multiple rocket systems; antiradar and stealth technology; and nuclear-propelled submarines. Should it be implemented, this would contribute significantly to transforming military postures into strictly defensive, *NOD*-type postures, as would the accompanying proposals for various *CSBMs*, such as reductions of armed forces in border areas [*Jane's Defence Weekly,* 3 August 1991, 179].

There has been very little debate on defensive strategies in LA. However, in Argentina a project (EURAL: Centro de Investigaciones Europa-Latinamericanos, i.e., Center for European-Latin American Research) has been under way under the direction of Thomas Scheetz to devise a concrete *NOD* posture for Argentina, taking into account mutual threat perceptions between Argentina, Brazil, and Chile, providing for democratic control of the armed forces, and allowing for reductions in defense expenditures.

LATVIA.   See *Baltic Region.*

LAWRENCE, THOMAS EDWARD.   See *Lawrence of Arabia.*

LAWRENCE OF ARABIA (1888–1935).   British agent who during *WWI* served as advisor to the Arab forces of Sharif Hussein and sons during their revolt (1916–1918) against Ottoman rule [Lawrence 1929; 1935; cf. Hart 1934; 1967a: 181–183; Beaufre 1972: 134–148; Shy & Coller 1986: 830–831].

The campaign took the form of *guerrilla warfare,* and both L and his biographers exploited the experience gathered from the revolt to develop a *guerrilla strategy.* According to *Liddell Hart,* L developed a strategy that was

> the antithesis of orthodox doctrine: Whereas normal armies seek to preserve contact, the Arabs sought to avoid it. Whereas normal armies seek to destroy the opposing forces, the Arabs sought purely to destroy materiel—and to seek it at points where there was no force. . . . Blows might induce them to concentrate, and simplify both their supply and security problems. Pin-pricks kept them spread out. Yet for all its unconventionality this strategy merely carried to its logical conclusion that of following the line of least resistance. (Hart 1967a: 181–182)

L himself described his strategy as the antidote to the "spirit of the offensive," which predominated during WWI (see *Plan 17, *Schlieffen Plan*). In contrast to the linear fighting between the opposing trenches along the western front, he likened the Arab forces to "an influence, an idea, a thing intangible, invulnerable, without front or back, drifting about like gas," which rendered the opponents "helpless without a target" (cf. *No-Target Principle*). Rather than concentrate their forces for large-scale encounters, the Arabs should maintain "elusiveness" by defending no specific points, the choice as to what would be decisive belonging to the defenders. In contrast to the traditional "wars of contact," L recommended waging "a war of detachment. We were to contain the enemy by the silent threat of a vast unknown desert, not disclosing ourselves till we attacked." The (tactical) attacks were to aim more against the enemy's *logistics* than against its forces.

L compared his strategy with naval strategy [1935: 337], both because of the similarities between the desert and the sea, and because of the importance of "commanding the desert" in a Mahanian sense (even though his strategy was more similar to French naval strategy; see *Guerre de Course*). Furthermore, the camel raiding parties employed by the Arabs were self-contained, like ships, and might cruise confidently around the enemy, sure of unhindered retreat into the depth of the desert. One of the most important determinants of success was the vast space available for the revolt, exploited through "the strategic principle of widest distribution of force" [1929: 130, 136].

L also formulated the principles of the unassailable base area and the importance of the support of the population, which have subsequently been associated with *Mao Zedong* [Lawrence 1935: 190–196].

LAWS OF WAR.   (Occasionally in singular.) Generic term for the unified body of the LOW, contained in a number of international treaties and rulings of international courts as well as implied by international customary law [Walzer 1977; Best 1980; De Lupis 1987; Schachter 1991; McCoubrey & White 1992].

1. **Status and Importance.** As all international law, the LOW differ from national legislation in that there is no corresponding supranational authority to enforce them. The closest approximations to such authority are the various international organizations, such as the *U.N.*, empowered to impose sanctions, etc.

Against this anarchical setting (and in light of numerous violations of the LOW) it remains disputed if, or to what extent, the LOW deserve to be taken seriously at all, although prohibition of specific types of weapons is nearly universally acknowledged as important. Those who maintain that the LOW are important expose themselves to accusations of naivete and of depreciating friction in war. The underlying *just war* theory has, moreover, been charged with actually legitimizing wars that in reality are never "just." Quite a strong case can nevertheless be made for the assertion that the prescriptions of the LOW do influence the calculations of military planners and field commanders, if occasionally only in the sense that they seek to conceal violations thereof.

- First, in most armed conflicts the two sides share an interest in averting an unbounded *escalation* that would be dysfunctional and indeed suicidal, hence in observing certain *thresholds*. Even though the parties may not always agree on the appropriate threshold (e.g., because one side enjoys escalation dominance at a certain level), they often do. In that case the LOW (or every code contained therein) may serve as a demarcation line, on the (at least temporary) observance of which the parties may agree, be it formally or tacitly. In either case, each side would be able to punish any violation by the other by simply reciprocating, either in a *Tit-for-Tat* mode or through escalation [cf. Schelling 1960; Axelrod 1984].
- Second, history seems to indicate that legal regulations on warfare do in

fact serve as guidelines for actual military conduct. Even though the LOW are often violated, even more frequently they are observed. Indeed, even in the case of violations, those responsible usually seek to legitimize their actions through legal arguments rather than by outrightly denying the validity of the LOW.

- Third, armed forces worldwide seem to take the LOW seriously, at least in peacetime: courses in the LOW are obligatory for officers in most countries; field manuals and textbooks on the subject are part of the curriculum for military education; military exercises tend to be in accordance with the regulations of the LOW; etc. To some extent, the lessons may be internalized and become an element in the professional *code d'honneur* of the officer corps, thus strengthening the instinctive aversion against obvious atrocities.
- Fourth, in most wars there are some prospects of sanctions in the form of court-martial and postwar trials of the Nuremberg Tribunal type. Although the prospect of sanctions may loom larger for the vanquished than for the victor (*vae victis!*), the loser usually has the strongest incentives to violate the LOW, whereas the prevailing side can afford magnanimity.
- Finally, international public opinion is a force in its own right. Neither states nor individual commanders take pleasure in public condemnation, as might well follow from violations of the LOW, widespread instances of which are unlikely to remain undetected in the age of television and instantaneous press coverage.

2. **Prohibited Weapons.** One might distinguish (somewhat arbitrarily) between, on the one hand, restrictions on the possession, use, or both of weapons, and, on the other hand, regulations on the actual waging of war. The restrictions in force on weapons include the following:

- *Chemical weapons* are regulated, inter alia, by the 1899 and 1907 Hague Conventions (prohibiting the use of "poisons and poisoned weapons") and the 1925 Geneva Protocol, which prohibited the use, albeit not the possession, of chemical weapons. Because a number of states explicitly reserved for themselves the right of retaliatory use, it was tantamount to little more than a first-use prohibition [SIPRI 1973]. In 1993, a new *chemical weapons convention* was signed that, upon ratification and entry into force, will prohibit the production and stockpiling of chemical weapons and make the destruction of existing stocks mandatory [Robinson & al. 1993].
- Biological weapons were likewise prohibited by the 1925 Geneva Protocol, even though their stockpiling remained unconstrained until the entry into force of the 1972 Biological Weapons Convention [De Lupis 1987: 219–222].
- "Weapons causing superfluous injury or unnecessary suffering" were prohibited in the 1977 Geneva Protocol 1, but the first steps in this direction had already been taken in the 1868 St. Petersburg Declaration and the 1899 Hague Convention. The concrete application of the principles, however, remains disputed [Röhling & Sukovic 1976; Lumsden 1978; SIPRI 1981]. In 1993 Switzerland (as repository of the protocol) launched an initiative to have antipersonnel landmines included among the prohibited weapons [*Pacific Research,* 6:2 (1993): 20].

All of these conventions contribute to a certain humanization of warfare, but they all suffer from serious defects: abstention by significant states, a number of reservations on the part of the signatories, and insufficient scope. The most criticized omission is undoubtedly that the use of nuclear weapons remains legal, in spite of their obviously highly indiscriminate effects.

3. **Regulations on the Waging of War.** Among the most important regulatory instruments are the Hague Conventions and the Geneva ("Red Cross") Convention [cf. Moreillon 1987; Rosas & Stenbäck 1987; Meurant 1987; Doswald-Beck 1987; O. Dürr 1987; Bring 1987; Best 1980; Walzer 1977]. The most general principle therein is that warfare is legal only against belligerents and proscribed against civilians, neutrals, or former combatants rendered *hors de combat* through surrender, incapacitation, or capture. A concrete application of this criterion has been sought in a number of conventions and treaties. Of special relevance for our present topic are the following:

- The Hague Conventions of 1899 and 1907, e.g., protected neutrals, including their shipping, and established the principles of *open towns.
- The Geneva ("Red Cross") Convention of 1949 facilitated the establishment, by special arrangement, of "neutralized *zones*" within individual states that would be exempt from (legal) attack. Furthermore, it regulated the rights and obligations of occupants relative to the indigenous population. And, it forbade reprisals as a means of warfare.
- The Additional Protocols of 1977 further improved the opportunities for *neutralization* of zones through unilateral declaration. It also provided a certain legal protection for participants in "internal warfare," i.e., insurgents and guerrillas, and it shifted the burden of proof from the defender to the offender, by defining civilians as people who are not combatants (rather than the other way around).

4. **Forbidden Targets.** In addition to the prohibitions on war against civilians, there are certain restrictions on war against material objects. Places of religious worship, cultural objects, and the like are "protected" even better than most human beings. Because the environment is already endangered and would be further devastated through war, it has been accorded some protection by the LOW. The natural environment is hence not a legitimate target, both by virtue of the explicit prohibition of warfare against the environment in the 1977 EnMod Convention, and implicitly through the general prohibition against indiscriminate warfare [Westing 1978].

5. **Alternative Defense and the LOW.** Both *NOD and *civilian-based defense proponents have occasionally been charged with presupposing an unrealistic degree of respect for the LOW on the part of an aggressor. This is a simplification. Alternative defense schemes have sought, above all, to render violations of the LOW unrewarding, inter alia, by giving an aggressor no opportunities for justifying transgressions as reprisals; hence also the emphasis placed on the defender's respect for the LOW [cf. Löser 1981; Nolte & Nolte 1984].

LEAGUE OF NATIONS.   International organization established in 1919 through the Treaty of Versailles and the Covenant of the LON in the immediate aftermath of *WWI, and with a view to preventing its repetition [Holsti 1991: 175–212; Baker 1923].

The basis for the LON was the ideas of U.S. President Woodrow Wilson, set out in his famous "14 Points" [Foley 1923; Levin 1972; Tucker 1993; cf. W. Johnston 1982; Cronon (ed.) 1965]. They envisioned a *collective security system that would provide effective sanctions (including collective armed intervention) against aggressors. The LON played an important role in the disarmament endeavors of the interwar years, e.g., the unsuccessful 1932 *World Disarmament Conference [Noel-Baker 1979; Borg 1992].

The LON suffered from a number of limitations and flaws, which might explain its obvious lack of success in preventing *WWII* [cf. Carr 1946: 22–62], which has discredited the idea of collective security until the present day:

- The ideological belief in the fundamental harmony of interest (i.e., utopianism or idealism) and denial of real conflicts, which hampered conflict resolution and crisis management.
- The temporary or permanent abstention of a number of great powers, including the *U.S.*, the *Soviet Union*, and Germany.
- The discrepancy between proclaimed intentions and actual capabilities with regard to sanctions and intervention.

LEBOW, RICHARD NED.　See *U.S.*

LEMKE, HANS-DIETER.　Serving Bundeswehr officer, temporarily affiliated with the Foundation for Science and Politics in Ebenhausen.

HDL on several occasions recommended a shift to *NOD* postures as the paramount objective of *arms control*. The goal should be a situation in which both sides would possess armed forces of an exclusively defensive utility, under which preconditions, even numerical disparities, would become rather irrelevant. More specifically, such a situation would be characterized by the following combination, reminiscent of the *SAS* proposal: "The distinguishing features of defensive superiority and strategical attack incapability would thus be a combination of strong main forces, which are capable of only defense and weaker . . . formations, which are also capable of operational and tactical *counteroffensives* within the boundaries of the defense" [1989: 15, 19; cf. 1990; 1990a; 1991; 1992].

LEVITE, ARIEL.　See *Israel.*

LIDDELL HART, BASIL H. (1895–1970).　British captain, strategist, and military historian whose name has been associated with both offensive, armored warfare and with defensive strategies [Bond 1977].

In the 1920s BLH followed the lead of *J. F. C. Fuller* by advocating a "new model army" capable of *blitzkrieg* operations by virtue of its complete mechanization. After *WWII*, BLH even asserted that he had been the main source of inspirations for the German generals (Guderian and others) and sought to substantiate the assertion with interviews that apparently were partly forged, or at least "doctored." His advocacy of armor notwithstanding, BLH consistently stressed the need for supporting *infantry* to a greater extent than did Fuller [Hart 1951; cf. Mearsheimer 1988a: 33–48, 184–201; Bond & Alexander 1986].

In the 1930s BLH recommended British abstention from any commitment to defense of the Continent, thereby providing a military reasoning in favor of the appeasement policy of Prime Minister Chamberlain. He argued this point with reference to the strength of the defensive, which would presumably allow for a fairly easy defense of Britain as well as of states on the Continent. His proposed "limited liability doctrine" was in line with technological trends, whereas a British continental commitment would have required offensive-capable forces that were at a technological disadvantage [Hart 1937; 1939; cf. Gibson 1941; Mearsheimer 1988a: 99–126]. Furthermore, BLH was a firm believer in the validity of the *three-to-one rule*, implying a wide margin of inherent defensive superiority.

As early as 1920 BLH had formulated "contracting-funnel" tactics as an appropriate way for infantry to thwart an attack, e.g., by armored forces. He envisioned a stationary, firepower-intensive *forward defense*, combined with mobile forces in the rear, and suggested luring the attacker into the rear where it would be easier to defeat [Hart 1921; cf. Mearsheimer 1988a: 26–33], a scheme similar to that of certain modern *NOD proponents, such as the *SAS or *Albrecht A. C. von Müller. BLH's advocacy of the defensive also influenced his advice to the British government with regard to the *World Disarmament Conference, namely, to seek qualitative *disarmament in general, and to curtail the production of *tanks in particular [Hart 1932].

Both blitzkrieg and defensive warfare were depicted as instances of the *indirect approach, to which conception BLH devoted a work, published in several editions (and with changing titles) since its first appearance in 1929. In Strategy: The Indirect Approach, BLH developed the concept of grand strategy [1967a: 321–322, 353–360; cf. Kennedy (ed.) 1991] and its relation to *strategy and tactics. He also formulated a set of "military axioms" (comparable to the *principles of war):

(1) Adjust your end to your means; (2) Keep your object always in mind; (3) Choose the line (or course) of least expectation; (4) Exploit the line of least resistance; (5) Take a line of operation which offers alternative objectives; (6) Ensure that both plan and dispositions are flexible—adaptable to circumstances; (7) Do not throw your weight into a stroke whilst your opponent is on guard; (8) Do not renew an attack along the same line (or in the same form) once it has failed. (Abridged from Hart 1967a: 335–336)

BLH was also, according to some analyses, instrumental in the development of nuclear strategy, particularly with his insistence that only *limited war thenceforth made sense, and that deterrence, in order to be credible, had to be "graduated" (pointing in the direction of *flexible response) [Freedman 1989: 97–100; cf. Hart 1960]. Above all, a conventional defense had to be devised that would render nuclear means nearly superfluous, and that would be unsusceptible to nuclear attack: a widely dispersed, in-depth, *militia-like posture [Hart 1960: 89–96, 165–173], in line with what was also proposed by *Ferdinand Otto Miksche.

BLH elaborated on his recommendation for a new kind of conventional defense with his analyses of *guerrilla strategy and even *civilian-based defense [Hart 1934; 1967]. The principles of battle avoidance, *dispersion, and hit-and-run were distinctive of *guerrilla warfare, which had previously been a defensive means but whose offensive utility was growing because of the aforementioned limited-war imperative imposed by nuclear deterrence. Whereas guerrilla warfare tended to be excessively violent and thereby demoralizing, *nonviolent resistance (e.g., *Fabian tactics and the *Good Soldier Schweick approach) promised most of the same advantages without the disadvantages [1967: 207]. BLH expected a full-scale commitment to civilian-based defense to be at most a long-term prospect but nevertheless urged governments to adopt it as a fallback, post-occupation, option [Hart 1967: 210].

The theoretical merits of BLH remain disputed. He is probably the most widely quoted of all military writers since *Clausewitz, yet some of the most prominent modern analysts regard him as an epigone whose contribution to strategic theory has consisted in popularizing Fuller [Dupuy 1980: 323; cf. Bellamy 1990: 305; Mearsheimer 1988a: ix–x & passim]. Nevertheless, he has often been quoted by *NOD proponents in support of their strategic views [e.g., Löser & al. 1982: 46–49; Unterseher 1989b: 157; Møller 1991c]. See also: *Concentration, *Principles of War, *Lawrence of Arabia, *Logistics.

LIGHT INFANTRY.   See *Infantry.

LIMITED WAR.   War wherein one or both sides deliberately refrain from using their entire disposable military power, i.e., seek to keep *escalation* within bounds. LW is not tantamount to a small or "harmless" war but is merely war that is less than total. A meaningful distinction is between classical LW, limited nuclear war, and limited subnuclear war.

1. **Limited Classical War.** Nearly all wars in the prenuclear era were limited in the sense that certain *thresholds* were observed: as a general rule, war was not waged against the civilian population, the use of poison was rare, etc. Part of the explanation for this may have been fear of retribution on the part of the victims, but cultural norms (inter alia, the Roman-Christian *just war* tradition) undoubtedly also played a restraining role [cf. J. T. Johnson 1981].

2. **Limited Nuclear War.** Whether a limited nuclear war is a possibility remains disputed, to say the least. Optimists have drawn parallels to the limitation of all other types of war and have pointed to the obvious incentive for both sides to avoid escalation to suicidal levels [Martin 1979; Kissinger 1969]. Pessimists have advised against the belief that war limitation would remain possible, once the nuclear threshold had been crossed [I. Clark 1982: 203–237; Brodie 1959: 305–357; Jervis 1984: 109–111]. First, there would be tremendous technical difficulties with maintaining the required degree of centralized political control after a breakdown of lines of communication [cf. Bracken 1983; Blair 1985; Slocombe 1987; Wohlstetter & Brody 1987], and subsequent to the unavoidable subdelegation of launch authority, to field commanders facing "use-them-or-lose-them" dilemmas [Kelleher 1987; 1988; Bracken 1987a]. Second, limitation of a nuclear war would presuppose tacit understandings between the contestants on the need for preventing escalation *in extremis*. It would further require obvious thresholds at which to halt, as well as a high (and probably unrealistic) degree of rationality on the part of the decisionmakers, who would inevitably be under severe stress and immense time pressure.

Even if it turned out to be possible to keep escalation within bounds, the damage caused by such a limited nuclear war, in the course of which thousands of tactical and short-range nuclear weapons would have been detonated on the European battlefield, would have been immense [Weizsäcker (ed.) 1971]. A 1955 *NATO* "map exercise" implied a dropping of 335 nuclear bombs in Europe, causing 1.5–1.7 million fatalities and 5 million wounded [Stromseth 1988: 18, 23; Mechtersheimer & Barth 1986: 111]. That such results would appear far from limited to the Europeans, and especially not to the Germans, explains why NATO has formally dismissed the notion of limited nuclear war. Nevertheless, it can be argued that the *flexible response* strategy presupposes precisely such a limitation because it would otherwise make no sense to envision the use of tactical nuclear weapons as an intermediary stage between direct conventional defense and all-out nuclear retaliation.

3. **Limited Subnuclear War.** The very fact that war may escalate to suicidal proportions is bound to exert a restraining influence on all belligerents, even below the nuclear threshold. Because of *existential deterrence*, it would, for instance, no longer be prudent to wage intense conventional countervalue warfare because of the risk of nuclear retaliation. Hence the inevitably limited character of war in a *subnuclear setting*, for which *NOD* proposals have usually been conceived [cf. Boserup 1986a; Joxe 1984]. See also *Totally Destructive Conventional War*, *Damage Limitation*.

LINEAR EQUATION.   See *Lanchester's Laws*.

LITHUANIA.   See *Baltic Region*.

LODGAARD, SVERRE.   Former researcher at *SIPRI*, subsequently director of the Peace Research Institute Oslo (PRIO) in *Norway*, since 1993 director of *UNI-DIR*. SL has for many years been a proponent of various alternative defense conceptions:

- *Common security*, politically as well as militarily [1984; 1985; 1985a];
- *No-first-use* of nuclear weapons [Blackaby & al. 1984; 1985a];
- *Nuclear-weapons-free zones*, both in the Nordic region and in central Europe [1982; 1983; 1988; idem & Berg 1983];
- Various forms of conventional *disengagement*, including a withdrawal of battle *tanks*, artillery, and other offensive-capable equipment from the line of contact between the blocs [idem & Berg 1983; 1985];
- Various *CBM*s and *CSBM*s, intended primarily to enhance *crisis stability* through deployment and exercise constraints [1986; 1987; 1989; 1991]; and
- *Maritime CBM*s and *naval arms control* [1989a–b; idem & Holdren 1990].

LOGISTICS.   Rear services that supply the weapons, ammunition, fuel, support equipment, and other goods needed for keeping the armed forces in fighting condition [Creveld 1977; 1991; J. Thompson 1991]. L are of importance for alternative defense and *NOD* in three respects.

- Through modifications of L, otherwise offensive-capable forces can be made strictly defensive, i.e., incapable of operating on enemy territory yet mobile on friendly territory. This has, e.g., been the guiding principle of the *spider-and-web* model of the *SAS*, which envisions a replacement of long-range transport capabilities with dispersed depots on the defended territory [SAS 1984; 1987; Unterseher 1984a; 1988b; 1988d; 1989b–d; 1990c; Borkenhagen 1984; 1984a].
- For Western Europe to have been able to fight a protracted war of defense, as envisioned by many NOD proponents, might have required supplies from the *U.S.*, hence some (preferably defensive) measures of *SLOC* (sea lines of communications) defense would have been essential [Møller 1987; 1989e; 1992: 187–194; cf. Grove 1990a; 29–39; 1990b; Mearsheimer 1986].
- NOD strategists have placed great emphasis on attacking an invader's logistic chains, as a partial alternative to engaging its forces directly. Accordingly, they have been following *Liddell Hart*, who advocated such tactics as an instance of his *indirect approach*: "To cut an army's lines of communication is to paralyse its physical organization" [Hart 1967a: 183].

LONDON DECLARATION.   Document issued by the North Atlantic Council 5–6 July 1990 [in Rotfeld & Stützle (eds.) 1991: 150–152], in which *NATO*'s heads of state and government reiterated previous assertions of the alliance's defensive intentions and further emphasized the *détente* and cooperation track of the 1967 Harmel Report: "The Atlantic Community must reach out to the countries of the East which were our adversaries in the Cold War, and extend to them the hand of friendship. . . . We will never in any circumstance be the first to use force." It was further proposed to sign a solemn statement to the effect that "we are no longer adversaries and reaffirm our intention to refrain from the threat or use of force."

With regard to military matters, NATO declared its readiness to intensify military contacts, including a sequel to the *Vienna Seminar* on military doctrine. NATO additionally proposed a sequel to the *CFE*, i.e., a "CFE-1A," on limitations on manpower in Europe, as well as a continuation of the conventional *arms con-

trol process, with the aim of further limiting offensive capabilities. Finally, the LD provided broad guidelines for a reform of NATO strategy, based on "smaller and restructured active forces" that would be highly mobile and versatile and partly multinational. The general state of readiness would be reduced, implying a heavier reliance on *mobilization*. Because of improved stability on the conventional level, the need for "sub-strategic nuclear systems of the shortest range" would diminish and negotiations on their reduction could commence. Thenceforth, nuclear forces would be "truly weapons of last resort" (see *no-first-use*, *no-early-first-use*). A strategic review was completed for the 1991 summit in Rome (see *Rome Declaration*).

LONDON NAVAL CONFERENCES.   Largely unsuccessful follow-up negotiations (1930 and 1935–1936) to the 1922 *Washington Naval Conference* [Kaufman 1990: 113–192; cf. Bull 1973; R. Hoover 1980; E. Müller 1986a].

For the 1930 LNC, the *U.S.* (under President Herbert Hoover) sought limitation of especially *offensive weapons* systems, a quest that was unsuccessful. First, it proved impossible to define "offensive weapons," and second, the major powers (the U.S., the *U.K.*, *France*, *Italy*, Germany, and *Japan*) were already engaged in a naval arms race. The only result was a U.S.–British–Japanese treaty defining ratios of certain types of vessels (especially cruisers) and prohibiting unrestricted submarine warfare.

The 1935–1936 negotiations were even more disappointing. Because both Germany and Japan were by now clandestinely preparing for war, they obstructed negotiations with entirely propagandistic initiatives, such as a Japanese proposal for a complete abolition of *aircraft carriers* and battleships [Kaufman 1990: 163]. Further, the U.K. preempted a treaty by signing a bilateral naval treaty with Germany in 1935, allowing for a German naval buildup. Essentially nobody really wanted an agreement; the negotiations were simply discontinued in 1936, when the 1930 treaty expired.

LÖSER, JOCHEN.   From 1936 to 1945 JL served as an officer of the Wehrmacht, and from 1956 to 1974 in the Bundeswehr, retiring with the rank of major-general. From 1977 to 1979 he collaborated with *Horst Afheldt* at the Max Planck Institute. Even though his principal works have been written in collaboration with others, JL's part in them is fairly clear.

JL's main publication was *Neither Red nor Dead: To Survive Without Nuclear War: An Alternative*, in which his particular *NOD* model was described [1981], and it has been only slightly modified since. The main purpose of reform was *damage limitation*, i.e., to devise a viable nonsuicidal form of defense, which implied, inter alia, abolition of battlefield nuclear weapons (combined with a *no-first-use* strategy) and abstention from a military defense of conurbations. Furthermore, *civil defense* would constitute an integral element of the country's *total defense*.

The military element would be an *area-covering defense* (*Raumdeckende Verteidigung*), constituting a mixture between an *area defense* and *forward defense*, and consisting of three *zones* with different levels of readiness: a frontier defense (almost exclusively consisting of standing forces); a (cadred) area-covering defense *web*; and an (even more skeletonized) rear area. The defense web would consist of light as well as heavier *infantry* (*Jäger*) formations, equipped above all with antitank and *air defense* weapons but also with some rocket launchers or mortars, short-range missiles, light armored vehicles, and antitank helicopters. The forces would be deployed in-depth, albeit with considerably greater mobility than envisioned by Afheldt. The "shield forces," furthermore, would be combined with "sword forces" capable of *counteroffensives* and comprising, e.g., *tanks* (see *Sword-and-Shield*). To minimize offensive capabilities, the forces would operate primarily in the rear zone, be multinational, and cadred to a large extent. In further contrast to Afheldt, JL also envisioned maintenance of the air force, assigned primarily to deep interdiction missions. His defense scheme would presumably lend itself to piecemeal implementation, e.g., in a manner resembling that envisioned by *gradualism*: through a process of *disengagement*, both sides would gradually withdraw offensive-capable forces from the frontier.

After publication of his original model, JL added little to its military essentials but elaborated on its ramifications, occasionally with rather odd results. He sought to make his proposal (appear) compatible with the U.S. Army's *AirLand Battle* doctrine, as well as with *NATO*'s *FOFA* doctrine and the Reagan administration's *SDI* [Löser & Anderson 1984: 45, 55; Löser 1985: 525–526, 531; 1986: 530; cf. for a later change of mind: 1987a: 59]. JL also took his disengagement proposal to the point of advocating *neutralization* of central Europe, which would presumably be facilitated by a shift to NOD on the part of the European NATO states [Löser & Schilling 1984, 145–166; Löser 1985a]. The two Germanys would become neutral, "along the Austrian example" and, in the fullness of time, form the nucleus of a central European "zone of peace": a confederation encompassing the two Germanys, the Benelux countries, Czechoslovakia, *Poland*, *Hungary*, *Austria*, Romania, and *Yugoslavia* [Löser & Schilling 1984: 24–37; 142; cf. Löser 1985a: 301–302; 1987; 1990: 142].

LOW-ALTITUDE FLYING.   See *Herman Hagena*.

LÜBKEMEIER, ECKHARD.   See *Friedrich Ebert Stiftung*.

LUTTWAK, EDWARD.   Senior fellow at the Center for Strategic and International Studies, and one of the most innovative strategic thinkers in the *U.S.*

Having dealt extensively with defense management [1984; 1986: 86–115], grand strategy (illustrated with the examples of the Roman and Soviet Empires) [1976; 1983], and operational art [1980] (cf. *military science*), EL in 1987 devoted a volume to a systematic analysis of strategy, constituting undoubtedly one the main works of the 1980s on the subject: *Strategy: The Logic of War and Peace* [1987; cf. 1986].

Strategy contained one of the most competent critiques of *NOD* in general and the *guerrilla*-like, in-depth *territorial defense* schemes of *Horst Afheldt* and others in particular (including EL's collaborator *Steven Canby*) [1987: 126–140; cf. 1986: 116–120. For an attempted rebuttal: Prins 1988a]. EL was far from blind to the tactical strength of such dispersed, agile forces (e.g., operating from protected "islands"); they would indeed present the stipulated Soviet invaders with enormous difficulties because they would "circumvent the Soviet Army's greatest strength, its ability to break through solid fronts, while exploiting its weakest point, its lack of small-unit flexibility" [Luttwak 1987: 130; cf. 1983: 103]. The weakness of NOD, according to EL, consisted in its lack of response to the obvious Soviet countermove: "to advance as quickly as possible in fewer, more concentrated columns aimed straight at the largest German cities . . . in order to induce the West German government to ask for an armistice" [1987: 137]. Although in principle it would be possible (indeed militarily advantageous) to prepare for urban warfare, EL (undoubtedly correctly) assessed as highly improbable that any German government would execute such a strategy for fear of its devastating effects on the civilian population.

LUTZ, DIETER S.   Director of the *IFSH* since the retirement of *Egon Bahr* in 1994. An extremely prolific writer, DSL has contributed to the alternative defense debate in several fields, albeit only rarely with genuinely new ideas. His main contribution has been in the conceptual field, linking the concepts of *NOD* and *common security* (CS) to other central security political concepts.

One of DSL's analytical points of departure was confidence-building, e.g., with a view to enhanced *crisis stability*. He recommended broadening the agenda for *CBM* negotiations (under the auspices of the *CSCE* and the *CDE*) to include restrictive CBMs, i.e., what are usually referred to as *CSBM*s, in which category he mentioned, e.g., *buffer zones* and defensive strategies [1981; 1982a].

Gradually, DSL came to concentrate on CS. Leaving the innovative thinking to other members of the IFSH (noteworthily, *Reinhard Mutz*), DSL surveyed the history and political status of the concept and linked it to other concepts. E.g., he found CS to be incompatible with military *alliances* as well as with the "deterrence system." Hence, the only legitimate form of *dissuasion* was presumably *deterrence by denial*, particularly in the shape of a *structural inability to launch an attack* (*StruNA*, i.e., NOD) [1986; 1986b; 1986d; 1986f; idem & Theilmann 1986].

In addition to a terminological innovation of questionable importance (the substitution of *StrUnA* for StruNA), DSL suggested broadening the concept to take into account a state's inability to attack for whatever reason, including a democratic form of government, i.e., "StruNa in the wider sense." Furthermore, he rejected the more permissive definitions of "offensiveness" (*Hannig*, *Erwin Müller* & al.) and upheld the demand for a complete "prohibition of cross-border capabilities for launching an attack," including aircraft or missile strikes against enemy territory [1987b–c; 1988; 1989]. On the basis of this "puritan" NOD definition, DSL published a study on aircraft, *Everything That Flies Must Go* [1989e; cf. 1989c; 1990d]. In it he advocated reductions of strategic and medium-range bombers as well as the eventual scrapping of all fighter-bombers in the entire *ATTU* area, along with prohibition of tactical aircraft for ground-attack capabilities (i.e., ordinary CAS: conventional air support). True to his general advocacy of withdrawal of all kinds of offensive-capable forces from forward positions [1988a; 1988c], he also recommended a rearward relocation of the remaining air fleets so as to create a 100-km-wide "no-aircraft corridor." In place of all these systems, DSL suggested the use of *SAM*s and other surface-to-air means.

DSL has been a longtime advocate of *collective security*, which he conceived of as an integral part of, as well as a preparatory step toward, what he has called the "New European Peace Order," a sequel to a "CS regime." Contrary to the latter, it would also entail provisions for collaboration in times of war, inter alia, in the form of collective sanctions. On the other hand, collective security would "merely" secure "negative *peace*" in the sense of an absence of war; the more ambitious peace order would also advance "positive peace" [1985; 1986; 1986b; 1986d; cf. idem (ed.) 1985; 1991].

Since the 1989 revolution and in light of the German unification debate, DSL has reiterated his proposals for collective security, explicating its institutional foundations (not unlike those created at the Paris *CSCE* summit): a "European Security Council"; a permanent secretariat; a general assembly; various institutions for peaceful conflict resolution; a multitude of mechanisms for mutual consultation, etc. Furthermore, while envisioning supranational armed forces to supersede national ones eventually, for the transitionary period DSL recommended a combination of the two, with the stipulation that the national armed forces should be strictly defensive. In a Europe with a grand total of 700,000 troops, no more than 105,000 (unmistakably defensive) German troops would be justified [Lutz 1989d; 1990: 42–45, 58–59; 1990a; 1990b; 1990c; 1993].

M

MACK, ANDREW. See *Australia.

MAGINOT LINE. French system of fixed, garrisoned fortifications along the northeastern border of *France, constructed in the interwar years under the direction of Minister of Defense André Maginot. The logic behind it was that a second world war would resemble *WWI, and that the defense would prevail, i.e., that the costs of breaking through fixed defenses would be prohibitive (as was the case with the WWI trenches).

The ML was a success in the sense that it was never overrun or penetrated. In another sense, it was a failure, first because it was never completed (i.e., taken all the way to the sea), and second, because of the accompanying neglect of mobile forces, which left Germany with the option of circumvention [Gibson 1941; Posen 1984b: 105-140; Bond & Alexander 1986; Bellamy 1990: 44–45; Pedroncini 1991]. An explanation of this failure has been offered: the paradox that the French did not trust the ML enough and "behaved as if it was not there," i.e., deployed the bulk of their forces in defense of a line that did not need it, thereby depleting the defense of the rear and facilitating a German *blitzkrieg [Dupuy 1990: 79–80].

"ML" and "ML mentality" [e.g., Keegan 1989: 60] have ever since been pejorative terms, used, inter alia, in critiques of *NOD, as well as of positional and *forward defense in general. The analogy is only superficial. A few NOD schemes [e.g. Afheldt 1983; or Hannig 1984] have resembled the ML in the sense of being stationary and positional (and most have shared with the ML the emphasis on *barriers and fortifications), but the more sophisticated NOD schemes have combined such stationary forces with mobile forces [SAS (eds.) 1984; 1989].

MANEUVER WARFARE. See *Land Forces.

MAO TSE-TUNG. See *Mao Zedong.

MAO ZEDONG. Chinese communist leader, guerrilla captain, and subsequently leader of *China until his death in 1974. In his theoretical writings on (as well as in his practice of) armed struggle M was at least as inspired by *Sun Tzu as by the Marxist classics. According to certain analysts, M is thus merely one exponent of "the Asian way of warfare," following the same strategy as employed during the *Taiping Rebellion [Bellamy 1990: 191–237; cf. Hsü 1975: 277–324], and as later used by guerrillas during the *Vietnam War.

1. **Background.** M's military strategy was developed in the context of the struggle against the Japanese invaders and the Kuomintang. After the breakdown of a temporary alliance with the Kuomintang, the Communists established themselves in Kiangsi Province (1928), whence they were forced to flee in 1934 by successive encirclement campaigns. After the famous Long March (1934–1935) [Snow 1972; Wilson 1977; Ch'ên 1967: 160-200], they established themselves in an extensive

base area at Yenan in Shensi Province. The Japanese aggression spurred yet another alliance with the Kuomintang, in the course of which the Communist guerrillas gradually surpassed the Nationalists. After the eventual defeat of *Japan* in 1945 (to which the Communists contributed substantially, along with the *U.S.* and the *Soviet Union*), they launched a large-scale offensive against the Kuomintang, with some Soviet assistance, e.g., in the form of weapons conquered from the Japanese occupants of Manchuria [Belden 1973; cf. Guillermaz 1972: 375–428]. Through this (relatively conventional) offensive the former guerrillas succeeded in evicting the nationalists to Taiwan and proceeded to establish the People's Republic of China in 1949.

After 1949 the defense policy of *China* was heavily influenced by the Maoist legacy, with Lin Piao standing out as the prospective theoretical (and power-political) heir to M, particularly during the Cultural Revolution. The Maoist idea of "encircling the cities" was elevated to a Chinese grand strategy (with presumed validity for the entire Third World), according to which the "world villages," i.e., the Third World, would gradually encircle and defeat the "world cities," i.e., the Western countries. Only after the fall of Lin Piao in 1971 and the subsequent ascendancy of Deng Xiao-ping, was Chinese defense policy "normalized" [Lin Piao 1965; cf. Baylis 1987a; Dellios 1989].

2. **Strategic Thought.** M's strategic thinking was elaborated in a number of quite sophisticated theoretical treatises written during the Yenan period [Mao Zedong 1936, 1938, 1938a, 1938b†]. His grand strategy was a revolutionary war against imperialism, which would automatically be a *just war* [1936: 183]. Concerning general strategy, M recommended protracted war, tantamount to an emphasis on piecemeal *attrition*. Nevertheless, he strongly advised against the positional battles most often associated with attrition warfare: "Oppose protracted campaigns and a strategy of quick decision, and uphold a strategy of protracted war and campaigns of quick decision. Oppose fixed battle lines and positional warfare, and favor fluid battle lines and mobile warfare" [1936: 199–200].

The Red Army should not strive to surpass its adversaries in terms of armaments and the like but capitalize on its inherent advantages, primarily in the realm of politics (i.e., adopt an asymmetrical or *indirect approach*). In a protracted war, the security of the rear is of paramount importance, hence the emphasis placed on base areas and on the collaboration of the local inhabitants. It might, furthermore, be advantageous to trade space for time by means of a strategic retreat, i.e., by "luring the enemy in the deep," as had happened (according to the official interpretation at least) through the Long March [1936: 215].

Despite its strategically defensive stand, the Red Army should adopt a mixture of defensive and offensive tactical postures [1938: 83–91] because the only defense worthy of the name is an active or "offensive defense" that constitutes merely a preparation for the *counteroffensive*; "*defensive defense*" is a spurious form of defense [1936: 207]. In such "offensive" operations, the aim should be annihilation of the enemy's forces, for which purpose an overwhelming superiority should be sought prior to any engagement [1936: 231, 238, 241], in addition to the advantage

---

†Mao's early writings were substantially revised for *Selected Works* in light of subsequent ideological developments, as demonstrated by Stuart Schram, *Mao Tse-tung* (Harmondsworth: Penguin, 1967), and idem, *The Political Thought of Mao Tse-tung* (same publisher). For our present purposes, however, none of the revisions seem to affect the gist of the argument, and the 86 million copies sold of *Selected Works* by 1968 [Snow 1972: 596] imply that they at least reflect the Chinese image of Mao.

of surprise. If such a favorable force ratio should prove unobtainable, a strategic retreat should be preferred to a risky battle [1936: 211]. The tactics of guerrillas advocated by M is very reminiscent of Sun Tzu. The guidelines were summed up in the famous "Sixteen Word Formula": "The enemy advances, we retreat; the enemy camps, we harass; the enemy tires, we attack; the enemy retreats, we pursue" [1936: 213].

M's writings and practice have inspired many Western NOD proponents, among whom are *Guy Brosollet* [1975], *Emilio Spannocchi* [1976], *Jochen Löser* [Löser & al. 1982: 207–218], *Alain Joxe* [1984; 1991], and their theoretical ancestors, such as *André Glucksman* [1967].

MARITIME C(S)BMS.   Confidence-building (*CBMs*) and confidence-and-security-building (*CSBMs*) measures applicable to naval forces. Because of *NATO*'s opposition, no genuine MC(S)BMs are in force, nor have negotiations been conducted on the subject.

1. **Taxonomy.** Just as is the case for the CBMs and CSBMs in force already, one might distinguish between transparency-enhancing (CBMs) and constraining (CSBMs) measures. MCBMs would merely be intended to improve the predictability of maritime operations, thus to prevent potentially dangerous misunderstandings. MCSBMs would constrain maritime activities, in the form of prohibition against operations in certain areas, in particular size and/or composition, or at special times. The purpose would be to create the basis for trust, e.g., by removing potentially offensive options.

2. **Approximations to MC(S)BMs.** The absence of MC(S)BMs aside, various regulations are in force that resemble them in various respects.

- In the mandate for the Stockholm *CDE* negotiations, it was stated that CBMs pertaining to "adjoining sea areas" were applicable to military activities "whenever these activities affect security in Europe as well as constitute a part of activities taking place within the whole of Europe" [Brauch (ed.) 1987: 337]. These modest goals notwithstanding, it proved impossible to include in the CDE final document more than a requirement for prior notification of, and invitation of observers to, activities such as major amphibious landings [Brauch 1987a; Ghebaldi 1989; Hill 1989: 70–71].
- The 1972 Prevention of Incidents at Sea agreement between the *U.S.* and the *Soviet Union* made peacetime naval practices somewhat less confrontational and provided a forum for regular consultations between naval professionals. Similar agreements were subsequently signed between the USSR and other Western naval powers [cf. Lynn-Jones 1988; 1990; Prawitz 1990].

3. **Soviet Proposals.** MC(S)BMs were a long-standing Soviet objective, but the USSR had to acquiesce in the Western recalcitrance concerning the topic in the CSCE process. It tabled its intention to raise the question of independent naval exercises at a later stage [in Brauch (ed.) 1987: 447–449]. Subsequently, the USSR included suggestions for MC(S)BMs, e.g., in the *Vladivostok* and *Murmansk Initiatives*, inter alia, proposals for prior notification of, and invitation of observers to, naval maneuvers, and mutual reductions of military activities in general and *ASW* (antisubmarine warfare) activities in particular [Gorbachev 1987; cf. Brundtland 1988; 1989a; *Tass* 8 June 1988; cf. Möttölä 1990; Purver 1988; 1988a; 1990].

4. **Relevant MC(S)BMs.** Independent analysts from the *NOD* community have also made a number of concrete proposals for MC(S)BMs [Borawski 1988a;

Grove 1988: 4–9; Børresen 1988; Haesken & al. 1988; Purver 1988; 1988a; Wiberg 1990; P. Howard 1990], among them the following:

- Simply extending terrestrial CBMs and CSBMs to the seas, i.e., invite observers to, and give advance notification of, major naval exercises.
- A "routinization" of naval activities, intended to avoid ambiguous shifts from normal patterns to intensified crisis patterns [Grove 1988]. However, although this might well improve *crisis stability*, it might, on the other hand, damage peacetime stability, say, if it were to imply a permanent peacetime deployment of offensive-capable warships in sensible areas.
- Limiting the total number of vessels participating in individual exercises; in major naval maneuvers in the proximity of the respective adversary; in exercises including amphibious operations or land-attack weaponry, etc. Naval exercises might also be made somewhat less provocative by being deprived of their combined-arms character. Naval aviation and surface warships might, e.g., be obliged to exercise separately rather than jointly because either type would be considerably less offensive in isolation than if supported by the other.
- Various *disengagement* schemes, e.g., pertaining to submarines and especially with a view to stabilizing deterrence by constraining strategic ASW. Most realistic might be some combinations of "stand-off" and/or "keep-out" *zones* for attack submarines. As long as the USSR (or Russia) adhered to a *bastion* strategy (i.e., maintained its strategic submarines in secluded seas), such seas might be acknowledged as "keep-out zones" for U.S. attack submarines. As a quid pro quo, the southern parts of the North Atlantic might be declared a "keep-out zone" for Soviet attack submarines. Furthermore, a "stand-off zone" for Soviet strategic submarines might be established off the U.S. coasts [Purver 1990; Møller 1987; 1989b].

MARITIME STRATEGY.   Also known as Forward MS, naval strategy of the U.S. Navy since the early 1980s [Watkins 1986; Kelley 1986; Brooks 1986; Friedman 1988].

The MS envisioned "taking the war to the enemy," i.e., threatening the Soviet naval forces in their home waters, above all the Barents Sea. On closer analysis, it was revealed as a rather amorphous bundle of strategic conceptions, intended also as a rationale for a naval expansion, i.e., a six-hundred-ship navy. Common to all of these conceptions was an emphasis on the offensive [Mearsheimer 1986; Tunander 1989]:

- Horizontal *escalation*, i.e., a strategy for the deliberate expansion of an initially localized conflict by retaliation in other spots of the globe.
- "Direct military impact" of sea-based forces on land battle in Europe, by means of forces such as the Marine Corps, carrier-based aircraft, and SLCMs.
- Strategic *ASW*, i.e., plans for hunting Soviet SSBNs in their *bastions* from the very start (i.e., during the conventional phase) of a future war.
- Counterforce coercion, i.e., nuclear war fighting.
- Protection of *NATO*'s transatlantic *SLOC*s (sea lines of communication) in a "close blockade mode."

The MS was criticized for its negative impact on *crisis stability* because it would give the USSR an incentive for a surge deployment of its naval forces during a crisis. It was, furthermore, particularly unwelcome from the point of view of the Nordic region because it tended to move these countries upward on the Soviet target list. However, the primary Soviet response to the development of the MS was a multitude of *arms control* initiatives, inter alia, the *Murmansk* and *Vladivostok Initiatives* [Gorbachev in *Tass,* 1 October 1987; idem 1987:133–148; Shaskolsky 1989; Churilin 1990; cf. Brundtland 1989a; Archer 1989].

The MS has been legitimized by the need for protection of NATO's SLOCs. An alternative means of SLOC protection was thus a precondition for abandoning the MS, lest NATO-Europe should be prepared to do without U.S. assistance in the event of a large-scale war, which would nearly be tantamount to disbanding NATO. Only very few *NOD* proponents have taken this problem seriously and sought to devise viable alternative means of SLOC defense [Møller 1987; 1989b; 1989e; cf. Mearsheimer 1986; Grove 1987; 1990a–b; Miller 1990a; Pott 1986]. Most have merely approached the problem with suggestions for *maritime* CBMs and similar modest measures [Arkin 1990; Børresen 1988; Fieldhouse (ed.) 1990; Grove 1988; Lodgaard 1989a–b; idem (ed.) 1990; Mack 1990; 1990d; Purver 1988; 1988a; 1990; Miller 1990; Wiberg 1990].

MARTINEZ-PUJALTE, ANTONIO-LUIS.   See *Spain.*

MAZING, VALERI.   See *ISKAN.*

MBFR.   Negotiations on mutual balanced force reductions in Europe, formally called MURFAAMCE (Mutual Reductions of Forces and Armaments and Associated Measures in Central Europe) [Dean 1985a; 1987: 153–184; Blacker 1988; Hopmann 1987; E. F. Jung 1988; M. Müller 1988a; Rühl 1982; A. Carter 1989: 230-257].

The MBFR were, in a certain sense, a quid pro quo for the *CSCE,* i.e., a Soviet concession to *NATO* in reciprocation of NATO's acceptance of the Soviet proposal for a CSCE. Besides NATO's official objective for the negotiations (to reduce the Soviet conventional superiority in Europe) the protracted process of negotiations (1973–1989) also served ulterior motives, namely, to preclude a unilateral (partial) U.S. withdrawal of forces from Europe. Regarding the *Soviet Union,* it may never really have desired an agreement. Apart from the lack of political support, there were other pitfalls in the MBFR as an approach to conventional *arms control*:

- The geographical scope (central Europe) was so narrow as to make little military sense.
- The focus was on active-duty personnel rather than on equipment.
- The data issue was not settled before the negotiations commenced, and it consumed considerable attention for most of the talks' duration. In addition to the (likely) USSR's underestimation of its troops levels, there were various objective counting difficulties involved, stemming, inter alia, from the different composition of forces.
- The proclaimed goals (equal ceilings for both sides of 700,000 ground force personnel out of a grand total of 900,000 troops) were so modest and insignificant as to attract very limited political attention.

Having accomplished no tangible results after fourteen years, the MBFR were formally concluded 2 February 1989, only a few weeks after agreement on the mandate for their successor, the *CFE* negotiations [*Atlantic News,* 10 February 1989].

MCCGWIRE, MICHAEL.   British naval analyst, subsequently senior research fellow at the Brookings Institution (Washington, D.C., *U.S.*), and upon retirement affiliated with the Global Security Programme at Cambridge University in the *U.K.*

MM's main field of research has been the military policy of the *Soviet Union.* He consistently pleaded for more realistic threat assessments that would also take intentions into account, i.e., empathize with the Soviet leaders. According to MM's reconstruction, their "hierarchy of objectives" included, as the first priority, the survival and well-being of the Soviet state. The second-order priorities were to avoid war as the primary goal but not to lose a war should it prove unavoidable. The main determinant of Soviet military planning was thus the assessment of the likelihood of war, which varied over time. The hierarchy was, furthermore, fraught with an intrinsic contradiction: the measures intended to avoid losing war (such as the offensive military strategy) might well make it more likely to occur [1987b: 40 & passim].

Against this background MM concluded, rather pessimistically, in 1987 that "NATO has no alternative to living with Soviet forces that are structured for offensive operations against the West" [1987b: 376]. Even so, the West was in a position to make Soviet leaders regard war as less likely, thereby reducing the actual threat: through *arms control* and/or defensive restructuring, and above all, by abandoning the notion of deterrence. What mattered most was not to deter the Soviet Union from a war it was eager to avoid anyway but to convince it that war was in fact avoidable [1984; 1986]. Taking into account the *Gorbachev* revolution, MM later arrived at the conclusion that the USSR had unilaterally abandoned previous fears of war, hence elevated war prevention to an unchallenged first priority. This explained why it now felt comfortable with a defensive posture vis-à-vis the West, which paved the way for an assertive arms control "offensive" [1987; 1987a; 1988a–b; 1991; 1991a]. MM urged the West to exploit this unique opportunity by proceeding to build a "mutual security regime" [1988; 1989; 1991: 413-415].

MCNAMARA, ROBERT S.   See *U.S.*

MEARSHEIMER, JOHN J.   Professor of political science at the University of Chicago.

One of the main preoccupations of JJM has been various aspects of conventional defense and deterrence. He has been a stern critic of the trends throughout the 1980s toward maneuver warfare, such as the *AirLand Battle* doctrine and *Samuel Huntington*'s proposals for *conventional retaliation* [1981; 1983a]. As an alternative, he has suggested a greater emphasis on *PGM*-type weapons for antitank defenses [1979], which might allow for quite an effective NATO *forward defense* against a Soviet attack. The main reason "Why the Soviets Can't Win Quickly in Central Europe" (the title of an article from 1982) was, however, the *three-to-one rule*, in the validity of which JJM has been a firm believer, even in the face of very competent and congenial critics [1988; 1989; idem & al. 1989; cf. for a critique: Epstein 1989].

Though he did not lack faith in the strength of conventional defenses, JJM advised against adoption of a *no-first-use* policy, which would mean that conventional defense should be measured against much higher standards [Mearsheimer 1986b; 1989a]. He was unpersuaded by the logic of *existential deterrence*, according to which a *subnuclear setting* was a given, almost regardless of who deployed how many nuclear weapons where and against whom. For deterrence to exert its benign effects, according to JJM, nuclear weapons had to be actually available, i.e., deployed. This reasoning led him in 1990 to advocate a German nuclear capability as a contribution to stability in post–Cold War Europe [1990]. The proposal was criticized widely [Russett, Risse-Kappen, & Mearsheimer 1990], but in 1993 he came out with similar argumentation in favor of Ukrainian retention of nuclear power [Mearsheimer 1993; cf. for a critique: Miller 1993].

Given the existence of nuclear weapons and a credible first-use option, however, war prevention could largely rest on conventional means. JJM demonstrated this in *Conventional Deterrence* [1983], where he analyzed historical instances of deterrence failures, finding them all to have been caused by the aggressor's belief in the possibility of achieving a swift and decisive victory.

Paradoxically, JJM's focus on land battle in central Europe was also what in 1986 motivated him to venture a critical analysis of the *Maritime Strategy* of the U.S. Navy. He critically scrutinized the official rationales for threatening the Soviet SSBN *bastions*, including that of thereby diverting forces from the central land battle, and found their credibility to be limited. The only genuine naval impact on the all-important land battle would be in the form of supplies, necessitating a naval emphasis on *SLOC* defense. This did not, however, presuppose offensive naval strategies but could be insured by seeking defensive sea control south of the GIN (Greenland-Iceland-Norway) perimeter [1986; cf. Møller 1987].

MECHTERSHEIMER, ALFRED.   Colonel in the Bundeswehr (ret.) and former member of the Bundestag for the *CDU. Under the influence of the *peace movements'* campaign against *INF* deployment, AM resigned from the armed forces and was expelled from the CDU, after which he joined the Green Party (cf. *Die Grünen*). He also founded and became the director of a private peace research institute in Starnberg.

AM's publications have primarily been critical analyses of *NATO* strategy in general and nuclear strategy in particular [Mechtersheimer & Barth (eds.) 1983], but he has also proposed several alternatives. First, a certain *Europeanization*, albeit more in the sense of independence from the *U.S. than of actual European defense cooperation. Second, a radical democratization of the military *alliances* as well as of defense decisionmaking. Third, as steps toward the long-term goal of demilitarization of Europe, he supported a shift to *NOD, albeit rather reservedly and accompanied by a warning to the peace movement against the lure of "technocratic panaceas." Furthermore, he recommended "not to waste too much energy on future defense conceptions, but first pave the way for their political implementation." AM nevertheless lent his support to a reform of the Bundeswehr, which should provide for area coverage combined with a modicum of *forward defense* and possibilities of reinforcements, and which should be based on locally conscripted forces. The troops of such a Bundeswehr should be equipped with, e.g., light antitank and *air defense* missiles, but nonmilitary means of defense should receive at least the same attention, particularly with regard to conurbations, which would be declared *open towns* [Mechtersheimer 1984; 1986; 1987].

MELMAN, SEYMOUR.   Professor emeritus of Industrial Engineering at Columbia University and a leading U.S. critic of the war economy and the *MIC* [1970; 1985], as well as a long-term advocate of *conversion* [1987; 1988].

MIC.   Military-industrial complex. Term coined by *U.S. President Eisenhower in his farewell address, signifying a coalition between social groups with vested interests in continued military spending: the armed forces and ministry of defense bureaucracies, the arms producers, and the scientific staff involved with military R&D (research and development); also known as the "Iron Triangle" [Pursell (ed.)

1972; Melman 1970; 1985; R. Barnett 1970; Kaldor 1981; Prins (ed.) 1983: 133–168; cf. for a critique: Sarkesian (ed.) 1972]. The MIC has been an important element in the internal ("autistic") *armament dynamics* [Albrecht 1990a; Thee 1990; Brauch 1990c; cf. Greenwood 1975; Allison 1983; idem & Morris 1975], and as such an obstacle to *disarmament*.

Various forms of alternative defense would have different effects on the MIC, even though this remains a rather unexplored field of research:

- The consequences of full-scale abolition of armed forces in favor of *civilian-based defense* would be the most immediate and drastic remedy because it would automatically abolish the MIC.

- The implications of a shift to *NOD* postures would be much less drastic, and primarily consist of a side effect, the modification of weapons profiles (see *Land Forces*). It would, inter alia, imply a shift of emphasis from weapons platforms (*tanks*, aircraft, warships, etc.) to weapons and munitions; and, within the latter category, from long-range to shorter-range weapons. This would, however, be tantamount to a shift of the industrial emphasis from the current main contractors (airframe plants, shipyards, etc.) toward other companies, which might facilitate industrial diversification. *Ceteris paribus,* this would tend to weaken the MIC. Furthermore, to the extent that NOD would be based (partly or completely) on reserve forces or *militias* [cf. Löser 1981; Bülow 1986], the hierarchies of the armed services would be weakened. Even though NOD has been charged with being "the last resort of militarism" [Tromp 1990: 171], it would thus rather tend to weaken the MIC.

MIDDLE EAST.   See *Israel*, *Gulf War*, *Commonwealth Institute*.

MIKSCHE, FERDINAND OTTO.   FOM started his military career as an officer in the armed forces of Czechoslovakia but fought in *WWII* as a member of De Gaulle's French forces. From 1945 until his retirement, he served on the French General Staff.

FOM wrote a great number of books on military affairs, including *Unconditional Surrender* (1952), in which he advocated a large-scale (thirty to fifty divisions) and unconstrained West German rearmament [1952: 319–329]. A later work, *Atomic Weapons and Armies,* was much more in line with alternative defense thinking, outlining a conventional force posture for a nuclear battlefield [1955: 160–204]. The ever-present threat of tactical nuclear attack called for great *dispersion* of forces, hence FOM suggested a division of the ground forces into five main elements:

- Tactical nuclear forces;
- Regular *infantry*, i.e., the "shield forces" that might have to fight according to guerrilla tactics;
- Mechanized forces, i.e., the "sword" (see *Sword-and-Shield*);
- Rear-area forces; and
- Paratroopers.

Although far from willing to forgo offensive options, FOM also mentioned the possibility that "for a power weak in standing armies, it might be of advantage in the early stages to avoid mobile warfare altogether, and to adopt a purely defensive attitude, warding off a hostile invasion from the secure shelters of a kind of *Maginot Line*" [1955: 181].

MILITARY DOCTRINE.   See *Military Science.*

MILITARY-INDUSTRIAL COMPLEX.   See *MIC.*

MILITARY PROFESSION.   Created by the social division of labor, implying that a special group performs the funtion of national defense (or attack), either for a short period or as a lifelong profession.

1. **Loyalty versus Dissent.** In modern societies the armed forces constitute one institution competing with others over distribution of the available resources, inter alia, as one component in the *MIC* (military-industrial complex). Furthermore, the different branches of the armed forces, i.e., the services, compete among themselves over distribution of the military's share: the well-known phenomenon of interservice rivalries, which many *NOD* proponents have recommended to eliminate through abolition of the services in favor of an integrated military system encompassing land, air, and maritime forces [e.g. Gerber 1984]. Rivalries notwithstanding, internal cohesion has been an important element in the military's bargaining power vis-à-vis the rest of society; defection has tended not to be taken lightly [Allison & Morris 1975; cf. Møller 1991: 135–137; 1992: 93–93, 205–206].

Military bargaining power has further depended on the prevailing assumptions about the respective strengths of offense and defense, which determine the payoffs on military investments. If offense is believed to be the stronger, the armed forces may seek to tempt decisionmakers with the prospects of advantageous offensive warfare. Or they may argue budget claims with reference to the need for dissuading adversaries from such wars of aggression. If defense is believed to be the stronger, fewer resources will be needed for military purposes: aggression will not be worthwhile, and defense will be easy and cheap. It therefore serves the vested interest of the armed forces to proclaim an offensive advantage [Evera 1984: 206–398; 1985; J. Snyder 1984; 1985; Posen 1984b, 179–219].

Notwithstanding the strong pressure on members of the armed forces to remain loyal to the common cause, inter alia, by denying the primacy of the defensive, numerous officers from all over the world (most of them retired, however) have voiced dissenting opinions, e.g., in favor of *no-first-use* and/or NOD. Examples include *Eckart Afheldt*, Hugh Beach [1985: 1986; 1988; 1990], *Bogislav von Bonin*, *Franz Borkenhagen*, Field Marshall Lord Carver [1981; 1982; 1983; 1983a], *Ludger Dünne*, *Johannes Gerber*, *Jens-Jørn Graabaek*, *Norbert Hannig*, Michael Harbottle [1988; 1988a; 1990], Gene La Rocque [1988; 1989], *Jochen Löser*, *Alfred Mechtersheimer*, Mikhail Milhstein, *Wilhelm Nolte*, Antoine Sanguinetti [1984; 1988], and *Franz Uhle-Wettler*). Many of these dissenters have joined forces across the East-West boundary, inter alia, in the international *Generals for Peace* organization [Kade 1981; Harbottle & al. 1984; cf. for examples: Prins (ed.) 1984].

In Germany, such dissent has been facilitated, e.g., by the notion of the armed forces as *citizens in uniform*, which has been furthered as an alternative to the Prussian "state within the state." Even though dissenting officers have thus enjoyed legal protection, the case of Commodore *Elmar Schmähling* nevertheless showed that there are limits to the freedom of expression [Baudissin 1986; cf. for a critique: Bald & al. 1981].

2. **Bellicosity?** It is sometimes alleged that the military profession is characterized by bellicosity, i.e., actually enjoys war for its own sake. If so, the only remedy would be a complete abolition of the armed forces, such as proposed by advocates

of *civilian-based defense*, only some of whom, however, have subscribed to the said view of the armed services. Furthermore, historical instances aside, such allegations seem unfounded as far as modern armies in industrial and postindustrial societies are concerned, regardless of whether they are professional, conscripted, or *militia*-type. As further hedges against society's being dragged into a war by the allegedly bellicose military, a greater *democratization* and political control of the armed forces have been suggested, e.g., by NOD proponents [Tatchell 1985].

   3. **Forms of Enlistment.** There are three basic modes of fielding armed forces (along with various combinations), each of which has its special strengths and weaknesses, also with regard to offensive and defensive purposes:

- Professionals who serve for long periods are very disciplined and loyal, wherefore they tend to be suitable for demanding, offensive-type operations. Because of professionals' high costs, their numbers are usually relatively low and the *mobilization* pool small, implying a greater suitability for interventionary purposes and the like than for defense against large-scale invasion.

- Universal male (with the exception of Israel) *conscription* is a relatively inexpensive way of fielding large forces of average quality and provides for a large mobilization pool. It is thus well-suited for defensive purposes, particularly when adequate strategic *warning* can be counted upon.

- Militias provide the greatest mobilization potential combined with the lowest number of standing forces. Both skills and equipment tend to be rather primitive, hence their unsuitability for offensive operations. Militias may, however, be quite adequate for national defense, provided timely warning is available [cf. Duic 1985; Fernau 1987; Haltiner 1987; Bald (ed.) 1987; idem & Klein (eds.) 1988; Klein (ed.) 1991].

Even though many NOD proposals have envisioned more extensive use of reserves or even a (wholesale or partial) shift to a militia system, the majority of NOD schemes would require no major changes in modes of enlistment. With a few exceptions [e.g. Booth 1983; cf. idem & Baylis 1988: 172-173], British and U.S. authors have tended to presuppose all-volunteer armies [ADC 1983: 138–140; Chalmers 1985: 158–173; Boston Study Group 1979: 274–276], whereas Germans have usually taken conscription as their point of departure, merely envisioning a better utilization of reserves [Bülow 1986; Funk 1988; 1988a; cf. for an exception: Gerber 1982: 71; 1984].

   4. **Employment.** The armed forces globally provide jobs for several millions of people. A shift to NOD might allow for reductions in military personnel, but many NOD schemes are so personnel-intensive that restructuring might partly offset the reductions in military expenditures that presumably would be brought about by NOD (cf. *armaments dynamics*). Even though this would imply a smaller budget share for investment and fewer jobs in the arms industry, available studies of *conversion* have indicated that job losses could be more than compensated for by civilian production (at least in the medium term, albeit not always immediately) [Melman 1988; Dumas (ed.) 1988; idem & Thee (eds.) 1989; Köllner & Huck (eds.) 1990; Renner 1992; Lall & Marlin 1992; Southwood 1991; Markusen & Yudken 1992; Paukert & Richards (eds.) 1991; cf. for a Third World perspective: N. Ball 1988: 332–334].

   5. **Internal Functions.** Armed forces are sometimes used against a state's own citizens, not merely by dictatorships and not always (albeit probably most often) in support of reactionary goals. A restructuring to NOD would seem to pose no particular problems in this respect because NOD-like armed forces might also be utilized

for internal (repressive or merely policing) functions, as demonstrated with horrifying clarity in the former *Yugoslavia*. Indeed, most of the weapons systems rendered superfluous and/or undesirable under an NOD regime (fighter-bombers, *aircraft carriers*, heavy artillery, *tanks,* and nuclear weapons, etc.) are generally not suitable for internal purposes in any event, whereas the light *infantry* emphasized in most NOD models would be. The only seeming incompatibility between NOD and internal use of armed force would be circumstantial, namely, the greater likelihood that democratic states, rather than dictatorships, would shift to NOD, combined with the reluctance of democracies to resort to arms in political struggles.

MILITARY SCIENCE.   Generic term for the science of preparing for and waging war.

1. **Hierarchy of MS.** The usual distinction in the West (adapted from *Clausewitz*, *Fuller*, *Liddell Hart*, and others) has been between the following branches, arranged in hierarchical order, with politics falling beyond the realm of MS, yet in the final analysis being what should determine the others because war is "a continuation of politics by other means" [Clausewitz 1980: 34/1984: 87 (Book I.1.24)]:

- Grand strategy, also known as total strategy [Beaufre 1963: 24–25], which was defined by Liddell Hart as "policy in execution," with the explanation that its object was to "co-ordinate and direct all the resources of a nation . . . towards the attainment of the political object of the war" [Hart 1967a: 322; cf. Kennedy (ed.) 1991; Luttwak 1976, 1983].
- *Strategy,* defined rather narrowly by Clausewitz as "the use of engagements for the objects of war" [1980: 84/1984: 128 (Book 2.1); cf. Luttwak 1987: 239–241]. It has been defined in slightly broader terms by modern strategists, as "the distribution and application of military means to fulfil the ends of politics" (by Liddell Hart [1967a: 321]), and as "the art of applying force in order to attain the ends of politics" by *André Beaufre* [1963: 16]).
- Operational art, also known as "grand tactics," which has been omitted from most taxonomies. When included, its object would be the totality of military activities within a theater of war [Luttwak 1980; 1987: 91–112; Simpkin 1986: 23–24].
- Tactics, defined by Clausewitz as "the use of armed forces in the engagement" [1980: 84/1984: 128 (Book II.1)].

In addition, there is the vague term "doctrine" (as in *FOFA* or *AirLand Battle* doctrine), the location of which tends to be somewhere between strategy and operational art.

The terminology of the former *Soviet Union* was significantly different, and more systematic than that of the West (see Table 25). It subdivided military science into the following hierarchically ordered categories:

- Doctrine: the primarily political assignment of military means to political ends (roughly comparable to grand strategy);
- Strategy: the application of military means to these objectives;
- Operational art: the science of major operations within a vast "theatre of military operations" (TVD); and
- Tactics: dealing with the activities of individual formations in single engagements [Lider 1983; Savkin 1972; Arbatov 1988].

Table 25. Western and Soviet Military Terminology

Western Terminology	Soviet Terminology
Politics	Politics
(Grand strategy)	Doctrine
Strategy	Strategy
(Operational art)	Operational art
Doctrine	
Tactics	Tactics

2. **Alternative Defense and MS.** Both traditional and alternative defense are hierarchically ordered, either implicitly or explicitly. Table 26 illustrates the hierarchy of alternative defense concepts, set up against traditional (and predominantly offensive) concepts of comparable scope.

Table 26. Hierarchy of Military Science

	Examples	
	Traditional	Alternative
Politics	Rollback	*Détente
Grand strategy	Imperialism	*Common security
Strategy	*Flexible response	*NOD
		*Civilian-based defense
Operational art	*Blitzkrieg	*Spider-and-web
Doctrine	*AirLand Battle	*Defense wall
Tactics	Armored assault	Ambushes, *barriers

MILITIA. Special form of recruitment of armed forces, based on very short compulsory military training followed by frequent refresher courses and maneuvers. Armed forces of the M type tend to be very small in terms of peacetime strength yet to have a very high *mobilization potential [Duic 1985; Fernau 1987; Haltiner 1987].

M forces have formed the backbone of the armed forces of *Switzerland for centuries [Cramer 1984; Fuhrer 1987], but they also play an important role (in the form of home guards) in the military organization of countries such as *Austria [cf. Bundeskanzleramt 1985; Spannochi 1990; Danzmayr 1990; 1991a; Fuhrer 1987; Schneider 1987; Tauschitz 1990], *Finland [Kanninen & Tervasmäki 1985; Visuri 1985; 1990; 1991; Väyrynen 1982; 1989; Ries 1989; 1989a: 175–375; Kemp (ed.) 1990] and the former *Yugoslavia [Roberts 1976:172–217; Lukic 1984; Bebler 1987; Acimovic 1987; Piziali 1989; Radinovic 1989], as well as (to a somewhat lesser extent) *Denmark [Heegaard-Poulsen 1980].

A greater reliance on M has been recommended by many *NOD advocates because M incorporate the combination of defensive strength and incapability of attack. The availability of local M forces throughout a country would ensure the ability to conduct an area-covering and in-depth *territorial defense*. Furthermore, the typical equipment of M (hand weapons, and antitank and air defense weapons, etc.) would be quite adequate for defensive purposes. On the other hand, the absence of large standing forces and heavy mechanized and armored forces would make an army composed largely of M relatively incapable of offensive operations and completely rule out the option of *surprise attack*. Finally, the special nature of M forces would make it unlikely that their mobilization in times of crisis could be perceived as provocative, a feature that would both improve *crisis stability* and minimize *self-deterrence*.

MINH, HO CHI (1890–1969).   Leader of the Communist Party of French Indochina, subsequently Democratic Republic of Vietnam. He presided as political leader over history's probably longest guerrilla war (see *Vietnam War*), even though the military direction of the North Vietnamese armed forces (as well as, indirectly, of the Vietcong), was delegated to *Vo Nguyen Giap*. HCM made several contributions to the development of *guerrilla strategy*, even though these have to be extracted from political declarations and orders to the armed forces [Minh 1973].

In 1944, in connection with the setting up of the "Armed Propaganda Brigade" (the nucleus of what was soon to become the National Liberation Army), HCM formulated the principles of the armed struggle, underlining, e.g., that "greater importance is attached to its political than to its military action. . . . Ours being a national resistance by the whole people we must mobilize and arm the whole people. . . . Concerning tactics we must apply *guerrilla warfare*, maintain secrecy, quickness of action and initiative: now in the east, now in the west, arriving unexpectedly and departing without leaving any traces" [Minh 1973: 47–48; cf. Giap 1970: 67–78].

HCM consistently maintained the need to abide by the principles of guerrilla warfare: to avoid big battles, maintain *self-sufficiency*, etc. At the same time, he emphasized the need to coordinate the struggle with that of the regular National Defense Army, a division of labor that he later specified as guerrilla warfare in enemy-occupied territory and regular warfare in the liberated zones. He furthermore predicted a very protracted war of resistance and rightly so, as the coming years were to show [Minh 1973: 78–80, 112–117, 146–152; cf. Giap 1970: 93, 103–104].

MINIMUM DETERRENCE.   Configuration of the nuclear arsenals intended to deter nuclear attack by means of as few offensive nuclear weapons as possible. MD thus presupposes *no-first-use* of nuclear weapons and may well be conceived of as merely a transitionary stage toward complete abolition of nuclear weapons. Numerous schemes have been proposed for the transition from the present levels to MD, either unilaterally or, more often, as a guideline for *arms control* negotiations such as *START* (Strategic Arms Reduction Talks). The thitherto prevailing East-West conflict meant that all MD schemes dealt merely with the two superpowers and disregarded minor nuclear powers entirely.

1. **Bipolar MD.** The size of the MD arsenals envisioned by different authors have ranged from a few thousand to a few hundred warheads [Szilard 1964; J. N. Miller 1988; Feiveson 1989; idem & Hippel 1990; 1990a; idem & al. 1985; idem & Hippel 1990a; Arbatov 1989; Bundy & al. 1993; Dean 1992b; idem & Gottfried 1991; Glaser 1992; Karp 1992; Kaysen & al. 1991; Rotblat & al. (eds.) 1993]. Most authors, however, have agreed that the composition of the arsenals at each consecutive stage of build-down would be of considerable importance. For the sake of stability and in order to allow for the smallest possible arsenals, weapons survivability would have to be maximized, e.g., through the following precautions:

- An abstention from targeting the other side's strategic forces.
- A build-down of counterforce and particularly hard-target-kill capabilities: single-warhead missiles with limited throw weight and moderate accuracies would be preferred to missiles with high throw weight and low CEPs (circular error probable). Also, single-warhead missiles would be preferable to missiles with MIRVs (multiple, independently targeted reentry vehicles), as acknowleded in the *Bush Initiative*.
- Survivable deployment of strategic missiles, most of them on submarines (as far as the *U.S.* is concerned) or mobile ICBMs in the case of the *Soviet Union* (or Russia).
- Maintenance of some heterogeneity, e.g., through a triad structure such as the present one, making the synchronization of a surprise attack impossible [cf. Scowcroft Commission 1983].

MD arsenals might thus comprise SLBMs (submarine-launched ballistic missiles), silo-based as well as mobile ICBMs (intercontinental ballistic missiles), and strategic bombers, with cruise missiles (CM) as penetration devices. The required total size of the MD arsenals might be in the range of five hundred warheads (WH), which might, as an example, be distributed as in Table 27 [cf. Møller 1992: 162–166; 189–190].

Table 27.   Minimum-Deterrence Postures

| | USA | | USSR/Russia | |
	Systems	WH	Systems	WH
ICBMs/warheads (WH)		50	50		250	250
Bombers/CMs		50	100		25	50
Submarines/SLBMs/WH	25	250	250	15	150	150
Sea-launched CMs		100	100		50	50
Total		450	500		475	500

Neither theater nor battlefield nuclear forces would be required for MD, wherefore a complete withdrawal of U.S. nuclear forces from Europe would be implied.

2. **Multipolar MD.** The bipolar framework characterizing the nuclear era until 1991 has been rendered obsolete by the dissolution of the *Soviet Union*, if not by the 1989 revolutions in Eastern Europe. The resultant *multipolarity* may significantly complicate the devising of sensible MD postures because it is an open question who should deter whom. On the other hand, it strains the imagination to conceive of, say, the U.S. seeking to deter a British nuclear strike, or vice versa. The only serious problems seem to be those caused by nuclear proliferation.

A certain political proliferation has already occurred as a result of the dismantling of the USSR, and the danger looms of a further proliferation, especially among Third World countries [Spector 1990; Barnaby 1993; Ham 1993; Fischer 1992]. Certain authors have advocated an "orderly proliferation" (e.g., by allowing Germany nuclear status) in the belief that nuclear status tends to instill caution in the state attaining it, hence that a multipolar nuclear world would be inherently more stable than conventional multipolarity [Waltz 1981; cf. Mearsheimer 1990; 1993; Garrett 1990: 63–72; cf. for a critique: Evera 1990a; 1990b]. However, taking into account that the possession of nuclear weapons may not in fact "teach" states to behave sensibly, and that proliferation is most unlikely to proceed in an orderly fashion, most authors have preferred a strengthening of the existing nonproliferation regime around the *NPT [Fischer 1992].

On the presumption, however, that the present nuclear powers are here to stay and that a limited proliferation may occur (and that a partial nuclear multipolarity will result), it might still be possible to devise an MD regime for this setting. It would, however, presuppose that both old and new nuclear powers would abstain from counterforce capabilities because these might jeopardize the survivability of the second-strike capabilities of potential adversaries, thus inducing them to expand their arsenals (thereby starting an *arms race). Should all nuclear powers, on the other hand, limit themselves to countervalue (or "pure deterrence") capabilities, the requirements for MD, i.e., "deterrence sufficiency," would remain the same: Even a state confronting, say, ten nuclear-armed opponents would need only "one assured destruction capability" (e.g., the size of the five hundred warheads suggested above) in order to deter them all because none could be sure of escaping retaliation.

MISSILE TECHNOLOGY CONTROL REGIME.   See *MTCR.

MISSING LINK.   The idea that a military posture might be deprived of its offensive capabilities by simply omitting one element. The synergies between weapons and the combined-arms nature of all modern armed forces would presumably mean that offensive or defensive capabilities would be a function of the totality (cf. *offense-defense distinction), wherefore minimal changes in the configuration might result in disproportionate changes in the overall capabilities. Naval power without land-attack weapons or amphibious forces would, e.g., not be truly offensive vis-à-vis land powers, *tanks would not be so without air cover, etc. The main proponent of this approach to creating *NOD postures has been *Erwin Müller [1984].

The ML approach might also be a prominent feature of the suggested combination of NOD with *collective security. The former would call for strictly defensive forces; the latter would presuppose offensive capabilities, inter alia, strategic mobility, without which states would be unable to fulfil their pledges of assistance to a party under attack. It might, however, be possible to create offensive-capable multinational task forces out of national armed forces that were strictly defensive by virtue of their MLs, if only the MLs of the various states differed [Møller 1993a]. See also *multinationality.

MIX CONCEPTS.   Generic term for a combination of *civilian-based defense and military defense, in most cases of an *NOD character. Proponents of such combinations have included the *ADC [1983], *Wilhelm Nolte [Nolte & Nolte 1984], *Johan Galtung [1983; 1984; 1984a], and Jack Salmon [1988].

MIXED-MOTIVE GAMES.   Term from *games theory signifying games in which cooperation and competition are intertwined. The best-known example is *prisoner's dilemma.

MOBILITY.   See *Land Forces.

MOBILIZATION.   Readying of the armed forces and the rest of society for war, either as a preparation for attack, or for defensive purposes if an attack is expected.

1. **M Potential.** Regardless of the contingency, all countries need to mobilize before they are ready to go to war: troops have to be in place; weapons have to be in

working order; command posts have to be staffed around-the-clock and surveillance intensified; and forces need to assume combat positions.

The M potential of states differs, inter alia, because of their different modes of recruitment. Countries with professional armed forces tend to have the highest peacetime readiness but the smallest M potential; the opposite is the case for countries with *militia armies.

2. **M Races.** In situations where war becomes conceivable, states tend to respond to one another's moves. One state's partial M might thus induce its adversaries likewise to mobilize, which might, in turn, make the speedy M of the first country's forces seem all the more urgent, etc. This might lead to a genuine M race, as was the case prior to *WWI [Tuchman 1962; Taylor 1969; cf. for a critique: Trachtenberg 1990]. The implication would be a low degree of *crisis stability because of a malign interaction between the two sides [cf. Bracken 1983]. Eventually, the point of full M might be reached, at which stage each side would have to decide between stepping down, i.e., initiating a demobilization, or going ahead, i.e., launching the attack, in preemption of the attack it fears the other is about to launch. Because the level of M is quite decisive for the force ratios between the two sides (cf. *Joshua Epstein, *Force Comparison), it may well seem wisest to go ahead.

3. **NOD and M.** Most *NOD advocates have recommended a low state of peacetime readiness of the armed forces because it would minimize capabilities for *surprise attack: a greater reliance on reserves, a withdrawal of forces from the forward line, etc. [e.g. Löser 1981; Dean 1989c: 232–242; cf. 1988b]. There may be serious drawbacks to this approach from a stability point of view. If the reserve forces were to include forces that possess significant offensive capabilities, their M in times of crisis might trigger the aforementioned malign interaction. Mobilizing them might well be mistaken for attack preparations. Moreover, if offensive-capable forces with rear peacetime deployment were to be brought forward, this might well be perceived as an actual attack in progress, etc. Counterintuitive though it may seem, it might be preferable to maintain offensive-capable forces at approximately their envisioned combat strength and as close to their combat stations as possible, while maintaining the more defensive forces at a low state of readiness. The M of the latter in times of crisis would not lend itself to misinterpretation but might still dissuade a prospective attacker by signaling defense readiness.

MOBILIZATION RACE.   See *Mobilization, *WWI.

MØLLER, BJØRN.   Danish peace researcher; Ph.D. in political science and M.A. in history; since 1985 research fellow (subsequently senior research fellow) at the *Centre for Peace and Conflict Research; and lecturer at the University of Copenhagen (*Denmark). BM has since 1985 been working on the project Non-Offensive Defence in Europe, and since 1991 also on the project *Conversion board. In 1993–1994, he initiated and has been in charge of the project Global Non-Offensive Defence Network, funded by the Ford Foundation. He is a board member of the *SAS.

BM's main contributions to the *NOD debate (apart from the present work) have been

- Publication of the international research newsletter *NOD: Non-Offensive Defence,* since 1991 *NOD and Conversion.*
- A comprehensive analytical survey of the German debate on NOD: *Resolving the Security Dilemma in Europe: The German Debate on Non-Offensive Defence* [Møller 1991; cf. 1990d–e; 1994].

- A theoretical analysis of the compatibility of NOD and *common security* with *Realism* and Neorealism, and the relationship between NOD and *arms control*, *alliance policies*, deterrence, and confidence-building: *Common Security and Non-Offensive Defense: A Neorealist Perspective* [1992; cf. 1988; 1989a; 1990c; 1990h; Møller & Wiberg 1989; 1990].
- Analyses of the application of NOD principles to the aerial and maritime environments [1987; 1989; 1989e; 1990b; 1990i; 1992: 183–194].
- Analyses of the regional application of NOD to the Nordic region [1985; 1987; 1987c; 1989b; 1989f; 1990b–c; 1991b; Gleditsch & al. 1990: 105–154].
- Analyses of how to combine NOD with *collective security* [1990; 1992a; 1992d; 1993a].

MTCR.   Missile Technology Control Regime. Agreement signed in 1987 between *Canada*, *France*, the *FRG*, *Italy*, *Japan*, the *U.K.* and the *U.S.*, which the *Soviet Union* later pledged its intention to observe, as did *China* [Anthony 1991a; Bailey 1991a; Karp 1991a; 1992; Kniest 1992; Shuey 1992].

The MTCR was envisioned as a supplement to the *NPT* (Nonproliferation Treaty), which prohibited transfer of nuclear weapons and technology by intending also to curb the proliferation of suitable means of delivery of nuclear weapons, especially in the Third World [cf. Carus 1990; Findlay (ed.) 1991; Levite 1992; Mack 1991a; 1991b; Navias 1993; Neuneck & Ischebeck (eds.) 1992; Nolan 1991; 1992; Karp 1991; 1991a; Singh 1992]. The MTCR, however, has been a very loose agreement, neither having institutional foundation nor containing any specific regulations. It merely obliged the signatories to include certain criteria in their national arms-transfer regulations. Furthermore, it stipulated to which technologies the criteria applied: above all rocket systems (or parts thereof) with payloads exceeding 500 kg and ranges above 300 km, on the implicit assumption that these were especially offensive. The MTCR thus reflected the growing realization that offensive capabilities should be minimized. See also *Arms Trade Regulations*, *Counterproliferation Initiative*.

MUELLER, JOHN.   See *Concert*, *U.S.*

MÜLLER, ALBRECHT A. C. VON.   Philosopher and peace researcher, former director of the Working Group Afheldt at the Max Planck Institute in Starnberg, FRG [Dürr & al. 1984], and subsequently director of *EUCIS* (European Center for International Security), consisting almost exclusively of himself. Besides the center, AvM's main basis has been the *Pugwash* movement, particularly the study group Conventional Forces in Europe under its auspices, for which he has served as convener, jointly with the late *Anders Boserup* [cf. Boserup & Neild (eds.) 1990].

Besides the philosophical *Art of Peace* [Müller 1984], AvM's production has consisted almost entirely of short articles (with an astounding degree of repetition). His chief contribution to the *NOD* debate has been the invention of the (awkward and linguistically incorrect) term "structural nonattack capability" (*Strukturelle Nichtangriffsfähigkeit* [*Frankfurter Rundschau*, 1 June 1983; cf. Müller 1987: 60].

AvM called his NOD model the "integrated *forward defense*." It was an eclectic model, combining features from the models of *Horst Afheldt*, *Hannig*, *Löser*, and the *SAS* into three successive *zones*: a 5-km-wide *fire barrier* (featuring MLRS, attack drones, and sophisticated mines combined with "close interdiction" onto enemy territory); a 75-km-wide "network zone" (with *PGM*-armed *infantry*); and a 60-km-wide "manoeuvre zone" with armored units possessing *counteroffensive* capabilities [1986b: 248–249; 1986: 216–220; 1989a: 129–130]. In a later version, the two forward zones were merged into a 25-km-deep '*web* zone" featuring light infantry and combat engineers [1988a; 1988c]. The model relied heavily on emerging technologies (see *E.T.*), such as sophisticated sensor arrays, combat drones, and intelligent passive munitions, presumably tending to benefit the defender [1985:276–279; 1986: 205–208; 1989a: 127–129].

What set AvM apart from genuine NOD proponents was his advocacy of *conditional offensive superiority*. Even though he presented this as merely a special case of *mutual defensive superiority*, it reflected, rather, a quest for inverting the offense-defense ratio: when the attacking forces had penetrated the two forward zones, they would presumably find themselves so outnumbered that the defenders might launch a counterinvasion against the home territory of the attacker [Müller 1986b: 253–254; 1987a: 519; 1988c: 92–93], an instance of *conventional retaliation* almost identical to that proposed by *Samuel Huntington* [1983]. AvM differed further from NOD proponents by supporting maintenance of land-based nuclear weapons in Europe. As far as their numbers were concerned, he even moved against the tide by first suggesting a maximum of two hundred warheads, later raising this to first four hundred and then five hundred [1987: 62; 1988; 1989: 340–341; Schmückle & Müller 1988; Müller & Karkoszka 1988].

MÜLLER, ERWIN.   Research fellow at the *IFSH* whose work has focused on confidence-building [1982], *NOD* [1984; 1989], and *conventional stability* [1988; 1989].

Through his studies of *CBM*s (confidence-building measures), EM arrived at the conclusion that NOD would be a viable approach, indeed would constitute the CBM par excellence [Müller 1982; cf. Møller & Wiberg 1990]. In his subsequent search for a viable form of NOD, EM's approach was to seek the shortest route, i.e., to achieve the sought-for inability to attack through minimal changes, to which the combined-arms character of modern warfare provided the key. It implied that offense was a function of a complex synergy of ground and air forces, *logistics*, and *C³I*. Hence the possibility of eliminating offensive capabilities through creation of a *missing link*, i.e., the deliberate omission of a vital component. Deprived of this, the armed forces might lack offensive capability yet retain an undiminished defensive capability. The component might be either mobile antitank or *air defense* or *bridge-building* equipment, without which the armed forces would possess no strategic mobility. On this precondition, EM found nothing objectionable in long-range striking power, e.g., in the form of conventional missiles. EM was no more "puritanical" concerning nuclear weapons, in which sphere he maintained that the defensive criterion merely required pure deterrence, based on strategic weapons that were "first-strike resistant, yet first-strike incapable," i.e., a form of *minimum deterrence* [1984; 1988; 1989].

In direct continuity with his analysis of NOD, EM proceeded (in light of the *CST* and *CFE* negotiations) with the analysis of conventional stability, concluding that stability presupposed that neither of two sides possessed the ability to win an offensive war (i.e., what others have called *mutual defensive superiority*). Contrary to most other NOD advocates, EM regarded this as unattainable through unilateral restructuring and hence recommended a bilateral implementation, either through a withdrawal of offensive-capable weapons systems (the preferred approach of the IFSH) or "asymmetrical missing links," as the ideal variety whereof EM pointed to a "dual missing link": no long-range weapons systems, and no cover (i.e., armor) on foreign territory [1988: 51-61; cf. 1988a; 1989].

MULLER, JEAN-MARIE.   See *France.*

MULTINATIONALITY.   In the military sphere, M signifies *integration* between sovereign states, which may take the form of joint command structures or (rarely) military formations, jointly operated military support facilities, or the joint defense of a particular front. Examples have been *NATO*'s integrated command structure (to which national forces have been envisioned to be subordinated in case of war or crisis); NATO's common air defense system, NADGE; the Franco-German brigade; and the "layer-cake" NATO posture along the central front (until 1990–1991).

It has often been asserted by NATO officials and others that NATO would be politically incapable of launching an attack, hence that structural changes (such as proposed by *NOD* advocates) would be superfluous. Besides the argument that *democracy* itself would be a strong inhibition against aggressive war, it has also been asserted that NATO's M would prevent it from aggression. First, politically, because it was inconceivable that sixteen states would agree to attack another state. Second, operationally, because no member state would be able to operate (e.g., by attacking another state) without the assistance (and by implication, consent) of its allies. Indeed, NATO has often been viewed as a hedge against a resurgent German expansionism, hence the total subordination of the Bundeswehr to NATO command and the separation of German sectors of the front by those of other nations. The validity of these arguments has been conceded by some NOD advocates, either wholly or at least partially [Senghaas 1986: 106; Löser 1981: 259–266], albeit usually with the caveat that structural safeguards might also have been required to meet Soviet concerns.

In the course of its 1990–1991 strategy review, NATO decided to increase M by promoting multinational (corps-size) military formations. That this might also be motivated by the wish to minimize offensive capabilities was indicated by Colonel David Miller (staff officer at the AFCENT NATO Headquarters): "It is to the Soviet Union's advantage for Germany to be a NATO member, where restraints . . . can be exercised by its allies. . . . Two further safeguards are possible within such an alliance. One is to designate national specializations such that no nation has an independent offensive capability. . . . Another approach is to create multinational formations [whereby] . . . the power of any one nation is limited yet further" [*Jane's Defence Weekly,* 5 May 1990, 838; cf. Miller 1991].

MULTIPOLARITY.   Structure of an international system (global, regional, or subregional) in which alignments between states (or, in principle, other actors) tend to cluster around at least three poles, as compared with unipolarity and *bipolarity.*

Because of several unattractive features of the bipolar system (such as its tendency to expand the systemic conflict to the entire globe, its nuclearism, etc.), certain alternative defense advocates have argued in favour of M. *Horst Afheldt,* e.g., asserted that a multipolar structure would be more stable than a bipolar one, while emphasizing that a precondition therefore would be "reciprocal structural defensive superiority," i.e., an implementation of *NOD* by all states [Afheldt 1989a: 92–97, 162–163]. Others have been somewhat less sanguine about the prospects of M, pointing to the problems of achieving stability when alignment patterns are unpredictable. The sought-for *mutual defensive superiority* may be impossible to define (much less to attain in practice) between M states of different sizes [cf. Huber 1990a; idem & Avenhaus (eds.) 1993; Møller 1992a].

To what extent the emergence of M is fraught with dangers hinges on assessment of the causes of the unprecedented long peace in Europe since WWII. Some have attributed this to bipolarity, with the implication that M would lead to a reemergence of a war system. Others have attributed it to the effect of nuclear weapons (cf. *existential deterrence*) which has made war for political purposes most unattractive. Still others have found the horrors of large-scale conventional war, as well as widespread war weariness, to be a sufficient explanation of the long peace [cf. Lynn-Jones (ed.) 1991, especially Mueller 1988 (also 1989); Gaddis 1986; Jervis 1988 (also 1989)]. Authors conceding that bipolarity may, at least, have contributed something to the long peace have made proposals for arrangements to accompany M, such as *concerts* or even *collective security* systems [Jervis 1985; Mueller 1990; 1991; Kupchan & Kupchan 1991; Senghaas 1990e; 1991; 1991a-b; cf. Møller 1990; 1991: 270–273; 1992: 71–75; Chalmers 1990; 1990a].

MURMANSK INITIATIVE.   Set of Soviet proposals concerning the northern oceans presented by *Mikhail Gorbachev* in a speech in Murmansk, 1 October 1987 [*Tass,* 1 October 1987; Archer 1989].

The MI was intended as a response to the *Maritime Strategy* of the U.S. Navy. Its purpose was to stabilize the northern region by *naval arms control* and *maritime C(S)BMs*, as well as by nonmilitary means. Besides proposals for *disengagement* *zones* and the like, the MI suggested possibilities of East-West cooperation on exploration of Arctic resources, inter alia, with a view to furthering *interdependence*. Even though the MI met with a negative response from the *U.S.*, the response of the Nordic states was somewhat more positive, albeit still cautious [cf. Brundtland 1988: 108–113; 1989a].

MUTUAL DEFENSIVE SUPERIORITY.   Situation of reciprocal *defense dominance* between two potential adversaries. MDS has been declared by *NOD* proponents to constitute a precondition of *stability*, hence a viable substitute for *balance* as the appropriate goal of *arms control* negotiations.

MDS implies that either side should be able to fend off an attack by the other, while itself being incapable of successfully launching one, under which circumstances neither side would have anything to gain by initiating war. The stabilizing effects of MDS were first described by *André Glucksman* [1967: 42–43] and *C. F. von Weizsäcker* [1976: 116, 150, 165–166], both of whom took as their point of departure *Clausewitz*'s observations on the inherent supremacy of the defense [1980:360–371/1984: 357–366 (Book VI.1–3); cf. Aron 1983:144–171; Gat 1988].

MDS has been expressed in mathematical formulae, e.g., by *Anders Boserup* [1985; 1990c], *Reiner Huber* [1988; 1988a; 1990] and *Bjørn Møller* [1987a; 1990f; 1992: 87–89, 152–157]. The simplest such formula is:
$$D^A > O^B \ \& \ D^B > O^A,$$
where O stands for offensive and D for defensive military strength. To operationalize these inequalities is very complicated, inter alia, because it would imply assigning two different values to weapons systems, formations, etc.

MUTUAL SECURITY.   Synonym for *common security* [Smoke & Kortunov (eds.) 1991].

MUTZ, REINHARD.   Earned his doctorate with a dissertation on the *MBFR* negotiations. A senior research fellow at the *IFSH*, RM was the author of one of the most theoretically stringent analyses of *common security*, understood as "theoretically more than an approach, but less than a full-fledged model" and "politically more than a postulate, but less than a ready program." RM refrained from simplistic and politicizing assertions (so often made by *Dieter Lutz* of the same institute) about the incompatibility of common security with deterrence, merely stating that the two were impossible to unite in the long run [1986: 103, 154-155]. Furthermore, he has contributed to the IFSH's exploratory work on *conventional stability* and the objectives of conventional *arms control* [1988; 1988a; 1990; idem & Neuneck 1990].

N

NACC. North Atlantic Cooperation Council. Consultative forum established by *NATO* in 1991, bringing together members of the Western alliance with former *Warsaw Pact* member states [Gerosa 1992; Linnenkamp 1993; Ondarza 1993]. Important preparatory steps were NATO's *London Declaration* (July 1990), and the statement "Partnership with the Countries of Central and Eastern Europe" (June 1991) [*NATO Review,* 39:3 (June 1991): 28–29]. However, the NACC was formally inaugurated only with the *Rome Declaration* of November 1991 [*ibid.* 39:6 (December 1991): 19–22].

The NACC "Statement on Dialogue, Partnership and Cooperation," signed 20 December 1991, declared that the focus would be placed on "consultations and cooperation . . . on security and related issues, such as defence planning, conceptual approaches to *arms control,* democratic concept of civilian-military relations, civil-military coordination of air traffic management and the *conversion* of defence production to civilian purposes" [*Atlantic News,* 21 December 1991, annex 11, 1–2]. The membership of NACC has since been expanded to include all the Soviet successor states. The agenda has also been concretized and expanded, to include, inter alia, a high-level seminar on defense policy and management, envisioned to deal with, among other things, a "discussion of concepts such as defensive sufficiency, stability, flexibility and crisis management," a formulation reminiscent of both the Soviet notion of *reasonable sufficiency,* and the concept of *NOD* [*Atlantic News,* 3 April 1992, annex, 1–3]. This was clarified later (18 December 1992) with the formulation that "we intend thus to contribute to achieving a pattern of democratically controlled and smaller military forces which are *structured with defensive intent,* at minimum levels consistent with legitimate security requirements" [*Atlantic News,* 22 December 1992, annex, 1].

An important additional function of NACC has all along been to provide the former Warsaw Pact states with some institutional affiliation with NATO, yet without formal membership and security guarantees. Unsatisfied with this solution, the countries of Eastern Europe tabled applications for NATO membership, which NATO was reluctant to accept. As yet another compromise solution, NATO in January 1994 inaugurated the *Partnership for Peace* program, launched under the auspices of NACC [*Atlantic Documents,* 83 (12 January 1994)].

NATIONAL SECURITY.   See *Security, *Common Security.*

NATO.   North Atlantic Treaty Organization. Military *alliance* founded in 1949 and currently consisting of the *U.S., *Canada,* and fourteen European states: the *U.K.,* the *FRG,* the *Netherlands, *Belgium,* Luxemburg, *Denmark, *Norway,* Iceland, *France, *Italy,* Portugal, *Spain,* Greece, and Turkey.

1. **NATO's Attitude Toward NOD.** NATO's attitude toward *NOD* has been somewhat contradictory. On the one hand, high-ranking officials such as then SACEUR (Supreme Allied Commander, Europe) Bernard Rogers and the late

General Secretary Wörner have commented negatively on NOD, stating, inter alia, that to forgo offensive options would weaken the alliance's defensive capability [*Adelphi Paper* 205: 13; Wörner 1985]. On the other hand, the objectives that NATO proposed for the **CFE* negotiations were clearly inspired by the NOD logic in their insistence that offensive capabilities were what had above all to be eliminated (cf. **Bruxelles Declaration*).

2. **Is NATO Defensive?** NATO's reluctance to endorse the notion of NOD seemed based on the assertion that the Western alliance was already completely defensive, an assertion that certainly had much to recommend it:

- NATO has consistently proclaimed its strictly defensive intentions.
- Ever since the Harmel Report of 1967, NATO has emphasized the need for combining deterrence with **détente* [NATO Information Service 1989: 402–404].
- The very fact that what has united NATO has been a shared commitment to **democracy*, and that the alliance (with a few occasional exceptions) has consisted exclusively of democracies has served as a powerful inhibition against aggression [cf. Kant 1795; Rummel 1979; 1985; 1991; Chan 1984; Weede 1984; Russett 1990].
- The democratic (i.e., unit-veto) structure of the alliance has implied that a NATO aggression would presuppose unanimity among all sixteen member states, something that was practically inconceivable.
- NATO's military strategy since the 1950s has been **forward defense*, which is a basically defensive conception.
- NATO's **logistics* and munitions supplies have made it incapable of long-range and/or protracted operations.
- The "layer-cake" military posture along the Central Front came close to constituting a material guarantee against aggression, since no state represented there would have been able to launch an aggression on its own (see **multinationality*).

3. **What Should Change?** NATO's general defensive orientation notwithstanding, NOD advocates (especially in the early years) addressed their proposals to the Western alliance rather than to the **Warsaw Pact*. The reason, first, may have been tactical: because the East seemed "petrified," it appeared more worthwhile to criticize the side most likely to take it seriously and perhaps comply, i.e., NATO. Many NOD advocates also hoped that a shift to NOD on the part of NATO might spur a corresponding reorientation on the Eastern side.

Second, and more substantially, NOD advocates were concerned about certain elements in the military strategy of NATO, as well as of certain member states:

- The nuclear emphasis, which had only been slightly reduced by the replacement of massive retaliation with **flexible response* [Stromseth 1988];
- The emphasis on a positional forward defense, which was unlikely to hold in the face of a concentrated Soviet attack;
- The growing emphasis on **deep strikes*, which might well appear provocative to the other side; and
- The emphasis on the part of the U.S. (and to a certain extent the **U.K.*) on maneuver warfare in the framework of the **AirLand Battle* doctrine.

4. **After the Cold War.** NATO's *raison d'être* ever since its founding had been the Soviet threat. Its disappearance in the years 1989–1990 hence left NATO somewhat in disarray. Nevertheless, the Western alliance managed rather swiftly to adjust to the new situation, both politically and militarily [Asmus & al. 1993].

Politically, NATO started by extending "its hand in friendship" to its former adversaries in the Warsaw Pact with the *London Declaration* of 1990 [*NATO Review*, 38:4 (August 1990): 32–33]. This was followed by the *Rome Declaration* of 1991, which launched the North Atlantic Cooperation Council (*NACC*) as a forum for consultation and cooperation with the former opponents [*ibid.*, 39:6 (December 1991): 19–22; cf. Gerosa 1992; Wittmann 1992; Linnenkamp 1993; Ondarza 1993]. When this did not satisfy the former Warsaw Pact member states, increasingly eager to join NATO, NATO in 1994 launched *Partnership for Peace*, envisioned as a halfway station toward membership [*Atlantic Documents*, 83 (12 January 1994)].

Militarily, NATO embarked upon a strategy review that was finalized in 1991, with implications for both conventional and nuclear forces. In "The Alliance's New Strategic Concept," adopted at the Rome Summit (7–8 November 1991) [*NATO Review*, 39:6 (December 1991): 25–32; Fortmann 1992; Leech 1991], the point of departure was an acknowledgment that "the threat of a simultaneous, full-scale attack on all of NATO's European fronts has effectively been removed," hence "minimum Allied warning time has increased accordingly." Furthermore, the main danger was no longer assumed to be that of premeditated aggression but "the adverse consequences of instabilities that may arise from the serious economic, social and political difficulties, including ethnic rivalries and territorial disputes."

As a consequence, NATO qualified, to the point of nearly abandoning, the principles of flexible response and forward defense:

- Nuclear weapons were relegated to means of the very "last resort";
- The diminished need for standing conventional forces in a high state of readiness was acknowledged;
- Conventional forces were divided into main defense forces, reaction forces (subdivided into "immediate" and "rapid"), and *augmentation forces;* and
- An intention was asserted to move toward multinational formations, inter alia, as a contribution to building "a European defense identity" [cf. D. Miller 1991; Lowe & Young 1991; Meyer & Sens 1992].

Even though this went some way toward meeting the demands of NOD proponents and other critics, other things—the increased emphasis on mobility, and the greater willingness to take part in operations beyond NATO's traditional area of responsibility—seemed to point in another, more offensive direction [Böge 1991a; Lübkemeier 1991; Borinski 1993; Bracken & Johnson 1993].

NAVAL ARMS CONTROL.   *Arms control* applied to naval forces.

1. **Taxonomy.** NAC may be subdivided into two main categories:

- Functional NAC, i.e., constraints on the operations of naval forces, usually in the form of *maritime C(S)BMs*).
- Structural NAC, limitations or reductions of naval forces, also known simply as NAC.

2. **History and Present Status.** The record of NAC until the present has been mixed, above all because of the natural reluctance of states enjoying naval supremacy (the *U.S.* and before that the *U.K.*) to place this in jeopardy [Hill 1989: 26–36; Ranft 1979; Prawitz 1990; Skogan 1991]. A few moderately successful agreements were reached before *WWII* that limited and/or reduced naval forces significantly. The 1922 and 1930 naval treaties managed to curb an ongoing arms race between the U.S., the U.K., and *Japan*, and achieved a certain build-down (see *Washington Naval Conference*) [Bull 1973; E. Müller 1986a; R.G. Kaufman 1990]. Since 1945 no major treaties have been signed, and very few negotiations

have been conducted. The only exceptions have been the postwar constraints on the German wartime allies [Järvenpää 1992; Poggiolini & Nuti 1992; Pülöp 1992; Vetschera 1985a; 1992], which also regulated naval holdings (cf. *Austria, *Finland), and a few treaties that have marginally affected naval forces (or rather: forces deployed at sea):

- The SALT treaties on strategic nuclear weapons also placed ceilings on the number of strategic submarines;
- The 1972 Prevention of Incidents at Sea agreement codified the "rules of the road" for U.S. and Soviet warships with a view to avoiding unintentional confrontations (it has subsequently been copied by other pairs of states [Lynn-Jones 1988; 1990]);
- The Stockholm *CDE Conference reached agreement on some limitations of naval forces operating in conjunction with *land forces, but only of marginal importance: a ceiling on the number of amphibious forces participating in exercises [Brauch 1987a; Ghebaldi 1989; Hill 1989: 70–71]; and
- The two treaties establishing *nuclear-weapons-free zones (the Tlatelolco and Raratonga Treaties) also apply (formally, albeit not effectively) to parts of the high seas.

3. **Soviet Initiatives.** The *Soviet Union was all along favorably inclined toward NAC, albeit notably less so in the period of Soviet naval expansion, i.e., the 1970s. Beginning in the mid-1980s, its interest grew, primarily in response to the new *Maritime Strategy of the U.S. Navy. E.g., it proposed extension of the *CBM regime to the oceans, and creation of areas of restricted submarine activities, inter alia, in the *Murmansk and *Vladivostok Initiatives [Gorbachev 1987; cf. Bomsdorf 1989a; Brundtland 1988; 1989a; Shaskolsky 1989].

Small NATO countries such as *Denmark and *Norway have had a somewhat more favorable attitude toward NAC in general and the Soviet proposals in particular than have the U.S. or the U.K. Furthermore, a growing number of (mostly retired, but also a few active) naval officers have expressed support, in principle, for the notion of NAC [Eberle 1990; Carroll 1990; Winnefeld 1989; McCoy 1990]. Many of the arguments against NAC have also been undermined by developments in other fields. Above all, the disappearance of the Soviet threat has made the Western alleged need for a counterweight redundant [cf. U.S. Navy Admiral Larson, quoted in Fieldhouse (ed.) 1990: 250]. Additionally, the arms control process has become so all-encompassing as to make the exemption of navies look increasingly conspicuous, and politically maybe untenable. There are reasons, accordingly, to expect that some forms of NAC may become a realistic option in the years ahead.

4. **Relevant Objectives.** A number of moderate (and thus presumably realistic) steps have been suggested that would contribute to making naval strategies and postures more defensive, thereby more stabilizing:

- A prohibition against tactical nuclear weapons at sea [cf. Paul Nitze in New York Times, 6 April 1988; Hill 1989: 116–123], which has become quite unproblematic (albeit at the same time perhaps less relevant) in view of the unilateral measures to the same effect by both superpowers, begun by the *Bush Initiative;
- A ban on nuclear SLCMs [cf. Goetemoeller 1987];
- Reductions in the number of *aircraft carriers, combined with limitations on the range of shore-based naval aviation;
- Reductions of nuclear-powered attack submarines [cf. Lodgaard 1989a–b; Eberle 1990: 330];

- Numerical ceilings on major surface vessels in general, and on amphibious ships in particular;
- As a first step, regional limitations on all of the above;
- Various maritime CSBMs of both global and regional scope [Borawski 1988; Grove 1988: 4–9; Børresen 1988; Haesken & al. 1988; Purver 1988; 1988a; Wiberg 1990; P. Howard 1990].

NAVAL STRATEGY.   See *Maritime Strategy*, *Naval Arms Control*, *Sea Power*.

NEGATIVE PEACE.   See *Peace*.

NEILD, ROBERT.   Professor emeritus of economics and praelector of strategic studies at Cambridge University, former director of *SIPRI*.

RN has all along been critical of *NATO* strategy in general and its reliance on *nuclear deterrence* in particular [1981; 1990: 29–45]. His main contributions to alternative defense theory, however, have been in the realms of *armament dynamics* and *stability*. He has been extremely critical toward the traditional approach to *arms control*, especially because of its emphasis on *balance*, which is impossible to define clearly, unattainable, and irrelevant. As an alternative, he recommended unilateral restructuring of forces [1986a; cf. 1984; 1990: 106–110].

Regarding *stability*, RN advocated *mutual defensive superiority* as a more appropriate yardstick than balance. Using O and D to represent offensive and defensive strength, and A and B to denote two adversaries, he identified three typical symmetrical stances, in addition to which there would be asymmetrical cases of one-sided dominance [1990: 19]:

- $D^a = O^a = D^b = B^b$: numerical parity with no distinction between offensive and defensive capabilities.
- $O^a > D^b$, $O^b > D^a$: a situation of "mutual offensive superiority," where either side would stand to gain from starting a war.
- $D^a > O^b$, $D^b > O^a$: a stabilizing mutual defensive superiority stance, where neither side would have anything to gain from starting a war; for obvious reasons the goal recommended by RN.

On the nuclear level, mutual defensive superiority would be tantamount to *existential deterrence* defined by a notion of "sufficiency." The latter might well be zero because what would deter (i.e., induce caution in) the other side would be the nuclear potential rather than specific weapons [1990: 29–45]. More than anything else, however, stability on the nuclear level would depend on stability on the subnuclear level. Concerning conventional forces, RN recommended a shift to *NOD* postures, which would also be advantageous because they would capitalize on developments in weapons technology. The increasing accuracies achievable by means of *PGM* technology, e.g., implied that *dispersion* of forces became advantageous, and that the *Lanchester Law* of *concentration* was no longer valid [1986; 1990: 48–52].

NEOREALISM.   See *Realism*, *Barry Buzan*, *Bjørn Møller*.

NETHERLANDS, THE.   Member of *NATO* and the WEU since their founding, and until the 1970s with a solid consensus backing for NATO strategy and policy.

1. **Crumbling Consensus.** In the course of the debate on the neutron bomb, and even more so during the *INF* debate, consensus began to erode. A strong *peace*

movement emerged, based on the IKV (*Interkerkkelijk Vredesberaad* [Interchurch Peace Council]), but with strong links to both the CDA (Christian Democratic Appeal) and the PvdA (*Partij van der Arbeid* [Labor Party]) [Siccima 1985; 1988a; Everts 1985; Koole & Lucardie 1986]. However, even though the PvdA eventually opposed INF deployment, the other political parties gave their support to it, including deployment of forty-eight Tomahawk ground-launched cruise missiles in the N.

2. **NOD Debate.** Since the early 1980s, there has been quite an extensive debate on NOD in the N, especially in the academic community and peace movement, but to some extent also in the CDA and the PvdA. The latter has thus frequently used such terms as *common security* and *structural inability to attack* [Commisie 1986; PvdA 1984; 1987; cf. N. Petersen 1984; Boeker 1990]. Central participants in the academic debate have included

- Interuniversity Network for Studies on Technology Assessment in Defense (*INSTEAD*), based in the universities of Twente and Amsterdam, and including Marlies ter Borg, *John Grin*, Henny van der Graaf, Wim Smit, and others [Borg 1989; 1992; idem & Barnaby 1986; idem & Grin 1986; idem & Smit (eds.) 1987; 1989; H. Graaf 1989; Grin 1990; 1990a–b; idem & Unterseher 1988; 1989; 1990; 1990a; idem & Graaf (eds.) 1990; Muijen 1990; Smit 1986; 1989; idem & al. 1987].
- Professor *Egbert Boeker* at the Free University in Amsterdam, who has published concrete proposals and analyses of alternative defense both individually and jointly with, e.g., *Frank Barnaby* and *Lutz Unterseher* [Boeker 1986; 1988; 1988a; 1990; 1990a; idem & Unterseher 1986; Barnaby & Boeker 1982; 1988; 1989].
- Sjef Orbons, a longtime member of the *SAS*, later employed by the Dutch Ministry of Defense in the research section; he had previously published an elaborate comparison of various NOD schemes, with a special view to their use of advanced military technologies [1986; 1989];
- Ben Dankbaar at the University of Limburg, who has written about the potentials of *PGMs for NOD [1982; 1984; 1986; 1989].

3. **Nonmilitary Defense.** There has also been considerable interest in *civilian-based defense* (CBD) in the N. The government in the 1970s even sponsored a major research project on civilian defense, directed by Hylke Tromp [1978]. Tromp was quite critical of NOD, pointing out that "in the perception of an adversary, the possibility of being confronted with a perfect defense is extremely threatening, because it . . . does not rule out an attack from behind impregnable defenses" [1989]. Contradicting the assertions of NOD advocates that NOD would facilitate disarmament, Tromp even charged that it might reveal itself as "a recipe to continue and expand the *arms race*" [1990: 171].

More recent research on nonmilitary forms of defense has included a major research program at the University of Leiden, which in 1985 published *Social Defence and Soviet Military Power.* On the basis of extensive case studies (including Lithuania in 1940; *GDR, 1953; *Hungary, 1956; and Czechoslovakia, 1968, the authors arrived at considerably less sanguine conclusions than had most proponents of CBD: it had often failed and would most likely also have done so in the cases when it was not attempted [Schmid & al. 1985].

Perhaps in recognition of the limitations of CBD, another study by Lt. Col. A. A. Klumper in 1983 advised against a full-fledged shift to CBD while recommending the inclusion of CBD methods alongside military defense. In the resultant "sub-defensive system," CBD would primarily serve as a fallback option to which the state might resort if military defense had to be abandoned [Klumper 1983].

NETWORK.   See *Web.*

NEUTRALISM.   Policy of impartiality (and occasionally equidistance) resembling the policies traditionally associated with *neutrality*, but neither a prerogative of neutral states nor a policy to which a neutral state is legally committed. N might thus signify a more autonomous and less *alliance*-conformist policy on the part of alliance members, such as has been proposed, e.g., for the *FRG* [Albrecht 1982; 1983].

NEUTRALITY.   Originally signifying merely the status of nonparticipation in a particular war, but gradually coming to mean a policy of impartiality, either with regard to an individual conflict or more generally.

1. **Legal Aspects of N.** The rights and obligations of neutral states have been codified in international law and the *laws of war*, such as the *Hague Conventions* [Hakovirta 1988; De Lupis 1987: 139–142; Windsor 1989]. On the one hand, N rules out direct participation in war as well as direct assistance to the belligerents, usually understood to mean also membership in an *alliance*. On the other hand, N neither requires ideological impartiality nor rules out membership in international organizations.

2. **Degrees of N.** Just as was the case with membership of various alliances, neutrality has assumed different forms and degrees [Harkovirta 1988: 9–10; Karsh 1988; Vukadinovic 1989; Widmer 1989; Windsor 1985]. The main categories have been the following:

- *Ad hoc* N, referring merely to a policy of nonparticipation but entailing neither legal obligations in peacetime nor any commitment to political impartiality; such, for instance, has been the status of *Sweden*, traditionally described as "nonalignment, pointing toward N in war."
- Freely chosen permanent N, exemplified by the status of *Switzerland.*
- Obligatory permanent N, to which a state may by committed by a peace treaty, such as was the case of *Finland* and (less formally) *Austria.*

Switzerland has chosen a very austere definition (implying, e.g., nonmembership in the *U.N.*), but most neutrals have felt free to pursue active foreign policies and to join numerous international organizations. Ireland is a member of the EC, and both Sweden and Austria have applied for membership. Finland was for most of the postwar period constrained by the FCMA (Friendship, Cooperation, and Mutual Assistance) Treaty with the Soviet Union and pursued a policy of near equidistance between the blocs [cf. Möttölä 1984; R. Allison 1985: 174–175; Joenniemi 1989; 1989a]. It has now abrogated this treaty and applied for EC membership.

3. **Functions of N.** It has often been argued that the Euro-neutrals had special roles to play in the Cold War environment [Binter 1986a; Birnbaum 1987; Luif 1986; Skuhra 1986a; Sundelius 1987; 1989; idem (ed.) 1987; Väyrynen 1987a].

- As active mediators, who could facilitate negotiations between the blocs—the traditional role of the neutrals. In the course of the *CSCE* process, the neutrals and nonaligned countries formed a special group (the so-called N&N Group), occasionally putting forward concrete proposals that both constituted attempts at mediation and reflected the special interests of the N&N states [Sundelius 1987; Freeman 1991: 165–169, 172–180].
- As elements of a *buffer zone*, shielding each side against *surprise attack* from the other. This function was not only a reflection of the traditional role of neutrals but also a function of geographical coincidence: the fact

that most of the Euro-neutrals happened to be located between the blocs. It was even proposed to expand this neutral *zone by obliging Germany to N as a corollary of unification [Kiessling 1989; 1989a; 1990a; Löser & Schilling 1984; cf. for a critique: Lutz 1982].

With the waning of the Cold War and the East-West conflict, the traditional roles of the neutrals have been rendered less meaningful. There is no longer a need for mediators between blocs but, at most, between states. And the need for buffers is no longer obvious. Furthermore, the very connotations of N as conceived through the postwar era have become unclear, at least as far as peacetime N is concerned [Auffermann & Joenniemi 1989; Andrén 1991; Binter 1991; Bitzinger 1991; Carton 1991; Kruzel 1989a; McSweeney 1990; Neuhold 1989].

4. **N and Defense Strategy.** N does not legally presuppose any particular defense strategy, only that the neutral state seek to prevent the use by any belligerent of its territory for purposes of war. In practice, the neutrals have sought to enhance the credibility of their intentions to remain neutral through a *tous azimut* defense strategy, i.e., through preparations for defense in all directions. To plan for a *forward defense* along only one border would presumably compromise N because one party to a potential war would be singled out as a likely aggressor. Most neutrals have therefore chosen a form of *territorial defense* whereby it is planned to defend the entire territory against an attack from any direction, thus meeting the legal duties of N [Bouchet 1989; Däniker 1990; 1990a; Danspeckgruber 1986; Kruzel 1989; Karsh 1988: 63–77; Wiberg 1994].

Even though these principles have tended to make neutral defenses strictly defensive (albeit only as a matter of fact rather than of logical implication), authors of a pacifist persuasion have suggested abandoning military defenses altogether in favor of *civilian-based defense*, just as proponents of these strategies have often seen the neutral states as the most likely to implement their schemes in practice [cf. Herz 1971; Ebert 1974a; Binter 1989; Stock & Krottmayer 1986]. The possible intrinsic merits of these conceptions notwithstanding, it remains disputable whether nonmilitary defenses would satisfy the requirement for credible N [cf. De Lupis 1987: 139-142].

NEUTRALIZATION.   Process of partial or complete *dealignment and/or *disengagement of either an entire country or its parts, perhaps, but not necessarily, leading to *neutrality [cf. De Lupis 1987: 141–142].

NEUTRALS AND NONALIGNED STATES.   So-called N&N Group that collaborated, inter alia, in the context of the *CSCE process. See *Neutrality.

NEW DEMOCRATS.   See *Canada.

NEW THINKING.   Generic term for various "new" conceptions in the foreign and defense policy of the *Soviet Union during the *Gorbachev era, which subsequently spread to the rest of the *Warsaw Pact [Gorbachev 1988; D. Klein 1988; Heininger 1989; Schmidt & Schwarz 1988; 1989a–d; S. M. Meyer 1988; cf. Adomeit 1989; Türpe 1989; Benjowski & Schwarz 1989]. The term apparently stems from the *Russel-Einstein Manifesto of 1955 [in Rotblat 1967: 77–79] repeated in a 1984 publication by Anatolii Gromyko (son of the Soviet foreign minister) and Vladimir Lemoiko: New Thinking in the Nuclear Age [cf. Holden 1991: 8]. NT has above all been the brainchild of the "*institutchiki," i.e., the civilian academics at the *IMEMO and the *ISKAN, but it required a receptive ear on the part of the political leadership, the Politburo of the Communist Party [Bialer 1988; Meyer

1988; Shenfield 1987; Bluth 1990: 22–35; Holden 1990b; 1991: 8–17, 52-58; Holloway 1989; 1989a; Wettig 1991: 68–88; Oudnaren 1991; Krakau & Diehl 1989a; Nichols & Krasik 1990; Conner & Poirier 1991; Lambeth 1992; Bellamy 1992; Diehl 1993].

NT has included the following interrelated elements:

- The acknowledgement that global problems (such as environmental disaster, resource depletion, world hunger, etc.) are the shared responsibility of capitalism and socialism rather than merely the result of imperialist exploitation, for which capitalism alone was to blame.
- The notion that *peaceful coexistence* might be a quasi-permanent stage in international relations rather than merely a temporary suspension of a particular form of conflict, namely, war. This was based on reassessment of the orthodox Leninist thesis that imperialism had an inbuilt proclivity toward war.
- The notion of *common security*, which had previously been rejected for its implication that even (presumably inherently peaceful) socialist states might constitute threats to the security of their neighbors, or at least be sincerely perceived as such [Zagladin 1989; G. Arbatov 1991; cf. Jahn 1986].
- A positive assessment of *interdependence* as a contribution to *peace* as well as development [Bezrukov & Kortunov 1991]. Previously, autarky had been the ideal, inter alia, because of the perceived need to prevail in the global "class struggle."
- The conception of the "*Common European House*," implying the dominance of shared interests between European states (including the *U.S.* and the USSR) over systemic conflicts [Gorbachev 1988: 194-195; Karaganov 1990; 1990a–b; Bezrukov & Davydov 1991; Lapins 1991; cf. Adomeit 1989].
- The notion that modern industrial (or postindustrial) societies were so vulnerable as to be incapable of waging any form of war, nuclear or conventional; what might be called the *Totally Destructive Conventional War* hypothesis. By implication, *Clausewitz*'s conception of "war as a continuation of policy by other means" was declared obsolete [Proektor 1988; cf. Clausewitz 1980: 34/1984: 87 (Book I.1.24)].
- The understanding that *security* was primarily to be built by political means, i.e., by reducing hostility and resolving conflicts.
- The new emphasis on war prevention, even at the expense of the ability to prevail in a possible war waged by capitalism against the socialist states [cf. MccGwire 1987; 1987a–b; 1988a–b; 1991].
- The overriding emphasis given to *disarmament*, both in order to reduce the foreign threat to socialism and with a view to free resources for internal modernization (*perestroika*) by way of *conversion* [Faramanzyan 1989; Izyumov 1991; Kireyev 1990; 1991; cf. Ballentine 1991; Deger & Sen 1991].
- The notion of *reasonable sufficiency* (rather than superiority or *balance*) as the appropriate guideline for military planning in both the conventional and the nuclear spheres [Zhurkin & al. 1989; cf. Fitzgerald 1990; Holoboff 1990; Odom 1991].
- The belief in the inherent *defensive dominance,* hence in the possibility of renouncing the option of large-scale *counteroffensives* in favor of an *NOD*-type defense [A. Arbatov 1988; 1989; 1989a–d; 1990; 1990a; idem, Amirov, & al. 1989; idem, Kishilov, & al. 1989; Arbatova & al. 1989;

Baranovsky & Streltsov 1988; Kokoshin 1988; 1988a; 1989; 1989a-b; 1990; 1991; idem & Larionov 1987; 1989; 1990; Konovalov 1989; 1990; 1990a; 1991; idem & Mazing 1989; Makarevsky 1988; 1989; Mazing 1989; Shaskolsky 1989; Streltsov 1989].

NEW ZEALAND.   NZ has been called "the most remote country in the world." It is an island state whose closest neighbor is the tiny island of New Caledonia, 1200 miles (2,000 kms.) away. The nearest major land masses are uninhabited and demilitarized Antarctica and *Australia, with which NZ has always had friendly relations [Graham 1989: 1]. NZ has never experienced any major security problems, its small population (3.3 million inhabitants) notwithstanding.

What nevertheless spurred a major debate among New Zealanders on security and defense issues in the 1980s was the attitude of the nuclear powers toward NZ, *in casu,* the *U.S. with whom NZ was aligned in the ANZUS Treaty. In particular the "neither-confirm-nor-deny" policy of the U.S. Navy with regard to its nuclear weapons appeared to constitute a violation of NZ's nuclear-free status. It had to be assumed that major vessels routinely carrying nuclear weapons entered NZ ports without unloading these arms. Port calls became an issue after the 1984 elections, which were won by the Labour Party. Under pressure from party left-wingers and the peace movement, the new prime minister David Lange denied access to U.S. Navy vessels that could not confirm to be nonnuclear. This resulted in a severe dispute with the U.S., which effectively expelled NZ from ANZUS and waged a worldwide press campaign against it. The U.S. was probably more concerned about possible "contagious" effects of NZ's move than about the denial of port access in itself [Clements 1989; Landais-Stamp & Rogers 1989; McMillan 1987]. Thanks, inter alia, to the (discreet) support of Australia as well as to the moral support of *peace movements* across the world, NZ abided by its decision. Further, NZ was instrumental in establishing a *nuclear-weapons-free zone* consisting of the entire Southern Pacific; the Raratonga Treaty of 1985 including, besides NZ, Australia and the numerous small island states in the area [*SIPRI Yearbook* 1986: 509–519; Fry 1986; Landais-Stamp & Rogers 1989].

NIEZING, JOHAN.   See *Belgium.

N&N.   Neutral and nonaligned states in the *CSCE process. See *Neutrality.

NOD.   Nonoffensive defense (occasionally non-offensive defence). Term coined in 1985 by *Bjørn Møller* and *Anders Boserup* [Boserup 1985; 1988a; 1989; Møller 1986a; 1987; 1987a–b; 1988; 1988a–b; 1989; 1989a–e; 1989g; 1990b, d, e–i; 1991; 1991a, c; 1992; Møller & Wiberg 1989; 1990; cf. Wiberg 1989c, e; 1990b–d]. Gradually, it came to dominate the debate in the *Soviet Union and the *Warsaw Pact [Kostko 1988; Makarevsky 1989; Prystom 1988; Sauerwein 1990; Schwarz 1988; Wieczorek 1990], but it has also played a role in the alternative defense debate elsewhere [Baudisch (ed.) 1988; M. Müller 1988; Auffermann (ed.) 1990; Brauch 1991a; Cheeseman 1990; Gates 1987; Gupta 1989; Martinez-Pujalte 1991; Neild 1989a; Niezing 1990; UNIDIR (ed.) 1990a; Tromp 1990a; Joxe 1990a; Flanagan 1988a], inter alia, as a result of the wide circulation of the international research newsletter *NOD: Non-Offensive Defence,* which has been issued since 1985 by the *Centre for Peace and Conflict Research,* edited by Bjørn Møller.

1. **Terminology.** "NOD" is a synonym for "nonprovocative defense" (NPD, the dominant term in the Anglo-Saxon countries and the Netherlands) and *"defensive defense"* (dominant in the popular debate everywhere), as well as of the German

terms *"*StruNA*" or *"*StrUnA*" ("structural inability to attack").2wes Attempts at distinguishing between conceptions corresponding to the several terms are misleading because the terms are almost completely interchangeable; the only difference is that they highlight different features of the same conception. "NOD" highlights the absence of offensive capabilities (while maintaining the need for strong defensive capabilities), as do *"StruNA"* and *"StrUnA,"* which merely emphasize that this is a matter of structure. "Defensive defense" highlights defensive strength, yet primarily as a way of rendering offensive capabilities superfluous [Afheldt 1983; 1984; 1987; 1987c; Boserup & Neild (eds.) 1990]. "NPD" stresses the confidence-building effects thereof, as does the term *"confidence-building defense"* proposed by the *SAS* [SAS (eds.) 1989]. Compared with its synonyms, however, "NOD" has an additional, grammatical, advantage: it lends itself to conjugation and other linguistic manipulation. It can be used in several modes:

- As an adjective: "noddy," "noddier," "noddiest," signifying different degrees of nonoffensiveness.
- As a verb: to "'noddify," with the derived noun "noddification," referring to a process of restructuring in the direction of NOD.
- As a noun: "noddiness," signifying the degree of nonoffensiveness.
- As an adverb: "noddily," referring to a specific way of defending a country.

2. **Definition.** Various definitions have been proposed for NOD (see also *offense/defense distinction*). The most elaborate structural definition has been the following: "The size, weapons, training, logistics, doctrine, operational manuals, war-games, manoeuvres, text-books used in military academies, etc. of the armed forces are such that they are seen in their totality to be capable of a credible defence without any reliance on the use of nuclear weapons, yet incapable of offence" [Barnaby & Boeker 1988: 137]. The need for an "etc." may, however, indicate that such structural definitions are inadequate. A functional definition would, instead, focus on options: "The armed forces should be seen in their totality to be capable of a credible defense, yet incapable of offense" [Møller 1992:1; cf. 1991: 8].

3. **Functional Scope.** Nearly all NOD proposals have focused on *land forces*, and some NOD proponents have even denied the relevance of naval and air forces for NOD. If only land forces were transformed from offensive or dual-purpose into strictly defensive forces, much, indeed, would have been accomplished. Without the ability to invade and occupy an opponent's territory, no state could any longer win wars in the traditional sense. To do so would require ground forces (albeit perhaps transported by air or sea), whereas neither navies nor air forces would be able to attain victory on their own. If a state (or an *alliance*) were therefore to lack ground forces capable of offensive utilization (i.e., of taking and holding ground) completely, the capabilities of its navy and/or air force would be irrelevant according to NOD criteria.

In the real world, however, all military operations are combined-arms operations, which often involve both air forces and navies. Furthermore, the synergies between the three military arms imply that their total offensive potential usually exceeds the sum of the constituent parts. Hence the need for taking the other services into account, at least regarding their contribution to the offensive capabilities of the ground forces. The contribution need not be defined narrowly because operations by an air force or a navy against those of the adversary may influence the land battle indirectly, even if conducted far from the front. Navies and air forces have to be taken into account, and the forces operating in these "dimensions" have to be, at the very least, compatible with the NOD requirements for the land forces and, ideally, supportive thereof. See also *Air Power*, *Sea Power*.

NO EARLY FIRST USE.   Partial *conventionalization*, entailing a raising of the nuclear threshold, hence a partial approximation to *no-first-use*.

NEFU has been proclaimed as *NATO*'s long-term goal since the early 1980s [Rogers 1982; 1983a; cf. Holst 1983; Harris 1988]. Indeed, in the early 1960s the *U.S.* had put pressure on its European allies to place greater emphasis on their conventional forces with a view to achieving NEFU, whence ensued *flexible response* as a compromise [cf. Stromseth 1988]. NEFU was initially attempted through a strengthening of conventional defenses, both quantitatively and qualitatively (see *FOFA*), to no great avail. However, it was greatly facilitated by the breakthrough in conventional *arms control* effected by the Soviet commitment to *reasonable sufficiency* and the ensuing *CFE* negotiations. The achievement of parity between NATO and the *Warsaw Pact* that would, as a minimum, have resulted from CFE, would probably have sufficed to make NATO comfortable with a shift to NEFU. *A fortiori* the complete collapse of both the Warsaw Pact and the *Soviet Union* itself made it quite uncontroversial for NATO to undertake such a commitment to NEFU in 1990–1991, when nuclear weapons were relegated to "means of the last resort." See *London Declaration* and *Rome Declaration*.

NO FIRST USE.   Policy according to which nuclear weapons will be used only in retaliation to a nuclear attack.

1. **History.** NFU was first proposed in the early 1950s by *George Kennan*, but a genuine debate on the subject did not result until the early 1980s, when Kennan reiterated the proposal jointly with Robert McNamara, McGeorge Bundy, and Gerard Smith (all three former prominent officials of the Kennedy administration) [Kennan 1982: 3-10; cf. Bundy & al. 1982]. The recommendations of this so-called "Gang of Four" were followed by an attempted rebuttal by four prominent West German mainstreamers who asserted that they spoke on behalf of Europe and advised against NFU with reference to the risk of eroding deterrence [Kaiser & al. 1982; cf. Brown & Farrar-Hockley 1985; Freedman 1984; Joffe 1985]. However, a number of their fellow countrymen, as well as other Europeans, supported NFU, at least as a medium-to-long-term goal, as did numerous Americans, including several generals and admirals and the entire Union of Concerned Scientists [Krell & al. 1983; Holst 1983; Kaufmann 1983; UCS 1983; Blackaby & al. 1984; Frei 1984; Warnke 1984; Arbess & Moravcsik 1988; Boeker 1990a; Lee 1988; Carver 1983a; Dettke 1984; Fisher 1986; Goldblat 1982; Hopmann 1982; Krause 1986; Lübkemeier 1988a; Ravenal 1983; Sigal 1983; 1983a; Weller 1983].

Whereas *NATO* and the *U.S.* have continued to oppose NFU, the *Soviet Union* in 1982 solemnly declared its commitment to the policy [in Blackaby & al. (eds.) 1984: 133–143; cf. Garthoff 1990: 80–89]. However, with promulgation of the military doctrine of the Russian Federation in 1994 (see *CIS*), the commitment was retracted [excerpts in *NOD & Conversion,* 28 (January 1994): 16–23].

2. **Implications.** NFU may signify at least four things, and much of the critique has been based on a (deliberate or unintentional) confusion of the four "levels":

- A mere declaration, such as that made by the Soviet Union. Critics have correctly pointed to its limited credibility, yet mostly without paying enough attention to its logical corollary: if such a declaration were without consequence, it could not possibly erode deterrence, even if made by a prospective defender, such as NATO.
- A functional implementation of NFU. This would necessitate a change of war plans, which would be visible in exercise patterns, etc. There was, in

fact, some evidence of such a Soviet shift, lending some credibility to the NFU declaration [cf. Shenfield 1984; Garthoff 1990: 80–89].

- A structural implementation of NFU. This would be a corollary of the functional implementation because the rationale for certain types of nuclear weapons would be removed by the revised war plans: forward-deployed and vulnerable tactical nuclear weapons would be incompatible with NFU because they could, realistically, be used only in a first-use mode [Blackaby & al. 1984].
- An abandonment of the first-use option. This would be the most extreme version of NFU. However, even though critics have assumed this to be the inescapable consequence of a NATO NFU policy (albeit not of the Soviet NFU declaration, hence the alleged erosion of deterrence), it strains the imagination to envision how the option could be abandoned without a complete abolition of nuclear weapons [cf. for a good example of this flawed logic: Mearsheimer 1989: 73, 85].

Presupposing a functional and/or structural implementation of NFU, both critics and the majority of the proponents would agree that NFU should be accompanied by an improvement of *conventional stability*, especially because NFU might be tantamount to an abandonment (or at least weakening) of U.S. *extended deterrence* [cf. Ravenal 1988; Sigal 1983: 7–23; Huth 1988; Cimbala 1990: 72–95; Dembinski 1991]. Most NFU proponents have therefore recommended a strengthening of conventional defenses, most often through (presumed) qualitative improvements, such as *FOFA* and other *E.T.* schemes. Others have pointed to the *arms control* approach as preferable, whereby conventional *balance* might be attained through a Soviet build-down rather than through NATO buildup.

3. **NFU and NOD.** The arms control logic lead some NFU proponents also to support a shift to *NOD*, through which nuclear first-use (and perhaps *nuclear deterrence* as such) might be rendered superfluous [Afheldt 1984b; Møller 1988a]. NFU and NOD are interrelated in other ways:

- The structural implementation of NFU might be tantamount to the de facto creation of a *nuclear-weapons-free zone* (or corridor), especially in central Europe. Proposals for such a *zone* have often also envisioned the withdrawal of offensive-capable conventional weapons [Lodgaard 1987; idem & Berg 1985; Bahr & al. 1988].
- An NOD posture characterized by great *dispersion* would tend to make the aggressor's *concentration* less rewarding, thus lessening the risks caused by NFU: whereas the aggressor might no longer have to assume a "nuclear-scared posture" (i.e., with relative dispersal) he would still have to operate in a similar configuration, albeit for entirely different reasons.

NOLTE, WILHELM. Lieutenant colonel in the Bundeswehr, assigned to the documentation department of the Bundeswehr Academy in Hamburg. At the end of 1993 WN left the Bundeswehr to set up his own information service.

His professional affiliation notwithstanding, WN is a radical antimilitarist and strict nuclear abolitionist who has published several extremely critical analyses of prevailing trends in *NATO* and Bundeswehr military planning [e.g., 1984; 1986a]. His own preferred alternative is what he called *autonomous dissuasion* (*Autonome Abhaltung*), and it was elaborated upon by WN and his brother Hans-Heinrich, a historian and expert on *civilian-based defense*. It is a combination of civilian resistance and elements of military defense intended to render nuclear weapons entirely redundant [Nolte & Nolte 1984; cf. W. Nolte 1985; 1986; 1986b; 1987; 1988b; 1989; 1990; 1990a; cf. Møller 1991: 127–131].

The envisioned *total defense* system would be strictly defensive, emphasizing "holding capability" and consisting of three components: military defense, *civil defense*, and *civilian resistance*. The division of labor among the three elements would primarily be determined by spatial factors: military defense would be acceptable (for reasons of *damage limitation*) only in areas of low population density; conurbations would be declared militarily undefended *open towns*, defended by civilian means. Cities would thus serve as "refugee fortifications" into which the population from the rural areas might be evacuated (as was common practice in the Middle Ages). The civilian-based defense element would prevent invaders from exploiting cities as secure bases and thus as stepping-stones in their piecemeal conquest [cf. Bracken 1976; Guertner 1982].

The military component was not described in any detail, even though the author(s) mentioned as important the exploitation of terrain advantages, *barrier-type* defenses, and an in-depth and *web*-like deployment. Later, however, WN opted for the *SAS* model, albeit with some reservations with regard to the envisioned maintenance of some *minimum-deterrence* nuclear capabilities [Nolte & Nolte 1984: 159, 221–225; cf. W. Nolte 1990; 1987: 221–223; 1987b: 70; 1986: 186; 1988: 465].

The author(s) went out of their way to emphasize the compatibility of their defense proposal with the existing social structure. The result would by no means be a "militarization of society" but, rather, a desirable "civilization of the military." On the basis of the *citizen-in-uniform* tradition of, e.g., Count von Baudissin, WH moreover appealed to the *military profession* for support, but not much was forthcoming [Nolte 1987a; cf. Baudissin 1986].

NONBATTLE.   Term used by *Guy Brossolet* [1975] for gradual *attrition* of an invader through a sequence of small encounters, as a substitute for the decisive battle of annihilation.

NONINTERVENTION REGIME.   Arrangement proposed by Randall Forsberg of the *IDDS* as a supplement to *NOD* [Forsberg 1987; cf. Evera 1990b].

The background for the recommended NIR was an apparent trend in U.S. attitudes toward the Third World in the 1980s, namely, a gradual erosion of the so-called Vietnam Syndrome, as the U.S. interventions in Grenada and Libya illustrated [cf. Klare 1981; Schraeder (ed.) 1992]. "Merely" to stabilize the military situation in Europe might thus not solve the security problems of the Third World but might even exacerbate them. The two superpowers would remain in possession of intervention-capable forces, such as *aircraft carriers*, amphibious and airlift capabilities, etc. They would, however, be somewhat relieved by the abating risk of large-scale (perhaps even nuclear) war in Europe and might therefore become less cautious in their behavior vis-à-vis each other and vis-à-vis third parties [cf. Feinberg 1983; Fukuyama 1986; Allison & Williams (eds.) 1990]. Hence the presumed advisability of combining NOD with an NIR pertaining to the Third World. Its purpose would be twofold: to delegitimize any use of armed force for other than defense purposes, and to render obviously superfluous such offensive-capable military equipment, which would have to be abolished worldwide anyway in order to make NOD credible.

The main problem with devising a viable NIR may be that foreign intervention has been only one rung on a ladder of increasingly militarized relations, and distinctions between rungs have been far from clear [see Levite & al. (eds.) 1992]. The problem has been compounded by the fact of interaction because most foreign interventions by one superpower have been legitimized with reference to the behavior of the other, such as in the following (entirely hypothetical) example:

- N experiences a political revolution (or counter revolution);
- A cuts off assistance (civilian as well as military) to N;
- B provides N with substantial military supplies and military advisors;
- A responds with covert operations or by supporting subversive activities in N [cf. Prados 1986; Ranelagh 1987; Woodward 1987];
- N steps up counterinsurgency operations with the assistance of B to the point of civil war;
- A intervenes militarily.

Such ladders notwithstanding (or indeed as a way of contravening a piecemeal involvement), an NIR would presuppose establishment of arbitrary limits. On the one hand, such limits (couched, e.g., in terms of troop ceilings) would help contain escalatory competitive involvement. On the other hand, if seen in isolation, such an NIR might be seen as *ipso facto* legitimizing everything beneath the intervention threshold; hence the rationale for combining the NIR with additional measures designed for crisis prevention and conflict resolution, as well as with *arms trade regulations*, etc.

NONMILITARY DEFENSE.   Synonym for *civilian-based defense*; suggested by *Johan Galtung* as best conveying the notion of functional equivalence: ND should be a viable substitute for military defense [Galtung 1968].

NONOFFENSIVE DEFENSE.   See *NOD*.

NONPROLIFERATION.   See *NPT*, *MTCR*, *Arms Trade Regulation*.

NONPROLIFERATION TREATY.   See *NPT*.

NONPROVOCATIVE DEFENSE.   See *NOD*.

NONVIOLENCE.   Generic term for a panoply of methods for internal political struggle as well as for national defense [Sharp 1973].

The defining criterion of N is the absence of (direct and physical) violence, an imperative that might be motivated by religious, ethical, or political considerations. N as advocated by *Gandhi*, *Tolstoy*, the *Quakers*, and others is tantamount to a religious obligation (based on, e.g., the *Sermon on the Mount* and Hindu scriptures). Hence, it is unconditional, irrespective of the consequences, be they personal suffering, capitulation to an aggressor, or whatever. The N advocated by most secular authors has been more pragmatic and might be based on the following assumptions (or a combination thereof):

- That violence corrupts, personally or societally, or both. Societal corruption might presumably be caused by the *military profession* and the *MIC*, both inevitable offsprings of the reliance on violence as a means of defense. Personal corruption might occur through the process of socialization in the armed forces, which allegedly might reach the point of a cult of violence. However, although there are well-documented instances of corrupt societies in which the military has played a predominant role (the South American dictatorships, for instance), as well as of special forces and the like infatuated with violence, these are clearly exceptions, rather than the rule [cf. LeShan 1992].

- That violence breeds violence, and hence is likely to be suicidal.
- That violence does not pay, i.e., that it is an ineffective (albeit in principle permissible) means of struggle.

The American civil rights movement, under the leadership of the Reverend *Martin Luther King*, seems to have been based on (besides deeply felt religious beliefs) a special variety of the latter assumption, focusing on the symbolic or communicative impact of N. It gave the impression of moral superiority by placing the onus of initiating violence on the opponent. It was realistically assumed that public opinion would tend to favor the injured party over the violent one.

Some of the same methods have been employed by the new *peace movements* in the 1980s, with the women's peace camps at Greenham Common and Comiso as two of the most outstanding examples. Rather than seeking to influence decision-makers directly, these activists emphasized the symbolic and novel content of their actions as a way of attracting the attention of the mass media [cf. Grosse 1987; Sharp 1973: 117-182; Rochon 1989].

NONVIOLENT DEFENSE.   National defense by means of *nonviolence*, usually in the form of *civilian-based defense*, including *nonviolent resistance*.

NONVIOLENT RESISTANCE.   Reactive application of the principles of *nonviolence* and *civilian-based defense* to situations such as invasions and internal *coups d'état*. Means would include noncooperation, strikes, demonstrations, etc. intended to hamper the attainment of genuine control by the usurpers.

NORDIC BALANCE.   During the Cold War, the Nordic region (*Denmark*, *Norway*, *Sweden*, and *Finland*, as well as, in certain respects, Iceland) stood out as an area of comparatively low tension. Not only were there no serious internal conflicts (hence the status as a *security community*), but the behavior of external powers toward the region and of the region's powers toward one another was also restrained. The low level of tension may have been due to a large extent to the emphasis placed on self-restraint by all Nordic countries. The various restraints were interrelated, constituting what has often been called the NB [Brundtland 1966; 1981; Heurlin 1984; cf. for a critique: Wiberg 1986].

The NB was no ordinary *balance of power*; on the contrary: the inconceivability of war between the Nordic states made the power distribution among them close to irrelevant. Neither was the NB tantamount to a status-quo-preserving balance. Rather, it was a dynamic way of manipulating changes in, and hence mitigating challenges emanating from, the region's surroundings, i.e., a balance of self-restraint and latent options [cf. Møller 1987; 1989b; 1989f; 1990b–c; 1991b].

This specific Nordic approach seems to have been based on an understanding of the *security dilemma*. It implied that some concern for the *security* of the adversary and abstention from actions that might be regarded as provocative would be in the best interest of all states, not least of small states such as the Nordic ones. Hence the reservations of Denmark and Norway with regard to their *NATO* obligations, allowing permanent stationing of neither foreign troops nor nuclear weapons on their territory in peacetime [Heurlin 1983; Boel 1988; Saeter 1983]. Furthermore, they observed various self-imposed constraints pertaining to military activities, intended to minimize provocative effects: no NATO maneuvers in the Far North of Norway and no NATO activities on the Danish island of Bornholm (in the middle of the Baltic Sea), etc. [Holst 1985: 109; Brundtland 1985; Ørvik 1986; Holbraad 1986; Petersen 1988; Møller 1990b].

The "semineutral" status of Denmark and Norway (somewhat less aligned with the West than might have been expected) was contrasted by Finland's almost exact opposite status: a neutral state that one might have expected to fall under Soviet domination, and which had to maintain a greater degree of equidistance between East and West than implied by *neutrality*. This was achieved by the so-called Paasikivi-Kekkonen line, which constituted a blend of *dissuasion* and nonprovocation, giving rise to the pejorative term "*Finlandization*" [cf. for a critique: Quester 1990]. Finland was obliged to forgo certain military capabilities and was committed to preventing the use of its territory for attacks against the USSR [Möttölä 1984; Allison 1985; Joenniemi 1989]. Furthermore, Finnish *territorial defense* policy always emphasized the dual tasks of efficiency and nonprovocation [Ries 1989; 1989a; Kanninen & Tervasmäki 1985; Visuri 1985; 1990].

In the center of the NB stood Sweden, whose "nonalignment in peace, aiming toward neutrality in war" was traditionally ensured by means of an impressive yet predominantly nonprovocative (and nuclear-free) defense effort [Agrell & Wiberg 1980; Roberts 1976: 62–123; Canby 1981; Andrén 1989; Agrell 1979; Jervas 1983; Wallin 1991], combined with a quite active foreign policy.

As a normative concept, the NB has remained latent during most of its (presumed) existence. Only once was it allowed to demonstrate more manifestly its utility for crisis management, namely, during the 1961 "Note Crisis" (when Soviet demands against Finland were met with discreet hints that Denmark and Norway might reconsider their NATO reservations). However, the absence of manifest applications during the remainder of the period cannot be taken as evidence of the concept's invalidity because it was precisely the purpose of the balance to prevent such crisis situations from ever emerging.

The NB might best be understood as a balance of latent options, or potential next steps, reciprocally deterring one another. It was tantamount to a special pattern of behavior, according to which policymakers took the likely reactions of neighboring states and allies into account. All showed considerable self-restraint and were able to persuade their respective allies to behave with greater circumspection, lest their policies with regard to the Nordic region should have undesirable repercussions for other Nordic states, hence for the system as a whole. Both superpowers took the concerns and policy constraints of their Nordic allies and "semiallies" (i.e., Finland) into some account, acquiescing in this (in some respects peculiar) state of affairs [Joenniemi 1986: 57–60].

The accomplishment was facilitated by the rather peripheral status of the region during most of the postwar period. This framework condition was seriously challenged as the region moved into the focus of the superpower contest in the 1980s, largely because of strategic trends and geostrategic factors entirely beyond the control of the regional states: the Soviet naval buildup and the U.S. *Maritime Strategy*, the interaction between which came close to upsetting the NB. Hence the renewed interest in the Nordic countries in conceptions such as *nuclear-weapons-free zones*, *disengagement*, and *CBMs*, which were intended, inter alia, to contravene the negative trends [Joenniemi 1986; cf. Møller 1989f; Gleditsch & al. 1990].

NORDIC NUCLEAR-WEAPONS-FREE ZONE. See *Nuclear-Weapons-Free Zones*.

NORTH ATLANTIC COOPERATION COUNCIL. See *NACC*.

NORTH ATLANTIC TREATY ORGANIZATION. See *NATO*.

NORWAY. Small democratic, capitalist state, since 1949 a member of *NATO*.

1. **Traditional Security Policy.** N's NATO membership was always circumscribed by certain reservations, albeit not quite as strong as those of *Denmark*. They were motivated by the intention to give the *Soviet Union* no cause for concern, and thereby to preserve the *Nordic balance*. First, N allowed no permanent stationing of foreign troops on its soil, while allowing for a prepositioning of heavy equipment for U.S. reinforcements. Second, N allowed no nuclear weapons on its territory in peacetime [Saeter 1983]. Third, N traditionally permitted no NATO maneuvers, especially none involving German forces, in its northernmost region, which borders on the USSR. Fourth, N observed a number of self-imposed constraints regarding military activities, intending thereby to minimize provocative effects. As a matter of deliberate policy, it deployed only modest and unmistakably defensive forces in the immediate vicinity of the USSR, once again in order not to allow the USSR to charge provocation. The armed forces of N have been rather nonoffensive in general, but particularly so in the forward areas [Holst 1985: 109; Brundtland 1985; Ørvik 1986a].

2. **New Trends in the 1980s.** The breakdown of the security political consensus in the 1980s was less drastic in N than in Denmark, and N never embarked on a "footnote policy" of overt dissent from the NATO line (as did Denmark). Nevertheless, the Social Democratic Party came to support a reduced role for nuclear weapons overall, and in particular the establishment of a *nuclear-weapons-free zone* in the Nordic region, probably as a reaction to the more confrontational policies of the *U.S.* under the Reagan administration [Evensen 1983; Alfsen 1982; Lodgaard & Thee (eds.) 1983; Møller 1985; Anon. 1991; Holst 1983]. Furthermore, N expressed (cautiously, yet still unmistakably) some interest in embarking on *naval arms control* and confidence-building. The purpose would be to mitigate the negative effects of the accelerating naval arms race in the Norwegian Sea, i.e., the interaction between the expansion of the Soviet Northern Fleet (based at Murmansk, close to Norway) and the U.S. Navy's Forward *Maritime Strategy* [Haesken & al. 1988; Børresen 1988; Holst & al. (eds.) 1985].

3. **Alternative Defense Debate.** Alternative defense has consistently had its proponents in N since *Johan Galtung* began advocating *nonmilitary defense* in the 1950s. In fact, N had a certain experience with this type of defense because the resistance against German occupation (1940–1945) had been largely nonviolent [Galtung 1959; 1964; 1965; 1968; 1971; 1981; 1983; 1984; 1984a].

The institute founded by Galtung, PRIO (Peace Research Institute Oslo), has all along hosted scholars involved with alternative defense, including the recent director, *Sverre Lodgaard*, and his successor, *Dan Smith*. Furthermore, the two journals published by the institute, *Journal of Peace Research* and *Bulletin of Peace Proposals* (subsequently *Security Dialogue*), have been among the most important fora for the international debate on *NOD* and *civilian-based defense*.

The more mainstream Norwegian Institute for Foreign Affairs, NUPI (Norske Utenrikspolitiske Institut), has also been involved in research on alternative defense and security (particularly in the framework of conventional arms control [Dalhaug 1989; 1990; Skjold 1989; Skogan & Brundtland (eds.) 1990], thanks, e.g., to its director, the late Johan Jørgen Holst. In addition to being one of the most prominent experts on strategy in Europe, he was also minister of defense for two periods and, until his untimely death in 1994, foreign minister. In the field of alternative defense, Holst's contributions were particularly noteworthy in the fields of *CBMs* and regional security [1985a; 1987; 1990; 1990c–e; 1991a].

Parts of the peace movement (especially "No to Nuclear Weapons") also took a certain interest in alternative defense, even though this was overshadowed by the preoccupation with nuclear disarmament [Furre & Teigene 1983]. Finally, a few military officers (most noteworthy, General Johan Christie) dissented openly from

NATO policy by supporting the positions of the peace movements and recommending a shift to NOD [Christie 1984; 1988].

NO-TARGET PRINCIPLE.   One of the central principles in *NOD strategy: to present no lucrative targets to enemy strikes, either by elimination of potential military targets or their *dispersion [Afheldt 1976: 218–220, 247; A. Müller 1989: 128–129; Unterseher 1989g: 201; 1989c: 327].

The rationale for the NTP was, first, to improve *crisis stability. This called for removing incentives for *preemptive attack or eliminating options of such strikes, or both. Removing vulnerable targets (such as airfields, missile sites, troop assembly points, and the like) would deprive an adversary of the option of preemption, i.e., make it less rewarding to launch a first strike. Because the said facilities tended to be the same as were associated with an offensive posture, the adversary would simultaneously be relieved of incentives to preempt. Since it would no longer have to fear an attack, it would be in an even better position to wait. Second, the rationale was to limit damage in the event of a war by removing rewarding military targets for enemy nuclear strikes, particularly nuclear weapons and delivery systems but also major concentrations of conventional forces. Certain NOD proponents (e.g., *Horst Afheldt) even recommended abandonment of *forward defense because the implied creation of a front would invite nuclear strikes [1976; 1983].

The NTP had strategic as well as tactical manifestations. On the strategic level, the initial preemptive *surprise attack would be made less rewarding and/or less urgent from the point of view of the respective opponent. On the tactical scale, analogous principles are presumably valid, albeit with the sole rationale of enhancing defense efficiency. This might imply replacing, as far as possible, lucrative (i.e., high-value and at the same time vulnerable) potential targets—such as tube artillery, major surface warships, aircraft, etc.—with a larger number of individually less important targets, i.e., to disperse defense assets. Hence the preference of NOD proponents for missile over tube artillery; for mobile land-based antiship missile batteries over frigates as missile platforms; or light, man-portable weapons over large and complex weapons systems, etc.

NPD.   Acronym for nonprovocative defense and synonym for *NOD.

NPT.   Treaty on nonproliferation of nuclear weapons, signed in 1968 and entered into force in 1970 [in Fischer 1992: 317–323]. Signatories fall into two categories: nuclear states that undertake not to transfer nuclear weapons to others, and nonnuclear states that pledge to remain so. The NPT could be seen as a logical companion of *nuclear-weapons-free zones.

The NPT was tantamount to freezing the situation in 1970, when there were five nuclear states (the *U.S., the *Soviet Union, *China, *France, and the *U.K.), hence it was clearly discriminatory—as has repeatedly been pointed out by the nonnuclear states at the review conferences. What has nevertheless strengthened the NPT regime has been the interest of nonnuclear states in ensuring that their adversaries remain nonnuclear, even at the expense of accepting the discriminatory treatment implicit in the NPT [Bailey 1991b: 8–49; Barnaby 1993; Carasales 1993; Carle 1992; Creveld 1993; Fischer 1992; Ham 1993; B. Roberts 1993; Quester 1992a; Spector 1984; 1990]. Only a few observers have been persuaded by the arguments of Kenneth Waltz that the possession of nuclear weapons induces caution and, accordingly, that proliferation should be furthered [Waltz 1981; Mearsheimer 1990; 1993; cf. Deudney 1993; Davis 1993; Chafetz 1993].

What may point to the need for further strengthening the NPT regime (for instance at the 1995 Review Conference) are, first, the fact that some of the most prominent "threshold states" remain outside the regime, including *Israel [Evron 1991; Cohen 1993], *India, and Pakistan. Second, the trend toward proliferation of precursor technologies and possible means of delivery (inter alia, ballistic missiles) [Neuneck & Ischebeck (eds.) 1992]. The trend may well be accelerated by the collapse of the Soviet Union, which has made many nuclear scientists and technicians redundant and desperate, perhaps willing to serve as "nuclear mercenaries" [Zagorski 1992; 1992a]. Third, the risk of chain reactions: should one country in a regional subsystem break out of the NPT regime, its adversaries might well follow suit, a risk that may be particularly topical in the Middle East, South Asia (the India-Pakistan conflict [Hagerty 1993; Singh 1992a; 1993]), and East Asia (the Korea-*Japan-*China triangle). Finally, the regime itself was unhinged by the dissolution of the Soviet Union, tantamount in certain respects to a *de jure* proliferation because four successor states inherited nuclear weapons, whereas only Russia proclaimed itself to be (and was accepted as) the nominal successor (see *CIS). Hence the efforts, inter alia, by the U.S., at persuading Kazakhstan, Belarus, and, above all, Ukraine to sign the NPT [Miller 1993a]. On the other hand, the world has seen an instance of "deproliferation," when South Africa abandoned its (thitherto clandestine) nuclear capability [*International Security Digest,* 1 (October 1993): 2].

NUCLEAR DETERRENCE.   The rationale of nuclear weapons, since their invention in 1945, has been their presumed deterrent, i.e., war-preventing, effect, according to the logic of Bernard Brodie: "Thus far the chief purpose of a military establishment has been to win wars. From now on its chief purpose must be to avert them. It can have no other useful purpose" [Brodie (ed.) 1946: 76].

1. **Deterrence Logic.** The presumed nuclear-weapons/war-prevention nexus has been partly based on a terminological confusion: The generic term "war prevention" tends to be equated with one particular species thereof, namely, deterrence by retaliation or even retaliation by nuclear means [cf. G. Snyder 1961: 14–16]. However, not only has the alleged synonymy between war prevention and ND been unfounded but deterrence itself has also rested on questionable premises. Inter alia, it presupposes that the adversary will seek a revision of the status quo and be prepared to use military means to attain that end. Furthermore, the adversary is presumed to be a rational actor, guided by cost-benefit considerations. The deterree will, accordingly, presumably be dissuaded from aggression by the prospect of nuclear retaliation, the costs of which would inevitably exceed any conceivable gains.

2. **Credibility Problems.** For all its apparent simplicity, the notion of punishing an aggressor by nuclear means encountered serious problems, particularly after the advent of nuclear *balance by the late 1950s or early 1960s; not in the sense of numerical parity but of mutual second-strike capabilities.

At the root of most of these problems lies awareness that ND would depend on the conceivability of carrying out the retaliatory threat; this has very little (if anything) to do with the size or composition of the nuclear arsenals but hinges on political deliberations about the use or nonuse of nuclear force, i.e., on political will. Furthermore, because ND is about making certain actions seem unattractive to another state presumably contemplating them, it is basically a psychological phenomenon. In consequence, ND presupposes both an unswerving will on the part of the deterrer to retaliate and the deterree's belief it will do so, hence the problem of credibility, i.e., of how to convince a would-be aggressor of one's determination to retaliate [cf. Kaufmann 1956].

3. **Contradictions and Dilemmas.** From the credibility problem have followed

a number of dilemmas [cf. Jervis 1984; 1987; 1989; 1989a; Lebow 1981; 1983; 1985; 1989; MccGwire 1986; Buzan 1987a: 163–196; Wiberg 1989], among which are the following:

- The deterrence-versus-*self-deterrence* dilemma, the gist of which is that the defender's deterrent threat is no more credible than the aggressor's implicit threat of counterretaliation; that the latter would thus tend to negate the former, i.e., induce self-deterrence.
- The *ex ante*-versus-*ex post* dilemma, according to which a retaliatory threat is rational only as a threat, whereas its implementation after a breakdown of deterrence would serve no rational purpose but could constitute merely revenge. Such revenge would, furthermore, violate all *Laws of War* and *just war* principles because the victims would primarily be civilians [see Shue (ed.) 1989; McCall & Ramsbotham (eds.) 1990].
- The rationality-versus-irrationality dilemma, stemming from the need to make contradictory assumptions about the rationality of the two sides. For deterrence to function at all, the prospective aggressor has to possess a considerable degree of rationality and be motivated by cost-benefit calculations that the deterrer would be in a position to influence. On the other hand, diametrically opposite assumptions must be made about the deterrer itself because its actual retaliation could never be rational.
- The limitation-versus-*escalation* dilemma, deriving from the first dilemma. An escape might presumably be found from self-deterrence if boundless escalation could somehow be avoided, e.g., through the acquisition of flexible nuclear options that would allow for calibrated responses to various levels of provocation. On the other hand, it would presumably be precisely the prospects of boundless escalation that would deter, hence the need for avoiding *thresholds* or "firebreaks" in the escalatory momentum that might allow a prospective aggressor to calculate its risks. Without thresholds escalation would presumably proceed unbounded once started, in which case to step onto even the first rungs of the escalation ladder would be tantamount to "slow-motion suicide," from which self-deterrence would inevitably follow [cf. Kissinger 1969; Martin 1979; Brodie 1959: 329; Clark 1982];
- The war prevention-versus-warfighting dilemma, combining elements of all of the above. If the deterrer were able only to threaten nuclear war, while incapable of actually waging it, his deterrent would be ineffective. Notwithstanding the solemn declarations by both superpowers that a nuclear war must never be fought, each felt compelled to prepare for such a contingency as well as to convey the impression that it would actually be prepared to fight, should the need arise. By so doing, it might conceivably have misled its respective adversary into believing that it was actually preparing for an offensive nuclear war, hence the conceivability of "reciprocal fears of surprise attack" [Schelling 1960: 207].

4. **Alternative Forms of Deterrence.** ND is far from the only possible form of war prevention. First, there are alternative forms of deterrence by retaliation, such as *conventional retaliation*. Second, a defender might opt for *deterrence by denial*, whereby the prospects of gaining the prize of aggression would be diminished, usually by means of a strong conventional defense. This has precisely been what *NOD* has sought to accomplish by means of an effective yet nonnuclear defense that possesses no offensive capabilities and has no means of retaliating against the aggressor (in a military sense at least).

5. **NOD and Nuclear Deterrence.** Most NOD proponents reject the notion of

ND altogether, whereby they may, however, be disregarding a certain rational nucleus that might be worth preserving as one element in a comprehensive strategy of war prevention. The empirical case for the efficiency of ND is very weak, but neither is there any empirical foundation against it, and certain of its tenets seem inherently plausible. On the presumption of approximate rationality on both sides, ND might well influence the cost-benefit calculations of any prospective attacker in the sense that there might be war plans that are never seriously contemplated because of ND but that might well have been contemplated in its absence.

Furthermore, ND has been accompanied by (and may have been a reason for) a "step-gently regime" in the conduct of international relations, at least concerning the superpowers and their immediate allies (cf. *Prevention of Nuclear War Agreement*) [Garthoff 1985: 334–344; cf. George 1988a; Allison & Williams (eds.) 1990, 269–271]. This regime, however, was probably not a function of any particular configuration of nuclear forces but, rather, of the availability, pure and simple, of nuclear weapons: what has been called "deterrence as a fact" or "*existential deterrence*" [Bundy 1986; cf. Freedman 1988]. It is even possible to conceive of a "blueprint deterrence," based exclusively on the potential of nuclear weapons and quite compatible with their physical abolition [Schell 1984: 119; cf. J. N. Miller 1988; Booth & Wheeler 1992]. In the shadow of weaponless ND, a prospective aggressor could not safely plan for escalating a conventional war beyond bounds because it would always have to reckon with the possibility of nuclear retaliation. Nor could wars any longer be won in the Clausewitzian sense of the victor's enforcing total submission on the vanquished. The losing side would always retain the last-resort option of negating victory by playing the nuclear trump card. In this sense, ND might indeed strengthen the defender and might thus be an appropriate supplement to, or perhaps even an integral element in, NOD [cf. Mearsheimer 1989a; 1990; Evera 1990a].

NUCLEAR EQUIVALENCE.   Possibility of achieving by conventional means (i.e., quasi-nuclear weapons) the same destructive effect as by means of nuclear weapons. The precise significance of equivalence is, however, dependent on the particular mission in question. Concerning blast effect, fuel-air explosives are, for instance, entirely comparable to tactical nuclear weapons, just as various forms of submunitions (such as Skeets) are with regard to tank-killing potential, and certain *chemical weapons* in terms of area contamination [Robinson 1983; cf. Cotter 1983]. NE has been portrayed as a means of *conventionalization* [ESECS 1983; cf. Harris 1988], yet it may not solve the problems associated with nuclear deterrence but even exacerbate them, unless accompanied by drastic reductions in nuclear holdings on both sides. Should one side, e.g., acquire capabilities for destroying the other's nuclear weapons by conventional means, *crisis stability* would deteriorate. The inhibitions against a counterforce strike would presumably be weaker if it could be carried out without crossing the nuclear threshold. The party fearing a disarming counterforce attack against its nuclear weapons would, however, have undiminished incentives for preemption.

NUCLEAR THRESHOLD.   See *Thresholds, *No Early First Use, *No First Use, *Damage Limitation, *Conventionalization.*

NUCLEAR-WEAPONS-FREE ZONE.   Political arrangements between a number of neighboring states for the purpose of making and/or maintaining a certain geographical region free of nuclear weapons.

1. **Rationale.** Contrary to the assertions of opponents, a NWFZ has never been conceived of as providing any defense against nuclear weapons (e.g., as an

area that could not be attacked with nuclear weapons: a truly absurd notion). A NWFZ has been intended as a contribution to curbing the nuclear *arms race* and improving *stability*, by stemming proliferation (in the case of a region with a significant risk thereof) and/or through *disengagement* (in the case of an already-nuclear region). The *Soviet Union* consistently favoured NWFZs; the *U.S.* and the other Western powers tended to oppose them [Kalyadin 1988; Meyer 1986].

2. **General Characteristics.** To be meaningful, a NWFZ would at least have to include the following elements:

- An obligation on the part of the participating states not to acquire nuclear weapons (and/or to destroy whatever nuclear weapons they might already possess) and not to allow other states to station nuclear weapons on their territory (usually, but not necessarily, including territorial waters and airspace).
- A commitment by other powers to respect the no-stationing clause.
- "Negative security guarantees" on the part of the nuclear powers to the states parties to the zone, i.e., commitments not to subject them to nuclear bombardment under any circumstances.
- Various "external limitations" on the part of the nuclear powers: as a minimum, a withdrawal of nuclear weapons specifically targeted (in terms of range and location) against the states party to the zone, i.e., which are in contravention of the aforementioned guarantee.
- An arrangement for monitoring all of the above provisions [Lodgaard 1982; 1983; SNU 1982; Rosas 1983a; Møller 1985; Kittel & al. 1991].

3. **Central European NWFZ.** Numerous proposals were made for a central European NWFZ in the 1950s: some of the *Germany Plans*, the *Rapacki Plan*, the *Gomulka Plan*, etc. In the 1980s proposals also abounded, among which were the *Palme Commission*'s proposal for a 300-km-wide nuclear-weapons-free corridor, the Polish *Jaruszelski Plan*, and the draft treaty negotiated by the *SPD* and the SED [Palme Commission 1982: 146–150; SPD & SED 1987].

The zonal schemes of the 1950s were conceived of as contributions to facilitating German (re)unification and as counters to the deployment in progress of tactical nuclear weapons; the rationales for the zones of the 1980s were more diverse. Some were embedded in schemes of German unification; others were envisioned as corollaries of a *no-first-use* policy; still others were embedded in disengagement schemes intended to improve *crisis stability* and/or to seal Europe off from the superpower confrontation. In many cases, NWFZ schemes were merely one element in broader zonal conceptions, which also encompassed conventional weaponry, particularly such as was traditionally associated with offensive warfare [Albrecht 1983a; Lodgaard & Thee (eds.) 1983; Lodgaard 1982; idem & Berg 1983; 1985; Møller 1988a; Brie 1988; Karkoszka 1988b; 1989c; Lutz 1988c].

4. **Nordic NWFZ.** Establishment of a NWFZ in the Nordic region (*Denmark*, *Norway*, *Sweden*, *Finland*, and perhaps Iceland) has been debated since the 1950s, when the Soviet Union made a proposal to this effect. It was followed in 1963 by the first *Kekkonen Plan* (proposed by the Finnish president), reiterated in 1978, in both cases without arousing any interest in the other Nordic countries.

A new wave in the NWFZ debate began in the early 1980s, when the social democratic parties endorsed the idea. Several concrete proposals were published, among which the most elaborate were those of *Jens Evensen* and of seven Nordic peace movements (*Nej til Atomvåben, Finlands Fredsförbund, De hundras Komitee, Folk fyri Fridi, Samtök Herstödsaanstaendiga, Nej til Atomvåpen,* and *Svenska Freds-och Skiljedomsföreningen*) ["Forslag om Norden som Atomvåbenfri

Zone" (1984), in Møller (ed.) 1985: 80–82; cf. Alfsen 1982; Evensen 1983; cf. Rosas 1983].

Decisions in favor of the NWFZ were taken by Nordic parliaments during the years after 1983, as a result of which an inter-Nordic civil service committee was appointed. It stalled the process until it finally reported back in 1991 [Anon. 1991]. Even though nearly all the obstacles that had thitherto stood in the way of a zone had now been removed (as a result of the developments in the East), no substantial steps had been taken by early 1994.

A Nordic NWFZ would be unique in several respects:

- It would be prophylactic rather than reactive, i.e., by seeking to preserve an already existing (and quite solid) nuclear-free status.
- It would include both *alliance members (Denmark and Norway) and neutrals (Sweden and Finland).
- It would be directed exclusively against external challenges (no Nordic state could conceivably want to build nuclear weapons, and all have formally renounced the right to do so).

It would thus be a considerably less radical departure from prevailing patterns of security policy than most other NWFZs under consideration [SNU 1982; Møller 1985; Gleditsch & al. 1990: 132–145; Wiberg 1986; Krohn 1989].

5. **Baltic NWFZ.** The inclusion or exclusion of the Baltic Sea in the Nordic zone has been disputed. Both critics and supporters of the Nordic NWFZ have pointed out that the Baltic Sea is beyond the jurisdiction of the states party to the envisioned Nordic NWFZ, hence could not possibly be included directly. However, in light of developments in Germany and *Poland, as well as in the Soviet Union in general, and the newly independent Baltic states in particular, there might be better opportunities than ever before for the establishment of a Baltic NWFZ comprising Poland, the *FRG, Denmark, Sweden, Finland, Estonia, Latvia, Lithuania, and parts of Russia. In addition to being directly linked to the Nordic zone, the Baltic zone would fit in directly with the proposed NWFZ in central Europe, along the lines of, say, the Palme Commission's proposal, the SPD-SED agreements, or the Jaruszelski Plan. Whereas the zone could not in any formal sense include the Baltic Sea (a *mare liberum*), by committing the littoral states (above all, Russia) to the withdrawal of nuclear weapons from their naval forces, it would remove all justification for other naval powers' navigation of the sea with nuclear-armed vessels [Krohn 1991c].

6. **Other European NWFZs.** The only other European subregion where there has been significant interest in a NWFZ has been the *Balkans [Behar & Nedev 1983; Drougos 1988; Platias & Rydell 1982]. Furthermore, the dissolution of the Soviet Union was accompanied by a rather curious variation on the NWFZ theme, in that former union republics (the Ukraine, Belarus, Kazakhstan) upon declaring themselves independent proclaimed it as their intention to constitute NWFZs, notwithstanding the fact that they had "inherited" Soviet nuclear weapons (cf. *CIS).

7. **Non-European NWFZs.** Two NWFZs already exist beyond the European orbit, namely, those established by the 1967 Tlatelolco Treaty, which covers most of *Latin America [SIPRI Yearbook 1969–1970: 237–256; Robles 1969; Calderon 1982], and the 1985 Raratonga Treaty, which covers the Southern Pacific [SIPRI Yearbook 1986: 509–519; Fry 1986; Landais-Stamp & Rogers 1989; Paul 1993]. Proposals have also been made for NWFZs covering Africa, the Indian subcontinent, the Middle East, the *Koreas, and other regions, most of which are, at best, realistic only as long-term perspectives [Kittel & al. 1991; Shim 1991].

O

ØBERG, JAN.   See *Sweden*.

OCA.   Offensive Counter-Air. Form of *air defense* through destruction of an enemy's aircraft on the ground.

OCA, e.g., played an important role in Germany's attacks in 1940 against the Low Countries and its 1941 attack against the *Soviet Union*, as well as in Japan's 1941 attack against the *U.S.* and the *U.K.* in the Pacific area. A more recent example is the destruction by the Israeli air force of the bulk of the Egyptian, Jordanian, and Syrian air forces, as well as a good part of the Iraqi air force, on 5 June 1967 [O'Ballance 1972: 49–84], which practically amounted to winning the entire June War (cf. *Israel*). Nearly the same happened in the opening phase of Operation Desert Storm (see *Gulf War*) in January 1991, when most of the Iraqi air force was literally wiped out in one of history's largest and most successful OCA operations [Friedman 1991: 147–196]. OCA may thus occasionally be decisive, but its effectiveness seems to hinge on the surprise element. Against well-prepared defenses, (such as those of the U.K. during the Battle of Britain) OCA tends to be much less effective.

Both the Soviet Union and the U.S. have traditionally paid considerable attention to OCA. However, OCA has serious detrimental effects on *crisis stability* because two sides subscribing to OCA conceptions tend to poise their offensive-capable air forces against each other, thus providing both with strong incentives for *preemptive attack*, and with clear risks of misunderstandings. Should one side, e.g., send its aircraft on airborne alert for fear of being taken out on the ground, the other side might misinterpret this as an OCA strike in progress and take off as well [cf. Møller 1987a; 1989; 1989b; Hagena 1994]. Further, doubts have been raised about the cost-efficiency of OCA, at least in an East-West context. First, runways might not be particularly lucrative as targets because prevention of their swift repair would require repeated strikes with very sophisticated, combined runway-cratering and area-denial munitions. Second, the chances of hitting the aircraft themselves may be rather slim because of countermeasures such as *dispersion*, hardened shelters, decoys, and various camouflage devices. Third, even if runways and additional support functions, as well as some aircraft, were to be effectively destroyed, such a strike would still leave undamaged what is perhaps the most important of all *air power* assets, namely, the skilled pilots [cf. Brown 1986: 178–190, 54–55].

OFFENSE.   See *Offense/Defense Distinction*; *Defense*, *Counteroffensive*.

OFFENSE/DEFENSE DISTINCTION.   The entire *NOD* discourse (including the assertions of improved *crisis stability* and *arms race stability* through *mutual defensive superiority*) hinges on the possibility of distinguishing between offensive and defensive postures and strategic conceptions. To be meaningful, distinctions have to be made at several levels of analysis (cf. *military science*), which should be kept analytically separate in order to avoid confusion [cf. Møller 1992: 147–151; Neild 1990: 15–16].

1. *__Politics__. It is possible to distinguish between defensive and offensive policies: to preserve the status quo is defensive; to seek a change is offensive (albeit sometimes justified). Politically defensive intentions, however, may well be compatible with plans for a defense by offensive military means, including *preventive war* and/or *preemptive attacks*.

2. *__Grand Strategy, Strategy, and Operational Art__. A defensive grand strategy would rule out preventive war and preemptive attacks and envisions only to defend a certain territory against attack, whereas an offensive grand strategy might, e.g., seek to improve future defensive capabilities through establishment of an extended *glacis* (as *Israel* did after the 1967 June War). A defensive *strategy*, in its turn, would merely envision the defense of existing territory. This might call for *counteroffensives* of different magnitudes: small (tactical-scale) counterattacks or operational or even strategic-scale counteroffensives through which the invader would not only be evicted but pursued beyond the border—perhaps all the way to its homeland in order to bring about capitulation, thereby preventing repetition of the attack.

3. __Tactics.__ The distinction between offensive and defensive tactics is simply a matter of who initiates each encounter: the tactical defender parrying the offensive party's attacks. This makes it obvious that wars could not possibly be waged by means of exclusively defensive tactics because doing so would imply relinquishing all initiative to the adversary. Indeed, a defensive strategy has often been combined with very offensive tactics because of the trade-off between strategic mobility and tactical agility, as well as because of the advantages to the defender of concealment. The attacker would have to move over long distances (in transport aircraft, amphibious vessels, or whatever), hence would be unable to bring along as many trucks and other vehicles as would be at the defender's immediate disposal. Further, the defender might fight in great *dispersion* and from shielded positions, whereas the attacker would have to move about, also across open fields, thereby often allowing the defender the first shot.

4. __Weapons.__ Even though occasional attempts have been made to distinguish between offensive and *defensive weapons*, close analysis makes it obvious that all weapons can, in principle, be used for offensive and defensive purposes. No absolute dichotomy can be arrived at. The 1930s quest proved futile and merely distracted attention from more fruitful approaches to qualitative disarmament (cf. *World Disarmament Conference*).

Nevertheless, in a particular context (defined both in space and time), weapons tend to benefit an attacker and a defender to different degrees. Hence the emphasis placed by most NOD proponents (referring, at least implicitly, to the context of Europe in the 1980s) on a build-down of, e.g., *tanks*, heavy artillery, fighter-bombers, and *aircraft carriers*. Hence also the selection of "treaty-limited items" in the *CFE* negotiations. Even though tanks are not per se offensive, a tank-heavy posture is almost inevitably more suitable for offensives than one emphasizing anti-tank weapons.

Even traditional military analysis has made distinctions between the value of a specific type of weapon in different contexts. The U.S. Army assigned offensive and defensive values to various categories of weapons, as set out in Table 28 [William Mako, quoted in Snyder 1987: 312]:

By implication, it would be possible to change the aggregate offensive capability through modifications in the composition of military formations, and in the equipment of individual units, as recommended by most NOD proponents.

Table 28.   U.S. Army Estimates of Weapon Strength, 1974

	Value	
	Offense	Defense
Tanks	64	55
Armored personnel carriers	13	6
Antitank weapons	27	46
Artillery	72	85
Mortars	37	47
Armed helicopters	33	44

5. **NOD and Traditional Strategy Compared.** On the basis of the above abstract distinctions, it is possible to clarify the difference between NOD and traditional strategic conceptions. In Table 29, a typical NOD model (that of the *SAS*) has been compared at each level of analysis with its counterparts in *NATO* and the *Soviet Union*. D and O stand for defensive and offensive; > and = signify emphasis; D > O means that the emphasis is on the defensive; and = approximate equality between the two.

Table 29.   NOD and Traditional Strategy, ca. 1985

	NOD	NATO	USSR
Politics	D	D	D
Grand strategy	D	D	D > O
Strategy	D	D	O > D
Operational art	D	D > O	O > D
Tactics	O = D	O = D	O = D
Posture	D	D > O	O > D

OFFENSIVE A L'OUTRANCE.   See *Plan 17.*

OFFENSIVE COUNTER-AIR.   See *OCA.*

OFFENSIVE-TURNED-DEFENSIVE.   According to *Clausewitz*, OTD was a logical corollary of attack, even though he described the introduction of defense as a "necessary evil" and even as an "original sin" and a "mortal disease" of the attack [1980: 584/1984: 424 (Book 7.2)].

The reasons for the "dialectical unity" of attack and defense were, first, that offensives are discontinuous and that every suspension of the forward momentum implies the introduction of defense. Second, a successful attack leaves behind it a rear area that needs to be protected (presupposing that the defender has means of threatening it, in a "defensive-turned-offensive" mode). Further, an attacker conquers ground that henceforth needs protection, i.e., defense. After the initial offensive thrust, the aggressor's forces therefore have to assume defensive positions in order to preserve their conquest; the "real" defender is forced to counterattack in order to restore the *status quo ante bellum* [cf. Mearsheimer 1983: 134–164; cf. Freedman 1987: 12–21], as was the case, e.g., during the *Gulf War* [Friedman 1991: 108–119; *International Defense Review,* 1991(9): 989–994].

Under such conditions, furthermore, the "intrinsic superiority of the defense" might be turned against the defender itself because the attacker would now be in a position to fortify his conquest, to construct defensive fighting positions, etc., whereas the defender would have to expose himself by attacking. Even though the attacker-turned-defender would only rarely enjoy the full defensive advantage, the force requirements for the defender-turned-attacker might nevertheless be quite high (cf. *Counteroffensive, *three-to-one rule, *Defense, *Defense Dominance).

OFFENSIVE WEAPON.   See *Offense/Defense Distinction.

OFFENSIVE-WEAPONS-FREE ZONE.   See *Zones, *Disengagement.

OLLENHAUER PLAN.   Precursor of the SPD's *Germany Plan, presented by the party's chairman, Erich Ollenhauer, 23 May 1957. It envisioned the integration of a reunited Germany into a comprehensive all-European security system (as a replacement of the *alliances), based, inter alia, on reciprocal obligations to refrain from aggression and on unconditional mutual assistance in case of attack, as well as on (unspecified) *arms control measures [Schramm & al. (eds.) 1972: 9–11; cf. Hacker 1989: 48–49].

OPEN TOWNS.   Conurbations that are militarily undefended yet protected by the *Laws of War, which proscribe attacks against cities declared to be OT [De Lupis 1987: 233–236]. Most *NOD proposals have placed a certain emphasis on *damage limitation, wherefore care is to be taken to avoid combat in densely populated areas, e.g., by declaring major cities OT [Löser 1981; Afheldt 1983; Møller 1989c–d]. However, because the high degree of urbanization, particularly in central Europe, might conceivably make it possible for an invader to use the militarily deserted cities as secure bases [Bracken 1976; cf. Guertner 1982], there might be trade-offs between damage limitation and military efficiency.To avoid demilitarizing an entire country, distinctions might have to be made between major and minor cities. These could be made quite arbitrarily because the establishment of OT would entirely be a matter of concrete declarations. Extensive "killing zones" between cities might thus be ensured, through which an invader would enjoy no free passage. An additional precaution might be to plan for *civilian-based defense in OT in order to make the occupation less rewarding for an attacker, as suggested by various NOD proponents [Nolte & Nolte 1984].

OPERATIONAL ART.   See *Military Science.

ORBONS, SJEF.   See the *Netherlands, *SAS.

OSGOOD, CHARLES.   American psychologist who in the early 1960s suggested a particular form of *gradualism as the best strategy for breaking the vicious circle of actions and reactions in U.S.–Soviet relations. The strategy was called *GRIT (Gradual Reciprocated Initiatives in Tension Reduction) [1962; cf. 1959; cf. Böge & Wilke 1984: 26–55; George 1988b; Goldstein & Freeman 1990; R. Fischer 1990: 196–202].

OSTPOLITIK.   See *Egon Bahr, *Change Through Rapprochement, *Détente, *FRG, *SPD.

OUT-OF-AREA.   Shorthand for the perennial dispute within *NATO on whether, or to what extent, the alliance should assume responsibilities beyond the Tropic of Cancer, stipulated as the southern boundary of NATO.

The *U.S. consistently pushed for an expanded role for NATO, or at the very least for an acknowledgment (accompanied by some reimbursement) of U.S. activities in the Third World in general, and in the Persian Gulf region in particular, as activities done on behalf of NATO. The allies, however, have been reluctant to acknowledge this, much less prepared to participate or reimburse the U.S., a disagreement that has loomed large in the long-standing dispute over burden sharing within NATO. In actual fact, NATO-Europe's reluctance to participate or sponsor OOA was a reflection not merely of "free riding" but also of disagreement about the appropriate means of solving extra-European conflicts [Bentinck 1986; cf. Conetta 1991].

The alternative defense community in Western Europe (including most of the socialist parties) has been almost unanimously opposed to OOA (cf. *Nonintervention Regime*), even though a certain reassessment seems to be under way as a consequence of the disappearance of the East-West conflict. This might open up opportunities for creating a *collective security* system, e.g., under the auspices of the *U.N., that might well have to involve OOA operations. Certain European NATO members (above all the *FRG) are, however, constitutionally debarred from participating in such operations [cf. Souchon 1990; Møller 1991: 256–259].

OXFORD RESEARCH GROUP.   See *U.K.

P

PACIFISM. Principled rejection of any use of arms, even for purposes of *self-defense*, for religious or other ethical reasons [J. T. Johnson 1987]. The term "pacifism" is sometimes also used by critics about the *peace movements* and others with (alleged) antidefense sentiments.

P is quite compatible with (indeed points directly toward) an advocacy of *civilian-based defense*, but it is hardly possible to combine P with proposals for *NOD*, which presupposes the use of arms. To the extent that NOD is conceived of merely as an intermediate step toward general and complete *disarmament*, however, it may receive the support of devoted pacifists, albeit only as the lesser evil.

PALME COMMISSION. Independent Commission on Disarmament and Security Issues, headed by the late Swedish Prime Minister Olof Palme until his assassination in 1986. The members included, inter alia, *Georgy Arbatov*, *Egon Bahr*, Gro Harlem Brundtland, Alfonso Garcia-Robles, David Owen, and Cyrus Vance. The PC published its major report, *Common Security; A Blueprint for Survival,* in 1982 and concluded its work with a smaller report, *A World at Peace: Common Security in the Twenty-First Century,* in 1989 [1982; 1989].

The PC is primarily known for lending authority to the concept of *common security*, first promoted by Bahr and other West Germans [Bahr 1983; 1985b; Komitee 1981: 18; cf. Lutz 1986a; 1992a; Rothschild 1992]. Further, the 1982 report included a number of concrete recommendations, including one for establishing a *nuclear-weapons-free zone* (or corridor) in central Europe, but most were rather cautious and reflected compromise between conflicting views. Even though the PC has been criticized by radicals for its conservatism, they have nevertheless often exploited the authority of the PC [Palme Commission 1982: 146–149; cf. Krippendorff & Stuckenbrock (eds.) 1983].

PARAMILITARY FORCES. Nonmilitary forces, such as nonuniformed, guerrilla-like armed forces, performing tasks that are usually the prerogative of the military. Use of PF has been recommended by a few *NOD* advocates, such as *Johan Galtung* [1983; 1984: 180-184; 1984a], but most have advised against this because a blurring of distinctions would weaken whatever protection the *Laws of War* might provide, both for combatants and civilians [cf. Møller 1989c-d].

PARIS CHARTER. The Paris Charter for a New Europe, signed in November 1990 by the (by then thirty-four) *CSCE* member states, solemnly declaring that "a new era of *democracy*, freedom and unity" was commencing in Europe. The signatories further decided to embark upon a certain institutionalization of the CSCE process, inter alia, through establishment of

- A council of foreign ministers that was to meet at least annually;
- A committee of high-ranking civil servants in permanent session;
- A secretariat, to be housed in Prague;

- A center for conflict prevention, to be housed in Vienna;
- A bureau for free elections, to be housed in Warsaw; and
- A parliamentary assembly [*SIPRI* 1991: 603–611].

PARIS TREATY. See *Finland.

PARITY. See *Balance.

PARTISAN WARFARE. See *Guerrilla Warfare.

PARTNERSHIP FOR PEACE. Program launched by *NATO in* January 1994 under the auspices of the North Atlantic Cooperation Council (see *NACC). The intention was to meet the wishes of former *Warsaw Pact member states for full membership in NATO, without thereby having to provide them with full security guarantees that might conceivably force NATO to become embroiled in conflicts from which it would prefer to stay aloof.

In the inaugural declaration from the summit meeting of the North Atlantic Council (10–11 January 1994) it was stated that "the Alliance remains open to the membership of other European countries." Rather than admitting new members, however, with PFP NATO intended to "invite Partners to join us in new political and military efforts." The purpose would be to "expand and intensify political and military cooperation throughout Europe, increase stability, diminish threats to peace, and build strengthened relationships by promoting the spirit of practical cooperation and commitment to democratic principles that underpin our Alliance." Rather than pledge military assistance, NATO promised to "consult with any active participant in the Partnership if that partner perceives a direct threat to its territorial integrity, political independence, or security." It was further envisioned that joint military exercises would be conducted, inter alia, with a view toward collaboration in *peacekeeping operations [*Atlantic Documents,* 83 (12 January 1994)]. Albeit somewhat less than enthusiastically, most former Warsaw Pact states had by mid-February announced interest in joining the PFP.

PEACE. In the past, P was often regarded as an anomaly, inter alia, by ultra-Realists believing in the *bellum omnium contra omnes* of Thomas Hobbes, and even as something deplorable (by bellicose authors, e.g., in Germany of the early twentieth century) [cf. Møller 1991: 40–43; 1992: 106–108, 121–122, 127–128]. Today the connotations of "P" are unambiguously positive. Nobody expecting to be taken seriously can proclaim himself to be against P, and wars are usually justified as means to attain (a lasting and/or just) P. This consensus on the desirability of P nothwithstanding, the meaning of the term remains disputed.

Various peace researchers have argued that the conception of P as merely the absence of war, i.e., large-scale interstate direct violence, is too narrow, that there is a need to distinguish between positive and negative P. This has led them to distinguish between two types of violence: direct and structural [Galtung 1969]. The world during the Cold War was characterized (at least as far as Europe was concerned) by negative P (i.e., an absence of war), but also by a perpetual risk of war and accompanying preparations for war (i.e., arms racing). Furthermore, societal and international inequalities, repression, exploitation, and other forms of structural violence and latent conflicts flourished, compounding the risks of an eventual conflagration. Positive P, on the other hand, would presuppose the elimination (or at least a substantial reduction) of both the potential for violence (inter alia, the military arsenals) and the latent (i.e., structural) violence. It required a just and equitable world order. A genuine "peace order" has accordingly been identified as the relevant goal, and as the only alternative to the prevailing "organized peaceless-ness" [Senghaas 1969; idem (ed.) 1971].

Following a similar chain of reasoning, certain proponents of the more modern concept of *common security have emphasized the need to combine hedges against war (*NOD included therein) with the promotion of positive P [Lutz 1986b; 1987e; Buro 1982: 85–86; 1983; 1983a; 1987; 1988]. Finally, the recent acknowledgment by others, *NATO among them, that *security has to be conceived in broader terms (to include, inter alia, economic and environmental security) has come close to an endorsement of the wider meaning of P.

PEACE-BUILDING.   One of the new tasks of the *U.N. mentioned by Secretary-General Boutros Boutros-Ghali in his 1992 report, *An Agenda for Peace, alongside *preventive diplomacy, *peacekeeping, and *peacemaking [Boutros-Ghali 1992: 475, 488–489]. It was defined as "action to identify and support structures which will tend to strengthen and solidify *peace in order to avoid a relapse into conflict." In this connection, he mentioned support for the building or strengthening of *democracy as a safeguard of peace, as well as the forging of cooperative relations between potential adversaries and the following military-related activities:

- Disarming of the warring parties and the destruction of weapons, as exemplified, inter alia, in the U.N. regulations pertaining to Iraq [see Molander 1992; Weller (ed.) 1993];
- Repatriation of refugees and the restoration of order;
- The removal of land mines, inter alia, with a view toward restoring agriculture; and
- Demilitarization.

PEACE DIVIDEND.   Presumed net gain (in terms of pecuniary and other resources) resulting from *disarmament, which might be "cashed in" through *conversion [cf. e.g., Kireyev 1991; Kirby 1991; Krause-Brewer 1991].

PEACE ENFORCEMENT.   *Ultima ratio* of a *collective security arrangement, tantamount to the use of military means to restore the *status quo ante bellum*. It is mentioned in Chapter VII of the *U.N. Charter, especially in Art. 42:

> Should the Security Council consider that measures provided for in Article 41 [dealing with sanctions and similar nonmilitary measures] would be inadequate or have proved inadequate, it may take such action by air, sea or land forces as may be necessary to maintain or restore international peace and security. Such action may include demonstrations, blockade, and other operations by air sea or land forces of Members of the United Nations. (in Roberts & Kingsbury [eds.] 1993: 511)

The action against Iraq undertaken by a coalition (see *Gulf War) was one instance of PE decided upon (with explicit reference to Chapter VII of the Charter) in Security Council Resolution 678 of 29 November 1990 [in Weller (ed.) 1993: 6]. It was against this background that PE was mentioned as one, albeit not the most important, form of *peacemaking in *An Agenda for Peace, presented by *U.N. Secretary-General Boutros Boutros-Ghali in 1992. It was stressed that "while such action should only be taken when all peaceful means have failed, the option of taking it is essential to the credibility of the United Nations as a guarantor of international security" [Boutros-Ghali 1992: 484].

Compared with the other military missions mentioned in *An Agenda for Peace* (**preventive deployment* and **peacekeeping*), PE would, as a general rule, be more militarily demanding. Moreover, it would usually require forces with considerable offensive capabilities because their task would be the forceful eviction of an aggressor from conquered territory (see **offensive-turned-defensive*). Hence the inherent difficulty with combining collective security with **NOD*, which might, according to some analysts, be overcome by way of **multinationality*, i.e., by creating offensive-capable task forces from defensive national components [Møller 1993a].

PEACEFUL COEXISTENCE.   Soviet conception dating back to Lenin but primarily associated with the Khrushchev and Brezhnev periods, when it was used almost synonymously with **détente* by the leadership of the **Soviet Union* [Jegorow 1972; F. Klein & al. (eds.) 1987; Schmidt & al. 1989; cf. Garthoff 1985: 36–53; Kremenyuk 1991].

1. **Soviet Conception of PC.** In the Soviet usage, PC signified a renunciation of war as a means of policy, albeit only within the special context of East-West relations. Accordingly, it meant a rejection of violent overthrow of capitalist governments ("export of the revolution") and a renunciation of war as a means of Soviet expansion. Even though it implied a certain reassessment of Western objectives, it still presupposed an innate capitalist hostility and war-proneness and attributed the actual state of nonwar to the military strength of socialism ("the favourable correlation of forces between the two world systems"), which deterred the West from attacking [cf. MccGwire 1987b: 13-66; 1991: 14-44, 80-114]. Two important limitations were embedded in this conception:

- PC tended to be conceived of as a mere suspension of hostilities (i.e., an extended cease-fire), wherefore no contradiction was experienced between PC and preparations for war, e.g., in terms of arms buildup and alliance formation.
- **Peace* was treated as divisible: the imperative of avoiding war between the superpowers and military blocs entailed neither any obligation to the resolution of outstanding conflicts nor any commitment to a suspension of other forms of conflict. The "class struggle" was merely waged by means short of war. Furthermore, beyond the European orbit, even armed conflict was regarded as permissible, if only to contravene intervention by the West (hence the involvement of the Soviet Union in the Somali-Ethiopian war and the Angolan civil war, as well as the invasion of **Afghanistan* [Garthoff 1985: 374–398, 502–537, 622–689, 887–938; Litwak 1990; A. George 1990].

2. **Western Attitude Toward PC.** The West was critical all along toward the notion of PC and generally sought to avoid using the term because of its connotations [cf. Jahn 1986; Wagenlehner 1987]. Nevertheless, in the process of negotiations with the East, and as a reflection of détente, the West also came to use the term "PC" rather extensively, e.g., in the 1972 U.S.–Soviet **Basic Principles Agreement*, and in the joint documents by the **SPD* and various ruling communist parties [Garthoff 1985: 290–298; cf. Bender 1986; H. Fischer 1987; Bruns 1990].

PEACEKEEPING.   During the Cold War, P constituted the main activity of the **U.N.* in the military realm, even though it was not mentioned in the **U.N. Charter*. Compared with the more ambitious missions mentioned in **An Agenda for Peace* [Boutros-Ghali 1992], such as **peacemaking* and **peace-building*, the purpose of P has been to solidify a *peace already achieved by other means, at least in the form of a cease-fire. In such a situation, P would usually consist of the interpositioning of U.N. forces between the two sides to a conflict; in the monitoring of a cease-fire; in various policing functions; and, occasionally, the monitoring of political activities, such as elections [Rikhye & Skjelsbaek (eds.) 1990; Goulding 1993; Morphet 1993; Roper & al. 1993; Berdal 1993]. P has presupposed the consent of (former and possibly future) belligerents, which has, in its turn, required impartiality. P has grown very considerably, especially since the end of the Cold War, in numbers as well as in scope, intensity, and costs [Hill 1993; Mackinlay 1993; Connaughton 1992; 1992a; Urquhart 1991; 1993; 1993a; Allen & al. 1993; Isaacs 1993; Berdal 1993; Kühne 1993; Renner 1993], to the point of nearly overstretching the means made available to the U.N.

PEACEMAKING.   New category of tasks of the *U.N. promulgated by Secretary-General Boutros Boutros-Ghali in his 1992 report, *An Agenda for Peace*, alongside *preventive diplomacy, *peacekeeping, and *peace-building [Boutros-Ghali 1992: 475, 488-489]. Here, P is defined as "action to bring hostile parties to agreement."

Among the tasks involved with P were mentioned mediation, negotiation, arbitration, and adjudication by the World Court of various disputes. Furthermore, the world community should assist in ameliorating the problems that might form the basis of conflict, or at least exacerbate it. Should such collaborative means prove insufficient, coercive means might be resorted to, such as economic sanctions (underpinned by military power) [see Hufbauer & al. 1990]. As a last resort, P might involve *peace enforcement by military means, undertaken under Chapter VII of the *U.N. Charter [in Roberts & Kingsbury (eds.) 1993: 511–513].

In addition to such direct responses "to outright aggression, imminent or actual" [Boutros-Ghali 1992: 484], the U.N. might also be called upon to restore and maintain a cease-fire, what has also come to be known as "peacekeeping plus," i.e., situations in which U.N. troops would be combatants, albeit not belligerents.

PEACE MOVEMENTS.   The history of PM goes back a very long time indeed, especially if one counts among PM (as seems entirely justified by virtue of their antimilitarist persuasion and agitation against war) the early workers' movements and parties as well as various religious groupings [cf. Kende 1989; Riesenberger 1987; Rajewsky 1987].

1. **PM of the 1980s.** It was not until the early 1980s that PM began to exert a significant influence on the political system. They did so, above all, by changing the popular mood and thereby transforming the framework conditions under which political parties had to wage their struggle for power [cf. Flynn & Rattinger (eds.) 1985]. All political parties had thenceforth to take into account the unpopularity of nuclear weapons and the *arms race, either by actually adopting the same viewpoints as the PM or by describing their policies in different and more accommodative terms. Nuclear buildup could be advocated only as a (deplorable but inevitable) precondition for build-down, and even programs such as the *SDI had to be couched in terms of making nuclear weapons impotent and obsolete.

What spurred the growth of the PM were the prospects of a renewed and accelerating nuclear arms race. This was heralded by two (partly interrelated) phenomena that brought to life two broad movements on partly separate tracks, albeit collaborating on a wide range of issues.

- The *INF issue, raised by NATO's "dual-track decision" of 1979, which envisioned deployment of 108 Pershing-2 missiles and 464 Tomahawk

GLCM (ground-launched cruise missiles) in Europe. This provided the impetus for the European PM, first of all in the *FRG, but also, e.g., in the *Netherlands and the *U.K., as well as Europe-wide, especially under the auspices of the END (European Nuclear Disarmament) movement [Rochon 1989; Wasmuht 1987; Carter 1992].

- The election of Ronald Reagan as U.S. president on a program of U.S. rearmament, not least as far as the nuclear arsenal was concerned. This led to the launch of the *Freeze movement: the largest of the PM in the history of the U.S. [Solo 1988; cf. Clotfelter 1986; Eichner 1990].

2. **West German PM.** The PM in the FRG were larger than in most other countries. One reason was that West Germany was envisioned as the stationing area for all the Pershings and a good part of the GLCMs, prospects that raised concerns about their becoming high-priority targets for Soviet *preemptive attacks. Concerns also about the global consequences of a nuclear war (e.g., the "nuclear winter" theory) were instrumental in swelling the ranks of the PM as well as the Green Party (see *Die Grünen), which paralleled the PM [Brauch 1983; Wasmuht 1987, 123-128; cf. Galtung 1986; 1988b; Szabo 1990: 35–39].

The PM continued to grow steadily up to 1983, when INF deployment commenced. Then the size of the demonstrations mushroomed; support from organizations such as churches and trade unions surged; more and more high-ranking politicians lent the movement their support; and new, more imaginative forms of protest were employed that yielded greater media attention [Bredthauer & Mannhardt (eds.) 1981: 157–180; Brauch 1983: 153–185; Wasmuht 1987: 101–172; Mushaben 1986; Rochon 1989: 3–52]. The PM remained rather amorphous, however, comprising devoted Christians, fanatical communists, eloquent antiauthoritarians, and many others. This proved to be a strength as well as a weakness. On the one hand, it prevented the PM from forcefully articulating a comprehensive set of demands because the political basis for joint activities had to be the highest common denominator, amounting to little more than "No to Cruise and Pershing." On the other hand, the diversity may have given the PM a longer lease on life. Although the size of demonstrations shrank rapidly after the peak in 1983, other activities continued, albeit on a much more modest scale, because the ideas had caught hold in new groups: among women, in the churches, in the trade union movement, etc. [Galtung 1984: 19-32]. Above all, the basic ideas promoted by the PM had taken root in the popular mood, in which the fear of war was significant and nuclear weapons were anathema.

The PM were tremendously successful in putting across their negative message but considerably less effective in formulating, and still less in promoting, concrete alternatives. The closest approximation thereto were the demands for *civilian-based defense (cf. *BOA) and *nuclear-weapons-free zones. The development of more comprehensive platforms was, however, the prerogative of individuals close to the PM, such as *Ulrich Albrecht, *Andreas Buro, *Joachim Wernicke, and *Ingrid Schöll [Albrecht 1982; 1983; Buro 1982; 1983; 1990a; Komitee 1981; Wernicke & Schöll 1985; Møller 1991: 180–188].

PEACE ORDER.   See *Peace.

PENINSULAR WAR.   War of the Spanish people against Napoleon, begun in 1808 with the rebellion launched in response to the emperor's instatement of his brother as king of *Spain. The PW was part of the English-led coalition's war against France, and the Spanish insurgents received a great deal of support from England. Large contingents of French forces were tied up in Spain, and the cumulated losses totaled around 500,000 French troops [Gates 1986]. The experience gathered from the PW formed the point of departure for the first theoretical analyses of *guerrilla strategy*. Besides *Jomini* and *Clausewitz*, two other analysts are noteworthy:

- Baron Georg Wilhelm Von Valentini, who analyzed both the Vendée and the PW. He was highly optimistic in his assessment of the future prospects of *guerrilla warfare* "in any defensive war in which the people play a purposeful role": "Even after he has won a battle, the enemy will never be able to gain a firm foothold in the country," with the result that "fought in this manner on a large scale, small wars become wars of extermination for the enemy armies" [extracts in Laqueur (ed.) 1978: 25–29].
- J. F. A. Lemiere de Corvay, who had fought in Spain himself, and who in 1823 presented a classic analysis of guerrilla strategy. The distinguishing features of guerrillas were that they "worked on the principle of avoiding any engagement in line with our armies," so that "perseverance in this plan thwarted all our schemes. By attacking small detachments, under-strength escorts, or any isolated men they met, they beat us point by point, undermined us. . . . Unless possessing superior forces, they never attacked soldiers travelling in bodies. Instead they fired on all isolated men. . . . They knew the country well and attacked violently in the most favourable spot" [J.F.A Lemiere de Corvey: "Des Partisans et des Corps Irreguliers" (Paris 1923) in Laqueur (ed.) 1978: 62–65].

PEOPLE'S WAR.   Popular term for *guerrilla warfare*, underlining its basis in, and considerable support from, the general population. The term was used, e.g. by *Clausewitz* and *Jomini* in connection with the *Peninsular War* [Clausewitz 1980:521–528/1984:479–483 (Book VI.26); cf. Hahlweg 1986].

PFLEIDERER PLAN.   Plan for German unification, presented in 1952 by MP for the *FDP* Georg Pfleiderer. Its logic was that because reunification was a politically offensive goal, it would spur Soviet fears if combined with Western arms buildup ("It has always been difficult to demonstrate defensive intentions"). As a solution, the PP envisioned taking Soviet concerns into account by allowing for a continued Soviet occupation of the areas east of the Oder-Neisse border (i.e., the parts of Germany ceded to *Poland*), and a Western "mirror occupation" of some of the westernmost areas. Between these, a reunited and neutral Germany should be allowed to emerge, the national armed forces of which should, however, be subjected to certain constraints [Albrecht (ed.) 1986: 60–65]. See also *Germany Plans*.

PGM.   Acronym for precision-guided munitions. PGM provided part of the original impetus for the development of *NOD* models because of their impressive results, inter alia, in the Vietnam and Yom Kippur Wars. PGMs appeared to have the potential to make the major weapons platforms traditionally associated with offensive warfare (*tanks*, aircraft, surface ships) impotent and obsolete, and thus to benefit the defender, according to James Digby, *John Mearsheimer*, and others [Digby 1975; Mearsheimer 1979]. Although some NOD advocates still cling to the belief in the small, cheap, and effective hi-tech weapons for antitank, *air defense*, and coastal defense purposes [Kaldor 1981; Hannig 1984; Barnaby 1986; idem & Boeker 1982], most have modified their initial optimism. PGMs continue to play an important role in most NOD schemes but no longer tend to be seen as panaceas, merely as one among several means to exploit the structural advantages of the defender.

PGRIT.   See **Progressive GRIT.*

PLAN 17.   French war plan developed prior to **WWI* and inspired by the *offensive a l'outrance* school (Foch, Joffre, and others). It envisioned a French attack against Germany through the Vosges, which would enable **France* to exploit the weakening of the left wing of the German front against it that would be created by the **Schlieffen Plan.* By thus poising an offensive-capable army with an offensive strategy against an equally offensive German one, the P17 made war all the more likely by providing both sides with powerful incentives for **preemptive attack* [J. Snyder 1984: 57–106; 1985; Evera 1985; Wallach 1972:137–164; Bellamy 1990: 69–73; Dupuy 1990: 226–227; cf. for a critique of these analyses: Trachtenberg 1990; Shimshoni 1990]. See also **Mobilization*, **Preemptive Attack*, **Preventive War.*

PNW.   See **Prevention of Nuclear War Agreement.*

POLAND.   Because of its exposed geopolitical situation (between two great powers: Germany and Russia) as well as its historical experience, and also because most conceivable large-scale wars in Europe would almost inevitably have involved P, P has always been very interested in **détente*, **disarmament*, and *demilitarization.*

1. **Before 1989.** Its tight integration with the **Warsaw Pact* notwithstanding, P took a number of independent initiatives during the Cold War period, e.g., the **Rapacki*, **Gomulka*, and **Jaruszelski Plans* [Stefanic 1988; Dobrosielski 1990; Karkoszka 1988b], the long-term perspective of which was a **disengagement* of Poland from the interbloc conflict. The same intentions might explain the Polish interest during the 1950s in **territorial defense*, manifested in the conception of the "Polish Front," the implementation of which would have constituted a partial **neutralization* of P [C. Jones 1981: 92–100; Holden 1989: 85–94; cf. A.R. Johnson & al. (eds.) 1982: 28–29; Walendowski 1988].

Poland was also among the first countries in the East to show an interest in **NOD*, albeit mostly from the traditional disengagement point of view. Leading academic proponents of these views included various researchers at the Polish Institute of International Affairs in Warsaw, such as Janusz Symonides [1989; cf. Polish Institute 1990], Janusz Prystrom [1988; 1988a; 1989; 1994], the present **SIPRI* director Adam Daniel Rotfeld [1983; 1986; 1987; 1988; 1989; 1989a; 1990; 1991], and **Andrzej Karkoszka* [1988; 1988a–b; 1989; 1989a–c; 1990].

In addition to the academic debate, the governing Polish United Workers Party (PUWP, i.e., the communist party) also conducted party-to-party negotiations with the West German **SPD.* They reached agreement on guidelines for a European security (including NOD and **common security*), as well as on a wide range of **CSBMs*, including some pertaining to the naval forces in the Baltic Sea [Ehmke 1986; SPD & PVAP 1985; 1988].

2. **German Question.** The immediate response of P to the revolutionary developments in 1989–1990 was mixed. On the one hand, they opened up opportunities for **democratization* and emancipation from Soviet hegemony. On the other hand, they raised concerns about German unification and possible irredentism, which were not allayed by Chancellor Kohl's reluctance, for several months, to endorse the Oder-Neisse border [*Europa-Archiv*, 1990(7): D179; cf. Møller 1991: 224–227]. Apart from the border question, P was apparently somewhat ambivalent about its preferences with regard to a united Germany. It tended to prefer a Germany tied down by its **NATO* membership to a more volatile and less predictable neutral

Germany, but it was also eager to accommodate Soviet concerns. The Polish solution to this dilemma was a quest to embed the German question as firmly as possible in an all-European framework, to forge subregional alliances, and to promote defensive restructuring.

The foreign minister, e.g., suggested the formation of mixed brigades (Polish-German, Czech-German, and/or Franco-German) to replace the Soviet troops stationed in the *GDR*, and insisted that whatever German units would be deployed there should be structured according to NOD principles [*Atlantic News*, 24 March 1990]. A clause to nearly the latter effect was actually included in the *Treaty on Good Neighbourship and Friendly Cooperation* that the two states signed in 1991, obliging them to "support a reduction of armed forces and armaments . . . to as low a level as possible which will suffice for defense but does not allow for attack" [*Bulletin*, 18 June 1991, 542].

3. **Post-1989 Developments.** Since 1989 P has been struggling with a complete revision of its security and defense policy. The most pressing issues have included

- A geopolitical reorientation away from Eastern Europe, to which P has never truly belonged, toward the central Europe that for centuries has been the "home" of P; furthermore, a growing interest in exploring cooperative ventures with the other states in the *Baltic Region* [Ilnicki & al. 1991].

- The severance of old and the forging of new alliance ties. P first, insisted on a Soviet troop withdrawal without delay that would not contribute to an international isolation of the Soviet Union. Second, P approached the Western institutions (*NATO*, WEU, EC), signaling its desire to join, albeit without making any formal application. Third, P promoted a triangular cooperation with Czechoslovakia and Hungary: the so-called Visegrad Triangle. These endeavors were weakened, however, by the fact that some of the participants (perhaps all) ordinarily saw the subregional collaboration as a vehicle for gaining admission to NATO [Spero 1992; 1992a; Bombík & Samson 1993; Dienstbier 1991; Hudak 1992; Kolar & al. 1993; Michta 1992; Valki 1992; Winkler 1992; Zielonka 1992].

- The development of a new, indigenous defense doctrine [H. M. 1990; Draus 1991; cf. Nowak 1990]. According to the testimony of the Polish representative to the *Vienna Seminar* in October 1991, this implied strengthening the strictly defensive character of the armed forces: "Our defense is based on the principle of 'sufficiency,' i.e., inability to wage aggressive operations." In conformity with this goal, heavy armored formations, missiles, *bridge-building* equipment, and various forms of mobile *logistics* were built down. Furthermore, the Polish Army has been reorganized into four military districts: two in the East and two in the West, as a manifestation of the *tout azimut* nature of Polish defense [*Jane's Defence Weekly*, 4 January 1992, 32; Schulz (ed.): 1992].

POSEN, BARRY.   See *U.S.*

POSITIVE PEACE.   See *Peace*.

POTT, ANDREAS.   Researcher at the *IFSH* whose interest has mainly been *naval arms control* and the application of *NOD* principles to naval forces [1985; 1986].

PRECISION-GUIDED MUNITIONS.   See *PGM*.

PREEMPTION.   See *Preemptive Attack*, *Crisis Stability*, *No-Target Principle*.

PREEMPTIVE ATTACK.   Attack launched by one side in the erroneous belief that its adversary has already launched an attack or has passed a point of no return toward an attack.

The determinant of a PA is not a choice between war or no war but between waiting for the attack (thereby allowing the opponent the advantages of *surprise attack*) or preempting it. The ensuing risk is that each side might erroneously fear an attack and seek to preempt the other's PA; war may spring out of "reciprocal fear of surprise attack" [Schelling 1960: 207–229]. The propensity for PA is a matter of *crisis stability* or instability, in its turn depending, e.g., on the deployment of forces. Hence the *no-target-principle* recommended by *NOD* advocates as a means of making PA less rewarding; and the principle of abstaining from offensive capabilities, inter alia, with a view to making a PA seem less urgent.

Even though some analysts have dismissed the risk of PA as entirely fictitious [Gray 1992; 1993; Trachtenberg 1990], according to recent revelations, the USSR actually contemplated a PA against the U.S./NATO in the early 1980s, in connection with the intelligence operation RYAN (Raketno-Yadernoye Napadenie [Nuclear Missile Attack]) [Andrew & Gordievsky 1990: 488–507].

PREVENTION OF INCIDENTS AT SEA AGREEMENT.   See *CBM*.

PREVENTION OF NUCLEAR WAR AGREEMENT.   Agreement on the Prevention of Nuclear War signed by the *U.S.* and the *Soviet Union* in 1973. The parties committed themselves to "act in such a manner as to prevent the development of situations capable of causing a dangerous exacerbation of their relations, so as to avoid military confrontations," and to consult immediately with each other should dangerous situations arise. The PNW can be seen as an elaboration on the *Basic Principles Agreement*, explicating the code of conduct appropriate for *détente* and making (at least partial) cooperation in crisis avoidance, management, and solution mandatory. In conformity with (albeit not necessarily because of) the PNW, the two superpowers consistently sought to avoid any direct contact between their armed forces [in Allison & Williams (eds.) 1990: 260–271; cf. Garthoff 1985: 334–355; Ury 1985: 53–54; Borawski (ed.) 1986; May & Harvey 1987].

PREVENTIVE DEPLOYMENT.   See *Preventive Diplomacy*, *Agenda for Peace*.

PREVENTIVE DIPLOMACY.   One of four main categories of *U.N.* activities for *peace* and *security* mentioned by Secretary-General Boutros Boutros-Ghali in his 1992 report *An Agenda for Peace* and defined as "action to prevent disputes from arising between parties, to prevent existing disputes from escalating into conflicts and to limit the spread of the latter when they occur" [1992: 475]. The following means were mentioned; the last two bear a striking resemblance to certain *NOD* schemes, above all, those that have envisioned *disengagement*.

- Measures to create confidence, including *CBM*s (confidence-building measures) in the narrow sense of the word, as well as the exchange of military missions and the establishment of risk-reduction centers.
- Fact-finding missions.
- Early *warning*, inter alia, in the form of means for synthesizing data gathered by national warning systems.
- Preventive deployment, either in internal crises or in interstate conflicts. In the latter, U.N. troops could be deployed on both sides of the border, which would presuppose the consent of both sides, or on one side only, at the request of the state feeling threatened by another. The latter eventuality was

an innovation because it implies singling one state out as a potential
aggressor.

- The creation of demilitarized *zones*, in which U.N. personnel would be
  deployed, preferably on both sides of the border but possibly on only one
  "for the purpose of removing any pretext for attack" [Boutros-Ghali 1992:
  476–480].

PREVENTIVE WAR.   War launched by one side in anticipation of a war at
some later stage under less favorable conditions. Under conditions of deteriorating
force ratios compared with its adversary or adversaries, a state might opt, e.g., for a
PW because of a preference for war now rather than later, even though it would prefer *peace* to war [Schelling & Halperin 1961: 9–17; Frei & Catrina 1983; Levy
1987; 1990; Lebow 1981; 1984].

NOD advocates have tended to take the risk of a PW (as well as of *preemptive
attacks*, i.e., "fast-pace PW") more seriously than traditional military thinkers. They
have rejected deterrence as an appropriate means of war prevention, with a few isolated exceptions. Regardless of how strong an inhibition *nuclear deterrence* might
be against premeditated wars of aggression, it might well be counterproductive with
regard to PW. To demonstrate strength and determination to use military power
might merely make war seem more inevitable, thereby strengthening motives for
PW. What would matter for the prevention of PW, rather, would be to make war
seem avoidable, inter alia, through abstention from offensive capabilities [cf.
MccGwire 1984; 1986].

PRINCIPLES OF WAR.   "Laws" of warfare of presumably perennial validity,
first formulated by *Jomini* in 1811 but also found in the writings of *Clausewitz*
and in the maxims of Napoleon. The authoritative codification into eight PW was
the work of *J. F. C. Fuller*, whose formulation has subsequently been revised only
slightly by modern armed forces [Jomini 1811, 4:275–286; cf. Dupuy 1990:
247–253]. With marginally different emphases, the PW are common to nearly all
present armies.

1. **Traditional PW.** The U.S. Army's nine PW have been formulated thus:

   1. Objective: Direct every military operation toward a clearly defined, deci-
      sive, and attainable objective.
   2. Offensive: Seize, retain, and exploit the initiative.
   3. Mass: Concentrate combat power at the decisive place and time.
   4. Economy of Force: Allocate minimum essential combat power to sec-
      ondary efforts.
   5. Maneuver: Place the enemy in a position of disadvantage through the flexi-
      ble application of combat power.
   6. Unity of Command: For every objective, insure unity of effort under one
      responsible commander.
   7. Security: Never permit the enemy to acquire an unexpected advantage.
   8. Surprise: Strike the enemy at a time and/or place and in a manner for which
      he is unprepared.
   9. Simplicity: Prepare clear, uncomplicated plans and clear concise orders to
      insure thorough understanding [HQ, Dep. of the Army 1982: B1–B5].

2. **NOD and the PW.** Certain *NOD* advocates have declared that the PW are
invalid and obsolete [Borg 1989]. Others have suggested a reformulation that might
be along the following lines (which underline continuity yet manage to include
most of the central NOD principles):

1. Direct every military operation toward a clearly defined, decisive, and attainable objective, i.e., toward the defense of national *security*, while assigning first priority to war prevention.
2. Seize, retain, and exploit the initiative by tactically taking the offensive, while remaining strategically on the defensive.
3. Prevent the enemy from defining the decisive place and time, inter alia, by waiting for the opposing forces to approach, by distributing forces across the entire territory, and by transforming the enemy's campaign into a protracted war of *attrition*.
4. Force the enemy, through being present everywhere, to allocate as much essential combat power as possible to secondary efforts.
5. Disadvantage the enemy through the flexible and asymmetrical application of combat power, i.e., by refraining from meeting like with like.
6. Ensure unity of effort under one responsible commander on a nationwide scale, while decentralizing command at the lover levels.
7. Never permit the enemy to acquire an unexpected advantage, inter alia, by avoiding the posing of lucrative targets that would invite preemption.
8. Strike the enemy at a time and/or place and in a manner for which he is unprepared, while guarding against surprise actions on his part.
9. Prepare clear, uncomplicated plans and clear, concise orders to ensure thorough understanding, but avoid rigidity and allow subordinate commanders a wide scope for initiative [Møller 1992: 138–141, 145–147].

PRINS, GWYN.   Anthropologist, former director of Studies in History, subsequently director of the Global Security Program, both at Cambridge University. (See *U.K.*)

GP's first major contribution to the defense debate was *Defended to Death*, written jointly with other Cambridge dons, which presented a powerful case against *nuclear deterrence* in general and the British independent deterrent in particular. While holding up general and complete *disarmament* as the ultimate goal, it also recommended immediate steps, including a citizens' reserve army equipped with "*defensive weapons*" [Prins (ed.) 1983: 269–270].

Through his participation in, e.g., the *Pugwash* Study Group on Conventional Forces in Europe, GP came to devote considerable attention to *NOD*, not least as an alternative to nuclear deterrence. Even though deterrence did not need an alternative ("One does not ask in the case of cancer: 'What will you put in its place?'") as GP formulated it [1986: 89]), *mutual security* recommended itself as the new paradigm to replace the deterrence system; GP urged the West to "catch up with Soviet thinking" in these respects [1990]. More recently, GP's role in the international NOD debate has come to resemble that of an *advocatus diaboli,* charging NOD proponents with "overstating . . . sometimes recklessly, what the adoption of their version of NOD could do," and of having fallen victim to the "military logic." Furthermore, he has criticized most NOD models for disregarding *logistics*, underestimating the importance of mobility, and exaggerating the potentials of emerging defensive technologies. These defects notwithstanding, GP assessed NOD as a valuable contribution to the transformation of Europe into a more mature system [1990a; cf. Tromp 1990; 1990a].

Concerning the emerging "new Europe," GP has devoted considerable attention to its "architecture" and the unique opportunities for building a genuine *collective security* system [1990a: 57–64; cf. 1990c]. The scope of the system should not be confined to Europe because the most acute challenges (both military and nonmilitary) require global action. He has envisioned an expanded role for the *U.N.* that should also include naval peacekeeping forces [1991; 1991a]. More generally, GP has advocated a refocusing of security studies that would reduce the saliency of NOD as well as of other military strategies. The real challenges facing humankind, according to GP, are no longer military but environmental. For security studies to escape marginalization, they should shift their focus accordingly [1990b; 1992].

PRIO. Peace Research Institute Oslo. See *Norway, *Johan Galtung, *Sverre Lodgaard.

PRISONER'S DILEMMA. Description of the *security dilemma in the terminology of *game theory. Two apprehended criminals (A and B) are depicted as being interrogated simultaneously but separately, i.e., unable to communicate with each other. Each is promised a light sentence in return for testifying against the other. It is understood that if neither testifies against the other, both will be acquitted.

The expected payoffs for the two prisoners are as follows. An acquittal may (quite arbitrarily) be given the value +10; a light sentence the value +5; and a harsh sentence a value of –10. "Cooperation" refers to the relationship between the two prisoners, i.e., implies keeping silent; "defection" means confessing and testifying against the other. Because A has no way of knowing whether B will give evidence, A has to assume the worst, namely, that B will testify against him, under which presumption he will be better off by testifying himself. Because B faces the same alternatives, each prisoner (according to game theory) will testify against the other and both will be convicted (value: –1), thus doing considerably worse than if they had cooperated (+10) [Rapoport 1989: 296–298, 556–560; idem (ed.) 1970; idem & Chammah 1970]. (See Table 30.)

Table 30.   Prisoner's Dilemma

		B	
		Cooperation	Defection
A			
	Cooperation	+10, +10	–10, –1
	Defection	–1, –10	–1, –1

It has been demonstrated that the PD has no "solution," i.e., that both sides invariably end up with suboptimal payoffs; accordingly, cooperation between partial adversaries will never be initiated, regardless of how mutually advantageous. However, in an iterated PD, one strategy has proved itself able to prevail and gradually to bring about cooperation: *Tit-for-Tat, originally suggested by *Anatol Rapoport and analyzed at length by *Robert Axelrod [1984; cf. Goldstein & Freeman 1990: 20–32, 130–138, 141–144; cf. for a critique: Stein 1990: 87–112].

PROGRESSIVE GRIT. Term coined by Joshua Goldstein and John Freeman [1990: 31–32, 130–136] for a special variety of *GRIT (Graduated Reciprocated Initiatives in Tension Reduction). The unilateral initiatives are envisioned to begin with rather insignificant measures of a largely symbolic nature but to be pursued over a long period of time and with increasing saliency and radicality. The results of simulation studies by the authors showed PGRIT to be somewhat more successful than GRIT: it might be slow to achieve cooperation, but over time it tended to achieve the greatest overall level of cooperation. On the other hand, its benign

effects tended to wear off rapidly after abandonment of the policy. See also
Extended GRIT, *Super-GRIT*.

PROJECT ON DEFENSE ALTERNATIVES.   See *Commonwealth Institute*.

PROLIFERATION.   See *Arms Trade Regulations*, *NPT*, *MTCR*.

PROTRACTED WAR.   See *Time Perspective*, *Guerrilla Warfare*.

PRYSTROM, JANUSZ.   See *Poland*.

PUGWASH.   International movement of scientists and other scholars inaugurated by the 1955 *Russell-Einstein Manifesto*. Since the inaugural conference in the
Canadian town Pugwash (hence the name) in 1957, conferences have been held
annually, the proceedings of which have been published regularly, in addition to
edited volumes containing the most important contributions [e.g., Rotblat (ed.)
1993; idem & d'Ambrosio (eds.) 1986; idem & Goldanski (eds.) 1989; idem &
Hellman (eds.) 1985; idem & Holdren (eds.) 1990].

Under the auspices of P, the Study Group on Conventional Forces in Europe was
founded in 1984 [Boserup & Neild (eds.) 1990], with *Anders Boserup* and
Albrecht A. C. von Müller serving as convenors. The participation of Soviet scholars from the *IMEMO* and *ISKAN* and prominent Western *NOD* experts facilitated the dissemination of Western ideas in the USSR (see *New Thinking*), thus laying the theoretical foundations for the endorsement, and subsequent partial
implementation, of a defensive doctrine by the *Soviet Union*. In 1993 another
study group under the auspices of P published an elaborate plan for a nuclear-
weapons-free world, in which they analyzed the technical, legal, and political possibilities of implementing such a scheme [Rotblat & al. (eds.): 1993].

PURVER, RONALD.   See *Canada*.

Q

QUAKERS. Christian sect that, inter alia, rejects the *just war* dogma of the Roman Catholic Church [Johnson 1987] in favor of a literal interpretation of the *Sermon on the Mount* and other relevant pieces of canonical scripture. Q regard the killing of other humans as impermissible under all circumstances, including *self-defense* [cf. Sharp 1979].

QUESTER, GEORGE. Professor of government at the University of Maryland, *U.S.*, who has for decades been one of the main U.S. proponents of *NOD*.

The main body of GQ's academic production consists of analyses of *nuclear deterrence* and its precursors, including strategic bombing during *WWII* [1966; 1986a–b; 1989; 1989c; 1991a]. Although agreeing with NOD proponents and other deterrence critics that counterforce targeting was problematic, GQ was nevertheless in favor of calibration and discrimination in target selection, inter alia, advocating economic targeting as an effective means of enforcing a Soviet capitulation in a future war [1986b].

GQ's main work in the NOD tradition was his *Offense and Defense in the International System* [1977; new ed. 1988], in which he traced the interaction between offense and defense through the ages with a special view to effects on the international system, especially its proclivity for war. His general conclusion was that the greater the superiority of the defense, the more stable and peaceful the system would be [cf. 1986: 27–57]. The precise criteria for distinguishing between offensive and defensive postures were elaborated upon later. Using the unfortunate term ""*defensive weapon*" (see *Offense/Defense Distinction*) GQ suggested distinguishing between, on the one hand, "weapons which reward, in terms of counterforce exchange, the side that moves forward, the side that takes the initiative to strike first" ("offensive" or "active" weapons), and, on the other hand, weapons "rewarding instead the side that elected to wait and let the attack come," i.e., "defensive" or "reactive" weapons [1986: 23; cf. 1991: 140–145].

In regard to his own country, GQ identified a strong defensive tradition in the U.S. approach to war and peace, dating back to the War of Independence, the Civil War, and the post–WWII notion of containment. Indeed, even the U.S. devotion to nuclear deterrence was presumably a reflection of the defensive commitment to the status quo [1989b].

GQ has also devoted some attention to the assessment of Soviet *New Thinking* and the different interpretation of *reasonable sufficiency* and NOD within the Soviet political system [1989a; 1991]. The inconsistencies and ambiguities notwithstanding, this reorientation presented the West with some unique opportunities for meaningful *arms control* settlement, for which the West should be willing to pay a certain price in terms of *disarmament*. Should a conventional *balance* be achieved in this way, "nuclear escalatory threats would fade from significance on either side." Until that juncture, GQ recommended their retention, not least as far as the *Soviet Union* was concerned [1991: 147].

R

RAGIONIERI, RODOLFO. Professor of applied mathematics at the University of Florence and cofounder of the *Forum on the Problems of Peace and War* (Forum per i problemi della pace e della guerra) in Florence, one of the main centers of the alternative defense debate in *Italy* [Ragionieri (ed.) 1989; Cerutti & Ragionieri (eds.) 1990].

RR's main work was *Common Security: A New Strategy of Peace Beyond Deterrence,* in which he analyzed the notion of *common security* as an attempted circumvention of the *security dilemma.* He furthermore identified *NOD* as the only military strategy compatible with this notion of *security* (albeit only for a transitionary period until the time was ripe for entirely nonmilitary solutions), and as the most reliable form of war prevention [Ragionieri 1989]. RR also dealt in more concrete terms with the security of Italy and the Mediterranean region, recommending a combination of a *nonintervention regime* with a range of *CBMs* (confidence-building measures) and *naval arms control* measures, the latter presumably facilitated through a shift to more defensive maritime forces [1989b; 1990; 1990a-b; 1991]. Relevant maritime objectives for *NATO* in the Mediterranean, according to RR, would be

- A defense of coastal and territorial waters, which could be done largely by land-based systems and small vessels;
- A defense of *SLOCs* (sea lines of communication), which might require a combination of "contiguous sea navies" [cf. Booth 1977: 121] and land-based naval aviation; and
- Sea denial in central parts of the Mediterranean, i.e., the prevention of their use by an aggressor (cf. *sea power*) [Ragionieri 1990a: 146–147].

RANDLE, MICHAEL. Former researcher at the Bradford School of Peace Studies (dealing, inter alia, with *civilian-based defense*) and former secretary of the *ADC*. MR was instrumental in writing and compiling ADC's two reports [Randle 1986; 1990; ADC 1983; 1987; Randle & Rogers (eds.) 1990].

RAPACKI PLAN. Plan for a central European *nuclear-weapons-free zone* (NWFZ), presented by the foreign minister of *Poland*, Adam Rapacki, in four consecutive, and slightly different, versions.

- Version 1 (2 October 1957) envisioned merely a NWFZ comprising the two German states, Poland, and Czechoslovakia.
- Version 2 (14 February 1958) supplemented Version 1 with negative security guarantees by calling upon the nuclear powers to accept an explicit prohibition of nuclear strikes against the zone.
- Version 3 (4 November 1958) envisioned a two-stage implementation: the first stage would essentially amount to a freeze by prohibiting the states parties from acquiring nuclear weapons or building facilities for them. A

simultaneous build-down of nuclear and conventional weapons within the zone was deferred until the second stage.
- Version 4 (28 March 1962) was largely identical in substance, and merely modified the details of subsequent stages.

The proposals were widely regarded as serious (i.e., as reflecting sincere, and perfectly understandable, Polish security concerns), albeit impossible for *NATO to agree to [Schramm & al. (eds.) 1972: 388–413; cf. Griffith 1978: 81–83; Stefanic 1988; Dobrosielski 1990]. They therefore gave rise to a number of Western proposals to partly the same effect, such as the *Gaitskell Plan and the various *Germany Plans. The subsequent Polish *Gomulka Plan and the *Jaruszelski Plan were in direct continuity with the RP.

RAPOPORT, ANATOLI.   Mathematician, psychologist, and peace researcher, currently professor at the University of Toronto.

One of AR's main contributions to the alternative defense debate has been his critique of strategic thinking in general and that of *Clausewitz and the neo-Clausewitzians in particular [1968; 1970a]. This also led him to criticize the prevailing trends in *game theory, even though he himself has been one of the most prominent game theorists. In particular, AR has written about the *prisoner's dilemma, both in mathematical terms and in the context of psychological experiments [1970; 1974: 173–225; 1989: idem & Chammah 1970]. It was he who submitted to the "prisoner's dilemma computer tournament" arranged by *Robert Axelrod the strategy of *Tit-for-Tat, which came out as the winning strategy [Axelrod 1984: 192–205; cf. Rapoport 1989: 273–274].

In 1989 AR published a monumental (620 pages) work, *The Origins of Violence: Approaches to the Study of Conflict*, in which he reiterated his critique of strategic thinking, while also commenting favorably on the "benign" versions, namely, *NOD strategy as well as "civilian," e.g., *civilian-based defense [1989: 468–476, cf. 247–343].

In 1992 AR published another attempted synthesis of peace research findings: *Peace: An Idea Whose Time Has Come* [1992]. Focusing above all on the "noosphere" (i.e., the realm of ideas, understood in a quasi-Platonic sense), he found the "germ" of the idea of peace to have been present for at least twenty-eight centuries. However, until the present it had lain dormant because the conditions for its spread and implementation were not yet ripe, hence the persistence of the war system. One factor promoting the replacement of this system with a peace system was the piecemeal evolution of cooperation, even between "egoists," as had been the guiding idea of tit-for-tat. AR found the urge for cooperation to be a (hitherto latent) inclination in human nature, which might gradually supplant competition, and do so in the same way as the latter had become dominant, namely, through reciprocity: "If . . . a state disarms and survives, and if one or more other states follow suit, the prospect of a disarmed world becomes more realistic. As it becomes more realistic, *disarmament becomes attractive. Eventually a threshold may be reached after which the momentum of disarmament carries the process to its conclusion" [1992: 78].

RARATONGA TREATY.   See *Nuclear-Weapons-Free Zones, *New Zealand.

REACTIVE DEFENSE.   Generic term for various *NOD models, particularly those envisioning the most stationary mode of deployment. A common feature of these models has been the inability of forces to go into action preemptively: they could become active only after the enemy (i.e., invader) had entered their circumscribed field of responsibility [Hofmann, Huber, & Steiger 1986; 1986a].

REALISM.   School in the study of international relations that has been predominant since 1945, above all in the *U.S.* Precursors of modern R include Thucydides, Machiavelli, and Hobbes.

1. **R versus Idealism.** R originated as a response (antithesis) to what the Realists themselves pejoratively called "idealism," portrayed as a preoccupation with how the world ought to be (e.g., norms, *regimes*, and international law) at the expense of reality. A prominent exponent of idealism was *Immanuel Kant*, who was concerned about "perpetual peace" and with the application of ethical principles to all fields of human activity, including that of international relations [Kant 1788; 1795; cf. Piepmeier 1987; Donaldson 1993].

Idealism had been the dominant approach to international politics in the interwar years. However, it experienced a conspicuous failure with regard to its main offspring, the ineffective *collective security* system built around the *League of Nations*, which proved capable of preventing neither fascist nor communist aggression [Carr 1946: 22–40]. Hence the quest for an alternative, "realist" paradigm, which did not, however, entirely displace idealism. The modern peace research tradition contains strong elements of idealism (e.g., the world order and federalist traditions), as does the literature about *common security*, which has occasionally been seen as the antithesis of R [Center for the Study of Democratic Institutions 1973; Galtung 1980. For a critical analysis, see Claude 1972; L. Neumann 1988; Suganami 1989].

2. **Fundamental Tenets of R.** What unites Realists and sets them apart from idealists are, inter alia, the following tenets [Carr 1946; Morgenthau 1960; cf. M.J. Smith 1986; Aron 1984]:

- The principal actors in international affairs are states;
- States and their leaders pursue national interests;
- Politics is fundamentally power politics;
- Moral principles are inapplicable to interstate relations;
- The international system is anarchical, i.e., lacks supranational authority over and above the individual states; and
- *Stability* and *peace* are attainable in international politics only under (rare) conditions of "legitimate international order" [Kissinger 1957: 1–3, 180–181; Bull 1977: 3–22] (where it might take the form of a *concert*) or as a result of the workings of *balance*-of-power mechanisms and other forms of reciprocity [Bull 1977: 117, 139].

Finally, Realists tend to assign primacy to military power, seeing most other factors as contributions to war-waging potentials. Military power, furthermore, according to R, has to be usable ("fungible") in order to perform its balancing functions [cf. Aron 1984: 28, 58–80, 133–156; Morgenthau 1960: 118, 211]. This has earned R (not quite justly) a certain reputation for war-proneness or even bellicosity. Most proponents of alternative defense and security have therefore, as a matter of principle, distanced themselves from R.

3. **Neorealism.** Neorealism could be seen as synthesis between classical R and elements of the critique raised against it, above all for focusing on the state to the exclusion of other actors. Neorealists have taken the trends toward complex global *interdependence* into account and focused on structures (hence the occasional term "structural R") [Waltz 1979; Gilpin 1981; Keohane (ed.) 1986; Buzan 1991; cf. for a precursor: Herz 1951: 95, 121, 132–133, 146, 226–227].

Certain Neorealists, most notably *Barry Buzan*, have also qualified the Realist explanation of international behavior as reflecting the pursuit of "national interest defined in terms of power" [Morgenthau 1960: 33] by focusing attention on the less ambitious concept of *security*. This has uncovered some of the paradoxes and dilemmas inherent in the Realist worldview, e.g., the *security dilemma* and the *defense dilemma*, both of which imply that the quest for power and/or security may be self-defeating [Buzan 1991: 270–327].

4. **Neorealism, the Security Dilemma, and Common Security.** In their search for an escape from these dilemmas a few Neorealists came to support *common security* and *NOD* [e.g. Buzan 1987; 1989; 1991: 15–16; Evera 1990]. Upon inspection, the three conceptions share a number of characteristics:

- A focus on the nation-state and the anarchical international system;
- An acknowledgment of real hostility and conflict, hence the possibility of deliberate aggression;
- A realization of the effects of the security dilemma;
- An acceptance of the need for military means for the preservation of national security;
- An emphasis on *arms control*; and
- An acknowledgment of the imperative of war prevention [Møller 1992: 9–47].

REASONABLE SUFFICIENCY.   A central element in the *New Thinking* of the *Soviet Union* during the *Gorbachev* era. The term's origins can be traced back to Robert McNamara and Richard Nixon [cf. Flynn 1989: 10–12; McNamara 1986: 40–47].

1. **Nuclear RS.** Just as had been the case for the *U.S.*, the Soviet Union first used the term "reasonable sufficiency" about nuclear weapons, e.g., in the program of the 1986 Communist Party convention, where it was stated that "our country is for eliminating weapons of mass destruction and limiting military potential to RS" [quoted in Garthoff 1990: 95]. The reasoning behind RS was the simple observation that both *balance* and superiority lost all meaning when applied to force ratios above a certain level of mutual assured destruction. RS might thus be defined as a secure second-strike capability of a magnitude sufficient for deterring an attack. This logic was subsequently pursued by civilian academics with concrete proposals for *minimum deterrence* postures [A. Arbatov 1988b–c; 1989e–f; 1990b–c; idem & Cochran 1991; idem & Lednev 1988], as well as by the political leadership itself. The latter's quest was reflected in the initial unilateral nuclear test moratorium and the subsequent *Year 2000 Plan* for a complete abolition of nuclear weapons by the turn of the millennium [Garthoff 1990; Goure 1990].

2. **RS in the Conventional Field.** The term "RS" was also applied to conventional forces, initially apparently without any clear conception of the implications. In the *Warsaw Pact*'s *Berlin Declaration*, it was, e.g., suggested to "keep to the limits sufficient for defense and repelling any possible aggression." However, Gorbachev personally defined it as "a structure in the military forces of countries that suffices for the prevention of possible aggression, but is insufficient for attack"—as good a definition of *NOD* as any.

One reason for the reorientation was a reassessment of the likelihood of war, *in casu* an authoritative endorsement of the view that war could be reliably avoided. Even if a small-scale war should erupt, it could presumably be contained. Hence a reordering of priorities with the objective of war prevention displacing that of being able to prevail in war as the first priority [cf. MccGwire 1987; 1987a–b; 1988a–b; 1991].

As a corollary of the new political doctrine (cf. *military science*), military doctrine stood in need of reformulation, with a new emphasis being placed on war prevention and a defensive strategy. Some thought was given to these matters, but the concrete implications remained a matter of dispute between the civilian advisors (*institutchiki*) at, e.g., the *IMEMO* and the *ISKAN* on the one side [A. Arbatov 1988; 1989; 1989a–d; 1990; 1990a; idem, Amirov, & al. 1989; idem, Kishilov, & al. 1989; Arbatova & al. 1989; Baranovsky & Streltsov 1988; Kokoshin 1988; 1988a; 1989; 1989a–b; 1990; 1991; idem & Larionov 1987; 1989; 1990; Konovalov 1989; 1990; 1990a; idem & Mazing 1989; Makarevsky 1988; 1989; Mazing 1989; Shaskolsky 1989; Streltsov 1989; Tschernyschew 1989; 1989a; Zhurkin & al. 1989; cf. Fitzgerald 1990; Arnett 1990a; Holoboff 1990; Hines 1990; Krakau & Diehl 1989a; Diehl 1993; Odom 1991; Conner & Poirier 1991; Bellamy 1992] and the armed forces on the other. The former tended to favor more-defensive strategic conceptions at the expense of *counteroffensive* capabilities; the latter (such as Defense Minister *Dmitri Yasov*) were unwilling to renounce the capabilities for a large-scale (strategic) counteroffensive. Hence the civilian-military dispute that was a contributory factor to the abortive 1991 coup d'état [Herspring 1987; 1989; Rice 1987; 1989; Larrabee 1988; Adragna 1989; Covington 1989; Davenport 1991; Green & Karaskik (eds.) 1990; Holden 1991b].

RECIPROCAL SECURITY (German: gegenseitige sicherheit).   Term first used by *Ken Booth*, but introduced into the political discourse in 1988 by West German Defense Minister *Rupert Scholz* of the *CDU*, apparently nearly synonymous with *common security* [*Bulletin der Bundesregierung,* 26 August 1988; cf. Lutz & Theilmann 1986: 224–228].

REGIME.   Set of written or tacit rules and norms within a certain issue area and/or between a number of parties that improves predictability by providing for a convergence of mutual expectations.

A R may but need not be formalized and involve supranational authorities with responsibility for enforcing the norms of which the R consists. A R may also be based entirely on reciprocity, i.e., on the recognition by each party that a violation on its own part is likely to result in similar behavior on the part of the rest, and hence would be dysfunctional. The conception of R has been associated with the idealist tradition in the study of international affairs, but it might also be compatible with *Realism*, inter alia, with the assumption that states tend to pursue national interests [cf. Krasner (ed.) 1982; Jervis 1982; Young 1986; George, Farley, & Dallin (eds.) 1988; Haggard & Simmons 1987; Rittberger & al. 1990].

The debate on *common security* and *NOD* has occasionally been couched in terms of forging a R, such as a "defensive R," "*common security* R," "*nonintervention R*," etc. [e.g., Gottfried & al. (eds.) 1989; Kokoshin 1989; Lutz 1991b; MccGwire 1988; Meulman 1991; Platias & Rydell 1982; Schmähling 1990; Schrogl 1990], albeit sometimes without much theoretical rigor. See also *MTCR*, *Security Regime*.

REINFORCEMENT.   See *Augmentation Forces*, *Mobilization*, *SLOC*.

REMACLE, ERIC.   See *Belgium*.

RESTRICTED MILITARY AREA.   See *CBM*.

RISK REDUCTION CENTERS.   See *CBM*.

ROBERTS, ADAM.   Professor of international relations at Oxford University.

AR's first contributions to alternative defense thinking were in the field of *civilian-based defense* (CBD). In 1967 he was the editor of, and main contributor to, one of the most influential works: *The Strategy of Civilian Defence: Non-Violent Resistance to Aggression* [Roberts (ed.) 1967], in which he elaborated on CBD as a functional equivalent of (and *ipso facto* viable substitute for) military defense. Besides this main function, CBD would also be applicable for other purposes, such as resistance to internal coups d'etat [Roberts 1967; cf. 1969; 1970].

Through his involvement in a study of the defense of *Sweden* with a special view to the inclusion of CBD methods [1978], AR came to devote more of his attention to military forms of defense, *in casu *territorial defense*, such as practiced by, e.g., the neutral and nonaligned states in Europe. On the basis of case studies of Sweden and *Yugoslavia*, AR in his 1976 work, *Nations in Arms,* recommended territorial defense as a viable strategy for *NATO*, as well as for the *U.K.* [1976: 267–277; cf. 1983; 1989]. It would be "a system of defense in depth," i.e., "the governmentally-organized defense of a state's own territory, conducted on its own territory and . . . aimed at creating a situation in which an invader . . . is harassed and attacked from all sides" [1989: 215].

AR did, however, still regard CBD as a viable means of national defense, albeit not relevant for all contingencies, wherefore he recommended a combination of military and CBD methods. Disregarding proposals for clear temporal and/or spatial separation of the two forms of defense (cf. *mix concepts*), he simply maintained that CBD might be a valuable "special option for special circumstances" [1978; cf. 1967a: 295–296; 1989: 217].

Most recently, AR has devoted himself to studies of the U.N. as a possible building block for a *collective security* system [1992; 1993; 1993a; idem & Kingsbury (eds.) 1993].

ROGERS PLAN.   A set of proposals by SACEUR (Supreme Allied Commander Europe) Bernard Rogers for a greater emphasis on conventional forces (cf. *CDI*) that led to NATO's adoption of the *FOFA* doctrine [Rogers 1982; 1983a; cf. Berg & Herolf 1984].

ROMANIA.   See *Balkans*.

ROMBERG, WALTER.   Professor of mathematics at the East German Academy of Sciences, and affiliated with the research association of the Evangelical Churches of the *GDR*. After the 1989 revolution, WR was appointed minister of finance for the East German *SPD*. Since his resignation and the all-German elections, WR has been working in the European Parliament, and since the mid-1980s, he has been affiliated with the *SAS* [cf. Møller 1991: 204–205, 216–217].

WR in 1986 issued a major study, *Crisis-Stable Military Security in Central Europe,* the first substantial work on *NOD* by an East European author. One point of departure was a number of declarations by the Protestant churches about *common security* and the need for "typically defensive conceptions of defense." Another was the Western literature on *crisis stability*, on the basis of which he proposed an alternative military structure that would minimize incentives for *preemptive attacks*. It would combine a structural incapability of "strategic offensives" with adequate defensive capabilities. The various NOD models were compared according to the same criteria, and only those complying with the criteria were found to be viable alternatives, above all that of the SAS [Romberg 1986: 2, 5–35, 71].

WR's own concrete proposal amounted to a *disengagement* scheme envisioning three stages, inspired by the *Palme Commission*'s proposal for a *nuclear-weapons-free zone*. In the first stage, battlefield nuclear weapons and heavy, mechanized forces would be withdrawn from a 2 x 50–100 km *zone* on either side of the border, which would thenceforth be defended by means of light *infantry* with various forms of *barriers*. Subsequently, the same weapons systems would be withdrawn a further 50 km. In the third and final stage, the remaining parts of the territories of the *FRG* and the GDR would be rid entirely of nuclear weapons, and the numbers of offensive-capable major weapons systems would be reduced substantially [1986: 53, 56–60; cf. Palme Commission 1982, 146–149].

WR's concrete proposal remained largely unchanged, albeit embedded more explicitly in a *gradualism* framework [Romberg 1988; 1989; 1989a], but in light of the unification process (and his own central role therein), WR in March 1990 published a detailed proposal for the military future of Germany. It envisioned the four Allied powers' maintaining 30,000 troops in Germany, whereas the German armed forces would number 100,000 on the territory of the FRG, and 30,000 in what used to be the GDR [quoted in Lutz 1990: 45].

ROME DECLARATION.   Declaration adopted by *NATO* (the North Atlantic Council) at its meeting in Rome, 7–8 November 1991, at the level of heads of state and government. The participants declared that "we are working toward a new European security architecture in which NATO, the *CSCE*, the European Community, the WEU and the Council of Europe complement each other" [*Atlantic Documents:* 76 (13 November 1991)].

Furthermore, the document *The Alliance's New Strategic Concept* was adopted, wherein it was acknowledged that "the threat of a simultaneous, full-scale attack on all of NATO's European fronts has effectively been removed," and that this enabled NATO to modify its military strategy, while reaffirming proven fundamental principles. One of the principles was that "the Alliance is purely defensive in purpose: none of its weapons will ever be used except in *self-defense*, and it does not consider itself to be anyone's adversary."

Regarding force structure, it was stated that both the overall size and the state of readiness of NATO's forces were to be reduced; and that "the maintenance of a comprehensive in-place linear defensive posture in the central region will no longer be required" (tantamount to an abandonment of *forward defense* as a general guideline). The combination of present forces and reserves, and the increased emphasis on the latter were confirmed (see *Augmentation Forces*, *Mobilization*), as was the goal of increasing the *multinationality* of military formations. Finally, a drastic reduction of substrategic nuclear forces was promised (see *No Early First Use*) [*Atlantic Documents:* 75 (9 November 1991)].

ROTFELD, ADAM DANIEL.   See *Poland*, *SIPRI*.

RPV.   Remotely piloted vehicle. Remotely guided, pilotless, usually miniature aircraft, used mostly for reconnaissance purposes, but with a significant combat potential as well. RPVs have had a considerable attraction for NOD proponents by virtue of their having (at least some of) the advantages of aircraft yet none of their problematic features. Like aircraft, they are highly mobile (albeit usually with rather limited range) and they exploit the third dimension, which makes them useful for antitank missions. At the same time they are not dependent on fixed airfields (cf. the *no-target principle*); they are relatively cheap; and their limited payload makes them rather harmless for ground-attack purposes [Herolf 1986; Barnaby 1986; Møller 1989].

RUSH-BAGOT TREATY.   See **Canada*.

RUSSELL-EINSTEIN MANIFESTO.   Drafted by Bertrand Russell in 1955 and signed by Albert Einstein and other prominent scientists [in Rotblat 1967: 77–79]. Its emphasis was on the overriding importance of preventing nuclear war, in view of which the REM urged people and governments to "learn to think in a new way," a phrase that has recently been taken over by the innovative thinkers in the Soviet foreign policy establishment under the term **New Thinking*. The signatories of the REM in 1957 convened a conference in **Pugwash*, Canada, hence the name of the movement inaugurated on this occasion.

RUSSIAN FEDERATION.   See **Soviet Union, *CIS*.

S

SALSS. *Sozialwissenschaftliche Analysen zur Leistungssteigerung Sozialer Systeme* (Social Science Analyses for the Performance Improvement of Social Systems). West German sociological institute that has, inter alia, undertaken critical analyses of official threat perceptions, Bundeswehr planning, and the pros and cons of various alternative defense proposals [1984; 1984a; 1985]. Through its codirector, *Lutz Unterseher*, the SALSS has been closely affiliated with the *SAS*.

SAM. Surface-to-air missile. *Air defense* missile capable of intercepting aircraft, as well as certain other flying targets [Perkins (ed.) 1987: 111–145; Elsam 1989: 57–66; N. Brown 1986: 123–136]. Among modern SAMs, one could mention, e.g., NATO's HAWK and IHAWK, and the Soviet SA-1 through 13. The most sophisticated modern SAM is Patriot, which possesses at least some ATM (antitactical missile defense) capabilities, in addition to the ability to track and engage several flying targets simultaneously, and which proved quite capable in Operation Desert Storm (cf. *Gulf War*). Small shoulder-fired SAMs (such as the Stinger) operated by *infantry* units have likewise proven themselves very capable (inter alia, in *Afghanistan*), not least against helicopters [*Jane's Defence Weekly,* 10 January 1987, 789; Smit 1986; Friedman 1991: 147–168, 341–342; Postol 1991].

In *NOD* postures, SAMs have usually been preferred to aircraft, even the most defensive ones, i.e., interceptor fighters [cf. Afheldt 1983: 64, 131; 1989a: 113–116; Hannig 1984: 108–116; Lutz 1989c; 1989e; 1990d; Boserup & Graabaek 1990:163–164; Unterseher 1989g; Møller 1989; 1992: 183–187]. The reason has been that SAMs, as opposed to fighters, would be unable to escort aircraft on offensive missions, wherefore they would be both tactically and operationally defensive. The only exceptions to this rule might be very-long-range SAMs with the potential of extending an air defense "envelope" over parts of enemy territory or mobile SAMs capable of accompanying mobile land and maritime forces on offensive missions. The latter is partly the case with SAMs on board offensive-capable warships or escort vessels in naval task forces [cf. Perkins (ed.) 1987: 132–145], as well as with the integrated battlefield air defense systems of army formations. The Soviets' reorganization of their air defense system (*PVO Strany*) in the early 1980s (implying a decentralization down to the level of *TVD* and army level) was accordingly interpreted as signifying a shift toward the offensive [*International Defense Review* 1987(7): 869–873]. See also *Extended Air Defense*, *Air Power*.

SANGUINETTI, ANTOINE. See *France*.

SAPERSTEIN, ALVIN M. Professor of physics at Wayne State University in Detroit. In the years 1983–1986, AMS developed a model of *disengagement* envisioning the creation of depletion *zones* in central Europe. The equipment to withdraw from 100-km-wide strips on each side of the border would include heavy *tanks*, artillery, mobile missile launchers, and *bridge-building equipment*, in addition to which major concentrations of troops, helicopters, and aircraft would be avoided. Within the depleted zone, forces would consist of local *militia*-like troops, which would engage in "Swiss style" defense, armed with antitank and *air-defense *PGMs [1985; 1986; 1986a; 1987; 1988].

SAS. *Studiengruppe Alternative Sicherheitspolitik* (Study Group on Alternative Security Policy). International group of *NOD experts headquartered in Bonn and chaired by *Lutz Unterseher. Other German members include *Hermann Scheer, *Wolfgang Vogt, *Franz Borkenhagen, and (until recently) *Hans Günter Brauch. The international membership includes *Bjørn Møller (Denmark), *John Grin, Sjef Orbons, and Henny van der Graaf (the *Netherlands), Carl Conetta and Charles Knight (*U.S., see *Commonwealth Institute), and *Malcolm Chalmers (U.K.). In addition to numerous books and articles by individual SAS members, the group has published two main collective works: *Structural Change in Defense* and *Confidence-Building Defense* [SAS (eds.) 1984; 1989; cf. Møller 1991: 97–111].

The SAS's NOD model(s) are more comprehensive than most others because they include concrete proposals for a restructuring of air forces and navies, and because they are applied concretely to regions other than Europe. Nevertheless, the focus has been on *land forces* in general, and those deployed in Germany and central Europe in particular. The SAS has been very critical toward *NATO and Bundeswehr planning, as well as their underlying assumptions, inter alia, the official threat assessments. Accordingly, NATO has been criticized for both its nuclear emphasis and the more recent *deep-strike and *AirLand Battle plans that threatened *crisis stability [SALSS 1984; 1984a; Unterseher 1984b, 111–115; 1987; 1987b; 1989a; 1989c: 299–308]. Closer analysis revealing the East-West *balance of forces to be roughly equal, neither conception was regarded as warranted [SALSS 1984; Unterseher 1990; Chalmers & Unterseher 1987; 1988; 1989]. The Bundeswehr was charged by the SAS with an infatuation with the offensive; with undermining the foundations of the West German welfare state through its excessive costs and manpower requirements; and with damaging *democracy by not taking popular attitudes toward defense and security into account [SALSS 1985; Grass 1984; 1990; Thimann 1989; Bebermeyer 1984; 1989; 1990; idem & Grass 1984; Voigt 1989; 1990].

The actual model of the SAS has been distinguished from others by its envisioning the employment of only available and tested technology; by its low manpower as well as investment costs; by its being a genuine *forward defense, combined with *territorial defense; and by its possessing substantial tactical, but no strategic, mobility that might be tantamount to (or be mistaken for) an offensive capability. Hence the nickname "*spider-and-web": within their "*web," the "spiders" (the mobile forces) would be very agile, but beyond it effectively immobile [Boeker 1986: 62–65; idem & Unterseher 1986; Grin & Unterseher 1988; Unterseher 1988b; 1989f].

In regard to (pre-unification) land forces, the SAS proposed a posture with three components :

- A stationary and weblike "containment force" consisting partly of reserves and armed with various antitank and *air defense weapons.
- A "rapid commitment force" (also called the "spiders" or the "fire brigade") intended, e.g., for reinforcement of exposed stationary units and deployed within the containment web. It would consist of three types of forces: *Jäger* battalions, cavalry regiments, and shock formations.
- A "rear protection force" comprising both stationary and mobile elements.

The three elements would collaborate closely. Because of its relative lightness, the containment force was not supposed, e.g., to engage the armored forces of the invader directly but, rather, to immobilize and fix them so that the "spider" forces could deliver the final blow [SAS 1984; 1984a; Dick & Unterseher 1986; Unterseher 1989b; 1989d].

Concerning air forces, the SAS recommended a combination of unilateral restructuring and *arms control* pointing in the direction of a build-down of offensive capabilities and a removal of lucrative targets. The defender would then abandon deep interdiction as well as *OCA* in favor of a greater emphasis on air defense, in its turn a joint task for the land forces with their *SAM*s and the air force's fighters. Finally, to the (limited) extent that conventional air support remained warranted, it would feature VTOL (vertical take-off landing) aircraft such as the Harrier [SAS 1984b; Unterseher 1988c; 1989g; cf. Møller 1989; 1992: 183–187].

With respect to the maritime component, the SAS called for abandonment of offensive ("blue water") ambitions, thereby criticizing both the U.S. Navy's *Maritime Strategy* and certain tendencies in the Bundeswehr. As an alternative, the Federal Navy should limit its ambitions to defending the Baltic Sea and parts of the North Sea, implying that both destroyers and frigates would become redundant, along with large submarines. The need, rather, would be for "an evidently defensively oriented coastal defense with ships, combat helicopters, and good blocking equipment," as well as land-based antiship missiles [Bebermeyer & Unterseher 1986; 1989; cf. Møller 1987; 1989e; 1990b; 1992: 187–194].

The SAS model would presumably be entirely compatible with the maintenance of an integrated *NATO* posture. A certain role specialization was nevertheless recommended, implying that allied forces would gradually assume the main responsibility for air defense and naval matters, whereas German forces would bear the brunt of land defense along the Central Front, albeit with an allied contribution (mostly for reasons of coupling). Concrete recommendations have also been issued (authored mainly by the U.S. members) for a restructuring of the U.S. Seventh Army, according to which most U.S. troops could be withdrawn from Europe because they would be made partly redundant by the "containment web," leaving most of the remaining U.S. divisions to serve as "spiders" [Knight 1988; Conetta & idem 1990; 1990a–b; Unterseher 1990b].

In light of approaching German unification, the SAS modified its proposals somewhat, particularly with regard to the land forces. On the one hand, Lutz Unterseher lectured East German officers on how the NVA (National People's Army) might be configured as "web forces" within the framework of a modified *Genscher Plan*. On the other hand, the SAS model was also adapted to suit the Bundeswehr of the united Germany. According to these analyses, its strength might immediately be reduced to 350,000, and that of the allied forces in West Germany to 150,000. Within the next eight to ten years, the armed forces of the united Germany might be further reduced to 250,000 and the allied forces to 100,000, without this causing insurmountable *conversion* problems (such as might well be the case with more radical proposals). Because of the lower *force-to-space* ratio resulting from having to defend a larger territory with fewer forces, however, the defense would have to be focused on central strongholds (i.e., *bastions*) to a larger extent than previously [Unterseher 1991a].

Some members of the SAS, in light of the new circumstances of 1989–1990, also began to analyze the opportunities for creating genuine *collective security* structures, as well as the role of NOD therein, with a particular focus on Europe [Brauch 1990; 1991; Chalmers 1990; 1990a; Møller 1990; 1991: 270–274; 1992: 71–75]. Finally, the SAS also accepted the "*out-of-area* challenge" posed by the 1990–1991 Gulf crisis by devising a possible alternative military posture for the

Persian Gulf region that would take into account the extremely low force-to-space ratios in the vast Arabian deserts, inter alia, by "stronghold defense" postures [Conetta, Knight, & Unterseher 1991; 1991a].

SCANDILUX.   Forum for regular consultations between the social democratic parties of the Scandinavian and Benelux countries, with the *SPD*, the Labour Party, and the French *Partie Socialiste* attending as observers. In this forum, the SPD representative, *Egon Bahr*, was quite influential, e.g., was instrumental in bringing about the general reorientation of security policies in the early 1980s. In fact, ideas traceable to him could be seen as gradually permeating into the sister parties, inter alia, through the cooperation of the S. Above all, three new common perspectives united these parties (with the partial exception of the PS): the concept of *common security*; the outspoken quest for greater (West) European independence; and alternative defense in general and *NOD* in particular [Petersen 1984].

SCHEER, HERMANN.   Parliamentary speaker for the *SPD* on *arms control* issues, and a longtime member of the *SAS* [Scheer 1986; cf. Møller 1991: 170–171, 239–240, 257–258]. In 1986 Scheer elaborated on a possible *peace* strategy in his book *Liberation from the Bomb;* his point of departure was an elaborate critique of prevailing strategic trends [1986a; cf. 1984]. He urged as an antidote that the two German states collaborate in building down tension and loosening the bloc division of the continent. A triangle of the two Germanies and *Austria* might presumably, in the fullness of time, become the center of a European peace order [1986a: 226, 237–238; cf. Bahr 1985]. To facilitate such a development, a piecemeal demilitarization should be pursued in the whole of Europe, alongside political and economic *integration*, up to the point of an all-European union, the prerogatives of which would also include security and defense policy.

Concerning arms control, HS supported the notion of *Egon Bahr*: to remove nuclear weapons from the territory of states not disposing of them themselves [Scheer 1986a: 231–233, 252–253, 279–285; cf. Bahr in Palme Commission 1982: 182]. This partial denuclearization would not necessitate a conventional buildup because "the conventional inferiority of NATO no longer exists," as he bluntly stated. He therefore supported an early *MBFR* agreement, to be followed by negotiations focusing on particularly offensive types of weaponry [1986a: 251, 268].

During the unification debate in 1990, HS and Heidemarie Wieczorek-Zeul (a member of the party presidium) issued a memorandum, "Germany and NATO," wherein they argued against membership of the united Germany in NATO in its present configuration. Likewise dismissing *neutrality*, their alternative was a loosened alliance, implying that NATO should abandon military integration along with the *flexible response* strategy. They further proposed a partial demilitarization of Germany, combined with a shift to a strictly defensive structure, i.e., *NOD*. The troops of the four powers should be drastically reduced and eventually withdrawn completely with the entry into force of a "European Peace and Security Arrangement," i.e., a *collective security* order [Wieczorek-Zeul & Scheer 1990; cf. 1990a].

SCHELL, JONATHAN.   American journalist, author of the 1982 bestseller *The Fate of the Earth* (on the horrors of nuclear war) and *The Abolition* [1984], in which he made a number of proposals for the reduction and eventual abolition of nuclear weapons.

Nuclear abolition to JS did not mean a "non-nuclear world" because "the possibility that nuclear arsenals will be used cannot be ruled out." Hence, "we shall always live in a nuclear world" [1984: 23–24]. As the best solution to this problem, JS pleaded for a "weaponless deterrence," i.e., a sort of *existential deterrence*, wherein "blueprint would deter blueprint, equation would deter equation." Even though *nuclear deterrence* would not rest on the actual deployment of any nuclear weapons, it would nevertheless work in the sense that war henceforth could not possibly be a rational instrument of national policy [1984: 119, 26. See also Booth & Wheeler 1992]: what others have called a *subnuclear setting* [Boserup 1985; 1986; 1989; 1990].

Pursuing the same logic, JS recommended *no-first-use*, albeit with the indispensable qualification "unless we start to lose." He thus apparently regarded its inherently limited credibility as speaking in favor of NFU, a paradoxical argument also made by other authors [1984: 53–55; cf. Møller 1986a; 1987a; 1988a. For a critique: Mearsheimer 1986b]. Nevertheless, JS agreed with both critics and proponents of NFU that conventional *balance* was an essential precondition for proceeding with abolition. For this purpose, he proposed what was tantamount to a shift to *NOD*, by emphasizing that forces should be configured with a view to being able to prevent invasions yet unable to undertake them. They should thus be "loaded down with anti-tank weapons but low on *tanks*, well equipped with anti-aircraft weapons but ill equipped with aircraft." JS (erroneously) compared such a posture with the example of *Switzerland*, apparently ignorant of the extreme tank-heaviness of the Swiss army [Schell 1984: 138–139].

SCHELLING, THOMAS C.   Professor of political economy at Harvard University, and one of the most influential strategic thinkers in the nuclear era.

In his 1960 work *The Strategy of Conflict*, which was largely couched in the terminology of *games theory*, TCS described and analyzed a number of cooperative and/or mixed-motive game situations from ordinary life as well as interstate relations. He thus contravened the almost exclusive focus on *zero-sum games* that dominated during the Cold War. Furthermore, TCS elaborated on the notion of *crisis stability*, e.g., with his scenario of "reciprocal fears of surprise attack." Two adversaries, neither of whom had to be expansionist or aggressive, might become locked in a vicious circle, whence a war desired by neither side might ensue: "A modest temptation on each side to sneak in a first blow, a temptation too small by itself to motivate an attack, might become compounded through a process of interacting expectations, with additional motives for attack being produced by successive cycles of 'He thinks we think he thinks we think . . . he thinks we think he'll attack; so he thinks we shall; so he will; so we must'" [Schelling 1960: 207]. Such scenarios have provided the point of departure for a number of *NOD* strategies, which have been designed especially to break such vicious circles.

TCS himself also dealt with at least one variety of alternative defense, namely, *civilian-based defense*. He treated it as a *strategy* in its own right (i.e., not just as "passive defense") and assessed it partly favorably, albeit with numerous reservations [1967]. In addition to dealing with military strategy, TCS was also among the first in the "strategic community" to theorize at length about *arms control* [Schelling & Halperin 1961]. The primary purpose thereof should not be *disarmament* but strategic *stability*, which might even require more, rather than less, weaponry. What mattered was not so much the quantities as qualitative features of (especially nuclear) armaments: vulnerability, precision, deployment patterns, etc. Evaluating retrospectively the record of arms control in 1986, TCS was therefore quite critical, particularly with regard to the actual focus on quantity to the detriment of qualitative arms control [Schelling 1986].

SCHLIEFFEN PLAN.   War plan developed prior to *WWI* by Count Alfred von Schlieffen and his associates (the younger Moltke and others) in the German General Staff. The purpose was to solve Germany's problem of a potential two-front war: against *France* and others in the West, and against Russia in the East. The SP envisioned a grand-scale envelopment of the French forces by means of an invasion through neutral *Belgium*. Germany was urged to deplete its eastern front (and hence fighting a holding battle against Russia) in order to allocate as many divisions as possible to the initial battle against France. After the defeat of France, the forces would be directed against Russia.

The highly offensive SP was matched by the equally offensive French *Plan 17*, with the result that by the spring of 1914, two offensive armies were poised against each other, both believing in the feasibility of a swift and decisive victory. In this situation, the interlocking *mobilization* schemes drove the two sides toward *preventive war* and/or *preemptive attack*, hence made WWI close to inevitable [Holborn 1941; Wallach 1967: 52–181; 1972:89–135; Howard 1985; Evera 1985; J. Snyder 1984: 107–156; Rothenberg 1986; Showalter 1991; Tuchman 1962: 33–161; Bellamy 1990: 69–73; Dupuy 1990: 226–227; cf. for a critique of these analyses: Trachtenberg 1990; Shimshoni 1990].

SCHMÄHLING, ELMAR.   Rear admiral (ret.) in the Bundeswehr. Having already been given an official reprimand by the Defense Ministry for his outspoken criticism of *NATO* strategy and his unequivocal *NOD* advocacy, ES was dismissed following publication of his book *The Impossible War* [1990e] in the spring of 1990. Since then, he has been a freelance writer with links to the *peace movement* [*Die Zeit,* 26 January 1990; *Atlantic News,* 19 January 1990; cf. Møller 1991: 142–147].

ES's contribution to the NOD debate has above all been in the field of naval strategy. He regarded surface warships as obsolete in a combat environment dominated by antiship missiles and submarines and hence recommended a shift toward "many small, possibly camouflaged and/or mobile weapons carriers at the coast or offshore" [1989b–c; 1990c; 1990f]. With respect to *land forces*, ES proposed greater reliance on *barrier*-type defenses (based on sophisticated mines) and locally conscripted *militia*-like light forces, fighting primarily at short distances. The defense would capitalize on prevailing trends in military technology (the potentials of long-range, accurate, indirect fire as a substitute for mobility) as well as on home ground advantages by laying fiber-optic networks and emplacing sensors underground and on high-rise buildings, and by storing mines, munitions, *bridge-building equipment*, etc. [1987; 1988; 1988a–c; 1990e: 181–183, 237–251, 284–290]. The result would be a stationary (Raumgebunden) posture incapable of large-scale offensive operations, whose size need not exceed 300,000–350,000 troops, as far as Germany was concerned. Because it would improve rather than weaken NATO's defensive capabilities, it would lend itself to unilateral implementation, even though a *Warsaw Pact* reciprocation would be most welcome. In the latter case, *mutual defensive superiority*, i.e., a high degree of *conventional stability*, would be the result, which would allow for deep cuts in the nuclear arsenals [1988a–c; 1990b; 1990e: 142–145, 172–183].

ES also challenged the hegemonic status enjoyed by the *U.S.* within NATO and urged the *FRG* to demand greater influence on strategy, in accordance with the "most affected principle." More generally, European collaboration in defense matters should be promoted [1989a; 1990e: 17-19, 99, 159, 268].

SCHMIDT, MAX.   Former director of the East German semiofficial political science institution *IPW, in which capacity he and his staff served as speechwriters for the SED leadership. He was, furthermore, influential in disseminating East German propaganda (including that pertaining to the new, defensive, *Warsaw Pact military doctrine, and the thesis of the *Totally Destructive Conventional War) to the East German population as well as to the West [1989; 1991; idem & al. 1989; idem & Schwarz 1989; 1989a–e; 1990; 1990a–b]. Soon after the 1989 revolution in the *GDR, he lost all his positions but retained a few abroad (as a relict of the past).

SCHMÜCKLE, GERD.   German general (ret.) and former deputy SACEUR (Supreme Allied Commander Europe). In 1988 GS published a proposal, *Concept of Stable Dissuasion,* jointly with *Albrecht A. C. von Müller*, wherein they recommended a withdrawal of offensive-capable weapons systems from the forward line, thenceforth to be defended by means of *barriers,* etc. The restructuring would be accompanied by a build-down of tactical nuclear weapons in Europe to "a few hundred," which would be assigned directly to SACEUR [Schmückle & Müller 1988].

SCHÖLL, INGRID.   Coauthor with *Joachim Wernicke* of *To Defend Rather Than to Annihilate: Escapes from the Nuclear Trap* [Wernicke & Schöll 1985].

SCHOLZ, RUPERT.   Former defense minister of the *FRG for the *CDU. While sharing the CDU's critical attitude toward the (seemingly social democratic) notions of *NOD [Scholz 1989] and *common security, RS introduced the (nearly synonymous) concept of *reciprocal security (gegenseitige sicherheit): "It requires willingness on the part of both parties to allow each other the same measure of security. It requires abstention from trying to solve political problems by military means. The concept of reciprocal security rests on the balance of mutually secured defense capability, on a system of reciprocal confidence-building measures, as well as on *détente and *disarmament" [*Bulletin der Bundesregierung,* 26 August 1988; cf. Lutz & Theilmann 1986: 224–228].

In light of unification, RS even went so far as to endorse explicitly a review of the principle of *forward defense. Contrary to the *SPD, however, he envisioned no in-depth or nonoffensive defense for its replacement but, rather, advocated a "mobile defense," which might well entail offensive capabilities. Moreover, he was skeptical about the viability of the *Genscher Plan because "a reunited Germany with two different, or even opposing, security zones is inconceivable" [*Atlantic News,* 4 April 1990; Scholz 1990, 244–246].

SCHUBERT, KLAUS VON.   Historian specializing in the history of West German rearmament [1970], member of the *SPD, and one of its leading ideologues, particularly with regard to *common security and *NOD [1988; 1989; 1990; 1992]. Until his death, KvS was leader of the Study Society of the Protestant Churches in the FRG [idem & al. 1988].

SCHULTE, LUDWIG.   West German military expert, specializing in weapons technology. His preoccupation with this subject led him to recommend both the *SDI and various *deep-strike conceptions, and *NOD-like employment of *PGMs for antitank and *air defense purposes [1985; 1990; Reinfried & Schulte 1987].

SCHÜTZE, WALTER.   See *France.

SCHWARZ, WOLFGANG.   East German political scientist on the staff of the *IPW and a close collaborator of *Max Schmidt [Schmidt & Schwarz 1988; 1989;

1989a–e; 1990; 1990a–b]. As a consequence, he lost his positions after the 1989 revolution.

WS was one of the first in the SED to study *NOD seriously. In his doctoral dissertation, "Structural Incapability of Attack in Europe: Requirements, Conceptions, Obstacles," he evaluated various West German NOD proposals [Schwarz 1989]. As his own alternative to prevailing trends, WS advocated a very radical and comprehensive form of NOD, including not merely the land forces but also navies and air forces. Regarding implementation, he dismissed *unilateralism because such a radical transformation called for a concerted collaborative effort by both sides. Furthermore, this *transarmament to NOD should be combined with actual *disarmament—i.e., require fewer troops and no major investments—wherefore WS was skeptical with regard to more demanding and high technology–dependent NOD schemes.

Besides advocating NOD, WS was a consistent proponent of the thesis of the *Totally Destructive Conventional War, which led him after the revolution to join the *SES.

SCHWEITZ OHNE ARMEE (Switzerland without an Army).   See *Switzerland.

SCHWEYK.   See *Good Soldier Schweyk.

SCORCHED EARTH STRATEGY.   Strategy whereby the defender embarks on a large-scale strategic withdrawal combined with destruction of whatever materials might be used by the invader. In this way, the enemy's logistical problems would be compounded, thus eventually forcing him to withdraw.

SES has usually been employed by the defender, but an aggressor might also under certain circumstances use it in order to starve out the defenders. The Soviet Army, e.g., occasionally used SES on a local and regional scale against the *Mujahedeen* in *Afghanistan [Dunnigan & Bay 1985: 101], and Iraq did so before leaving Kuwait in 1991 (cf. *Gulf War) [*International Defense Review*, 1991(9): 992–994].

SDI.   Strategic Defense Initiative launched by President Ronald Reagan in 1983. See *Strategic Defense.

SDP.   See *U.K.

SEA COMMAND.   See *Sea Power.

SEA CONTROL.   See *Sea Power.

SEA LINES OF COMMUNICATION.   See *SLOC.

SEA POWER.   Generic term signifying the ability to perform a panoply of naval (or better, maritime) missions while denying this ability to adversaries. Its military connotations notwithstanding, SP presupposes civilian use of the sea and has its fundamental *raison d'être* in such use [Mahan 1894; cf. Sprout 1941; Crowl 1986; Støren 1987: 36–47].

1. **Elements of SP.** The various military uses of the sea require the ability to perform a number of naval missions. These might (somewhat artificially) be divided into assertive (i.e., "offensive") and protective ("defensive") (Table 31).

Table 31. Defensive and Offensive Sea Power

Protective Use	Defensive Missions	Assertive Use	Offensive Missions
Supplying population and armed forces	Defense of SLOCs (sea lines of communication)	Naval blockade	SLOC severance *Guerre de course*
Defense of • home territory • off shore and overseas possessions	Costal defense Forward defense	Power projection	Amphibious operations Bombardment
Intrawar deterrence (stalemating)	Fleet-in-being Nuclear deterrence	War-winning	Decisive battles Strategic ASW (antisubmarine warfare)
Assertion of sovereignty	Patrolling Showing the flag	Offensive naval diplomacy	Gunboat diplomacy

The distinctions between assertive/offensive on the one side and protective/defensive on the other are, however, not entirely straightforward and should certainly not be understood to signify a good-versus-bad dichotomy (cf. *offense-defense distinction*). The *Gulf War* (in response to Iraq's aggression against Kuwait), e.g., featured primarily offensive operations on the part of the coalition and defensive ones on the part of Iraq, a classical example of the *offensive-turned-defensive* phenomenon.

2. **Sea Control and Command.** SP is inherently flexible and general purpose because naval forces can, within certain (wide) limits, be redeployed and reassigned at will to different theaters and missions. This flexibility has many advantages, including those of parsimony: Because the same forces can simply be used for many purposes simultaneously "use" need not be tantamount to engagement in combat but might include deterrence and "showing the flag." The attraction of this flexibility has undoubtedly been one reason that the Western powers have been reluctant to negotiate *naval arms control* and/or *maritime C(S)BMs*; these might limit their freedom of navigation as codified, e.g., in the "*mare-liberum*" principle.

To maximize naval flexibility and maritime freedom has been the essence of "sea command," conceived of as the appropriate goal for great sea powers and implying complete control in the sense that other states are able to navigate only in conformity with the will of the holder of sea command [Mahan 1894]. Sea command is thus much more ambitious and exclusive than "sea control," signifying a functionally and/or geographically circumscribed ability to navigate at will while denying this ability to an adversary. The quest for command is thus more liable than control to lead to conflict because it tends to transform the naval contest into a *zero-sum game*. For sea power to be compatible with broader security political principles such as those of *common security* and *NOD*, it would therefore have to be disaggregated and each element subjected to a critical assessment [cf. Møller 1987; 1989e; 1990b; 1992: 187–194].

3. **Technological Trends.** The offense/defense balance for maritime forces has shifted back and forth in the past. There are various indications that the present is a

period of defensive supremacy, due, inter alia, to the development of technologies related to firepower and *C^3I. *Concentration is becoming more a function of weapons and C^3I and less of platform mobility through naval maneuvering. A premium is, furthermore, being placed on concealment, to the detriment of mobility and platform concentration, and the survivability of large surface ships is being challenged by the proliferation of guided antiship missiles, etc. [Schmähling 1990c; FRAG 1986; Klippenberg 1989].

4. **Nonoffensive Sea Power.** At a minimum, a nonoffensive maritime defense would have to include the ability to perform the following missions:

- Defense against seaborne invasions, by means of a combination of land-based naval aviation and antiship missiles (assigned to the coastal artillery), small surface vessels, minefields, and submarines. Dependent on the concrete circumstances, it should, in principle, be a *forward defense, yet preferably not so forward as to intersect with an adversary's forward defenses. The appropriate defense perimeter would depend on geographical particularities, but *disengagement zones should, as far as possible, be created by opting for a perimeter less than half the distance to the adversary.
- Defense of vital *SLOCs, e.g., through a combination of *barrier-type defenses and a convoy-with-escorts system.
- *Minimum deterrence, for the immediate future by means of, e.g., seaborne nuclear forces, based either in defended *bastions or in the high seas.

In none of the three contexts should the defender present his forces for any "decisive battle" but, rather, he should opt for a modern "fleet-in-being" strategy [cf. Till 1982: 111–121; Støren 1987: 84–92]. It should, however, differ from the traditional one by being more of a combined-arms strategy in which land-based weapons would play an important role. Furthermore, it should be as elusive as possible through maximum dispersal, and its operational concepts should resemble "naval *guerrilla warfare" (cf. *guerre de course).

A posture designed with the three missions in mind would need no *aircraft carriers, battleships, land-attack cruise missiles, or tactical naval nuclear weapons. The average size as well as the required number of major surface warships (frigate size and above) would therefore be substantially reduced from their present postures. On the other hand, it would place a greater emphasis on minelayers and sweepers, attack submarines (most of which with diesel-electric propulsion), small missile-carrying vessels, land-based naval aviation and helicopters, and mobile, land-based antiship missile batteries [cf. Gerber 1982a: 78–83; Hannig 1984: 116–120; Bebermeyer & Unterseher 1986; 1989; Prins 1988; Boserup 1987; 1990a; Møller 1987; 1989e; Booth 1994].

Such naval strategies and postures would imply abandonment of certain missions, such as foreign intervention and coercive gunboat diplomacy. Because the need for foreign intervention would probably not disappear entirely (as shown by the Iraqi invasion of Kuwait), it had better be left to such international bodies as the *U.N., whereas a *nonintervention regime might be established for individual states. Some of the thenceforth redundant major warships might therefore be assigned to U.N. naval task forces for peacekeeping and (as a last resort) interventionary missions [Forsberg 1987; Evera 1990; Prins 1991; 1991a; Staley 1992].

5. **Implementation.** A restructuring to such naval postures could, in principle, be undertaken unilaterally, but there would be many advantages to reciprocity, e.g., in the framework of *naval arms control.

SECOND WORLD WAR.   See *WWII.

SECURITY.   The object of security policy, and *ipso facto* also of defense policy, yet without any authoritative meaning. A good definition is that of Arnold Wolfers: "Security, in any objective sense, measures the absence of threats to acquired values, in a subjective sense, the absence of fear that such values will be attacked" [Wolfers 1965: 150; cf. Buzan 1991: 16–17 & passim; Krell 1981].

Other than the distinction between its objective and subjective senses, S is multidimensional in several senses, a fact that has been disregarded by the traditional paradigm of international relations, that of *Realism*, with its preoccupation with military state-to-state threats [Morgenthau 1960; cf. for a critique: Booth 1991; 1991b]. The competing paradigm of *interdependence* [Keohane & Nye 1977; Tromp 1988], which to some extent has been incorporated in Neorealism [Buzan 1991; Gilpin 1981; Keohane (ed.) 1986], has highlighted a range of other threats and insecurities. Not only states and/or nations but also persons as well as humankind as a whole may be either secure or insecure, even though "security" may mean different things to different referent objects (see Table 32) [cf. Waever & al. 1993].

Table 32.   Security

Referent Object	Content
The State	Sovereignty, power
Collectivities	Identity
Individuals	Survival, well-being

Furthermore, not only states but also individuals, other states and entities, or indeed the structure of the international system may be sources of threats on all levels of analysis. Finally, threats may appear in many forms, and those of direct violence and aggression are not necessarily the most serious [cf. Galtung 1969; Brundtland Commission 1987; Mische 1989; Westing (ed.) 1989; Møller 1993; Prins 1992]; Table 33 captures only a few of these dimensions and types of threats. On the basis of this panoply of threats, a distinction might be derived between various forms of S such as personal, societal, cultural, economic, ecological, political, and military.

Table 33.   Security Threats: From Whom or What?

| | **Source** | | | | |
Object	Individuals and Groups	The State	Other States	International System	Structural
Individuals and groups Society	Crime Exploitation Discrimination	Oppression Structural violence Militarization	Psychological warfare Migration	Cultural dependency	Pollution Ozone depletion Resource scarcity Population growth
The state	Revolution		Attack		
International system			Arms racing Economic warfare Militarization		
The World			Nuclear war		

It furthermore becomes obvious that contradictions may arise between the quests for S on each level of analysis:

- Individual, societal, and global S may, for instance, be damaged by a preoccupation with military security, e.g., because of the opportunity costs of military preparations (cf. the *defense dilemma*).
- International S, and thereby also national S, may be jeopardized by the very quest for national S in the form of military preparations because of the *security dilemma*.

The alternative defense and S discourse has tended to adopt a broad conception of S. To the extent that continued attention to military S has at all been acknowledged as imperative, it has been underlined that the defense and security dilemmas have to be resolved, as implied by the *common security* approach. This, in turn, would presuppose a shift to *NOD* postures and would presumably facilitate disarmament, thereby making resources available for the handling of other, nonmilitary threats to security (cf. *conversion*).

SECURITY COMMUNITY.   A central concept of *integration* theory [Deutsch & al. 1957], which has been applied to the "Atlantic Community" as well as to western Europe integration. The first stage in the integration process might be a "pluralistic SC," i.e., a situation wherein war had ceased to be a genuine option but national sovereignty would remain unlimited. The second stage might be an "amalgamated SC," i.e., a confederation of states or even a federation, in which sovereignty would be relinquished (partly or completely) to international or supranational authorities.

Certain theorists of *common security* have regarded the transformation of the whole of Europe into a pluralistic SC as a viable goal, leaving the possibility of eventual amalgamation open [Schlotter 1982; Møller 1992: 20, 35]. It is obvious, however, that such a development would presuppose (or at least be facilitated by) a neutralization of the military factor, e.g., through elimination of reciprocal offensive options by means of a shift to *NOD*. A full-fledged SC would be almost synonymous with a *common security regime* [cf. Lutz 1987a].

SECURITY DILEMMA.   Intrinsic feature of international relations in a setting such as that described by *Realism*. The SD has, e.g., been described by the earliest proponents of this view, such as Thucydides [*The Peloponnesian War,* e.g., Book I:75 (speech of the Athenian representatives)] and Thomas Hobbes [e.g., *Leviathan* I:13], as well as by the founders of postwar Realism such as Morgenthau [1960: 54–67] or Kissinger [1957: 144–145]. The theory of the SD has, however, above all been associated with the names of Herz, *Jervis*, and *Buzan* [Herz 1950; 1951; Jervis 1976: 58–93; 1978; Buzan 1991: 294–327].

The gist of the SD is that in any dyadic relationship containing at least some conflictual elements, each side tends to perceive the other as a potential threat, against whom a certain defense is called for. If State A, however, were to respond to a perceived threat from State B, the latter (similarly regarding A as a potential threat) would tend to see A's response as an increased threat, calling for a response. The latter response may be interpreted by A as both confirmation of the original threat perception and an addition to the material threat, calling for a response in turn. And so on. Even in the absence of any malevolent intentions on either side, i.e., between two saturated and *security*-minded states, a malign interactive spiral might ensue, in the course of which the security of both sides would deteriorate. At one horn of the SD is a self-defeating quest for security; at the opposite horn are the risks attendant upon not responding to the perceived growing threat.

The SD manifests itself primarily in the arms race, in competitive *alliance* formation, and in crisis scenarios. It is thus at the core of the action-reaction phenomenon characterizing (at least one aspect of) the *armaments dynamics*. It is, furthermore, a factor promoting bipolarity in an initially polycentric or multipolar international system. When all states regard one another as potential enemies, States A and B might, e.g., be in a position to overwhelm State C if they pool their forces, which gives C an incentive to align itself with either A or B against the respective other, preferably preemptively. Hence, the tendency toward a continuous struggle for "preemptive alignment," leading to an increasing polarization of international relations. Finally, the SD may manifest itself in a low degree of *crisis stability*, evidenced in vicious circles of "reciprocal fears of surprise attack." In the midst of a political crisis, urges toward *preemptive attack* might acquire a momentum of their own, which might escape control and ultimately lead to war because of the "additional motives for attack being produced by successive cycles of 'He thinks we think he thinks we think . . . he thinks we think he'll attack; so he thinks we shall; so he will; so we must'" [Schelling 1960: 207].

Alternative defense and security proponents have asserted that they have found an escape from the SD in the conceptions of *common security* and *NOD*: By taking into account the security concerns of their opponents, adversarial states might simultaneously achieve a lasting improvement of their security vis-à-vis one another. This, in turn, would presuppose a distinction between offensive and defensive means of defense, and a clear emphasis on the latter, i.e., means of defense that do not threaten others [Jervis 1978; Väyrynen 1985a; Afheldt 1987c; E. Müller 1986: 164; Wheeler & Booth 1987; Wiseman 1989; Møller 1989b; 1992: 24–28]. This would presumably facilitate *disarmament* and contribute to "mellowing" bloc confrontation (or even pave the way for *neutralization*) as well as improve crisis stability.

SECURITY PARTNERSHIP.   See *Common Security*, *SPD*.

SECURITY REGIME.   *Regime* defined by *Robert Jervis* as consisting of "principles, rules and norms that permit nations to be restrained in the belief that others will reciprocate" [Jervis 1982; cf. Wheeler & Booth 1987: 316–317. On regimes in general, see Krasner (ed.) 1982]. A SR would not have to be formalized but might consist entirely of tacit understandings. It need not even require initial agreement between the two sides but could develop gradually on the basis of a coordinated and calibrated *unilateralism*. Such has, inter alia, been the rationale for the *GRIT* and *Tit-for-Tat* strategies [Møller 1992: 39–40; cf. Osgood 1962; Axelrod 1984; Goldstein & Freeman 1990].

Elements of a SR have continuously been present between the superpowers, and were almost codified in both the *Basic Principles Agreement* of 1972 and the *Prevention of Nuclear War Agreement* of 1973, in which the two sides committed themselves to show restraint and respect each other's vital interests [George 1989; George, Farley, & Dallin (eds.) 1988; Kanet & Kolodziej (eds.) 1991]. Furthermore, it has been present between the two sides of the East-West conflict on a regional scale, for instance in the so-called *Nordic balance* [Møller 1985; 1989f; 1990b–c; 1991b; Gleditsch & al. 1990; Bomsdorf 1989].

A SC might thus be an element in (or a precondition for) a *common security regime*. In this context, the SC might, e.g., be manifested in a deliberate abstention by both sides to a given adversarial dyad of states from offensive capabilities.

SED.   *Sozialistische Einheitspartei Deutschlands* (Socialist Unity Party of Germany). See *GDR*.

SELF-DEFENSE.   The only permissible justification of a recourse to war, i.e., *casus belli,* according to, inter alia, the 1928 Kellogg-Briand Pact and the **U.N. Charter.*

A certain ambiguity has been created by the fact that the U.N. Charter acknowledged the legitimacy of collective defense arrangements such as **alliances*, implying that a state under attack would be entitled to the help of other states who were not themselves under attack, hence not acting in direct SD. This might, furthermore, have implications for force structures because such allies would require some offensive capabilities (long-range mobility, means of forced entry, etc.) in order to meet their reinforcement obligations, whereas strictly defensive capabilities would suffice only for a state committed to "pure SD," i.e., a neutral state. The geographical scope of self-defense thus remains controversial.

There has also been some disagreement about the temporal scope of SD. It remains disputed whether a state is obliged to wait for an attack to occur (thereby permitting an attacker to exploit the surprise factor), or whether "anticipatory SD" might be legitimate. It might, for instance, take the form of a **preemptive attack*, if an attack was imminent. The latter has been the official position of **Israel*, and it may (according to certain analyses) have been that of the **Soviet Union*. The problem is that anticipatory SD would be practically indistinguishable from attack, and that the very fact of anticipation might render it impossible to ascertain (even *ex post facto*) whether an attack was actually imminent. Anticipatory SD raises the question within whose authority the judgment about the legitimacy of anticipation should fall. The **U.N.* has been unwilling to extend the scope of SD to include anticipation [Schachter 1991].

SELF-DETERRENCE.   Possible result of the MAD (mutual assured destruction) stance of reciprocal deterrence of the nuclear powers, hence an element in the several dilemmas of **nuclear deterrence* [cf. Jervis 1984; 1987; 1989; 1989a; Lebow 1981; 1983; 1985; 1989; MccGwire 1986; Buzan 1987a: 163-196; Wiberg 1989; Kissinger 1969; Martin 1979; Brodie 1959: 329; Clark 1982].

If State A were, for instance, to launch a limited attack against State B, deterrence theory would suppose retaliation by B, the prospects of which would presumably deter A from attacking in the first place. B might, however, be self-deterred from retaliation by the prospect of A's counterretaliation. If A were aware of this SD effect, the deterrent threat would lose credibility. SD could also operate on the conventional level of armed conflict, particularly if sliding scales of **escalation* from direct conventional response, from tactical to strategic nuclear weapons, were envisioned, as under **NATO*'s **flexible response* strategy. If B would be self-deterred from retaliation by means of strategic nuclear weapons, this SD might lead to SD at each successive step leading toward this level, unless the rungs on the escalation ladder were clearly separated (which is precisely what they should not be for the sake of deterrence). A state might thus be self-deterred from even the smallest first-use of nuclear weapons. To the extent that conventional and nuclear forces were intermingled (as was the case for both NATO and the **Warsaw Pact* during the Cold War), states might even be self-deterred from using their conventional forces out of fear that, for example, the takeoff of (conventionally armed but nuclear-capable) aircraft might be mistaken for a nuclear strike and hence incur retaliation. Even **mobilization* of conventional forces in anticipation of an attack might be liable to SD because states might fear that defense preparations could be misinterpreted as preparation for an attack.

One advantage of **NOD* would be the elimination of the SD factor: if the military measures needed for parrying an attack were to be clearly and unmistakably defensive (and if nuclear and conventional forces were clearly separated), they could not possibly provoke an otherwise avoidable attack, wherefore there would be no reason not to take them in anticipation of an attack.

SELF-SUFFICIENCY.   See *Invulnerability.*

SEMIALIGNMENT.   See *Neutralization.*

SENGHAAS, DIETER.   Professor at Bremen University, and one of the founding fathers of German peace research. His main contributions have been in the fields of development theory and *armaments dynamics* research [1972; 1990f]. Since the mid-1980s, DS has paid considerable attention to the potentials of *disengagement* and *NOD* [1985; 1986; 1987; 1988; 1988a; cf. Møller 1991: 82–83, 254–255].

Taking in 1986 the assumption of a maintenance of Europe's division into two opposing *alliances* as his point of departure, DS advocated disengagement and NOD as contributions to confidence-building and coexistence. Contrary to the conceptions of *nuclear-weapons-free zones*, DS emphasized that disengagement should commence with a disentanglement of heavy conventional equipment and large manpower concentrations. By allaying concerns about conventional aggression, this would deprive nuclear weapons of their main rationale and thus pave the way for a *no-first-use* strategy and deep cuts in the nuclear inventories. The proposed disengagement zone should stretch from 50 km west of to 100 km east of the dividing line, so as to reflect geographical asymmetries. Offensive-capable major weapons systems and items of *logistics* such as *tanks*, combat helicopters, and bridging equipment would be withdrawn from the zone, and each side would be encouraged to construct tank *barriers* of various kinds in its zone [1986: 103–107; 1987; 1988: 78–87].

Disengagement should not be seen as an escape route from the alliances, the very *multinationality* of which (particularly of *NATO*) constituted a political guarantee against attack. In the long-term perspective, however, an all-European system of *collective security* should be established, in its turn, as a further step toward what DS called "cooperative security," requiring as a minimum a demilitarization of the East-West conflict [1986: 106, 114–116; 1988a].

By 1990 DS had abandoned the notion of demilitarization as too conservative and had come to regard the entire East-West conflict as obsolete, hence the unique opportunities to create a "structure of lasting *peace*" in Europe within the span of a decade [1990d: 7–8, 11–14, 21; 1990a]. The foundations of the structure would be the *CSCE* and the EC; the military alliances would soon be rendered superfluous and outdated and would be replaced by a collective security system. A "European Security Council" should be established, having at its disposal a standing peacekeeping force of Blue Helmets, capable of military intervention, albeit only as a very last resort; the emphasis would be placed on "therapeutic conflict intervention."

With respect to united Germany DS suggested a build-down to no more than 100,000 troops, a force that would be structurally incapable of major offensives, besides being balanced by around 175,000 troops in the neighboring countries, as well as by the maintenance of 65,000 foreign troops stationed in Germany [1990d: 23–31, 46, 54–55, 87, 91; cf. 1990b; 1990c; idem in *Frankfurter Rundschau,* 22 June 1990].

SERMON ON THE MOUNT.   The theological point of departure for most of the religiously based *pacifism* to be found in Christianity [Matthew 5–7]. The admonition "love thine enemies" and the unqualified proscription of killing [Matthew 5:44, 5:21] have been taken literally by, e.g., the *Quakers* and *Tolstoy*, i.e., interpreted as ruling out any recourse to arms, even in *self-defense* [cf. J. T. Johnson 1987].

SES. *Studiengruppe Entmilitarisierung der Sicherheit* (Study Group on the Demilitarization of Security). Group established in postrevolutionary *GDR* in January 1990, primarily by former supporters of the regime. The members included *Wolfgang Schwarz* (of the *IPW*), sea captain *Siegfried Fischer* (of the NVA and the Military Academy), and André Brie from the Diplomatic Academy, subsequently speaker for the PDS (the reformed successor of the SED). After German unification, the group was dissolved [cf. Møller 1991: 208–209, 246].

Against this background, the conceptions promoted by the SES were surprisingly radical. E.g., reviewing the *Vienna Seminar* on military doctrines in January 1990, the SES stated that "the discussion now in progress on military doctrines should not take the defensive nature of these doctrines as its yardstick for comparison, but rather their conformity with the tendency toward demilitarization." The SES, furthermore, made quite radical proposals with regard to unification, e.g., by advocating *neutralization* combined with a complete demilitarization of the GDR. It rejected the notion of *NOD* on the grounds that "even the preference for the defense eliminates neither offensive capabilities, nor the threat emanating from there." The SES was, finally, a firm believer in the futility and utter destructiveness of any type of war on the European continent, even a strictly conventional one (cf. *Totally Destructive Conventional War*) [*Frieden und Abrüstung*, 1 (1990): 23–30; S. Fischer 1990; 1990a; cf. Schwarz 1989a].

SHADOW OF THE FUTURE. The value that two contestants assign to future and long-term gains relative to immediate and short-term gains, sometimes expressed as a "discount factor." As a general rule, cooperation is more likely to be the rational strategy of both sides if the SF is long (i.e., the discount factor small) because the short-term gains from defection (say, from not complying with an *arms control* agreement) would tend to be outweighed by lost opportunities of future cooperation (say, in terms of new arms control agreements). Couched in the terms of *games theory*, an iterated *prisoner's dilemma* might have a solution, but only on the precondition of a long SF [Axelrod 1984; cf. for a critique: Stein 1990: 86–112].

The *time perspective* would thus be an important determinant of cooperation in international relations. In its turn, it would depend on whether (at any given moment) the existing configuration of the international system is viewed as transitory or of indefinite duration. In the latter case, *common security* might become the guiding principle (manifested, e.g., in a shift to *NOD*-type military forces) and *interdependence* might be promoted. In the former case, both sides would tend to prepare for a final showdown, hence to emphasize relative over absolute advantages and tending to view their mutual relations as a *zero-sum game* [Møller 1992: 37–39]. The *Soviet Union* until the mid-1980s tended to adopt the former perspective (e.g., by conceiving of *peaceful coexistence* as a transitory strategy); its apparent shift to the latter perspective facilitated the shift to a defensive strategy and thus paved the way for the dismantling of the Cold War [cf., e.g., Shenfield 1987; 1988; Zagladin 1986; 1988].

SHARP, GENE. Professor emeritus of political science at Southeastern Massachusetts University, president of the Albert Einstein Institution, and the leading U.S. expert on *civilian-based defense* (CBD) and *nonviolence*. GS served for a number of years as director of the Program of Nonviolent Sanctions at Harvard University.

GS's main work is his monumental, three-volume *The Politics of Non-violent Action,* in which he analyzed 198 methods of nonviolent struggle for both national defense purposes and civil rights movements, etc. [1973; cf. 1967]. Among his main sources of inspiration were *Gandhi* (to whose teachings and practical experience he devoted an entire book [1979]), and the black civil rights movement in the United States led by *Martin Luther King*, which also accounted for many examples. Unlike the approaches of both men, however, GS's approach to nonviolence was more pragmatically than ethically or religiously motivated. Furthermore, whereas many CBD proponents have been revolutionaries (in the broad sense of the term), GS has emphatically denied any revolutionary implications of his defense scheme, realizing that any such function would jeopardize CBD's chances of implementation [1985: 79-82].

GS's assumptions about the efficacy of CBD presupposed that government would be possible only by consent, hence that widespread noncooperation would render government impossible, to such an extent that usurpers would eventually be forced to relinquish power: "It is control of the ruler's power by withdrawal of consent. It is control . . . by the subjects' declining to supply the power-holder with the sources of his power, by cutting off his power at the roots" [1973: 47, 8–31].

GS has also dealt more specifically with the use of nonviolence for national-defense purposes, e.g., in *Making Europe Unconquerable* [1985; cf. 1985a] and *National Security Through Civilian-Based Defense* [Sharp with Jenkins 1992]. He sought, inter alia, to dissipate the image of CBD as "passive defense" by promoting the alternative image of CBD as a "functional substitute for war": "Nonviolent action is a means of combat, as is war. It involves the matching of forces and the waging of 'battles,' requires wise strategy and tactics, and demands of its 'soldiers' courage, discipline and sacrifice" [Sharp 1973: 67].

While meticulously recording innumerable instances of, and recommending the use of, CBD methods for a whole panoply of hypothetical instances, GS failed to address the question of relevance. From the success of a strategy against indigenous usurpers more than half a century ago, it hardly follows that it would be effective against foreign invaders today. Furthermore, he tended to disregard more sophisticated and strategically minded CD advocacies, such as those of the late *Anders Boserup* [Boserup & Mack 1974], Johan Niezing [1987; 1990], and Jean-Marie Muller and associates [Muller 1984; Mellon & al. 1985]. Finally, even though GS emphasized the need for thinking CBD strategically, i.e., for relating specific CBD tactics to the stipulated objectives of the aggressor, he never took into account aggressions for "strategic purposes," i.e., such as would seek only to establish a bridgehead, conquer an airfield, or the like, and for which CD would seem, at best, of dubious relevance [Sharp & Jenkins 1990]. GS did, however, assert that CBD might be a potential equivalent of *forward defense*, namely, in the form of what he called a "non-violent *blitzkrieg*," tantamount to an intense outburst of nonviolent resistance, such as mass demonstrations, general strikes, etc. [Sharp 1985: 114–116].

His lifelong experience in the field notwithstanding, GS has tended to focus his attention (with regard to nonviolent action as well as CBD) on a score of thoroughly researched classical cases from the 1920s, the *WWII* resistance movements, France 1961, and the 1968 Soviet invasion of Czechoslovakia. He has almost completely disregarded instances where CBD either conspicuously failed or was never attempted, perhaps because it was doomed to failure [cf. Schmid 1985]. Furthermore, even though in his 1990 work he did mention the use of NV and CBD methods in China as well as Eastern Europe, he dealt with these instances only in passing and omitted entirely what might be the most interesting case of all, namely, the *Intifada* of the Palestinians in *Israel* [Sharp & Jenkins 1990].

SHARP, JANE M. O.   Senior research fellow at the Centre for Defense Studies, Kings College, London; previously senior researcher at *SIPRI* (1987–1991). JMOS's two main fields of research have been *alliance* policies and conventional *arms control*.

JMOS was director of a research program on the *Warsaw Pact* at Cornell University in the early 1980s [Holloway & Sharp (eds.) 1984] but subsequently devoted most of her attention to *NATO*'s problems with maintaining cohesion. Maintaining cohesion was presumably difficult because of the "alliance *security dilemma*," manifested in the oscillation between the two poles of entrapment and abandonment. On the one hand, fear of being trapped in conflicts because of alliance ties; on the other, a fear of being abandoned by allies in a crisis. This perennial problem of alliances, in JMOS's analysis, was at the root of the contradictory attitudes of NATO-Europe in the *INF* and, e.g., other arms control issues [1985; 1987; 1987a; 1991a; cf. G. Snyder 1984; idem & Diesing 1977].

The same dilemma influenced European attitudes toward the U.S. military presence, a topic to which SIPRI devoted a research project, with JMOS as leader. Taking into account the more favorable Soviet attitude to the U.S. presence, as well as the apparent U.S. pressure for a certain withdrawal, JMOS concluded by predicting a certain residual U.S. presence, combined with a gradual displacement of NATO as the main guarantee of security and a troika consisting of a modified NATO, an institutionalized *CSCE*, and a wider and deeper EC [1990a: 25; idem (ed.) 1990].

At SIPRI, JMOS was also responsible for keeping track of the negotiations on conventional arms control, in particular the *CFE* [1988; 1989; 1990b; 1991; 1991a; 1992; 1992a; 1993]. Assessing the eventual outcome as positive, she nevertheless cautioned against complacency because the CFE, rather than achieving symmetry, merely replaced old asymmetries with new ones. Furthermore, nothing had been accomplished with regard to naval forces, which were of greater concern to many countries than the type of forces on the CFE agenda [1991: 457-460]. She therefore warned that "the danger for the 1990's might be that traditional arms control policies could hold back progress that might otherwise be expected from the new thinking in political, economic and cultural relations" [1991a: 136].

SHENFIELD, STEVEN.   British Sovietologist with close links to the alternative defense community (e.g., the *Foundation for International Security*).

SS was among the first to analyze Soviet *New Thinking*, both insofar as its ideological foundations were concerned [1987; 1988] and with regard to its concrete security political manifestations, including *no-first-use* and *NOD*, etc. [1984]. Viewing Soviet foreign, security, and defense policy as partly responses to Western moves (i.e., as governed by the *security dilemma* [1985]), SS urged the West to develop a complex response strategy based on an acknowledgment of shared interests. Should the West, on the other hand, reject Soviet overtures for collaboration merely because of suspicion about Soviet motives, "the USSR would then become, by Western default, the main standard-bearer for 'all-human values.' . . . Conditions would be optimal for building up a broad united front of 'enlightened' world opinion looking to Moscow as its centre" [1987: 104].

SHEVARDNADZE, EDUARD.   Foreign minister of the *Soviet Union*, 1985–1990. After having resigned from his post in protest against what he saw as an approaching dictatorship, and after having supported Boris Yeltsin during the August 1991 coup-cum-revolution, S was persuaded to return as foreign minister in November 1991. Because of the subsequent dissolution of the Soviet Union, his tenure lasted for only about a month. He later became president of Georgia.

During his tenure as foreign minister, ES was one of the main promoters of the Soviet *arms control* offensive in general and of *NOD* in particular. In his opening address to the *CFE* negotiations in March 1989, ES presented a Soviet three-stage plan for a reduction of armed forces in Europe "down to levels sufficient exclusively for defense." During the first phase (two to three years), imbalances and asymmetries were to be eliminated with regard to troop numbers and armaments through a build-down to equal ceilings 10–15 percent below the lowest alliance holdings. Furthermore, "along the line of contact of the two politico-military *alliances*, strips [*zones*] with lower levels of arms would be set up, in which the most dangerous destabilizing kinds of arms would be subject to withdrawal, reduction, or limitation, and limitations would be imposed on military activities." In the second phase (also lasting two to three years) further cuts of 25 percent would be carried out, along with "further steps to restructure the armed forces on the basis of the principles of sufficiency for defense." In the third phase, "the armed forces should be given a strictly defensive character" [*Tass,* 6 March 1989].

SIA.   Acronym for *structural inability to attack* [Lutz 1988a].

SIEGFRIED LINE.   Fortified line along Germany's entire western border during *WWII*, around 50 km deep and with thousands of bunkers, in front of which was a deep minefield. Germany's faith in the SL (i.e., in the strength of the defensive) allowed it to plan for wide-ranging offensives (thus illustrating the inherent ambivalence of *defense*), as became apparent, e.g., during the negotiations with the USSR in September 1939. In a presentation on 27 September, Foreign Minister Rippentropp asserted that "Germany felt absolutely secure behind that line" and that "we were in a position to secure victory with mathematical accuracy" [*International Affairs* (Moscow) 1991(8): 117].

SIMPKIN, RICHARD.   Former tank regiment officer, and one of the most renowned Western analysts of mechanized warfare [1983]; from 1971 until his death, free-lance author.

In his main work, *Race to the Swift: Thoughts on 21st Century Warfare,* RS devoted a chapter to what he called "small-force manoeuvre strategy," i.e., *guerrilla strategy* [1986: 311–321]. He acknowledged that "the span of military techniques covered by the term 'revolutionary warfare' may actually represent a more effective way of waging war than operations by organised forces." This effectiveness was due to the application in extremis by *guerrilla warfare* of the *principles of war* concerning surprise and economy of force. RS saw the perspectives of small-scale warfare entering the strategic competition between the blocs, e.g., by way of special forces of the "*spetznaz*-type." Indeed, this type of unconventional warfare by means of guerrilla tactics might soon become the most likely form of warfare in the industrialized world, as well as the only available means of intervention in the Third World [1986: 316–317]. RS hence urged governments to "'shed their hidebound attitudes to peace and war. . . . Established armed forces need to . . . structure, equip and train themselves to employ the techniques of revolutionary warfare—to beat the opposition at their own game on their own ground" [1986: 320].

SINGH, JASJIT.   Air commodore, subsequently director of the Institute for Defense Studies and Analyses in New Delhi, *India.* JS is the author of numerous articles on *NOD*, wherein he has analyzed the concept as such as well as sought to apply it to the Indian subcontinent [e.g., 1989; 1989a–b; 1990; 1992b; 1994].

In his attempts to clarify the conception of NOD, JS has, e.g., suggested to define it in a number of "paradigms," the gist of which was the following:

- NOD is a politico-military doctrine, the aim of which is to ensure a durable *peace* and equal *security* for all states.
- War can no longer serve as an instrument of policy.
- Neither deterrence by punishment nor nuclear weapons are compatible with NOD.
- NOD must be applied globally and take into account the problems of foreign intervention.
- NOD presupposes greater transparency and efforts to curtail the development of destabilizing weapons technologies; in this connection, JS has especially pointed to the risks stemming from the proliferation of ballistic missiles [1992; 1992a].
- NOD should be strictly counterforce and not directed against the civilian parts of society.
- NOD requires reciprocity and universal application.
- NOD is merely a transitionary stage on the way to general and complete *disarmament* [Singh 1990].

SIPRI.   Stockholm International Peace Research Institute. Founded in 1966 and ever since one of the leading institutions in peace research. Many of the most prominent alternative defense advocates have, at one time or another, worked at SIPRI (e.g., *Anders Boserup*, *Randall Forsberg*, *Mary Kaldor*, *Sverre Lodgaard*, *Andrzej Karkoska*, and *Jane Sharp*), and three of its former directors have been in the forefront of the *NOD* debate: *Robert Neild*, *Frank Barnaby*, and Frank Blackaby.

Besides including several articles on NOD-related subjects in its yearbooks [Berg & Herolf 1984; Canby 1987; Fry 1986; Herolf 1983; 1986a; Karp 1991; Krohn 1991b; Lodgaard 1982; 1986; Rotfeld 1991; J. M. O. Sharp 1991; 1992; 1993] (including the annual Olof Palme Memorial Lectures [Brandt 1988]) and various monographs [Anthony 1990; Deger & Sen 1990; Harkavy 1989], SIPRI has held symposia and/or published anthologies on topics such as *common security*, *no-first-use*, *disengagement* and *nuclear-weapons-free zones*, and *arms trade regulations* [Anthony (ed.) 1991; Laurence & al. 1993; Blackaby & al. (eds.) 1984; Jacobsen (ed.) 1987; Lodgaard & Thee (eds.) 1983; Rotfeld & Stützle (eds.) 1991; SIPRI (ed.) 1978; Trapp (ed.) 1987; Väyrynen (ed.) 1985; Wachter & Krohn (eds.) 1989]. SIPRI has thus been a very important vehicle for the NOD debate, and it is likely to remain so because of the appointment of its present director, Adam Daniel Rotfeld, who was among the first in his home country, *Poland*, to take an interest in alternative defense.

SI VIS PACEM PARA BELLUM (If you want peace, prepare for war).   Roman proverb, usually attributed to Flavius Renatus Vegetius, whose actual formulation was "*Qui desiderat pacem, praeperat bellum*" [*De Rei Militari* (On military matters), Book 3, Prologue, in Phillips (ed.) 1985: 123; cf. Luttwak 1991: 3, 11]. It has occasionally been paraphrased by peace researchers as *Si vis pacem para pacem* (If you want peace, prepare for peace) or *Si pacem vis, para aequilibrium* (If you want peace, build a stable balance) [Webber 1990: 217].

SLOAN, STANLEY.   See *U.S.*

SLOC.   Sea lines of communication in general, but often used more specifically about *NATO*'s transatlantic SLOCs, via which the *U.S.* and *Canada* would supply and reinforce their European allies in the event of a war in Europe.

1. **Need for SLOC Defense.** NATO may have exaggerated the threat to its SLOCs posed by the Soviet Northern Fleet. First, *force comparisons* probably overstated the potentials of the latter by not taking qualitative and "generational" factors into account. Second, there was never any solid evidence to support the allegation that SLOC-severance was ever high on the Soviet list of naval priorities [cf. MccGwire 1979; 1981; Rohwer 1990; 1990a; Bathurst 1990; Daniel 1990; Harrick 1990]. Finally, the very need for transatlantic reinforcement of Europe may have been inflated by flawed force comparisons between the *land forces* of NATO-Europe and those of the *Warsaw Pact. A fortiori,* the combined result of Soviet and pact unilateral build-down, *CFE*-related reductions, and the subsequent dissolution of both the pact and the USSR itself (i.e., the replacement of an Eastern with a Western superiority) must have lessened the need for SLOC defense very considerably.

However, SLOC protection might still be of some importance, for several reasons. First, because the ability to transform any attempted *blitzkrieg* into a protracted *attrition* contest would be a viable alternative to the threat of nuclear *escalation.* In any such war of attrition overseas reinforcements and supplies might well matter, and access to an almost inexhaustible fountain of military and economic power such as the U.S. would almost inevitably prove decisive. Second, the cutback in U.S. forces stationed in Western Europe would be less controversial if the option of their return in an emergency were to be left open. The U.S. forces (and in more general terms, mobilizable military potential) would constitute well nigh invulnerable strategic reserves, deployed in what might be reckoned as NATO-Europe's rear. Seen in this light, it will have to be acknowledged that the West (contrary to the prevalent views) does possess a considerable strategic depth, at least comparable to that of the USSR or Russia. Both of these requirements would, however, presuppose that NATO's transatlantic SLOCs would remain passable in times of war. Only on this precondition would it be possible to sea-lift reinforcements and to provision Western Europe with civilian commodities such as oil [Grove 1987; 1990b: cf. Møller 1987; 1989e].

2. **Maritime Strategy.** The need for SLOC protection was used as a rationale for offensive naval strategies such as the U.S. Navy's Forward *Maritime Strategy.* Presumably, the envisioned strategic *ASW* (antisubmarine warfare) operations and forward-deployed carrier groups would tie up Soviet general-purpose naval forces in the defense of the *bastions* of their strategic submarines, thus making them unavailable for SLOC-severance warfare. To render this problematic strategy (from *NOD* and *crisis stability* points of view) redundant, alternative forms of SLOC protection would be desirable.

3. **Defensive Sea Control.** An element in such an alternative would be defensive sea control south of a GIUK (Greenland-Iceland-U.K.) perimeter (or perhaps GIN: Greenland, Iceland, Norway). If the Soviet (Russian) attack submarines could be prevented from passing this perimeter in any great numbers, NATO's SLOCs would be secure. Neither the GIUK nor the GIN perimeter would intersect with what might be a quite sufficient Soviet (Russian) defense perimeter, approximately on the dividing line between the Barents and Norwegian Seas. The Norwegian Sea might thus come to constitute a kind of *disengagement* zone between U.S. (NATO) and Soviet (Russian) naval forces [Mearsheimer 1986: 86–89; Møller 1987a; cf. Lodgaard 1989a; S. E. Miller 1990a]. The perimeter would, furthermore, be a highly convenient defense perimeter for NATO, thanks to the availability of bases ashore, allowing land-based aircraft, helicopters, and missiles to contribute significantly to the *barrier*-type defense, in which minefields would also play a central role. However, because such an ASW barrier would inevitably be penetra-

ble, it would need the backup of a convoy-with-escorts system, with escort vessels optimized for ASW purposes, i.e., consisting mostly of frigates, destroyers, small ASW carriers with helicopters, and (conventionally powered and armed) attack submarines [Grove 1990b; cf. Turner & Thibault 1982].

SMITH, DAN. Director of *PRIO in Oslo, formerly fellow at the Transatlantic Institute in Amsterdam, and a leading figure in the CND (Campaign for Nuclear Disarmament) and *END (European Nuclear Disarmament) movements, as well as a member of the *ADC (Alternative Defence Commission) throughout its existence [Thompson & Smith (eds.) 1980; Kaldor & Smith (eds.) 1982].

DS has been very critical of *NATO, which he has portrayed as characterized by U.S. hegemony; as the only alternative to which he has recommended *dealignment [Smith 1983; 1987; 1989]. Concerning military strategy, the attitude of DS was that of a nuclear abolitionist. E.g., he rejected the conceivability of "pure" deterrence [Krass & Smith 1982] and advocated a strictly conventional defense of both the *U.K. and the West in general [1980: 347–355; 1981; 1990]. This would be a "defensive deterrent" with no long-range-strike aircraft, nuclear weapons, or ocean-going naval forces. The purpose of the envisioned *territorial defense would merely be to defend British territory by means of "relatively small naval craft, . . . surface-to-air missiles, a limited air force for *air defense and possibly striking naval targets, and a highly mobile *infantry supported by helicopters" [1980: 261]. Such postures would presumably enhance *stability because they would create "a particular kind of imbalance: when each side's defensive capabilities out-weigh the other's offensive strength" [1990: 12], i.e., what others have called a situation of *mutual defensive superiority.

SMITH, PETER. Secretary of *Just Defence until his death in 1991 [1990; 1991]. In 1984 PS presented a concrete proposal for a defensive British military strategy, written jointly with research officer for the Labour Party Mike Gapes [Smith & Gapes 1984]. The two authors proposed a shift to "defensive deterrence," based upon an air force posture emphasizing *air defense by means of *SAMs; a coastal defense based on small patrol vessels and diesel-power submarines; and *land forces structured along the model of *Sweden but also including contingency plans for *civilian-based defense.

SNYDER, JACK. Associate professor of political science at Columbia University.

One of JS's main fields of research has been civil-military relations, which he initially studied from a historical perspective, resulting in a major work *The Ideology of the Offensive: Military Decision Making and the Disasters of 1914* [1984; cf. 1985]. Demonstrating how the military predilection for the offensive (cf. *Schlieffen Plan, *Plan 17) was a contributory cause of *WWI, he drew a parallel to the *Soviet Union of the 1980s, warning that it might produce "an extreme variant of the military's endemic offensive bias" [1985: 146].

A few years later, impressed by the "*Gorbachev revolution," JS made predictions about the Soviet Union that were considerably more optimistic, above all because civil-military relations had been transformed as a corollary of *glasnost* and *perestroika*. The availability of alternative sources of military advice (the *Institutchiki) allowed the political leadership to find a way out of the impasse of Soviet security and defense policy, above all because "the taming of the military creates the possibility for the change from conventional offensive to defense." The reorientation presumably presented the West with a unique opportunity for making an advantageous *arms control* bargain by trading limits on SDI and emerging technologies for a Soviet build-down of its *tank forces* [1986: 95 & passim; cf. 1987].

In 1990 JS returned to the topic of civil-military relations by presenting three alternative scenarios for the internal (and by implication also external) development of the new states in the collapsed Soviet bloc. One of these was "praetorianism," under which system "competing groups face each other nakedly, their struggle unmediated by any effective institutional framework" [1990:19]. Because the *MIC* (military-industrial complex) would inevitably be one of these groups, and the one with the most abundant resources, this boded ill for stability. JS's recommended remedy was the forging of a pan-European security system, including a modicum of *collective security* arrangements but also based on the reduction of national offensive capabilities [1990; 1990a].

SOCIAL DEFENSE.   Synonym for *civilian-based defense* (CBD). The term is intended to highlight a distinguishing feature of CBD, namely, its being a defense *of* society (rather than of territory or state) *by* society, rather than by a special group, *in casu* the armed forces. The term has been most favored by proponents of civilian-based defense in the German-language area [Ebert 1981a–b; 1983; 1987; 1987a; 1990; Jahn 1977; Bruckmann 1984; Nolte 1985; Sternstein 1988; Senghaas 1990g; Grünen im Bundestag 1987; Stock & Krottmayer 1986], as well as to some extent in the Scandinavian [Galtung 1964; T. Olson 1985] and Dutch ones [Klumper 1983; Geeraerts 1978; 1986; Niezing 1987; 1990; Schmid & al. 1985].

SOCIALDEMOKRATIET.   Social Democratic Party of *Denmark*, and the largest of the parties in the Danish parliament, *Folketinget.* S has held government power for extended periods since the 1930s and assumed leadership of a coalition government once again in 1993, after having been in opposition since the early 1980s. Although S until 1979 supported traditional security and defense policy, including the low-key Danish *NATO* membership and NATO's nuclear strategy [cf. Heurlin 1983; Boel 1988], the party underwent a metamorphosis in the early 1980s that was partly inspired by the *Scandilux* group [cf. Petersen 1984]. This landed it in support of nearly all of the *peace movement*'s demands and conceptions: *common security*, *nuclear-weapons-free zones*, *NOD*, and *conversion* [cf. S. M. Christensen (ed.) 1984; Møller 1990c; idem (ed.) 1985; 1986]. However, after its defeat in elections called over the issue of the right of nuclear-armed ships to visit Danish ports, S gradually returned to a more consensual and compromise-seeking security policy [Nordic Study Group 1990; cf. *Information,* 12–16 December 1991].

SOMMER, MARK.   See *U.S.*

SØRENSEN, BENT.   See *Denmark.*

SOVIET UNION.   Established in 1922 after the 1917 communist revolution as a successor to the Russian Empire. It was dissolved in 1991 by the founding republics Russia, Ukraine, and Belarus, who proceeded to establish a partial, yet much looser, successor, the *CIS*.

1. **Constant Factors.** Geopolitically, the SU both enjoyed advantages and suffered from vulnerabilities with regard to defensive strength and *security*. On the one hand, the enormous size of the country made it difficult to conquer and impossible to occupy, as became apparent during Napoleon's 1812 invasion, as well as the German attack in 1941. On the other hand, the length and diversity of the Soviet

borders (until 1991 the USSR bordered on *Norway, *Finland, *Poland, Czechoslovakia, *Hungary, Romania, Turkey, Iran, *Afghanistan, *China, Mongolia, and North *Korea) implied a great number of potential enemies and a plenitude of possible fronts to defend. This also conferred certain advantages in terms of offensive strength: in a war against NATO-Europe, the SU would have enjoyed the advantages of *interior lines of communication; NATO would have relied on *SLOCs (sea lines of communication).

The large population of the SU allowed for the raising of huge armed forces; the standing forces amounted to around five million men, and the military manpower base was occasionally estimated to be as many as fifty-five million men [U.S. Department of Defense, *Soviet Military Power 1986:* 103]. It was, however, always questionable whether such a *mobilization would be possible.

2. **Political System.** Until 1991 the SU was led by the Communist Party (CPSU) by means of a *nomenklatura* system. The power of the military and the *MIC (military-industrial complex) remained disputed: they were undoubtedly influential, but most analysts agreed that the CPSU was in command, inter alia, by virtue of the political control of the armed forces through a system of political officers [see Nichols 1993].

3. **Military Science.** Soviet *military science was very elaborate and hierarchically organized, identifying, inter alia, the prerogatives of the Communist Party and the armed forces. Soviet military science was divided into main levels:

- Doctrine, the (primarily political) assignment of military means to political ends;
- Strategy, the science of applying military means to these objectives;
- Operational art, the science of major operations within a TVD (theater of military action); and
- Tactics, the science of how to use military units for single engagements [Arbatov 1988; cf. MccGwire 1987b: 37–42; 1991: 81].

Although the formulation of military doctrine was a prerogative of the political leadership, the armed forces were largely at liberty to formulate conceptions on the three lower rungs of the hierarchy. Among the most important determinants of Soviet military policies was assessment of the likelihood and nature of a future war:

- Inevitable; likely; possible but unlikely; or inconceivable?
- Global and unlimited, i. e., thermonuclear; containable and possible to limit, i.e., largely conventional (yet with a nuclear option); or easily containable?

The more inevitable and the more total Soviet leaders assumed war to be, the greater apparently was their propensity for offensive operations, and vice versa [MccGwire 1987a; 1987b; 1991]. These assessments varied over time, accounting for the shifts in Soviet security and defense policy [Nation 1992].

4. **Stalin Era.** Josef Stalin in the early 1930s launched the notion of "socialism in one country." This pointed toward a defensive doctrine, with consolidation of Soviet rule as the highest priority, albeit with an option of exploiting emerging opportunities. The Hitler-Stalin pact of 23 August 1939 might thus have been defensively intended, but it involved, e.g., aggression against *Finland, the Baltic states, and *Poland. Even though the USSR at one stage contemplated a preventive war against Germany [Magenheimer 1994], the German attack on 22 June 1941 came as a surprise to Stalin and resulted in the temporary occupation of large parts of Soviet territory. After the Battle of Stalingrad in 1943, the Red Army shifted from strictly defensive operations to an operational and strategic *counteroffensive,

largely by means of *tanks and artillery, yet with important defensive elements, e.g., in the *Battle of Kursk [Nation 1992: 74–157].

After the war, Stalin seemed determined to prevent a repetition, inter alia, by occupying a *buffer zone in *Eastern Europe, which in 1955 was formalized as the *Warsaw Pact. Some demobilization notwithstanding, the Red Army retained preponderance in numerical strength, and its strategic conceptions envisioned offensive conventional warfare with extensive use of tanks and artillery. These policies seem to have been guided by the assumption that world war was inevitable in the long run, but that the USSR might be able to postpone the eventual showdown, e.g., through *peaceful coexistence and occasional *alliances, even with capitalist states.

5. **Khrushchev Era.** Nikita Khrushchev in the early 1960s supported the "revolution in military affairs" by shifting to a clear emphasis on *nuclear deterrence, accompanied by a substantial demobilization. The resultant conflict with the military may serve as a partial explanation of his ouster in October 1964 [Nichols 1993: 57–90]. Khrushchev's policy seems to have been guided by the assumption that world war was no longer inevitable, and that its likelihood depended on Soviet deterrence. Should war come, however, it was presumed that it would inevitably escalate into full-scale thermonuclear war.

6. **Brezhnev Era.** Leonid Brezhnev and his collective CPSU leadership maintained the emphasis of their predecessors on nuclear weapons, launching a major buildup and thereby attaining approximate nuclear parity with the U.S. by the early 1970s. The buildup was accompanied by a renewed interest in *arms control, whence developed the SALT treaties as well as the unilateral Soviet commitment to *no-first-use of nuclear weapons. At the same time, however, the SU resumed the buildup and modernization of its conventional forces in all categories. By the early 1980s the Soviet ground forces seem to have shifted to even more offensive strategic conceptions, envisioning deep thrusts into NATO's rear by means, e.g., of operational maneuver groups. The Soviet Navy was expanded and, under the leadership of Admiral Gorschkov, began to emphasize global presence and power-projection capabilities. The Soviet buildup, however, was accompanied by what was presumed to be an interest in conventional arms control, manifested, inter alia, in the *MBFR negotiations from 1973 to 1989, which yielded no tangible results.

These policies seem to have been guided by the assumption that world war was no longer inevitable, and that if it materialized, it would be possible to prevent it from escalating to nuclear war. On the other hand, horizontal *escalation, say from a local conflict to world war, was still regarded as inescapable, hence the need for a full-fledged offensive conventional war-fighting capability. The purpose thereof was to prevent the U.S. from establishing a bridgehead in Europe, hence the Continent would have to be conquered with the utmost speed [Alimurzayev 1989; Dick 1985; Donnelly 1982; Erickson & al. 1986; Hamilton-Eddy 1988; Hines 1982; idem & Petersen 1984; Lebow 1985a; Leites 1981; Luttwak 1983; Mazing 1989; MccGwire 1987b; Petersen 1987; Smith & Meier 1987; Snyder 1986; Vigor 1983].

7. **Gorbachev Era.** The accession to power of *Mikhail Gorbachev, and with him a whole new generation of political leaders (including Foreign Minister *Eduard Shevardnadze), inaugurated a complete reorientation of Soviet policy. First, glasnost paved the way for a reform of civil-military relations, including a breach of the armed forces' monopoly of military expertise (cf. *Institutchiki, *IMEMO, *ISKAN) [Bellamy 1992; Diehl 1993; Nichols 1993: 130–204]. Second, *perestroika necessitated a reordering of priorities, in casu a shift of emphasis toward domestic reform, at the inevitable expense of external assertiveness. *Disarmament and *conversion thus became high-priority imperatives, hence the innovative approach to arms control (see, e.g., *Year 2000 Plan, *Murmansk

Initiative, **Vladivostok Initiative*) and military doctrine (cf. **Reasonable Sufficiency*), as well as the rethinking of security policy in general, subsumed under the concept **New Thinking*.

The reorientation was facilitated by (indeed probably preceded by) an authoritative reassessment of the likelihood of war. Whereas it had previously been presumed that local crises (say, a U.S.-Soviet conflict in the Persian Gulf) would automatically escalate to world war, the pessimistic assumption was now abandoned in favor of a belief that conflict might be contained. As a consequence, the imperative of offensive operations vis-à-vis NATO in Europe disappeared, opening the door to a recognition of the advantages of a defensive strategy and posture that might contribute to war prevention as well as facilitate disarmament [cf. MccGwire 1987; 1987a; 1991]. The reform process, however, also implied a conflict with the armed forces, the rest of the military-industrial (or military-bureaucratic) complex [cf. Herspring 1987; 1989; Rice 1987; 1989; Larrabee 1988; Adragna 1989; Covington 1989; Davenport 1991; Green & Karasik (eds.) 1990; Holden 1991; S. Meyer 1991; Diehl 1993; Nichols 1993: 205–236], and the party bureaucracy, a struggle that culminated in the abortive August 1991 coup d'état (cf. **Dmitri Yasov*).

8. **Post-Gorbachev.** Developments since the reinstatement of Gorbachev can be characterized only as a democratic revolution, in the course of which domestic problems completely overshadowed foreign ones. The processes set in motion before August 1991 continued, but at a low key and without attracting much attention. The main problem became not whether but how to implement disarmament and withdrawal from empire, i.e., how to provide returning forces with housing and other necessities, in the face of increasing scarcity in practically all fields. (See **CIS*.)

SOZIALDEMOKRATISCHE PARTEI DEUTSCHLANDS.   See **SPD*.

SPAIN.   S, in a certain sense, has a defensive tradition dating back to its experience with **guerrilla warfare* during the **Peninsular War* against Napoleon [Gates 1986] and some similar experience from the Civil War (1936–1939) [Brouhe & Temines 1968; Jackson 1965]. Because of the outcome of the latter (the dictatorship of Generalissimo Franco), however, the debate on the military in S for decades concentrated on the domestic function of the armed forces. The period 1976–1980 was characterized by the attempt to subordinate the armed forces to civilian control as part of the general process of **democratization* [Viñas 1988: 153–157].

1. **S and NATO.** Because of its location far from the dividing line between the military blocs of the Cold War era, S only at a late stage took much interest in the East-West conflict and played only an insignificant role therein. To the extent that S has faced any external threats, these have come from North Africa (the *Mahgreb*) rather than from the East. S thus became a member of **NATO* only in 1982, after having confined itself to bilateral relations with the **U.S.* for decades. Spanish NATO membership was controversial (and, e.g., opposed unconditionally by **peace movements* and partly the PSOE [Spanish Workers' Socialist Party]), and its modalities were settled only after a referendum in 1986. The terms of membership (as defined in a memorandum from the Spanish government to NATO) also reflected the controversialism of the subject and resembled those of **France*. Even though S would participate fully in the alliance's political work, "the Spanish contribution to collective defense would take place outside the integrated military structure." Since then, some integration has nevertheless occurred in terms of procurement, **logistics, **air defense*, etc. [Viñas 1988; Duke 1989: 251–272].

2. **Alternative Defense Debate.** It is understandable that the alternative defense

debate in S has been less intense than in most other European countries. Nevertheless, there has been a certain interest and at least one fully developed *NOD* proposal.

- The CIP (*Centro de Investigación para la Paz* [Peace Research Center]) in Madrid in its journal *Papeles para la Paz* has followed the international NOD debate [e.g., Armengol 1990b; Ragionieri 1990].
- At the University of Valencia, Antonio-Luis Martinez-Pujalte has advocated *common security*, NOD combined with *civilian-based defense* (CBD), and steps toward a *collective security* system [1991].
- At the CIDOB in Catalonia (subsequently at the CIP), Vicenç Fisas Armengol in 1985 presented a detailed proposal, *An Alternative to the Defense Policy of Spain*, in which he analyzed all the essential principles of NOD (including CBD) [1985: 27–126]. In 1990 he followed this up with a detailed proposal for a nonoffensive defense of S, *Defense 2001*.

The author's analytical point of departure was the sanguine assessment that Spain "is not under threat from anyone," hence much smaller forces would suffice than were fielded or "'in the pipeline." The armed forces might presumably be reduced to a quarter of their present size, i.e., to between 80,000 and 100,000 men (all of whom should be professionals): 57 percent to the army, 18 percent to the navy, and 14 percent to the air force. This would allow for considerable financial benefit, even if (as demanded by the author) the new professional armed forces were to receive "fair" wages: a total cost of about 566,000 pesetas, just a little above one percent of the Spanish GDP, and 300,000 pesetas less than the actual defense budget.

The means to these desired ends would be a defensive restructuring, according to which Spain would renounce the ability to project force beyond its borders, inter alia, through abandonment of *aircraft carriers*, frigates, and fighter planes, and through reductions in the numbers of *tanks* and other armored vehicles, amphibious vessels, and aircraft in favor of larger holdings of short-range missiles and patrol boats. Even this defensive military structure would be viewed merely as a transitional stage on the way toward reliance on nonmilitary means of defense [1990: 89–129].

SPANNOCCHI, EMILIO.   General and former commander in chief of the federal army of *Austria*. In 1975 he published *Defense Without Self-Destruction*, which was a major source of inspiration for *Horst Afheldt* [Spannocchi 1976; cf. Afheldt 1976]. S's ideas have largely been implemented, inter alia, in the *Plan of National Defense* of 1985 [Bundeskanzleramt 1985; cf. Mayer 1992], which has been regarded by many analysts as the closest real-life approximation to an *NOD* posture.

Central to S's thinking was the "principle of proportionality," which might be exploited by small states in general and by neutral countries in particular. They could count on being attacked only as a sideshow to a larger scheme, i.e. as a means to a more ambitious end, hence the possibility of making it not worth while. Proportionality, however, also required a degree of *damage limitation* from the defender's point of view, lest the defense should result in disproportionate destruction (cf. the *Defense Dilemma*).

In formulating his strategic ideas, S drew heavily on the experience of *guerrilla warfare*, and he explicitly acknowledged inspiration from great guerrilla captains such as *Mao Zedong*, *Vo Nguyen Giap*, and *Yosip Broz Tito*, perhaps particularly the last because of the similarity between Austria's and *Yugoslavia*'s strategic position. From this body of historical experience S distilled a number of strategic principles, which would presumably be applicable to small states such as Austria. However, he took exception to the final offensive operations envisioned by all three captains. As an alternative means of *war termination and of restoration of the *status quo ante bellum,* S suggested relying on the *time factor:* "Time is a strategic concept. It can decide the war, if only because it demonstrates to the aggressor the lack of proportion between its expensive waste and the value of the political goal aimed for." A precondition for exploiting the time factor was that the defender did not risk his forces in any decisive battle.

To avoid such an eventuality, the defender should renounce operational and strategic (but of course not tactical) mobility yet still cover the entire territory rather than any front line or other fixed positions. The envisioned in-depth *territorial defense* should consist primarily of small *infantry* units of ten to fifty men, who should fight according to guerrilla principles because "this forces the attacker into a multitude of tactically unimportant skirmishes, of which he may successively win every single one, but not the war" [1976: 56–61]. To take advantage of their familiarity with the terrain, the defenders should be assigned to specific localities in advance, pointing toward a *militia*-like home guard (*Landwehr*) as the core of the defense posture. S did, however, picture supplementing these forces with mobile formations for reinforcement purposes [1976: 68, 82–85].

The civilian population was expected to contribute to the defense effort through both *nonviolent defense*, and various semimilitary activities such as sabotage. S nevertheless acknowledged the protection of the civilian population to be imperative, for which purpose he recommended extensive *civil defense* measures, as well as abstention from armed combat in urban areas (see *Open Towns) [1976:77, 81].

SPD. Sozialdemokratische Partei Deutschlands (Social Democratic Party of Germany). Founded in 1875 by trade unions and parties inspired on the one hand, by Karl Marx and Friedrich Engels and on the other hand, by Ferdinand Lasalle.

Regardless of internal disagreements on other matters, the SPD has a very long tradition of antimilitarism, a feature common to both lines in the labor movement. Demands for the abolition of standing armies in favor of a *militia* were to be found in the party platforms adopted in Eisenach (1869), Gotha (1875), and Erfurt (1891). Furthermore, since 1891 the party has supported international arbitration and conciliation. An aberration from traditional antimilitarism occurred during *WWI,* when the SPD lent its support to the German war effort, only to return to largely its previous positions after the war [Dowe & Klotzbach (eds.) 1984: 175, 181, 190, 213, 224; cf. Møller 1991: 160–179].

Immediately after *WWII,* the SPD remained neutralist and supported *disengagement* (cf. its *Germany Plan), without, however, being consistently defensive. It occasionally advocated an offensive "roll-back" strategy, including plans for border-crossing *counteroffensives.* Finally, in 1959 the party swung around to a wholesale endorsement of the security policy of the *CDU* government, including *NATO* membership, nuclear weapons on German territory, and a strong Bundeswehr [Rosolowsky 1987: 37–47; Krell 1989: 4; 1990: 5–6].

The present positions of the SPD emerged in the early 1980s, under the influence of the growing *peace movements* and the opposition to *INF* deployment in the *FRG.* From support for NATO's 1979 dual-track decision, the party by 1983 had landed on an unconditional rejection of INF deployment, against the opposition of former chancellor Helmut Schmidt and others [H. Schmidt 1984; Brauch 1983]. Along with this reassessment of the INF issue went second thoughts on *nuclear deterrence* as such, in which connection the SPD came out in support of *no-first-use, *nuclear-weapons-free zones,* a "third zero," etc. [cf. Kaiser & al. 1982; Krell & al. 1983; Krause 1986; Joffe 1987: 45–92]. The swing to an antinuclear attitude was accompanied by a growing interest in collaborative political strategies, such as "security partnership" (its preferred term for *common security*), that were, however, in perfect continuity with the SPD's traditional support for *détente* in general, and *Ostpolitik* and *change through rapprochement* in particular [Griffith 1978; Rosolowsky 1987; Ehmke & al. (eds.) 1986; Krell 1988; Bahr 1991].

In the 1980s *Ostpolitik* was revived, this time from the opposition benches, in the form of an unofficial diplomacy vis-à-vis the governing parties (as political parties, rather than as state institutions) of *Eastern Europe*. Draft treaties were negotiated with the SED of the *GDR* on both a *chemical-weapons-free zone* and a nuclear-weapons-free zone for central Europe. In 1987 the two parties even signed a joint document, on "The Ideological Struggle and Common Security," wherein the imperative of *peaceful coexistence* was acknowledged alongside a commitment to a "controversial dialogue." Furthermore, defensive military doctrines and the lack of capabilities for attack were mentioned as valuable contributions to bringing about a *common security regime* [SPD & SED 1985; 1987; 1987a; Voigt 1985; 1987; Hahn 1987; T. Meyer 1987; Birkenbach 1989; Seidelmann 1987; Eppler 1988]. Alongside these documents came an agreement with the Polish Unified Worker's Party (PUWP), "Criteria and Measures for Confidence-Building Security Structures in Europe," and a sequel, "Confidence- and Security-Building Measures and Arms Control in the Baltic Sea," both of which contained concrete proposals for less offensive postures and strategies [SPD & PVAP 1985; 1988; cf. Ehmke 1986].

The general security political reorientation in the 1980s was accompanied by an increasingly unequivocal support for alternative conventional defense structures [cf. Borkenhagen 1987; Szabo 1990: 72-78; Møller 1991: 164–178]. Central figures in this reorientation included *Egon Bahr*, *Andreas von Bülow*, *Hermann Scheer*, and *Karsten Voigt*. In 1983 the SPD executive appointed a working group on new strategies, initially chaired by Bahr and subsequently by von Bülow. The group recommended that a conventional strategy should be strictly defensive (inter alia, with a view to *disarmament*, not least of nuclear weapons, and confidence-building) but remained somewhat ambivalent toward *deep strike* [Bahr 1984: 281, 283]. As a further element in its exploratory work, the SPD arranged a hearing on alternative strategies in the Bundestag in 1983–1984, to which *NOD* theorists such as *Horst Afheldt*, *Lutz Unterseher*, and *Carl Friedrich von Weizsäcker* were invited [Biehle (ed.) 1985; Schuster & Wasmuht 1984; Horn & Buch 1984].

At the 1984 party conference in Essen, the SPD officially endorsed the objective of *StruNA* (i.e., NOD) by stating that "a new security concept must . . . gradually establish a defensive conventional structure, so that in the long run a structural incapability of attack is achieved" [Parteivorstand: *Für eine Neue Strategie des Bündnisses,* quoted in Lutz 1989: 20]. At the 1986 Nuremberg party conference, the SPD furthermore unanimously adopted a resolution, "Peace and Security Policy," wherein its views on NOD were elaborated upon, e.g., with regard to its implication for Bundeswehr planning, which should strengthen *forward defense* in general and "antitank defense, obstacle-construction, and *air defense*" in particular [SPD 1986].

At its 1988 Münster convention, the SPD adopted a document, "Peace and Disarmament in Europe," wherein the focus had shifted to *arms control* questions in general and the implications of the INF Treaty in particular. The party emphatically rejected any modernization of the theater nuclear arsenal and proposed a new set of negotiations on battlefield and short-range nuclear weapons. Furthermore, the SPD reiterated its demand for "a reduction of armed forces to such an extent that they remain capable of defense but become structurally incapable of attack." For the impending first round of the *CFE* negotiations, the SPD recommended a target of equal ceilings around 50 percent of present NATO holdings of tanks, heavy artillery systems, combat aircraft, and helicopters, combined with density limits intended to rule out force concentrations in preparation of attacks [SPD 1988].

The proposal was followed up in July 1989 by an elaborate plan, *European Security 2000,* authored by Bahr, von Bülow, and Voigt (see under Bahr), which stirred controversy. Hermann Scheer and Heidemarie Wieczorek-Zeul published an elaborate criticism; their dissent concerned above all the implicit postponement of the withdrawal of tactical nuclear weapons until the achievement of NOD structures, and the unwarranted emphasis on negotiations over unilateral steps (for which purpose the two authors recommended the *SAS model of a "*spider-and-web defense" [Bahr & al. 1989; cf. Wieczorek-Zeul & Scheer 1990a].

After the East German revolution, the SPD committed itself unreservedly to unification, e.g., in the document passed by the Berlin conference in December 1989: "The Germans in Europe." In it the party emphasized that "the unification of Europe and the uniting of the Germans are closely interrelated. One cannot be obtained at the expense of the other." Further, the SPD urged both German states to do their utmost to accelerate *disarmament, for which purpose it recommended a follow-up to the CFE negotiations "with the aim-point of structural incapability to attack" (i.e., NOD). Regarding the more distant future, the SPD underlined that "our goal is to replace the military alliances with a European *peace order."

Since unification, the SPD has toned down its opposition to the government's defense policy, focusing instead on the *out-of-area question: whether to allow German soldiers to operate beyond the borders of NATO, and if so under what conditions. From a more principled opposition, the party gradually came around to supporting limited engagements under the auspices of the *U.N., while demanding some clarification of the Constitution's apparent prohibitions against this [see, e.g., Matthies (ed.) 1993; Lutz (ed.) 1993].

SPIDER-AND-WEB.   The central idea of the *SAS model, according to which the defense posture should consist of two main elements: a stationary, area-covering *attrition net (the web), and mobile forces (the spiders), including *tanks, armored personnel carriers, tactical aircraft, and helicopters. The mobile elements would be rendered both more effective and less offensive through their synergies with the web. Within the web they would enjoy support from the dispersed *infantry squads and area-covering logistical and *C^3I structures; beyond it they would be handicapped by virtue of their shortage of organic air support, mobile *logistics, etc. As a further hedge against offensiveness, the number of mobile units should be built down substantially. The weapons profile should emphasize well-tested technologies and give ruggedness and numbers higher priority than sophistication. The defense possess no border-crossing *counteroffensive capabilities but should have substantial internal reinforcement and tactical counterattack capabilities because of the enhanced tactical mobility and the shorter distances to cover [SAS 1984, 1989; Grin & Unterseher 1988; Unterseher 1989c].

SQUARE LAW.   See *Lanchester's Laws.

STABILITY.   Quality of a system, e.g., an international system of states, that allows it to regain a stage of equilibrium after a disturbance [cf. Axelrod 1990].

S has acquired positive, but conservative, connotations, inter alia, because it has become nearly synonymous with an absence of violent change. However, S need not be tantamount to a maintenance of the status quo, and there need be no contradiction between S and change. It is possible, e.g., to distinguish between static and dynamic S: the former signifies a situation of immobility; the latter refers to a structure of the system allowing for development and movement without the collapse of the system as such.

In the field of international relations, two of the most central forms of S have been *arms-race* S and *crisis* S. Arms-race stability would rule out neither weapons innovation (say, a gradual substitution of defensive for offensive armaments) nor *disarmament*. Crisis stability might be quite compatible with political crisis, which might often be a precondition for necessary and beneficial change. See also *Stability-Instability Paradox*.

STABILITY-INSTABILITY PARADOX.   State of affairs wherein a situation may be stable in one respect yet unstable in another, the instability perhaps precisely a consequence of the stability [cf. Axelrod 1990].

The military situation in Europe during the Cold War was stable in the sense that borders were treated as inviolable, that states refrained from overt military activities directed against other states, and that no wars occurred between the two blocs. It was at the same time unstable in the sense that if something had gone wrong, the situation might well have escaped control because of the high degree of militarization, the tight integration within the blocs, the intermingling of nuclear with conventional weapons, etc. Even a small perturbation might conceivably have resulted in a total conflagration, which may have been part of the explanation why all states tended to behave with great caution in their relations with adversaries. Conversely, a stabilization of the military situation (such as sought for by means of *NOD*, *disengagement*, *no-first-use*, *disarmament*, etc.) might have lifted some of the previous constraints and emboldened states to pursue their objectives more assertively and with less circumspection [cf. Snyder 1990]. Hence the need for comprehensive approaches to the improvement of international stability that would take into account both *arms-race stability* and *crisis stability*. This has been attempted through the *common security* approach.

STALEMATE.   Defensive strategy for crisis resolution proposed by William Ury [1985: 115–125], particularly with regard to the nuclear superpowers, and presumably with a special relevance for "spin-off crises" [cf. Lebow 1981: 41–56]. In such instances, the reactive side (A) was advised to act as though the crisis situation derived from miscalculation on the part of the instigator (B). A should hence minimize provocation, i.e., refrain from escalatory steps, while making it clear to B that he intended to stand firm. A military way of conveying this message might be the laying of land or sea mines, thus "drawing a line," but presenting no immediate threat to B. The military stalemate should be accompanied by political consultations with a view to restoring the status quo.

As a strategy of war prevention, S in many ways resembled *NOD*, which likewise seeks no victory or unconditional surrender but merely aims at restoration of the status quo, and which would also do so by reliance on the *time factor*.

STAND-OFF ZONE.   See *Maritime C(S)BM*.

START.   Strategic Arms Reduction Treaty signed by the *U.S.* and the *Soviet Union* on 29 July 1991 after negotiating since June 1982 [Brooks 1991]. The treaty limited each side's strategic nuclear weapons holdings (see Table 34) [*Survival* 33:5 (September/October 1991): 466–470]. The ratification of the START Treaty was postponed by the dissolution of the Soviet Union (and the subsequent vacillations of Ukraine over whether to retain its inherited nuclear arsenal [see*CIS*]), but a follow-up agreement was reached in June 1992 between the U.S. and the Russian

Table 34.   Start I Treaty

	Ceiling	Reduction	
		U.S. (%)	USSR (%)
Delivery vehicles	1,600	29	36
Accountable warheads on ICBMs and SLBMs			
Total	6,000	43	41
ICBMs and SLBMs	4,900	40	48
Heavy ICBMs	1,540	—	50
Mobile ICBMs	1,100	—	—
Aggregate throw weight of ICBMs and SLBMs	—	—	54

Federation. It envisioned a further build-down by the year 2003 to levels of 3,000–3,500 warheads on strategic ballistic missiles, and an abolition of MIRV'ed ICBMs, such as envisioned in the *Bush Initiative* [Lockwood 1993; Umbach 1994; Sorokin 1994].

STAR WARS.   Popular nickname for *SDI.* See *Strategic Defense.*

STAYING POWER.   See *Time Perspective.*

STEINBRUNER, JOHN D.   See *U.S.*

STOCKHOLM CONFERENCE.   See *CDE.*

STRATEGIC ARMS REDUCTION TREATY.   See *START.*

STRATEGIC DEFENSE.   Generic term for means of defense against strategic nuclear weapons, i.e., an opponent's long-range and/or forward-deployed nuclear forces capable of hitting the home territory of the defender.

1. **Categorization.** In principle there are several forms of SD. First, a distinction might be made according to what is supposed to be defended, yielding two broad categories: point defenses of particular objects such as military forces; and much more demanding area defenses, intended to protect the entire territory, particularly population centers. Second, a distinction might be made according to what the defense is supposed to be against:

- BMD (ballistic missile defense), i.e., a defense against nuclear-armed, long-range ballistic missiles, i.e., ICBMs (intercontinental ballistic missiles) and SLBMs (submarine-launched ballistic missiles).
- *Air defense,* which in this connection refers to defense against long-range, nuclear-capable bombers.
- *Civil defense,* which protects the population directly against all types of nuclear threats.
- Counterforce warfare, the ultimate form of SD, which constitutes attempts at neutralizing an opponent's strategic nuclear forces before launching, either by means of air or missile strikes against airfields or missile silos, or through strategic *ASW* (antisubmarine warfare), directed against the oppo-

nent's SSBNs (strategic submarine, ballistic, nuclear) [for a post–Cold War proposal to this effect, see Powers & Muckerman 1994].

Usually, however, the term is taken to signify merely the first (occasionally also the second) of these categories.

2. **History.** Both superpowers devoted considerable attention to SD throughout the entire nuclear era [Garthoff 1984; Schwartz 1984]. The *Soviet Union* always placed great emphasis on air defense and in the 1960s deployed a primitive ABM (antiballistic missile) system. Even though in 1972 through the *ABM Treaty* and subsequent amendments this was limited to a defense of Moscow (i.e., the NCA [national command authority]), it has been continuously modernized ever since [Stevens 1984; Pike 1985].

The *U.S.* in the 1950s sought a defense against Soviet strategic bombers but subsequently abandoned the search. The Eisenhower, Kennedy, Johnson, and Nixon administrations, furthermore, explored the technical opportunities for an ABM defense based on sophisticated *SAMs*. Due to the technological obstacles, however, the level of ambition was gradually scaled down from an area defense (the Sentinel program) to point defenses of missile sites and the NCA (the Safeguard program of the Nixon administration) [Newhouse 1973]. However, it was increasingly realized that the very notion of SD was problematic, hence the interest in an ABM treaty, which was signed in 1972 [Drell & al. 1986; Schneiter 1984; Garthoff 1985: 127–198].

From 1972 to 1983 research on SD continued, but deployment was limited to one site for each superpower. The Soviet Union chose to maintain the Galosh system around Moscow; the U.S. dismantled its embryonic ABM defense of one missile site.

3. **SDI.** In March 1983 President Reagan launched the *SDI* (Strategic Defense Initiative) in a televised speech [in Miller & Evera (eds.) 1986: 257–258]. It was conceived of as an astrodome-type area defense of the entire U.S. The basis for these high ambitions was technological innovations and especially *E.T.* (emerging technologies). They presumably raised the prospect of overcoming the obstacles to a meaningful population defense that had haunted the vintage-1970 ABM program:

- Laser and particle-beam weapons for the interception of incoming missiles; by being significantly faster than their targets, these weapons would presumably possess a higher kill probability than their predecessors.
- The option of deploying such weapons on orbiting battle stations, thus allowing for interception during the boost and midcourse phases of missile trajectories. This would facilitate destruction because interception might occur before the launch of the MIRVs (multiple independently targeted reentry vehicles) against fewer targets.
- A quantum jump in data gathering and processing technologies, presumably allowing for effective (i.e., largely computerized), high-speed, and close to real-time battle management within the maximum thirty minutes that would be available from the launch of an enemy ICBM attack to its impact [Carter 1984; OTA 1985].

It was pointed out by critics and skeptics that numerous countermeasures would be able to thwart the defense [Garwin 1986a-b; Tirman (ed.) 1984: 46–152; Barnaby 1986a: 96–101], hence that it might not be "cost-effective at the margin," at least not in its original shape [cf. Nitze 1986]. The problems were a function of the immense destructive power embedded in nuclear weapons, implying that the standards to which SD should live up to in order to be meaningful would be very high indeed. Even, say, a 1 percent leakage (out of an enemy strategic arsenal of around 10,000 warheads) would be tantamount to an unprecedented disaster with perhaps tens of millions of casualties. Furthermore, the adversary would be able to negate any improvement of the accomplished efficiency by simply adding to its offensive nuclear forces in order to saturate the defenses, or it might resort to technological countermeasures.

SDI's level of ambition was therefore gradually scaled down from the originally envisioned area-covering population defense to a point defense system protecting the U.S. offensive strategic forces, i.e., a "defense of the deterrent" [Clausen 1986]. Furthermore, in light of the *START negotiations and the international thaw, the rationale has gradually shifted further to defense against small-scale attacks from minor nuclear powers (See *GPALS [global protection against limited strikes]).

4. **Implications of SD.** The logic behind SD might appear flawless. SD would presumably enable a nuclear power to defend itself against an enemy first strike, thereby minimizing (perhaps to the point of eliminating) the need for retaliatory capabilities. SD would thus presumably facilitate a nuclear build-down, or even abolition, which would, paradoxically, render SD superfluous. However, notwithstanding the attractions of SD as an *alternative* to *nuclear deterrence*, SD would be inherently ambivalent if deployed as a *supplement* to deterrence. By means of SD, a state might escape its opponent's retaliation, which would be tantamount to achievement of a disarming first-strike (hence war-winning) capability [McNamara & al. 1985; Glaser 1986; Stockton 1986].

Accordingly, SD might be not merely infeasible but also potentially destabilizing because it might signal the intention to break the stalemate of mutual deterrence. It would therefore probably elicit a response from the adversary, whereupon an *arms race* might well ensue. The only setting in which SD might be feasible and unproblematic would be under a *minimum deterrence* regime. Here, however, the attempted *damage limitation* would also be attainable through a simple build-down of the arsenals. Furthermore, the only political circumstances that might allow for a "safe" transition to an SD regime would tend to be the very same that would allow such truly deep cuts. Nevertheless, in early 1992, in light of the disappearance of the East-West conflict, Russia indicated its interest in a collaborative SD endeavor with the U.S., pointing toward jointly developed defenses against third parties [*Atlantic News*, 31 January 1992, 3].

STRATEGIC DEFENSE INITIATIVE.   See *Strategic Defense.*

STRATEGIC WARNING.   See *Warning.*

STRATEGY.   Defined by *Liddell Hart* as the science of the proper "distribution and application of military means to fulfil the ends of politics" [Hart 1967a: 321]. See *Military Science.*

STRONGHOLD DEFENSE.   See *Bastions.*

STRUCTURAL INABILITY TO ATTACK.   English translation of the German term *Strukturelle Angriffsunfähigkeit* [Lutz 1988a].

STRUKTURELLE ANGRIFFSUNFÄHIGKEIT.   *Structural Inability to Attack.*   Term coined by *Dieter S. Lutz* as a substitute for *Strukturelle Nichtangriffsfähigkeit* [Lutz 1989: 27–28, 31–97]. Lutz further proposed a distinction between SA in a narrow (i.e., almost strictly military) sense, synonymous to *NOD, and a wider sense, encompassing nonmilitary factors, such as *democracy* (see *StrUnA i.e.S. and *StrUnA i.w.S., respectively).

STRUKTURELLE NICHTANGRIFFSFÄHIGKEIT.   *Structural Nonattack Capability.*   Term coined by *Albrecht von Müller* and/or *Hans Peter Dürr* [*Frankfurter Rundschau,* 1 July 1983; cf. Lutz 1989: 13–16], which is just as poor German as English. Nevertheless, the term became very prominent in the German debate, inter alia, through the *SPD*'s advocacy. It has been suggested to replace the term with *Strukturelle Angriffsunfähigkeit,* i.e., *StrUnA* [Lutz 1989: 27–28, 31–97].

SA is simply a synonym for *NOD* (and the numerous other terms for the same concept), but it underlines the importance of structure. What makes a defense posture and/or strategy offensive or defensive is neither declaratory policy nor the presence or absence of specific weapons but the configuration and composition of the arsenal, i.e., its structure.

StrUnA.   Acronym for *Strukturelle Angriffsunfähigkeit* (notice the position of the capital letters).

StruNA.   Acronym for *Strukturelle Nichtangriffsfähigkeit* (notice the position of capital letters).

StrUnA i. e. S.   Acronym for Strukturelle Angriffsunfähigkeit in engeren Sinne (*Structural Inability to Attack* in a narrower sense). Synonym for *NOD* proposed by *Dieter S. Lutz* [1989: 35–37].

StrUnA i. w. S.   Acronym for Strukturelle Angriffsunfähigkeit im weiteren Sinne (*Structural Inability to Attack* in a wider sense). Somewhat broader conception of defensiveness than implied by *NOD* or *StrUnA i.e.S.,* proposed by *Dieter S. Lutz.* S signifies a more permanent state of defensiveness, resting on the internal structure of the state in question as well as on the structure of the international system. S was thus conceived of as an element in a "European *peace* order" [Lutz 1989: 27, 37–39].

Many analysts, including *NATO* "traditionalists," have asserted that *democracy* and *multinationality* would be elements of S, albeit without using this term. Democratic states presumably rarely go to war and never against one another, which would presumably make NATO structurally defensive. Moreover, the multinational structure of NATO would presumably render offensive operations on the part of individual states inconceivable [cf. Senghaas 1986: 106; 1990a; 1990d; 1990e]. Certain authors close to the *peace movement* (e.g., *Andreas Buro*) have pursued the same logic with demands for further *democratization* in general and curtailment of the influence of the *MIC* (military-industrial complex) in particular as necessary elements of S [Buro 1983; 1983a; 1987, 1988].

STUDIENGRUPPE ALTERNATIVE SICHERHEITSPOLITIK.   See *SAS.*

STUDIENGRUPPE ENTMILITARISIERUNG DER SICHERHEITSPOLITIK.   See *SES.*

STUDY GROUP ON ALTERNATIVE SECURITY POLICY.   See *SAS.*

STUDY GROUP ON THE DEMILITARIZATION OF SECURITY POLICY.   See *SES.*

STÜLPNAGEL, JOACHIM VON.   Chief of the Department of the Army in the German General Staff after the Versailles Treaty. In 1924–1925 he developed a defensive strategy intended to make the best of the German situation of clear inferiority vis-à-vis its wartime (and most likely future) enemies. Rather than seeking an unattainable decisive victory, Germany should merely seek to prevent the occupation of German territory. In a future war, the first stage would be strategically defensive, and no offensive should be attempted until the second stage of the war. At the time such a stage was regarded as a utopia, hence the emphasis was placed on the first stage.

The tactical and operational principles advocated by JvS were in many respects similar to the ones advocated by modern *NOD* proponents. The defender should concentrate on *attrition*, combining *forward defense* with protracted in-depth defense on its own territory. It was emphasized that the enemy "should be able to find no target, such as a fortified line or a concentrated group of artillery, but must encounter individual assault groups, machine gun nests, artillery squads and positions of mixed formations across the terrain, the targeting of which is hardly possible, and not cost-effective at all" [Stülpnagel 1924; 1924a; Bald 1987]. Several NOD principles were discernible in embryo in these plans: the *no-target principle*, the *territorial defense* posture, and that of forward defense.

SUBMARINES.   See *ASW*, *Sea Power*.

SUBNUCLEAR SETTING.   Term used by *Anders Boserup* and others for the special circumstances created by *existential deterrence*, as opposed to the classical setting.

In a SS, the possibility of *escalation* to suicidal levels of warfare will influence a war, thus making it impossible for military power (i.e., victory on the battlefield in the usual sense) to decide the outcome. The prospective loser might at any time up the ante by escalating, thus denying victory to the prospective victor (without itself winning, needless to say). Under such circumstances, exploitation of the *time factor* (or even the deliberate *stalemating* of the war) might make perfect sense because it might allow for *war termination* by political means. Thanks to the inherent superiority of the defense, the defender would, moreover, be in a position to enforce precisely this [Boserup 1985; 1986; 1989; 1990].

SUN TZU.   Also known as Sun Wu or Sun Pin, who lived circa 400–320 B.C. [cf. Bellamy 1990: 311; Sawyer (ed.) 1993: 149–156], author of the oldest known work on strategy, *The Art of War* [Sun Tzu 500 B.C.]. ST might count as the founding father of the *indirect strategy* today associated with the names of *Liddell Hart* [1967], *Beaufre* [1963, 1972], *Luttwak* [1987], and others. More directly, ST has provided a source of inspiration for the strategists of *guerrilla warfare*, particularly the Chinese, foremost among whom is *Mao Zedong* [1936; 1938; 1938a; 1938b]. Via this route, ST has indirectly influenced several of the precursors and founders of *NOD*, such as *Glucksmann* [1967], *Brossollet* [1975], *Spannocchi* [1976], and *Joxe* [1991:150–165]. Even today ST is often described as the greatest military thinker ever [Creveld 1991: 126; cf. Simpkin 1986: 3; Gray 1990: 44].

The main element in war for ST was not battle but the thwarting of the enemy's plans, i.e., maintenance of the initiative:

> Warfare is the Way [Tao] of deception. Thus although [you are] capable, display incapability to them. When committed to employing your forces, feign inactivity. When [your objective] is nearby, make it appear as if distant; when far away, create the illusion of being nearby. . . . Create disorder [in their forces] and take them. If they are substantial, prepare for them; if they are strong, avoid them. . . . If they are united, force them to

be separated. Attack where they are unprepared. Go forth where they will not expect it. (Sun Tzu 500 B.C.: 158)

ST recommended evading the enemy's main forces in order to attack its weak points, i.e., accept battle only on favorable terms (as in guerrilla warfare). Further, he advised commanders to adapt to the terrain, e.g., by garrisoning narrow passes (analogous to NOD's emphasis on fortifications and *barriers). In the realm of grand strategy as well, ST predated NOD's emphasis on a counter-*blitzkrieg strategy of protracted war with his warning that "no country has ever profited from protracted warfare" [ibid.: 159].

SUPER-GRIT.   Strategy for inducing lasting cooperation between adversaries by means of an ongoing sequence of sporadic, unilateral, cooperative initiatives. S-GRIT was invented by *Joshua Goldstein and *John R. Freeman [1990: 153–156], according to whom the strategy should be guided by three rules: "1. Use sustainable cooperative initiatives. . . . 2. Strengthen reciprocity. Try to respond in kind. . . . 3. Value the long-term future and seek to change psychological images and expectations. . . ." The authors even interpreted the Soviet "peace offensive" since 1985 as a (partly successful) instance of S-GRIT. See also *GRIT, *Extended GRIT, *Progressive GRIT.

SURFACE COMBATANT.   See *Sea Power.

SURPRISE ATTACK.   Attack providing the victim with little or no *warning [Betts 1982; 1985].

   1. **Taxonomy.** There are degrees of surprise in SA, and it takes different forms.

- Complete bolt-out-of-the-blue SA, preceded by neither strategic nor tactical warning.
- Partial SA, which is preceded by strategic warning pointing to a likely attack sometime in the indefinite future, but by tactical warning about neither the precise timing nor the direction of the attack.
- Tactical (or operational) SA, an attack during a war in progress, which is unexpected with regard to time, location, and/or magnitude.

Military history abounds with examples of the second and third varieties, but there have been only a few instances of complete SA [cf. R. Wohlstetter 1962]. Nevertheless, this eventuality formed the point of departure for *NATO's defense planning for decades, as well as for U.S. nuclear planning [cf. A. Wohlstetter 1959; Bracken 1983; Blair 1985].

   2. **Problem of SA.** Although a prepared defense would enjoy a number of advantages over the offensive form of combat (cf. *defense dominance, *three-to-one rule), a SA would inevitably benefit the attacker, precisely because the defender would be caught unprepared.

- On the ground, the attacker would be able to concentrate his forces against the weak points in the defense and thereby may be able to effect a breakthrough of the defender's forward line.
- With respect to air forces, a SA in the form of an *OCA (offensive counterair) strike might catch the defender's aircraft unprotected on the runways, as *Israel did in 1967 [O'Ballance 1972: 49–84].
- In the nuclear sphere, a SA might take the form of a counterforce strike against the enemy's missile silos and strategic bomber bases, coordinated

with a strategic *ASW* campaign directed against its sea-based deterrent. However, worst-case analyses notwithstanding, a disarming first-strike would be well nigh impossible to synchronize because of benign synergies inherent in the triad structure of strategic forces [Scowcroft Commission 1983]. Nevertheless, "reciprocal fears of SA" [Schelling 1960: 207–229] might conceivably lead to a *preemptive attack* by one side, acting under the erroneous impression that the other side might be contemplating a SA.

3. **Reducing Risks of a SA.** The reduction of risks of a SA would be tantamount to an improvement of *crisis stability*. Measures belonging to the following categories would be relevant in this respect:

- Deliberate abstention from first-strike capabilities, such as strategic nuclear weapons with high accuracies (i.e., low CEP [circular error probable]); forward basing of nuclear forces (providing the opponent with only short warning); and abstention from strategic ASW [cf. Harvard Nuclear Study Group 1983].
- *Disengagement* of forces, whereby the time required to prepare for attack and move forces into position would be prolonged, and the surprise factor thus minimized. The resultant *zones* would constitute "early warning devices" because the introduction of troops or prohibited weapons systems into the zone might safely be taken as evidence of a pending attack.
- Avoidance of high-priority (i.e., lucrative) vulnerable targets for a SA: the *no-target principle.*
- Improvement of surveillance and warning capabilities.
- Improved transparency of military activities through various *CBM*s and *CSBM*s.
- Finally, and most decisively, the elimination of offensive capabilities envisioned in *NOD* schemes would *a fortiori* also eliminate capabilities of a SA.

The last measure was acknowledged by both East and West when they reached agreement on the mandate for the *CFE* talks, the main goal of which was "to strengthen stability and security in Europe through . . . the elimination, as a matter of priority, of the capability for launching surprise attack and for initiating large-scale offensive action" [*Atlantic News,* 8 March 1989].

SURVEILLANCE.   See *Warning, *C³I, *Surprise Attack.*

SWEDEN.   Neutral, Western country belonging to the Nordic community scheduled to join the EU in 1995.

1. **Neutrality.** The *neutrality* of S has been a matter of political decision, not of international law. It is usually described as a policy of nonalignment, pointing toward neutrality in war, and is thus far less binding than, say, the neutrality of *Switzerland, *Austria,* or *Finland.* [Agrell 1983a; Andrén 1987; 1987a; 1988; 1991; Jervas 1983; 1987; Kruzel 1988; Ries 1989; Salicath & Støren 1988; cf. Hakovirta 1988: 8–14]. As a reflection thereof, S has traditionally pursued a policy of active neutrality that has merely excluded formal membership in military *alliances.* The main dimensions in Swedish foreign relations have included:

- The Nordic dimension, manifested in the cordial relations with the other Nordic countries, as well as in the *Nordic balance* [cf. Brundtland 1966; 1981; Wiberg 1986; Møller 1989f; 1991b].

- The Western dimension, which has included certain low-key military relations with *NATO and the *U.S. but has been dominated by political and economical relations with the EC.
- The "*N&N dimension, i.e., cooperation with the other N&N (neutral and nonaligned) states in Europe, above all in the context of the *CSCE process.
- The Eastern dimension, i.e. relations with the *Soviet Union, which were never friendly but usually normal and peaceful, even though the submarine "incidents" (Soviet incursions into Swedish territorial waters) of the 1980s resulted in temporary crises [Agrell 1986; 1986a; 1987; Leitenberg 1986; Lindberg 1987; Oldberg 1987; Amundsen 1985; Salicath & Støren 1988].
- The global dimension, which has been manifested in the high priority given to the *U.N., the relatively high level of development aid, etc.

2. **Defense Posture and Strategy.** The defense posture of S has occasionally been described as a model of *NOD. In reality, it contains far more offensive components than, say, that of Finland [Agrell 1979; Andrén 1989; Canby 1981; Jervas 1988; Ries 1989; Roberts 1976: 62–123]. S even for a period gave serious consideration to becoming a nuclear power [Jervas 1983; Agrell 1985; Wallin 1991] yet abandoned its plans in favor of a strong conventional defense.

Swedish defense planning has traditionally been guided by the logic of "marginal utility." On the assumption that S would never be the main target of (Soviet) aggression, but that an attack would be a sideshow in a war between NATO and the *Warsaw Pact, S was confident that it needed to confront only a small part of Soviet forces. This would have been done mainly by means of a "peripheral defense" (skalförsvar), i.e., a form of *forward defense backed up by large mobilizable forces in a *territorial defense mode and capable of waging protracted war against an aggressor.

S's military expenditures and the size of its mobilizable forces have always been higher than those of most NATO countries [Kruzel 1989: 147–148]. As a reflection of that circumstance and of an emphasis on self-sufficiency, S has always had a large arms-manufacturing industry. This has presented S with a dilemma that has often given rise to political controversies: whether to export arms, and if so, to whom and on what conditions. To keep down the costs of the weapons produced for the Swedish armed forces through longer production lines, S ranked as the thirteenth-largest arms exporter in the 1980s. However, exporting weapons to the Third World, including military regimes and/or warring states (as has happened on more than one occasion), seemed to contradict the traditional Swedish policy of furthering *disarmament and development [Hagelin 1982; Sigmund 1985; Lindén 1991; Gullikstad 1991; Westander 1988]. In its search for an escape from its dilemma, S has also given some attention to the *conversion of industry from military to civilian production [Thorsson 1985].

3. **Alternative Defense Debate.** S has always had a quite intense debate on its defense, as well as on NOD and *civilian-based defense in a European context.

- Members of the Center Party have advocated a shift to a more decentralized posture and fallback plans for guerrilla-like resistance [Johannson & Nordmark 1979].
- *Wilhelm Agrell at Lund University was one of the very first to discover the potentials of NOD [1979; 1983; 1984a–b; 1987; 1989].
- Håkan Wiberg (director of the *Center for Peace and Conflict Research in *Denmark), formerly at LUPRI (Lund University Peace Research Institute), did research on civilian-based defense in the 1970s [1977]. Since then he has devoted his attention to peace research in general [1990f (1st

ed. 1976)] and to the critique of *nuclear deterrence* [1983; 1989; 1989b; 1989d-e; idem (ed.) 1989], *CBMs* [1989a; 1990], and NOD [1989c; 1990b–d; 1994].

- Jan Øberg (originally a Dane) was formerly at LUPRI, then subsequently became director of the independent Transnational Foundation for Peace and Future Research (TFF) in Lund. He started his career with a dissertation on Danish defense policy [1980] but abandoned this interest in favor of a philosophical critique (much inspired by *Johan Galtung*) of "occidental" concepts of *peace* and *security*, pleading for a "paradigm shift" to a more "oriental," i.e., holistic view of the problems [1983; 1989a; Fischer & al. 1987]. His critique of technological fixes notwithstanding, JØ has remained a (qualified) supporter of *transarmament* [1983a; 1989].

- The Life and Peace Institute in Uppsala, under the directorship of Bernt Jonsson, has maintained close links to the Swedish churches and the Ecumenical Council, in particular. Its approach to defense alternatives has been primarily ethical. With a view to achieving savings on the defense budget, as well as to averting nuclear holocaust, the institute recommended a shift to NOD, albeit only for a transitionary phase until the time would be ripe for civilian-based defense [Life and Peace Institute 1990].

- Members of the peace movement (especially *Svenska Freds-och Skiljedomsföreningen* [Swedish Peace and Arbitration Society]) have also criticized what they have regarded as Swedish overspending on national defense as well as the excessive offensive potential of the Swedish armed forces. As an alternative, they have suggested a transarmament to NOD, combined with elements of civilian-based defense [M. Adler & al. 1989; Englén & Hatt 1989].

- This has also been the general approach of Lennart Bergfeldt, who has been involved, e.g., in a semigovernmental exploratory study on civilian-based defense as a complement to armed defense [1984; 1985; 1985a; 1990; cf. Roberts 1977]. The study was not entirely disregarded in subsequent defense planning.

Provisions for *nonviolent resistance* are included in official planning, as demonstrated in a leaflet published by the Directorate for Psychological Defense (under the auspices of the Defense Ministry): "Nonmilitary resistance complements the military resistance and is likewise based on a strong determination to resist. This constitutes an additional threat that a prospective attacker has to take into consideration" [quoted in Bergfeldt 1990: 81].

SWITZERLAND.   Small neutral country in the heart of Europe, enjoying what is, from the point of view of defense, probably the world's most favorable terrain.

1. **Neutrality.** S has the world's oldest neutral stance, codified after the Napoleonic Wars in 1815, and confirmed in the 1919 Versailles Treaty. It has never been seriously challenged militarily or politically, not even during the Cold War, above all because the defense capabilities of S have been beyond reproach [Fleury 1990; Freymond 1990; Schwok 1990; Spillmann 1989; Duic 1985; Aichinger & Maiwald 1989; Schiemann 1991]. S has traditionally had a very restrictive view of *neutrality* that connotes a rather passive and isolationist international stance (including an abstention from *U.N.* membership). It is debatable whether S has ever had an *arms control* policy at all [F. Tanner 1990; 1991]. Its austere approach to neutrality made it possible for S to stay aloof from the East-West conflict, its undivided sympathy for the West notwithstanding. However it has come increasing-

ly into question whether isolation will remain feasible [Dardel 1990; Däniker 1990; Gasteyger 1990; Krummenacher 1990].

2. **Defense Posture and Strategy.** Swiss defense strategy and posture have above all been conceived of in terms of *dissuasion*, i.e., as a strictly nonoffensive *deterrence by denial* through which a prospective aggressor would be persuaded that an attack would not be worthwhile because of S's ability and will to defend itself. By effectively disengaging from the military confrontation in Europe, S has almost (in combination with *Austria*) constituted a *buffer zone* [Däniker 1983; 1987; 1990a].

S's armed forces have been only one element in a general posture, i.e. *total defense* posture, wherein *civil defense* especially has played an important role. S has the most extensive shelter program in the world, with more than 5.5 million places available; it is scheduled to reach complete coverage (6.5 million) by the year 2000 [D. Fischer 1982; Galtung 1989c; Cramer 1984: 23–28; Spillmann 1989: 172; 1990]. The armed forces have been organized as a *militia*, implying very minuscule standing forces (around 3,000 cadres), yet S has the greatest per capita *mobilization* potential in the whole of Europe (more than 600,000 out of a population of 6.5 million) [Duic 1985; Fernau 1987; Haltiner 1986; 1987; Cramer 1984].

The Swiss conception of *territorial defense* was developed around 1973, yet on the basis of a much older tradition. It envisions extensive use of *barriers* and prepares for the demolition of communication lines (e.g., mountain tunnels) in order to block access to an invader. It also entails a modicum of *forward defense* in the form of the frontier brigades (*Grenzbrigaden*). Furthermore, the field army (three corps) is to be responsible for a mobile defense of most of S, with special emphasis on areas of particular strategic and/or political importance, i.e., a form of *bastion* defense. The last is even more manifest in the case of the *redoubt* brigades, tasked with continuing resistance from the inaccessible Alps even after a successful invasion and partial occupation [Cramer 1984: 21–23, 34–38; Däniker 1987; Weck & Maurer 1990; Fuhrer 1987: 113–123].

Even though this defense strategy is very much in line with those proposed by *NOD* advocates for the rest of Europe (indeed has inspired many of these), the arms posture of S is not particularly defensive but quite *tank*-heavy, with heavy artillery, long-range aircraft, etc. [Cramer 1984: 31–34; Fuhrer 1987; Seethaler 1988; Miliovejic 1990]. However, thanks to the location of S and its defense strategy, its arms posture seems to have been of no concern to other states.

3. **Alternative Defense Debate.** There has not been much of a debate on military alternatives in S. Paradoxically, leading Swiss officers (such as General Gustav Däniker) have even expressed themselves quite negatively about the NOD proposals pertaining to other countries [Däniker 1987: 94–103; 1987a], thereby creating the impression that S would prefer the rest of Europe to remain as offensive as possible.

On the other hand, those in S who have supported a defensive restructuring in the rest of Europe (e.g., peace researchers such as the late Daniel Frei [1984; 1989; idem & Catrina 1983] and Günther Bächler [1990; 1990a–b]) have either disregarded the military policy of S or advocated a complete abolition of the armed forces.

S has had quite powerful *peace movements*, including the Gruppe Schweitz ohne Armee (Group in Favor of Switzerland without an Army). In 1989 they took advantage of the Swiss Constitution to enforce a referendum on the abolition of the armed forces. Even though the motion was eventually turned down by the electorate, as many as 35.6 percent of the ballots supported *disarmament* [Gross & al. (eds.) 1989; cf. Herz 1971; Kelly & Bastian 1989].

SWORD AND SHIELD.   Metaphor used about two different combinations:

- **NATO*'s combination of nuclear (sword) and conventional (shield) forces, particularly until the abandonment in 1967 of massive retaliation in favor of **flexible response*.
- The combination recommended by certain **NOD* proponents of defensive (shield) and offensive-capable (mechanized, armored, with high firepower [sword]) forces, the latter usually deployed in the rear as operational reserves [Löser 1981; 1985; 1986; 1987a; Bülow 1986; 1986a; 1988; 1988a–c; idem & Funk 1987; 1988; 1988a].

SWORDS INTO PLOWSHARES.   The "motto" of **conversion* [originally from Isiah 2:4].

SYMONIDES, JANUSZ.   See **Poland.*

T

TACTICAL WARNING. See *Warning*.

TACTICS. See *Military Science*.

TAIPING REBELLION. Large-scale civil war in *China* (1851–1864) that cost at least 11 million lives. Both the *guerrilla strategy* applied in the TR and its use of large-scale evasive maneuvers foreshadowed and probably influenced twentieth-century guerrilla strategists such as *Mao Zedong*. The TR also showed signs of inspiration from *Sun Tzu*, thus, according to some military historians, constituting merely an instance of a particular Chinese (or more generally, Asian) form of warfare [Bellamy 1990: 218–228; cf. Hsü 1975: 277–321].

TANK-FREE ZONE. See *Zones*.

TANKS. One of the main means of offensive warfare in the twentieth century, primarily by virtue of their combination of mobility, protection (through armor), and high firepower.

T made their first appearance during *WWI* as offensive means of fighting entrenched opponents. During the interwar years, strategic thinkers such as *J. F. C. Fuller*, *Liddell Hart*, and others developed strategic conceptions of offensive *maneuver warfare* based on the combination of T and aircraft [Bond & Alexander 1986]. This was tried in practice with overwhelming success by the German armies in the *blitzkriegs* in the beginning of *WWII*. Since WWII, most armies have placed a great emphasis on T, for both defensive (e.g., in the armed forces of *Switzerland*) and offensive purposes. The Red Army of the *Soviet Union* was extremely tank-heavy, and Soviet strategy and operational art were to a large extent based on high-speed maneuver warfare by means of T [Simpkin 1983].

Because of this predominantly offensive utilization, most *NOD* proponents have sought to make T obsolete, inter alia, by stationary defense schemes. In such models, area coverage and a *web*-like configuration should make tanks, and the mobility they provide, superfluous for defensive purposes, above all for antitank defense [Afheldt 1983]. Other NOD proponents have devised schemes for various forms of antitank *barriers*, such as tank ditches, "killing zones" saturated with mines, or even *terrain modifications*, etc.

The potentials of advanced antitank weapons of the *PGM* type have led many NOD advocates to predict the obsolescence of T [Barnaby 1986: 67–75; cf. P. Walker 1981; 1985; 1986; Herolf 1983; Simpkin 1987]. Others have taken into account the wide range of countermeasures available to the designers of T (composite and reactive armor, etc.) and concluded that it would be premature to write off T altogether. Certain NOD schemes have therefore included T, albeit in configurations that would deprive them of most of their offensive (while preserving their defensive) capabilities: either in a *spider-and-web* mode [SAS (eds.) 1984; 1989; Unterseher 1989b–d; 1994; cf. E. Müller 1984] or in rear positions from which they would at least pose no threat of *surprise attack* [Löser 1981; Bülow 1986; 1986a; 1988c; 1989a; A. Müller 1986].

TARGETS.   See *No-Target Principle.*

TATCHELL, PETER.   Affiliated with the Labour Party of the *U.K.* as well as the CND. PT was the author of *Democratic Defence* [1985], in which he dealt critically with both the organizational and the strategic and political aspects of the present defense system.

With respect to organizational aspects, PT delivered a fervent critique of the authoritarian forms of leadership prevalent in the armed forces and advocated a radical *democratization* of the military system. Without recommending the introduction of general *conscription*, he nevertheless suggested a greater emphasis on home-guard structures [1985: 73–92, 157–176]. Regarding strategic aspects, PT rejected the dominant accent on nuclear and offensive-capable forces, advocating instead a strictly defensive form of defense, i.e., *NOD*, with inspiration drawn from three sources: the Continental NOD debate (*Afheldt*, *Brossollet*, *Spannochi*, *Uhle-Wettler*, and others); the experience of the European neutrals with *territorial defense* structures (*Sweden*, *Switzerland*, *Yugoslavia*); and the legacy of the socialist classics and the worker's movements in Britain and elsewhere, particularly his near-forgotten fellow countryman *Tom Wintringham* [1985: 93–156].

PT's concrete proposals were couched in terms of a ten-year reform of Britain's defense posture, by the end of which it would have been transformed into a strictly conventional defense based on a combination of frontier, territorial, and *civilian-based defense.* The regular forces in general and the Royal Navy and the BAOR (British Army of the Rhine) in particular would be gradually reduced and the territorial army would be greatly expanded. The number of large weapons platforms (*aircraft carriers*, destroyers, etc., as well as aircraft) would be reduced in favor of the weapons themselves: small, sophisticated (and preferably easy to handle) "anti-weapons" for *air defense* and for antitank and coastal defense purposes. Overall, the reform would lead to a reduction in total force levels (from regular forces numbering 321,000 down to 167,000, and from a mobilizable strength of 600,000 down to 457,000), as well as to a cut in defense expenditure (from £16 to £10 billion) [1985:177–199].

Judging such a reform to be inconceivable within the framework of *NATO*, PT recommended unilateral withdrawal, albeit not in favor of a "Little England" policy but accompanied by attempts to create a "self-defensive alliance of European nations," conceived of as "a non-provocative rival to NATO" [1985: 200–201].

TECHNO-COMMANDO.   Term used by *Horst Afheldt* for his light *infantry* forces deployed in a defense *web* and armed with high-technology antitank and *air defense* weapons [H. Afheldt 1983; cf. Weizsäcker (ed.) 1984; Carton 1984]. Also known as "techno-guerrillas."

TECHNO-GUERRILLA.   See *Techo-Commando.*

TERRAIN MODIFICATION.   TM might in certain cases be a very effective means of enhancing a country's defensive capabilities. Furthermore, it need not be tantamount to any conversion of the countryside from civilian to military use (which would probably be unacceptable on other accounts) but could take the form of provisions for dual-purpose utility [Møller 1991d; Booth & Baylis 1988: 122; Webber 1990: 197, 211; cf. Garrett 1989: 213–230]. Forests, plantations, canals, and artificial lakes and ponds would be obvious instances because a terrain replete therewith would both have enhanced recreational value in peacetime and be distinctly tank-hostile (i.e., more easily defensible against an armored thrust) yet strictly defensive.

TERRITORIAL DEFENSE.   TD may mean any form of defense of a nation's territory, or a special form thereof.

1. **Territoriality and National Defense.** In the first sense, TD has become nearly synonymous with national defense and is in any case its central element. First, land possesses intrinsic value (as opposed to the air and the sea) because people inhabit it and often depend on it for their livelihoods. Second, with the partial exception of small nomadic tribes, human groupings are tied to the land, in the sense that their identity is defined by their "belonging" to a particular territory, either the land they actually inhabit or some kind of "promised land." Nationhood is thus basically territorial. Third, the modern state is by its very nature territorial, i.e., is tantamount to holding sovereignty over certain tracts of land.

Territoriality is fundamentally an exclusive notion, and shared ownership is nearly unknown, with a few partial exceptions, such as Spitsbergen, the Antarctic, the mandate territories of the former League of Nations and the U.N., and Germany and Austria under the postwar occupation regimes. Regimes such as these are fragile and ephemeral, constituting merely transitory stages, i.e., the first steps of a territoriality in the making; or they may reflect the fact that the territory in question is of no great interest to anybody.

Territorial disputes have caused many wars throughout history, but the postwar period is unprecedented for having delegitimized war as a means of territorial expansion. TD, on the other hand, is a sine qua non of statehood.

2. **Taxonomy of TD.** In principle, there are four main types of TD, as well as various combinations thereof:

- Defense of the entire territory under a state's sovereignty, i.e., TD in the narrow sense of the term;
- *Bastion-type TD, whereby central parts of the country are defended at almost any cost and the invader is allowed to establish control over other parts of the territory temporarily;
- *Forward-defense*–type TD, whereby any incursion into the territory to be defended is to be avoided; and
- Deterrence-type TD, whereby the territory is protected by dissuading aggression in the first place.

The choice of TD type is not entirely free but determined by political factors as well as by size and location of the country. Small countries have a genuine option of TD in the narrow sense, which is not really available to huge states such as Russia, *Canada*, or *Australia*. Forward defense primarily appeals to states with predictable enemies and not excessively long borders, such as the *FRG* (before unification). Bastion defense seems appropriate for countries such as the *Soviet Union*, whereas countries such as *Israel* have tended to rely on the deterrent mode [cf. Levite 1990].

3. **TD Strategy and Posture.** The most straightforward way of achieving TD is through a TD strategy (in the narrow sense), such as those currently adhered to by the neutral and nonaligned states of Europe [cf. Roberts 1976; Lutz & Grosse-Jütte (eds.) 1982; Fuhrer 1987; Danspeckgruber 1986; Kruzel 1989; Wiberg 1994], some of which, however, have postures constituting mixtures between TD and bastion-type defenses. This is the case for *Austria* [Bundeskanzleramt 1985; Rotter 1984;

Vetschera 1985], *Switzerland* [Cramer 1984; Däniker 1987; Spillmann 1989], and it was the case for *Yugoslavia* [Lukic 1984; Bebler 1987], whereas the defense of *Sweden* rather resembles a compromise between a forward ("peripheral" or "perimeter") defense and TD [Agrell 1979; Andrén 1989]. *Finland*'s defense posture may in fact be the closest approximation to the ideal of TD, even though the uneven distribution of population has also prevented complete area coverage [Ries 1989; 1989a; FCMH (eds.) 1985].

A TD strategy would imply defending (in principle) the entire country by means of a rather even distribution of forces, something that might be not only militarily advantageous (if possible) but also politically advisable, especially for neutrals. A pure TD is, by its very nature, *tous azimuts,* i.e., not directed against any specified aggressor, and would hence signal equidistance rather than unidirectional hostility. TD would also signal peaceful intentions because the envisioned territorial coverage would diminish the need for mobility and would *ipso facto* allow a state to relinquish offensive capabilities. Likewise, TD would envision no pursuit of an invader across the border but rest content with evicting it from the defended territory, i.e., with restoration of the *status quo ante bellum.*

That TD implies defending the entire country is, of course, a simplification because most countries contain (sometimes vast) stretches of more or less inaccessible land: mountains, marshes, lakes, and forests, as well as areas shielded or surrounded by such features. Needless to say, the distribution of forces should therefore not be uniform but cover only actually exposed territory. Most countries in Europe would, in principle at least, be in a position to establish such a TD by virtue of high population density, even though the actual distribution of the population might match poorly the ideal distribution of the armed forces in, say, countries such as the Scandinavian.

THOREAU, HENRY DAVID.   American poet and anarchist writer (1817–1862), author of, inter alia, the essay *Resistance to Civil Government,* written in protest against the Mexican-American War [1849]. HDT advocated civil disobedience, i.e., *nonviolent resistance* (he served as a source of inspiration for, e.g., *Tolstoy* and *Gandhi*) and conscientious objection to military service.

THREE-TO-ONE RULE.   Numerical expression of *defense dominance*, i.e., of *Clausewitz*'s assumption that the defense is intrinsically superior to the offense [Clausewitz 1980:361/1984:358 (Book VI.1.2); cf. Dupuy 1979:12–15; 1980: 326–333].

According to the TTOR, the defender should be able to prevail against an attacker up to three times his own numerical strength, implying that a successful attacker would need a superiority exceeding 3:1. However, the status of the TTOR (rule-of-thumb, distillation of experience, or superstition) remains disputed, as does its validity [see, e.g., the debate in *International Security,* 12:4, 13:4, and 14:1; Epstein 1988; 1989; Mearsheimer 1988; 1989; idem & al. 1989; Posen 1988; Dupuy 1989]. Most analysts would agree that it would be valid for tactical engagements in a linear mode, i.e., breakthrough battles in which two opposing forces meet head-on: the defender would be able to exploit fully his advantages of prepared combat positions and the attacker would be forced to expose himself.

Entire wars might theoretically be little more than such linear engagements writ large, such as was almost the case for the trench warfare on the western front in *WWI*. Were maneuver warfare to predominate, however, the defender's stipulated tactical 3:1 advantage might be of small avail, inter alia, because the attacker might concentrate swiftly, thus achieving overwhelming (i.e., exceeding 3:1) tactical superiority at the decisive points.

It is not a settled matter whether, or to what extent, the TTOR would also be applicable to peacetime *balance assessments, such as the *force comparisons between *NATO and the *Warsaw Pact. *John J. Mearsheimer has stood out as one of the most prominent advocates of the TTOR, his faith in which allowed him to make very optimistic force assessments [1982; 1983: 58–60; 1988; 1988a: 119–120; 1989; idem & al. 1989: 128–144; cf. Posen 1984a]. He has, however, been criticized, both by equally optimistic analysts [Epstein 1989] and by "qualified pessimists" [Cohen 1988; cf. idem in Mearsheimer & al. 1989: 160–179]. Most European *NOD advocates would tend to agree with the assumption of the validity of the TTOR, applied on a theater scale, with the caveat that this might well require deliberate modifications of defensive postures and strategies [Boserup 1990c]. The U.S. Army in the 1976 edition of its basic field manual accepted the TTOR (cf. *Active Defense) but was more ambiguous in the 1982 edition, which inaugurated the offensive *AirLand Battle Doctrine [HQ Dep. of the Army 1976: 5.3; 1982: 8.5–6, 10.3–4; cf. Posen 1984a: 84]. This might be taken as evidence of a negative correlation between confidence in the defensive advantages and propensity for offensive tactics.

THRESHOLDS.   Breaks or discontinuities in *escalation of a conflict, the crossing of which would signify a qualitative change. Even though some T have been codified in the *laws of war, most have remained matters of convention and tacit understandings, hence inherently contested. That T would be observed at all in a war would reflect the shared interest of the two conflicting sides in avoiding undesirable and irrational escalation. It might also presuppose some degree of communication, albeit often in the form of deeds rather than words [Schelling 1960: 53–80].

All recorded wars until the present have been *limited wars in the sense that both sides have withheld some of their destructive power; *chemical weapons, e.g., were not used during *WII, even though both sides had them at their disposal [Moon 1984; McFarland 1986; Brauch 1982a: 76–92]. Nuclear weapons have not been used since 1945, not even in cases where a nuclear power lost a war against a nonnuclear state, as the U.S. did in the *Vietnam War and the *Soviet Union in *Afghanistan.

Even beneath the nuclear threshold, some limitations tend to be observed. The U.S. showed some restraint during the Korean War, inter alia, by not seeking to roll back the communists but opting for a stabilization along the 38th parallel [Carver 1981: 151–171; cf. Gaddis 1982: 109–117]. The same was the case during the Vietnam War, when the U.S. refrained from bombing the dike system of North Vietnam [Gibson 1988: 319–382; cf. Clodfelter 1989]. The USSR likewise behaved with some restraint in Afghanistan, inter alia, by not significantly expanding the level of troops, by largely refraining from counter-value bombing (although some of the area bombing was rather indiscriminate), and by not employing chemical weapons (in spite of repeated, but undocumented, allegations to this effect) [Urban 1988: 56–57, 211–215].

Some such constraints might, of course, be preconditioned by a democratic system of government in which the aversion of the population (i.e., electorate) against indiscriminate violence would be a factor that decisionmakers have to take into account. All nations, however, would be somewhat susceptible to world opinion, as was the USSR regarding the Afghan War, which seriously (and perhaps irreparably) damaged the Soviet image as a defender of the developing nations against "imperialism."

TIME FACTOR.   There are at least two senses in which the TF influences war.

First, the time to be anticipated between a state's receipt of *warning of an impend-ing attack would determine the required state of readiness: the shorter the warning time, the higher the state of readiness, and vice versa. Second, the expected as well as the actual duration of a war will be a determinant of its probability as well as its destructiveness.

1. **The TF and War Prevention.** The hope of attaining a swift and decisive vic-tory over an opponent has often been a determining factor in war-versus-no war deliberations. The prevailing (yet surprisingly erroneous) short-war expectations prior to *WWI might accordingly go a long way toward explaining its outbreak [Tuchman 1962; Howard 1985; Evera 1985; 1985a; J. Snyder 1984; 1985. For a cri-tique, see Trachtenberg 1990]. At least in a *subnuclear setting, protracted war would be utterly unattractive, inter alia, because the risk of something going wrong would tend to increase over time, perhaps making unwanted *escalation more like-ly. Furthermore, even though an aggressor might well believe the hoped-for prize would outweigh the expected (finite) losses over a finite period, it would probably be dissuaded by the prospect of (1) not attaining its objectives and (2) becoming embroiled in a protracted war of *attrition and incurring steadily cumulating losses. As a general rule, war prevention would therefore be strengthened by the ability of the defender to transform an intended *blitzkrieg into a drawn-out and indecisive war [Mearsheimer 1982; 1983].

2. **Defense and the TF.** A further contribution to war prevention, as well as to *war termination, is the fact that the defender, as a general rule, stands to benefit from a prolongation of war. As expressed by *Clausewitz: "The time that passes is lost to the aggressor. Time lost is always a disadvantage that is bound in some way to weaken him who loses it' [Clausewitz 1980: 391/1984: 383 (Book VI.8)]. Furthermore, surprise is the attacker's main advantage, but its effects tend to wear off rapidly, or even to be inverted to benefit the defender: "The one advantage the attacker possesses is that he is free to strike at any point along the whole line of defence, and in full force: the defender, on the other hand, is able to surprise his opponent constantly throughout the engagement by the strength and direction of his counterattacks" [Clausewitz 1980: 364/1984: 360 (Book VI.2)]. More than four-fifths of the engagements during the *Vietnam War were thus initiated by the Vietnamese guerrillas, i.e., the "defenders," who took the U.S. "aggressors" by sur-prise [Gibson 1988: 11, 103–110]. Finally, the importance of *logistics would tend to grow over time, in which regard the defender waging war on its own territory, enjoying *interior lines of communication, would have obvious advantages over the aggressor.

The defender might prolong the war by denying battle and suspending fighting, inter alia, through a general *dispersion of its forces, thereby denying the aggressor the opportunity to fight a decisive battle. Or the defender might withdraw, thus buy-ing time with space and exploiting the advantages of withdrawal over pursuit [A. Jones 1988; Posen 1986]. Numerous recommendations to these effects were to be found in some of the first works on *NOD [Spannocchi 1984; 1988; Brossolet 1975; Afheldt 1976; 1983].

3. **The TF and the East-West Confrontation.** The general observations above applied fully to both sides of the confrontation between *NATO and the *Warsaw Pact. The only war that either side stood any chance of "winning" was a blitzkrieg; each side would have been certain to lose an attritional contest on the opponent's territory, such as would have ensued if one had attacked the other [cf. Stephen Van Evera in Feldman (ed.) 1990: 102]. The USSR/Russia had demonstrated its strength in protracted defensive war during Napoleon's 1812 invasion, as well as during the 1941–1945 "Great Patriotic War." In both cases, the aggressor's logistical chains

broke down because of the enormous distances to traverse and the harsh climate. The defender exploited these by means of a grand-scale strategic withdrawal; the aggressor was lured into the depth of the country and starved out by means of *scorched-earth* tactics. The West, likewise, would have had to have been nearly certain to prevail in a protracted defensive war against the USSR by virtue of its vastly superior mobilizable potential, in its turn reflecting both economic superiority and a larger manpower pool.

TIT-FOR-TAT.   Presumed solution to the *prisoner's dilemma* whereby "cooperation can emerge between egoists without central authority." The strategy was suggested by *Anatol Rapoport* and analyzed at length by *Robert Axelrod* in *The Evolution of Cooperation* [1984; cf. Rapoport 1989: 273–274, 558–559; Brams & Kilgour 1988: 113–118; Jervis 1988; Stein 1990: 87–112; Goldstein & Freeman 1990: 29–32, 130–138, 141–144].

TFT presupposes a sequence of "games" (i.e., interactions) between two "players," who are able to learn and to "invest," i.e., to strive for long-term rather than immediate gains, hence the importance of the *shadow of the future*. Presuming that the shadow is long enough for it to influence the preferences of the players significantly, TFT envisions always commencing a "round" with cooperation but hedging against exploitation by consistently retaliating symmetrically, i.e., without escalating or de-escalating. It is a strategy of perfect reciprocity, i.e., of conditional cooperation, which is presumably suitable for de-escalating a conflict that supposedly neither side desires. One example would be a vicious circle reflecting the *security dilemma*, manifested, e.g., in an *arms race* or in malign interaction during a crisis. The strength of the strategy was tested by Axelrod in a computer tournament, a simulation game, but it presumably also has numerous real-life manifestations: in biological systems (where even bacteria, assertedly, tend to "play" according to TFT) and in social interaction, even in such confrontational forms as trench warfare. It has, however, been demonstrated that TFT tends to "echo," i.e., to lead to a continuing repetition of noncooperation [Goldstein & Freeman 1990].

TITO, JOSIP BROZ.   Yugoslav communist partisan leader and subsequent head of state in *Yugoslavia*. Even though he was the author of no major work on strategy, his concrete orders and proclamations directed to the guerrillas and the general population [Tito 1966] have made him a central figure in the development of *guerrilla strategy*. He influenced, among others, *Emilio Spannocchi* and other modern *NOD* proponents.

JBT immediately after the Italian invasion directed his followers to support the patriotic struggle, initially under the auspices of the regular army and later in the framework of People's Liberation Partisan Detachments. They were tasked with sabotage against railways, bridges, factories, etc. of utility for the invaders, and with the defense of villages and towns. The partisans were explicitly urged to "be ruthless in annihilating" their enemies [1966: 42–49]. JBT gradually sought to elevate the patriotic struggle above the level of partisan warfare, inter alia, through creation of the People's Liberation Army, where shock and partisan brigades were merged into an "almost-regular" army capable of major offensive operations and based on compulsory *conscription* in the liberated areas.

JBT advised against abandoning partisan features altogether:

> This means that we must be wary of inflexible fronts and avoid them, that we must not allow the enemy to use the tactic of imposing on us the defense of long, sprawling fronts, the mending of breaks in the front, and so on. On the contrary, our army must be imbued exclusively with the

spirit of offensive, not only when we are on the offensive, but when we are on the defensive as well. The spirit of offensive is reflected, during enemy offensives, in bold powerful strokes of Partisan tactics, the tactics of breaking through to the enemy rear, destroying not only communications but also enemy supply depots and bases which are either poorly protected during enemy offensives or left unguarded altogether. [1966: 87–88]

These guerrilla characteristics notwithstanding, JBT emphasized that it was not to be an "episodic" war but, rather, "a permanent war . . . of extermination against the invaders and domestic traitors" [1966: 207].

TLATELOLCO TREATY. See *Latin America.*

TOLSTOY, LEO. Nineteenth-century Russian landowner, high-ranking officer during the Crimean War, novelist (author of, inter alia, the monumental *War and Peace*), and pacifist of a Christian and anarchist persuasion. In the last capacity he has served as a source of inspiration for, among others, *Gandhi* and the proponents of *nonviolence* and *civilian-based defense.* Proponents of the latter were especially influenced by *The Kingdom of God Is Within You,* in which he advocated conscientious objection to military service [1893; cf. Kessler 1987]. The reasons were two: first, killing was in itself impermissible, even in *self-defense,* and second, the arming by one state tended to spur reciprocal action on the part of its enemies, resulting in an *arms race,* and perhaps war.

TOTAL DEFENSE. Term used in certain countries (e.g., *Denmark*) for a national defense posture consisting of three elements. These are almost without exception present in all countries, albeit with different emphases [Svensson 1980; Boelsen (ed.) 1982; Bundeskanzleramt 1985]:

- Military defense;
- Civilian preparedness, various precautions intended to enhance society's *survivability* in the face of aggression; and
- *Civil defense.*

Certain alternative-defense proponents have recommended a similar division of national defense, often emphasizing the equal importance of the three components:

- A strictly defensive, i.e., *NOD*-type, military defense;
- A broader conception of civilian preparedness, including training of the population in *civilian-based defense* and *nonviolent resistance,* along with measures to render society as invulnerable as possible; and
- Civil defense [e.g., Löser 1981: 89–95, 195–201; Nolte & Nolte 1984: 143–145; D. Fischer 1982; 1984; idem & al. 1987; Øberg 1983: 125–142; 1983a].

TOTALLY DESTRUCTIVE CONVENTIONAL WAR HYPOTHESIS. Hypothesis that gained growing support during the latter half of the 1980s, particularly in the *Warsaw Pact* countries, as well as in the peace-research and alternative-defense communities [see, e.g., M. Schmidt 1989; idem & Schwarz 1988; Schwarz 1989: 15–21; Benjowski & Schwarz 1988: 13–20; 1989; *S+F,* 1989:4. For a critique, see Møller 1989c–d].

1. **TDCWH.** The gist of the TDCWH was enunciated by *Mikhail Gorbachev,* according to whom, "even a conventional war . . . would be disastrous for Europe

today" to the extent that "conventional hostilities [could] . . . make the continent uninhabitable" [Gorbachev 1988: 195–196]. The implications of these views were, first, that the difference between nuclear and conventional war was shrinking because of the increasingly destructive potential of conventional weapons. Second, that even victorious conventional wars would be suicidal because the destruction would be so enormous as to make postwar recovery impossible. Third, that attempts at *damage limitation* were therefore futile because the only damage limitation worthy of its name would be war prevention. Paradoxically, *NOD* has often been advocated as the best means of war prevention, even though it is, on inspection, logically incompatible with the TDCWH. Fourth, that the Clausewitzian conception of war as a continuation of politics has been rendered obsolete because a totally (and mutually) destructive war would serve no political purpose.

2. **Evidence.** Vast amounts of empirical evidence were used in support of the TDCWH, the validity of which was not always matched by relevance. The most often heard arguments included the following:

- The European countries are replete with stored nuclear and chemical munitions whose destruction might well contaminate the atmosphere were the depots to be hit by nuclear or conventional explosives [see, e.g., Arkin & Fieldhouse 1985; Mechtersheimer & Barth 1986: 122–137].

- Nuclear power plants are abundant (more than three hundred), as are chemical factories and other installations with highly toxic contents. Not merely nuclear but also conventional attacks against these civilian targets could cause damage to the nearby population centers that might dwarf the consequences of the Chernobyl accident [see, e.g., Smit 1989a; Ramberg 1986].

- Industrial and postindustrial societies are increasingly vulnerable to breakdowns in power supply, data processing systems, etc.

- Finally, the lethality of certain conventional weapons has grown to such an extent as to partly blur the distinction between conventional and nuclear weapons, in the sense that the most lethal conventional weapons surpass the smallest nuclear ones in destructive power: fuel-air explosives, certain container bombs with miniature bomblets (e.g., Skeets), and similar quasi-nuclear weapons provide *nuclear equivalence* [Robinson 1983; Cotter 1983: 228–233; Bellamy 1987: 186–187; J. R. Walker 1987: 52–53; Barnaby 1986: 84–85; Lumsden 1978: 171–175].

Quite a convincing case can be made for expecting that an unlimited conventional war would devastate beyond recovery the battlefield upon which it would be waged.

3. **Implications of the TDCWH.** The TDCWH had some logical implications that were often overlooked by its proponents. Above all, it would tend to weaken the case for nuclear disarmament as well as exacerbate the so-called *defense dilemma.*

- If conventional and nuclear war are equally (because totally) destructive, then the only factor of importance is their respective likelihood. A smaller risk of nuclear war would thus have to be acknowledged as preferable to a larger (even by a narrow margin) risk of conventional war. Unless one is prepared to deny completely that *nuclear deterrence* might contribute to war prevention, one would have to dismiss nuclear disarmament, or even build-down, as dysfunctional—except as one element in general and complete *disarmament.* This would contradict both the rest of the argument made by the same authors and established facts as well as strong hypotheses: Even if the consequences of tactical nuclear war might not exceed those of large-scale conventional war (a big if), those of global strategic nuclear war certainly would, especially if the theory of the "nuclear winter" were to hold true [Sagan 1983; Ehrlich & al. 1984].

- Because defense against aggression would logically imply war, then the assumption that such a war would inevitably be suicidal (as implied by the TDCWH) would inescapably lead to the conclusion that states had better refrain from defending themselves: an extreme version of the *defense dilemma* [cf. Buzan 1991: 270–293]. In that occupation would not necessarily be irreversible (and its costs hence less than total), preemptive surrender would be the most sensible course to follow in the face of threats from aggressive states. Were such a defeatist attitude to become dominant, the world would be safe for aggressive war and the most ruthless states would be at liberty to expand beyond bounds.

4. **Validity of TDCWH.** The TDCWH seems to have been based on questionable assumptions about a future war. It presupposed that a future war (in Europe at least) would inevitably be total in the sense that each side would use all means at its disposal and observe no *thresholds* to escalation, a logically flawed premise:

- Proponents of the TDCWH seem implicitly to assume that the nuclear threshold would be observed; if not, it would have to be acknowledged that conventional war was inconceivable rather than totally destructive.
- At the same time, they presume that no other thresholds would be observed, even though nearly all previous wars have been limited (i.e., less than total) in one way or another, and particularly so in the nuclear era: The widespread nonuse of chemical weapons (even by states possessing large stockpiles), the abstention from indiscriminate bombings of cities in most (yet far from all) wars, and the traditional respect for the legal rights of neutrals and POWs are merely a few examples. That these thresholds have sometimes been crossed does not mean that they do not exist, and should not be allowed to conceal their even more frequent observance.
- Finally, it strains the imagination to envision a large-scale war in Europe in which the aggressor would not seek some degree of war limitation because it would hardly make sense to invade another country only in order to lay it in ruins. The aggressor would hence seek to avoid deliberate destruction of the civilian society, the industrial potential, and other aspects of which it (according to the hypothesis) would be seeking to conquer.
- Finally, what would tend to keep conventional escalation within bounds would be the fear of retribution. The side embarking on, say, terrorist bombardments would have to anticipate retaliation, either in the same coinage or in an escalatory mode, e.g., by means of nuclear weapons. The resultant upward spiral would have no obvious limit because of the infinite and absolute destructive power of thermonuclear weapons (the so-called *existential deterrence* effect).

If both sides would have an obvious incentive to contain escalation, it would not be unreasonable to assume that they might seek to avoid attacks on nuclear power plants and the like (i.e., countervalue warfare) in a future war (in fact, such agreements exist between *India* and Pakistan, and Argentina and Brazil), reducing this problem to that of collateral (unintentional) damage caused by counterforce warfare. The extent of the collateral damage would depend on the way in which a future war were to be waged, and the weapons employed. The quasi-nuclear character of certain conventional weapons notwithstanding, there is also an opposite trend at work toward weapons with very high hit probabilities and limited destructive yields (see *PGMs*), which would allow for more limited collateral damage [Herolf

1983; 1986; 1988; 1988a; Boyer 1985; P. F. Walker 1981; 1985; 1986; Canby 1987].

5. **Evaluation.** The TDCWH seems to foster defeatism with regard to not merely the appropriate response to aggression but also efforts at *damage limitation*. If war was inevitably suicidal, damage limitation would logically have to be acknowledged as futile. Unless the TDCWH leads to general and complete disarmament, it could well turn into a self-fulfilling prophecy. The fact notwithstanding that it has often been advocated by NOD proponents, the TDCWH is logically incompatible with NOD for the following reasons:

- One of the main purposes of NOD is damage limitation, which according to the logic of the TDCWH would be a futile endeavor;
- NOD is envisioned as a contribution to nuclear disarmament, which would make no sense according to the TDCWH; and
- NOD is intended to provide insurance against aggression through adequate defensive capabilities, whereas the TDCWH would lead to a rejection of the sensibility of national defense under all circumstances.

TRANSARMAMENT.   Term used by *Johan Galtung* and others to signify a restructuring of the armed forces, occasionally in two stages, both of them referred to as T: the transformation of military defense from dual-purpose to strictly defensive armed forces without offensive capabilities, i.e., *NOD*; and the transformation of the defense posture from military to *nonmilitary*, i.e., *civilian-based defense* [Galtung 1984a; 1988; cf. 1981; 1983; 1984].

TRANSPARENCY REVOLUTION.   See *C^3I*.

TREATY OF GOOD NEIGHBORLY RELATIONS, PARTNERSHIP, AND COOPERATION.   Treaty signed by the *FRG* and the *Soviet Union*, 13 September 1990, as a corollary of the agreement on German unification, accomplished through the *Two-Plus-Four negotiations*. Besides a general commitment to cooperation in various fields, the treaty was tantamount to a nonaggression treaty, the scope of which extended to all of Europe because both states solemnly committed themselves to respect the territorial integrity of all European states. Further, the treaty contained what might be called an *NOD* clause: "The Federal Republic of Germany and the Union of Socialist Soviet Republics will seek to ensure that armed forces and armaments are substantially reduced . . . in order to achieve . . . a stable balance at a lower level . . . which will suffice for defense, but not for attack" [in Rotfeld & Stützle (eds.) 1991: 190–194, §4].

TROMP, HYLKE.   See the *Netherlands*.

TURKEY.   See *Balkans*.

TWO-PLUS-FOUR.   Unofficial name for the 1990 negotiations on German unification between the four occupation powers (i.e., the victors in *WWII*: the *U.S.*, the *U.K.*, *France*, and the *Soviet Union*) and the two German states, with *Poland* participating as an observer at some stages [Oeser 1990; Albrecht 1992; cf. Møller 1991: 222–232; Rotfeld & Stützle (eds.) 1991]. The proposals made in the course of the negotiations to some extent were inspired by the concepts that had originated in the alternative defense and security debate and that had by then become commonplace. The stipulations about the posture of the former *GDR* hence reflected the notions of *disengagement* that had become prominent in previous years. The final agreement was signed on 12 September 1990 and among other things stipulated the following for the new, enlarged *FRG:*

- That the Oder-Neisse border was final and unalterable (§1.1);
- That Germany would abstain from anything that might jeopardize peace, particularly preparation for aggressive war (§2), a stipulation already in the FRG Constitution [Lutz 1987d; 1989a; 1990a; 1992; 1993; 1993a–b];
- That Germany renounced its right to produce or dispose of ABC weapons (§3.1);
- That Germany's armed forces would not exceed 370,000 troops, to be formalized in the *CFE* treaty (§3.2);
- That Soviet forces in the former GDR would be withdrawn by the end of 1994, and that until that date the Western powers would retain a contingent of forces in Berlin (§4.1, §5.2);
- That until the end of 1994, no German forces integrated in *NATO* command structures or other NATO forces would be stationed in the former GDR (§5.1) (cf. *Genscher Plan*);
- That after 1994 Germany would be free to station NATO-integrated German forces in the former GDR, but without nuclear weapons or means of delivery; allied forces would, however remain prohibited (§5.3) [Rotfeld and Stützle (eds.) 1991].

The German commitment to the 370,000 troop ceiling was formalized in connection with the signature of the CFE Treaty, upon the entry into force of which it was made conditional [Rotfeld and Stützle (eds.) 1991].

U

UCS. Union of Concerned Scientists. U.S. organization founded in 1969 by scientists concerned about the nuclear *arms race*, numbering around 100,000 members. Its staff has included some of the most prominent *arms control* proponents, such as *Jonathan Dean.

The UCS has played an important role in the debate on a nuclear *freeze* [Ford, Kendall, & Nadis 1982]; the *SDI (the Strategic Defense Initiative, i.e., "Star Wars") [Tirman (ed.) 1984; 1986]; *no-first-use of nuclear weapons [UCS 1983], and nuclear disarmament to *minimum deterrence standards [Dean & Gottfried 1991]. Even though its focus has been on nuclear arms control, the UCS has also devoted some attention to conventional forces, both as a corollary of the recommended abandonment of first-use and as a topic in its own right [Dean 1987; 1989c; 1994].

U.K. United Kingdom. Because of its insular position, the U.K. has traditionally not been particularly concerned about national defense in the same sense as, e.g., the *FRG. There has rarely been any serious threat of invasion but only minor challenges (compared with the "existential" ones faced by Germany), mostly directed against its overseas possessions (e.g., in the Falklands/Malvinas War) and its standing in continental Europe.

1. **Issues.** The major issues in the British national-defense debate in recent years have been the following:

- *NATO strategy in general and the role of nuclear weapons in particular, inter alia, in connection with the international *no-first-use debate [Carver 1983a; cf. Brown & Farrar-Hockley 1985; Kelly 1988; Dunn 1991; Reid & Dewar (eds.) 1991; McInnes 1991; 1991a].
- The appropriate strategy for *arms control in general and nuclear and *naval arms control in particular [Laird & Robertson 1990; Dunn 1990; Hoffman 1991; Hill 1989].
- Threat assessment, including the interpretation of developments in the *Soviet Union, as a determinant of defense and security policy [Holden 1989; 1990a; 1991; Booth 1990a; Wheeler 1991].
- The role of the independent British deterrent in general and of the Trident program in particular [McInnes 1991b].
- The role (if any) of the U.K. in NATO, including the questions of the stationing of U.S. nuclear weapons in Britain and of the future of the BAOR (British Army of the Rhine). A small minority have advocated *neutrality; a larger number have favored a loosening of the ties to NATO; and even more have supported withdrawal of the U.S. cruise missiles from the U.K. [Rees 1991; P. Johnson 1985; Hill-Norton 1983; Greene 1983].
- The *out-of-area problem, and in particular whether to renounce overseas commitments entirely, and the related question of establishing priorities between the Royal Navy and the rest of the armed forces.

- The question of how much to pay for defense, including the problem of *conscription* and of the structure (and possible *conversion*) of the defense industry [Booth 1983; Chalmers 1985; 1991; 1992; D. Smith 1980: 156–180; Kaldor 1983; Croft & Dunns 1990; Laurent 1991; Southwood 1991].
- The metaquestion of who should decide defense and security issues, and how [Tatchell 1985; Elworthy & Ingram 1991; Oxford Research Group 1990].

2. **Political Parties.** The parliamentary defense debate became increasingly politicized in the 1980s and took place under greater public scrutiny than had previously been the case. This was, inter alia, due to the influence of the *peace movements*, especially the large CND (Campaign for Nuclear Disarmament) [Taylor & Young (eds.) 1987]. Because of the electoral system, however, it would always take a very long time before such changes in public opinion would be reflected in the composition of Parliament.

- The attitude of the Conservative Party (the Tories) under Prime Minister Margaret Thatcher was consistently negative toward alternative defense (including nuclear disarmament, no-first-use, etc.) and only moderately in favor of conventional arms control. The replacement of Mrs. Thatcher with John Major led to a moderation of the rhetoric, but there have been no signs of a fundamental change of attitude. However, because the leadership change coincided with the international thaw as well as with towering economic problems, the Tory government was forced to embark on a certain disarmament. The force levels were lowered, and the anticipated saving more substantial in the government's *Statement on the Defence Estimates for 1991* than envisioned in the *Options for Change* plan of July 1990. Particularly, the U.K.'s commitment to the defense of the European mainland was to be affected by the planned cuts [Ramsbotham 1991].
- The party that has taken the greatest interest in alternative defense and *NOD* has, beyond comparison, been the Labour Party, whose shift in the early 1980s to a policy of unilateral nuclear disarmament (subsequently abandoned in favor of the arms control path to nuclear build-down) necessitated a review of conventional defense policy as well. Even though the party in 1984 committed itself to a policy of "*defensive deterrence*" (i.e., nuclear disarmament and *NOD*-type conventional forces), the electoral defeats apparently made Labour uneasy about its "courageous" defense policy. Because of its vacillations, Labour became the main addressee of the independent alternative defense proponents. Even though the policy of unilateral nuclear abolition was subsequently abandoned in favor of support for a negotiated build-down to zero, Labour has continued to support no-first-use, an immediate scrapping of tactical nuclear weapons, and, as a long-term objective, complete nuclear abolition. The attitude of Labour toward Europe has reflected a division of the party between "Atlanticists" and "Europeans," the latter constituting the "left," who have tended to think not merely in terms of the EC and/or the WEU but also of pan-European forms of cooperation [Labour Party 1984; cf. Gapes 1988; 1988a; George & al. 1991; Ramsbotham 1991].
- The Liberal Democrats (and before them the SDP–Liberal Alliance [SDP & Liberal Party 1986]) have taken some interest in the alternative defense options and frequently used terms such as *common security* and *non-provocative defense*, just as they supported nuclear build-down to *minimum-deterrence* standards. The LDP has, furthermore, probably been the

party most favorably inclined toward *Europeanization*, based on the WEU as well as the EC [Ramsbotham 1991: 30–31].

- The Green Party has remained even more marginalized than the LDP, notwithstanding a percentage of the votes for the European elections of 15 percent. It has been somewhat less pacifist than most of its continental sister parties and has unreservedly acknowledged the need for an adequate defense, albeit nonprovocative and nonnuclear. It shares with the other *Green parties* the conviction that the military *alliances* should be dismantled in favor of a pan-European security system [Ramsbotham 1991: 31].

3. **Independent Groups.** Besides the political parties, and the peace movement(s) as such, a large number of research groups, foundations, and the like have taken an interest in alternative defense or aspects thereof:

- Most influential was probably the Alternative Defence Commission (see *ADC*) in the early 1980s [ADC 1983; 1987; Randle & Rogers (eds.) 1990]. It devoted considerable attention to, and made concrete recommendations for, a shift to NOD, as far as the U.K. was concerned, as well as with regard to Europe as a whole, in both cases with the Labour Party as the main addressee. The ADC was succeeded by the Alternative Security Working Group, which included *Frank Barnaby*, Frank Blackaby, and others [Alternative Security Working Group 1989; Blackaby & al. 1989]. It was considerably less influential than its predecessor.
- *Just Defence* (featuring Frank Barnaby as chairman and, until his death in 1991, *Peter Smith* as secretary) has also appealed primarily to the Labour Party, with a certain but far from decisive success [Barnaby 1986; 1990; idem & Boeker 1982; 1986; 1988; idem & Borg 1986; various issues of the *Just Defence Newsletter;* P. Smith 1990; 1991; cf. idem & Gapes 1984].
- The *Foundation for International Security* has been significantly less "alternative" than *Just Defence* and has appealed more broadly to the other parties, not least the SDP and the Alliance, subsequently the LDP [Grove 1987; 1988; 1990; 1990a–b; Windass 1985; 1985a–b; 1987; 1987a; idem (ed.) 1985; idem & Grove (eds.) 1987; 1988].

Besides the above independent groups whose work has been devoted mainly to alternative defense, at least two other groups have influenced the debate, albeit more indirectly:

- The Oxford Research Group, featuring among others Scilla Elworthy (formerly McLean) and Oliver Ramsbotham, has dealt extensively with decisionmaking processes, primarily pertaining to nuclear weapons but including attitudes toward NOD [McLean (ed.) 1986; Oxford Research Group 1990; Elworthy & Ingram 1991; Ramsbotham 1989; 1991].
- The Saferworld Foundation has dealt with subjects such as European and global security after the Cold War, the structure of the defense industry, British defense requirements for the coming decade, and the implications for Western security of Soviet reform. Most recently, it has devoted considerable attention to *arms trade regulations* [Saferworld Foundation 1990; 1990a–e; 1991; Anthony, Eavis, & Greene 1991].

The *peace movements,* including SANE, the CND, and the END, have also periodically employed researchers (albeit sometimes with the title "information officers"). They have included specialists in alternative defense such as Sheena Philips [1990], *Philip Webber* [1990], and *Steven Shenfield* [1984; 1985; 1987; 1988].

4. **University Research.** In addition to the above (partly university-based) research groups, a few British universities might claim a certain research tradition in the field of alternative defense and security.

- The University College of Wales in Aberystwyth, where particularly *Ken Booth* and John Baylis have taken an interest in defense alternatives [Booth 1983; 1985; 1987; 1987a; 1990; 1990a; 1991; 1991a–c; idem & Baylis 1988; Baylis 1987; idem (ed.) 1983].
- The Bradford School of Peace Studies, which has employed, at one time or another, e.g., Alan Bloomgarden, *Malcolm Chalmers*, Paul Rogers, *Michael Randle*, Oliver Ramsbotham, and Owen Greene [Bloomgarden 1989; 1990; Chalmers 1984; 1985; 1989; 1990; 1990a; idem & Stevenson 1989; idem & Unterseher 1987; 1988; 1989; Greene 1988; 1990; Randle 1986; 1990; Rogers 1988; 1990; 1991].
- The University of Sussex, in particular the Science Policy Research Unit (SPRU), which has employed, e.g., *Mary Kaldor* [1981: 226–230; 1983; 1985; 1986]; and *Gerard Holden* (subsequently at the Peace Research Institute Frankfurt) [1989; 1990; 1990a–b; 1991; 1991a–b]. The Arms and Disarmament Information Unit at the same university for a number of years issued the *ADIU Reports,* in which a number of concrete NOD proposals were published for the first time.
- The University of Cambridge has fostered the Cambridge University Disarmament Seminar [Prins (ed.) 1983]. A partial successor to this has been the Global Security Programme, directed by *Gwyn Prins* [1984; 1986; 1988; 1988a; 1990; 1990a–c; 1991; 1991a], but involving also the former Brookings Institution fellow *Michael MccGwire* [1984; 1986; 1987; 1987a–b; 1988; 1988a-b; 1989; 1991; 1991a]. The former *SIPRI* director, and one of the most prominent theoreticians of alternative defense, *Robert Neild* also has the status of professor emeritus of Cambridge [1981; 1984; 1986; 1986a; 1989; 1989a].
- In addition to its forming the basis for the aforementioned Oxford Research Group, Oxford University has also provided the institutional basis for the former *civilian-based defense* and subsequently *territorial-defense* proponent *Adam Roberts* [1967; 1967a; 1969; 1970; 1976; 1977; 1978; 1983; 1983a; 1989; 1991]. Furthermore, it has held a chair for the prominent military historian Michael Howard, who has studied both *disengagement* and the dangers of offensive strategies but has remained skeptical with regard to NOD [1958; 1985; 1986; 1990; 1990a].
- Finally, King's College in London includes the Department of War Studies under the leadership of Professor Lawrence Freedman, whose studies on the *offense/defense distinction* and nuclear strategy rank among the most important [1984; 1987; 1988; 1989; 1991]. Also on the staff is former SIPRI researcher *Jane Sharp*, a long-standing NOD proponent [1984; 1985; 1987; 1987a; 1988; 1989; 1989a; 1990; 1990a–b; 1991; 1991a].

Besides the persons already mentioned, there have been quite a few (more or less wholehearted) supporters of NOD at various academic institutions all over the U.K., including

- Ian Bellany at the University of Lanchester [1990; 1991];
- *Barry Buzan* at the University of Warwick (and the *Centre for Peace and Conflict Research* in *Denmark*) [1987; 1987a; 1989];
- Michael Clarke at the Defence Studies Centre, London University [1985a; 1988; 1989; 1991]; and
- *Adrian Hyde-Price* at the University of Southhampton [1991; 1991a].

5. **The Military.** The armed forces as an institution have taken only little interest in defense alternatives, even though the Royal Military Academy at Sandhurst has fostered a few able critics of NOD, such as Charles Dick [1986; idem & Unterseher 1986]. Ministry of Defence lecturer at the University of Aberdeen, *David Gates*, has authored the most detailed critique of NOD to date [1991; cf. 1987; 1987a; 1990]. Among retired officers, however, there have been quite a few instances of support for NOD, no-first-use, and nuclear disarmament, most notably Field Marshall Lord Michael Carver [1982; 1983; 1983a], Michael Harbottle [1988; 1988a; 1990], Lord Cameron [1983], and General Hugh Beach [1985; 1986; 1988; 1990].

UHLE-WETTLER, FRANZ.   Lieutenant general (ret.) who ended his career as commandant of the NATO Defence College in Rome.

FUW has been one of the most prominent *infantry* advocates within the German armed forces. After having proposed a greater reliance on infantry in 1966, he elaborated upon this theme in 1980 in *Battlefield Central Europe* [Uhle-Wettler 1980; cf. 1966]. What drove FUW (with a background in the armored forces) to advocate the use of light infantry was primarily his concern over the deteriorating "teeth-to-tail ratio," i.e., the declining number of fighters in the armed forces compared with the growth of the rear services. This had adverse effects on mobility and resulted in exacerbated vulnerability because entire formations might be rapidly put out of action through attacks on the logistical chains [1980: 19–21, 27–36]. The "overtechnologizing" (*Übertechnifizierung*) of the armed forces also hampered adaptation to the specific German geography: extensive forests and a high degree of urbanization (and an even swifter growth in suburban sprawl) meant that a war in Germany would to a large extent take place in villages, woods, or mountains. For such an eventuality, mechanized forces were ill-adapted; only light infantry were really suitable—and ideal for taking full advantage of the potentials of antitank missiles [1980: 25, 59–62, 78, 87; 1990: 128–129; cf. Bracken 1976]. That the Bundeswehr had not exploited these obvious opportunities, FUW attributed to the "legacy of Guderian," i.e., to a predilection for the offensive and the ensuing insistence on the complete mechanization of the army, intended to provide the armed forces with "capability of wide-range offensives and large-scale maneuver warfare."

According to FUW, the defense needed both mechanized forces for combat in open, i.e., tank-friendly, terrain, and light infantry divisions for closed terrain. The light forces would compensate for their lack of mobility by *dispersion* and, furthermore, seek to force the invading forces to dismount and engage them on favorable terms, i.e., where the aggressor's long-range fire and mobility would be of small avail [1980: 61, 72–73, 93–94, 103, 107; cf. 1990: 126–127]. The appropriate tactics for such combat resembled *guerrilla warfare* (as well as the ideas of *Horst Afheldt*): the infantry would withdraw in the face of a superior opponent, "luring him into the depth" of the defended territory only to strike at his *logistics*. For the sake of (tactical) mobility, the infantry should be genuinely light, both in terms of weaponry and means of transportation, even though trucks, jeeps, and helicopters would be needed for some purposes [1980: 116–118, 120–123].

The envisioned posture would be personnel intensive and hence might require an increase in troop levels, but thanks to the diminished offensive capabilities, FUW did not expect this to cause serious security political problems [1980: 143–146].

UKRAINE.   See *CIS*.

ULLMAN, RICHARD H.   See *U.S.*

U.N.   The United Nations, founded in 1945 as the successor to the *League of Nations,* with the primary purpose of preventing a repetition of the two world wars. Because of the more realistic distribution of influence in the U.N. than in its predecessor, its accomplishments have been more in line with its ambitions.

The roles of the U.N. with relevance for alternative defense and security have included the following:

- Providing a setting for an international policy debate that may have contributed to greater empathy among nations, even though it has also implied the (ab)use of the U.N. General Assembly as a forum for propaganda.
- Commissioning of several studies that have explored central concepts on the security political agenda, such as North-South relations [Brandt Commission 1980], *common security* [Palme Commission 1982], environmental security [Brundtland Commission 1987], and *defensive security.*
- Contributing to conflict resolution through the provision of good services: mediation, arbitration, etc.
- *Peacekeeping*, through the dispatch of forces assigned by member countries to the U.N. They have been used for monitoring and implementing armistice agreements between warring parties, thereby paving the way for a more durable peace settlement, such as has happened in, e.g., Congo, the Middle East, Cyprus, and, most recently, *Yugoslavia* [Rikhye & Skjelsbaek (eds.) 1990; Goulding 1993; Morphet 1993; Roper & al. 1993; Berdal 1993].
- Imposing sanctions (e.g., against South Africa, Rhodesia, and, most recently, Iraq and Serbia) [Hufbauer & al. 1990].
- In a few cases (the Korean War and the *Gulf War*) armed forces have actually fought a war on behalf of the U.N., or at least under a U.N. flag.

Especially in the post–Cold War period, there have been numerous proposals for expanding the role of the U.N. to a full-fledged *collective security* system [e.g., Prins 1991; 1991a; cf. Gati 1991]. Central in this debate has been the report by Secretary-General Boutros Boutros-Ghali, *An Agenda for Peace* [1992], which introduced new terms into the U.N. political discourse such as *peace-building*, *preventive diplomacy*, and *peacemaking*.

A precondition for the last, especially the envisaged element of *peace enforcement*, might imply a more or less permanent assignment of armed forces to U.N. command (as envisioned in the *U.N. Charter* and proposed in *An Agenda for Peace*) and creation of a functioning military staff (the currently largely ineffective Military Staff Committee) [cf. Boulden 1993; Grove 1993]. Furthermore, it might necessitate creation of U.N. naval task forces as well as air forces at the disposal of the U.N. [cf. Mackinlay 1993; Connaughton 1992; 1992a; Urquhart 1991; 1993; 1993a; Allen & al. 1993; Isaacs 1993; Berdal 1993; Kühne 1993; Renner 1993; Prins 1991; 1991a; Staley 1992].

U.N. CHARTER.   Charter of the United Nations (see *U.N.*), adopted in 1945 [in Roberts & Kingsbury (eds.) 1993: 499–529]. In additon to determining the basic structure of the organization, the UNC has also served as one of the most authoritative international legal documents on the *laws of war* in general, and the prerequisites of *just war* in particular. The UNC proscribes recourse to force in international relations (as did the Kellogg-Briand Treaty) with a few explicit exceptions. Recourse to force is thus legal only for (individual or collective) *self-defense*, or in connection with attempts at restoring peace under U.N. auspices.

The UNC describes a set of actions to secure peace and security, ranging from nonmilitary, noncoercive means through sanctions to *peace enforcement*. The last is described in Chapter VII, especially Article 42: "Should the Security Council consider that measures provided for in Article 41 [sanctions, etc., BM] would be inadequate or have proved inadequate, it may take such action by air, sea or land forces as may be necessary to maintain or restore international peace and security. Such action may include demonstrations, blockade, and other operations by air, sea or land forces of Members of the United Nations" [in Roberts & Kingsbury (eds.) 1993: 511]. Because of the deadlock in the Security Council during the Cold War, "Article 42 operations" were very rarely resorted to. The U.N.–authorized action by the coalition against Iraq in the *Gulf War was, however, explicitly undertaken with reference to Article 42 [SR Resolution 678 (29 November 1990), in Weller (ed.) 1993: 6]. This presumably set a precedent for *collective-security actions under the auspices of the U.N., as originally conceived. The prospects and implications were set out in Secretary-General Boutros Boutros-Ghali's report *An Agenda for Peace [1992].

UNIDIR.   The United Nations Institute of Disarmament Research, located in Geneva, which has played a prominent role in sponsoring and coordinating research on *NOD [Graham 1989; UNIDIR (ed.) 1990a]; conventional *arms control [Banner & al. 1990; Ghebaldi 1989; Kokoyev & Androsov 1990; Scherbak 1991; Sur (ed.) 1991; UNIDIR (ed.) 1989; 1990]; *conversion [UNIDIR (ed.) 1991]; and the role of the *U.N. [Dhanapala (ed.) 1991]. Its present director is *Sverre Lodgaard, a long-standing advocate of NOD.

UNILATERALISM.   Mode of implementation, inter alia, of the envisioned *transarmament to *NOD, and as such an alternative to negotiated, bilateral implementation.

Many NOD proponents, especially in the early years, saw U as the only realistic escape from the cul-de-sac of traditional *arms control. Through a unilateral shift to NOD, a nation would presumably improve its own *security, regardless of whether the adversary maintained an offensive stance, simply because NOD would presumably be more effective for defense purposes [e.g., Afheldt 1983]. However, assessed according to conservative standards, taking all conceivable countermeasures on the part of an opponent into account, most NOD models failed the test [cf. Hofmann, Huber, & Steiger 1986]. More often, therefore, recommendations for unilateral transarmament were accompanied by hopes that the opponent would reciprocate. If the adversary (in casu the *Soviet Union) were likewise trapped in a *security dilemma, then unilateral defensive restructuring would remove unfounded fears, thereby permitting the other side to embark upon a restructuring as well. The result might be a benign spiral of transarmament-cum-*disarmament, similar to the processes envisioned by *gradualism. Because this would be a two-sided process, it would take the form of "coordinated unilateralism," whereby the pitfalls of arms control negotiations could, it is hoped, be avoided [Neild 1986a; cf. George 1988b; Ramberg (ed.) 1993].

Since the breakthrough in East-West relations accompanying the "*Gorbachev revolution," however, many NOD proponents, including some former unilateralists, have come to assess the arms control approach as more promising. Accordingly, they have abandoned their quest for unilaterally implementable models in favor of concrete (and not always very radical or alternative) arms control proposals [e.g., A. Müller 1990a; 1990c; idem & Karkoszka 1988; 1990].

UNION OF CONCERNED SCIENTISTS.   See *UCS.

UNITED KINGDOM.   See *U.K.

UNITED NATIONS.   See **U.N.*

UNITED STATES.   See **U.S.*

UNTERSEHER, LUTZ.   Sociologist, codirector of **SALSS,* chairman of the **SAS,* military consultant to the **SPD,* and personal advisor to **Hermann Scheer.*

LU has written extensively on all aspects of **NOD,* with a special emphasis on the details of postural changes: weapons profile, numbers, deployment patterns, costs, etc. [1987c; 1988c; 1989b, d-e; 1990; 1994; Bebermeyer & Unterseher 1986; 1989; Chalmers & Unterseher 1987; 1988; 1989]. He has also been the main "architect" of the SAS's **spider-and-web* principle [1984; 1984a-b; 1986; 1987; 1987a-b; 1987d; 1988d; 1989b-c, f; 1990b; 1991b; Boeker & Unterseher 1986; Brauch & Unterseher 1984; Grin & Unterseher 1988; 1989; 1990; 1990a]. Whereas he had previously focused on the **FRG,* LU has since the end of the Cold War begun to pay greater attention to the possible application of the defensive principles to other regions, e.g., **Italy* and **Denmark* [1988a; Sørensen & Unterseher 1989; 1990], as well as to extra-European areas of conflict such as the Middle East, **Korea,* **India*/Pakistan, and South Africa [Conetta, Knight, & Unterseher 1991; 1991a].

U.S.[†]   The strongest and perhaps most secure country in the world, at least as far as external threats are concerned. Because of its size and inaccessibility, and thanks to the fact that the only neighboring states are weak and/or peaceful, the U.S. had no need to fear attack before the advent of the nuclear era, whereupon it became vulnerable to a Soviet missile attack [Nordlinger 1991; Evera 1989]. That a joint congressional committee in 1946 could describe the Japanese attack on Pearl Harbor (7 December 1941) as "the greatest military and naval disaster in our Nation's history" demonstrates that the security of the U.S. has been quite unprecedented [Mueller 1991].

1. **Defensive Tradition.** Because of its near-ideal degree of **security,* the U.S. has all along had a different attitude toward war than most other states. Whereas most other states have had war imposed upon them, the U.S. has usually been able to chose which wars to participate in, and hence has had a more voluntaristic attitude toward war.

There has, nevertheless, been a strong defensive tradition in U.S. military thinking and practice ever since the War of Independence (1775–1783) [Quester 1989b]. That war and the Civil War (1861–1865) were characterized by a predominance of defensive operations and defensive postures [Hagerman 1988; A. Jones 1988: 409–418]. On the other hand, neither the war with Mexico over Texas (in reality over California) in 1845–1846 nor the overt imperialism in the last decades of the nineteenth century were defensive at all.

The voluntaristic attitude toward war also furthered the belief that war could be avoided, presupposing only the will to do so, hence the strong tradition of idealism. The best-known manifestation of this was the Wilsonian idealism that gave rise to the **League of Nations.* Another manifestation was the **disarmament* attempts of President Hoover in the interwar years, including the quest for banning **offensive weapons* in the contexts of the **Washington Naval Treaty* and the **World Disarmament Conference* [Baker 1923; Cronon (ed.) 1965; Foley 1923; Johnston

---

†Because of the size of the United States, and the richness of its security political debate, the following survey is even more sketchy than those of most other countries.

1982; Levin 1972; Hoover & Gibson 1943; cf. Buckley 1970; Bull 1973; R. Hoover 1980; Kaufman 1990; E. Müller 1986; Noel-Baker 1979].

2. **Issues.** In the post-WWII period, U.S. strategic debate has been characterized primarily by the new and unprecedented vulnerability stemming from *nuclear deterrence*. Since the failure of the initial attempts at achieving a complete ban on nuclear weapons (the Baruch Plan of 1946 [Myrdal 1976: 72–78; Newhouse 1989: 63–64]), the previous idealism has largely been superseded by *Realism*. This implied, inter alia, a preoccupation with managing rather than transcending deterrence (the *SDI* constituted a noteworthy exception) [Miller & Evera (eds.) 1986; Tirman (ed.) 1984; 1986; C. Snyder (ed.) 1986; McNamara & al. 1985; Stockton 1986].

Prominent issues throughout the postwar period with some relevance for the alternative defense debate have included the following:

- The relative importance of Europe and the Third World [Evera 1990a; David 1989; Walt 1989; Desch 1989; Johnson 1989; Art 1991], one manifestation of which has been the debate on the pros and cons of interventionism [Klare 1981; Epstein 1981; 1983; Forsberg 1987; Evera 1990b; Schraeder (ed.) 1992]. Another has been the controversy over a possible *disengagement* of the U.S. from Europe [Krauss 1986; Ravenal 1978].

- The appropriate priority between land and naval power, as manifested in the debate on the *Maritime Strategy* [Watkins 1986; Kelley 1986; Brooks 1986; Friedman 1988; cf. Mearsheimer 1986].

- The best strategy for a land battle, with the debate on the *AirLand Battle* doctrine as the best example from the 1980s [FM 100-5 1982; cf. Czege 1984: 105–108; Lindner 1984; Mallin 1990: 234–239; Garrett 1989: 45–47; Bellamy 1990: 131–138; Sutton & al. 1984; Strachan 1984: 34–38; Richardson 1986].

- The possibility of substituting technological cunning for muscle and manpower, as illustrated by the *FOFA* doctrine [Rogers 1983; 1985; Berg & Herolf 1984; OTA 1987; Cardwell 1986; Farndale 1988; Wijk 1986; 1986a; 1989; Flanagan 1986; 1987; Abshire 1987; cf. Tegnelia 1985; Canby 1985; 1985a; Herolf 1986a; 1988; OTA 1987; Burgess 1986; Dugan 1990; Sharfman 1991; Skingsley 1991], the *CDI* and other debates on *E.T.* (emerging technologies) [Flanagan 1986; 1987; Abshire 1987; cf. Tegnelia 1985; Canby 1985; 1985a; Berg & Herolf 1984; Herolf 1986a; 1988; OTA 1987]. In all of these a certain faith in technological fixes has been noticeable.

- Since the end of the Cold War, a prominent theme has been that of a New World Order [cf. Buzan 1994] and the role of the U.S. therein [Haas 1993; Herrmann 1991; Keohane & al. (eds.) 1993; Mead 1993; Rochester 1993; Rubinstein 1991; Singer & Wildawsky 1993; Steinbruner 1991; Art 1991].

3. **Political Debate.** There has been direct support from neither of the two major political parties for any of the alternative defense conceptions. However, there has been piecemeal endorsement of central elements by prominent politicians and/or former political advisors such as Les Aspin [1987] and Zbigniew Brzezinski [1988]. Most prominent among such political (or "ex-political") personalities has been the so-called 'Gang of Four': former SALT negotiator Gerard Smith, former national security advisor McGeorge Bundy, former secretary of defense Robert S. McNamara, and former ambassador to the Soviet Union *George Kennan*. Jointly as well as individually, the "gang" published a very influential appeal for *no-first-use* of nuclear weapons as well as various critiques of the direction in which nuclear

strategy seemed to be heading, including an influential critique of the SDI program [Bundy 1986; 1990; idem & al. 1982; 1993; Kennan 1982; McNamara 1986; 1989; idem & al. 1985; Kaysen & al. 1991].

Furthermore, some of the research entities serving Congress have taken quite seriously some of the themes characterizing the European *NOD debate, above all the Congressional Research Service [Sloan 1986]. In 1988, at the initiative of Representative Solarz, it prepared a draft proposal for the *CST talks, wherein it envisioned the creation of "exclusion *zones" from which particularly offensive equipment such as *tanks, self-propelled artillery, *bridge-building equipment, and major ammunition and fuel storage depots would be removed by both sides [*Arms Control Reporter,* 407.C.1–2].

Throughout the 1980s, the political debate was influenced by the activities of quite strong *peace movements, above all *Freeze and SANE [Solo 1988; Conetta (ed.) 1988].

4. **Alternative Defense Doves.** Among the most direct supporters of alternative defense have been groups and institutions with their roots in, or otherwise close to, the peace movements:

- One of the first examples is the Boston Study Group in the late 1970s, the members of which included the directors of the two following institutes, Randall Forsberg and Paul Walker [Boston Study Group 1979].
- The *IDDS has tended to regard itself almost as a research and information resource of the peace movements, inter alia, because its director has been Randall Forsberg, who launched the Freeze movement [1984; 1985; 1986; 1987; 1987a–b; 1988; 1988a; idem & al. 1983; 1991].
- The *IPIS was founded by two other members of the Boston Study Group, Pam Solo and Paul F. Walker, and it has likewise emphasized links with the peace movements [Solo 1988; Walker 1981; 1985; 1986].
- The *Commonwealth Institute, in its turn, was in a certain sense an off-spring of the IDDS, where both Charles Knight and Carl Conetta had previously worked [Conetta 1991; idem & Knight 1990; 1990a–b; 1991; 1991a; 1992; 1993; idem & al. 1991; 1991a; Knight 1988].
- The Union of Concerned Scientists (*UCS) is actually a peace movement in its own right, with over 100,000 members. In addition to its critical analyses of the SDI and related subjects, the UCS has also made several NOD-inspired sets of recommendations for conventional restructuring [1983; Ford & al. 1982; Tirman (ed.) 1984; 1986], not least thanks to its arms control advisor *Jonathan Dean [1985; 1985a; 1987; 1988; 1988a–b; 1989; 1989a–f; 1990; 1990a–k; 1991; 1992; 1992a; 1993; 1994; idem & Resor 1990].
- The *Alternative Defense Project has also, at least ideologically, been very close to the peace movement, as was its predecessor EXPRO (Exploratory Project on the Conditions of Peace) [Hollins & al. 1989; cf. Hollins 1982; Sommer 1983; 1984; 1985; 1986; 1988; 1989]. Its leading members have included Harry Hollins, Averill Powers, and Mark Sommer, but persons like Randall Forsberg and *Dietrich Fischer [1982; 1984; 1984a; 1988; 1988a; idem & al. 1987] have also been involved at some stage.
- Also partly affiliated with the peace movement (at least ideologically) have been various convinced nuclear abolitionists, such as *Jonathan Schell [1984] and *Freeman Dyson [1984].
- Finally, the U.S. (like most other countries) has always had its *civilian-based defense constituency, the most prominent figure in which has been *Gene Sharp, residing in his own institute, the Albert Einstein Institution,

in Cambridge, Massachusetts [1967; 1973; 1978; 1979; 1985; 1985a; idem & Jenkins 1990; 1992].

5. **Nuclear Strategy "Owls."** Whereas all of the above could be ranked as doves in opposition to the dominant hawks of the strategic studies community, the intellectual field in the U.S. has also encompassed a large category of "owls," i.e., moderate critics and skeptics demanding no radical changes in existing strategies.

- Ever since the 1960s, there have been advocates of *minimum deterrence* within the scientific community [Szilard 1964]. Contemporary advocates of this approach to reversing the nuclear arms race have included Harold Feiveson and Frank von Hippel, who have also been involved in cooperative projects with their Soviet counterparts. Von Hippel has also participated in the *Pugwash* Study Group on Conventional Forces in Europe [Feiveson 1989; idem & Hippel 1990; 1990a; idem, Ullman, & Hippel 1985; Hippel & Sagdeev (eds.) 1990].
- A more moderate version of the same quest for stability has been represented by the self-proclaimed "owls" at Harvard (Graham Allison, Albert Carnesale, Joseph Nye, and others), whose main preoccupation has been with *crisis stability* and *arms race stability*, inter alia, with a view to preventing inadvertent war. In this connection they have assessed conceptions such as NOD for their possible effects [Harvard Nuclear Study Group 1983; Allison & al. (eds.) 1985; Nye & al. (eds.) 1988; Allison & Ury (eds.) 1989].
- The prevention of inadvertent nuclear war has also been the preoccupation of various research projects on crisis stability, such as those directed by Desmond Ball (now in *Australia*) and others [Ball & al. 1987], Paul Bracken [1976; 1983; 1983a; 1987; 1987a; Gottfried & Bracken (eds.) 1990], and William Ury [1985], as well as, of course, the founding father of this "discipline," *Thomas Schelling* [1960; 1966; 1967; 1986; idem & Halperin 1961]. To some extent, all of these authors have also taken into account the risks of conventional instability spilling over into the nuclear realm. Furthermore, it was to a large extent the results from this type of research that inspired the first European NOD models.

6. **Conventional Deterrence "Owls."** The continuous debate on conventional deterrence reflects the same interest in stability.

- In the forefront in this debate has been the journal *International Security,* the editorial staff and/or panel of which have included such analysts as *Stephen Van Evera* [1984; 1985; 1985a; 1990; 1990a–b; idem in Feldman (ed.) 1990: 95–103], *Jack Snyder* [1984; 1985; 1986; 1987; 1990; 1990a; 1991]; Sean Lynn-Jones; *John J. Mearsheimer* [1979; 1981; 1982; 1983; 1983a; 1986; 1986a–b; 1988; 1988a; 1989; 1989a; idem & al. 1989]; and Barry Posen [1982; 1984a–b; 1988; 1991; 1992; idem & Evera 1983]—all of whom have dealt at length with the implications of offensive and defensive strategies, both from a historical perspective and with regard to the East-West conflict. The same holds true for former editor Steven Miller, who was reappointed to the post in 1992 after a tenure at *SIPRI*. Besides conventional stabilization generally, he has also devoted some attention to naval stabilization, pointing to the potential of defensive strategies, yet with the caveat that their implementation is unlikely [1990; 1990a; 1991].
- The Brookings Institution in Washington, D.C., has also given considerable attention to conventional forces and their potential for supplanting, wholly or partially, nuclear deterrence. Brooking's no-first-use study from 1983

(by John Steinbruner, Leon Sigal, William Kaufmann, and others) was quite influential in the debate, also by virtue of its drawing attention to the possibilities of improving NATO's conventional defenses by means of *barriers,* etc. [Steinbruner & Sigal (eds.) 1983; Sigal 1983; 1983a–b; Steinbruner 1983; Kaufmann 1983]. Further, Richard Betts has dealt extensively with the risks of, and possible ways of averting, *surprise attacks,* in which connection he has pointed to defensive restructuring [1982; 1985; 1985a–b]. In light of the waning of the Cold War and Soviet *New Thinking,* Brookings analysts such as Kaufmann, Steinbruner, *Joshua Epstein,* and *Michael MccGwire* have explored the opportunities of forging a new security system, thereby allowing for a substantial build-down of U.S. forces and military expenditures [Epstein 1990; cf. 1985; 1988; 1989; Kaufmann 1990; 1990a; MccGwire 1987; 1987a–b; 1988; 1988a–b; 1989; 1991; 1991a; Sigal 1987; Steinbruner 1988; 1990]. Finally, resident fellow *Raj Gupta* has argued the case for NOD on the basis of "military geometry" [1993].

- Besides hosting prominent researchers from the East such as *Andrzej Karkoszka* and *Pal Dunay,* the Institute for East-West Security Studies in New York has focused on conventional arms control in general and the opportunities provided by Soviet rethinking in particular [Blackwill & Larrabee (eds.) 1989; Larrabee 1984; 1988; idem & Lynch 1987].

- Stephen Flanagan is among the critics of the nuclear emphasis in NATO strategy and has on some occasions written about NOD, albeit from a critical vantage point [1986; 1987; 1988; 1988a; idem & Hamilton 1988; idem & Hampson (eds.) 1986].

7. **Political Science.** The U.S. political science community is by far the largest in the world. Even though it has traditionally been dominated by "hard-nosed" adherents of Realism, to a growing extent its members have begun to focus on cooperative security political strategies, in which connection they have also begun to discover the potentials of NOD and *common security.*

- Numerous political scientists have taken an interest in the (partly hidden and/or implicit) elements of cooperation in the Cold War conflict, e.g., in the form of *regimes* [Krasner (ed.) 1982; Kanet & Kolodziej (eds.) 1991; cf. Keohane & Nye 1977]. The most noteworthy study in this regard is the monumental volume edited by Alexander George, Philip Farley, and Alexander Dallin, *U.S.–Soviet Security Cooperation* [1988; idem (eds.) 1988; cf. George 1988a–c; 1989; 1990].

- Others have more directly focused on how to achieve cooperation under anarchy and between states that are fundamentally selfish, i.e., how to solve or somehow transcend the *security dilemma.* Notable among these scholars are the "discoverer" of *Tit-for-Tat,* *Robert Axelrod* [1979; 1984; 1990; idem & Keohane 1985], and the classical proponents of *gradualism,* *Charles Osgood* [1959; 1962; 1986] and *Amitai Etzioni,* as well as their more recent successors *Joshua Goldstein* and *John Freeman* [Goldstein & Freeman 1990].

- Other political scientists have arrived at partly the same conclusions through a study of perceptual biases and analyses of the inherent contradictions in deterrence. Most prominent in this category have been *George Quester* [1977 (2d ed. 1988); 1985; 1986; 1986a–b; 1989; 1989a–c; 1990; 1990a; 1991; 1991a]; *Robert Jervis* [1976; 1978; 1982; 1985; 1988a; 1991; 1991a–b; idem & Snyder (eds.) 1991]; and Richard Ned Lebow [1981; 1983; 1984; 1985; 1985a; 1989].

- Some, from the vantage point of Realism or neorealism, have sought to sketch different futures for Europe and the world in an "architectural mode," so to speak. Such analyses have included those of John Mueller (also known as a proponent of the *Totally Destructive Conventional War hypotheis* albeit from a different vantage point than that of the eastern proponents), who has recommended establishment of a *concert* structure for post–Cold War Europe [1988; 1989; 1990; 1991] and Richard Ullman's *Securing Europe,* which contained explicit recommendations for shifting to an NOD structure [1991: 72–74 & passim; cf. 1985].

- Finally, there is the most "eccentric" part of the political studies community, namely, those who have rejected Realism as unrealistic and subscribed to some version of idealism. Most of these scholars have also regarded themselves (rightly) as peace researchers. The group has included proponents of World Order, such as the offspring of the old Federalist movement; the World Policy Institute with, e.g., Richard Falk [Falk 1975; 1987; 1989; 1991; Lifton & Falk 1982; Falk & Mendlowitz (eds.) 1973], Michael Lucas [1988; 1990], and Bruce Weston [1990]; Robert Johansen of the Institute for Peace Studies, University of Notre Dame [1990; 1991a–c]; Michael Klare of the Five College Program in Peace and World Security Studies [1984; idem & Thomas (eds.) 1991]; and Richard Smoke and his colleagues of the Center for Foreign Policy Development at Brown University [Smoke & Kortunov (eds.) 1991]. All of these scholars have envisioned the development of cooperation to the point of nearly substituting a rule of law for international conflict, and many of them have deemed alternative defense strategies useful for the transitionary period. What has distinguished this group has been its broad conception of *security* as encompassing economic, cultural, and ecological security.

8. **Strategic Critics.** In addition to the authors mentioned so far, who are primarily political critics of (or skeptics with regard to) military policies, there have been numerous genuine military critics, both civilians and military officers.

- The U.S. is one of the few countries in the world that has a large civilian strategic studies community as well as private consultants; the latter tend to be temporarily on the Pentagon's payroll. Some of these have recommended military reforms that incorporate certain elements from NOD schemes. This group has included, among others, *Steven Canby* [1972; 1975; 1979; 1980; 1981; 1982; 1983; 1984; 1985; 1985a; 1986; 1986a; 1987; 1989; 1990; 1994], who has also advocated NOD-like reforms for broader security political reasons, and *Edward Luttwak,* who has argued against a wholesale shift to a defensive strategy, while recommending a greater use of NOD-like tactics [1980; 1984; 1987; 1991]. The same has been the case for *James Garrett* [1989; 1990] and Gary Guertner of the U.S. Army War College [1982; 1990]. Some "armchair strategists" such as Colin S. Gray (now back in England) have, on the other hand, criticized all defensive ideas, both the alternative ones and official strategy, as being far too cautious [1979; 1990; 1991],

- To an even larger extent than most other countries (except perhaps the *FRG*) the U.S. has seen a great number of retired officers, including both generals and admirals, coming out against official strategy upon retirement. Some of these have been affiliated with the international association *Generals for Peace and Disarmament* [cf. Harbottle & al. 1984]; others have been temporarily engaged by the peace movements. The latter include Vice Admiral John Marshall Lee, who was *primus motor* behind the UCS's no-first-use report [1988; cf. UCS 1983]. Still others formed the Washington-based Center for Defense Information, which has been issuing critical analyses of U.S. defense policy as well as making concrete recommendations for alternative postures [La Rocque 1988; 1989].

USSR. See *Soviet Union*.

V

VALKI, LAZLO.   See **Hungary*.

VAN CREVELD, MARTIN.   See **Creveld, Martin Van*.

VAN EVERA, STEPHEN W.   See **Evera, Stephen W. Van*.

VÄYRYNEN, RAIMO.   See **Finland*.

VERIFICATION.   See **CBM*.

VIENNA SEMINAR(S).   High-level consultations on military doctrines held under the auspices of the **CSCE* as an element of the CSBM negotiations. The origin of the VS was the **Warsaw Pact*'s 1987 **Berlin Declaration*, in which **NATO* was challenged to such consultations. The first VS was held in January–February 1990; the second in October 1991 under the auspices of the newly established Conflict Prevention Centre. In the course of the two VS, all participating states sought to portray their military conceptions as entirely defensive [Hamm & Pohlman 1990; 1990a; Krohn 1991b].

VIETNAM WAR.   Probably the most protracted **guerrilla war* in history, lasting from **WWII* to 1973, proceeding through several stages.

1. **Stages.** The first stage (1941–1946) commenced with the Vietminh's resistance against interchanging Japanese and French rule in Indochina during WWII, in the course of which they enjoyed some assistance from the **U.S.* [Maclear 1981: 1–26]. The second stage (1946–1954) featured an anticolonial struggle against **France*, until the French withdrew and Vietnam was temporarily divided at the Geneva conference in 1954. Particularly during the last years, the U.S. became increasingly involved through its assistance to France. The third stage (1961–1973) was characterized by increasingly intense combined guerrilla and regular warfare against the Americans and their allies in South Vietnam waged by the Vietcong (National Liberation Front) with the assistance of North Vietnam (Democratic Republic of Vietnam). The final stage (1973–1975) lasted from the U.S. withdrawal in 1973 until the victorious offensive in South Vietnam.

2. **Characteristic Features.** Throughout most of the war, guerrilla warfare was combined with more regular military operations waged by the North, which served as a large-scale guerrilla base area. It was connected with the South by a long series of tracks, the famous Ho Chi Minh Trail. **Logistics* generally favored the indigenous forces, the guerrillas. The guerrillas succeeded repeatedly in breaking the enemy's supply trains, and the Americans found it impossible to close the Ho Chi Minh Trail, part of which ran through Thailand [J. Thompson 1991: 133–219].

The Vietnamese forces undertaking these missions grew from small guerrilla bands into an army composed of different branches and based on compulsory military service (in the North) [Giap 1970: 114–115, 172–173, 177, 216–219, 301–302]. Its undertakings included major offensives (such as the multipronged 1968 Tet Offensive) as well as sizable siege operations against enemy strongpoints, such as that of the French at Dien Bien Phu in 1954 and that of the U.S. at Khe Sanh. All three instances were conceived of as decisive battles to break the will of the opponent, and all were nearly successful. On several occasions the "guerrillas" used heavy artillery barrages in preparation for their assaults, departing from traditional guerrilla techniques. The final stage of the struggle (after U.S. withdrawal) also included a large-scale offensive land campaign, employing *tanks, artillery, and aircraft, even though guerrilla warfare continued to play an important part [Maclear 1981: 43–62, 255–273, 427–451; Giap 1970: 117–160; Gibson 1988: 155–224].

3. **Vietcong Strategy.** The guerrillas in South Vietnam for the most part used traditional guerrilla tactics: ambushes, harassment raids against enemy outposts, interdiction of supply lines, minelaying, etc. Their tactics (as opposed to the strategy) were predominantly offensive, and over 80 percent of the firefights were initiated by the Vietcong [Gibson 1988: 11, 103–110; Maclear 1981: 217–222]. They generally enjoyed the support of the rural population, as well as (as the Tet Offensive showed) of substantial parts of the urban population, which allowed the guerrillas to "swim in the countryside like a fish in water." Their tactics also "forced" the adversary to wage the war in an indiscriminate manner, causing numerous civilian casualties, and thus eventually politically to undermine its ability to wage war [Bellamy 1990: 230–237; cf. Beaufre 1972; Carver 1981: 101–120, 172–202].

4. **U.S. Strategy.** The U.S. applied numerous strategic conceptions of counterinsurgency warfare in the VW, nearly all unsuccessfully. First, it sought to sever the organic links between the guerrillas and the rural population by constructing fortified villages ("strategic hamlets"), inspired by similar strategies used by the *U.K. in Malaya. In actual fact, the scheme restricted the movement of the civilian villagers more than that of the guerrillas, thus allowing the Vietcong to pose as liberators and the enemy to be cast as oppressors. Furthermore, the scheme created de facto free-fire zones outside the hamlets whereby it contributed to the indiscriminate butchering of civilians and hence to the popularity of the guerrillas [Gibson 1988: 83–85, 135–151, 186–187, 226–233; cf. Carver 1981: 12–27].

Second, the Americans sought to exploit their supremacy in *air power, supposedly giving them an edge in terms of surveillance as well as powerful means of air-to-ground combat. This approach was rendered futile by the elusiveness of the guerrillas, who took advantage of the jungle and other terrain features, and who possessed quite sophisticated *air defense systems. The technological sophistication of U.S. aircraft proved of little avail because actual combat situations very rarely allowed its utilization (supersonic aircraft, e.g., practically never flew at supersonic speeds) [N. Brown 1986: 241; Gibson 1988: 377–378].

Third, the Americans trained and deployed special forces such as the Green Berets (designed especially for such contingencies) and developed search-and-destroy-tactics for the ordinary forces, etc. However, the dispatched search-and-destroy units tended to find themselves entrapped and annihilated, rather than the other way around, because the guerrillas refused to expose themselves by taking the bait before having attained crushing tactical superiority [Gibson 1988: 103–110]. Furthermore, the intermingling of combatants with civilians meant that the actual targets were often the latter, whence a trend toward a perversion of the war to indiscriminate slaughter, of which the My Lai massacre is the best-documented example. The overall circumstances had disastrous effects on morale (demonstrated by increasing drug abuse in the armed forces) and on political support for the war [Maclear 1981: 373–382; Walzer 1977: 309–315].

Fourth, the U.S. sought to exploit its military production base by developing a panoply of advanced antipersonnel weapons, some to be delivered from the air, others for use by the land forces, and still others for installation in a booby-trap mode: napalm, fragmentation weapons, small-caliber/high-velocity munitions, etc. Such weaponry inflicted extremely severe suffering on the civilian population but proved incapable of solving the problem of the elusiveness of the targets [Gibson 1988: 357-382].

Fifth, the U.S. sought to defeat the guerrilla in ordinary *attrition terms by steadily building pressure with huge numbers of soldiers. It was an approach that did not take into account the expansive character of the guerrilla movement. In the last resort the foreigners almost invariably ended up outnumbered [Gibson 1988: 154–173]. Furthermore, reinstatement of the draft and the heavy troop toll tended to move the political struggle back to the foreigners' homeland. After having taken over from the French (who had lost 95,000 troops), the U.S. ended up with 550,000 fighting troops in Vietnam, and lost around 60,000 [Gibson 1988: 9, 67, 95].

Finally, the U.S. unavailingly pursued strategic bombing in an attempt to coerce the Vietnamese into submission. Notwithstanding the more than 8 million tons of explosives dropped from 1965 to 1973 (2 million were spent during the whole of WWII), the Vietnamese refused to surrender [Gibson 1988: 319–382; Clodfelter 1989].

5. **Lessons.** The VW had far-reaching repercussions in the entire strategic community, not merely in the U.S. but across the world. It seemed to provide some important lessons for future wars, technologically, strategically, and (most significant) politically.

Technologically, it seemed to show the limits of offensive military technologies. If B-52 bombers could be shot down by fairly inexpensive *SAMs wielded by "primitive" armed forces, then were the investments perhaps misguided? It also led to a certain interest in what was later to become *NOD by apparently showing that alternative tactics were capable of defeating even the most advanced military technologies. Strategically as well, the VW seemed to convey the message that "technowar" was a flawed conception because "the enemy" was far from a mirror image of the Americans, for instance, by having an entirely different economic structure that was largely resistant to counterindustrial warfare [Gibson 1988: 23]. Furthermore, the VW seemed to demonstrate that the defense has gained an edge over offensive forms of combat. Politically, the war created a "Vietnam Syndrome" in the U.S., i.e., a distinct reluctance to become involved in costly wars (of questionable moral value) for less than vital stakes. Through the combination of its lessons, the VW was an important precondition for the emerging interest in *NOD, and many of the NOD designers explicitly referred to the Vietnamese example.

VISURI, PEKKA.   See *Finland.

VLADIVOSTOK INITIATIVE.   Arms control initiative pertaining to the Asia-Pacific region, announced by *Gorbachev in a speech in Vladivostok in 1986 and apparently addressed primarily to *Japan. The VI contained concrete proposals for a *nuclear-weapons-free zone on the Korean Peninsula and in the southern Pacific, various *maritime CSBMs for the Pacific Ocean, and "a reduction of armed forces and conventional armaments to limits of *reasonable sufficiency" [in Gorbachev 1987:133–148; cf. Findlay 1990; Garett & Glaser 1990; A. I. Johnston 1990; Langdon & Ross 1990; Mack 1990].

VOGT, WOLFGANG R.   Professor of sociology at the Leadership Academy of the Bundeswehr in Hamburg, member of the *SAS.

WRV's main contribution to the *NOD debate in the *FRG has been his repeated warning that a continuation of a defense policy resting on *nuclear deterrence and conventional arms buildup would undermine popular support for, and indeed delegitimize, national defense: a special variety of the *defense dilemma [Vogt 1989; 1989a; 1990; 1990b; 1991]. He has also analyzed the effects of the deterrence system on ways of thinking, with a special emphasis on "enemy images," and urged a paradigm shift based on the principles of *common security [1990a; idem & Rubber-Vogt 1989].

VOIGT, KARSTEN.   The *SPD's security political speaker in the Bundestag, a longtime supporter of *NOD (without specific attachment to any particular model), and coauthor of the *European Security 2000* proposal of 1989 (see under *Bahr) [Bahr & al. 1989].

KV has consistently advocated *conventional stability and identified this as a situation where neither side would be capable of attack. As a consequence, he criticized (inter alia, as the SPD representative in the North Atlantic Assembly) *NATO's *deep-strike plans, such as *FOFA [Voigt 1984; 1987a], while actively supporting *disengagement conceptions, such as chemical- and *nuclear-weapons-free zones, inter alia, through his participation in the SPD's negotiations with the SED on these matters [1985; 1987].

In 1988 KV published a detailed comparison of the various proposals of *Afheldt, *Hannig, *Löser, *von Müller, *von Bülow, and the *SAS. According to his findings, Afheldt's model was simply too weak, and that of Hannig too vulnerable because of its numerous lucrative targets. Although von Müller's and von Bülow's models were presumably strong enough to withstand an attack, their strictly defensive character was disputable, wherefore they failed to pass the stability criterion. On balance, KV found the SAS model to be preferable, albeit presumably lacking in deterrent effect because of its inability to hold an aggressor's homeland at risk [1988].

In light of the new developments, KV in 1990 was very optimistic with regard to the prospects of conventional arms control and urged NATO to adopt the principles of NOD as its *arms control guideline: "The concept of reciprocal structural inability to attack does, on the one hand, envision a maintenance of *forward defense. On the other hand, it makes abandonment of both *flexible response and the previous form of forward defense imperative. Such an abandonment is urgent because without it, NATO would end up in the political cul-de-sac of a structural inability to disarm" [Voigt 1990a: 149, 152].

In the years 1990–1991 KV devoted most of his energies to devising political schemes for the combination of German unification with Soviet and East European reform and pacification, as well as with West European *integration (seen in a pan-European perspective). He did not envision a dissolution of NATO but consistently emphasized the major role to be played by the EC and the *CSCE in the transformation of the European landscape [1989; 1989a; 1989c; 1990; 1990a–f].

VO NGUYEN GIAP. See *Giap, Vo Nguyen.

W

WALKER, PAUL F.   See **IPIS.*

WAR LIMITATION.   See **Limited War.*

WARNING.   Prior knowledge by a state of an impending attack or an attack in progress. Adequate W is a precondition for defense preparations, and thus for actually benefiting from the intrinsic **defense dominance* that pertains only to a prepared defense (cf. **surprise attack, *three-to-one rule*). It is common to distinguish between two general forms of W, for each of which different sources of information may be relevant:

- Strategic W, i.e., information about an adversary's political intentions (gathered, e.g., through spying) and longer-term material preparations for an attack, in terms of diplomacy, military deployments, **mobilization*, etc. (cf. **force comparison*).
- Tactical W, i.e., information about a prospective attacker's military preparations, usually gathered by national technical means such as surveillance aircraft, intelligence-gathering satellites, etc. [Stares 1991: 82–125; Karas 1984: 94–147; Jasani & Lee 1984: 6–57; Blair 1985].

The degree of effective W depends on several factors, among which technology is not the most important. Most surprise attacks are actually preceded by indicators that in retrospect are seen to have constituted decisive proof of an impending attack only were not recognized as such [Betts 1982; 1983; 1985; cf. R. Wohlstetter 1962; Magenheimer 1994]. Accordingly, the main problem may not be the gathering of but the interpretation of information and the available political options of taking appropriate action. The interpretation of data has, for instance, often been determined by vested interests: air forces have tended to exaggerate aerial threats; navies, the maritime challenges; etc. [cf. Prados 1982; Freedman 1986].

Two aspects of **NOD*-type defense restructuring might have an effect on W:

- Various tank-free, "offensive-weapons-free," or otherwise depleted forward **zones* (i.e., a *disengagement of forces) would constitute valuable early warning devices because an attack would inevitably be preceded by a deployment of forces within the zones.
- Renouncement of offensive capabilities would eliminate the element of **self-deterrence* that might prevent a state from responding to (nearly always ambiguous) early W.

WAR PREVENTION.   See **Deterrence by Denial, *Nuclear Deterrence, *Preventive War, *Preemptive Attack, *Time Factor, *Crisis Stability.*

WARSAW PACT.   Popular term for the Warsaw Treaty Organization, WTO

(1955–1991). Military alliance between the *Soviet Union*, Bulgaria, *Poland*, the *GDR*, *Hungary*, Czechoslovakia, Romania, and, until the early 1960s, Albania.

1. **Functions.** The WP was established in 1955, officially in response to West German membership in *NATO* but in actual fact probably for other reasons:

- The WP made it possible to couch Soviet diplomacy vis-à-vis NATO in multilateral terms.
- It gave the USSR a modicum of justification for interference in the internal affairs of other communist countries, ultimately for armed intervention.
- It provided a convenient institutional framework for the continued stationing of Soviet forces in *Eastern Europe*.

2. **Organization.** All legal paraphernalia notwithstanding, the WP was never an egalitarian alliance but all along was dominated by the Soviet Union. The level of *integration* and uniformity was high in all respects yet unidirectional. Whereas the armed forces of the smaller WTO states were rendered incapable of independent action, the Soviet armed forces remained unconstrained.

Certain Western analysts, most notably *Christopher Jones*, have gone so far as to see this as the main rationale for the WP: to preserve Soviet regional hegemony by the implicit threat of armed intervention. This, in turn, ruled out a shift to *territorial defense* postures on the part of the minor WP countries because a shift would have allowed them to stand up to Soviet pressure [C. Jones 1981; 1981a; 1984; 1992; cf. for a critique: Holden 1989: 20–28]. If this interpretation is accurate, it would also explain why burden sharing was so disproportionate in the WP, even compared with NATO: a ratio between Soviet and non-Soviet manpower as uneven as 4:1 [Rice 1984; Holden 1989: 46–51, 164–167].

3. **Offensive or Defensive?** Geopolitically, according to most Western analyses, the WP served either as a protective *buffer zone* or as a stationing area for Soviet armed forces poised for an offensive, or both.

In the "defensive" buffer-zone reading, the USSR may in fact have feared a Western threat on the basis of, inter alia, the Russian experience with Western invasions through the centuries [cf. e.g., Steele 1985: 70–79; MccGwire 1987b: 13–35; Dibb 1988: 109–114]. This *security* imperative might also explain why the WP's "Northern Tier" enjoyed a lesser freedom of action than the rest of the alliance: any NATO invasion would have passed through East Germany and Poland or Czechoslovakia, whereas invasions through Hungary, Bulgaria, or Romania were hardly feasible.

The actual stationing pattern of Soviet forces was also explicable in offensive terms: the placing of groups of Soviet forces well forward, above all in East Germany [cf. Lippert 1987], could facilitate a *surprise attack* against Western Europe. Some of the planning documents discovered in East Germany after unification did, indeed, seem to support this interpretation [Rühl 1991c–d]. On the other hand, it would even be possible to view the Soviet quest for an invasion capability as defensively motivated, i.e., as an attempt to deter the *U.S.* from aggressive ventures by holding Western Europe hostage with an implicit invasion threat [cf. MccGwire 1987b: 48–49].

It has also been assumed in the West that the non-Soviet Pact forces were intended to reinforce the Red Army in an attack against the West, in conformity with the Soviet conception of coalition warfare. However, skeptics all along pointed to the questionable reliability of Eastern European armed forces, particularly in a war of aggression against the West, with whom the sympathy of the overwhelming majority of the population lay [cf. Johnson & al. 1982; Volgyes 1982; Herspring & Volgyes 1980; MacGregor 1986; MacCausland 1986; Walendowski 1988]. In retrospect, these doubts seem to have been confirmed.

4. **Post–Cold War WP.** By 1987 the entire WP had followed the Soviet example by proclaiming *NOD* as official doctrine. The first signs were appearing, furthermore, of an emerging interest in formulating indigenous military doctrines along territorial defense lines (see *Berlin Declaration, *Budapest Appeal*) [Holden 1989: 97–114; J. Krause 1988; Dunay 1990]. This development was facilitated by revocation of the Brezhnev Doctrine in favor of official endorsement of the principle of national self-determination. This, in turn, paved the way for the 1989 revolutions that overthrew the communist regimes in the whole of Eastern Europe [Dawisha & Valdez 1987; Kramer 1990; cf. Summerscale 1981; Larrabee 1984]. Subsequently, in 1991 the WP was dissolved, in two stages. Prior to this, the *GDR* had already left the organization (in fact, disappeared as a state) as a corollary of German unification.

WARSAW TREATY ORGANIZATION.   See *Warsaw Pact.*

WAR TERMINATION.   The topic of WT has often been neglected in strategic analyses [see, however, Iklé 1971; Cimbala & Dunn (eds.) 1987] including those emanating from the alternative defense community.

In the prenuclear era, wars were as a rule (as least theoretically) terminated by the eventual exhaustion of the belligerents, which made the more-worn-down party surrender to its more-enduring opponent. Hence *Clausewitz*'s conception of war as a boundless contest of wills that could be decided (i.e., terminated) only by the weaker side's bending to the will of the stronger party [1980: 19–20/1984: 77 (Book I.1.4)]. Since the appearance of nuclear weapons, however, strategies of total war have become absurd, at least between nuclear powers and states enjoying the protection of their *extended deterrence.* Because the prospective loser could at any time inflict a practically infinite amount of destruction on the near-victor, wars can no longer be won, particularly not nuclear wars, as acknowledged by nearly everybody [for an exception, see Gray 1979]. Of course, a major war in the nuclear age might be terminated through a full-scale nuclear exchange, something that would not merely terminate war but also annihilate both of the warring states, as well as most likely extinguish life on the planet Earth [cf. Kahn 1965: 50–51, 194–195; Bracken 1987; Quester 1986; Abt 1985: 39–57]. This utterly unattractive prospect has put total war (in both a nuclear and a *subnuclear setting*) as a means of politics beyond the pale. To the extent that politics might still use military means, it would have to be in the form of *limited war*, where neither side would expend its entire strength.

Because wars can no longer be won through exhaustion, they can be terminated only by politics as the ultimate arbiter. Political, as opposed to military, decisionmaking would presuppose a suspension of the act of fighting (see *stalemate*), so that, paradoxically, the defender's ability to transform a would-be swift and conclusive war into a protracted war might become the decisive factor in its termination. To thwart *blitzkrieg* attempts by neither seeking nor presenting one's forces for decisive battles but, rather, opting for a relentless sequel of "*nonbattles*" was precisely what most *NOD* advocates recommended, albeit primarily for the sake of war prevention [Brossollet 1975; Afheldt 1976; 1983; cf. Mearsheimer 1983].

To allow politics to reign would not, unfortunately, automatically ensure rational decisionmaking and hence timely war termination. Political decisionmakers in the warring states would have to decide whether a continuation of war would be preferable to a termination on perhaps unfavorable terms. However, what may have appeared rational before war would not automatically remain rational. Whereas decisionmakers might well have been deterred by the total costs of war, had they anticipated them correctly, the decisive variable in midcourse decisionmaking would be not aggregate but marginal costs. Couched in such terms, it might well appear rational to continue a war well beyond the point where the cumulated costs would exceed any conceivable gains. The only answer to this problem may be compromise, i.e., to allow the prospective loser (even if this is the aggressor) an honorable exit that would minimize marginal losses.

WASHINGTON NAVAL CONFERENCE.   Conference held in Washington 1921–1922 at the initiative of the *U.S.* for the purpose of preventing and/or halting the naval *arms race* in progress between, above all, the U.S. and *Japan* [Buckley 1970; Bull 1973; R. Hoover 1980; Kaufman 1990: E. Müller 1986].

The original intentions of the U.S. were (in the formulation of Secretary of State Charles Evans Hughes) to "limit provocative armaments that threatened aggression, bred mistrust, stimulated competition in arms, and led to war" [Kaufman 1990: 44–48]. These almost *NOD*-like objectives notwithstanding, U.S. objectives soon translated into simple numerical ratios for "capital ships" (i.e., battleships) between the U.S., the *U.K.*, and Japan, as well as, at a later stage, *France* and *Italy*. Agreement on the 5:5:3 ratio for battleships was accompanied by (and in fact presupposed) agreements on a limitation of fortifications and naval bases, technical definitions of the distinction between cruisers and battleships, and tonnage limitations for *aircraft carriers* (about which nobody really cared much at this stage).

WEB.   Dominant structural principle in most *NOD* models, especially those of a *territorial-defense* type [e.g. Afheldt 1983; SAS (eds.) 1989]. A web-like configuration of the defense forces would imply that forces were distributed across a large area (for instance the entire territory to be defended) in small units, each of which would be responsible for merely a small area. The web would ensure that the defender could maintain the cohesion of its forces in the absence of long-range mobility because each unit would be able to assist neighboring units. The individual units might operate under a high degree of autonomy, and hence the command structure might be decentralized to a considerable extent, thereby becoming less vulnerable (see *C³I*). The entire web would be reactive in the sense that each unit could go into action only when the aggressor had reached its designated area of operations. This would underline the defensive nature of the armed forces as well as help contain *escalation*. To the extent that forces with an inherent mobility are included in the defense posture (e.g., for internal reinforcement) these might be immobilized beyond the state's own territory (i.e., made strictly defensive) by being made dependent on the stationary web, as envisioned in the *spider-and-web* model of the *SAS* [Unterseher: 1988b; 1989b–c, f; 1990b; Boeker & Unterseher 1986; Grin & Unterseher 1988; 1989; 1990: 1990a].

WEBBER, PHILIP.   Political scientist affiliated with the Bradford School of Peace Studies and the Centre for Defense and Disarmament Studies, University of Hull.

In 1990 PW published one of the most elaborate studies on NOD in the Anglo-Saxon world: *New Defence Strategies for the 1990s: From Confrontation to Coexistence,* which was intended as a rebuttal of Vegetius's aphorism, *"Si vis pacem, para bellum."* PW suggested as a replacement, "Si *pacem vis, para aequilibrium"* (If you want peace, build a stable balance) [1990: 217]. The work featured an elaborate critique of present strategies, such as *flexible response,* the U.S. Navy's *Maritime Strategy,* and the various *deep-strike* conceptions, all of which he charged with negative effects on *crisis stability.*

PW's proposed alternative was *defensive deterrence,* i.e., a conventional and self-sustainable *NOD* posture, presumably made possible by trends in the development of military technology that tended to benefit the defender. Concretely, PW proposed a zonal arrangement for Europe, inspired by the various West German NOD models:

- The first 4 km behind the front would be a "sensor *zone" devoid of troops.
- Behind this would be a 30-km-deep "*forward defense zone" featuring troops with antitank weapons and various barriers.
- This would followed by a 70–100-km-wide "combined defense zone," apparently inspired by the *spider-and-web principle of the *SAS group.
- The remaining territory would be designated a "reserve and maneuver zone" [1990: 155–174].

According to Webber's calculations, the new posture would be entirely affordable, even if implemented unilaterally.

WEINBERGER INITIATIVE.   See *CDI.

WEIZSÄCKER, CARL FRIEDRICH VON.   Professor of physics and philosophy at the Max Planck Institute in Starnberg, brother of *Richard von Weizsäcker, a former president of the *FRG.

CFW has been a longtime collaborator of *Horst Afheldt and has edited and/or contributed to the most important works on *NOD in the FRG, thus serving as a "mentor" of the NOD debate [Weizsäcker (ed.) 1971; (ed.) 1976; (ed.) 1984; (ed.) 1990; Afheldt 1976]. CFW's main individual contribution to NOD theory was *Ways in Peril* [Weizsäcker 1979], wherein he addressed the problem of *stability. A precondition for *balance-of-power politics to further war prevention, according to CFW, was that "the defense would be so militarily superior to the offense that everybody could be secure against everybody else, including any conceivable coalition" [1979: 116]. No state would feel truly secure unless its defensive capability exceeded that of the offensive capability of all possible opponents. The balance-of-power equation between states with identical requirements would, however, be without solution unless "Clausewitzian weapons" were to predominate, i.e., such weaponry as would ensure the supremacy of the defense: "If only Clausewitzian weapons are available, there are no immanent reasons for an unlimited *arms race. At equal force levels, each party is superior in the defense to the other's attack, and therefore need not fear such an attack, even if it were to be undertaken" [Weizsäcker 1979: 165–166, 150]. CFW was thus the first (along with *André Glucksmann) to "discover" the concept of *mutual defensive superiority, which has usually been attributed to the late *Anders Boserup [Boserup 1985].

WEIZSÄCKER, RICHARD VON.   President of the FRG (1984–1994), mayor of Berlin (1981–1984), *CDU parliamentarian, and brother of *Carl Friedrich von Weizsäcker.

On more than one occasion, RvW made very favorable statements about *NOD, inter alia, during his visit to Moscow in 1987: "A balanced defense without the capability of attack is of importance" [*Bulletin:* 73 (15 July)]. During the official visit of Erich Honecker in September 1987, RvW even used the term "structural inability to attack" (*Strukturelle Nicht-Angriffsfähigkeit*), thus deliberately disregarding the fact that it had been so closely associated with the *SPD [quoted in Lutz 1988e: 26].

WERNICKE, JOACHIM.   Author (with *Ingrid Schöll) of *To Defend Rather Than Annihilate: Escapes from the Nuclear Trap,* which could be seen as part of the debate in the *peace movement. Criticizing *NATO strategy for its nuclear emphasis and for being suicidal should deterrence fail, the two authors recommended a genuinely *defensive defense* (of the *forward defense* type), combined with a greater emphasis on *civil defense* [Wernicke & Schöll 1985].

WESTERN EUROPEAN UNION.   See *Europeanization.*

WEU.   Western European Union. See *Europeanization.*

WEYGAND LINE.   French defense line established during the German attack in 1940; conceived by General Maxime Weygand as a supplement to the *Maginot Line*. The WL ran from the Channel coast along the Somme and Aisne Rivers and consisted of defended forest areas and villages configured in a "hedgehog" mode, the defense of which emphasized antitank weapons. The WL was too light and possessed insufficient depth to prevent a swift armored breakthrough [Keegan 1989: 83–85].

WIBERG, HÅKAN.   See *Centre for Peace and Conflict Research*, *Sweden.*

WINDASS, STAN.   See *Foundation for International Security.*

WINTER WAR.   See *Finland.*

WINTRINGHAM, TOM.   Left-wing journalist and founding member of the British Communist Party (which he left in 1938). TW fought on the Republican side in the Spanish Civil War and upon return to Britain became one of the most outspoken advocates of the creation of a "Home Guard." In the years 1939–1942, he published several books and pamphlets in which he advocated a new "anti-*blitzkrieg* strategy." This was remarkably similar to the *NOD* ideas of the 1980s [1940; 1940a; 1941; 1941a; 1942; cf. Tatchell 1985: 129–156], bringing to mind above all the schemes of *Horst Afheldt*, *Steven Canby*, and the *SAS*.

The proposed defense posture would be built around the Home Guard, which would be raised and deployed locally and would constitute "an army that is so thick a line that it fills the whole map." The result would be a *territorial defense* that would be strategically static yet tactically mobile [quoted in Tatchell 1985: 139]. TW likened this posture to "a sieve that lets some parts of the attack through . . . but holds the other part of the attack which cannot penetrate the sieve," and to "a *web* in which the mobile forces are caught and netted until they break themselves to pieces thrashing about within the web." TW thus proposed "a defense web made up of tank-proof islands of resistance," joined by minefields and protecting one another by their fire.

Among the more baroque ideas of TW was his advocacy of urban *guerrilla warfare*, to be found in a book coauthored with "Yank" Levy [Levy 1941] and in a *Picture Post* feature from 24 January 1942. He recommended street fighting: "*Tanks* are of little use in streets against a vigorous resistance . . . *infantry* opposing the *tanks* can always be hidden, on either side, at very short ranges. Streets are, for tanks, long and dangerous 'defiles,' like passes through hills where the enemy can lurk, not only close on each side, but also above the vehicles" [quoted in Tatchell 1985: 135].

WISEMAN, GEOFFREY.   See *Australia.*

WORLD COMMISSION ON ENVIRONMENT AND DEVELOPMENT. See *Brundtland Commission.*

WORLD DISARMAMENT CONFERENCE.   Major disarmament conference held in Geneva from 1932 to 1934 under the auspices of the *League of Nations*; it achieved no tangible results [Noel-Baker 1979; Neild 1990: 137–144; Wiseman 1989: 47–50; Borg 1992; Zubok & Kokoshin 1989; Hall 1987; Tronnberg 1985; cf. Kaufman 1990]. The WDC was the most ambitious attempt ever at quantitative as well as qualitative *arms control*, featuring numerous proposals for general and complete *disarmament*, alongside more modest proposals for partial disarmament. The main reason for its complete failure may have been a simple lack of political will; both Germany and *Japan* were embarking on major rearmament.

The WDC was noteworthy, inter alia, because it constituted the first attempt by the international community to ban "offensive weapons" (or "aggressive weapons"), as envisioned, e.g., under the Hoover Plan proposed by the *U.S.* president in 1932. It proved impossible to define such weapons to everybody's satisfaction, even though the finest military minds, such as *Liddell Hart* and *J. F. C. Fuller*, took part in the public debate on the topic [cf. Hart 1932; Fuller 1932].

WORLD WAR I.   See *WWI*.

WORLD WAR II.   See *WWII*.

WÖRNER, MANFRED.   From 1982 to 1988 West German defense minister for the *CDU*; from 1988 until his death in 1994 general secretary of *NATO*.

During his tenure as defense minister, MW repeatedly spoke against *NOD*, both in person [1985] and through the ministry. The ministry's 1985 *White Book* included an official critique of NOD, undoubtedly reflecting the CDU's and MW's viewpoints. Nevertheless, the defensive orientation of NATO was emphasized as imperative because "neither preemptive war nor offensive and preventive operations onto the territory of the opponent . . . are politically conceivable or militarily feasible options for NATO" [*Bundesministerium der Verteidigung* 1985: 30, 80–81].

WTO.   Warsaw Treaty Organization. See *Warsaw Pact*.

WWI.   The First World War (1914–1918). In a certain sense the war constituted the analytical point of departure for *NOD* by virtue of the lessons derived from it. These included the following:

- That the defensive form of combat is inherently the stronger, as the fighting in the trenches on the western front seemed to demonstrate; attacks were horrendously costly in terms of manpower because of the combined defensive strength of machine guns, barbed wire, and surveillance aircraft [cf. A. Jones 1988: 434–488].
- That "military expertise" is deceptive and not a prerogative of professionals: the most accurate prediction of the actual course of WWI was that of the Polish banker *Ivan Bloch* [1889], whereas the officer corps on all sides believed in the possibility of swift and bloodless victories but were proven completely wrong [cf. Howard 1985].
- That beliefs in the strength of the offensive are dangerous, regardless of whether they are correct or erroneous: The "cult of the offensive" in both Germany and France (manifested in the *Schlieffen Plan* and *Plan 17*) practically compelled each side to attack the other side preemptively because each poised highly offensive-capable forces against the other [Evera 1984; 1985; 1985a; J. Snyder 1984; 1985].
- That *alliances* may be a mixed blessing; they might not only dissuade an

aggressor from war but also contribute to making war unavoidable, as well as expand its scope.

- That wars might occur without premeditation, and that risks of "inadvertant" war merit serious attention [Levy 1990; cf. for a critique: Trachtenberg 1990; Gray 1992].

Because these lessons differ significantly from those derived from *WWII*, a central element in the NOD-versus-traditional strategy debate has been whether a future war would resemble one or the other world war.

WWII. Second World War (1939–1945). WWII has constituted the paradigm case of war for both East and West throughout the postwar period, even though each has derived different lessons from it.

The *Soviet Union* seems to have derived the following lessons from its experience in the Great Patriotic War (1941–1945), horrendous losses in the range of 20 million casualties):

- That the West could not be trusted, and that the Soviet Union was, in the final analysis, on its own.
- That there was no satisfactory substitute for military strength.
- That the offensive form of combat was the stronger, and, in particular, capabilities for armored warfare were decisive.
- That war on Soviet soil was not to be tolerated; any future war should be carried to the aggressor's home territory as swiftly as possible.

This interpretation strengthened the postwar Soviet predilection for the offensive, with accompanying emphases on forward deployment of forces, on superiority in conventional ground forces, *buffer zones*, etc. [cf. Alimurzayev 1989; Dick 1985; Donnelly 1982; Erickson & al. 1986; Hamilton-Eddy 1988; Hines 1982; idem & Petersen 1984; Lebow 1985a; Leites 1981; Luttwak 1983; Mazing 1989; MccGwire 1987b; P. Petersen 1987; Smith & Meier 1987; Snyder 1986; Vigor 1983].

However, the Soviet Union might just as well have derived alternative lessons, some of which have been debated as a reflection of *New Thinking*:

- That defensive operations are sometimes decisive, as they were in the *Battle of Kursk* [Kokoshin & Larionov 1987].
- That forward deployment and buffer zones are not necessarily useful, as might have been learned from the experience with the partition of *Poland* in 1939.

The West, in its turn, seems to have derived partly different lessons from WWII:

- That appeasement (such as attempted by British Prime Minister Chamberlain) is bound to fail [cf. Lebow 1984].
- That deterrence is the best means of *war prevention*.
- That deterrence requires offensive capabilities and preponderant destructive power, *in casu* nuclear weapons.
- That *alliances* should be strengthened and that, in particular, the participation of the *U.S.* in any future war in Europe should be ensured in advance, i.e., that alliances should include the U.S.

Those are the lessons that alternative defense proponents have challenged, inter alia, by pointing to the lessons derived from *WWI* as an alternative "paradigm" [cf. Afheldt 1987b: 34–49].

Y

YASOV, DMITRI. Defense minister of the *Soviet Union* until the abortive coup in August 1991, in which he was a central participant. After the (politically enforced) Soviet adoption of *NOD* as a central element in military doctrine, the newly appointed DY outlined this doctrine in the Soviet journal *International Affairs* (intended primarily for foreign consumption). Even though he expressed approval of the new doctrine in general terms, he remained equivocal as to whether *Warsaw Pact* doctrine *was* defensive in all respects (meaning that it could remain unchanged) or whether it *should become* strictly defensive, implying a readiness to abandon offensive features on the strategic, operational, and tactical levels [Yasov 1987].

This apparent disagreement (or at least difference of emphasis) may have reflected more profound controversies about civil-military relations, with the armed forces seeking to prevent an encroachment on their customary prerogatives, namely, formulation of the military-technical level of military doctrine (see *Military Science*) [cf. Holden 1991b; 1991; Herspring 1987; 1989; Holloway 1989; Rice 1987; 1989; Larrabee 1988; *Jane's Defence Weekly*, 31 October 1987: 1010–1012; Green & Karasik (eds.) 1990; Nichols 1993; idem & Krasik 1990].

YEAR 2000 PLAN. Soviet set of proposals for nuclear disarmament, presented by *Mikhail Gorbachev*, 15 January 1986 [in Gorbachev 1987: 7–22; cf. Garthoff 1990: 94–148]. The goal was a complete abolition of nuclear weapons by the year 2000, to be approached through three overlapping stages:

- In the first stage (lasting six to eight years) both the *U.S.* and the *Soviet Union* would reduce their strategic nuclear weapons by half and build down the number of delivery vehicles to a maximum of six thousand each; the development of strategic defenses (i.e., *SDI*) would be discontinued; U.S. and Soviet *INF* weapons would be eliminated from Europe; and the *U.K.* and *France* would commit themselves not to expand their arsenals.
- In the second stage (five to seven years) the two "small" nuclear powers would participate fully in the process, and all tactical nuclear weapons would be eliminated.
- In the final stage (from 1995 to 2000) all remaining nuclear weapons would be eliminated.

YUGOSLAVIA. Federal state established in 1945 and consisting until 1991 of the six republics Serbia, Croatia, Slovenia, Bosnia-Herzegovina, Macedonia, and Montenegro, and two autonomous provinces, Kosovi and Wojwodina.

1. **Partisan Movement.** The resistance of the Yugoslav partisans against the Italian and German invaders during *WWII* was among the most successful instances of *guerrilla warfare* ever, effectively liberating the country virtually without any direct foreign assistance. The largest resistance movement was led by

the Communist Party (under the leadership of *Josip Broz Tito*), which at a very early state joined the armed struggle against the Italian invasion. Initially, party members were ordered to join the regular army rather than forming special formations ["Report to the Comintern 1941," in Tito 1966: 42–43]. However, what they regarded as defeatism on the part of the government soon led the communists to form separate People's Liberation Partisan Detachments, which were conceived of as patriotic rather than communist forces. During the initial stage of the resistance their activities consisted mostly of sabotage (intended to deprive the invaders of the use of the infrastructure, i.e., railways, bridges, factories, workshops, ammunition and arms dumps, etc.) and a rather extensive killing of enemy troops and collaborators [cf. Tito 1966: 48–49].

At a subsequent stage, political bodies were set up to supervise (or legitimize) the armed forces, in their turn merged into the People's Liberation Army [Tito 1966: 75, 86–88, 145, 154]: almost a regular army based on compulsory *conscription* in the liberated areas and with considerable offensive capabilities, albeit not without residual partisan features. It was recommended to exploit the enemy as the main source of weapons supplies, even though this was supplemented with a flow of weapons from the Allies as well as with indigenous arms production in the liberated territories. Furthermore, hit-and-run tactics and ambushes were employed on a wide scale, as were attacks against the infrastructure, which might otherwise have been exploited by the enemy. A total of 1,684 km of rail, 830 railway stations, 1,099 trains with 1,918 locomotives and 19,759 railway cars, and more than 3,000 bridges were destroyed [Winkler 1986: 25, 28; Gschwendtner 1986: 39]. At the same time brigade-size shock units capable of substantial offensive strokes were formed. Even a naval section was set up under the auspices of the People's Liberation Army for the purpose of carrying the fighting into the Adriatic Sea [Tito 1966: 222–223, 240–244, 260–261].

The total size of the resistance amounted to more than 350,000 active fighters. It pinned down sixteen to twenty Italian divisions throughout the war and continued unabated after Germany had taken over Italy's role [Trgo 1966: 332; Gschwendtner 1986; Winkler 1986; Tito 1966: 147]. Even though the Red Army eventually participated in the liberation of Y, it was well understood that the Yugoslav partisans would have been quite capable of effecting liberation on their own, a fact that gave considerable basis for the development of an independent stand vis-à-vis the *Soviet Union* in the years to come [C. Jones 1981: 79–83; 1984; Roberts 1976: 124–171; Dedijer 1952].

2. **General People's Defense.** The postwar structure of Y's defense bore the imprint of the liberation struggle, particularly after adoption in 1968 of the new doctrine, General People's Defense (codified in legislation in 1969), which was revised in 1974 [Roberts 1976: 172–217; Lukic 1984; Bebler 1987; 1991; Acimovic 1987; Piziali 1989; Radinovic 1989; M. Brown 1991; Gow 1992]. The General People's Defense was conceived as a *territorial defense* posture and strategy that envisioned a protracted struggle against any invader. It would encompass the entire country and emphasize light antitank and *air defense* weapons yet also comprise significant mechanized forces. The guiding principle of the defense structure would be *decentralization* in terms of the delegation of authority from the federation to authorities in the republics and in terms of local autonomy, which would imply that arms would be distributed to a large number of local constituencies, factories, etc. In addition to the *militia*-like armed forces, Y also maintained a centralized field army, as well as naval and air forces under centralized command.

The armed forces of Y were credited with great defensive strength, and it was widely assumed, by NOD proponents and others, that their specific nature might have been what dissuaded the Soviet Union from ever attacking [C. Jones 1981a: 79-83].

3. **Impact of "New Thinking."** Even though the defense of Y was, as a matter of principle, *tous azimuts,* i.e., directed against any possible attacker, in reality the only serious threat with which the political leadership ever reckoned was that of a Soviet invasion. The apparent disappearance of this threat, inter alia, as a result of the repudiation of the Brezhnev Doctrine and other elements of Soviet *New Thinking,* therefore gave rise to some Yugoslav reassessment of military requirements versus domestic needs [Bebler 1989]. Studies on *conversion* of the (quite large) arms-producing industry were initiated in 1989–1990 [Milosavljevic (ed.) 1991].

4. **From Civil to Interstate War.** What prevented the impending reassessment of defense needs was, above all, the outbreak of open civil war in 1991, after repeated secessionist attempts through the 1980s (and earlier) [Bischof 1991; Bebler 1991; 1991a; M. Brown 1991; CSS 1991; Furkes & Schlarp (eds.) 1991; Ramet 1992; Wiberg 1992]. The recognition granted to the secessionist republics transformed the war into one between states. The course of the civil war has been determined to a large extent by the previous emphasis on autonomy and decentralization: there was an abundance of weapons at the disposal of local communities and the secessionist republics. This allowed for Slovenia to assert its independence with relatively moderate bloodshed [Bebler 1992; 1993; 1993a; Gow 1992], whereas the secession of Croatia resulted in bloody war with the federal (i.e., Serbian) government, followed by an even bloodier war in Bosnia-Herzegovina [see Arbucle 1992].

The Yugoslav tragedy has occasionally been utilized for a critique of NOD in general, but the argument is, at most, justified only insofar as particular models are concerned. Even though the defense posture of Y resembled *some* NOD models in *some* respects, nobody has ever suggested copying it. On the contrary, NOD advocates have consistently emphasized that defense should be adapted to each country's special circumstances, politically, geographically, and otherwise.

Z

ZERO-SUM GAME.   Game in which the gains of one player equal the losses of the other, i.e., a game of pure conflict (see *Game Theory*, *Chicken*). The superpower competition, particularly in the Third World, was often seen as analogous to a ZSG by one or both sides, implying that the two sides tended to focus on relative gains and losses: even though both might "lose" the *arms race* (in the sense of expending resources without achieving a usable preponderance of strength), it might still make sense to race if the adversary stood to lose more—this would be tantamount to a relative gain [Stein 1990: 113-145].

The time perspective adopted by the players in a ZSG is an important determinant of whether a conflict is viewed as zero-sum or variable-sum, i.e., of whether cooperation is regarded as possible. The longer the *shadow of the future*, the less likely it is that a conflict is to be seen as zero-sum, and the better the opportunity for cooperation.

ZONES.   Distinguishing feature of many *NOD* models as well as of various other *arms control* and *disarmament* arrangements. The logic behind Z is that of *disengagement*, i.e., spatial separation of the forces of two adversaries, whereby the risk of their ending up in a shooting war presumably would be diminished. The available *warning* time for both sides is prolonged, and the inherent superiority of the defense thereby amplified because the side intending to attack would first have to redeploy forces into the depleted zone, thereby alerting the envisioned target of aggression. Furthermore, the risk of misunderstandings leading to war would be reduced by clear and unambiguous regulations on permitted and prohibited forces and/or military activities in the Z.

Various types of Z have been proposed, by NOD proponents and others, including

- Demilitarized Z;
- *Nuclear-weapons-free* Z, e.g., in the Nordic region, *Latin America*, Africa, the southern Pacific, the Middle East, and the Korean Peninsula;
- *Chemical-weapons*-free Z [SPD & SED 1985; cf. Voigt 1985; Trapp (ed.) 1987];
- *Tank*-free Z [Brzezinski 1988];
- Offensive-weapons-free Z (also known as "defensive Z"), i.e., zones along a border where only strictly defensive forces would be permitted [e.g. Bahr 1987a; idem & al. 1988; Lutz 1988c; 1989b];
- Maneuver-free Z, e.g., as proposed by Romania during the *CDE* negotiations [cf. Blackwill & Legro 1989]; and
- Naval disengagement Z, for example, as a combination of "attack submarine keep-out Z" and "strategic submarine stand-off Z" [Tunander 1989; 1990; cf. Purver 1988; 1988a; 1990; Møller 1987].

ZOPFAN. Zone of Peace, Freedom, and Neutrality. Project launched by the states of the Association of South-East Asian Nations (ASEAN) in 1971 that envisioned a certain *neutralization* of Southeast Asia [Hänggi 1993].

Bibliography

This is a multilingual bibliography, covering literature in English, German, French, Dutch, Italian, Spanish, Danish, Swedish, Norwegian, and Latin.

All works quoted in the main body of the dictionary have been included, as has a wide range of additional literature on the subjects covered. The "density" (in terms of exhaustiveness) is highest at the center, i.e., nonoffensive and other types of genuinely alternative defense. It decreases (i.e., becomes increasingly selective) toward the periphery, containing various forms of alternative security arrangements and instruments as well as "not-all-that-alternative" defense proposals.

As a general rule, only one version of each item has been included: either the most recent or the most complete. The inclusion of "gray" publications, such as working papers and unpublished conference papers, is very selective. Translations into languages other than English have usually been omitted.

Items are ordered alphabetically by author and chronologically by year, but the markings a, b, c, etc. are entirely arbitrary. Individually authored items precede joint products, and authored items precede edited ones. The order might thus be: NN 1990a; NN 1990b; NN & al. 1989; NN (ed.) 1989; NN & al. (eds.) 1988. The German letters with umlaut, i.e., ä, ö, ü (Ä, Ö, Ü) have been alphabetized as the corresponding letters without umlaut.

To save space, cross-references have been used extensively with regard to anthologies, and publishers' names have been abridged by omitting most "suffixes" (Press, Publishers, Incorporated, Verlag, etc.). The following abbreviations have been used for periodicals:

AFP	*AFES-Papers* (Mosbach: AFES-PRESS)
AFR	*AFES-PRESS Reports* (same publisher)
AP	*Adelphi Papers*
APZ	*Aus Politik und Zeitgeschhichte.* Beilage zur Wochenzeitung Das Parlament
AuP	*Aussenpolitik*
BAS	*Bulletin of the Atomic Scientists*
Berichte	*Berichte des Bundesinstituts für ostwissenschaftliche und internationale Studien* (Köln)
Blätter	*Blätter für deutsche und internationale Politik* (Köln: Pahl-Rugenstein Verlag)
BPP	*Bulletin of Peace Proposals* (Oslo: Peace Research Institute, Oslo, PRIO)
Bulletin	(Bonn, Presse- und Informationsamt der Bundesregering).
CES	*Cahiers d'études strategiques* (Paris: CIRPES)
CRPV	*Current Research in Peace and Violence* (Tampere, Finland: TAPRI)
DA	*Deutschland Archiv. Zeitschrift für Fragen der DDR und der Deutschlandpolitik*
DDA	*Defense and Disarmament Alternatives* (Brookline, Mass.: Institute for Defense and Disarmament)

Dialog	*Dialog: Beiträge zur Friedensforschung* (Stadtschlaining: Österreichisches Institut für Friedensforschung und Friedenserziehung)
EA	*Europa-Archiv: Zeitschrift für internationale Politik* (Bonn: Deutsche Gesellschaft für auswärtige Politik)
ES	*European Security*
EuS	*Europäische Sicherheit*
EW	*Europäische Wehrkunde: Wehrwissenschaftliche Rundschau*
FA	*Foreign Affairs*
F&A	*Frieden und Abrüstung* (Bonn: Initiative für Frieden, internationalen Ausgleich und Sicherheit)
HB	*Hamburger Beiträge zur Friedens- und Konfliktforschung* (Hamburg: Institut für Friedensforscung und Sicherheitspolitik)
HSFK-B	*HSFK-Berichte* (Frankfurt a.M.: Hessische Stiftung für Friedens- und Konfliktforschung)
HSFK-R	*HSFK-Reports* (same publisher)
IA	*International Affairs* (London)
IA(M)	*International Affairs* (Moscow)
IDR	*International Defense Review*
IPW-B	*IPW-Berichte* (Berlin, GDR: Institut für internationale Politik und Wirtschaft)
IPW-F	*IPW-Forschungshefte* (same publisher)
IPW-P	*IPW-Papers* (same publisher)
IR	*Internationale Wehrrevue*
IS	*International Security*
JPR	*Journal of Peace Research* (same publishers as BPP)
JSMS	*Journal of Soviet Military Studies*
JSS	*Journal of Strategic Studies*
KAAD	*Kurzpapier der Abteilung Aussenpolitik und DDR-Forschung* (Bonn: Friedrich-Ebert-Stiftung)
MR	*Military Review*
NGFH	*Die Neue Gesellschaft: Frankfurter Hefte* (Bonn: Verlag J. H. W. Dietz Nachf./Friedrich-Ebert-Stiftung)
NSN	*NATO's Sixteen Nations*
NUPIN	*NUPI Notat* (Oslo: Norwegian Institute of International Affairs)
NUPIP	*NUPI Paper* (same publisher as NUPIN)
NUPIR	*NUPI Report* (same publisher as NUPIN)
OE	*Osteuropa: Zeitschrift für Gegenwartsfragen des Ostens*
ÖMZ	*Österreichische Militärische Zeitschrift* (Bundesministerium für Landesverteidigung, Vienna)
POC	*Problems of Communism*
PRIF-R	*PRIF-Reports* (Frankfurt: Peace Research Institute Frankfurt)
Proc	*U.S. Naval Institute Proceedings*
RJ	*RUSI Journal*
S+F	*S+F: Vierteljahresschrift für Sicherheit und Frieden* (Baden-Baden: Nomos)
SIPRI	*World Armaments and Disarmament: SIPRI Yearbook*
SMR	*Soviet Military Review*
SS	*Security Studies*
Studien	*Studien der Abteilung Aussenpolitik und DDR-Forschung* (Bonn: Friedrich-Ebert-Stiftung)
SWB	*Summary of World Broadcasts, BBC Monitoring.*
WP	*World Politics*

WP-CPCR   *Working Papers* (Copenhagen: Centre for Peace and Conflict
Research, Copenhagen)
WPJ   *World Policy Journal*
WQ   *Washington Quarterly*

Aaron, Henry J. (ed.) 1990: *Setting National Priorities: Policy for the Nineties* (Washington, D.C.: Brookings).

Abenheim, Donald 1990: "The Citizen in Uniform: Reform and Its Critics in the Bundeswehr," in Szabo (ed.): 31–51.

Abshire, Donald M. 1987: "NATO's Conventional Defense Improvement Effort: An Ongoing Imperative," *WQ* (Spring): 49–60.

Abt, Clark C. 1985: *A Strategy for Terminating a Nuclear War* (Boulder, Colo.: Westview).

Acimovic, Ljubivoje 1987: "The Case of Yugoslavia," in Lodgaard & Birnbaum (eds.): 154–173.

Acheson, Neil 1981: *The Polish August: The Self-limiting Revolution* (Harmondsworth: Penguin).

Acker, Alexander 1984: "Einsatz von Raketenartillerie im Verteidigungsnetz," in Weizsäcker (ed.): 89–120.

Ackland, Len & Steven McGuire (eds.) 1986: *Assessing the Nuclear Age: Selections from the Bulletin of the Atomic Scientists* (Chicago: Educational Foundation for Nuclear Science).

Adam, Bernard 1990: "Les années 90: Décennie du désarmement et de la nouvelle sécurité européenne," in GRIP (ed.), *Memento Defense Desarmament, 1990* (Bruxelles: GRIP): 13–22.

Adams, F. Gerard (ed.) 1992: *The Macroeconomic Dimensions of Arms Reduction* (Boulder, Colo.: Westview).

Adams, Gordon & Stephen Alexis Cain 1989: "Defense Dilemmas in the 1990s" (*IS*, 13:4) in Miller & Lynn-Jones (eds.): 84–94.

ADC (Alternative Defence Commission) 1983: *Defence Without the Bomb* (New York: Taylor & Francis).

——— 1987: *The Politics of Alternative Defence: A Role for a Non-Nuclear Britain* (London: Paladin).

Adler, David Jens 1984: "Fodnoteoprøret: Socialdemokratiets sikkerhedspolitik fra Hedtoft og H.C. til Anker og Ritt," in Christensen (ed.): 49–63.

Adler, Emanuel 1992: "Arms Control, Disarmament, and National Security: A Thirty Year Retrospective and a New Set of Anticipations," in Adler (ed.): 1–20.

——— (ed.) 1992: *The International Practice of Arms Control* (Baltimore: Johns Hopkins U.P.).

Adler, Mats, Cilla Lundström & Lars Ångström 1989: *I stället för upprustning* (Malmö: Pax).

Adomeit Hannes 1988: "Gorbatschows aussen- und sicherheitspolitische Initiativen: Chancen für Westeuropa?" *S+F*, 8:2, 65–69.

——— 1989: "The 'Common House': Gorbachev's Policies Towards Western Europe," in Clesse & Schelling (eds.): 212–224.

——— 1990: "Gorbachev and German Unification: Revision of Thinking, Realignment of Power," *POC*, 39:4 (July/August): 1–23.

——— 1990a: "The Impact of Perestroika on Soviet European Policy," in Hasegawa & Pravda (eds.): 242–266.

Adragna, Steven P. 1989: "A New Soviet Military? Doctrine and Strategy," *Orbis*, 33:2 (Spring): 165–180.

Adrets, André 1986: "Franco-German Relations and the Nuclear Factor in a Divided Europe," in Robin F. Laird (ed.), *French Security Policy: From Independence to Interdependence* (Boulder, Colo.: Westview): 105–120.

Afheldt, Eckhardt 1984: "Verteidigen, ohne Selbstmord zu Begehen," in Schröder (ed.): 115–132.

————— 1984a: "Verteidigung ohne Selbstmord. Vorschlag für den Einsatz einer leichten Infanterie," in Weizsäcker (ed.): 41–88.

————— Horst Afheldt, Andreas von Bülow, Walter Kons & Helmut Willmann 1985: "Streitgespräch: Alternative Strategien," *NGFH*, 2: 112–121.

Afheldt, Horst 1971: "Analyse der Sicherheitspolitik durch Untersuchung der kritischen Parameter," in Weizsäcker (ed.): 23–74.

————— 1971a: "Entwicklungstendenzen der Sicherheitspolitik in Europa und umfassendere Ansätze zur Friedenssicherung," in Weizsäcker (ed.): 417–456.

————— 1976: *Verteidigung und Frieden: Politik mit militärischen Mitteln* (München: Carl Hanser).

————— 1977: "Das Problem der Sicherheitspolitik für die Bundesrepublik heute: Rationale Einsatzoptionen für den Ernstfall?" in Schwartz (ed.): 443–458.

————— 1978: "Friedenspolitik mit militärischen Mitteln in den neunziger Jahren," in Schwartz (ed.): 639–656.

————— 1978a: "Tactical Nuclear Weapons and European Security," in SIPRI (eds.): 262–295.

————— 1983: *Defensive Verteidigung* (Reinbek: Rowohlt).

————— 1984: "Defensive Verteidigung. Warum?" in Weizsäcker (ed.): 13–28.

————— 1984a: "Neue Ansätze in der Sicherheitpolitik: Isolierte oder umfassende Problemlösungen," in SAS (eds.): 87–102.

————— 1984b: "The Necessity, Preconditions and Consequences of a No-First-Use Policy," in Blackaby & al. (eds.): 57–66.

————— 1985: "Abrüstung der Staaten und die kollektive Organisation des Weltfriedens?" in Lutz (ed.): 64–81.

————— 1985a: "Frieden durch Überlegenheit oder Frieden durch Stabilität: Vorstufen zur kollektiven Sicherheit?" in Lutz (ed.): 245–259.

————— 1985b: *Pour une défence non suicidaire en Europe,* preface by Jean Klein, postface by Georges Buis (Paris: Découverte).

————— 1987: "Defensive Verteidigung als politisch-strategische Konzeption," in Bühl (ed.): 155–172.

————— 1987a: "Was heisst strukturelle Nichtangriffsfähigkeit?" *S+F*, 7:1, 59.

————— 1987b: *Atomkrieg. Das Verhängnis einer Politik mit militärischen Mitteln,* 2d ed. (München: dtv).

————— 1987c: "Der Weg aus dem Dilemma: Rein defensive konventionelle Verteidigung in Europa," *Dialog,* 10: 101–108.

————— 1988: "Konventionelle Stabilität, Zivilschutz, Defensivüberlegenheit: Redebeiträge," in Bahr & Lutz (eds.): 402–408.

————— 1988a: "New Policies, Old Fears," *BAS,* 44:7, 24–27.

————— 1989: "Alternative Defence Postures for NATO and the WTO," in Tromp (ed.): 187–198.

————— 1989a: *Der Konsens: Argumente für die Politik der Wiedervereinigung Europas* (Baden-Baden: Nomos).

————— 1990: "European Security and German Unity," in Brauch & al.: 9–20.

————— 1990a: "Überlegungen zu einer defensiven Verteidigung in Theorie und Praxis," in Heisenberg & Lutz (eds.): I:259–271.

————— 1992: "Conflict of Interests, Nuclear Weapons, and Flexible Response: A Critical German View," in Brauch & Kennedy (eds.): 121–146.

—————, Hans Günter Brauch, Malcolm Chalmers & Jonathan Dean 1990: "German Unity and the Future European Order of Peace and Security," *AFR,* 38.

————— & Hellmuth Roth 1971: "Verteidigung und Abschreckung in Europa," in Weizsäcker (ed.): 285–302.

———— & Phillip Sonntag 1971: "Stabilität und Abschreckung durch strategische Kernwaffen, eine Systemanalyse," in Weizsäcker (ed.): 303–416.

Agrell, Wilhelm 1979: *Om kriget inte kommer: En debattbok om Sveriges framtida försvars- och säkerhetspolitik* (Stockholm: Liber).

———— 1982: *De yderste våben: Om kernevåben og kernevåbenkrig* (Copenhagen: Mellemfolkeligt Samvirke).

———— 1983: *Et tredje standpunkt i forsvarsspørgsmålet* (Copenhagen: SNU).

———— 1983a: "Spelet om neutraliteten," *Försvar i Nutid,* 2.

———— 1984: "Icke-provokativt försvar: Diskussionen kring ett freds- och försvarsstrategiskt alternativ" (Kalmar: Världspolitikens dagsfrågor).

———— 1984a: "Small But Not Beautiful," *JPR,* 21:2, 157–167.

———— 1984b: *Sveriges civila säkerhet* (Stockholm: Liber).

———— 1985: *Alliansfrihet och atombomber: Kontinuitet och förandring i den svenska Forsvarsdoktrinen, 1945–1982* (Stockholm: Liber).

———— 1986: *Bakom Ubåtskrisen* (Stockholm: Liber).

———— 1986a: "The Baltic and Scandinavia: The Implications of Changes in Military Strategy," *CRPV,* 12: 43–53.

———— 1987: "Offensive versus Defensive: Military Strategy and Alternative Defence," *JPR,* 24:1, 75–86.

———— 1987a: *Hvorfor ubådskrænkelserne* (Copenhagen: SNU).

———— 1989: "Avskräckning och icke-provokation: En europeisk och några nordiske erfarenheter," in Wiberg (ed.): 117–145.

———— & Håkan Wiberg 1980: "Säkerhet och nedrustning," *Försvar i Nutid,* 4/5.

Ahto, Sampo 1985: "Finnish Tactics in the Second World War," in FCMH (eds.): 177–188.

Aichinger, Wilfried & Arthur Friedrich Maiwald 1989: "Die Grossmächte und die Europäischen Neutralen, Teil 1: Die UdSSR," *ÖMZ,* 2: 105–115.

Akavia, Gideon Y. 1991: "Defensive Defense and the Nature of Armed Conflict," *JSS,* 14:1 (March): 27–48.

Albrecht, Ulrich 1982: "Western European Neutralism," in Kaldor & Smith (eds.): 143–161.

———— 1983: "Neutralismus und Disengagement: Eine Alternative für die Bundesrepublik," in DGFK (eds.): 479–504.

———— 1983a: "The Political Background of the Rapacki Plan of 1957 and Its Current Significance," in Steinke & Vale (eds.): 117–135.

———— 1985: "Alternative Designs of European Security, the Palme Commission Report and the Conventionalization of Forces," in Paul (ed.): 141–183.

———— 1990: "The Role of Germany in Perspective," in Randle & Rogers (eds.): 121–140.

———— 1990a: "The Role of Military R&D in Arms Build-Ups," in Gleditsch & Njølstad (eds.): 87–104.

———— 1992: *Die Abwicklung der DDR: Die "2+4-Verhandlungen": Ein Insider-Bericht* (Opladen: Westdeutscher).

———— (ed.) 1986: *Europa Atomwaffenfrei: Vorschläge, Plane, Perspektiven. Mit einem Essay von Berthold Meyer* (Köln: Pahl-Rugenstein).

————, Burkhard Auffermann & Pertti Joenniemi 1988: "Neutrality: The Need for Conceptual Revision," *Occasional Papers,* 35 (Tampere: TAPRI).

———— & al. 1992: "Europa danach: Aussen- und Sicherheitspolitik nach dem Zerfall der Sowjetunion und dem Wiederaufstieg Deutschlands," *Blätter,* 37:4 (April): 405–425.

Albrecht-Carrié, René 1968: *The Concert of Europe, 1815–1914* (New York: Harper & Row).

Alexandrova, Olga 1992: "Die Ukraine in Ostmitteleuropa: Bindeglied zwischen Ost und West?" *EA*, 47:18 (October 25): 535–544.

——— 1993: "Divergent Russian Foreign Policy Concepts," *AuP*, 44:4 (Fall): 363–372.

——— 1994: "Russia as a Factor in Ukrainian Security Concepts," *AuP*, 45:1 (1st Quarter): 68–78.

Alexandrow, Valentin 1993: "Entsowjetisierung und nationale Traditionen. Der Militärreform in Bulgarien," *EuS*, 42:1 (January): 32–36.

Alford, Jonathan 1979: "Confidence-Building Measures in Europe: The Military Aspects," *AP*, 149: 4–13.

——— 1983: "Perspectives on Strategy," in Steinbruner & Sigal (eds.): 91–105.

——— 1986: "No Early Use," in Paskins (ed.): 34–48.

——— & Kenneth Hunt (eds.) 1988: *Europe in the Western Alliance: Toward a European Defence Identity?* (New York: St. Martin's, IISS).

Alfsen, Erik 1982: "For a Nordic Nuclear-Weapon-Free Zone," in Kaldor & Smith (eds.): 135–142.

Alger, Chadwick & Michael Stohl (eds.) 1988: *A Just Peace Through Transformation: Cultural, Economic, and Political Foundations for Change* (Boulder, Colo.: Westview, IPRA).

Alimurzayev, Grigori 1989: "A Shield or a Sword? History of Soviet Military Doctrine," *IA(M)*, 5: 100–109.

Allebeck, Agnes Courades 1989: "Arms Trade Regulations," in *SIPRI*, 1989: 319–338.

Allen, William W., Antoine D. Johnson & John T. Nelsen, II 1993: "Peacekeeping and Peace Enforcement Operations," *MR*, 73:10 (October): 53–61.

Allgeier, Karl-Heinz 1992: "Key TBM Technologies and Consequences for Proliferation," in Neuneck & Ischebeck (eds.): 115–124.

Allison, Graham T. 1983: "What Fuels the Arms Race?" in Reichart & Sturm (eds.): 463–480.

——— 1989: "Conclusion: Windows of Opportunity," in Allison & Ury (eds.): 309–332.

——— & Robert Blackwill 1991: "America's Stake in the Soviet Future," *FA*, 70:3 (Summer): 77–97.

——— & Frederic A. Morris 1975: "Armaments and Arms Control: Exploring the Determinants of Military Weapons," *Dædalus*, Summer: 99–129.

———, Albert Carnesale & Joseph S. Nye, Jr. 1988: "Conclusion," in Nye, Allison & Carnesale (eds.): 215–235.

———, Carnesale & Nye (eds.) 1985: *Hawks, Doves and Owls: An Agenda for Avoiding Nuclear War* (New York: Norton).

——— & William Ury (with Bruce Allyn) (eds.) 1989: *Windows of Opportunity: From Cold War to Peaceful Competition in U.S.–Soviet Relations* (Cambridge, Mass.: Ballinger).

Allison, Roy 1985: *Finland's Relations with the Soviet Union, 1944–84* (New York: St. Martin's).

——— 1991: "New Thinking About Defence in the Soviet Union," in Booth (ed.): 215–243.

——— 1993: "Military Forces in the Soviet Successor States," *AP*, 280.

——— & Phil Williams 1990: "Crisis Prevention: Patterns and Prospects," in Allison & Williams (eds.): 246–264.

——— & Phil Williams 1990a: "Superpower Competition and Crisis Prevention in the Third World," in Allison & Williams (eds.): 1–25.

——— & Phil Williams (eds.) 1990: *Superpower Competition and Crisis Prevention in the Third World* (Cambridge: Cambridge U.P.).

Alpher, Joseph 1992: "Security Arrangements for a Palestinian Settlement," *Survival,* 34:4 (Winter): 49–67.

Alt, Franz 1983: *Frieden ist möglich* (München: Piper).

—— 1986: "Die geistige Atombombe: Vertrauen," in Gohl & Niesporek (eds.): 21–28.

Altenburg, Wolfgang 1990: "Die Zukunft der NATO," in Weizsäcker (ed.): 303–316.

Altermann, Eric A. 1985: "Central Europe: Misperceived Threats and Unforeseen Dangers," *WPJ,* 2:4 (Fall): 681–709.

Alternative Defence Commission: see ADC.

Alternative Security Working Group 1989: "The Next Steps in Europe" (London: Alternative Security Working Group).

Altmann, Jürgen 1991: "Abwehrsysteme gegen Kurz- und Mittelstreckenraketen," in Müller & Neuneck (eds.): 249–263.

—— & Bernhard Gonsior 1989: "Nahsensoren für die kooperative Verifikation der Abrüstung von konventionellen Waffen," *S+F,* 7:2, 77–81. Also in Müller, E. & Neuneck (eds.) 1990: 147–160.

——, Benoit Morel, Theodore Postol & Thomas Risse-Kappen 1987: "Anti-tactical Missile Defenses and West European Security," *PRIF-R,* 3.

—— & Joseph Rotblat (eds.) 1989: *Verification of Arms Reductions: Nuclear, Conventional and Chemical* (Berlin: Springer).

Aly, Abdel Mohem Said 1992: "Quality vs. Quantity: The Arab Perspective of the Arms Race in the Middle East," in Stahl & Kemp (eds.): 61–74.

—— 1992a: "Arms Control and the Resolution of the Arab-Israeli Conflict: An Arab Perspective," in Spiegel (ed.): 151–158.

Amirov, Oleg, Nikolai Kishilov, Vadim Makarevsky & Yuri Usachev 1988: "'Conventional War': Strategic Concepts," in IMEMO (eds.): 347–370.

——, Nikolai Kishilov, Vadim Makarevsky & Yuri Vsachev 1988a: "Problems of Reducing Military Confrontation," in IMEMO (eds.): 371–400.

Ammon, Herbert 1990: "The Uncertainties of Détente and the Necessity of a Political Alternative to Bloc Confrontation in Europe," in Randle & Rogers (eds.): 109–120.

—— & Peter Brandt (eds.) 1981: *Die Linke und die nationale Frage* (Reinbek: Rowohlt).

Amundsen, Kirsten 1985: "Soviet Submarines in Scandinavian Waters: Implications for Western Security," *Atlantic Community Quarterly,* 23:3, 251–261.

Andersen, F. Søgaard (ed.) 1975: *Ikkevold: Idé og virkelighed* (Copenhagen: Rhodos).

Andersen, Inger Bjørn 1990: "Norden som fredszone," in Hansen (ed.): 136–145.

Andrén, Nils 1987: "Looking at the Superpowers," in Sundelius (ed.): 160–181.

—— 1987a: "The Case of Sweden," in Lodgaard & Birnbaum (eds.): 134–154.

—— 1988: "Svensk neutralitetspolitik inför nya utmaningar," *Internasjonal Politikk,* 1: 57–72.

—— 1989: "Swedish Defence: Traditions, Perceptions and Policies," in Kruzel & Haltzel (eds.): 175–199.

—— 1990: "Thinking About a New European Structure," in Clesse & Rühl (eds.): 89–102.

—— 1991: "On the Meaning and Uses of Neutrality," *Cooperation and Conflict,* 26:2 (June): 67–84.

Andrew, Christopher & Oleg Gordievsky 1990: *KGB: The Inside Story of Its Foreign Operations from Lenin to Gorbachev* (London: Hodder & Stoughton).

Anon. 1991: "Kjernevåpenfri sone i nordisk område: Rapport fra den nordiske embetsmannsgruppe for undersøkelse av forutsetningene for en kjernevåpenfri sone i nordisk område," *Aktuelle utenriksspörsmål*, 1 (Oslo: Utenriksdepartementet).

——— (ed.) 1969: *Civilian Defence—Gewaltloser Widerstand als Form der Verteidigungspolitik (Tagungsbericht)* (Biehlefeld: Bertelsman).

——— (ed.) 1986: *The U.S.S.R. Proposes Disarmament (1920s–1980s)* (Moscow: Progress).

——— (ed.) 1993: "Wehrpflicht im Umbruch," *F&A*, 1.

Ansprenger, Franz 1993: "Blauhelme—Hoffnung und Alpdruck der Vereinten Nationen," *Blätter*, 38:11 (November): 1321–1332.

Antal, John F. 1992: "Maneuver versus Attrition: A Historical Perspective," *MR*, 72:10 (October): 21–33.

Anthony, Ian 1990: *The Naval Arms Trade* (London: Oxford U.P., SIPRI).

——— 1991: "Introduction," in idem (ed.): 1–13.

——— 1991a: "The Missile Technology Control Regime," in Anthony (ed.): 219–227.

——— 1993: "Assessing the UN Register of Conventional Arms," *Survival*, 35:4 (Winter): 113–129.

——— (ed.) 1991: *Arms Export Regulations* (Oxford: Oxford U.P., SIPRI).

———, Agnes Courades Allebeck, Gerd Hagmeyer-Gaverus, Paolo Miggiano & Herbert Wulf 1991: "The Trade in Major Conventional Weapons," in *SIPRI*, 1991: 197–231 (with appendices).

———, Paul Eavis & Owen Greene 1991: *Regulating Arms Exports: A Programme for the European Community* (Bristol: Saferworld).

Arbatov, Alexei G. 1988: "Military Doctrines," in IMEMO (eds.): 201–224.

——— 1988a: "The ABM Treaty and the Prevention of the Arms Race in Space," in IMEMO (eds.): 67–108.

——— 1988b: "The Problem of Reducing Strategic Offensive Forces," in IMEMO (eds.): 109–130.

——— 1988c: "Criteria of Correlation of Strategic Offensive Forces," in IMEMO (eds.): 225–238.

——— 1989: "Defence Dilemmas," *SMR*, 9: 22–23.

——— 1989a: "Defence Sufficiency and the Restructuring of the Armed Forces," in IMEMO (eds.): 211–224.

——— 1989b: "How Much Defence is Sufficient?" *IA(M)*, 4: 31–44.

——— 1989c: "Problems of Reducing the Military Confrontation," in Blackwill & Larrabee (eds.): 258–279.

——— 1989d: "Concepts of Conventional Stability and Reductions of Arms in Europe," in Altmann & Rotblat (eds.): 126–130.

——— 1989e: "Reduction of Strategic Nuclear Arms," in IMEMO (eds.): 27–60.

——— 1989f: "Space-Based ABM Weapons and the ABM Treaty," in IMEMO (eds.): 61–100.

——— 1990: "Defence Dilemmas," in Boserup & Neild (eds.): 12–18.

——— 1990a "Problems of Defensive Defence," in Cerutti & Ragionieri (eds.): 163–168.

——— 1990b: "Rethinking Nuclear Deterrence: In Search of a New Basis for European Security," in Cerutti & Ragionieri (eds.): 39–43.

——— 1990c: "Prospects for Tactical Nuclear Arms Control in Europe," in Clesse & Rühl (eds.): 254–263.

——— 1992: "Dilemmas of Strategic Defence and START II," in Ball & Horner (eds.): 154–177.

———— 1992a: "The Soviet Union, Naval Arms Control and the Norwegian Sea," in Fürst & al. (eds.): 44–66.

———— & A. Kalladin 1992: "2000: Prospects of Control over Arms and Conversion," *Ways to Security,* 1/7 (Moscow: Peace Research Institute): 3–26.

————, Oleg Amirov & Nikolai Kishilov 1989: "Assessing the NATO-WTO Military Balance in Europe," in Blackwill & Larrabee (eds.): 66–89.

———— & Thomas Cochran 1991: "Finale Is Important, Not START," *New Times,* 45: 18–20.

————, Nikolai Kishilov, Oleg Amirov & Yuri Streltsov 1989: "Negotiations on Conventional Armed Forces in Europe," in IMEMO (eds.): 275–318.

———— & Gennadi Lednev 1988: "Strategic Equilibrium and Stability," in IMEMO (eds.): 239–264.

———— & Alexander Savelyev 1988: "Command, Control, Communications and Intelligence System as a Factor of Strategic Balance," in IMEMO (eds.): 265–287.

———— & Alexander Savelyev 1989: "Military Doctrines: Discussions in the East," in IMEMO (eds.): 197–210.

Arbatov, Georgy 1990: "Glasnost, Talks and Disarmament," in Boserup & Neild (eds.): 187–194.

———— 1991: Preface to Smoke & Kortunov (eds.): xiii–xxiii.

———— & Willem Oltmans 1983: *Cold War or Détente? The Soviet Viewpoint* (London: Zed).

Arbatova, Nadezhda, Vladimir Baranovsky, Gennadi Kolosov, Alexander Kokeyev & Andrei Chervyakov 1989: "Security from a West European Point of View," in IMEMO (eds.): 351–384.

Arbess, Daniel J. & Andrew M. Moravcsik 1988: "Lengthening the Fuse: No First Use and Disengagement," in Nye, Allison & Carnesale (eds.): 69–92.

Arbucle, Tammy 1992: "Yugoslavia. Strategy and Tactics of Ethnic Warfare," *IDR,* 25:1 (January): 19–22.

Archer, Clive 1989: "Western Reponses to the Murmansk Initiative," *Centre Piece,* 14 (Aberdeen: Centre for Defence Studies).

———— (ed.) 1988: *The Soviet Union and Northern Waters* (London: Routledge).

Archibugi, Daniele 1989: "Peace and Democracy: Why Such an Unhappy Marriage?" in Harle (ed.): 33–86.

———— 1993: "The Reform of the UN and Cosmopolitican Democracy: A Critical Review," *JPR,* 30:3 (August): 301–315.

Arkin, William M. 1990: "Controlling Superpower Naval Operations," in Fieldhouse (ed.): 144–157.

————, Damian Durrant & Hans Kristensen 1991: "Nuclear Weapons Headed for the Trash," *BAS,* 47:10 (December): 15–19.

———— & Richard Fieldhouse 1985: *Nuclear Battlefields: Global Links in the Arms Race* (Cambridge, Mass.: Ballinger).

Armengol, Vicenç Fisas 1985: *Una alternativa a la política de defensa en España* (Barcelona: Editorial Fontamara).

———— 1990: "Capacidades e intenciones: El debate sobre las armas offensivas/defensivas," *Politica Exteriori,* 16: 124–130.

———— 1990a: *Defensa 2001: Una proposta de defensa no offensiva per a Espana* (Barcelona: Centre UNESCO de Catalunya).

———— 1990b: "El CIP en el Congreso II: Modelos de fuerzas armadas en el contexto de la actual situation internacional," *Papeles para la Paz,* 39–40 (Madrid: Centro de Investigation para la Paz): 24–34.

———— 1994: "NOD in the Western Mediterranean," in Møller & Wiberg (eds.): 166–175.

Armitage, M. J. & R. A. Mason 1983: *Air Power in the Nuclear Age, 1945–82* (London: Macmillan).

Armstrong, John A. 1964: Introduction to idem (ed.): 3–72.

———— (ed.) 1964: *Soviet Partisans in World War II* (Madison: University of Wisconsin Press).

———— & Kurt DeWitt 1964: "Organization and Control of the Partisan Movement," in Armstrong (ed.): 73–140.

Arnett, Robert L. 1990: "Arms Control in Soviet National Security Policy Under Gorbachev," in Hudson (ed.): 277–298.

———— 1990a: "Perestroika in Decision-Making in Soviet National Security Policy," *JSMS*, 3:1 (March): 32–45.

———— 1994: "Russia After the Crisis: Can Civilians Control the Military?" *Orbis*, 38:1 (Winter): 41–57.

Arnold, Klaus 1990: "KSE-Verifikation: Vertrauenskredit-nicht ohne Bürgschaft," *EW*, 9: 539–544.

Arnswald, Ulrich 1992: "Ende der Sowjetunion—Rüstungskontrolle am Ende?" in Birkenback & al. (eds.): 138–148.

Aron, Raymond 1983: *Clausewitz: Philosopher of War* (London: Routledge).

———— 1984: *Paix et guerre entre les nations,* 8th ed. (Paris: Calmann-Levy).

Art, Robert J. 1991: "A Defensible Defense: America's Grand Strategy After the Cold War," *IS*, 15:1 (Spring): 5–53.

———— & Kenneth N. Waltz (eds.) 1965: *The Use of Force: International Politics and Foreign Policy* (Boston: Little Brown).

Asada, Mashiko 1990: "Revived Soviet Interest in Asia: A New Approach," in Langdon & Ross (eds.): 35–71.

Asmus, Ronald D., Richard L. Kugler & F. Stephen Larrabee 1993: "Building a New NATO," *FA*, 72:4 (September–October): 28–40.

Aspin, Les 1987: "Conventional Weapons in Europe: Unilateral Moves for Stability," *BAS*, 10: 12–16.

———— & William Dickinson 1992: *Defense for a New Era: Lessons of the Persian Gulf War* (Washington, D.C.: Brassey's).

Atkeson, Edward B. 1993: "Theatre Forces in the Commonwealth of Independent States," in Goodby & Morel (eds.): 113–149.

———— 1993a: "US Theatre Forces in the Year 2000," in Goodby & Morel (eds.): 150–168.

Auffermann, Burkhard 1988: "Neutralism as an Alternative Security Concept," *CRPV*, 11:3, 83–86.

———— 1990: "Introduction: Non-Offensive Defence—A Concept Wins Political Acceptance," in idem (ed.): 5–12.

———— & Pertti Joenniemi 1989: "Neutrality in the Changing Europe: Stagnation or New Thinking," in Karl & Auffermann (eds.): 42–48.

———— (ed.) 1990: "NOD or Disarmament in the Changing Europe?" *Research Reports*, 40 (Tampere: Tampere Peace Research Institute).

Aupers, Gerard J. 1988: "Consequences of Conventional Warfare: Essentials of Research," *BPP*, 19:2, 191–195.

Avenhaus, Rudolf, Reinar K. Huber & John D. Kettelle (eds.) 1986: *Modelling and Analysis in Arms Control* (Berlin: Springer).

Avenhaus, Rudolf, Hassane Karkar & Michel Rudnianski (eds.) 1991: *Defense Decision Making: Analytical Support and Crisis Management.* Proceedings of the First ARESAD International Conference on Decision Making and Defense, Paris, 22–23 November 1989 (Berlin: Springer).

Awad, Mubarak L. 1985: "Palestine: Pour une résistance nonviolente," *Alternatives Non Violentes*, 55: 29–39.

Axelrod, Robert 1979: "Rational Timing of Surprise," *WP*, 31:1 (January): 228–246.

———— 1984: *The Evolution of Cooperation* (New York: Basic Books).

———— 1990: "The Concept of Stability in the Context of Conventional War in Europe," *JPR*, 27:3, 247–254.

———— & Robert O. Keohane 1985: "Achieving Cooperation Under Anarchy: Strategies and Institutions," *WP*, 38:1 (October): 226–254.

Axt, Heinz-Jürgen 1993: "Kooperation unter Konkurrenten: Das Regime als Theorie der aussenpolitischen Zusammenarbeit der EG-Staaten," *Zeitschrift für Politik*, 40:3 (September): 241–259.

Ayoob, Mohammed 1992: "The Security Predicament of the Third World State: Reflections on State Making in a Comparative Perspective," in Job (ed.): 63–80.

———— 1993: "Squaring the Circle: Collective Security in a System of States," in Weiss (ed.): 45–62.

Babbage, Ross 1990: "Trends in Conventional Warfare and Technologies," in Ball & Downes (ed.): 108–128.

Babst, Stefanie & Heribert Schaller 1993: "Die Ukraine—ein nukleares Sicherheitsrisiko," *EuS*, 42:9 (September): 446–449.

Bacarrere-Becane, Catherine 1991: "La politique de défense de l'URSS et des pays membres de l'Organization du Traité de Varsovie," in Buffotot (ed.): 161–174.

Bächler, Günther 1990: "Gewaltverzicht durch Demokratisierung: Dimensionen der Demokratisierung der friedens- und sicherheitspolitischen Diskussion," in Vogt (ed.): 176–187.

———— 1990a: "Ein Funke springt über: Die Schweizer Kampagne zur Abschaffung der Armee," in Birkenbach & al. (eds.): 218–227.

———— 1990b: "Friedensgestaltung durch Demokratisierung," in Calliess (ed.): 792–810.

———— 1992: "'Ein Beitrag an die internationale Stabilität: Vornehmlich in Europa': Demokratische Gestaltung des Friedens," in Bächler (ed.): 195–257.

———— 1992a: "Friedensgestaltung durch Demokratisierung," in Jopp (ed.): 125–142.

———— (ed.) 1992: Friede und Freiheit: Die Schweitz in Europa (Chur, Switzerland: Rüegger).

———— (ed.) 1992a: *Perspektiven: Friedens- und Konfliktforschung* in Zeiten des Umbruchs (Chur, Switzerland: Rüegger).

———— & Christiane Rix 1987: "Sicherheitspolitische Differenzierungsprozesse, Wandel der Rolle des Militärischen in den 'Ost-Ost-Beziehungen'," *HB*, 21.

Backerra, Manfred (ed.) 1993: *NVA: Ein Rückblick für die Zukunft* (Köln: Markus).

Bader, Rolf 1991: "Strukturelle Nichtverteidigbarkeit hochindustrialisierter Staaten und Gemeinsamer Frieden," in Lutz (ed.): 33–48.

Baev, Pavel K. 1993: "'Zero Options' for Cooperative Denuclearization," *JPR*, 30:4 (November): 455–460.

———— 1993: "Peace-keeping as a Challenge to European Borders," *Security Dialogue*, 24:2 (June): 137–150.

Bagnall, Nigel 1991: "The Central Region: A Strategic Overview," in Reid & Dewar (eds.): 239–242.

Bahr, Egon 1963: "Wandel durch Annäherung," in Bahr 1991: 11–17.

———— 1983: "Peace: A State of Emergency," in Steinke & Vale (eds.): 141–154.

———— 1984: "Bericht der Arbeitsgruppe 'Neue Strategien' beim SPD-Parteivorstand vom Juli 1983," in Brauch (ed.): 275–290.

———— 1985: "Die Chancen der Geschichte in der Teilung suchen," *S+F*, 3:2, 70–72.

———— 1985a: "Gemeinsame Sicherheit: Voraussetzung für Kollektive Sicherheit," in Lutz (ed.): 98–106.

———— 1985b: "Observations on the Principle of Common Security," in Väyrynen (ed.): 31–36.

———— 1986: "Gemeinsame Sicherheit: Einführende Überlegungen," in Bahr & Lutz (eds.): 15–27.

———— 1986a: "Von der Strategie der Abschrekung zur gemeinsamen Sicherheit," in Ehmke & al. (eds.): 95–102.

———— 1987: "Gemeinsame Sicherheit: Ein Weg aus der Sackgasse," in Bahr & Lutz (eds.): 19–28.

———— 1987a: "Korridor ohne Waffen für den Angriff," *EW*, 3: 140–143.

———— 1987b: "Strukturelle Nichtangriffsfähigkeit: Aktuelle Anmerkungen," *S+F*, 5:1, 3–5.

———— 1988: "Militärische Konfrontation: Konventionelle Stabilität. Strukturelle Nichtangriffsfähigkeit," in Bahr & Lutz (eds.): 13–24.

———— 1988a: *Zum Europäischen Frieden: Eine Antwort auf Gorbatschow* (Berlin: Siedler).

———— 1989: "Das Bündnis und deutsche Interessen" (manuscript for a speech to the III International Wehrkundetagung 1989 in Munich), *F&A*, 29 (1/89): 29–36.

———— 1989a: "Perspectives pour la sécurité européenne," *Memento Defense— Desarmament 1989* (Brussels: GRIP): 117–128.

———— 1990: "Für den Westen und den Osten: Die Zeit der grossen Schritte," in Heisenberg & Lutz (eds.): II:328–339.

———— 1991: *Sicherheit für und vor Deutschland: Vom Wandel durch Annäherung zur Europäischen Sicherheitsgemeinschaft* (München: Carl Hanser).

———— 1991a: "Europa nach dem Golfkrieg," *S+F*, 9:3, 116–118.

———— 1993: "Die Verfassung steht über dem Bündnis," in Lutz (ed.): 53–62.

———— & Dieter S. Lutz (eds.) 1986: *Gemeinsame Sicherheit: Idee und Konzept, Bd. I: Zu den Ausgangsüberlegungen, Grundlagen und Strukturmerkmalen Gemeinsamer Sicherheit* (Baden-Baden: Nomos).

———— & Dieter S. Lutz (eds.) 1987: *Gemeinsame Sicherheit: Dimensionen und Disziplinen. Bd. II: Zu rechtlichen, ökonomischen, psychologischen und militärischen Aspekten Gemeinsamer Sicherheit* (Baden-Baden: Nomos).

———— & Dieter S. Lutz (eds.) 1988: *Gemeinsame Sicherheit: Konventionelle Stabilität. Bd. III: Zu den militärischen Aspekten Struktureller Nichtangriffsfähigkeit im Rahmen Gemeinsamer Sicherheit* (Baden-Baden: Nomos).

————, Dieter S. Lutz, Erwin Müller & Reinhard Mutz 1988: "Defensive Zonen: Stellungnahme der IFSH zur Strukturellen Nichtangriffsfähigkeit und Konventionellen Stabilität in Europa," in Bahr & Lutz (eds.): 483–492.

————, Andreas von Bülow & Karsten Voigt 1989: "Europäische Sicherheit 2000: Überlegungen zu einem Gesamtkonzept für die Sicherheit Europas aus Sozialdemokratischer Sicht," *F&A*, 30 (2/89).

Bahr, Hans Eckehard 1990: "Von der Armee zur europäischen Friedenstruppe," in idem (ed.): 9–48.

———— (ed.) 1990: *Von der Armee zur europäischen Friedenstruppe* (München: Knaur).

Bailey, Kathleen C. 1991: "Arms Control for the Middle East," *IDR*, 4: 311–314.

———— 1991a: "Ballistic Missile Proliferation: Can It Be Reversed?" *Orbis*, 35:1 (Winter): 5–14.

———— 1991b: *Doomsday Weapons in the Hands of Many* (Urbana: University of Illinois Press).

Bajusz, William D. & Lisa D. Shaw 1990: "The Forthcoming 'SNF Negotiations'," *Survival,* 32:4, 333–347.

Baker, Ray Stannard 1923: *Woodrow Wilson and World Settlement,* 3 vols. (New York: Doubleday, Page).

Bak-Nielsen, Aase 1990: "Danmarks bedste forsvar," in Hansen (ed.): 15–23.

Balazs, Judit & Håkan Wiberg (eds.) 1993: *Peace Research for the 1990s* (Budapest: Akadémiai Kiadó).

Bald, Detlev 1987: "Die Lage Deutschlands und der Krieg der Zukunft: Sicherheitspolitische und militärische Planungen des deutschen Generalstabs in den Jahren 1924/25," in idem (ed.): 15–52.

———— 1987a: "Erfahrungen des Weltkrieges und Ansätze zu einem alternativen Verteidigungskonzept in den fünfziger Jahren, die Miliz," in idem (ed.): 71–82.

———— 1989: "Traditions in Military-Strategic Thought in Germany and the Problem of Deterrence." Paper presented to the 30th ISA Convention.

———— 1992: "Staat, Militär und die Erhaltung des Friedens," in Jopp (ed.): 143–164.

———— (ed.) 1987: *Militz als Vorbild? Zum Reservistenkonzept der Bundeswehr* (Baden-Baden: Nomos).

———— (ed.) 1992: *Die Nationale Volksarmee: Beiträge zu Selbstverständnis und Geschichte des deutschen Militärs von 1945–1990* (Baden-Baden: Nomos).

———— & Paul Klein (eds.) 1988: *Wehrstruktur der neunziger Jahre: Reservistenarmee, Miliz oder . . . ?* (Baden-Baden: Nomos).

————, Thomas Krämer-Badoni & Roland Wakenhut 1981: "Innere Führung und Sozialisation: Ein Beitrag zur Sozio-Psychologie des Militärs," in Steinweg (ed.): 134–166.

Ball, Desmond 1993: "Strategic Culture in the Asia-Pacific Region," *SS,* 3:1 (Autumn): 44–74.

———— & al. 1987: *Crisis Stability and Nuclear War* (Ithaca, N.Y.: AAAS & Cornell University Peace Studies Program).

———— & Cathy Downes (eds.) 1990: *Security and Defence: Pacific and Global Perspectives* (Sydney: Allan & Unwin).

———— & David Horner (eds.) 1992: *Strategic Studies in a Changing World: Global, Regional and Australian Perspectives* (Canberra: Australian National University, Research School of Pacific Studies, Strategic and Defence Studies Centre).

———— & Jeffrey Richelson (eds.) 1986: *Strategic Nuclear Targeting* (Ithaca: Cornell U.P.).

Ball, Nichole 1988: *Security and Economy in the Third World* (Princeton, N.J.: Princeton U.P.).

———— 1989: "International Experience of Conversion," in Dumas & Thee (eds.): 17–22.

Ballentine, Karen 1991: "Soviet Defence Industry Reform: The Problems of Conversion in an Unconverted Economy," *Background Paper,* 36 (Ottawa: Canadian Institute for International Peace and Security).

Bamberg, Sebastian & al. 1991: "Konversion: Eine Strategie für soziale Bewegungen zur Veränderung der Industriegesellschaft? Projekte, Ansätze, Perspektiven," *F&A,* 39.

Banerjee, D. 1989: "Non-provocative Defence: Conceptual Framework and Parameters," in Singh & Vekaric (eds.): 111–134.

Banner, Allen V., Andrew J. Young & Keith W. Hall 1990: *Aerial Reconnaissance for Verification of Arms Limitation Agreements: An Introduction* (New York: UNIDIR).

Baranovsky, Vladimir G. 1990: "Europe: Emergence of a New International Political Structure," *IMEMO Discussion Papers,* no. 5 (Moscow: IMEMO).

—— 1993: "Post-Soviet Conflict Heritage and Risks," *SIPRI:* 131–170.

—— & Yuri Streltsov 1988: "Multilateral Negotiations Within the Framework of the Conference of Security and Cooperation in Europe: Military-Political Questions," in IMEMO (eds.): 331–346.

Barany, Zoltan D. 1992: "East European Armed Forces in Transitions and Beyond," *East European Quarterly,* 26:1 (March): 1–30.

Bardehle, Peter 1993: "Kooperative Denuklearisierung: Ein neues Konzept der amerikanischen Sicherheitspolitik und seine Probleme," *EA,* 48:5 (10 March): 140–148.

Barfoed, Niels, Thomas Bredsdorff, Leif Christensen & Ove Nathan (eds.) 1989: *The Challenge of an Open World: Essays Dedicated to Niels Bohr* (Copenhagen: Munksgaard).

Barnaby, Frank 1984: *Future War: Armed Conflict in the Next Decade* (London: Michael Joseph).

—— 1985: "Must the Nuclear Arms Race Lead to War?" in Huzzard & Meredith (eds.): 17–33.

—— 1986: *The Automated Battlefield* (New York: Free Press).

—— 1986a: *What on Earth Is Star Wars?* (London: Fourth Estate).

—— 1990: "New Military Technologies Suit the Needs of a New European Security System," *Discussion Paper* (Appleby, Cumbria, U.K.: Just Defence).

—— 1992: *The Role and Control of Weapons in the 1990s* (London: Routledge).

—— 1993: *How Nuclear Weapons Spread: Nuclear-Weapon Proliferation in the 1990s* (London: Routledge).

—— (ed.) 1991: *Building a More Democratic United Nations* (London: Frank Cass).

—— & Egbert Boeker 1982: "Defence Without Offence: Non-nuclear Defence for Europe," *Peace Studies Papers,* 8 (Bradford: School of Peace Studies).

—— & Egbert Boeker 1988: "Non-Nuclear, Non-Provocative Defence for Europe," in Hopmann & Barnaby (eds.): 135–145.

—— & Egbert Boeker 1989: "Non-Provocative, Non-Nuclear Defence of Western Europe," in Tromp (ed.): 199–212.

—— & Marlies ter Borg 1986: "Problems Facing NATO," in Barnaby & ter Borg (eds.): 3–17.

—— & Marlies ter Borg (eds.) 1986: *Emerging Technologies and Military Doctrine: A Political Assessment* (London: Macmillan).

—— & Stan Windass 1983: "What Is Just Defence?" (Oxford: Just Defence).

Barnett, Richard 1970: *The Economy of Death: A Hard Look at the Defense Budget, the Military Industrial Complex, and What You Can Do About Them* (New York: Atheneum).

Barnett, Robert W. 1984: *Beyond War: Japan's Concept of Comprehensive National Security* (Washington, D.C.: Brassey's).

Barth, Peter 1988: "Raumverteidigung," in Lippert & Wachtler (eds.): 314–329.

—— 1990: "Die Architektur einer neuen europäischen Friedensordnung," in Köllner & Huck (eds.): 117–149.

—— 1990a: "Konventionelle Abrüstung und Sicherheit in Europa," in Köllner & Huck (eds.): 59–116.

Bartke, Matthias 1991: "Entstehung und Grundprinzipien der kollektiven Sicherheit," in Lutz (ed.): 115–132.

—— & Margret Johannsen (eds.) 1990: "Zur Zukunft Deutschlands: Reden vom 2. internationalen Ost-West-Workshop über Gemeinsame Sicherheit," *HB,* 44.

Barylski, Robert V. 1992: "The Soviet Military Before and After the August Coup:

Departmentalization and Decentralization," *Armed Forces and Society,* 19:1 (Fall): 27–46.

Basler, Gerhard 1990: "Die 'Herbstrevolution' und die Ost-West-Beziehungen der DDR," *EA,* 1: 13–18.

Bastian, Till 1993: *Frieden schaffen mit deutschen Waffen: Krieg als Mittel der Politik? Plädoyer für ein ziviles Deutschland* (Köln: PapyRossa).

Batani, Dimitri 1989: "Il problema dei sistemi ATBM e antiaerei in Italia," in Ragionieri (ed.): 183–209.

Bathurst, Robert B. 1990: "The Soviet Navy Through Western Eyes," in Gillette & Frank (eds.): 59–80.

Batiouk, Victor 1992: "Ukraine's Non-Nuclear Option," *Research Paper,* 14 (New York: United Nations; Geneva: UNIDIR).

Baudisch, Kurt (ed.) 1988: *European Security and Non-Offensive Defence* (Berlin: WFSW).

Baudissin, Wolf Graf von 1982: "Vertrauensbildende Massnahmen—Ihre Bedingungen und Perspektiven," in Lutz & Müller (eds.): 11–22.

———— 1986: "Dreissig Jahre Bundeswehr: Licht und Schatten," in Borkenhagen (ed.): 15–32.

———— (ed.) 1983: *From Distrust to Confidence: Concepts, Experiences and Dimensions of Confidence-Building Measures,* vol. II (Baden-Baden: Nomos).

Bay, Christian 1982: "Vad bör vi försvara?" in Ofstad & Öhberg (eds.): 180–208.

Baylis, John 1983: "Introduction: Defence Policy for the 1980s and 1990s," in Baylis (ed.): 1–30.

———— 1987: "Revolutionary Warfare," in Baylis & al.: I:209–232.

———— 1987a: "Chinese Defence Policy," in Baylis & al.: II:113–141.

———— 1993: "Britain and the Future of NATO," in Clarke & Sabin (eds.): 3–20.

———— (ed.) 1983: *Alternative Approaches to British Defence Policy* (London: Macmillan).

————, Ken Booth, John Garnett & Phil Williams 1987: *Contemporary Strategy,* 2d ed., vol. I, *Theories and Assumptions;* vol. II, *The Nuclear Powers* (London: Croom Helm).

Beach, Hugh 1985: "Comment on Chapter 3," in Roper (ed.): 57–60.

———— 1986: "On Improving NATO Strategy," in Pierre (ed.): 152–185.

———— 1988: "Flexible Response and Nuclear Weapons: A British View," in Kaiser & Roper (eds.): 126–136.

———— 1990: "Conventional Forces in Europe: Deep Cuts and Security," in Boserup & Neild (eds.): 204–214.

Beaufre, André 1963: *Introduction á la Strategie* (Paris: Collin).

———— 1965: *Deterrence and Strategy* (London: Faber & Faber).

———— 1972: *La guerre revolutionnaire: Les nouvelles formes de la guerre* (Paris: Fayard).

———— 1974: *Strategy for Tomorrow* (London: Jane's).

Beazley, Kim 1990: "Response," in Cheeseman & Kettle (eds.): 209–221.

Bebermeyer, Hartmut 1984: "Verteidigungsökonomie am Scheideweg," in SAS (eds.): 9–34.

———— 1989: "Die Sicherheit und ihre Kosten," in SAS (eds.): 118–136.

———— 1990: "The Fiscal Crisis of the Bundeswehr," in Brauch & Kennedy (eds.): 128–147.

———— & Bernd Grass 1984: "Unsere Streitkräfte auf dem Weg in die Ressourcenkrise," in Brauch (ed.): 176–189.

———— & Lutz Unterseher 1986: *Eine künftige Bundesmarine im Rahmen einer defensiven Verteidigungskonzeption* (Bonn: SAS).

———— & Lutz Unterseher 1989: "Wider die Grossmannssucht zur See: Das Profil einer defensiven Marine," in SAS (eds.): 165–187.

———— & Lutz Unterseher 1992: "Zur Bestimmung der Friedensdividende: Fiskalische Konsequenzen von Alternativen deutscher Verteidigungsplanung," in Hartwig (ed.): 53–66.

Bebler, Anton 1984: "Konflikte zwischen sozialistischen Ländern," *ÖMZ*, 2: 112–120.

———— 1987: "Jugoslawiens nationale Verteidigung," parts 1, 2, *ÖMZ*, 4, 5: 301–310, 414–422.

———— 1989: "La nouvelle politique soviétique et la sécurite de la Yougoslavie," *Defense nationale*, 45 (July): 95–110.

———— 1991: "Staat im Staate: Zur Rolle des Militärs," in Furkes & Schlarp (eds.): 106–133.

———— 1991a: "The Armed Forces in the Yugoslav Conflict," *IDR*, 4: 306–309.

———— 1992: "Der Krieg in Jugoslawien, 1991–1992," *ÖMZ*, 30:5 (September–October): 397–411.

———— 1992a: "The Neutral and Non-Aligned States in the New European Security Architecture," *ES*, 1:2 (Summer): 133–143.

———— 1993: "The Yugoslav People's Army and the Fragmentation of a Nation," *MR*, 73:8 (August): 38–51.

———— 1993a: "Slovenia's Territorial Defense," *IDR*, 26:1 (January): 65–67.

Becher, Klaus 1990: "Security Co-operation in Europe: Domestic Challenges and Limitations," in Clesse & Rühl (eds.): 63–79.

Bechler, Rosemary (ed.) 1989: *Conventional and Non-Nuclear Disarmament: A Question for Peace Movements* (Helsinki: World Peace Council).

Behar, Nansen & I. Nedev 1983: "A Nuclear Weapon-Free Zone in the Balkans," in Lodgaard & Thee (eds.): 93–99.

Behrmann, Jörn 1990: "Temporalisierung interstruktureller Interaktion: Die Zeitpolitische Dimension des europäischen Einigungsprozesses," in Calliess (ed.): 811–854.

Bejar, Hector 1982: "The Failure of the Guerillas in Peru," in Chaliand (ed.): 282–299.

Belden, Jack 1973: *China Shakes the World* (Harmondsworth: Penguin Books).

Bell, Raymond E. 1990: "Die Zukunft der Infanteriedivision ('leicht')," *ÖMZ*, 1: 27–36.

Bellamy, Chris 1987: *The Future of Land Warfare* (London: Croom Helm).

———— 1990: *The Evolution of Modern Land Warfare: Theory and Practice* (London: Routledge).

———— 1992: "'Civilian Experts' and Russian Defence Thinking: The Renewed Relevance of Jan Bloch," *RJ*, 137:2 (April): 50–56.

Bellany, Ian 1990: "The Distinction Between Offensive and Defensive Armaments," *Arès*, 12 (1990/91): 64–68.

———— 1991: *A Basis for Arms Control* (Aldershot: Dartmouth).

Bender, Peter 1986: "Sicherheitspartnerschaft und friedliche Koexistenz: Zum Dialog zwischen SPD und SED," *NGFH*, 4: 342–348.

———— 1992: "Wandel durch Annäherung," in Lutz (ed.): 197–204.

Benjowski, Klaus & Wolfgang Schwarz 1988: "Military Issues of Security in Europe—Theoretical Aspects," *CRPV*, 11:3, 95–103.

———— & Wolfgang Schwartz 1989: "Militärische Aspekte der Sicherheit—gemeinsame Sicherheit statt Abschreckung," in Schmidt & al.: 106–122.

Benjowski, Klaus, Bertel Heurlin, Gregor Putensen & Håkan Wiberg (eds.) 1990: *Endzeit für Deutschlands Zweistaatlichkeit: Konsequenzen für Europa: Deutsch-dänische Reflxionen europäischer Sicherheitsprobleme*. Protokolle eines Symposiums über die deutsch-nordischen Beziehungen und die europäische Sicherheit in Kopenhagen, 3–4 Mai 1990.

Bennet, Jeremy 1967: "The Resistance Against the German Occupation of Denmark, 1940–1945," in Roberts (ed.): 154–172.

Bennett, G. C.: *The New Abolitionists: The Story of Nuclear-Free Zones* (Elgin, Ill.: Brethren).

Bennike, V. 1946: "Jernbanesabotagen i Jylland," in La Cour (ed.): III:81–96.

Bentinck, Marc 1986: "NATO's Out-of-Area Problem," *AP*, 211.

Berdal, Mats R. 1993: "Whither UN Peacekeeping: An Analysis of the Changing Military Requirements of UN Peacekeeping with Proposals for Its Enhancement," *AP*, 281.

————— 1994: "Peacekeeping in Europe," *AP*, 284 ("European Security after the Cold War. Part I"), pp. 60–78.

Berdennikov, Mikhail V. 1992: "Russia and Her Security Policies," *RJ*, 137:6 (December): 5–9.

Berg, Per & Gunilla Herolf 1984: "'Deep Strike': New Technologies for Conventional Interdiction," *SIPRI*, 1984: 291–318.

Berg, Rolf 1986: "Military Confidence-Building in Europe" in Lynch (ed.): 9–68.

Berger, Thomas U. 1993: "From Sword to Chrysanthemum. Japan's Culture of Anti-Militarism," *IS*, 17:4 (Spring): 119–150.

Bergfeldt, Lennart 1984: "Kompletteranda Motståndsformer," *Statens Offentliga Utretninger*, 1984 (Stockholm: Försvarsdepartementet): 10.

————— 1985: "Civilian Defence as a Complement," *Cooperation and Conflict*, 20:4, 279–286.

————— 1985a: "Complementary Forms of Defence: Report of the Swedish Commission on Resistance," *BPP*, 16:1, 21–33.

————— 1986: "Experiences of Civilian Resistance: The Case of Denmark 1940–1945, Report no. 3: Introduction, Analysis, Conclusions," (Uppsala: Uppsala University, Dept. of Peace and Conflict Research).

————— 1990: *Civilmotstånd: Vision ock verklighet* (Malmö: Pax).

Bergstein, Svend 1985: "Kritik af tekno-forsvaret," *Forsvaret Idag*, 5: 8–9.

Bertele, Manfred 1982: "Struktur und Dislozierung von Landstreitkräften in Europa: Mögliche Funktionen vereinbarter Beschränkungen bei der Schaffung militärischen Gleichgewichts," in Nehrlich (ed.): 375–420.

Bertram, Christoph 1987: "Gemeinsame Sicherheit und nukleare Abschreckung," in Bahr & Lutz (eds.): 277–281.

Bertrand, Maurice 1993: "The Historical Development of Efforts to Reform the UN," in Roberts & Kingsbury (eds.): 420–436.

Best, Geoffrey 1980: *Humanity in Warfare: The Modern History of the International Law of Armed Conflicts* (London: Methuen).

————— 1986: "Irregular Options for Western European Defence," in Paskins (ed.): 84–101.

Betts, Richard K. 1982: *Surprise Attack: Lessons for Defense Planning* (Washington, D.C.: Brookings).

————— 1983: "Intelligence 'Failures'," in Reichart & Sturm (eds.): 679–693.

————— 1985: "Surprise Attack and Preemption," in Allison, Carnesale & Nye (eds.): 54–79.

————— 1985a: *Preventing Nuclear War: A Realistic Approach* (Bloomington: Indiana U.P.).

————— 1985b: "Conventional Deterrence: Predictive Uncertainty and Policy Confidence," *WP*, 37:2 (January): 153–179.

————— 1992: "Systems for Peace or Causes of War? Collective Security, Arms Control, and the New Europe," *IS*, 17:1 (Summer): 5–43.

Bezrukov, Mikhail & Yuriy Davydov 1991: "The Common European Home and Mutual Security," in Smoke & Kortunov (eds.): 152–160.

—————— & Andrei Kortunov 1991: "Interdependence: A Perspective on Mutual Security," in Smoke & Kortunov (eds.): 10–18.

Bialer, Seweryn 1988: "'New Thinking' and Soviet Foreign Policy," *Survival,* 30:4, 291–309.

—————— 1989: "Where Should We Go from Here? An American View," in Allison & Ury (eds.): 265–283.

Biddle, Stephen D. 1988: "The European Conventional Balance: A Reinterpretation of the Debate," *Survival,* 30:2, 99–121.

Bidwai, Praful & Achin Vanaik 1991: "India and Pakistan," in Karp (ed.): 258–276.

Biedenkopf, Kurt H. 1986: "Die Akzeptanz einer Friedenssicherung mit Waffen," in Gohl & Niesporek (eds.): 28–44.

Biehle, Alfred (ed.) 1985: *Alternative Strategien: Das Hearing im Verteidigungsausschuss des Deutschen Bundestages: Die schriftlichen Gutachten und Stelllungnahmen* (Koblenz: Bernard & Graefe).

Bienen, Henry (ed.) 1992: *Power, Economics, and Security: The United States and Japan in Focus* (Boulder, Colo.: Westview).

Bills, Scott L. 1979: "The United States, NATO and the Colonial World," in Lawrence S. Kaplan & Robert W. Clawson (eds.), *NATO After Thirty Years* (Wilmington, Del.: Scholarly Resources): 149–164.

Binnendijk, Hans 1989: "NATO's Nuclear Modernization Dilemma," *Survival,* 31:2 (March–April): 137–155.

—————— 1990: "European Security Alternatives," in Clesse & Rühl (eds.): 175–182.

Binter, Josef 1986: "Österreichische Neutralität und Friedenspolitik: Fragestellungen, Folgerungen und Vorschläge," *Dialog,* 7: 307–340.

—————— 1986a: "Neutralität, alternative Sicherheit und Friedenspolitik," *Dialog,* 6: 181–242.

—————— 1987: "Mehr Sicherheit durch Aufrüstung? Probleme und Perspektiven der österreichischen Landesverteidigung," *Dialog,* 10: 297–312.

—————— 1989: "Neutralität und Defensivkonzept: Politische Aspekte 'Alternativer Sicherheitskonzeptionen'—Anmerkungen zur militärischen Sicherheit Österreichs," in Karl & Auffermann (eds.): 49–55.

—————— 1991: "Die Zukunft der Neutralität in einem Europa des Wandels," *Arbeitspapier,* 2 (Stadtschlaining: Austrian Institute for Peace Research and Peace Education).

—————— 1992: "Neutrality in a Changing World: End or Renaissance of a Concept?" *BPP,* 23:2 (June): 213–218.

Birkenbach, Hanne-Margret 1989: "Die SPD-SED-Vereinbarungen über den 'Streit der Ideologien und die gemeinsame Sicherheit': Eine Fallstudie zu den Chancen und Schwierigkeiten kommunikativer Friedensstrategien," *HB,* 37.

—————— 1990: "Frieden durch Streit? Politisch-psychologische Rahmenbedingungen für die Überwindung von Feindbildern," in Steinweg & Wellmann (eds.): 151–188.

—————— & al. 1986: "Transanlantic Crisis—A Framework for an Alternative West European Peace Policy?" in Kaldor & Falk (eds.): 113–142.

——————, Uli Jäger & Christian Wellmann (eds.) 1989: *Jahrbuch Frieden 1990: Ereignisse, Entwicklungen, Analysen* (München: Beck).

——————, Uli Jäger & Christian Wellmann (eds.) 1990: *Jahrbuch Frieden 1991: Ereignisse, Entwicklungen, Analysen* (München: Beck).

——————, Uli Jäger & Christian Wellmann (eds.) 1991: *Jahrbuch Frieden 1992: Ereignisse, Entwicklungen, Analysen* (München: Beck).

————, Uli Jäger & Christian Wellmann (eds.) 1992: *Jahrbuch Frieden 1993: Konflikte, Abrüstung, Friedensarbeit* (München: Beck).

————, Uli Jäger & Christian Wellmann (eds.) 1993: *Jahrbuch Frieden 1994: Konflikte, Abrüstung, Friedensarbeit* (München: Beck).

Birnbaum, Karl E. 1987: "The Neutral and Non-Aligned States in the CSCE Process," in Sundelius (ed.): 148–159.

———— 1987a: "Threat Perceptions and National Security," in Lodgaard & Birnbaum (eds.): 39–67.

———— 1987b: "Das Problem des Vertrauens in den Ost-West-Beziehungen," in Brauch (ed.): 66–73.

Bischof, Henrik 1991: "Systemkrise in Jugoslavien," *Studie*, February.

————, Wilhelm Bruns, Christian Krause, Wulf Lapins, Eckhard Lübkemeier & Oliver Thränert 1990: "Der künftige sicherheitspolitische Status Deutschlands: Probleme und Perspektiven der '2+4-Gespräche'," *KAAD*, 42.

Bitzinger, Richard A. 1991: "Neutrality for Eastern Europe: Problems and Prospects," *BPP*, 22:3 (September): 281–289.

Blackaby, Frank & al. 1989: *The 'Comprehensive Concept' of Defence and Disarmament for NATO: From Flexible Response to Mutual Defensive Superiority* (British American Security Information Council, Committee for National Security, and Alternative Security Working Group).

Blackaby, Frank, Josef Goldblatt & Sverre Lodgaard 1984: "No-First-Use of Nuclear Weapons—An Overview," in Blackaby, Goldblatt & Lodgaard (eds.): 3–26.

————, Josef Goldblatt & Sverre Lodgaard (eds.) 1984: *No-First-Use* (London: Taylor & Francis, SIPRI).

Blacker, Coit D. 1988: "The MBFR Experience," in George, Farley & Dallin (eds.): 123–143.

Blackwill, Robert D. 1986: "European Influences and Constraints on U.S. Policy Toward the Soviet Union," in Horelick (ed.): 127–151.

———— 1988: "The Security Implications of Conventional Arms Control," in Nehrlich & Thomson (eds.): 123–135.

———— 1989: "Conceptual Problems of Conventional Arms Control" (*IS*, 12:4) in Miller & Lynn-Jones (eds.): 272–291.

———— 1990: "Militärische Aspekte der europäischen Sicherheit nach INF," in Heisenberg & Lutz (eds.): I:314–332.

———— & F. Stephen Larrabee 1989: "What Next for Conventional Forces in Europe?" in Blackwill & Larrabee (eds.): 462–472.

———— & F. Stephen Larrabee (eds.) 1989: *Conventional Arms Control and East-West Security* (Durham: Duke U.P.).

———— & Jeffrey W. Legro 1989: "Constraining Ground Force Exercises of NATO and the Warsaw Pact," *IS*, 14:3, 68–98.

Blagovolin, Sergei E. 1990: "Soviet Military Power: How Much, What Kind, What For?" *IMEMO Discussion Papers*, 4 (Moscow: IMEMO).

Blair, Bruce 1985: *Strategic Command and Control: Redefining the Nuclear Threat* (Washington, D.C.: Brookings).

———— 1992: *The Logic of Accidental Nuclear War* (Washington, D.C.: Brookings).

Blaker, Jim 1987: "Transparancy, Inspections and Surprise Attacks," in Windass & Grove (eds.): II (no pagination).

Blank, Stephen & Thomas Durrell-Young 1992: "Challenges to Eastern European Security in the 1990s," *ES*, 1:3 (Autumn): 381–411.

Blechman, Barry M. 1983: "Do Negotiated Arms Limitations Have a Future?" in Reichart & Sturm (eds.): 408–419.

———, William J. Durch & Kevin P. O'Prey 1990: *NATO's Stake in the New Talks on Conventional Armed Forces in Europe* (London: Macmillan).

———, William J. Durch, W. Philip Ellis & Cathleen S. Fischer 1991: *Naval Arms Control: A Strategic Assessment* (New York: St. Martin's).

——— & Cathleen Fischer 1990: "West German Arms Control Policy," in Laird (ed.): 71–122.

Bloch, Jean [Ivan] de 1889: *La Guerre* (Traduction de l'ouvrage russe La Guerre Future aux points de vue Technique, Économique et Politique), 6 vols. (Paris: Paul Dupont).

Blomberg, Jaako 1993: "Finland's Evolving Security Policy," *NATO Review,* 41:1 (February): 12–16.

Bloomfield, Lincoln P. 1993: "Collective Security and US Interests," in Weiss (ed.): 189–208.

Bloomgarden, Alan H. 1989: "Aircraft and Arms Control: A Match Made in Vienna?" *Peace Studies Briefing,* no. 39 (Bradford: University of Bradford, Department of Peace Studies).

——— 1990: "Low Flying and Conventional Arms Reductions" (Bradford: University of Bradford, Department of Peace Studies).

Bluhm, Georg 1990: "Deutschland: Entspannung und der Frieden in Europa," in Weizsäcker (ed.): 54–76.

Blumenwitz, Dieter & Gottfried Zieger (eds.) 1989: *Die deutsche Frage im Spiegel der Parteien* (Bonn: Wissenschaft und Politik).

Bluth, Cristoph 1990: *New Thinking in Soviet Military Policy* (London: Pinter).

——— 1991: "Military and Security Issues," in Pravda (ed.): 227–260.

Boden, Dieter 1991: "Der KSE-Vertrag," *NGFH,* 38:2 (February): 107–110.

Bodie, William C. 1992: "Strategy and Successor States: Report from Kiev," *World Affairs,* 154:3 (Winter): 107–114.

Boeker, Egbert 1986: *Europese veiligheid: Alternatieven voor de huidige veiligheidspolitiek* (Amsterdam: VU Uitgeverij).

——— 1988: "Strukturelle Nichtangriffsfähigkeit aus niederländischer Sicht," in Bahr & Lutz (eds.): 149–154.

——— 1988a: "Was heisst Konventionelle Stabilität?" in Bahr & Lutz (eds.): 409–411.

——— 1990: "'Gemeinsame Sicherheit' und 'strukturelle Nichtangriffsfähigkeit'. Die Diskussion in den Niederländen," in Lutz & Schmähling (eds.): 245–252.

——— 1990a: "A Nonprovocative Conventional Posture in Europe with No First Use of Nuclear Weapons," in Schmähling (ed.): 117–136.

——— & Lutz Unterseher 1986: "Emphasizing Defence," in Barnaby & Borg (eds.): 89–109.

Boel, Erik 1988: *Socialdemokratiets atomvåbenpolitik, 1945–88* (Copenhagen: Akademisk Forlag).

Boelsen, Bjørn Watt (ed.) 1985: *Totalforsvar: Sikkerhed og Samfundsøkonomi* (Copenhagen: FOV).

Boëne, Bernard & Michel Louis Martin (eds.) 1991: *Conscription et Armée de Métier* (Paris: Fondation pour les études de défense nationale).

Bogdanov, Radomir & Andrei Kortunov 1989: "On the Balance of Power," *IA(M),* 8: 313.

Böge, Volker 1986: "Grüne und sozialdemokratische Friedenspolitik im Vergleich," *S+F,* 4:4, 199–202.

——— 1988: "Common Security—An Alternative?" *Green Peace Policy,* 13: 11–12.

——— 1990: "Rüstungssteuerung in der Sackgasse—mit einseitig-unabhängiger

Abrüstung einen neuen Anfang machen!" in Heisenberg & Lutz (eds.): I:369–384.

———— 1991: "Entmilitarisierung Deutschlands, Auflösung der Blöcke, Eurokollektive Sicherheit: Grundpfeiler einer neuen europäischen Friedensordnung," in Lutz (ed.): 147–160.

———— 1991a: "WVO-Auflösung und NATO-Modernisierung: Neue Militärordnung für Europa," in Birkenbach & al. (eds.): 137–151.

———— 1991b: "Mut zur Selbstbeschränkung: Deutsche Machtpolitik, das Grundgesetz und die UNO," *Blätter,* 36:7 (July): 818–831.

———— & Matthias Pack 1990: "Verhandeln statt handeln: Konventionelle Rüstungskontrolle in Wien," in Birkenbach & al. (eds.): 129–141.

———— & Irene Schulert 1986: "Abrüstung und Umrüstung bei Andreas von Bülow: Zur Kritik des 'Bülow-Papiers' aus Grüner Sicht," *S+F,* 4:3, 162–164.

———— & Peter Wilke 1984: *Sicherheitspolitische Alternativen: Bestandaufnahme und Vorschläge zur Diskussion* (Baden-Baden: Nomos).

Boggs, Marion W. 1941: *Attempts to Define and Limit "Aggressive" Armaments in Diplomacy and Strategy* (Columbia: University of Missouri Press).

Bok, Sissela 1985: "Common Insecurity," in Väyrynen (ed.): 19–30.

———— 1989: *A Strategy for Peace: Human Values and the Threat of War* (New York: Pantheon).

Bölkow, Ludwig 1990: "Ist Verteidigungsdominanz technisch möglich?" in Weizsäcker (ed.): 113–122.

Bollmann, Kaj 1990: "Et nyt trusselsbillede i Europa," in Hansen (ed.): 61–73.

Bolt, Richard 1990: "The New Australian Militarism," in Cheeseman & Kettle (eds.): 25–72.

Bolyatko, Anatoly V. 1989: "The Military Dimension of European Security," in Clesse & Schelling (eds.): 342–348.

Bolz, Klaus 1987: "Gemeinsame Sicherheit und Ost-West Wirtschaftsbeziehungen," in Bahr & Lutz (eds.): 129–142.

Bombík, Svetoslav & Ivo Samson 1993: "Security Perspectives for Central Europe: A Slovak View," *Perspectives: Review of Central European Affairs,* 1:1, 37–43.

Bomsdorf, Falk 1989: "Zur Nordinitiative der Sowjetunion," *AuP,* 40:1, 59–69.

———— 1989a: *Sicherheit im Norden Europas: Die Sicherheitspolitik der fünf nordischen Staaten und die Nordeuropapolitik der Sowjetunion* (Baden-Baden: Nomos).

Bond, Brian 1977: *Liddell Hart: A Study of His Military Thought* (reprint, London: Gregg Revivals, 1992).

———— & Martin Alexander 1986: "Liddell Hart and De Gaulle: The Doctrines of Limited Liability and Mobile Defense," in Paret (ed.): 598–623.

Bond, Douglas G. 1988: "The Nature and Meanings of Nonviolent Direct Action: An Exploratory Study," *JPR,* 25:1, 81–89.

Bonesteel, Ronald M. 1992: "Viability of a CIS Security System," *MR,* 72:12 (December): 36–47.

———— 1993: "The CIS Security System: Stagnating, in Transition, or on the Way Out?" *ES,* 2:1 (Spring): 115–138.

Bonin, Bogislav von 1954: "Juli-Studie," in Brill (ed.) 1989: 69–101 (with appendices).

———— 1955: "Angst vor Neutralität?" in Brill (ed.) 1976: 53–57.

———— 1955a: "Das Gleichgewicht der Kräfte," in Brill (ed.) 1976: 49–52.

———— 1955b: "Die militärische Bedeutung atomarer Waffen," in Brill (ed.) 1976: 32–52.

———— 1955c: "Wiederbewaffnung und gleichzeitig Sicherheit: Ein

Lösungsvorschlag für Regierung und Opposition," in Brill (ed.) 1989: 132–141.

———— 1956: "Januar-Denkschrift," in Brill (ed.) 1989: 255–266.

———— 1958: "Befehl im Widerstreit?" in Brill (ed.) 1976: 71–84.

———— 1958a: "Unfähigkeit oder böser Wille? Eine kritische Betrachtung der Militärpolitik der Bundesregierung," in Brill (ed.) 1976: 58–70.

———— 1966: "Die Schlacht von Kursk: Ein Modell für die Verteidigung der Bundesrepublik," in Brill (ed.) 1976: 94–105.

———— & Arthur Stegner 1955: "Stegner-Bonin-Plan," in Brill (ed.) 1989: 248–253.

————, Wend von Wietersheim & Smilo Freiherr von Lüttwitz 1955: "Wiedervereinigung und Wiederbewafnung—kein Gegensatz," in Brill (ed.) 1976: 21–26.

Booth, Ken 1977: "Warships and Political Influence," in Michael MccGwire & James McDonnel (eds.), *Soviet Naval Influence: Domestic and Foreign Dimensions* (New York: Praeger): 459–482.

———— 1983: "Strategy and Conscription," in Baylis (ed.): 154–190.

———— 1985: "The Case for Non-nuclear Defence," in Roper (ed.): 32–56.

———— 1987: "Disarmament and Arms Control," in Baylis & al. (eds.): I:140–186.

———— 1987a: "The Evolution of Strategic Thinking," in Baylis & al. (eds.): I:30–70.

———— 1987b: "New Challenges and Old Mindsets: Ten Rules for Empirical Realists," in Jacobsen (ed.): 39–66.

———— 1990: "Steps Towards Stable Peace in Europe: A Theory and Practice of Coexistence," *IA*, 66:1, 17–45.

———— 1990a: "The Foreign Policy Context: Brezhnev's Legacy and Gorbachev's Talents," in Gillette & Frank (eds.): 19–56.

———— 1991: Preface to idem (ed.): ix–xvi.

———— 1991a: "Introduction: The Interregnum: World Politics in Transition," in Booth (ed.): 1–28.

———— 1991b: "War, Security and Strategy: Towards a Doctrine For Stable Peace," in Booth (ed.): 335–376.

———— 1991c: "Security in Anarchy: Utopian Realism in Theory and Practice," *IA*, 67:3, 527–545.

———— 1991d: "Security in the New Europe: Ten Debates About 'Whose Security?' and 'Which Europe?'" Paper presented to the conference Sicherheit im neuen Europa, Evangelische Akademie Loccum and SAS, 27–29 September 1991.

———— 1993: "The Role of Navies in Peacetime: The Influence of Future History on Sea Power," in Smith & Bergin (eds.): 145–162.

———— 1993a: "Liberal Democracy, Global Order and the Future of Transatlantic Relations," in Center for Defence Studies (ed.): 353–366.

———— 1994: "NOD at Sea," in Møller & Wiberg (eds.): 98–114.

———— (ed.) 1991: *New Thinking About Strategy and International Security* (London: HarperCollins).

———— & John Baylis 1988: *Britain, NATO and Nuclear Weapons: Alternative Defence versus Alliance Reform* (London: Macmillan).

———— & Eric Herring 1994: *Keyguide to Information Sources in Strategic Studies* (London: Mansell).

———— & Nicholas J. Wheeler 1992: "Beyond Nuclearism," in Karp (ed.): 21–55.

Boothby, Derek 1990: "Maritime Change in Developing Countries: The Implications for Naval Arms Control," in Fieldhouse (ed.): 77–90.

Borawski, John 1986: "The Potential Contribution of Arms Control," in Flanagan & Hampson (eds.): 163–190.

——— 1986: "The World of CBMs," in Borawski (ed.): 9–42.

——— 1988: *From the Atlantic to the Urals: Negotiating Arms Control at the Stockholm Conference* (Washington, D.C.: Brassey's).

——— 1988a: "Next Steps in Confidence and Security Building Measures" in Windass & Grove (eds.): no pagination.

——— 1989: "From Stockholm to Vienna: Confidence and Security-Building Measures in Europe," *Arms Control Today,* May: 11–15.

——— 1991: "Conventional Arms Control in Europe in 1989," in Danspeckgruber (ed.): 135–145.

——— 1992: *Security for a New Europe: The Vienna Negotiations on Confidence and Security-Building Measures, 1989–90, and Beyond* (London: Brassey's).

——— (ed.) 1986: *Avoiding War in the Nuclear Age: Confidence-Building Measures for Crisis Stability* (Boulder, Colo.: Westview).

Borg, Marlies ter 1989: "From Offence as a Principle of War to Defence as a Principle of War Prevention," in Borg & Smit (eds.): 15–34.

——— 1992: "Reducing Offensive Capabilities—the Attempt of 1932," *JPR,* 29:2 (May): 145–160.

——— & Frank Barnaby 1986: "Paving the Way to European Security," in Barnaby & Borg (eds.): 267–275.

——— & John Grin 1986: "The Military Relevance of Recent Cooperative ET Projects," in Barnaby & Borg (eds.): 177–196.

——— & Marianne Tulp 1987: "Defence Technology Assessment: Improving Defence Decision-Making" (Den Haag: NOTA).

——— & Wim Smit (eds.) 1987: *Tactical Ballistic Missile Defence in Europe: Feasible, Affordable, Desirable?* (Amsterdam: Free U.P.).

——— & Wim Smit (eds.) 1989: *Non-provocative Defence as a Principle of Arms Control and Its Implications for Assessing Defence Technologies* (Amsterdam: Free U.P.).

Borinski, Phillip 1993: "Die neue NATO-Strategie: Perspektiven militärischer Sicherheitspolitik in Europa," *HSFK-R,* 1 (Frankfurt: HSFK).

Borkenhagen, Franz H. U. 1984: "Anmerkungen eines Logistikers," in SAS (eds.): 174–175.

——— 1984a: "Grundzüge einer effizienten Logistik in einem militärischen Defensivkonzept," in Weizsäcker (ed.): 29–40.

——— 1985: "Militärische Defensivkonzepte in einem System der Kollektiven Sicherheit," in Lutz (ed.): 260–274.

——— 1985a, "Gefahren für eine defensive Vorneverteidigung," *ÖMZ,* 1: 47–52.

——— 1987: "Aspekte der sicherheitspolitischen Diskussion in der Bundesrepublik Deutschland: Aus dem Blickpunkt der SPD," *ÖMZ,* 2: 138–145.

——— 1987a: "Strukturwandel im Einklang mit dem Bündnis: Mögliche Rahmenbedingungen für eine Strukturelle Nichtangriffsfähigkeit," *NGFH,* 5: 448–451.

——— 1988: "Der europäische Pfeiler der westlichen Allianz," *APZ,* B 18/88 (29 April): 35–45.

——— 1988a: "Antworten zu Fragen im Rahmen der Anhörung 'Konventionelle Stabilität'," in Bahr & Lutz (eds.): 412–418.

——— 1991: "Eine europäische Streitkraft für Europa," in Lutz (ed.) 1991: 217–233.

——— (ed.) 1986: *Bundeswehrdemokratie in Oliv: Streiträfte im Wandel* (Berlin: Dietz).

Børresen, Jacob 1988: "Maritime CBMs in Nordic Waters—Potential and Limitations from a Naval Point of View," in Möttölä (ed.): 235–246.

Boserup, Anders 1981: "Deterrence and Defence," *BAS,* December: 11–13.

———— 1982: "Nuclear Disarmament: Non-Nuclear Defence," in Kaldor & Smith (eds.): 185–192.

———— 1985: "Non-offensive Defence in Europe," in Paul (ed.): 194–209.

———— 1986: "Staten, samfundet og krigen hos Clausewitz," in Niels Berg (ed.), *Carl von Clausewitz: Om Krig,* 3 vols. (Copenhagen: Rhodos): III:911–930.

———— 1986a: "Escaping from the Nuclear Trap," in Boserup, Christensen & Nathan (eds.): 81–90.

———— 1987: "Two Papers on Maritime Strategy," *WP-CPCR,* 1987:1.

———— 1988: "Non-Provocative Defence," in Hellman (ed.): 109–114.

———— 1988a: "Common Security and the Concept of Non-Offensive Defense," in Alger & Stohl (eds.): 455–469.

———— 1988b: "A Way to Undermine Hostility," *BAS,* 44:7 (September): 16–19.

———— 1989: "Common Security and the Concept of Non-Offensive Defence," in Tromp (ed.): 173–186.

———— 1990: "Krieg, Staat und Frieden: Eine Weiterführung der Gedanken von Clausewitz," in Weizsäcker (ed.): 244–263.

———— 1990a: "Maritime Defence Without Naval Threat: The Case of the Baltic," in idem & Neild (eds.): 179–184.

———— 1990b: "Deterrence and Defence," in Boserup & Neild (eds.): 7–11.

———— 1990c: "Military Stability and Defence Dominance at the Tactical Level," in idem & Neild (eds.): 79–108.

———— 1990d: "Mutual Defensive Superiority and the Problem of Mobility Along an Extended Front," in idem & Neild (eds.): 63–78.

———— 1990e: "Approaches to Defensive Defence," in Randle & Rogers (eds.): 21–30.

———— & Michael Clemmesen 1984: "Et militært spor i en ny østpolitik," in SNU (ed.), *Øst-Vest forholdet: Vestlige muligheder i østpolitikken* (Copenhagen: SNU).

———— & Jens Jørn Graabaek 1990: "A Zonal Approach to the Neutralisation of Airpower in Europe," in Boserup & Neild (eds.): 159–165.

———— & Andrew Mack 1974: *War Without Weapons: Non-Violence in National Defence* (London: Pinter).

———— & Robert Neild (eds.) 1990: *The Foundations of Defensive Defence* (London: Macmillan).

————, Ludvig Christensen & Ove Nathan (eds.) 1986: *The Challenge of Nuclear Armaments: Essays Dedicated to Niels Bohr and His Appeal for an Open World* (Copenhagen: Rhodos).

Boskma, Peter & Frans-Bauke van der Meer 1986: "Trends in Military Technology," in Barnaby & Borg (eds.): 21–26.

Bosserelle, Daniel 1989: "The Role of Military Forces within a European Security System," in Wachter & Krohn (eds.): 101–113.

Boston Study Group 1979: *The Price of Defense: A New Strategy for Military Spending* (New York: Times).

Bouchet, Benoit 1989: "Neutralité et défense de l'Europe," *Défense Nationale,* May: 53–68.

Boudreau, Thomas E. 1987: "Geopolitical Demarche: Establishing a Non-Nuclear 'Corridor of Confidence' in Europe" (Limerick: Irish Peace Institute).

———— 1987a: "The Next Step After INF: Establishing a Corridor of Confidence in Europe," *Occasional Paper,* 1 (Cambridge, Mass.: Institute for Resource and Security Studies).

———— 1990: "The Berlin Plan: A Central-European Peace Proposal," *Prism Project Report* (Collegeville, Minn.: St. John's University).

Boulden, Jane 1993: "Prometheus Unborn: The History of the Military Staff Committee," *Aurora Papers,* 19 (Ottawa: Canadian Centre for Global Security).

———— & David Cox 1991: *The Guide to Canadian Policies on Arms Control, Disarmament, Defence and Conflict Resolution, 1991* (Ottawa: Canadian Institute for International Peace and Security).

Boulding, Elise (ed.) 1992: *New Agendas for Peace Research: Conflict and Security Reexamined* (Boulder, Colo.: Lynne Rienner).

Boulding, Kenneth 1962: *Conflict and Defense* (New York: Harper).

———— 1978: *Stable Peace* (Austin: University of Texas Press).

———— 1988: "Moving from Unstable to Stable Peace," in Gromyko & Hellman (eds.): 157–167.

———— 1989: *Three Faces of Power* (London: Sage).

Bourdet, Claude 1984: "The Rebirth of the Peace Movement," in Howorth & Chilton (eds.): 190–201.

Boutros-Ghali, Boutros 1992: "An Agenda for Peace: Preventive Diplomacy, Peacemaking and Peace-Keeping. Report of the Secretary-General Pursuant to the Statement Adopted by the Summit Meeting of the Security Council on 31 January 1992," in Roberts & Kingsbury (eds.) 1993: 468–498.

———— 1992a: "Empowering the United Nations," *FA,* 72:5 (Winter): 89–102.

———— 1993: "An Agenda for Peace: One Year Later," *Orbis,* 37:3 (Summer): 323–332.

Boutwell, Jeffrey 1986: "The German Search for Security," in Flanagan & Hampson (eds.): 113–136.

———— 1990: "Party Politics and Security Policies in the FRG," in Szabo (ed.): 123–148.

Bowie, Robert R. 1992: "Arms Control in the 1990s," in Adler (ed.): 53–67.

Boyer, Yver 1985: *Les nouvelles technologies en matière d'armes conventionelles: Leurs implications stratégiques et politiques* (Geneva: Institut des Nations Unies pour la recherche sur le désarmament).

Bracken, Paul 1976: "Urban Sprawl and NATO Defence," *Survival,* 18:6, 254–260.

———— 1983: *The Command and Control of Nuclear Forces* (New Haven: Yale U.P.).

———— 1983a: "The NATO Defence Problem," *Orbis,* Spring: 83–105.

———— 1987: "War Termination," in Carter & al. (eds.): 197–214.

———— 1987a: "Delegation of Nuclear Command Authority," in Carter & al. (eds.): 352–372.

———— 1993: "Nuclear Weapons and State Survival in North Korea," *Survival,* 35:3 (Autumn): 137–153.

———— & Stuart E. Johnson 1993: "Beyond NATO: Complementary Militaries," *Orbis,* 37:2 (Spring): 205–221.

———— & Kurt Gottfried 1990: "The New Strategic Environment," in Gottfried & Bracken (eds.): 3–81.

Bradley, Bill, Hans-Dietrich Genscher, Harry Ott & John C. Whitehead 1987: "Implications of Soviet New Thinking," *Special Report* (New York: Institute for East-West Security Studies).

Bradsher, Henry S. 1985: *Afghanistan and the Soviet Union,* expanded ed. (Durham: Duke U.P.).

Brahm, Hans & al. 1991: *Sowjetpolitik unter Gorbatschow: Die Innen- und Aussenpolitik der UdSSR, 1985–1990* (Berlin: Duncker & Humblot).

Brams, Steven J. & D. Marc Kilgour 1988: *Game Theory and National Security* (New York: Basil Blackwell).

Brandt, Børge & Kaj Christensen 1968: *BOPA Sabotage: Brudstykker af en Beretning fra Kampen mod de Tyske Besættelsesstyrker* (Copenhagen: Sirius).

Brandt Commission 1980: *North-South: A Programme for Survival: Report of the Independent Commission on International Development Issues* (Cambridge, Mass.: MIT).

Brandt, Willy 1985: *Der organisierte Wahnsinn: Wettrüsten und Welthunger* (Köln: Kiepenheuer & Witsch).

———— 1988: "The SIPRI 1987 Olof Palme Memorial Lecture: 'Security and Disarmament, Change and Vision'," in *SIPRI*, 1988: 539–547.

Brans, J. P. (ed.) 1984: *Operational Research '84* (Amsterdam: Elsevier).

Brauch, Hans Günter 1982: "Disarmament-Supporting Measures: Conceptual Innovation and Institutional Reform of the National Arms Control and Disarmaments Machineries," in Carlton & Schaerf (eds.): 82–104.

———— 1982a: *Der chemische Alptraum: Oder gibt es einen C-Waffen-Krig in Europa?* (Bonn: Dietz).

———— 1983: *Die Raketen kommen: Vom NATO-Dobbelbeschluss bis zur Stationierung* (Köln: Bund).

———— 1984: "Sicherheitspolitik im Umbruch: Aussenpolitische Rahmenbedingungen und Entwicklungschancen sicherheitspolitischer Alternativen," in Vogt (ed.): 145–159.

———— 1985: "Perspektiven einer europäischen Sicherheitspolitik," in Strübel (ed.): 27–46.

———— 1987: "Vertrauensbildende Massnahmen (VBM) und Konferenz über Vertrauens- und Sicherheitsbildende Massnahmen und Abrüstung in Europa (KVAE): Eine Einführung," in idem (ed.): 21–64.

———— 1987a: "Vertrauensbildung und Vertrauens- und Sicherheitsbildende Massnahmen zwischen Rüstungsdynamik und Rüstungskontrolle, Überlegungen zur Weiterentwicklung des V(S)BM-Konzepts," in idem (ed.): 401–421.

———— 1987b: "From SDI to EDI—Elements of a European Defence Architecture," in idem (ed.), *Star Wars and European Defence: Implications for Europe: Perceptions and Assessments* (London: Macmillan): 436–499.

———— 1988: "The Federal Republic of Germany, Searching for Alternatives," in Rudney & Reuchler (eds.): 79–102.

———— 1988a: "West German Alternatives for Reducing Reliance on Nuclear Weapons," in Hopmann & Barnaby (eds.): 146–182.

———— 1989: "Nuclear Weapons in Europe after the INF Agreement," in Harle & Sivonen (eds.): 155–178.

———— 1990: "Europa-Haus des Friedens: Ein Fundament, drei Pfeiler und ein Baustein für eine europäische Friedensordnung," in Vogt (ed.): 52–70.

———— 1990a: "German Unity, Conventional Disarmament, Confidence Building Defense and a New European Order of Peace and Security," in idem (ed.): 21–74.

———— 1990b: "Proposal for a Third Generation of Confidence- and Security-Building and Risk-Reduction Measures: A Supplementary Element of Confidence-Building," in Cerutti & Ragionieri (eds.): 68–73.

———— 1990c: "Weapons Innovation and US Strategic Weapons Systems: Learning from Case Studies?" in Gleditsch & Njølstad (eds.): 175–219.

———— 1991: "From Collective Self-Defence to a Collective Security System in Europe," *Disarmament,* 14:1, 1–20.

———— 1991a: "The New Europe and Non-Offensive Defense Concepts: Implications for Military Force Planning of United Germany," *AFR,* 15.

—— 1991b: "Nuclear Disarmament Prospects for Europe in the 1990s: Political Change and the Future Role of Nuclear Weapons," in Harle & Sivonen (eds.): 159–202.

—— 1992: "German Unity, Conventional Disarmament, Confidence-Building Defence, and a New European Order of Peace and Security," in Brauch & Kennedy (eds.): 3–44.

—— 1992a: "Institutionelle Bausteine einer gesamteuropäischen Sicherheitsarchitektur," *AFR*, 23.

—— (ed.) 1984: *Sicherheitspolitik am Ende: Eine Bestandsaufnahme, Perspektiven und neue Ansätze* (Gerlingen: Bleicher).

—— (ed.) 1987: *Vertrauensbildende Massnahmen und Europäische Abrüstungskonferenz: Analysen, Dokumente und Vorschlage* (Gerlingen: Bleicher).

—— (ed.) 1990: "German Unity and the Future European Order of Peace and Security," *AFR*, 38.

—— (ed.) 1990a: "Military Doctrine and Arms Control," *AFR*, 29.

—— (ed.) 1990b: "Verification and Arms Control: Implications for European Security," *AFR*, 36.

—— (cd.) 1991: "Weapons Technology, Disarmament and Verification: IPRA Defense and Disarmaments Study Group Paper 1," *AFR*, 41.

—— & Lutz Unterseher 1984: "Getting Rid of Nuclear Weapons," *JPR*, 21:2, 193–200.

—— & Robert Kennedy (eds.) 1990: *Alternative Conventional Defense Postures for the European Theater*, vol. I: *The Military Balance and Domestic Constraint* (New York: Crane Russak).

—— & Robert Kennedy (eds.) 1992: *Alternative Conventional Defense Postures in the European Theatre*, vol. II: *The Impact of Political Change on Strategy, Technology, and Arms Control* (New York: Crane Russak).

—— & Gerd Neuwirth (eds.) 1992: "Confidence and Security Building Measures in Europe: From the Stockholm to the Vienna Document," *AFR*, 24.

——, Henny J. van der Graaf, John Grin & Wim A. Smit (eds.) 1992: *Controlling the Development and Spread of Military Technology: Lessons from the Past and Challenges for the 1990s* (Amsterdam: VU U. P.).

Bredow, Wilfred von 1989: "Die deutsch-deutschen Beziehungen, ein Sonderfall," in Haberl & Hecker (eds.): 207–225.

—— 1992: *Der KSZE-Prozess: Von der Zähmung zur Auflösung des Ost-West-Konflikts* (Darmstadt: Wissenschaftliche Buchgesellschaft).

Bredthauer, Karl D. & Klaus Mannhardt (eds.) 1981: *Es geht ums Überleben: Warum wir die Atomraketen ablehnen* (Köln: Pahl-Rugenstein).

Brie, André 1988: "European Nuclear-Weapon-Free Zone Facilitates Structural Incapacity to Attack on Both Sides," in Baudisch (ed.): 89–94.

—— 1989: "Reducing the Most Offensive Weapon Systems," in Borg & Smit (eds.): 221–230.

—— & Manfred Müller 1988: *Europa: Wieviel Waffen reichen aus?* (Berlin: Staatsverlag der DDR).

Brigot, André 1989: "A Neighbour's Fears: Enduring Issues in Franco-German Relations," in Prestre (ed.): 85–104.

Brill, Heinz 1987: *Bogislaw von Bonin im Spannungsfeld zwischen Wiederbewaffnung, Westintegration, Wiedervereinigung*, Band 1: *Ein Beitrag zur Entstehungsgeschichte der Bundeswehr, 1952–1955* (Baden-Baden: Nomos).

—— (ed.) 1976: *Opposition gegen Adenauers Sicherheitspolitik: Eine Dokumentation zusammengestellt von Heinz Brill* (Hamburg: Neue Politik).

———— (ed.) 1989: *Bogislaw von Bonin im Spannungsfeld zwischen Wiederbewaffnung, Westintegration, Wiedervereinigung,* Band 2, *Dokumente und Materialien* (Baden-Baden, Nomos).

Brinckley, Douglas 1987: "Kennan-Acheson: The Disengagement Debate," *Atlantic Community Quarterly,* Winter: 413–425.

Bring, Ove E. 1987: "Humanitarian Law or Arms Control," *JPR,* 24:3 (September): 275–286.

———— 1990: "Naval Arms Control and the Convention on the Law of the Sea," in Fieldhouse (ed.): 137–143.

———— 1990a: "International Law and Arms Restraint at Sea," in Lodgaard (ed.): 187–197.

Brock, Lothar 1993: "Die 'Agenda für den Frieden': Chancen für eine neue UNO?" in Birkenback & al. (eds.): 49–60.

————, Mathias Jopp, Berthold Meyer, Norbert Ropers & Peter Schlotter 1988: "France, Great Britain and West Germany: Dissenting Promoters of Security Cooperation in Western Europe," *PRIF-R,* 4.

Brodie, Bernard (ed.) 1946: *The Absolute Weapon* (New York: Harcourt Brace).

———— 1959: *Strategy in the Missile Age* (Princeton: Princeton U.P., RAND).

Bromke, Adam 1993: "Post-Communist Countries: Challenges and Problems," in Karp (ed.): 17–44.

Brooks, Linton F. 1986: "Naval Power and National Security: The Case for the Maritime Strategy" (*IS,* Fall) in Miller & Evera (eds.) 1988: 16–46.

———— 1991: "The Strategic Arms Reduction Treaty: Reducing the Risk of War," *NATO Review* 39:5 (October): 7–10.

Brossollet, Guy 1975: *Essai sur la Nonbattaille* (Paris: Éditions Belin). Also as "Das Ende der Schlacht," in Weizsäcker (ed.) 1976: 93–223.

Brost, Wolfgang 1988: "'Strukturelle Nichtangriffsfähigkeit'—ein irreführendes Schlagwort in der sicherheitspolitischen Debatte," in Bahr & Lutz (eds.): 155–164.

Brouée, Pierre & Emile Temines 1968: *Revolution und Krieg in Spanien* (Frankfurt: Suhrkamp).

Brower, Kenneth S. 1990: "Technology and the Future Battlefield: The Impact on Force Structure, Procurement and Arms Control," *RJ,* 135:1, 53–60.

Brown, Chris 1992: *International Relations Theory: New Normative Approaches* (Hemel Hempstead: Harvester Wheatsheaf).

Brown, Frederic J. 1992: *The U.S. Army in Transition, II: Landpower in the Information Age* (Washington, D.C.: Brassey's).

Brown, James F. 1986: "The Impact of NATO Doctrinal Choices on the Policies and Strategic Choices of Warsaw Pact States: Part I," *AP,* 296: 11–19.

———— 1992: *Nationalism, Democracy and Security in the Balkans* (Aldershot: Dartmouth).

Brown, Michael 1991: "Yugoslavia's Armed Forces—Order of Battle," *Jane's Intelligence Review,* 3:8 (August): 366–373.

Brown, Michael L. & Thomas J. Leney 1984: "Conventional Defense: Technology, Doctrine and Force Structure," in Golden & al. (eds.): 163–176.

Brown, Neville 1986: *The Future of Air Power* (London: Croom Helm).

———— 1992: *The Strategic Revolution: Thoughts for the Twenty-First Century* (London: Brassey's).

———— & Anthony Farrar-Hockley 1985: *Nuclear First Use* (London: Buchan & Enright, RUSI).

Brown, Sheryl & Kimber M. Schraub (eds.) 1993: *Resolving Third World Conflict: Challenges for a New Era* (Washington, D.C.: United States Institute of Peace).

BRSF (Bremische Stiftung für Rüstungskonversion und Friedensforschung) (ed.) 1990: *Chancen für Rüstungskonversion* (Bremen: BSRF).

——— (ed.) 1991: *Rüstungskonversion vor Ort: Probleme, Alternativen, Perspektiven* (Bremen: BSRF).

Bruce, Maxwell, Horst Fischer & Thomas Mensah 1993: "A NWFW Regime: Treaty for the Abolition of Nuclear Weapons," in Rotblat & al. (eds.): 119–131.

Bruckmann, Wolfgang 1984: "Bericht vom Hearing 'Militärische Rüstung abbauen, Soziale Verteidigung aufbauen! Perspektiven einer neuen europaischen Friedensordnung jenseits von NATO und Warschauer Pakt'," *Mediatus,* 10: 10–12.

——— 1990: "Sicherheitspolitische Grundsätze und Abrüstungsszenarien—Deutsche Fragen, europäische Antworten," in Calliess (ed.): 604–626.

Brundtland, Arne Olav 1966: "Nordisk Balanse før og Nå," *Internasjonal Politikk,* 5: 491–541.

——— 1981: "Den klassiske, den omsnudde og den fremtidige nordiske balance," in Ole Nørgaard & Per Carlsen (eds.), *Sovjetunionen, Østeuropa og dansk Sikkerhedspolitik* (Esbjerg: Sydjysk Universitetsforlag): 83–99.

——— 1985: "Norwegian Security Policy: Defence and Nonprovocation in a Changing Context," in Flynn (ed.): 171–223.

——— 1988: "Den nye sovjetiske nordpolitiken og mulige norske svar," *Internasjonal Politikk,* 23: 95–133.

——— 1989: "Factual Groundwork," in UNIDIR (eds.): 7–15.

——— 1989a: "Sovjetisk utfordring og vestlige svar: Norsk sikkerhetspolitikk på nye vilkår?" *Internasjonal Politikk,* 3: 105–142.

Brundtland Commission (World Commission on Environment and Development) 1987: *Our Common Future* (Oxford: Oxford U.P.).

Brunn, A., L. Baehr & H.-J. Karpe (eds.) 1992: *Conversion—Opportunities for Development and Environment* (Berlin: Springer).

Brünner, Henning 1988: "Nichtoffensive Verteidigung und die Umstrukturierung und Umrüstung der konventionellen Kräfte: Zum Verhältnis von Technologie und Doktrin," in Calliess (ed.): 283–288.

Bruns, Wilhelm 1989: "Normalisierung oder Wiedervereinigung: Wie geht es weiter in den deutsch-deutschen Beziehungen?" (Bonn: Friedrich-Ebert-Stiftung).

——— 1990: "Von der Koexistenz über die Vertragsgemeinschaft zur Konföderation? Thesen," *KAAD,* 35.

——— 1990a: "Die äusseren Aspekte der deutschen Einigung," *Studien,* 40.

——— 1990b: "Zur Diskussion über den sicherheitspolitischen Status des deutschen Bundesstaates," *KAAD,* 37.

——— Christian Krause & Eckhard Lübkemeier 1984: *Sicherheit durch Abrüstung: Orientierende Beiträge zum Imperativ unserer Zeit* (Bonn: Neue Gesellschaft).

——— & Wulf Lapins (eds.) 1991: "Die Bundeswehr vor neuen Aufgaben und Herausforderungen," *Studien,* 44.

Brzezinski, Ian J. 1993: "Polish-Ukrainian Relations: Europe's Neglected Strategic Axis," *Survival,* 35:3 (Autumn): 26–37.

Brzezinski, Zbigniev 1988: "Peaceful Change in a Divided Europe," in Nehrlich & Thomson (eds.): 15–27.

——— 1989: "Post-Communist Nationalism," *FA,* 68(5): 1–25.

Brzoska, Michael 1992: "Assessing Proposals for a Conventional Arms Export Control Regime," in Brauch & al. (eds.): 301–311.

Buchbender, Ortwin, Hartmut Bühl & Heinrich Quaden (eds.) 1987: *Sicherheit und Frieden. Handbuch der weltweiten sicherheitspolitischen Verflechtungen:*

Militärbündnisse, Rüstungen, Strategien. Analysen zu den globalen und regionalen Bedingungen unserer Sicherheit, 3d, expanded ed. (Herford: Mittler).

Buckley, Thomas 1970: *The United States and the Washington Conference* (Knoxville: University of Tennessee Press).

Buehring, Guenter 1989: "New Negotiations on Conventional Forces and Confidence and Security-building Measures in Europe," *Disarmament,* 12:2 (Summer): 11–17.

Buffotot, Pierre (ed.) 1991: "Politiques de Défense," *ARES,* 12:3.

Bull, Hedley 1973: "Strategic Arms Limitation: The Precedent of the Washington and London Naval Treaties," in Kaplan (ed.): 26–52.

———— 1977: *The Anarchical Society: A Study of Order in World Politics* (London: Macmillan).

———— 1987: *Hedley Bull on Arms Control,* selected and introduced by Robert O'Neill and David N. Schwartz (London: Macmillan).

Bülow, Andreas von 1984: "Gedanken zur Weiterentwicklung der Verteidigungsstrategie in West und Ost," in Brauch (ed.): 223–244.

———— 1984a: *Alpträume West gegen Alpträume Ost: Ein Beitrag zur Bedrohungsanalyse* (Bonn: SPD).

———— 1985: "Strategie vertrauensschaffender Sicherheitsstrukturen in Europa: Wege zur Sicherheitspartnerschaft," *Blätter,* 10: 1267–1274. Also in Lutz (ed.) 1989: 101–121.

———— 1986: "Defensive Entanglement, an Alternative Strategy for NATO," in Pierre (ed.): 112–151.

———— 1986a: "Vorschlag für eine neue Bundeswehrstruktur der 90er Jahre: Einstieg in die strukturelle Nichtangriffsfähigkeit," *EW,* 11: 636–638.

———— 1987: "Stellungnahme zu Christian Krauses Kritik," *S+F,* 5:3, 195–197.

———— 1988: "Friedensordnung und Verteidigung in Europe: Technik, Taktik, Strategie und Politik, Gestaltungsraum und Sachzwänge," in Bülow, Funk & Müller: 9–46.

———— 1988a: "Neues Denken auch in den Militärdoktrinen," in Bülow, Funk & Müller: 195–207.

———— 1988b: "Thesen zur strukturellen Nichtangriffsfähigkeit," *Blätter,* 11: 1376–1379.

———— 1988c: "Vorschlag für eine Bundeswehrstruktur der 90er Jahre: Auf dem Weg zur konventionellen Stabilität," in Bülow, Funk & Müller, 95–110.

———— 1989: "Conventional Stability: An Overall Concept," in Borg & Smit (eds.): 45–48.

———— 1989a: "Restructuring the Ground Forces," in Borg & Smit (eds.): 161–174.

———— & Helmut Funk 1987: "Zum Problem der Strukturellen Nichtangriffsfähigkeit," *S+F,* 7:1, 14–16.

———— & Helmut Funk 1988: "Statements zur Konventionellen Stabilität," in Bahr & Lutz (eds.): 422–428.

———— & Helmut Funk 1988a: "The Achievement of Mutual Conventional Forces Defender Superiority in Central Europe from the Urals to the Atlantic," *Pugwash Newsletter,* 25:3, 108–110.

————, Helmut Funk & Albrecht A. C. von Müller 1988: *Sicherheit für Europa* (Koblenz: Bernard & Graefe).

Bundeskanzleramt 1985: *Landesverteidigungsplan* (Wien: Österreichische Staatsdruckerei).

Bundesministerium der Verteidigung 1985: *Weissbuch 1985: Zur Lage und Entwicklung der Bundeswehr* (Bonn: Ministry of Defense).

Bundy, McGeorge 1986: "Existential Deterrence and Its Consequences," in Douglas Maclean (ed.), *The Security Gamble* (Totowa, N.J.: Rowman & Allanheld): 3–13.

—— 1990: "From Cold War Toward Trusting Peace," *FA*, 69:1 (Winter): 197–212.

—— 1991: "Nuclear Weapons and the Gulf," *FA*, 70:4 (Fall): 83–94.

——, William J. Crowe, Jr. & Sidney Drell 1993: "Reducing Nuclear Danger," *FA*, 72:2 (Spring): 140–155.

——, George F. Kennan, Robert S. McNamara & Gerard Smith 1982: "Nuclear Weapons and the Atlantic Alliance" (*FA*, Spring) in W. Bundy (ed.) 1985: 23–40.

Bundy, William P. (ed.) 1985: *The Nuclear Controversy: A Foreign Affairs Reader* (New York: New American Library).

Burgelin, Henri 1991: "L'Europe et le Guerre de Golfe," *Stratégique*, 51/52: 161–178.

Burgess, John A. 1986: "Emerging Technologies and the Security of Western Europe," in Flanagan & Hampson (eds.): 64–84.

Burgess, Michael 1989: *Federalism and European Union: Political Ideas, Influences and Strategies in the European Community, 1972–1987* (London: Routledge).

Burke, David P. 1982: "The Defense Policy of Romania," in Murray & Viotti (eds.): 323–341.

Buro, Andreas 1982: *Zwischen sozialliberalem Zerfall und konservativer Herrschaft: Zur Situation der Friedens und Protestbewegung in dieser Zeit* (Offenbach: 2000).

—— 1983: "Principles and Problems of Alternative Peace Strategies," in Steinke & Vale (eds.): 173–186.

—— 1983a: "Striktes Defensivkonzept als Alternative," *Antimilitarismus*, 3.

—— 1987: "Strukturelle Nichtangriffsfähigkeit unter friedens-politischer Perspektive," *S+F*, 5:1, 16–21.

—— 1988: "Bemerkungen zur Diskussion über konventionelle Stabilität," in Bahr & Lutz (eds.): 429–432.

—— 1990: "Die Zivilisierung Europas steht noch voran: Zur Begründung der Kampagne 'Bundesrepublik ohne Armee' (BoA)," *Blätter*, 9: 1106–1115.

—— 1990a: "BoA—Eine Kampagne entsteht: Thesen zum Beitrag der bundesrepublikanischen Friedensbewegung," in Calliess (ed.): 648–662.

Burt, Gordon 1988: "Introduction to the Crisis," in Burt (ed.): 1–11.

—— 1988a: "A Strategic Conclusion," in Burt (ed.): 134–149.

—— (ed.) 1988: *Alternative Defence Policy* (London: Croom Helm).

Burt, Richard 1988: "An Increased Emphasis on Conventional Defence: A U.S. View," in Nehrlich & Thomson (eds.): 28–34.

Busch, Marc L. & Eric R. Reinhardt 1993: "Nice Strategies in a World of Relative Gains: The Problem of Cooperation Under Anarchy," *Journal of Conflict Resolution*, 37:3 (September): 427–445.

Butkevicius, Audrius 1993: "The Baltic Region in the New Europe," in Center for Defence Studies (ed.): 169–178.

Butterwege, Christoph 1989: "Die Opposition auf der Suche nach einer friedenspolitischen Konzeption: Sozialdemokratische und Grüne Alternativen zur Sicherheitspolitik der Bundesregierung," in IMSF (eds.): 278–291.

—— & Eva Senghaas-Knobloch (eds.) 1992: *Von der Blockkonfrontation zur Rüstungskonversion? Die Neuordnung der internationalen Beziehungen, Abrüstung und Regionalentwicklung nach dem Kalten Krieg* (Münster: LIT).

Buzan, Barry 1987: "Common Security, Non-provocative Defence, and the Future of Western Europe," *Review of International Studies,* 13: 265–279.

———— 1987a: *An Introduction to Strategic Studies: Military Technology and International Relations* (London: Macmillan).

———— 1989: "The Future of Western European Security," in Wæver & al. (eds.): 16–45.

———— 1989a: "Common Security as a Policy Option for Japan and Western Europe," in Nakarada & Øberg (eds.): 71–82.

———— 1991: *People, States and Fear: An Agenda for International Security Studies in the Post–Cold War Era,* 2d ed. (London: Harvester Wheatsheaf; Boulder, Colo.: Lynne Rienner).

———— 1991a: "Is International Security Possible?" in Booth (ed.): 31–55.

———— 1991b: "New Patterns of Global Security in the Twenty-First Century," *IA,* 67:3 (July): 431–451.

———— 1992: "Third World Regional Security in Structural and Historical Perspective," in Job (ed.): 167–189.

———— 1994: "Does NOD Have a Future in the Post–Cold War World?" in Møller & Wiberg (eds.): 11–24.

————, Morten Kelstrup, Pierre Lemaitre, Elzbieta Tromer & Ole Wæver 1990: *The European Security Order Recast: Scenarios for the Post–Cold War Era* (London: Pinter).

Byrd, Peter (ed.) 1991: *British Defence Policy: Thatcher and Beyond* (London: Harvester Wheatsheaf).

Byrne, Paul 1988: *The Campaign for Nuclear Disarmament* (London: Croom Helm).

Cable, James 1981: *Gunboat Diplomacy, 1919–1979: Political Applications of Limited Naval Force* (New York: St. Martin's).

Cahen, Alfred 1986: "Relaunching the Western European Union: Implications for the Atlantic Alliance," *NATO Review,* 4: 6–12.

———— 1986a: "Unity Through Common Defence: Western European Union," *NSN* (June): 38–71.

———— 1989: "The Construction of Europe in the Context of a Changing Transatlantic Relationship: Developments in East-West Relations and in the Overall Situation," in Clesse & Schelling (eds.): 273–282.

Calderon, Felix 1982: "Nuclear-Weapon-Free Zones: The Latin American Experiment," in Carlton & Schaerf (eds.): 252–272.

Calleo, David P. 1987: *Beyond American Hegemony: The Future of the Western Alliance* (New York: Basic Books).

———— 1987a: "NATO's Middle Course," *Foreign Policy,* 69 (Winter): 135–147.

Callesen, Gerd 1989: "'Wenn Dänemark seine Land- und Seebefestigungen schleift und seine Flotte abschafft . . .' Vom Vorschlag der dänischen Sozialdemokratie, 1908 vollständig abzurüsten und der Reaktion führender europäischer Sozialisten," in Gross & al. (eds.): 21–42.

Calliess, Jörg (ed.) 1988: "Konventionelle Stabilität und Vertrauensbildende Verteidigungskonzepte," *Loccumer Protokolle,* 19/1988.

———— (ed.) 1989: "Abrüstung und Strukturwandel: Wege zu einer Neuordnung des europäischen Sicherheitssystems," *Loccumer Protokolle,* 16/1989.

———— (ed.) 1990: "Der Neubau Europas," *Loccumer Protokolle,* 19/1990.

Calogero, Francesco 1972: "Some Remarks and a Proposal Concerning the Limitations of Strategic Armaments," *Proceedings of the 22nd Pugwash Conference:* 305–317.

———— 1977: "A Novel Approach to Arms Control Negotiations," *Proceedings of the 22nd Pugwash Conference:* 1–10.

———— 1983: "A Flexible Approach to a Nordic NWFZ," in Lodgaard & Thee (eds.): 247–254.

———— 1993: "An Asymptotic Approach to a NWFW," in Rotblat & al. (eds.): 191–200.

————, Marco De Andreis, Gianluca Devoto & Paolo Farinella 1989: "Le Idee di 'Difesa Alternativa' ed il Ruolo dell'Italia" (research paper commissioned by the Centro Militare di Studi Strategici [CeMiSS], Rome).

Calvocoressi, Peter 1992: "A Problem and Its Dimensions," in Rodley (ed.): 1–13.

Cameron, Lord 1983: "Alternative Strategies: Strategy, Tactics and New Technology," in Baylis (ed.): 92–116.

Camilleri, J. A. & Jim Falk 1992: *The End of Sovereignty? The Politics of a Shrinking and Fragmenting World* (London: Edward Elgar).

Campbell, Edwina 1990: *Germany's Past and Europe's Future* (London: Brassey's).

Canby, Steven L. 1972: "NATO Muscle: More Shadow Than Substance," *FA*, 8 (Fall): 38–49.

———— 1975: "Damping Nuclear Counterforce Initiatives: Correcting NATO's Inferiority in Conventional Military Strength," *Orbis*, 19:1, 47–71.

———— 1979: "NATO Strategy: Political-Military Problems and Divergent Interests and Operational Concepts," *MR*, 59:4 (April): 50–58.

———— 1980: "Territorial Defense in Central Europe," *Armed Forces and Society*, Fall, 51–67.

———— 1981: "Swedish Defence," *Survival*, 23(3), 116–123.

———— 1982: "Classical Light Infantry and the Defense of Europe" (Washington, D.C.: C & L Associates).

———— 1983: "Military Reform and the Art of War," *Survival*, 25:3, 120–127 (abridged version of Canby 1984).

———— 1984: "Military Reform and the Art of War," in Clark & al. (eds.): 126–146.

———— 1985: "New Conventional Force Technology and the NATO–Warsaw Pact Balance: Part 1," *AP*, 188: 7–24.

———— 1985a: "The Operational Limits of Emerging Technology," *IDR*, 6: 875–880.

———— 1986: "Can Non-Provocative Defence Provide Atlantic Security?" in Barnaby & Borg (eds.): 215–219.

———— 1986a: "Light Infantry Perspective," in Tromp (ed.): 149–164.

———— 1987: "Conventional Weapon Technologies," in *SIPRI*, 1987: 85–95.

———— 1989: "The Quest for Technological Superiority: A Misunderstanding of War?" *AP*, 237: 26–40.

———— 1990: "Leichte Infanterie in der Vorneverteidigung," in Weizsäcker (ed.): 230–243.

———— 1994: "Weapons for Land Warfare," in Møller & Wiberg (eds.): 74–84.

Carasales, Julio 1993: "Nuclear Non-Proliferation After 1995," in Rotblat (ed.): 54–62.

Cardwell, Thomas E. 1984: "One Step Beyond: AirLand Battle, Doctrine not Dogma," *MR*, 64:4, 45–54.

———— 1985: "AirLand Battle Revisited," *MR*, 65:9, 413

———— 1986: "Follow-on Forces Attack: Joint Interdiction by Another Name," *MR*, 66:2, 4–11.

Carle, Christophe 1992: "Mayhem or Deterrence? Regional and Global Security from Non-Proliferation to Post-Proliferation," in Stahl & Kemp (eds.): 45–57.

Carlson, Don & Craig Cromstock (eds.) 1986a: *Securing Our Planet: How to*

Succeed When Threats Are Too Risky and There's Really No Defense (New York: St. Martin's).

———— & Craig Cromstock (eds.) 1986b: *Citizen Summitry: Keeping the Peace When It Matters Too Much to Be Left to Politicians* (New York: St. Martin's).

Carlton, David & Carlo Schaerf (eds.) 1982: *The Arms Race in the 1980s* (London: Macmillan).

———— & Carlo Schaerf (eds.) 1989: *Perspectives on the Arms Race* (London: Macmillan).

Carpenter, Ted Galen 1992: "Direct Military Intervention," in Schraeder (ed.): 153–172.

Carr, Edward Hallett 1946: *The Twenty Years' Crisis, 1919–1939: An Introduction to the Study of International Relations,* 2d ed. (New York: Harper, 1964).

Carroll, Eugene J. 1984: "Nuclear Weapons and Deterrence," in Prins (ed.): 3–14.

———— 1990: "US-Soviet Naval Competition: Dangers and Risks," in Fieldhouse (ed.): 64–76.

Carter, April 1967: "Political Conditions for Civilian Defence," in Roberts (ed.): 274–290.

———— 1987: "The Peace Movement Alternative," in Smith & Thompson (eds.): 103–123.

———— 1989: *Success and Failure in Arms Control Negotiations* (Oxford: Oxford U.P.).

———— 1992: *Peace Movements: International Protest and World Politics Since 1945* (London: Longman).

Carter, Ashton B. 1984: "BMD Applications: Performance and Limitations," in Carter & Schwartz (eds.): 98–181.

———— & David N. Schwartz (eds.) 1984: *Ballistic Missile Defense* (Washington, D.C.: Brookings).

————, John D. Steinbruner & Charles A. Zraket (eds.) 1987: *Managing Nuclear Operations* (Washington, D.C.: Brookings).

Carton, Allan 1984: "Dissuasion Infra-nucleaire: L'Ecole allemande de techno-guerilla," *CES,* 3.

———— 1991: *Les neutres, la neutralité et l'Europe* (Paris: Fondation pour les études de défense nationale).

Carus, W. Seth 1990: "Ballistic Missiles in the Third World: Threat and Response," *Washington Papers,* 146 (New York: Praeger).

———— 1992: "Weapons Technology and Regional Stability," in Stahl & Kemp (eds.): 9–16.

Carver, Michael (Field Marshall Lord) 1981: *War Since 1945* (New York: Putnam's).

———— 1982: *A Policy for Peace* (London: Faber).

———— 1983: "Getting Defence Priorities Right," in Baylis (ed.): 76–91.

———— 1983a: "No First Use: A View from Europe," *BAS,* March: 22–26.

Catrina, Christian 1988: *Arms Transfers and Dependence* (New York: Taylor & Francis).

———— 1990: "International Arms Transfers: Supplier Policies and Recipient Dependence," *Disarmament,* 13:4, 113–130.

Catudal, Honoré M. 1988: *Soviet Nuclear Strategy from Stalin to Gorbachev: A Revolution in Soviet Military and Political Thinking* (Berlin: Spitz).

Ceadel, Martin 1989: *Thinking About Peace and War* (London: Oxford U.P.).

Ceillier, C. F. 1990: "Les idées stratégiques en France de 1870 a 1914: La Jeune École," in Hervé Coutau-Bégarie (ed.), *L'évolution de la pensée navale* (Dossier, 41) (Paris: Fondation pour les études de défense nationale/ Commission française d'histoire maritime): 195–231.

Cekota, Jaromir 1991: "The Political Economy of Disarmament in the 1990s," *HB*, 58.

Center for the Study of Democratic Institutions 1973: "Preliminary Draft of a World Constitution," in Falk & Mendlowitz (eds.): 455–468.

Centre for Defence Studies (ed.) 1993: *Brassey's Defence Yearbook, 1993* (London: Brassey's).

Cernat, Iulian & Emanoil Stanislav (eds.) 1976: *National Defence: The Romanian View* (Bucharest: Military Publishing House).

Cerutti, Furio & Rodolfo Ragionieri (eds.) 1990: *Rethinking European Security* (New York: Crane Russak).

Chace, James 1992: *The Consequences of the Peace: The New Internationalism and American Foreign Policy* (Oxford: Oxford U.P.).

Chafetz, Glenn 1993: "The End of the Cold War and the Future of Nuclear Proliferation: An Alternative to the Neorealist Perspective," *SS*, 2:3/4 (Spring/Summer): 127–158.

Chaliand, Gerard 1977: *Revolution in the Third World* (Harmondsworth: Penguin).

———— 1981: *Rapport sur la Resistance Afghane* (Paris: Biblioteque Berger-Levrault).

———— 1982: Introduction to idem (ed.): 1–34.

———— 1982a: "With the Guerillas in 'Portuguese' Guinea," in idem (ed.): 186–215.

———— 1991: "Strategie indirecte et guerilla," *Stratégique*, 1: 177–184.

———— (ed.) 1982: *Guerilla Strategies: An Historical Anthology from the Long March to Afghanistan* (Berkeley: University of California Press).

Chalmers, Malcolm 1984: "Can NATO Afford a Non-Suicidal Strategy?" in Prins (ed.): 218–233.

———— 1985: *Paying for Defence: Military Spending and British Decline* (London: Pluto).

———— 1989: "Reciprocal Unilateralism as a Complement to CFE: Two Studies," *AFR*, 37.

———— 1990: "Beyond the Alliance System," *WPJ*, 7:2, 215–250.

———— 1990a: "The European Security System After the Cold War: Beyond the Alliances," in Brauch (ed.): 75–126.

———— 1991: "Britain's Defences in a New Europe: Forces and Budgets," in Unterseher & al. 1991b (no pagination).

———— 1992: *Biting the Bullet: European Defence Options for Britain* (London: Institute for Public Policy Research).

———— 1993: "Developing a Security Regime for Eastern Europe," *JPR*, 30:4 (November): 427–444.

———— 1993a: "Britain and Alliance Burden-Sharing," in Clarke & Sabin (eds.): 102–129.

———— & Owen Greene 1993: "Implementing and Developing the United Nations Register of Conventional Arms," *Peace Research Report*, 32 (Bradford: University of Bradford, School of Peace Studies).

————, Bjørn Møller & David Stevenson 1989: "Alternative Conventional Defense Structures for Europe: British and Danish Perspectives," *AFR*, 22.

———— & David Stevenson 1989: "Mutual Restructuring of Conventional Forces in Europe: A British Perspective," in Chalmers & al.: 67–94.

———— & Lutz Unterseher 1987: "Is There a Tank Gap? A Comparative Assessment of the Tank Fleets of NATO and the Warsaw Pact," *Peace Research Report*, 19 (Bradford: University of Bradford, School of Peace Studies).

——— & Lutz Unterseher 1988, "Is There a Tank Gap? Comparing NATO and Warsaw Pact Tank Fleets," *IS,* 13:1, 5–49.

——— & Lutz Unterseher 1989: "Correspondence: The Tank Gap Data Flap: The Authors Reply," *IS,* 13:4, 187–194.

Chan, Stephen 1984: "Mirror, Mirror on the Wall . . . Are Freer Countries More Pacific?" *Journal of Conflict Resolution,* 28:4 (December): 617–648.

Charlier, J. 1990: "Military and Strategic Consequences of the Evolution of the European Security Context," in Clesse & Rühl (eds.): 572–581.

Chatterji, Manas & Linda Rennie Forcey (eds.) 1992: *Disarmament, Economic Conversion, and Management of Peace* (New York: Praeger).

Checkel, Jeff 1993: "Ideas, Institutions, and the Gorbachev Foreign Policy Revolution," *WP,* 45:2 (January): 271–300.

Cheeseman, Graeme 1988: "Alternative Defence Strategies and Australia's Defence," *Working Paper,* 51 (Canberra: Research School of Pacific Studies).

——— 1989: "An Alternative Defence Posture for Australia," *Working Paper,* 59 (Canberra: Research School of Pacific Studies).

——— 1990: "The Application of the Principles of Non-Offensive Defence Beyond Europe: Some Preliminary Observations," *Working Paper,* 78 (Canberra: Peace Research Centre).

——— 1990a: "Over-reach in Australia's Regional Military Policy," in Cheeseman & Kettle (eds.): 73–92.

——— 1991: "An Alternative Approach for Structuring the Army," in David Horner (ed.), *Reshaping the Australian Army: Challenges for the 1990s,* Canberra Papers on Strategy and Defence, no. 77 (Canberra: Australian National University, Strategic and Defence Studies Centre): 141–170.

——— 1993: *The Search for Self-Reliance: Australian Defence Since Vietnam* (Melbourne: Longman Cheshire).

——— and Michael McKinley 1991: "Australia's Regional Security Policies, 1970–1990: Some Critical Reflections," *Working Paper,* 101 (Canberra: Australian National University, Peace Research Centre).

——— & St. John Kettle (eds.) 1990: *The New Australian Militarism: Undermining Our Future Security* (Leichardt, Australia: Pluto).

Ch'ên, Jerome 1967: *Mao and the Chinese Revolution* (Oxford: Oxford U.P.).

Cheon, Seong W. 1993: "North Korea's Nuclear Problem: Current State and Future Prospects," *Korean Journal of National Unification,* 2: 85–104.

Cherkasov, Pyotr, Fyodor Rybinsev, Anatoly Yoryev, and G. Murphy Donovan 1989: "On Alexei Arbatov's Article 'How Much Defence Is Sufficient'," *IA(M),* 8: 133–143.

Chernaga, Vladimir: "Russian Concept of National Security—Political and Military Aspects." Paper presented to the workshop "Soviet, Russian and Baltic Security," Svaneke, Denmark, 25–27 September 1991.

Chervov, Nikolai 1989: "Limitation and Reduction of Conventional Arms: Objectives and Methods," in UNIDIR (eds.): 45–48.

Chichester, Michael & John Wilkinson 1983: "Flexibility, Mobility and Involvement," in Baylis (ed.): 138–153.

Child, Jack 1992: *The Central American Peace Process, 1983–1991: Sheathing Swords, Building Confidence* (Boulder, Colo.: Lynne Rienner).

Chinworth, Michael W. 1992: *Inside Japan's Defense: Technology, Economics and Strategy* (Washington, D.C.: Brassey's).

Chipman, John 1988: "Allies in the Mediterranean: Legacy of Fragmentation," in idem (ed.): 53–85.

——— (ed.) 1988: *NATO's Southern Allies: Internal and External Challenges* (London: Routledge).

Christensen, Søren Møller (ed.) 1984: *Man har et standpunkt . . . Socialdemokratiet og sikkerhedspolitikken i 80'erne: En debatbog* (Copenhagen: Eirene).

Christensen, Thomas J. & Jack Snyder 1990: "Chain Gangs and Passed Bucks: Predicting Alliance Patterns in Multipolarity," *International Organization,* 44:2, 137–168.

Christie, Johan K. 1984: "Alternativ Sikkerhed" (Egtved: Fredsbevaegelsens Forlag).

———— 1988: "From Power Confrontation to Co-Operation," in Fedorenko (ed.): 99–112.

Chubin, Shahram 1991: "Post-War Gulf Security," *Survival,* 33:2 (March/April): 140–157.

Churilin, Alexander 1990: "Soviet Approaches to the Limitation of Naval Activities and Naval Armaments," in Lodgaard (ed.): 130–137.

Cimbala, Stephen J. 1989: *NATO Strategies and Nuclear Weapons* (London: Pinter).

———— 1990: "Military Strategy: Soviet Threat Perceptions and Escalation Control," *JSMS,* 3:4 (December): 545–585.

———— 1991: *Uncertainty and Control: Future Soviet and American Strategy* (London: Pinter).

———— 1993: "Nonoffensive Defense and Strategic Choices: Russia and Europe After the Cold War," *Journal of Slavic Military Studies,* 6:2 (June): 166–202.

———— & Keith A. Dunn (eds.) 1987: *Conflict Termination and Military Strategy: Coercion, Persuasion, and War* (Boulder, Colo.: Westview).

Claesson, Paul 1983: *Grønland: Middelhavets Perle* (Copenhagen: Eirene).

Clark, Asa A., IV, Peter W. Chiarelli, Jeffrey S. McKitrick & James W. Reed (eds.) 1984: *The Defense Reform Debate: Issues and Analysis* (Baltimore: Johns Hopkins U.P.).

Clark, Ian 1982: *Limited Nuclear War: Political Theory and War Conventions* (Oxford: Martin Robertson).

Clark, Michael T. & Simon Serfaty (eds.) 1991: *New Thinking and Old Realities: America, Europe, and Russia* (Washington, D.C.: Seven Locks Press, Johns Hopkins Foreign Policy Institute).

Clarke, Douglas 1992: "A Guide to Europe's New Security Architecture," *ES,* 1:2 (Summer): 126–132.

Clarke, Jonathan 1993: "Replacing NATO," *Foreign Policy,* 93 (Winter): 22–40.

Clarke, Michael 1985: "American Reactions to Shifts in European Policy: The Changing Context," in Roper (ed.): 78–99.

———— 1985a: "The Alternative Defence Debate: Non-Nuclear Defence Policies for Europe," *ADIU Occasional Paper,* 3.

———— 1988: "Public Attitudes and the Defence Debate in Britain," in Kaiser & Roper (eds.): 60–74.

———— 1989: "The Debate on European Security in the United Kingdom," in Waever & al. (eds.): 121–140.

———— 1991: "Alternative Defence: The New Reality," in Booth (ed.): 189–212.

———— (ed.) 1993: *New Perspectives on Security* (London: Brassey's).

———— & Rod Hague (eds.) 1990: *European Defence Cooperation: America, Britain and NATO* (Manchester: Manchester U.P.).

———— & Philip Sabin (eds.) 1993: *British Defence Choices for the Twenty-First Century* (London: Brassey's).

Claude, Inis L., Jr. 1972: "Appraisal of the Case for World Government," in Robert L. Pfaltzgraff, Jr. (ed.), *Politics and the International System,* 2d ed. (Philadelphia: Lippincott): 585–596.

Clausen, Peter A. 1986: "Limited Defense: The Unspoken Goal," in Tirman (ed.): 147–160.

Clausewitz, Carl von 1980: *Vom Kriege, Ungekürzter Text nach der Erstauflage* (1832–1834) (Frankfurt: Ullstein).

———— 1984: *On War,* edited and translated by Miclear Howard and Peter Paret (Princeton: Princeton U.P.).

Clemens, Clay 1988: "Beyond INF, West Germany's Centre-Right Party and Arms Control in the 1990s," *IA,* 65:1, 55–74.

Clements, Kevin 1989: *Back from the Brink: The Creation of a Nuclear-Free New Zealand* (London: Allen & Unwin, Port Nicholson Press).

Cleminson, F. Ronald 1989: "Conventional Arms Reductions in Europe: A Verification Model," in Altman & Rotblat (eds.): 158–166.

Clemmesen, Michael H. 1986: *Om defensivt forsvar* (Copenhagen: SNU).

———— 1986a: "Force Postures and European Security," in Rotblat & d'Ambrosio (eds.): 88–92.

———— 1988: *Styrkeforholdet mellem øst og vest* (Copenhagen: SNU).

———— 1990: "En militær kritik af ikke-offensivt forsvar," in Sørensen (ed.): 154–168.

Clesse, Armand & Lothar Rühl (eds.) 1990: *Beyond East-West Confrontation: Searching for a New Security Structure in Europe* (Baden-Baden: Nomos).

Clesse, Armand & Thomas C. Schelling (eds.) 1989: *The Western Community and the Gorbachev Challenge* (Baden-Baden: Nomos).

Clesse, Armand & Rudolf Tókés (eds.) 1992: *Preventing a New East-West Divide: The Economic and Social Imperatives of the Future Europe* (Baden-Baden: Nomos).

Clodfelter, Mark 1989: *The Limits of Air Power: The American Bombing of North Vietnam* (New York: Free Press).

Close, Robert 1976: *L'Europe sans défence: 48 heures qui pourrait changer la face du monde* (Brussels: Editions Arts et Voyages).

———— 1981: *Encore un effort et nous aurons définitivement perdu la troisième guerre mondiale* (Paris: Pierre Belfond). Quoted from the German translation: *Das Ungleichgewicht des Schreckens: Führt der Rüstungswettlauf zwischen Ost und West zum Dritten Weltkrieg?* (Vienna: Molden, 1982).

———— 1981a: "Abschliessende Betrachtungen: Weder nuclearer Holocaust noch Capitulation," in Löser (ed.): 218–232.

Clotfelter, James 1986: "Disarmament Movements in the United States," *JPR,* 23:2 (June): 97–102.

Cockburn, Andrew 1984: *The Threat: Inside the Soviet Military Machine,* updated and rev. ed. (New York: Vintage).

Coffey, Joseph I. & Klaus von Schubert (eds.) 1989: *Defense and Detente: American and West German Perspectives on Defense Policy* (Boulder, Colo.: Westview).

Cohen, Avner 1993: "Nuclear Weapons, Opacity, and Israeli Democracy," in Yaniv (ed.): 197–226.

Cohen, Eliot A. 1988: "Toward Better Net Assessment: Rethinking the European Conventional Balance," *IS,* 13:1 (Summer): 50–89.

Cohen, Raymond 1991: *Negotiating Across Cultures: Communication Obstacles in International Diplomacy* (Washington, D.C.: United States Institute for Peace).

Cohen, Richard & Peter A. Wilson 1990: *Superpowers in Economic Decline: U.S. Strategy for the Transcentury Era* (New York: Crane Russak).

Collins, A.S. 1984: "Current NATO Strategy: A Recipe for Disaster," in Prins (ed.): 29–41.

Colson, Bruno 1991: "La culture stratégique américaine et le Golfe," *Stratégique,* 51/52: 113–132.

Coman, Ion & al. 1984: *The Romanian National Defence Concept* (Bucharest: Military Publishing House).

Commisie van het Wetenschappelijk Instituut voor het CDA 1986: *Vrede wegen: Alternatieven voor het veiligheidsbelei beoordeeld* (Den Haag: CDA).

Committee for Basic Rights and Democracy 1983: "Five Proposals for a New Security Policy," in Steinke & Vale (eds.): 179–186.

Conetta, Carl 1991: "European 'Out-of-Area' Operations: An Alternative Defense Perspective." Paper presented to the conference "Sicherheit im neuen Europa," Evangelische Akademie Loccum and SAS, 27–29 September.

———— & Charles Knight 1990: "After Conventional Cuts: New Options for NATO Ground Defense," *Working Paper,* 7 (Brookline, Mass.: Institute for Defense and Disarmament Studies).

———— & Charles Knight 1990a: "How Low Can NATO Go?" *DDA,* 3:2, 1–3.

———— & Charles Knight 1990b: "In Defense of Europe: Rethinking US Army Procurement Priorities," *Briefing Paper,* 3 (Brookline, Mass.: Institute for Defense and Disarmament Studies).

———— & Charles Knight 1991: "Defense Procurement Policy for the 1990s: Selected Army and Air Force Systems," *Project on Defense Alternatives Briefing Report* (Cambridge, Mass.: Commonwealth Institute).

———— & Charles Knight 1991a: "After Desert Storm: Rethinking US Defense Requirements," *Project on Defense Alternatives Briefing Report,* 2 (Cambridge, Mass.: Commonwealth Institute).

———— & Charles Knight 1992: "Reasonable Force: Adapting the US Army and Marine Corps to the New Era. Part I. Threat Environment and Force Size Requirement," *Project on Defense Alternatives Briefing Report,* 3 (Cambridge, Mass.: Commonwealth Institute).

———— & Charles Knight 1993: "Rand's *New Calculus* and the Impasse of US Defense Restructuring," *Project on Defense Alternatives Briefing Report,* 4 (Cambridge, Mass.: Commonwealth Institute).

————, Charles Knight & Lutz Unterseher 1991: "Toward Defensive Restructuring in the Middle East," *BPP,* 22:2 (June): 115–134.

————, Chalres Knight & Lutz Unterseher 1991a: "Toward Defensive Restructuring in the Middle East," *Project on Defense Alternatives Research Monograph* (Cambridge, Mass.: Commonwealth Institute).

———— (ed.) 1988: *Peace Resource Book: A Comprehensive Guide to the Issues, Organizations, and Literature, 1988–1989* (Cambridge, Mass.: Ballinger).

Connaughton, Richard 1992: *Military Intervention in the 1990s: A New Logic of War* (London: Routledge).

———— 1992a: "Military Intervention and UN Peacekeeping," in Rodley (ed.): 165–197.

Conner, Albert Z. & Robert G. Poirier 1991: "The Institutions of the USSR Academy of Sciences: An Examination of Their Roles in Soviet Doctrine and Strategy," *JSMS,* 4:1 (March): 62–86.

Cooley, John K. 1991: "Pre-war Gulf Diplomacy," *Survival,* 33:2 (March–April): 125–139.

Cooney, James A. & al. (eds.) 1984: *The Federal Republic of Germany and the United States: Changing Political, Social, and Economic Relations* (Boulder, Colo.: Westview).

Coopieters, Bruno 1990: "Die pazifistischen Sekten, die Bolschewiki und das Recht auf Wehrdienstverweigerung," in Steinweg (ed.): 308–360.

Copel, Etienne 1984: *Vaincre la guerre: C'est possible!* (Paris: Lieu Commun; quoted from the paperback edition, 1985).

——— 1986: "To Defend Our Territory by All Available Means," in Tromp (ed.): 233–238.

Corcoran, Edward 1988: "Operational Countermobility," in Windass & Grove (eds.): no pagination.

Cordesman, Anthony H. 1988: "Alliance Requirements and the Need for Conventional Force Improvements," in Nehrlich & Thomson (eds.): 77–102.

——— & Abraham R. Wagner 1990: *The Lessons of Modern War,* 3 vols. (London: Mansell).

Cordier, Sherwood S. 1992: "Scandinavia and Finland: Security Policies and Military Capabilities in the 1990s," *AFR,* 47.

Corradini, Allessandro 1990: "Consideration of the Question of International Arms Transfers by the United Nations," *Disarmament,* 13:4, 131–143.

Cotter, Donald R. 1983: "Potential Future Roles for Conventional and Nuclear Forces in Defense of Western Europe," in ESECS: 209–253.

Coulmy, Daniel 1991: "Le Japon et sa défense," *Dossiers* (Paris: Fondation pour les études de défense nationale): 43.

Coutau-Bégarie, Hervé (ed.) 1991/1992: *L'Evolution de la pensé navale,* 2 vols. (*Dossiers,* 41, 46) (Paris: Fondation pour les études de défense nationale).

Covington, Stephen R. 1989: "NATO and Soviet Military Doctrine," *WQ,* 12:4 (Autumn): 73–82.

——— 1989a: "Soviet Military Doctrine: The Soviet Military: Prospects for Change," *JSMS,* 2:2 (June): 241–265.

Cox, David 1993: "Exploring *An Agenda for Peace:* Issues Arising from the Report of the Secretary-General," *Aurora Papers,* 20 (Ottawa: Canadian Centre for Global Security).

Crain, Fitzgerald 1989: "Eine sozialpsychologische Kritik helvetischer Igelmentalität," in Gross & al. (eds.): 109–131.

Cramer, Benedict 1984: "Dissuassion infranuclaire: L'armee de milice suisse: mythes et realites strategiques," *CES,* 4.

Crenzien, Bodo 1990: *Stabilitet og sikkerhed: Militære aspekter af konventionel nedrustning* (Copenhagen: SNU).

Creveld, Martin Van 1977: *Supplying War: Logistics from Wallenstein to Patton* (Cambridge: Cambridge U.P.).

——— 1985: *Command in War* (Cambridge: Harvard U.P.).

——— 1989: *Technology and War from 2000 B.C. to the Present* (New York: Free Press).

——— 1991: *The Transformation of War* (New York: Free Press).

——— 1991a: "Logistics Since 1945: From Complexity to Paralysis," in Reid & Dewar (eds.): 219–226.

——— 1993: *Nuclear Proliferation and the Future of Conflict* (New York: Free Press).

Croft, Stuart 1990: "The Future of Nuclear Deterrence in Europe: NATO and Nuclear Strategy," in Clesse & Rühl (eds.): 245–253.

——— 1990a: "Nuclear Arms Control and the UK," in Hoffman (ed.): 73–92.

——— 1991: "Stock-Taking After the Cold War," in idem (ed.): 201–216.

——— 1993: "Cooperative Security in Europe," in Centre for Defence Studies (ed.): 101–116.

——— (ed.) 1991: *British Security Policy: The Thatcher Years and the End of the Cold War* (London: HarperCollins).

—— & Phil Williams (eds.) 1992: *European Security Without the Soviet Union* (London: Frank Cass).

—— & David H. Dunns 1990: "The Impact of the Defence Budget on Arms Control Policy," in Hoffman (ed.): 53–70.

Cronin, Audrey K. 1990: "Changing American Views of European Security," in Cerutti & Ragionieri (eds.): 79–86.

Cronon, E. David (ed.) 1965: *The Political Thought of Woodrow Wilson* (Indianapolis: Bobbs-Merrill).

Crowe, William, Phillip Karber, William Odom, Leon Gouré, Edward Luttwak, Lutz Unterseher, David Holloway, Michael MccGwire & Ted Warner 1990: "The Red Army on the Defense? Reporting New Doctrine and the Changing Soviet Threat," *Deadline,* 5:2 (New York: Center for War, Peace and the News Media).

Crowl, Phillip A. 1986: "Mahan: The Naval Historian," in Paret (ed.): 444–480.

CSS (Center for Strategic Studies) 1991: "Anatomy of the Yugoslav Crisis," *CSS Papers,* 6 (Belgrade: CSS).

Cuthbertson, Ian M. 1989: "The Political Objectives of Conventional Arms Control," in Blackwill & Larrabee (eds.): 90–115.

—— & David Robertson 1990: *Enhancing European Security: Living in a Less Nuclear World* (London: Macmillan).

Cyr, Arthur 1990: "Forces and Politics in Europe," in Sarkesian (ed.): 139–151.

Czege, Huba Wass, de 1984: "Army Doctrinal Reform," in Clark & al. (eds.): 101–120.

Czempiel, Ernst-Otto 1990: "Die Modernisierung der Atlantischen Gemeinschaft," *EA,* 8: 275–286.

—— 1992: "Demokratie und Frieden: Theoretische und politikstrukturelle Aspekte einer europäischen Friedensordnung," in Lutz (ed.): 451–460.

——, Liparit Kiuzadjan & Zdenek Masopust 1990: *Non-Violence in International Crises.* Proceedings of the First International Symposium on Non-Violent Solutions of International Crises and Regional Conflicts, Frankfurt am Main, February 1989 (Vienna: European Coordination Centre for Research and Documentation in Social Sciences).

Daalder, Ivo H. 1987: "A Tactical Defence Initiative for Western Europe," *BAS,* May: 34–39.

—— 1991: *The Nature and Practice of Flexible Response: NATO Strategy and Theater Nuclear Forces Since 1967* (New York: Columbia U.P.).

—— 1992: "Cooperative Arms Control: A New Agenda for the Post–Cold War Era," *CISSM Papers,* 1 (College Park: University of Maryland, CISS).

—— 1992a: "The Future of Arms Control," *Survival,* 34:1 (Spring): 51–73.

—— & Jeoffrey Boutwell 1988: "TBMs and ATBMs: Arms Control Considerations," in Hafner & Roper (eds.): 179–208.

Dae-Jung, Kim 1993: "Korean Reunification: A Rejoinder," *Security Dialogue,* 24:44 (December): 409–414.

Dahlitz, Julie 1992: "Legal Issues Concerning the Feasibility of Nuclear Weapons Elimination," in Karp (ed.): 80–102.

Dalby, Simon 1991: "Rethinking Security: Ambiguities in Policy and Theory," *International Studies* (Burnaby, BC: Simon Fraser University, Department of Political Science).

Dalhaug, Arne Bård 1989: "Konvensjonell nedrustning i Europa og konsekvenser for Norge," *NUPIR,* 135.

—— 1990: "Konvensjonell nedrustning," *Internasjonal Politikk,* 48:1, 51–66.

Dallin, Alexander, Ralph Mavrogordato & Wilhelm Moll 1964: "Partisan Psychological Warfare and Popular Attitudes," in Armstrong (ed.): 197–337.

Damgaard, Knud 1984: "Forsvaret år 2001: Teknologi og struktur i fremtiden," in Christensen (ed.): 136–157.

Daniel, Donald C. F. 1990: "Alternate Models of Soviet Naval Behaviour," in Gillette & Frank (eds.): 237–248.

Däniker, Gustav 1983: "Die Armee des neutralen Kleinstaates als friedenssichernde Kraft," *ÖMZ*, 1: 19–26.

—— 1987: *Dissuasion: Schweizerische Abhaltestrategie Heute und Morgen* (Frauenfeld: Huber).

—— 1987a: "'Defensive Verteidigung': Ein gefährlicher Pleonasmus," *Allgemeine Schweizerische Militärzeitschrift*, 11: 701–702.

—— 1989: "Rüstung als Friedenssicherungsfaktor in der Schweiz," in Wagemann (ed.): 57–64.

—— 1990: "Der neutrale Kleinstaat in Europa," *ÖMZ*, 6: 461–465.

—— 1990a: "Die neutralen Länder Europas und die Strategie der Abhaltung," in Weizsäcker (ed.): 204–211.

—— 1992: *Wende Golfkrieg: Vom Wesen und Gebrauch künftiger Streitkräfte* (Frauenfeld: Huber).

Danilovich, A. A. 1992: "On New Military Doctrines of the CIS and Russia," *JSMS*, 5:4 (December): 517–538.

Dankbaar, Ben 1982: "Alternative Defence Policies and Modern Weapon Technology," in Kaldor & Smith (eds.): 163–184.

—— 1984: "Alternative Defence Policies and the Peace Movement," *JPR*, 21:4, 141–155.

—— 1986: "Emerging Technologies and the Politics of Doctrinal Debate," in Barnaby & Borg (eds.): 127–140.

—— 1989: "Alternative Defence Concepts and European Security," in Karl & Auffermann (eds.): 89–97.

Danspeckgruber, Wolfgang F. 1986: "Armed Neutrality: Its Application and Future," in Flanagan & Hampson (eds.): 242–279.

—— 1991: "Neutrality and the Emerging Europe," in idem (ed.): 265–288.

—— (ed.) 1991: *Emerging Dimensions of European Security Policy* (Boulder, Colo.: Westview).

Danzmayr, Heinz 1990: "Österreichs Landesverteidigungsplan," *Allgemeine Schweizerische Militärzeitschrift*, 1: 613.

—— 1991: "Neue Herausforderungen an Rüstungskontrolle und Aufrüstung in Europa," *Dialog*, 20: 167–174.

—— 1991a: "Streitkräfte im neuen Europa: Leitideen, Aufträge, Strukturen, Optionen." Paper presented to the conference "Sicherheit im neuen Europa," Evangelische Akademie Loccum and SAS, 27–29 September 1991.

Dardel, Jean-Jacques de 1990: "New Challenges Facing Swiss Foreign Policy," in Miliovejic & Maurer (eds.): 122–136.

Darilek, Richard E. 1986: "The Stockholm CDE Negotiations," in Borawski (ed.): 90–102.

—— & John K. Setar 1989: "Constraints," in Blackwill & Larrabee (eds.): 379–404.

Däubler, Wolfgang 1987: "Rechtliche Aspekte Gemeinsamer Sicherheit—Die UN-Charta als Ausgangspunkt," in Bahr & Lutz (eds.): 33–46.

Davenport, Brian A. 1991: "The Ogarkov Ouster: The Development of Soviet Military Doctrine and Civil/Military Relations in the 1980s," *JSS*, 14:2 (June): 129–147.

David, Dominique (ed.) 1989: *Le politique de défense de la France: Textes et documents* (Paris: Fondation pour les Études de la Défense nationale).

David, Steven R. 1989: "Why the Third World Matters," *IS*, 14:1 (Summer): 50–85.

Davis, Jacquelyn K. 1994: "Restructuring Military Forces in Europe," *AP*, 284 ("European Security after the Cold War. Part I"): 79–96.

———— & Robert L. Pfaltzgraff, Jr. 1989: "The West German Left: Implications for U.S. Policy," in Griffith & al.: 111–126.

Davis, Paul K. 1990: "Central Region Stability at Low Force Levels," in Huber (ed.): 143–163.

Davis, Warren F. 1990: "Technology and Alternative Security: A Cherished Myth Expires," in Weston (ed.): 43–77.

Davis, Zachary S. 1993: "The Realist Nuclear Regime," *SS*, 2:3/4 (Spring/Summer): 79–99.

Davy, Richard (ed.) 1992: *European Detente: A Reappraisal* (London: Sage).

Dawisha, Karen & Jonathan Valdez 1987: "Socialist Internationalism in Eastern Europe," *POC*, March–April: 1–14.

———— & Philip Hanson (eds.) 1981: *Soviet-East European Dilemmas: Coercion, Competition, and Consent* (New York: Holmes & Meier).

Dawydow, Jurij P. & Dmitrij W. Trenin 1992: "Die Zukunft der Streitkräfte Russlands: Die Auflösung eines Imperiums und das Problem einer Militärreform," *EA*, 47:13 (10 July): 353–364.

Deak, Peter 1989: "Restructuring Conventional Forces in Defensive Modes," in Rotblat & Goldanskii (eds.): 86–92.

———— 1993: "Review of Hungarian Security Policy Developments, 1945–1990," in "International Security and East-Central Europe," *Occasional Papers*, 1993 (Budapest: Hungarian Academy of Sciences, Centre for Security Studies): 3–33.

Dean, Jonathan 1985: "Directions in Inter-German Relations," *Orbis*, Fall: 609–632.

———— 1985a: "MBFR: Past and Future," in Paul (ed.): 82–91.

———— 1987: *Watershed in Europe: Dismantling the East-West Military Confrontation* (Lexington, Mass.: Lexington).

———— 1987a: "Alternative Defence: Answer to the NATO's Central Front Problems?" *IA*, 64:1 (Winter).

———— 1988: "Negotiated Force Reductions and Constraints on NATO and Warsaw Pact Forces in Europe," in Windass & Grove (eds.): no pagination.

———— 1988a: "How Can Arms Control Raise the Nuclear Threshold in Europe?" in Hopmann & Barnaby (eds.): 291–301.

———— 1988b: "The Exclusionary Zone: An American Approach," in Kirk (ed.): 5–8.

———— 1988c: "The INF Treaty Negotiations," in *SIPRI*: 375–394.

———— 1989: "Conventional Talks: A Good First Round," *BAS*: 26–31.

———— 1989a: "Cutting Back the NATO-Warsaw Treaty Confrontation," *Disarmament*, 12:2, 18–25.

———— 1989b: "Verifying Conventional Force Reductions and Limitations," in Blackwill & Larrabee (eds.): 280–306.

———— 1989c: *Meeting Gorbachev's Challenge: How to Build Down the NATO–Warsaw Pact Confrontation* (London: Macmillan).

———— 1989d: "Political Connotations of New Military Strategies," in Tromp (ed.): 319–330.

———— 1989e: "How to Reduce NATO and Warsaw Pact Forces," *Survival*, March–April: 109–122.

———— 1989f: "Disarmament and European Security," in Calliess (ed.): 27–41.

———— 1990: "Components of a New Security System for Europe," *IA(M)*, 12: 25–32.

———— 1990a: "Components of a Post–Cold War Security System for Europe," in Brauch (ed.): 127–162.

———— 1990b: "Defining Long-Term Western Objectives in CFE," *WQ*, 13:1 (Winter): 169–184.

———— 1990c: "The 1990 CSCE Summit: Step Toward a Treaty?" *DDA*, 3:6–7: 3–10.

———— 1990d: "The CFE Negotiations, Present and Future," *Survival*, 31:4 (July/August): 313–324.

———— 1990e: "Alternative Defence: Answer to the Bundeswehr's Post-INF Problems?" in Szabo (ed.): 188–216.

———— 1990f: "Immer noch vonnöten: Ein gemeinsames Zentrum zur Risikominderung von NATO und Warschauer Pakt," in E. Müller & Neuneck (eds.): 135–146.

———— 1990g: "Foreword: Main Elements of a Security System for Europe," in Huber (ed.): 11–18.

———— 1990h: "The Vienna Force Reduction Talks: Moving Towards Deep Cuts," in Boserup & Neild (eds.): 194–203.

———— 1990i: "Nicht-offensive Verteidigung bei den laufenden Wiener Abrüstungsverhandlungen," in Weizsäcker (ed.): 293–299.

———— 1990j: "The Vienna Force Reduction Talks: Resolved and Unresolved Issues," in Rotblat & Holdren (eds.): 50–65.

———— 1990k: "Components of a Post–Cold War Security System for Europe," in Calliess (ed.): 366–399.

———— 1991: "Europe's New Security Blanket," *Pacific Research*, 4:1, 7–10.

———— 1992: "Components of a Post–Cold War Security System for Europe," in Brauch & Kennedy (eds.): 45–68.

———— 1992a: "Constraining Technological Weapons Innovation in the Post–Cold War Environment," in Brauch & al. (eds.): 35–41.

———— 1992b: "Breakthrough in World Nuclear Security," in Lutz (ed.): 337–343.

———— 1993: "Unter welchen Bedingungen kann die Europäische Sicherheitsgemeinschaft erreicht werden?" *S+F*, 11(3): 126–128.

———— 1994: "Using Arms Control to Promote NOD in Europe," in Møller & Wiberg (eds.): 37–50.

———— & Kurt Gottfried 1991: *Nuclear Security in a Transformed World* (Washington, D.C.: Union of Concerned Scientists).

———— & Kurt Gottfried 1992: *A Program for World Nuclear Security,* new ed. (Cambridge, Mass.: Union of Concerned Scientists).

———— & Stanley R. Resor 1990: "Constructing a European Security System," in Gottfried & Bracken (eds.): 85–135.

———— & Randall Watson Forsberg 1992: "CFE and Beyond: The Future of Conventional Arms Control," *IS*, 17:1 (Summer): 76–121.

Dedijer, Vladimir 1952: "The Yugoslav Partisans," in Chaliand (ed.) 1982: 63–87.

Defence Agency 1990: *Defence of Japan* (Tokyo: Japan Times).

Defence Research Trust 1986: "Common Security Programme I: Common Security in Europe" (Adderbury, Oxfordshire: Defence Research Trust).

Deger, Sadat 1991: "World Military Expenditure," in *SIPRI*, 1991: 115–163.

———— & Somnath Sen 1990: *Military Expenditure: The Political Economy of International Security* (Oxford: Oxford U.P., SIPRI: Strategic Issue Papers).

———— & Somnath Sen 1991: "The Economics of Conversion: A Case Study of the Soviet Union." Paper presented at the Centre for Peace and Conflict Research at the University of Copenhagen, 16 October 1991.

Deittert, Albert 1986: "Von der Gefahrengemeinschaft zur Gemeinsamen Sicherheit," in Gohl & Niesporek (eds.): 113–126.

DeLauer, Richard D. 1986: "Emerging Technologies and Their Impact on the Conventional Deterrent," in Pierre (ed.): 40–70.

Dellios, Rosita 1989: *Modern Chinese Defence Strategy: Present Developments, Future Directions* (London: Macmillan).

De Lupis, Ingrid Detter 1987: *The Law of War* (Cambridge: Cambridge U.P.).

Dembinski, Matthias 1990: "Europa ohne taktische Nuklearwaffen? Konzeptionelle Überlegungen zur Minimierung der erweiterten Abschreckung," *HSFK-R,* 4.

—— 1991: "Tactical Nuclear Weapons After the End of the East-West Conflict: The Concept of Minimized Extended Deterrence," in Harle & Sivonen (eds.): 98–128.

—— 1993: "Mit START zum Ziel der allgemeinen und vollständigen Abrüstung? Stand und Perspektiven der Bemühungen um 'kooperative Denuklearisierung'," *HSFK-R,* 3.

—— & al. 1989: "No End to Modernization? Short-range Missile Modernization and the Deficiencies in the NATO Security Debate," *PRIF-R,* 67.

Deng, Francis M. 1993: *Protecting the Dispossessed: A Challenge for the International Community* (Washington, D.C.: Brookings).

Desch, Michael C. 1989: "The Keys that Lock Up the World: Identifying American Interests in the Periphery," *IS,* 14:1 (Summer): 86–121.

Dettke, Dieter 1984: "No First Use: Ein Beitrag zur Konventionalisierung der Verteidigung Europas," in SAS (eds.): 103–107.

Deudney, Daniel 1993: "Dividing Realism: Structural Realism versus Security Materialism on Nuclear Security and Proliferation," *SS,* 2:3/4 (Spring/Summer): 7–36.

Deutsch, Karl W. & al. 1957: *Political Community and the North Atlantic Area: International Organization in the Light of Historical Experience* (Princeton: Princeton U.P.).

DeWitt, Kurt 1964: "The Partisans in Soviet Intelligence," in Armstrong (ed.): 338–360.

Deyanov, Radoslav 1990: "The Role and Security Objectives of Confidence-building Measures at Sea," *Disarmament,* 13:4, 77–97.

DGFK (Deutsche Geselschaft für Friedens- und Konfliktforschung) (eds.) 1983: *DFGK Jahrbuch, 1982/83: Zur Lage Europas im globalen Spannungsfeld* (Baden-Baden: Nomos).

Dhanapala, Jayantha (ed.) 1991: *The United Nations, Disarmament and Security: Evolution and Prospects* (New York: United Nations; Geneva: UNIDIR).

Diab, M.Z. 1992: "A Proposed Security Regime for an Arab-Israeli Settlement," in Spiegel (ed.): 159–172.

Dibb, Paul 1988: *The Soviet Union: The Incomplete Superpower,* 2nd ed. (London: Macmillan).

Dick, Charles J. 1985: "Soviet Operational Concepts," 1, 2, *MR,* 65:9, 65:10 (September, October): 29–45, 4–19.

—— 1986: "Soviet Responses to Emerging Technology Weapons and New Defensive Concepts," in Barnaby & Borg (eds.): 220–238.

—— 1992: "Initial Thoughts on Russia's Draft Military Doctrine," *JSMS,* 5:4 (December): 552–566.

—— & Lutz Unterseher 1986: "Dialogue on the Military Effectiveness of Non-provocative Defence," in Barnaby & Borg (eds.): 239–250.

Diehl, Ole 1993: *Die Strategiediskussion in der Sowjetunion: Zum Wandel der sowjetischen Kriegsführungskonzeption in den achtziger Jahren* (Wiesbaden: Deutscher Universitätsverlag).

————— & Anton Krakau 1989: "Das Militärpotential des Warschauer Paktes in Europa," *Berichte,* 36.

Dienstbier, Jiri 1987: "Pax Europeana: A View from the East," in Smith & Thompson (eds.): 180–189.

————— 1991: "Central Europe's Security," *Foreign Policy,* 83 (Summer): 119–127.

Digby, James 1975: "Precision-Guided Munitions," *AP,* 118.

Din, Allan M. 1990: "A Dual Technology Approach to Stability," in Huber (ed.): 203–211.

Djiwandono, Soedjati 1990: "Nonprovocative Defence Strategy," in UNIDIR (eds.): 109–118.

Dobriansky, Paula J. & David B. Ricvkin, Jr. 1990: "Changes in Soviet Military Thinking: How Do They Add Up and What Do They Mean for Western Security?" in Green & Karasik (eds.): 147–184.

Dobrosielski, Marian 1990: "Vom Rapacki- zum Jaruzelski-Plan: Ein polnischer Weg," in Lutz & Schmähling (eds.): 113–142.

Domange, J. M. 1985: *Le Réarmement du Japon* (Paris: Fondation pour les études de défense nationale).

Donaldson, Thomas 1993: "Kant's Global Rationalism," in Nardin & Mapel (eds.): 136–157.

Donnelly, Christopher N. 1982: "The Soviet Operational Manoeuvre Group: A New Challenge for NATO," *IDR,* 9: 1177–1186.

————— 1990: "The Development of Soviet Military Policy in the 1990s," *RJ,* 135:1 (Spring): 11–16.

————— 1991: "The Central Front in Soviet Strategy," in Reid & Dewar (eds.): 35–43.

————— 1992: "Evolutionary Problems in the Former Soviet Armed Forces," *Survival,* 34:3 (Autumn): 28–42.

Doran, Charles F. 1992: "The Globalist-Regionalist Debate," in Schraeder (ed.): 55–71.

Dörfer, Ingemar & Johan Tunberger 1990: "Aspects of Stability in a Less Nuclear Europe," in Schmähling (ed.): 63–72.

DoswaldBeck, Louise 1987: "The Civilian in Crossfire," *JPR,* 24:3 (September): 251–262.

Dowe, Dieter & Kurt Klotzbach (eds.) 1984: *Programmatische Dokumente der deutschen Sozialdemokratie* (Bonn: Dietz).

Downing, Wayne J. 1986: "Light Infantry Integration in Central Europe," *MR,* 66:9, 18–29.

Downs, George W., David M. Rocke & Randolph M. Siverson 1985: "Arms Races and Cooperation," *WP,* 38:1 (October): 118–146.

Doyle, Michael 1983: "Kant, Liberal Legacies, and Foreign Affairs," *Philosophy and Public Affairs,* 12(3): 205–35, and 12(4): 323–353.

Dragsdahl, Jørgen 1989: "How Peace Research Has Reshaped the European Arms Dialogue," *Annual Review of Peace Activism* (Boston: Winston Foundation for World Peace): 39–45.

Draus, Franciszek 1991: "La ligne Oder-Neisse et évolution des rapport germano-polonais: La nouvelle doctrine militaire de la Pologne," *Dossier,* 36 (Paris: Fondation pour les études de défence nationale).

Dregger, Alfred 1988: "Disarmament with Security: A German View of Current Alliance Developments," *Atlantic Community Quarterly,* Winter: 404–412.

————— 1989: "Sicherheit für Europa und das Atlantische Bündnis," in Wagemann (ed.): 97–106.

Drell, Sidney, Philip J. Farley & David Holloway 1986: "Preserving the ABM

Treaty: A Critique of the Reagan Strategic Defense Initiative," in Miller & Evera (eds.): 57–97.

Dreyer, June Teufel 1982: "The Chinese Militia: Citizen Soldiers and Civil-Military Relations in the People's Republic of China," *Armed Forces and Society,* 9:1 (Fall): 63–82.

Drifte, Reinhard 1990: *Japan's Rise to International Responsibilities: The Case of Arms Control* (London: Athlone).

Driscoll, R. F. 1993: "European Security: Theoretical and Historical Lessons for Emerging Structures," *ES,* 2:1 (Spring): 15–22.

Dross, Armin (ed.) 1980: *Polen: Freie Gewerkschaften im Kommunismus?* (Reinbek: Rowohlt).

Drougos, Athanassios 1988: "A Nuclear-Weapon-Free Zone in the Balkans," in Baudisch (ed.): 95–101.

Druckman, Daniel 1993: "The Psychology of Arms Control and Reciprocation," in Ramberg (ed.): 21–41.

Dugan, Michael J. 1990: "The Future of FOFA: An Operational Perspective," *NSN,* September: 41–48.

Duic, Mario 1985: "Milizen: Begriffe, Modelle, Vergleiche, Versuch eines Überblicks," *ÖMZ,* 2: 104–112.

—— 1985a: "Die Schweiz 1939–1945: Erfahrungen in Sicherheitspolitik und umfassender Landesverteidigung," *ÖMZ,* 5: 407–416.

Duke, Simon 1989: *United States Military Forces and Installations in Europe* (Oxford: Oxford U.P., SIPRI).

Düllfer, Jost 1990: "Internationales System, Friedensgefährdung und Kriegsvermeidung: Das Beispiel der Haager Friedenskonferenzen 1899 und 1907," in Steinweg (ed.): 95–116.

Dumas, Lloyd J. 1988: "Turning Obstacles into Opportunities: Overcoming the Economic Damage of the Arms Race," in Hellman (ed.): 29–37.

—— 1989: "Economic Conversion: The Critical Link," in Dumas & Thee (eds.): 3–16.

—— 1990: "Economics and Alternative Security: Toward a Peacekeeping International Economy," in Weston (ed.): 137–175.

—— (ed.) 1988: "Swords into Plowshares: Conversion of Military Production to Socially Useful Purposes," *BPP,* 19: 1.

—— & Marek Thee (eds.) 1989: *Making Peace Possible: The Promise of Economic Conversion* (Oxford: Pergamon).

Dumoulin, André 1989: "De la doctrine de dissuasion minimale au concept de suffi-cance raissonable," *Memento Defense—Desarmament,* 1989 (Brussels: GRIP): 63–70.

—— 1991: "Du syndrome Scud à la renaissance des ATBM," *Stratégique,* 51/52: 319–344.

Dunay, Pál 1987: "Hungary's Security Policy," *HB,* 17.

—— 1988: "The Role of Allies in Helping to Structure Arms Control Issues: A View from Hungary," in Kirk (ed.): 25–30.

—— 1990: "Military Doctrine: Change in the East?" *Occasional Paper Series,* 15 (New York: Institute for East-West Security Studies).

—— 1991: "The CFE Treaty: History, Achievements and Shortcomings," *PRIF-R,* 24.

—— 1993: "CFE and the CIS: The Difficult Road to Ratification," *PRIF-R,* 29.

—— 1993a: "The Prospects of Arms Control Beyond the East-West Conflict," in "International Security and East-Central Europe," *Occasional Papers* (Budapest: Hungarian Academy of Sciences, Centre for Security Studies): 34–90.

―――― 1993b: "Das alte Ungarn im neuen Europa? Ungarische Sicherheitspolitik nach dem Systemwandel," *HSFK-R*, 2.

―――― 1993c: "Hungary: Defining the Boundaries of Security," in Karp (ed.): 122–154.

―――― & Laszlo Valki 1990: "Gemeinsame Sicherheit und Strukturelle Nichtangriffsfähigkeit: Eine ost-mitteleuropäische Sicht," in Lutz & Schmähling (eds.): 143–182.

Dunleavy, Patrick 1988: "A Non-Nuclear, Non-NATO Britain: Is There an Electoral Pathway?" in Burt (ed.): 106–133.

Dunn, David H. 1990: "Conventional Arms Control, Theatre Nuclear Weapons and European Security," in Hoffman (ed.): 93–114.

―――― 1991: "Challenges to the Nuclear Orthodoxy," in Croft (ed.): 9–28.

Dunn, Keith A. & William O. Staudenmaier 1985: "The Retaliatory Offensive and Operational Realities in NATO," *Survival*, 27:3, 108–118.

Dünne, Ludger 1988: "'Strukturelle Nichtangriffsfähigkeit' und konventionelle Landstreitkräfte," in Bahr & Lutz (eds.): 100–107.

―――― 1988a: "Zur Frage der Realisierung 'Struktureller Nichtangriffsfähigkeit' im Rahmen konventionell bewaffneter Landstreitkräfte," *HB*, 34.

Dunnigan, James F. 1983: *How to Make War: A Comprehensive Guide to Modern Warfare*, updated ed. (New York: Quill).

―――― & Austin Bay 1985: *A Quick and Dirty Guide to War: Briefings on Present and Potential Wars* (New York: Morrow).

Dupree, Louis 1989: "Post-Withdrawal Afghanistan: Light at the End of the Tunnel," in Saikal & Maley (eds.): 29–51.

Dupuy, Trevor N. 1979: *Numbers, Predictions and War* (New York: Bobbs-Merrill).

―――― 1980: *The Evolution of Weapons and Warfare* (London: Jane's).

―――― 1989: "Combat Data and the 3:1 Rule," *IS*, 14:1 (Summer 1989): 195–210.

―――― 1990: *Understanding Defeat: How to Recover from Loss in Battle to Gain Victory in War* (New York: Paragon).

―――― (ed.) 1993: *International Military and Defense Encyclopedia*, 6 vols. (Washington, D.C.: Brassey's).

Dupuy, William E. 1985: "The Light Infantry: Indispensable Element of a Balanced Force," *Army*, June: 26–41.

Dürr, Hans-Peter 1987: "Stabilitätsorientierte Sicherheitspolitik," *S+F*, 5:1, 22–23.

―――― 1988: "Kooperation statt Konfrontation: Wie wir die globalen Herausforderungen gemeinsam anpacken konnen," *Wissenschaftiche Welt*, 32:1, 13–19.

――――, Horst Afheldt & Albrecht A. C. von Müller 1984: *Exposé zu dem Forschungsprojekt "Stabilitätsorientierte Sicherheitspolitik"* (München: Max Planck Institute).

Dürr, Olivier 1987: "Applicability of Humanitarian Law of Armed Conflict," *JPR*, 24:3, 263–274.

Duus, Henning 1983: "En nation under våben," *Grus*, 11: 90–110.

Dyson, Freeman 1984: *Weapons and Hope* (New York: Harper & Row).

Earle, Edward Mead (ed.) 1941: *Makers of Modern Strategy: Military Thought from Machiavelli to Hitler* (New York: Atheneum, 1970).

Eavis, Paul & Owen Greene 1992: "Regulating Arms Export: A Programme for the European Community," in Brauch & al. (eds.): 283–299.

Eberhardt, Hans 1986: "Zur Militär- und Rüstungspolitik der Schweiz," *ÖMZ*, 5: 405–411.

―――― 1988: "Die Wehrpolitik der Schweiz zwischen Realität und Ideologie," *ÖMZ*, 6: 496–501.

Eberle, James 1988: "A European SACEUR?" in Alford & Hunt (eds.): 221–229.
——— 1990: "Global Security and Naval Arms Control," *Survival,* 32:4 (July/August): 325–332.
Ebert, Theodor 1967: "Die Organisation und Leitung der Sozialen Verteidigung," in idem 1981a: 90–119.
——— 1967a: "Gewaltloser Widerstand gegen stalinistische Regimen? Der Juni-Aufstand in der DDR," in idem 1981a: 52–77.
——— 1967b: "Non-violent Resistance Against Communist Regimes," in Roberts (ed.): 173–194.
——— 1967c: "Organization in Civilian Defence," in Roberts (ed.): 255–273.
——— 1968: "Der zivile Widerstand in der Tschechoslowakei: Eine Analyse seiner Bedingungen und Kampftechniken," in idem 1981a: 78–89.
——— 1968a: "Soziale Verteidigung: Eine Alternative zur 'Vorwärts-verteidigung'? Ein Forschungsbericht," in Ebert 1981a: 9–29.
——— 1968b: *Gewaltfreier Aufstand: Alternative zum Bürgerkrieg* (Waldkircher).
——— 1969: "Der zivile Widerstand in der Tschechoslowakei und die Zukunft demokratischer Wehrpolitik," in Anon (ed.): 191–208.
——— 1969a: "Die Entwicklung der sozialen Verteidigung [Civilian Defence]," in Anon (ed.): 33–46.
——— 1969b: "Grundzüge der Strategie der sozialen Verteidigung," in idem 1981a: 30–51.
——— 1970: "Der zivile Widerstand in der Tschechoslowakei 1968," in idem (ed.): 294–306.
——— 1970a: "Von aggressiver Drohung zu defensiver Warnung: Das Konzept der Sozialen Verteidigung," in idem 1981a: 141–191.
——— 1971: "Vermutungen über den dialektischen Prozess zur Sozialen Verteidigung," in idem 1981b: 9–28.
——— 1972: "Die Wehrpolitische Lücke im Programm antikapitalistischer Strukturreformen," in idem 1981b: 29–43.
——— 1974: "Widerstandsmöglichkeiten gegen innenpolitisches Eingreifen mit bewaffneten Einheiten," in idem 1981b: 44–72.
——— 1974a: "Ziviler Widerstand-mix oder pur? Zur Kontroverse zwischen Miitärstrategen und Friedensforschern über die Zukunft wehrhafter Neutralität," in idem 1981b: 73–103.
——— 1976: "Wie das Volk lernt, die soziale Verteidigung zu wollen-oder: Der lange March zu einer politischen Kultur der Widerstandsbereitschaft," in Ebert 1981b: 104–113. (English version: "Disarmament as a Non-violent Initiative: How People Learn to Want Civilian Defence, or the Long March Towards a Political Culture of Preparedness to Resistance," in Geeraerts (ed.) 1977: 66–76.)
——— 1977: "Soziale Verteidigung als Forschungsaufgabe: Schriftliche und mündliche Ausführungen bei der Konsultation der Regierung der Niederlande am 12 September 1977 in Den Haag," in idem 1981b: 114–128.
——— 1977a: "Über die Grenzen militärischen Denkens: Besprechung von Spannochi/de Brossolet 'Verteidigung ohne Schlacht'," in idem 1981b: 148–150.
——— 1978: "Die neue Frauenbewegung und die gewaltfreie Konfliktaustragung," in idem 1981b: 145–147.
——— 1978a: "Die Soziale Verteidigung im Bezugsfeld alternativer Sicherheitskonzepte," in idem 1981b: 151–163.
——— 1978b: "Verteidigungspolitik aus der Sicht der Ökologiebewegung," in idem 1981b: 129–144.
——— 1980: "Energieversorgung im Verteidigungsfall," in idem 1981b: 164–170.

———— 1980a: "Gewaltfreier Widerstand als Mittel der Verteidigungspolitik: Das Konzept der Sozialen Verteidigung in der Argumentation von Kriegsdienstverweigerer," in idem 1981b: 177–192.

———— 1981a: *Soziale Verteidigung,* vol. I: *Historische Erfahrungen und Grundzüge der Strategie* (Waldkirch: Waldkircher).

———— 1981b: *Soziale Verteidigung,* vol. II: *Formen und Bedingungen des zivilen Widerstandes* (Waldkirch: Waldkircher).

———— 1983: "Techno-Commando Units or Social Defense? A Comparison," in Steinke & Vale (eds.): 197–203.

———— 1984: "Eröffnungsreferat zur Anhörung 'Alternative Strategien'," *S+F,* 2:3, 57–60.

———— 1984a: "Mit Raketen kann man gegenüber gewaltfreiem Wiederstand nicht viel anfangen: Das Konzept der sozialen Verteidigung," in Schröder (ed.): 65–79.

———— 1984b: "Ziviler Wiederstand im besetzten Gebiet," in Weizsäcker (ed.): 231–264.

———— 1987: "Soziale Verteidigungeine Alternative? Von der Notwendigkeit und Möglichkeit einer gewaltfreien Alternative zur militärischen Abschreckung," in Heisenberg & Lutz (eds.): 602–621.

———— 1987a: "Soziale Verteidigung als politisch-operatives Konzept?" in Bühl (ed.): 197–218.

———— 1987b: "Die Diskussion um militärische und nichtmilitärische Defensivkonzepte aus deutscher Sicht und in Blick auf Österreich," *Dialog,* 10: 115–140.

———— 1990: "Soziale Verteidigung: Sicherheitspolitik mit gewaltfreien Mitteln nach der Öffnung der deutsch-deutschen Grenze," in Heisenberg & Lutz (eds.): II:234–259.

———— (ed.) 1970: *Ziviler Widerstand: Fallstudien aus der Innenpolitischen Friedens- und Konfliktforschung* (Düsseldorf: Bertelsmann).

———— (ed.) 1971: *Soziale Verteidigung: Friedens- und Sicherheitspolitik in den 80er Jahren* (Berlin: Burchhardthaus).

Eberwein, Wolf-Dieter 1992: "Demokratie und Krieg," *S+F,* 10:4, 196–202.

———— 1993: "Ewiger Frieden oder Anarchie? Demokratie und Krieg," in Forndran & Pohlman (eds.): 139–166.

Eden, Lynn 1992: "The End of Superpower Nuclear Arms Control?" in Karp (ed.): 164–201.

Edmonds, Martin 1991: "The Political-Military Dimension of East-West Relations: The Quest for Security in Europe," in Jordan (ed.): 235–252.

Eekelen 1990: "WEU and the Gulf Crisis," *Survival,* 32:6 (November–December): 519–532.

Ehmke, Horst 1986: "Wege zur Sicherheitspartnerschaft: Aus der Tätigkeit der gemeinsamen Arbeitsgruppe von SPD-Bundestagsfraktion und PVAP," *Blätter,* 6: 669–674.

————, Karlheinz Koppe & Herbert Wehner (eds.) 1986: *Zwanzig Jahre Ostpolitik: Bilanz und Perspektiven* (Bonn: Neue Gesellschaft).

Ehrhart, Hans-Georg 1987: *Französische Verteidigungspolitik im Wandel? Zur Problematik konventioneller Verteidigungskooperation zwischen Paris und Bonn* (Bonn: Friedrich-Ebert-Stiftung).

———— 1989: "Konventionelle Abrüstung als Gegenstand deutsch-französischer Sicherheitskooperation: Entwicklungen, Interesen, Perspektiven," *Studien,* 34.

———— 1991: "Zur Funktionsweise der UNO als System kollektiver Sicherheit: Politische Erfahrungen," in Lutz (ed.): 291–302.

———— 1991a: "Zur Rolle der EG beim Aufbau einer europäischen Ordnung des Gemeinsamen Friedens," in Lutz (ed.): 319–338.

———— 1992: "Der 'Westen' und die Sicherheit des 'Ostens': Lage, Risikeen, Optionen," in Lutz (ed.): 395–408.

————, Anna Kreikemeyer & Andrei V. Zagorski (eds.) 1993: *The Former Soviet Union and European Security: Between Integration and Re-Nationalization* (Baden-Baden: Nomos).

Ehrlich, Paul, Carl Sagan, Donald Kennedy & Walter Orr Roberts 1984: *The Cold and the Dark: The World After Nuclear War* (London: Sidgwick & Jackson).

Eichner, Klaus 1990: "The U.S. and West German Peace Movements of the Eighties: An Empirical Comparison," in Kodoma & Vesa (eds.): 193–202.

Eikenberg, Katrin 1993: "Somalia: Vom Krieg der Clans zum Krieg der UNO?" in Matthies (ed.): 183–198.

Einhorn, Hans 1990: "Abrüstung als ökonomische Aufgabe: Erkenntnisse— dargestellt am Beispiel der ehemaligen DDR," *Beiträge zur Konfliktforschung,* 20:4, 59–79.

———— & al. 1990: *Kosten und Gewinn: Ökonomie der Abrüstung* (Berlin: Brandenburgisches Verlagshaus).

Eiselt, Gerhard 1988: "Die Neutralisierung Europas als Weg zur Einheit Deutschlands," *DA,* 21:11, 1163–1166.

Ekéus, R. 1983: "How to Proceed Towards a Nordic Nuclear Weapon-Free Zone," in Lodgaard & Thee (eds.): 239–246.

Elsam, M. B. 1989: *Air Defence,* Brassey's Air Power: Aircraft, Weapons Systems and Technology Series, 7 (London: Brassey's).

El-Shafei, Omran 1991: "Denuclearization of Africa: Overview and Update," in U.N. Department for Disarmament Affairs (eds.), *Disarmament and Security in Africa* (New York: United Nations): 76–84.

Elshtain, Jean Bethke (ed.) 1992: *Just War Theory* (Oxford: Blackwell).

Elster, Ellen 1983: "Ikkevoldsforsvarets muligheder," in Thoft & Jacobsen (eds.): 4–17.

Elworthy, Scilla & Paul Ingram 1991: "Defence and Security in the New Europe: Who Will Decide?" *Current Decisions Report,* 7 (Oxford: Oxford Research Group).

Ember, Carol R., Melvin Ember & Bruce M. Russett 1992: "Peace Between Participatory Polities: A Cross-Cultural Test of the 'Democracies Rarely Fight Each Other' Hypothesis," *WP,* 44:4 (July): 573–599.

Enders, Thomas 1986: "ATM: Europas Verteidigung auch gegen Flugkörper," *EW,* 10: 560–568.

———— 1988: "Nukleare Abschreckung in und für Europa: Einige grundsätzliche Überlegungen nach Unterzeichnung des INF-Vertrags," *Interne Studien,* 9 (Sankt Augustin: Konrad-Adenauer-Stiftung).

———— 1990: "Militärische Herausforderungen Europas in den neunziger Jahren," *EA,* 10: 321–329.

———— 1990a: "The Role of Nuclear Deterrence in Non-Provocative Defence," *Disarmament,* 13:2, 24–38.

———— & Peter Siebenmorgen 1988: "Überlegungen zu einem sicherheitspolitischen Gesamtkonzept der Bundesrepublik Deutschland," *EA,* 14: 385–392.

————, Peter Siebenmorgan & Ulrich Weisser 1990: *Schlüssel zum Frieden— Sicherheitspolitik in einer neuen Zeit* (Bonn: Bouvier).

Engell, Hans 1986: *Ja, var der ingen fare* (Værløse: Kontrast).

———— 1986a: *Defensivt forsvar—et forsvar uden troværdighed* (Copenhagen: Det konservative Folkeparti).

Englén, Peter & Greger Hatt 1989: *Försvar utan hot: Hur icke-provokativ försvars-politik ger Europas stater större säkerhet* (Malmö: Pax).

Eppler, Erhard 1981: *Wege aus der Gefahr* (Reinbek: Rowolt).

———— 1983: *Die tödliche Utopie der Sicherheit* (Reinbek: Rowohlt).

———— 1988: *Wie Feuer und Wasser: Sind Ost und West Friedensfähig?* (Reinbek: Rowohlt).

Epstein, Joshua M. 1981: "Soviet Vulnerabilities in Iran and the RDF Deterrent" (*IS*, 6:2) in Miller (ed.) 1986: 309–341.

———— 1983: "Horizontal Escalation: Sour Notes of a Recurrent Theme," (*IS*, 8:3) in Miller & Evera (eds.) 1988: 102–114.

———— 1985: *The Calculus of Conventional War: Dynamic Analysis Without Lanchester Theory* (Washington, D.C.: Brookings).

———— 1987: *Strategy and Force Planning: The Case of the Persian Gulf* (Washington, D.C.: Brookings).

———— 1988: "Dynamic Analysis and the Conventional Balance in Europe," *IS*, 12:4 (Summer): 154–165.

———— 1989: "The 3:1 Rule, the Adaptive Dynamic Model, and the Future of Security Studies," *IS*, 13:4 (Summer): 90–127.

———— 1990: *Conventional Force Reductions: A Dynamic Assessment* (Washington, D.C.: Brookings).

Erickson, John 1991: "The Future of Soviet Military Doctrine," in Hemsley (ed.): 105–121.

———— 1993: "Fallen from Grace: The New Russian Military," *WPJ*, 10:2 (Summer): 19–24.

————, Lynn Hansen & William Schneider 1986: *Soviet Ground Forces: An Operational Assessment* (Boulder, Colo.: Westview).

ESECS (European Security Study Group) 1983: *Strengthening Conventional Deterrence in Europe: Proposal for the 1980s: Report of the European Security Study* (London: Macmillan).

———— 1985: *Strengthening Conventional Deterrence in Europe: A Program for the 1980s: European Security Study: Report of the Special Panel* (Boulder, Colo.: Westview).

ESG (European Strategy Group) 1988: *The Gorbachev Challenge and European Security: A Report from the European Strategy Group* (Baden-Baden: Nomos).

Etzioni, Amitai 1962: *The Hard Way to Peace: A New Strategy* (New York: Collier).

European Strategy Group: See ESG.

Evangelista, Matthew A. 1990: "Cooperation Theory and Disarmament Negotiations in the 1950s," *WP*, 42:4 (July): 502–528.

———— 1990a: "Why the Soviets Buy the Weapons They Do," in Gleditsch & Njølstad (eds.): 295–313.

Evans, Gareth 1993: *Cooperating for Peace: The Global Agenda for the 1990s and Beyond* (St. Leonard's, NSW: Allen & Unwin).

Evensen, Jens 1983: "The Establishment of Nuclear Weapon-Free Zones in Europe: Proposal on a Treaty Text," in Lodgaard & Thee (eds.): 167–189.

Evera, Stephen W. Van 1984: "Causes of War." Ph.D. diss., University of California, Berkeley.

———— 1985: "The Cult of the Offensive and the Origins of the First World War," in Miller (ed.): 58–107.

———— 1985a: "Why Cooperation Failed in 1914," *WP*, 38:1 (October): 80–117.

———— 1989: "Offense, Defense, and Strategy: When Is Offense Best?" in Feldman (ed.): 95–102.

———— 1990: "Primed for Peace: Europe After the Cold War," *IS*, 15:3, 7–57.

———— 1990a: "Why Europe Matters, Why the Third World Doesn't: American Grand Strategy After the Cold War," *JSS,* 13:2, 1–51.

———— 1990b: "Wars of Intervention: Why They Shouldn't Have a Future, Why They Do," *DDA,* 3:3, 1–8.

Everts, Philip P. 1985: "Public Opinion on Nuclear Weapons, Defence, and Security: The Case of the Netherlands," in Flynn & Rattinger (eds.): 221–274.

Evron, Yair 1991: "Israel," in Karp (ed.): 277–297.

Eyal, Jonathan 1991: "Central and Eastern Europe," in Pravda (ed.): 84–121.

———— 1992: "Military Relations," in Pravda (ed.): 35–72.

Ezz, Esmat A. 1992: "The Role of New Technology and the Interests of Developing Countries," in Neuneck & Ischebeck (eds.): 215–220.

Faber, Mient Jan 1990: "Detente in Europe," in Kaldor, Holden & Kalk (eds.): 25–51.

Fadin, Andrei 1993: "Spasmen der Gewalt: Der soziale Sinn der postsowjetischen Kriege," *Blätter,* 38:7 (July): 844–858.

Fairmichael, Rob 1983: "An Alternative Defence for Ireland. Some Considerations and a Model of Defence without Arms for the Irish People," *Dawn,* 95–96: 9–16.

Faivre, Maurice 1987: "Débat sur les défenses alternatives," *Défense nationale,* 43, January: 41–54.

Falk, Richard A. 1975: *A Study of Future Worlds* (New York: Free Press).

———— 1987: "Superseding Yalta: A Plea for Regional Self-Determination," in Kaldor & Falk (eds.): 28–51.

———— 1989: "The Superpowers and a Sustainable Detente in Europe," in Kaldor, Holden & Falk (eds.): 133–153.

———— 1991: "Theory, Realism, and World Security," in Klare & Thomas (eds.): 6–24.

———— 1992: "Recycling Interventionism," *JPR,* 29:2 (May): 129–134.

———— & Saul Mendlowitz (eds.) 1973: *Regional Politics and World Order* (San Francisco: Freeman).

Faramanzyan, Rachik 1989: "The Problem of Conversion: Theoretical and Practical Aspects," in IMEMO (eds.): 587–594.

Farer, Tom J. 1993: "The Role of Regional Collective Security Arrangements," in Weiss (ed.): 153–186.

———— & Felice Gaer 1993: "The UN and Human Rights: At the End of the Beginning," in Roberts & Kingsbury (eds.): 240–296.

Farinella, Paolo 1989: "Difesa non nucleare e non offensiva sul fronte nordorientale italiano," in Ragionieri (ed.): 141–162.

Farley, Philip J. 1988: "Arms Control and U.S.-Soviet Security Cooperation," in George, Farley & Dallin (eds.): 618–640.

———— 1988a: "Managing the Risks of Cooperation," in George, Farley & Dallin (eds.): 679–691.

Farndale, Martin 1988: "Follow on Forces Attack," *NSN,* 3:2, 42–50.

———— 1990: "Could NATO Cope Without US Forces?" in Sharp (ed.): 431–455.

Farwick, Dieter 1990: "Modelle der konventionellen Sicherheit," in Heisenberg & Lutz (eds.): II:352–363.

FCMH (Finnish Commission of Military History) (eds.) 1985: *Aspects of Security: The Case of Independent Finland* (Vasaa: FCMH).

Fedorenko, Sergei (ed.) 1988: *Reflections on Security in the Nuclear Age: A Dialogue Between Generals East and West* (Moscow: Progress).

Feinberg, Richard E. 1983: *The Intemperate Zone: The Third Wold Challenge to U.S. Foreign Policy* (New York: Norton).

Feiveson, Harold 1989: "Finite Deterrence," in Shue (ed.): 271–292.

———— & John Duffield 1984: "Stopping the Sea-Based Counterforce Threat," in Miller & Evera (eds.) 1988: 285–300.

———— & Frank N. von Hippel 1990: "Beyond Start: How to Make Much Deeper Cuts," *IS,* 15:1, 154–180.

———— & Frank N. von Hippel 1990a: "Stability of the Nuclear Balance After Deep Reductions, II," in Hippel & Sagdeev (eds.): 23–50.

————, Richard H. Ullman & Frank von Hippel 1985: "Reducing US and Soviet Arsenals," *BAS,* 41:7, 144–150.

Feld, Werner J. 1993: *The Future of European Security and Defense Policy* (Boulder, Colo.: Lynne Rienner).

Feldman, Shai 1992: "Security and Arms Control in the Middle East: An Israeli Perspective," in Stahl & Kemp (eds.): 75–91.

———— (ed.) 1990: *Technology and Strategy: Future Trends* (Boulder, Colo.: Westview).

Fels, Gerhard, Reiner K. Huber, Werner Kaltefleiter, Rolf F. Pauls & Franz-Joseph Schulze (eds.) 1990: *Strategie-Handbuch,* 2 vols. (Herford: Mittler).

Fernau, Heribert 1987: "Der Milizbegriff," *ÖMZ,* 6: 495–504.

Fieldhouse, Richard 1989: "Naval Nuclear Weapons: Status and Implications," in Fieldhouse & Taoka: 83–155.

———— 1989a: "Naval Forces and Arms Control: A Look to the Future," in Fieldhouse & Taoka: 155–170.

———— 1990: "Naval Nuclear Arms Control," in Fieldhouse (ed.): 158–186.

———— 1990a: "Naval Forces and Arms Control," in Fieldhouse (ed.): 3–42.

———— 1992: "Nuclear Weapons Developments and Unilateral Reduction Initiatives," *SIPRI:* 66–84.

———— (ed.) 1990: *Security at Sea: Naval Forces and Arms Control* (Oxford: Oxford U.P.).

———— & Shunji Taoka 1985: *Superpowers at Sea: An Assessment of the Naval Arms Race* (Oxford: Oxford U.P.).

Fifoot, Paul 1992: "Functions and Powers, and Inventions: UN Action in Respect of Human Rights and Humanitarian Intervention," in Rodley (ed.): 133–164.

Filchev, Ivan 1988: "The Defensive Character of the Military Doctrine of the Warsaw Treaty States," in Baudisch (ed.): 13–17.

Finan, J. S. 1992: "The Republic of South Africa: Nonoffensive Defence," *Occasional Paper,* 15 (Winnipeg: University of Manitoba, Programme in Strategic Studies).

Findlay, Trevor 1990: "North Pacific Confidence-Building: The Helsinki-Stockholm Model," in Langdon & Ross (eds.): 72–86.

———— (ed.) 1991: *Chemical Weapon and Missile Proliferation: With Implications for the Asia/Pacific Region* (Boulder, Colo.: Lynne Rienner).

Fischer, David 1992: *Stopping the Spread of Nuclear Weapons: The Past of the Prospects* (London: Routledge).

Fischer, Dietrich 1982: "Invulnerability Without Threat: The Swiss Concept of General Defense," *JPR,* 19:3, 205–225.

———— 1984: *Preventing War in the Nuclear Age* (Totowa, N.J.: Rowman & Allanheld).

———— 1984a: "Peace as a Road to Disarmament," *Disarmament,* 7:3, 81–87.

———— 1988: "No First Use of Nuclear Weapons," in Krieger & Kelly (eds.): 125–142.

———— 1988a: "Eight Points of Convergence," in Baudisch (ed.): 18–25.

———— 1993: *Nonmilitary Aspects of Security: A Systems Approach* (Aldershot: Dartmouth).

————, Wilhelm Nolte & Jan Øberg 1987: *Frieden gewinnen: Mit autonomen Initiativen aus dem Teufelskreis ausbrechen* (Freiburg: Dreisam).

Fischer, Horst 1987: "Koexistenz und Kooperation im modernen Völkerrecht: Gemeinsame Sicherheit und die Struktur des Rechts der Friedenssicherung," in Bahr & Lutz (eds.): 55–84

———— 1993: "Der Schutz von Menschen im Krieg: Humanitäres Völkerrecht und Humanitäre Intervention," in Matthies (ed.): 87–104.

Fischer, Johannes (ed.) 1975: *Verteidigung im Bündnis: Planung, Aufbau und Bewährung der Bundeswehr, 1950–1972* (München: Bernard & Graefe, Militärgeschichtliches Forschungsamt).

Fischer, Ronald J. 1990: *The Social Psychology of Intergroup and International Conflict Resolution* (New York: Springer).

Fischer, Siegfried 1990: "Militärdoktrinen und internationale Sicherheit: Gedanken über die Zukunft der NVA," *S+F*, 1: 13–17.

———— 1990: "Die Sicherheitspolitik der DDR im Übergang," in Benjowski & al. (eds.): 14–21.

———— 1994: "Is War Possible in Europe?" in Møller & Wiberg (eds.): 115–126.

———— & Otfried Nassauer (eds.) 1992: *Satansfaust—Das nukleare Erbe der Sowjetunion* (Berlin & Weimar: Aufbau).

Fisera, Vladimir Claude 1984: "The New Left and Defence: Out of the Ghetto?" in Howorth & Chilton (eds.): 233–247.

Fisher, David 1986: "No First Use," in Paskins (ed.): 21–33.

Fitzgerald, Mary C. 1989: "Defensive Defense: Marshal Ogarkov's View," *IDR*, 9: 1143–1147.

———— 1990: "Gorbachev's Concept of Reasonable Sufficiency in National Defense," in Hudson (ed.): 175–196.

———— 1992: "The Evolving Post-Soviet Military Doctrine," *IDR*, 25:5 (May): 410.

———— 1992a: "Russia's New Military Doctrine," *RJ*, 137:5 (October): 40–48.

Flanagan, Stephen J. 1986: "NATO's Conventional Defense Choices in the 1980s," in Flanagan & Hampson (eds.): 85–112.

———— 1987: "Emerging Tensions over NATO's Conventional Forces," *IDR*, 1: 31–39.

———— 1988: "Nonprovocative and Civilian-Based Defenses," in Nye, Allison & Carnesale (eds.): 93–110.

———— 1988a: "Nonoffensive Defense Is Overrated," *BAS*, 44:7 (September): 46–48.

———— & Andrew Hamilton 1988: "Arms Control and Stability in Europe: Reductions Are Not Enough," *Survival*, September–October: 448–463.

———— & Fen Osler Hampson (eds.) 1986: *Securing Europe's Future*. A Research Volume from the Center for Science and International Affairs, Harvard University (London: Croom Helm).

Fleury, Antoine 1990: "Switzerland and the Organisation of Europe: An Historical Perspective," in Miliovejic & Maurer (eds.): 97–109.

Flor, Roland 1986: "Der bewaffnete Konflikt in Afghanistan," *ÖMZ*, 2: 147–154.

———— 1990: "Thesen zur künftigen Rüstungsentwicklung: Neue Dimensionen des Kriegs- und Gefechtsbildes durch Hochtechnologien," *ÖMZ*, 5: 407–412.

Flory, Maurice 1993: "L'ONU et les opérations de maintien et de rétablissement de la paix," *Politique Étrangere*, 58:3 (Autumn): 633–640.

Flowerree, Charles C. 1986: "CBMs in the U.N. Setting," in Borawski (ed.): 103–115.

Flynn, Gregory 1989: "Doctrine, Images, and the East-West Military Relationship," in Flynn (ed.): 1–26.

———— (ed.) 1985: *NATO's Northern Allies: The National Security Policies of*

Belgium, Denmark, the Netherlands, and Norway (London: Rowman & Allanheld).

——— (ed.) 1989: *Soviet Military Doctrine and Western Policy* (London: Routledge).

——— & David Scheffer 1990: "Limited Collective Security," *Foreign Policy,* 80, Fall, 77–101.

——— & Hans Rattinger (eds.) 1985: *The Public and Atlantic Defense* (London: Rowman & Allanheld).

FM 1005 1976: U.S. Department of the Army, *HQ: Field Manual 100-5: Operations* (Washington, D.C.: GPO).

——— 1982: U.S. Department of the Army, *HQ: Field Manual 100-5: Operations* (Washington, D.C.: GPO).

——— 1986: U.S. Department of the Army, *HQ: Field Manual 100-5: Operations* (Washington, D.C.: GPO).

Fochepoth, Josef (ed.) 1988: *Adenauer und die deutsche Frage* (Göttingen: Vandenhoech & Ruprecht).

Foley, Hamilton 1923: *Woodrow Wilson's Case for the League of Nations* (Port Washington, N.Y.: Kennicott).

Folke, Steen 1983: "Socialisterne og militæret," in Thoft & Jacobsen (eds.): 70–77.

Ford, Daniel, Henry Kendall & Steven Nadis 1982: *Beyond the Freeze: The Road to Nuclear Sanity* (Boston: Beacon).

Forndran, Erhard 1990: "Abrüsting und Rüstungskontrolle—Chancen und Risiken nach dem Vertrag über die Mittelstreckenraketen," in Heisenberg & Lutz (eds.): I:348–368.

——— & Hartmut Pohlman (eds.) 1993: *Europäische Sicherheit nach dem Ende des Warschauer Paktes* (Baden-Baden: Nomos).

Forsberg, Randall 1984: "The Freeze and Beyond: Confining the Military to Defense as a Route to Disarmament," *WPJ,* 1:2, 285–318.

——— 1985: "Putting the Arms Race in Reverse," in Huzzard & Meredith (eds.): 186–200.

——— 1986: "Parallel Cuts in Nuclear and Conventional Forces," in Ackland & McGuire (eds.): 325–334.

——— 1987: "The Case for a Third-World Nonintervention Regime" (Brookline, Mass.: IDDS).

——— 1987a: "A Global Approach to Nonprovocative Defense," *Alternative Defense Working Paper,* 6 (Brookline, Mass.: IDDS).

——— 1987b: "Abolishing Ballistic Missiles: Pros and Cons," *IS,* 12:1 (Summer): 190–196.

——— 1988: "Toward a Nonaggressive World," *BAS,* 44:7 (September): 49–55.

——— 1988a: "Conventional Stability and Confidence-Building Structure: Criteria for the Change of Defense Plans," in Calliess (ed.): 94–107.

——— 1992: "Keep Peace by Pooling Armies," *BAS,* 48:4 (May): 41–42.

——— Steven Van Evera, Hayward R. Alker, Jr., Jonathan Dean, Carl Kaysen, Joanne Landy, Steven Miller & Jane M. O. Sharp 1992: "After the War: A Debate on Cooperative Security," *Boston Review,* 17:6 (November–December): 7–19.

———, Richard L. Garwin, Paul C, Warnke & Robert W. Dean 1983: *Seeds of Promise: The First Real Hearings on the Nuclear Arms Freeze* (Andover, Mass.: Brick).

———, Rob Leavitt, Matthew Goodman & Carl Conetta 1988: "Guide to Peace Issues and Strategies," in Conetta (ed.): 3–31.

———, Rob Leavitt & Steve Lilly-Webber 1991: "Conventional Forces Treaty Buries Cold War," *BAS,* 47:1 (January–February): 32–37.

Fortmann, Michel 1992: "NATO Defence Planning in a Post-CFE Environment: Assessing the Alliance Strategy Review (1990–1991)," in Haglund & Mager (eds.): 41–62.

———— & Thierry Gongora (with the assistance of Stéphane Lefebvre) 1993: "Modern Land Warfare Doctrines and Non-Offensive Defence: 'An Interesting Idea'," *Research Note,* 93/06 (Ottawa: Department of National Defence, Canada, Operational Research and Analysis, Directorate of Strategic Analysis).

Fought, Stephen 1991: "Translating the Lessons and Methods of Nuclear Arms Control to the Conventional Arena," in Smoke & Kortunov (eds.): 317–333.

Foundation for International Security 1989: *Beyond the Cold War: Building Common Security in Europe* (Adderbury, Oxfordshire: Foundation for International Security).

FRAG (Forsvarsministerens Rådgivnings- og Analysegruppe) 1986: "Mulig anvendelse af landbaserede sømålsmissiler i Danmark," *Report,* RAG 04/1986 (Copenhagen: Ministry of Defence).

Franck, Thomas M. & Georg Nolte 1993: "The Good Offices Function of the UN Secretary-General," in Roberts & Kingsbury (eds.): 143–182.

Freedman, Lawrence 1984: "The No-First-Use Debate and the Theory of Thresholds," in Blackaby & al. (eds.): 67–78.

———— 1986: *U.S. Intelligence and the Soviet Strategic Threat,* 2d ed. (Princeton: Princeton U.P.).

———— 1987: "Strategic Defence in the Nuclear Age," *AP,* 224.

———— 1988: "I Exist, Therefore I Deter?" *IS,* 13:1 (Summer): 177–195.

———— 1989: "The Politics of Conventional Arms Control," *Survival,* 31:5, 387–396.

———— 1989a: *The Evolution of Nuclear Strategy,* 2d ed. (London: Macmillan, IISS).

———— 1990: "Arms Control at Sea," in Pay & Till (eds.): 232–240.

———— 1991: "Arms Control: Thirty Years On," *Dædalus,* Winter: 69–82.

———— 1991a: "Whither Nuclear Strategy?" in Booth (ed.): 75–89.

———— 1992: "The End of Formal Arms Control," in Adler (ed.): 69–83.

———— 1994: "European Security after the Cold War. Part I: Strategic Studies and the New Europe," *AP,* 284: 15–27.

———— (ed.) 1983: *The Troubled Alliance: Atlantic Relations in the 1980s* (London: Heinemann).

———— (ed.) 1990: *Europe Transformed: Documents on the End of the Cold War* (London: Tri-Service Press).

———— & David Boren 1992: "'Safe Havens' for Kurds in Post-War Iraq," in Rodley (ed.): 43–92.

———— & Efraim Karsh 1991: "How Kuwait Was Won: Strategy in the Gulf War," *IS,* 16:2 (Fall): 3–41.

Freeman, John 1991: *Security and the CSCE Process: The Stockholm Conference and Beyond* (London: Macmillan).

Frei, Daniel 1984: "Alternatives to the First Use of Nuclear Weapons," in Blackaby & al. (eds.): 79–88.

———— 1989: "Sicherheit und Vertrauensbildung—Theoretische Aspekte," in Calliess (ed.): 347–369.

———— (with Christian Catrina) 1983: *Risks of Unintentional Nuclear War* (Totowa, N.J.: Allanheld).

Freier, Shalheveth 1993: "International Security in a NWFW," in Rotblat & al. (eds.): 145–152.

Freundl, Siegfried 1984: "Überlegungen zu einem Informations- und Führungsnetz für eine rein defensive Verteidigung," in Weizsäcker (ed.): 167–178.

Freymond, Jean F. 1990: "Neutrality and Security Policy as Components of the Swiss Model," in Marko & Maurer (eds.): 175–193.

Friedman, Norman 1988: *The US Maritime Strategy* (London: Jane's).

———— 1991: *Desert Victory: The War for Kuwait* (Annapolis, Md.: Naval Institute Press).

Fritsch-Bournazel, Renata 1988: *Confronting the German Question: Germans on the East-West Divide* (Oxford: Berg).

———— 1989: "The German Factor in Future East-West Relations," in Clesse & Schelling (eds.): 225–235.

———— 1990: *Europa und die deutsche Einheit* (Bonn: Aktuell).

———— 1991: "German Unification: Views from Germany's Neighbours," in Heisenberg (ed.): 70–87.

Fritzsche, Helmut 1989: "The Security Debate in the Evangelical Church of the GDR," in Wæver & al. (eds.): 46–60.

Frost, Gerald (ed.) 1991: *Europe in Turmoil: The Struggle for Pluralism.* European Defence and Strategic Studies Annual, 1990–1991 (London: Adamantine).

Fry, Greg E. 1986: "The South Pacific Nuclear-Free Zone," in *SIPRI,* 1986: 499–522.

Fry, Michael D. 1989: "Some Thoughts on the Role of Military Forces Within a European Security System," in Wachter & Krohn (eds.): 79–86.

Fry, Michael G. 1989: "Historians and Deterrence," in Stern & al. (eds.): 84–97.

Frye, Alton 1990: "Engagement and Restraint: Dilemmas of European Security," in Clesse & Rühl (eds.): 80–88.

———— 1992: "Zero Ballistic Missiles," *Foreign Policy,* 88, Fall, 3–20.

Fuchs, Katrin 1985: "Frieden ist möglich—durch Abrüstung und Entspannung," in Fuchs & al. (eds.): 52–74.

————, Hajo Hoffmann & Horst Klaus (eds.) 1985: *Konzepte zum Frieden: Vorschläge für eine neue Abrüstungs- und Entspannungspolitik der SPD* (Berlin: SPW).

Fuhrer, H. R. 1987: "Austria and Switzerland: The Defense Systems of Two Minor Powers," in Gann (ed.): 95–125.

Fukuyama, Francis 1985: "Escalation in the Middle East and Persian Gulf," in Allison & al. (eds.): 115–147.

———— 1986: "U.S.-Soviet Interactions in the Third World," in Horelick (ed.): 198–224.

———— 1991: "Democratization and International Security," *AP,* 266: 14–24.

———— 1992: *The End of History and the Last Man* (New York: Free Press).

Fuller, Graham E. 1991: "Moscow and the Gulf War," *FA,* 70:3 (Summer): 55–76.

Fuller, J. F. C. 1932: "What Is an Aggressive Weapon?" *English Review,* 54 (June): 601–605.

———— 1944: "Armour and Counter Armour, Part 3: Defence Against Armoured Attack," *Infantry Journal,* May: 39–44.

Funabashi, Yoichi 1991: "Japan and the New World Order," *FA,* 70:5 (Winter): 58–74.

Funk, Helmut 1988: "Die Integrierte Vorneverteidigung Mitteleuropas," in Bülow, Funk & Müller: 111–193.

———— 1988a: "Von der Doktrin zur Struktur," in Bülow, Funk & Müller: 208–219.

Furkes, Josip & Karl-Heinz Schlarp (eds.) 1991: *Jugoslawien: Ein Staat zerfällt* (Reinbek: Rowohlt).

Furre, Berge & Ingolf Håkon Teigene 1983: *Forsvar for Fred* (Oslo: Pax).

────── & Ingolf Håkon Teigene 1983a: "På vej mod et alternativ," in Thoft & Jacobsen (eds.): 28–39.

Fürst, Andreas, Volker Heise & Steven E. Miller (eds.) 1992: *Europe and Naval Arms Control in the Gorbachev Era* (Oxford: Oxford U.P., SIPRI).

Fyodorov, Y. E. 1990: "Transition to Defence-Oriented Configurations," in UNIDIR (eds.): 167–171.

Gabriel, Jürg Martin 1990: "Switzerland and Economic Sanctions: The Dilemma of a Neutral," in Miliovejic & Maurer (eds.): 232–245.

Gaddis, John Lewis 1982: *Strategies of Containment: A Critical Appraisal of Postwar American National Security Policy* (New York: Oxford U.P.).

────── 1986: "The Long Peace: Elements of Stability in the Postwar International System," *IS,* 10:4, 99–142.

Gaer, Felice D. 1993: "The United Nations and the CSCE: Cooperation, Competition, or Confusion?" in Lucas (ed.): 161–206.

Galtung, Johan 1959: *Forsvar uten Militærvesen* (Oslo: Fanum).

────── 1964: "Two Concepts of Defense," in Galtung 1976: 280–340.

────── 1965: "On the Meaning of Nonviolence," in Galtung 1976: 341–377.

────── 1968: "On the Strategy of Nonmilitary Defense: Some Proposals and Problems," in Galtung 1976: 378–426.

────── 1969: "Violence, Peace and Peace Research," in Galtung 1976: 109–134.

────── 1971: "Nichtmilitärische Verteidigungsmassnahmen," in Ebert (ed.): 84–93.

────── 1973: *The European Community: A Superpower in the Making* (London: Allen & Unwin).

────── 1975: *Essays in Peace Research,* vol. I: *Peace: Research, Education, Action* (Copenhagen: Ejlers).

────── 1976: *Essays in Peace Research,* vol. II: *Peace: War and Defense* (Copenhagen: Ejlers).

────── 1980: *The True Worlds: A Transnational Perspective* (New York: Free Press).

────── 1981: "What Kind of Defence Should We Have?" in Myrdal & al.: 157–168.

────── 1982: "Nya perspektiv på det säkerhetspolitiska samarbetat mellan de nordlige staterne," in Ofstad & Öhberg (eds.): 150–163.

────── 1983: "Et defensivt kombi-forsvar," in Thoft & Jacobsen (eds.): 18–27.

────── 1984: *There Are Alternatives: Four Roads to Peace and Security* (Nottingham: Spokesman).

────── 1984a: "Transarmament: From Offensive to Defensive Defense," *JPR,* 21:2, 127–140.

────── 1986: "The Green Movement: A Socio-Historical Exploration," in 1988a: 343–356.

────── 1987: "Peace and the World as Inter-civilizational Interaction," in Väyrynen (ed.): 330–347.

────── 1987a: "What Is Meant by Peace and Security? Some Options for the 1990s," *Dialog,* 10: 11–31.

────── 1987b: "The Real Star Wars Threat," *Nation,* 28 February: 248–249.

────── 1988: "From Disarmament to Transarmament: Evolving Trends in the Study of Disarmament and Security," in Galtung 1988a: 27–41.

────── 1988a: *Essays in Peace Research,* vol. VI: *Transarmament and the Cold War: Peace Research and the Peace Movement* (Copenhagen: Ejlers).

———— 1988b: "The Peace Movement: A Structural-Functional Exploration," in Galtung 1988a, 322–342.

———— 1989: "Non controllo degli armamenti, ma controllo delle doctrine," in Ragionieri (ed.): 7–14.

———— 1989a: *Nonviolence and Israel/Palestine* (Honolulu: University of Hawai, Institute for Peace).

———— 1989b: *Europe in the Making* (New York: Crane Russak).

———— 1989c: "Die Schweiz als Vorbild für andere?" in Gross & al. (eds.): 181–188.

———— 1990: "Stabilität oder Unsicherheit: Einige Überlegungen zur Friedens- und Sicherheitspolitik der neunziger Jahre," in Heisenberg & Lutz (eds.): I:50–59.

———— 1990a: *60 Speeches on War and Peace* (Oslo: PRIO).

———— 1990b: "Die Zukunft der Armeen?" *Dialog,* 19: 15–28.

———— 1991: "What Would Peace in the Middle East Be Like—and Is It Possible?" *BPP,* 22:3 (September): 243–247.

———— 1992: "Arms Control and Disarmament in Europe," in Øberg (ed.): 121–133.

———— & Jan Øberg 1992: "Experimental Politics After the Cold War: A Peace Zone in Cooperation with the United Nations," in Øberg (ed.): 255–280.

———— & Arne Næss 1955: *Gandhis politiske etikk* (Oslo: Fanum).

————, D. Kinkelbur & M. Nieder (eds.) 1993: *Gewalt im Alltag und in der Weltpolitik* (Münster: Agenda).

Galvin, John R. 1989: "Problems of Conventional Arms Control," *Survival,* 31:2 (March–April): 99–107.

Gamba-Stonehouse, Virginia 1991: "Argentina and Brazil," in Karp (ed.): 229–257.

Gann, L. H. (ed.) 1987: *The Defence of Western Europe* (London: Croom Helm).

Ganser, Helmut 1988: "Konventionelle Stabilität," in Bahr & Lutz (eds.): 433–437.

Gapes, Mike 1988: "The Evolution of Labour's Defence and Security Policy," in Burt (ed.): 82–105.

———— 1988a: "Labour's Defence and Security Policy," in Hopmann & Barnaby (eds.): 341–355.

———— 1990: "After the Cold War: Building on the Alliances," *Fabian Tracts,* no. 540 (London: Fabian Society).

Gardner, Lloyd C. 1992: "The Evolution of the Interventionist Impulse," in Schraeder (ed.): 23–36.

Gareev, M. A. 1992: "On Military Doctrine and Military Reform in Russia," *JSMS,* 5:4 (December): 539–551.

Garett, Banning N. & Bonnie S. Glaser 1990: "Arms Control in the Management of US Security Interests in the Asia-Pacific Region," in Langdon & Ross (eds.): 154–169.

Garnett, John 1987: "Technology and Strategy," in Baylis & al.: I:91–109.

Garrett, B. N. 1991: "Ending the United States–Soviet Cold War in East Asia: Prospects for Changing Military Strategies," *WQ,* 14:2 (Spring): 163–177.

Garrett, James M. 1989: *The Tenuous Balance: Conventional Forces in Central Europe* (Boulder, Colo.: Westview).

———— 1990: *Central Europe's Fragile Deterrent Structure* (London: Wheatsheaf).

———— 1992: "Confidence-Building Measures: Foundation for Stability in Europe," *JSS,* 15:3 (September): 281–303.

Garthoff, Raymond L. 1984: "BMD and East-West Relations," in Carter & Schwartz (eds.): 275–329.

———— 1985: *Détente and Confrontation: American-Soviet Relations from Nixon to Reagan* (Washington, D.C.: Brookings).

———— 1988: "Gorbachev and Soviet Military Power," *WQ,* 11:3, 131–158.

———— 1989: "Military Strategy: Lectures from the Voroshilov General Staff Academy," *JSMS,* 2:2 (June): 157–205.

———— 1990: *Deterrence and the Revolution in Soviet Military Doctrine* (Washington, D.C.: Brookings).

———— 1992: "The Tightening Frame: Mutual Security and the Future of Strategic Arms Limitation," in Leebaert & Dickinson (eds.): 57–78.

Gärtner, Heinz 1987: "Alternative Sicherheitskonzepte und Disengagement," *Dialog,* 10: 141–172.

———— 1989: "Internationale Sicherheit durch vertrauensbildende Massnahmen im Bereich der Militär- und Rüstungskontrollpolitik," in Calliess (ed.): 392–398.

———— 1990: "Wird Europa sicherer? Thesen zur Rüstungskontrolle und Vertrauensbildung," in Schweizerische Friedensstiftung (eds.): 143–151.

———— 1991: "Kollektive und nationale Sicherheit," *Dialog,* 20: 214–224.

———— 1992: *Wird Europa sicherer? Zwischen kollektiver und nationaler Sicherheit* (Laxenburg: Braumüller).

Garwin, Richard 1986a: "Countermeasures and Costs," in Snyder (ed.): 149–156.

———— 1986b: "The Soviet Response: New Missiles and Countermeasures," in Tirman (ed.) 129–146.

———— 1989: "Deep Cuts in Strategic Nuclear Weapons: Possible? Desirable?" in Rotblat & Goldanskii (eds.): 2–13.

———— 1993: "Nuclear Weapons for the United Nations?" in Rotblat & al. (eds.): 169–180.

Gasteyger, Curt 1990: "Swiss Neutrality: Obsolete or Obstinate?—The Challenges of the Future," in Miliovejic & Maurer (eds.): 194–206.

Gat, Azar 1988: "Clausewitz on Defence and Attack," *JSS,* 11:1, 20–26.

———— 1989: *The Origins of Military Thought from the Enlightenment to Clausewitz* (Oxford: Clarendon).

Gates, David 1986: *The Spanish Ulcer: A History of the Peninsular War* (New York: Norton).

———— 1987: *Non-Offensive Defence: A Strategic Contradiction?* (London: Alliance).

———— 1987a: "Area Defence Concepts: The West German Debate," *Survival,* 29:4 (July/August): 301–317.

———— 1989: *Light Divisions in Europe: Forces of the Future* (London: Alliance, Institute for Defence and Strategic Studies).

———— 1990: "An Interim Structure for NATO Forces: Proposals and Problems," in Clesse & Rühl (eds.): 235–244.

———— 1991: *Non-Offensive Defence: An Alternative Strategy for NATO?* (London: Macmillan).

Gati, Toby Trister 1991: "The UN Rediscovered: Soviet and American Policy in the United Nations of the 1990s," in Jervis & Bialer (eds.): 197–223.

Gazit, Shlomo 1992: "After the Gulf War: The Arab World and the Peace Process," in Spiegel (ed.): 17–26.

Geeraerts, Gustav 1977: "Two Approaches of Civilian Defence: Observations on the Development of the Civilian Defence Concept," in idem (ed.): 5–19.

———— 1978: "Two Approaches to Civilian Defence: Instrumentalists and Structuralists," *BPP,* 9:4, 316–320.

———— 1986: "Voortgang en Vooruitgang in het Denken over Sociale Verdediging: Een constructieve evaluatie," *Polemological Papers,* 3 (Bruxelles: Vrije Universiteit, Centrum voor Polemologie).

———— (ed.) 1977: *Possibilities of Civilian Defence in Western Europe, Proceedings of the 2nd International Working Conference on Violence and*

Non-Violent Action in Industrialized Societies, Brussels, March 24–26th, 1976 (Amsterdam & Lisse: Swets & Zeitlinger).

Geisenmeyer, Stefan 1971: "A Defensive Weapons Mix for Europe: Pandora, Medusa, Dragon Seed," *Survival,* 9: 307–309.

Geissler, Erhard 1984: "Implications of Genetic Engineering for Chemical and Biological Warfare," in *SIPRI,* 1984: 421–454.

——— 1985: "B-Waffen und Geningenieure," in Werner Dosch & Peter Herrlich (eds.), *Ächtung der Giftwaffen: Naturwissenschaftler warnen vor Chemischen und Biologischen Waffen* (Frankfurt: Fischer): 160–166.

Gelman, Harry 1986: "The Rise and Fall of Détente: Causes and Consequences," in Horelick (ed.): 55–85.

Genscher, Hans Dietrich 1987: "Chancen einer zukunftsfähigen Gestaltung der West-Ost-Beziehiungen" (speech in Davos, 1 February 1987), *Blätter,* 3: 439–446.

——— 1989: "Kooperative Strukturen der Sicherheit: Eine Perspektive für die Friedensordnung in Europa," *S+F,* 7:2, 132–136.

——— 1991: "German Responsibility for a Peaceful Order in Europe," in Rotfeld & Stützle (eds.): 20–29.

George, Alexander L. 1988: "Factors Influencing Security Cooperation," in idem & al. (eds.): 655–678.

——— 1988a: "Incentives for U.S.-Soviet Cooperation and Mutual Adjustment," in idem & al. (eds.): 641–654.

——— 1988b: "Strategies for Facilitating Cooperation," in idem & al. (eds.): 692–711.

——— 1988c: "U.S.-Soviet Efforts to Cooperate in Crisis Management and Crisis Avoidance," in idem & al. (eds.): 581–599.

——— 1989: "The Search for Agreed Norms," in Allison & Ury (eds.): 45–66.

——— 1990: "Superpower Interests in Third Areas," in Allison & Williams (eds.): 107–120.

——— 1991: "U.S.-Soviet Relations: Evolution and Prospects," in Smoke & Kortunov (eds.): 19–41.

———, Philip J. Farley & Alexander Dallin 1988: "Epilogue: Perspectives," in idem, Farley & Dallin (eds.): 712–722.

———, Philip J. Farley & Alexander Dallin (eds.) 1988: *U.S.–Soviet Security Cooperation: Achievements, Failures, Lessons* (New York: Oxford U.P.).

George, Bruce & al. 1991: "The British Labour Party and Defense," *Washington Papers* (New York: Praeger): 153.

George, James L. 1990: *The New Nuclear Rules: Strategy and Arms Control After INF and START* (London: Pinter).

——— 1990a: "Naval Arms Control: Mediums and Messages," *Naval War College Review,* 43:3, 32–42.

George of Bohemia, King 1464: "Tractatus pacis toti christianitati fiendae," in Vanecek (ed.) 1964: 69–80. (English translation, "Treaty on the Establishment of Peace Throughout Christendom": 81–90.)

Gerber, Johannes 1973: "Zur Problematik des sozialen Ertrages bei Aufwendungen für die Verteidigung," in idem 1989: 54–62.

——— 1982: "Ist unsere Verteidigung noch finanzierbar?" in idem 1989: 63–75.

——— 1982a: "Das Milliardending," in idem 1989: 76–89.

——— 1982b: "Die Bundeswehr in der Strukturkrise," in idem 1989: 90–97.

——— 1984: "Fordert die Wirtschaftlichkeit eine neue Struktur des Heeres?" in Jacobsen & Lemm (eds.): 39–52.

——— 1985: *Die Bundeswehr im Nordatlantischen Bündnis* (Regensburg: Walhalla & Praetoria).

——— 1987: "Die konventionelle Nichtangriffsfähigheit als militärische Dimension der Sicherheit in Mittel-und Nordeuropa," *S+F,* 1: 23–30.

——— 1988: "Militärökonomische Aspekte zur militärischen Stabilität in Mitteleuropa," in Bahr & Lutz (eds.): 438–444.

——— 1989: *Beiträge zur Praxis der alternativen Verteidigung, herausgegeben von Reinhard Meyers* (Münster: LIT).

——— 1989a: "Ökonomie und Streitkräfte," in idem 1989: 186–203.

——— 1989b: "Landkriegführung: Entwurf eines Beitrages für die International Military and Defence Encyclopedia, London–New York 1991," in Gerber 1989: 204–250.

Géré, Francois 1991: "La stratégie intégrale des États-Unis ou une leçon de guerre démocratique," *Stratégique,* 51/52: 83–112.

Gerhard, Wilfried 1989: "Angst vor Gorbatschow? Neues Denken in der Sowjetunion und politische Bewusstseinsstrukturen in der Bundesrepublik Deutschland," in Vogt (ed.): 47–66.

Gerhold, Dankward 1992: "Armaments Control of Germany: Protocol III of the Modified Brussels Treaty," in Tanner (ed.): 71–100.

Gerlach, Jeffrey R. 1993: "A U.N. Army for the New World Order?" *Orbis,* 37:2 (Spring): 223–236.

Gerosa, Guido 1992: "The North Atlantic Cooperation Council," *ES,* 1:3 (Autumn): 273–294.

Gersdorff, Gero von 1984: "Sozialdemokratie und Sicherheitspolitik seit 1945," in Brauch (ed.): 37–46.

Gerster, Florian 1990: "Zwei-Zonen-Staat unter alliierter Kontrolle? Der Sicherheitspolitische Rahmen der deutschen Einheit," *S+F,* 8:2, 69–73.

Gervasi, Tom 1987: *The Myth of Soviet Military Supremacy* (New York: Harper & Row).

Geschwendtner, Jürg 1986: *Deutsche Antipartisanen-kriegsführung: Gegenoffensive deutscher Verbände in Slowenien und Zentralkroatien September bis November 1943* (Frauenfeld: Huber).

Geyer, Michael 1986: "German Strategy in the Age of Machine Warfare, 1914–1945," in Paret (ed.): 527–597.

Ghebaldi, Victor-Yves 1989: "Confidence-Building Measures Within the CSCE Process: Paragraph-by-Paragraph Analysis of the Helsinki and Stockholm Regimes," *Research Paper,* 3 (New York: UNIDIR).

——— 1990: "Les négotiations de Vienne sur les forces armées conventionelles et les mésures de confiance," *Arès,* 12: 32–63.

——— 1990a: "CFE: First Year Status Report," *IDR,* 4: 367–369.

——— 1990b: "Switzerland and the Issue of Peaceful Settlement of Disputes Within the CSCE Process," in Miliovejic & Maurer (eds.): 162–172.

——— 1993: "The CSCE Forum for Security Cooperation: The Opening Gambits," *NATO Review,* 41:3 (June): 23–27.

——— 1993a: "Towards a CSCE in the Mediterranean: The CSCM," in Lucas (ed.): 335–343.

Giap, Vo Nguyen 1970: *The Military Art of People's War: Selected Writings of General Vo Nguyen Giap,* ed. Russell Stetler (New York: Monthly Review).

Gibson, Christian Lea 1986: "The Soviet Invasion of Afghanistan: Pursuing Stability and Security," in Jonathan R. Adelman (ed.), *Superpowers and Revolution* (New York: Praeger): 266–285.

Gibson, Irving M. 1941: "Maginot and Liddell Hart: The Doctrine of Defense," in Earle (ed.): 365–387.

Gibson, James William 1988: *The Perfect War: The War We Couldn't Lose and How We Did* (New York: Vintage).

Giessmann, Hans-Joachim 1991: "Einhegung militärischer Technologie—neues Bewährungsfeld der Rüstungskontrolle?" in Müller & Neuneck (eds.): 353–360.

———— 1992: *Das unliebsame Erbe: Die Auflösung der Militärstruktur der DDR* (Baden-Baden: Nomos).

———— 1992a: "Technology Proliferation Control Regimes: Challenges, Risks, and Policy Options in a Changing Environment," in Neuneck & Ischebeck (eds.): 207–214.

———— 1993: "Wozu noch Rüstung? Ein Plädoyer für Abrüstung und Konversion," in Lutz (ed.): 115–136.

———— (ed.) 1992: *Konversion im vereinten Deutschland: Ein Land—zwei Perspektiven?* (Baden-Baden: Nomos).

———— & Josef Nietzsch 1989: "Qualitative Aspects of Balance Assessments— Voting for an Extended Approach to Negotiations (Theses)," in Borg & Smit (eds.): 107–114.

Gilges, Konrad 1985: *Frieden ohne NATO: Perspektiven einer linken Friedenspolitik* (Hamburg: VSA).

———— 1985a: "Essener Parteitag und wie weiter? Zur Friedenspolitischen Programmdiskussion in der SPD," in Fuch & al. (eds.): 44–51.

Giller, Joachim 1989: "Insel der Seligen? Bedrohungsbewusstsein und Friedenskultur im neutralen Kleinstaat Österreich," in Vogt (ed.): 154–166.

Gillette, Philip S. & William C. Frank, Jr. (eds.) 1990: *The Sources of Soviet Naval Conduct* (Lexington, Mass.: Lexington).

Gilpin, Robert 1981: *War and Change in World Politics* (Cambridge: Cambridge U.P.).

Glad, Betty (ed.) 1990: *Psychological Dimensions of War* (London: Sage).

Gladkov, Petr 1991: "Domestic Preconditions for the Transition to Mutual Security: A Soviet Perspective," in Smoke & Kortunov (eds.): 42–58.

———— 1991a: "Emerging European Architecture." Paper presented to the conference "Sicherheit im neuen Europa," Evangelische Akademie Loccum and SAS, 27–29 September 1991.

Glantz, David M. 1991: "Future Directions in Soviet Military Strategy," in Hemsley (ed.): 123–144.

———— 1991a: "Military Strategy: Soviet Military Art: Challenges and Change in the 1990s," *JSMS*, 4:4 (December): 547–593.

———— 1992: *The Military Strategy of the Soviet Union: A History* (London: Frank Cass).

Glaser, Charles L. 1986: "Why Even Good Defenses May be Bad," in Miller & Evera (eds.): 25–56.

———— 1992: "Nuclear Policy Without an Adversary: U.S. Planning for the Post-Soviet Era," *IS*, 16:4 (Spring): 34–78.

———— 1993: "Why NATO Is Still Best: Future Security Arrangements for Europe," *IS*, 18:1 (Summer): 5–50.

Gleditsch, Nils Petter 1990: "Research on Arms Races," in Gleditsch & Njølstad (eds.): 87–104.

———— 1992: "Democracy and Peace," *JPR*, 29:4 (November): 369–376.

———— Håkan Wiberg & Dan Smith 1992: "The Nordic Countries: Peace Dividend or Security Dilemma," *Cooperation and Conflict*, 27:4 (December): 323–348.

————, Bjørn Møller, Håkan Wiberg & Ole Wæver 1990: *Svaner på vildveje? Nordens sikkerhed mellem supermagtsflåder og europæisk opbrud* (Copenhagen: Vindrose).

———— & Olav Njølstad (eds.) 1989: *Arms Races: Technological and Political Dynamics* (London: Sage).

Glicksman, Alex 1987: "NATO After INF: Toward a New Alliance Consensus," *National Defense*, 433, December, 51–55.

———— 1988: "Observations on New Military Technologies and Conventional Arms Control," in Calliess (ed.): 82–92.

———— 1990: "Defensive Defence in the Middle East," in UNIDIR (eds.): 145–151.

Gloannec, Anne-Marie, Le 1986: "Die Deutschen und die französische Abschreckungsstreitmacht: Deutsch-französische Zweideutigkeiten," in Kaiser & Lellouche (eds.): 90–100.

Gloeckner, Eduard 1990: "Der Minister, die Armee und die deutsche Einheit," *EW*, 7: 384–387.

Glotz, Peter 1985: "'Common Security' und 'Strukturelle Nichtangriffsfähigkeit': Plädoyer für eine Politisierung des historischen Begriffs Kollektiver Sicherheit," in Lutz (ed.): 107–114.

———— 1987: "Mehr Sicherheit durch FOFA?" *EW*, 4: 195–199.

———— 1990: "Gesamteuropa—Skizze für einen schwierigen Weg," *EA*, 2: 41–50.

———— 1990a: "Sicherheitspartnerschaft zwischen ideologischen Blöcken in Europa," in Weizsäcker (ed.): 347 358.

Glucksman, André 1967: *Le Discours de la Guerre* (Paris: l'Herne).

———— 1983: *La force du vertige* (Paris: Grasset).

Glynn, Patrick 1992: *Closing Pandora's Box: Arms Races, Arms Control and the History of the Cold War* (New York: Basic Books).

Goblirsch, Josef 1984: "Neue Leitidee für die Verteidigung: Technotaktik," in Weizsäcker (ed.): 121–166.

Godballe, Mogens 1987: *Selvforsvar i atomalderen: En debatbog om defensivt forsvar* (Copenhagen: Politisk Revy).

Goetemoeller, Rose E. 1987: "Land-attack Cruise Missiles," *AP*, 226.

———— 1993: "Unilateralism in Soviet and Russian Arms Control," in Ramberg (ed.): 43–70.

Gohl, Diethelm 1986: "Schritte zur Abrüstung: Plädoyer für das Gradualismus-Konzept," in Gohl & Niesporek (eds.): 148–164.

———— & Heinrich Niesporek (eds.) 1986: *Sicher auf neuen Wegen: Mit Beiträgen von Franz Alt, Kurt Biedenkopf, Gustav Fehrenbach und anderen* (Warendorf: Gohl).

Goldanskii, Vitalii & Stanislav Rodionov 1993: "An International Nuclear Security Force," in Rotblat & al. (eds.): 181–190.

Goldblat, Jozef 1982: "The Nuclear Non-Proliferation Treaty and No First Use of Nuclear Weapons," *BPP*, 13:4, 315–321.

———— 1993: "Making Nuclear Weapons Illegal," in Rotblat & al. (eds.): 153–168.

Golden, James R. 1984: "Conventional Deterrence: Conclusions and Policy Implications," in Golden & al. (eds.): 221–230.

————, Asa A. Clark & Bruce E. Arlinghaus (eds.) 1984: *Conventional Deterrence: Alternatives for European Defense* (Lexington, Mass.: Lexington).

Goldmann, Kjell 1988: *Change and Stability in Foreign Policy: The Problems and Possibilities of Detente* (Princeton: Princeton U.P.).

Goldstein, Joshua S. & John R. Freeman 1990: *Three-Way Street: Strategic Reciprocity in World Politics* (Chicago: University of Chicago).

Goodby, James E. 1988: "The Stockholm Conference: Negotiating a Cooperative Security System for Europe," in George, Farley & Dallin (eds.): 144–172.

———— & Benoit Morel (eds.) 1993: *The Limited Partnership: Building a Russian-US Security Community* (Oxford: Oxford U.P.).

Gorbachev, Mikhail 1986: *Political Report of the CPSU Central Committee to the 27th Party Congress* (Moscow: Novosti).

———— 1987: *For a Nuclear-Free World: Speeches and Statements by the General Secretary of the CPSU Central Committee on Nuclear Disarmament Problems, January 1986–January 1987* (Moscow: Novosti).

———— 1988: *Perestroika: New Thinking for Our Country and the World,* 2d ed. (London: Fontana).

———— 1988a: "Address to the 43rd Session of the U.N. General Assembly," *Tass* press release, 7 December.

Gordenker, Leon & Thomas G. Weiss 1993: "The Collective Security Idea and Changing World Politics," in Weiss (ed.): 3–18.

———— & Thomas G. Weiss 1993: "Whither Collective Security? An Unsettled Idea in a Changing World," in Weiss (ed.): 209–219.

Gordon, Michael R. 1984: "Technology and NATO Defence: Weighing the Options," in Golden & al. (eds.): 149–162.

Gormley, Dennis M. 1986: "The Impact of NATO Doctrinal Choices on the Policies and Strategic Choices of Warsaw Pact States: Part II," *AP,* 296: 20–34.

Gottfried, Kurt & al. (eds.) 1989: *Towards a Cooperative Security Regime in Europe* (Ithaca: Cornell University, Peace Studies Program).

———— & Paul Bracken (eds.) 1990: *Reforging European Security: From Confrontation to Cooperation* (Boulder, Colo.: Westview).

Goulding, Marrack 1993: "The Evolution of United Nations Peacekeeping," *IA,* 69:3 (July): 451–464.

Goure, Daniel 1990: "Soviet Doctrine and Nuclear Forces into the Twenty-First Century," in Green & Karasik (eds.): 75–107.

Gow, James 1992: *Legitimacy and the Military: The Yugoslav Crisis* (London: Pinter).

———— (ed.) 1993: *Iraq, the Gulf Conflict and the World Community* (London: Brassey's).

———— & Lawrence Freedman 1992: "Intervention in a Fragmenting State: The Case of Yugoslavia," in Rodley (ed.): 93–132.

Graabæk, Jens Jørn 1979: "'Drukner' kampflyet i 80'ernes og 90'ernes teknologi?" *Militært Tidsskrift,* November/December, 349–369.

———— 1986: "Tanker om Fremtiden" (Copenhagen: Hjemmevaernsbladet).

———— 1989: "Skitse til strukturaendringer 1992–95," in Forsvarskommisionen 1989, *Forsvaret i 90'erne: Beretning fra Forsvarskommissionen af 1988* (Copenhagen: Statens Informationstjeneste): appendix 22.

Graaf, Henny van der 1989: "Verification of a Conventional Armed Forces Agreements in Europe," in Borg & Smit (eds.): 245–256.

Grachev, Pavel S. 1992: "The Defence Policy of the Russian Federation," *RJ,* 137:5 (October): 5–7.

Graf, Wilfried 1986: "Zur Glaubwürdigkeit der Neutralität: Ideologische Dimensionen österreichischer Sicherheits- und Friedenspolitik," *Dialog,* 6: 371–436.

Graham, Kennedy 1989: *National Security Concepts of States: New Zealand* (New York: Taylor & Francis, UNIDIR).

———— 1992: "Global Restructuring: Arms Control and Disarmament," in Tehranian & Tehranian (eds.): 62–74.

Granberg, Donald 1978: "GRIT in the Final Quarter: Reversing the Arms Race Through Unilateral Initiatives," *BPP,* 9:3, 210–221.

Grass, Bernd 1984: "Das Wehrpotential der 90er Jahren: Eine demographisch-statistische Analyse," in SAS (eds.): 35–46.

———— 1990: "The Personnel Shortage of the Bundeswehr Until the Year 2000," in Brauch & Kennedy (eds.): 97–112.

Gray, Colin S. 1979: "Nuclear Strategy: A Case for a Theory of Victory" (*IS*, Summer) in Miller (ed.) 1984: 23–56.

———— 1990: *War, Peace and Victory: Strategy and Statecraft for the Next Century* (New York: Simon & Schuster).

———— 1991: "Do the Changes Within the Soviet Union Provide a Basis for Eased Soviet-American Relations? A Skeptical View," in Jervis & Bialer (eds.): 61–76.

———— 1992: *House of Cards: Why Arms Control Must Fail* (Ithaca: Cornell U.P.).

———— 1993: "Arms Control Does Not Control Arms," *Orbis,* 37:3 (Summer): 333–348.

Gray, Richard B. 1969: *International Security Systems: Concepts and Models of World Order* (Itasca, Ill.: Peacock).

Graziani, Antonietta 1986: "Ipotesi di difesa alternativa in Europa," *Working Papers,* 4 (Rome: Forum Humanum Project).

Green, William C. 1990: "Assessing Soviet Military Literature: Attempts to Broker the Western Debate," in Green & Karasik (eds.): 13–27.

———— & Theodor Karasik (eds.) 1990: *Gorbachev and His Generals: The Reform of Soviet Military Doctrine* (Boulder, Colo.: Westview).

Greene, Owen 1983: *Europe's Folly: The Facts and Arguments About Cruise* (London: CND).

———— 1988: "Common Security: Priorities for Reforms in the NATO-WTO Confrontation," in Baudisch (ed.): 26–31.

———— 1990: "Reforming NATO Policy: Prospects for the Early 1990s," in Randle & Rogers (eds.): 45–70.

Greenwood, David 1983: "Economic Constraints and Political Preferences," in Baylis (ed.): 31–61.

———— 1985: "Economic Implications: Finding the Resources for an Effective Conventional Deterrence," in Windass (ed.): 77–86.

———— 1993: "Concentration of Effort and Complementarity," in Clarke & Sabin (eds.): 175–197.

Greenwood, Ted 1975: *Making the MIRV: A Study in Defense Decision Making* (Cambridge, Mass.: Ballinger).

———— 1993: "NATO's Future," *ES,* 2:1 (Spring): 1–14.

Gregg, Robert W. 1993: *About Face? The United States and the United Nations* (Boulder, Colo.: Lynne Rienner).

Greiner, Gottfried 1987: "The German Territorial Army in NATO Defence: An Evolving Role," in Windass & Grove (eds.) 1987: I (no pagination).

Griffith, Ivelaw L. (ed.) 1991: *Strategy and Security in the Caribbean* (New York: Praeger).

Griffith, John 1984: "Nuclear Weapons and International Law," in Prins (ed.): 154–171.

Griffith, William E. 1978: *The Ostpolitik of the Federal Republic of Germany* (Cambridge: MIT).

———— : "The American View," in Moreton (ed.): 49–63.

———— 1989: "The Security Policies of the Social Democrats and the Greens in the Federal Republic of Germany," in Griffith & al.: 1–20.

———— & al. 1989: *Security Perspectives of the West German Left: The SPD and the Greens in Opposition* (Washington, D.C.: Pergamon-Brassey's).

Grin, John 1990: "Options for the Communication System of NATO's Land Forces," *AFR*, 31.

——— 1990a: *Military-Technological Choices and Political Implications: Command and Control in Established NATO Posture and a Non-Provocative Defence* (Amsterdam: Free U.P.).

——— 1990b: "Eliminating Offensive Capabilities: Exploring Multilateral Mechanisms Beyond Arms Reduction: Implications for Verification," in Grin & Graaf (eds.): 71–91.

——— 1992: "Assessing Military Aircraft Innovation and R&D Paths," in Smit & al. (eds.): 215–240.

——— & Hans Günter Brauch 1992: "Introduction: Controlling the Development and Spread of Destabilizing Technologies—Perspectives and Challenges for Arms Control and Disarmament in the 1990s," in Brauch & al. (eds.): 3–16.

——— & Lutz Unterseher 1988: "The Spiderweb Defense," *BAS*, 44:7, 28–31.

——— & Lutz Unterseher 1989: "Spezialisierung auf die Defensive: Einige Zusammenhänge," in SAS (eds.): 139–148.

——— & Lutz Unterseher 1990: "Military Restructuring for Conventional Disarmament," in Auffermann (ed.): 103–114.

——— & Lutz Unterseher 1990a: ". . den Bedrohungszirkel unterbrechen: Spinnennetz: Ein militärtheoretischer Beitrag zur Um- und Abrüstung," in Vogt (ed.): 243–262.

——— & Henny van der Graaf (eds.) 1990: *Unconventional Approaches to Conventional Arms Control Verification: An Exploratory Assessment* (Amsterdam: Free U.P.).

Grinevsky, Oleg A. 1989: "East-West: Problems of Security, Confidence and Disarmament," in Clesse & Schelling (eds.): 304–313.

Gromyko, Anatoly & Martin Hellman (eds.) 1988: *Breakthrough: Emerging New Thinking* (New York: Walker).

Gross, Andreas, Fitzgerald Crain & Bruno Kaufmann (eds.) 1989: *Frieden mit Europa: Eine Schweitz ohne Armee als Beitrag zur Zivilisiereung der Weltinnenpolitik* (Zürich: Realutopia).

Grosse, Heinrich W. 1987: "Martin Luther King (1929–1968): Die Stimme derer, die keine Stimme haben," in Rajewsky & Riesenberger (eds.): 317–325.

Grosser, Dieter (ed.) 1992: *German Unification: The Unexpected Challenge* (Oxford: Berg).

Grove, Eric 1987: "The Northern Flank and the Atlantic Link," in Windass & Grove (eds.): 1 (no pagination).

——— 1988: "Naval Cooperative Security," in Windass & Grove (eds.): no pagination.

——— 1988a: Introduction to Julian S. Corbett, *Some Principles of Maritime Strategy* (1911; new ed., London: Brassey's): xi–xiv.

——— 1989: "The Norwegian Sea—NATO's First Line of Defence," in idem (ed.): 1–19.

——— 1990: "The Law of the Sea: Ocean Management and Confidence-Building," *Disarmament*, 13:4, 98–104.

——— 1990a: *Maritime Strategy and European Security, Common Security Studies*, 2 (London: Brassey's).

——— 1990b: *The Future of Sea Power* (London: Routledge).

——— 1992: "Naval Technology and Stability," in Smit & al. (eds.): 197–214.

——— 1993: "UN Armed Forces and the Military Staff Committee: A Look Back," *IS*, 17:4 (Spring): 172–182.

——— (ed.) 1989: *NATO's Defence of the North* (London: Brassey's).

———— & Stan Windass (eds.) 1988: *The Crucible of Peace: Common Security in Europe, Common Security Studies,* 1 (London: Brassey's).

Grünen, die 1981: *Das Friedensmanifest der Grünen* (Bonn: Grünen).

———— 1985: *Frieden und Freiheit durch einseitige Abrüstung: Bei uns anfangen: Diskussionsbeiträge zum Ratschlag der Friedensbewegung, 1985* (Bonn: Grünen).

———— (eds.) 1985: *Euromilitarismus: Zur Bedeutung der "Europäisierung der Sicherheitspolitik"* (Bonn: Grünen).

Grünen im Bundestag, die 1984: *Angriff als Verteidigung: AirLand Battle, AirLand Battle 2000, Rogersplan: Eine Dokumentation* (Bonn: Grünen im Bundestag).

———— 1987: *Ohne Waffen, aber nicht Wehrlos: Das Konzept der Sozialen Verteidigung* (Bonn: Grünen im Bundestag).

———— 1988: *Militärblock West: Die NATO-Broschüre der Grünen* (Bonn: Grünen im Bundestag).

Guertner, Gary L. 1982: "Nuclear War in Suburbia," *Orbis,* Spring, 49–69.

———— 1990: *Deterrence and Defence in a Post-Nuclear World* (London: Macmillan).

Guevara, Ernesto Che 1959: "Le role social de l'armee rebelle," in idem 1976: 136–146.

———— 1959a: "Qu'estce qu'un guerillero?" in idem 1976: 132–135.

———— 1960: "La guerre de guerilla," in idem 1976: 23–129.

———— 1963: "La guerre de guerilla: une methode," in idem 1976: 147–165.

———— 1968: *Reminiscences of the Cuban Revolutionary War* (London: Allen & Unwin).

———— 1976: *Oeuvres 1: Textes Militaires* (Paris: Francois Maspero).

Gugel, Günthert & Uli Jäger (eds.) 1989: *Die Sicherheitspolitik der Sowjetunion im Wandel* (Tübingen: Verein für Friedenspädagogik).

Guillen, Abraham 1973: "Urban Guerilla Strategy," in Chaliand (ed.) 1982: 317–323.

Guillermaz, Jacques 1972: *A History of the Chinese Communist Party, 1921–1949* (London: Methuen).

Gullikstad, Espen 1991: "Sweden," in Anthony (ed.): 146–155.

Gummett, Philip 1988: "New Conventional Technology and Alternative Defence: Help or Hindrance?" in Burt (ed.): 70–81.

Gupta, Narendra L. 1989: "Non-offensive Defence—Sustaining the Doctrine," in Singh & Vekaric (eds.): 135–146.

Gupta, Raj 1993: *Defense Positioning and Geometry: Rules for a World with Low Force Levels* (Washington, D.C.: Brookings).

Gupta, Ronjon Das 1988: "Anregungen zur Panzerrüstung" (Bonn: SAS).

Güth, Werner & Eric van Damme 1991: "Gorby-Games: A Game Theoretical Analysis of Disarmament Campaigns," in Avenhaus & al. (eds.): 215–240.

Haas, Ernst B. 1966: *International Political Communities* (New York: Anchor).

———— 1970: "The Study of Regional Integration: Reflections on the Joy and Anguish of Pretheorizing," in Falk & Mendlowitz (eds.) 1973: 103–130.

———— 1972: "The Balance of Power: Prescription, Concept or Propaganda?" in Robert L. Pfaltzgraff, Jr. (ed.), *Politics and the International System,* 2d ed. (Philadelphia: Lippincott): 452–480.

———— 1987: "War, Interdependence and Functionalism," in Väyrynen (ed.): 108–127.

———— 1993: "Collective Conflict Management: Evidence for a New World Order?" in Weiss (ed.): 63–117.

Haas, Richard N. 1988: "Future Options for Arms Control in Europe," in Nehrlich & Thomson (eds.): 136–153.

———— 1988a: "Arms Control at Sea: The United States and the Soviet Union in the Indian Ocean, 1977–1978," in George, Farley & Dallin (eds.): 524–539.

Haberl, Othmar Nikola & Hans Hecker (eds.) 1989: *Unfertige Nachbarschaften: Die Staaten Osteuropas und die Bundesrepublik Deutschland* (Essen: Reimar Hobbing).

Hacker, Jens 1989: "Die deutsche Frage aus der Sicht der SPD," in Blumenwitz & Zieger (eds.): 39–66.

Haeberlin, Friedrich, Hans-Joachim Hoffmann, Volker Kröning, Peter von Oertzen, Hermann Scheer & Lutz Unterseher 1985: "Konservative Verteidigungsplanung erhöht die Kriegsgefahr in Europa," *Vorwärts Dokumentation,* 21 September.

Haeffner, Peter 1987: "Drohnen für die Aufklärung: Sehen, deuten und melden vom rechten Ort zur rechten Zeit," *EW,* 10: 574–578.

Haesken, Ole & al. 1988: "Confidence Building Measures at Sea," *FFI/Rapport,* 88/5002 (Kjeller, Norwegen: Forsvarets Forskningsinstitut).

Hæstrup, Jørgen 1954: *Kontakt med England, 1940–1943* (Copenhagen: Trajan, 1968).

———— 1959: *Hemmelig Alliance,* 2 vols. (Copenhagen: Forlaget Trajan, 1968).

Hafner, Donald L. & John Roper (eds.) 1988: *ATBM and Western Security: Missile Defenses for Europe* (Cambridge, Mass.: Ballinger).

Haftendorn, Helga 1986: "Wurzeln der Ost- und Entspannungspolitik der sozial-liberalen Koalition," in Ehmke & al. (eds.): 17–28.

Hagelin, Björn 1982: "Grenzen der Sicherheit: Politik und Wirtschaft in Schweden," in Lutz & Grosse-Jütte (eds.): 107–139.

Hagena, Hermann 1990: "Reform für die NVA: Vom Parteisoldaten zum Staatsbürger in Uniform," *EW,* 3: 172–175.

———— 1990a: "Vorschlag für die Volkskammer: Die Radikale Wende der Nationalen Volksarmee," *EW,* 4: 206–210.

———— 1990b: *Tiefflug in Mitteleuropa: Chancen und Risiken offensiver Luftkriegsoperationen* (Baden-Baden: Nomos).

———— 1990c: "Konventionelle Kräftevergleiche in Europa: Grundlage von Bedrohungsanalyse, Streitkräfteplanung und konventioneller Rüstungskontrolle," in E. Müller & Neuneck (eds.): 63–74.

———— 1994: "NOD in the Air," in Møller & Wiberg (eds.): 85–97.

Hagerman, Edward 1988: *The American Civil War and the Origins of Modern Warfare* (Bloomington: Indiana University Press).

Hagerty, Devin T. 1993: "The Power of Suggestion: Opaque Proliferation, Existential Deterrence, and the South Asian Nuclear Arms Competition," *SS,* 2:3/4 (Spring/Summer): 256–383.

Haggard, Stephan & Beth A. Simmons 1987: "Theories of International Regimes," *International Organization,* 41:3, 491–517.

Haglund, David G. & Olaf Mager 1992: "Homeward Bound?" in Haglund & Mager (eds.): 273–285.

———— & Olaf Mager (eds.) 1992: *Homeward Bound? Allied Forces in the New Germany* (Boulder, Colo.: Westview).

Hahlweg, Werner 1986: "Clausewitz and Guerilla Warfare," in Handel (ed.): 127–133.

Hahn, Erich 1987: "Überlegungen zum Dokument 'Der Streit der Ideologien und die gemeinsame Sicherheit'," *Blätter,* 11: 1486–1493.

Hakovirta, Harto 1987: "East-West Tensions and Soviet Policies on European Neutrality," in Sundelius (ed.): 198–217.

———— 1988: *East-West Conflict and European Neutrality* (Oxford: Clarendon).

Hall, Cristopher 1987: *Britain, America and Arms Control, 1921–1937* (London: Macmillan).

Haltiner, Karl 1986: "Struktur, Tradition und Integration der schweizerischen Miliz," *S+F*, 3: 126–133.

———— 1987: "Integration der schweizerischen Miliz in der Gesellschaft," in Bald (ed.): 135–146.

Ham, Peter van 1993: *Managing Non-Proliferation Regimes in the 1990s: Power, Politics and Policies* (London: Pinter).

Hamann, Rudolf & Volker Matthies (eds.) 1991: *Sowjetische Aussenpolitik im Wandel: Eine Zwischenbilanz der Jahre 1985–1990* (Baden-Baden: Nomos).

Hamilton, Andrew 1985: "Redressing the Conventional Balance: NATO's Reserve Military Manpower," *IS*, 10:1 (Summer): 111–136.

Hamilton-Eddy, Jane 1988: "Recent Developments in the Soviet Conventional Threat to Europe," in Nehrlich & Thomson (eds.): 67–76.

Hamlin, Douglas 1990: "Conventional Arms Control and Disarmament in Europe: Canadian Objectives," *Working Papers*, no. 20 (Ottawa: Canadian Institute for International Peace and Security).

Hamm, Manfred R. & Hartmut Pohlman 1990: "Military Doctrines and Strategies. The Missing Keys to Success in Conventional Arms Control?" *AuP*, 1: 52–72.

———— & Hartmut Pohlman 1990a: "Military Strategy and Doctrine: Why They Matter to Conventional Arms Control," *WQ*, 13:1, 185–198.

Hammick, Murray & al. 1991: "The Gulf War in Review: Report from the Front," *IDR*, 9: 979–1001.

Hammond, Paul Y. 1990: "What Future for the US Military Presence in Europe?" in Clesse & Rühl (eds.): 492–507.

Hampson, Fen Osler 1985: "Escalation in Europe," in Allison, Carnesale & Nye (eds.): 80–114.

———— 1986: "Is There an Alternative to NATO?" in Flanagan & Hampson (eds.): 191–217.

————, Harald von Riekhof & John Roper (eds.) 1992: *The Allies and Arms Control* (Baltimore: Johns Hopkins U.P.).

Handel, Michael I. 1986: Introduction to Handel (ed.): 1–34.

———— 1986a: "Clausewitz in the Age of Technology," in Handel (ed.): 51–94.

———— 1992: *Masters of War: Sun Tzu, Clausewitz and Jomini* (London: Frank Cass).

———— (ed.) 1986: "Special Issue on Clausewitz and Modern Strategy," *JSS*, 9: 2–3.

Hänggi, Heiner 1993: *Neutralität in Südostasien: Das Project einer Zone des Friedens, der Freiheit und der Neutralität* (Bern: Haupt).

Hanley, David 1984: "The Parties and the Nuclear Consensus," in Howorth & Chilton (eds.): 75–93.

Hannig, Norbert 1979: "Lässt sich Westeuropa konventionell verteidigen?" *IW*, 1: 1237–1241.

———— 1981: "The Defence of Western Europe with Conventional Weapons," *IDR*, 14:11, 1439–1443.

———— 1981a: "NATO-Nachrüstungsbeschluss: Militärisch notwendig?" *Armada International*, 4: 34–46.

———— 1984: *Abschreckung durch konventionelle Waffen: Das David-und-Goliath Prinzip* (Berlin: Spitz).

———— 1985: "Fiasko der NATO-Verteidigung und Wege zu einer Neuorientierung," *Armada International*, 1: 51–66.

———— 1986: "Das DEWA-Konzept: Eine Alternative zu FOFA," *IW*, 7: 897–898.

———— 1986a: "NATO's Defense: Conventional Options Beyond FOFA," *IDR*, 7.

————— 1986b: "Verteidigen ohne zu bedrohen: Die DEWAKonzeption als Ersatz der NATO-FOFA," *AFP*, 5.

————— 1987: "Die 'Strukturelle Nichtangriffsfähigkeit' im weiteren und engerem Sinne," *S+F*, 1: 30–32.

————— 1988: "Antworten auf Fragen zur Konventionellen Stabilität," in Bahr & Lutz (eds.): 445–451.

————— 1989: "Deterrence by Means of Conventional Weapons: The David-Goliath Principle," in Tromp (ed.): 223–236.

————— 1990: "Die DEWA-Konzeption: Verteidigen ohne zu bedrohen," in Weizsäcker (ed.): 150–157.

Hanolka, Harro 1984: "Soziale Voraussetzungen und Konsequenzen einer Netzverteidigung der Bundesrepublik," in Weizsäcker (ed.): 199–230.

Hanrieder, Wolfram 1990: "The Federal Republic, the United States, and the New Europe: A Historical Perspective," in Jordan (ed.): 67–86.

Hansen, Lynn M. 1990: "Towards an Agreement on Reducing Conventional Forces in Europe," *Disarmament*, 13:3, 61–76.

————— 1991: "Arms Control and Limitations," in Hemsley (ed.): 257–271.

Hansen, Peter Kragh 1990: "Et Danmark uden militær," in Hansen (ed.): 146–155.

————— (ed.) 1990: *Danmark uden våben* (Copenhagen: Aldrig Mere Krig).

Hansen, Roger 1972: "Regional Integration: Reflections on a Decade of Theoretical Efforts," in Michael Hodges (ed.), *European Integration* (Harmondsworth: Penguin): 184–199.

Harbottle, Michael 1988: "Security for All," in Fedorenko (ed.): 20–39.

————— 1988a: "European Security and the Concept of Non-Offensive Defence," in Baudisch (ed.): 32–35.

————— 1990: "Just Defence Peacekeeping in Europe," *Discussion Paper* (Appleby: Just Defence).

————— & al. 1984: *Generals for Peace and Disarmament: A Challenge US/NATO Strategy* (New York: Universe).

Harden, Sheila (ed.) 1994: *Neutral States and the European Community* (London: Brassey's).

Hardin, Russell, John J. Mearsheimer, Gerald Dworkin & Robert E. Goodin (eds.) 1985: *Nuclear Deterrence: Ethics and Strategy* (Chicago: University of Chicago Press).

Harkabi, Yehoshafat 1988: *Israel's Fateful Decisions* (London: Tauris).

Harkavy, Robert E. 1989: *Bases Abroad: The Global Foreign Military Presence* (Oxford: Oxford U.P., SIPRI).

Harle, Vilho (ed.) 1989: "Studies in the History of European Peace Ideas," *Research Reports*, 37 (Tampere: TAPRI).

————— & Jyrki Iivonen (eds.) 1990: *Gorbachev and Europe* (London: Pinter).

————— & Pekka Sivonen (eds.) 1989: *Europe in Transition: Politics and Nuclear Security* (London: Pinter).

————— & Pekka Sivonen (eds.) 1991: *Nuclear Weapons in a Changing Europe* (London: Pinter).

Harman, Chris 1974: *Bureaucracy and Revolution in Eastern Europe* (London: Pluto).

Harrick, Robert W. 1990: "Soviet Naval Strategy and Missions, 1946–1960," in Gillette & Frank (eds.): 165–191.

Harris, John B. 1988: "From Flexible Reponse to No Early First Use," in Hopmann & Barnaby (eds.): 86–114.

Harris, Sally 1991: "Alternative Security Architectures," in Kirby & Hooper (eds.): 57–80.

Hart, Basil Lidell: see Lidell Hart, Basil.

Hart, Douglas & Barry Blechman 1989: "East-West Arms Control and Soviet Military Doctrine," in Flynn (ed.): 331–382.

Hartley, Keith & Todd Sandler (eds.) 1990: *The Economics of Defence Spending: An International Survey* (London: Routledge).

Hartung, William D. 1990: "Breaking the Arms-Sales Addiction," *WPJ*, 8:1, 1–26.

———— 1992: "Curbing the Arms Trade: From Rhetoric to Restraint," *WPJ*, 9:2 (Spring): 219–247.

Hartwig, Jürgen (ed.) 1992: *Rüstungskonversionspolitik: Ein Länderbericht: Sicherheits- und verteidigungspolitische, regional- und strukturwirtschaftliche sowie völkerrechtliche Aspekte der Abrüstung und der Truppenreduzierung* (Bremen: Steintor).

Harvard Nuclear Study Group (Albert Carnesale & al.) 1983: *Living with Nuclear Weapons* (New York: Bantam Books).

Harvey, Hal 1988: "Defense Without Aggression," *BAS*, 44:7 (September): 12–15.

Hasegawa, Tsuyoshi & Alex Pravda (eds.) 1990: *Perestroika: Soviet Domestic and Foreign Policies* (London: Sage, Royal Institute of International Affairs).

Hassner, Pierre 1990: "Europe Beyond Partition and Unity: Disintegration or Reconstitution," *IA*, 66:3, 461–475.

Hatchell, Ron 1989: "Restructuring the Air Forces for Non-Provocative Defence," in Borg & Smit (eds.): 177–187.

Hatschikjan, Magarditsch A. & Wolfgang Pfeiler 1989: "Deutsch-sowjetische Beziehungen in einer Periode der Ost-West-Annäherung," *DA*, 22:8, 883–889.

Haugaard, Svend 1990: "Danmark udem militært forsvar," in Hansen (ed.): 74–79.

Haukijärvi, Lilli-Rakel 1989: "Auf dem Weg zu einem Finnland ohne Armee," in Gross & al. (eds.): 233–240.

Hausleitner, Mariana 1989: "'Gemeinsame Sicherheit' und das 'Europäische Haus'—Konzeptionelle Gemeinsamkeiten und Unterschiede," in Karl & Auffermann (eds.): 21–27.

Heckrotte, Warren 1986: "A Soviet View of Verification," *BAS*, 41:9, 12–15.

Heegaard-Poulsen, P. A. 1980: *Hjemmeværnet* (Copenhagen: FOV).

Heimann, Bernhard 1989: "On Political and Military Aspects of a European Security System," in Wachter & Krohn (eds.): 87–100.

Heimann, Gerhard 1990: "Die Auflösung der Blöcke und die Europäisierung Deutschlands," *EA*, 5: 167–172.

Heinemann-Grueder, Andreas 1994: "The Russian Military and the Crisis of State," *AuP*, 45:1 (1st quarter): 79–89.

Heininger, Horst 1989: "Aggresivität und Friedensfähigkeit des heutigen Kapitalismus: Methodologische Aspekte," *IPW-B*, 4: 1–7.

Heisbourg, Francois 1986: "Conventional Defense: Europe's Constraints and Opportunities," in Pierre (ed.): 71–111.

Heisenberg, Wolfgang 1991: "European Security After German Unification," in Heisenberg (ed.): 106–130.

———— (ed.) 1991: *German Unification in European Perspective* (London: Brassey's).

———— & Dieter S. Lutz (eds.) 1987: *Sicherheitspolitik kontrovers: Auf dem Weg in die neunziger Jahre* (Baden-Baden: Nomos).

———— & Dieter S. Lutz (eds.) 1990: *Sicherheitspolitik kontrovers*, 3 vols. (*Schriftenreihe: Studien zur Geschichte und Politik*, 291/3 vols.) (Bonn: Bundeszentrale für politische Bildung).

Heisler, Martin O. (ed.) 1990: "The Nordic Region: Changing Perspectives in International Relations," *Annals of the American Academy of Political and Social Science*, 512.

Held, Finn: "Et forsvar er et forsvar uden militær," in Hansen (ed.): 34–60.

Heller, Mark A. 1992: "Middle East Security and Arms Control," in Spiegel (ed.): 129–138.

Hellman, Sven (ed.) 1988: *Disarmament—But How?* (Stockholm: Swedish Professionals Against Nuclear Arms).

Hemsley, John 1991: "The Influence of Technology on Soviet Doctrine," in Hemsley (ed.): 161–176.

———— (ed.) 1991: *The Lost Empire: Perceptions of Soviet Policy Shifts in the 1990s* (London: Brassey's).

Hennes, Michael 1992: "Weltweiter Schutz gegen begrenzte Raketenangriffe," *ÖMZ*, 30:6 (November–December): 527–533.

Herf, Jeffrey 1991: *War by Other Means: Soviet Power, West German Resistance, and the Battle of the Euromissiles* (New York: Free Press).

Hermann, Charles F., Charles W. Kegley & James N. Rosenau (eds.) 1987: *New Directions in the Study of Foreign Policy* (Boston: Allen & Unwin).

Herolf, Gunilla 1983: "Anti-tank Missiles," in *SIPRI*, 1983: 245–266.

———— 1986: "The Future of Unmanned Aircraft," in Barnaby & Borg (eds.): 147–155.

———— 1986a: "Emerging Technology," in *SIPRI*, 1986: 193–208.

———— 1988: "From Deep Strike to Alternative Defense: The Role of Technology," in Alger & Stohl (eds.): 470–474.

———— 1988a: "New Technology Favors Defense," *BAS*, 44:7 (September): 42–45.

Herrera-Lasso, Luis 1990: "Nonoffensive Defence Strategies in Central America," in UNIDIR (eds.): 152–164.

Herrmann, Richard K. 1991: "The Middle East and the New World Order: Rethinking U.S. Political Strategy After the Gulf War," *IS*, 16:2 (Fall): 42–75.

Herspring, Dale R. 1987: "On Perestroyka: Gorbachev, Yasov, and the Military," *POC*, July–August, 99–107.

———— 1989: "The Soviet High Command Looks at Gorbachev," *AP*, 235: 48–62.

———— 1993: "The Process of Change and Democratization in Eastern Europe: The Case of the Military," in Lampe & Nelson (eds.): 55–74.

———— & Ivan Volgyes 1980: "How Reliable Are Eastern European Armies?" *Survival*, 22:5 (September–October): 208–218.

Herz, John M. 1950: "Idealist Internationalism and the Security Dilemma," *WP*, 2: 157–180.

———— 1951: *Political Realism and Political Idealism: A Study in Theories and Realities* (Chicago: University of Chicago Press).

———— 1981: "Political Realism Revisited," *International Studies Quarterly*, 25:2, 182–197.

Herz, Ulrich 1971: "Umstellung auf soziale Verteidigung aus der Sicht der neutralen Staaten Schweden, Schweiz und Österreich," in Ebert (ed.): 94–102.

Hesslein, Bernd C. 1990: "Abschied von Himmerod oder: Die Bundeswehr am Ende des Kalten Krieges-ratlos," *S+F*, 8:2, 107–109.

Hettne, Björn, Jyrki Käkönen, Sverre Lodgaard, Peter Wallensteen & Håkan Wiberg 1991: "The Directors of Nordic Peace Research Institutes: Norden, Europe and the Future," *PRIO Report*, 3 (Oslo: PRIO).

Hetzer, Christine & Roland Hetzer 1990: "Aktuelle Herausforderungen einer Militärreform in der DDR vor dem Hintergrund der Wiener KVAE-Verhandlungen und des Verhältnisses zwischen der DDR und der BRD," in Benjowski & al. (eds.): 22–29.

Heurlin, Bertel 1983: "Dansk Kernevåbenpolitik," in idem (ed.): 91–113.

———— 1984: "Indledning," in idem (ed.): 9–19.

———— (ed.) 1983: *Kernevåbenpolitik i Norden* (Copenhagen: SNU).

———— (ed.) 1984: *Nordiske Sikkerhedsproblemer* (Copenhagen: SNU).

Hewett, A. & Victor H. Winston (eds.) 1991: *Milestones in Glasnost and Perestroika: Politics and People* (Washington, D.C.: Brookings).
——— & Victor H. Winston (eds.) 1991a: *Milestones in Glasnost and Perestroika: The Economy* (Washington, D.C.: Brookings).
Hewish, Mark, Robert Salvy, Bill Sweetman & Gerard Turbe 1987: "Reconnaissance and Combat Drones," *IDR*, 9: 1197–1204.
Heydrich, Wolfgang, Joachim Krause, Uwe Nehrlich, Jürgen Nötzold & Renhardt Rummel (eds.) 1992: *Sicherheitspolitik Deutschlands: Neue Konstellationen, Risiken, Instrumente* (Baden-Baden: Nomos).
Higgins, Ronald 1990: *Plotting Peace: The Owl's Reply to Hawks and Doves* (London: Brassey's).
High Frontier Europe 1986: "The European Defence Initiative, EDI" (Rotterdam: HFE).
Hill, J. R. 1989: *Arms Control at Sea* (London: Routledge).
——— 1990: "Superpower Naval Arms Control: Practical Considerations and Possibilities," in Fieldhouse (ed.): 118–134.
Hill, Roger 1993: "Preventive Diplomacy, Peace-Making and Peace-Keeping," *SIPRI*: 45–65.
Hill-Norton, Lord 1983: "Return to a National Strategy," in Baylis (ed.): 117–137.
Hinde, Robert A. (ed.) 1992: *The Institution of War* (New York: St. Martin's).
Hinds, Jim E. 1986: "The Limits of Confidence," in Borawski (ed.): 184–198.
Hines, John G. 1982: "The Principle of Mass in Soviet Tactics Today," *MR*, 62:8 (August): 13–23.
——— 1990: "Calculating War, Calculating Peace: Soviet Military Determinants of Sufficiency in Europe," in Huber (ed.): 185–199.
——— & Phillip A. Petersen 1984: "The Soviet Conventional Offensive in Europe," *MR*, 64:4 (April): 2–29.
Hinrichs, Hans 1977: "Vorneverteidigung oder Abwehr in der Tiefe?" *EW*, 3: 131–138.
Hinteregger, Gerald 1989: "Some Misconceptions About Austrian Neutrality," in Kruzel & Haltzel (eds.): 269–276.
Hintermann, Eric 1988: "European Defence: A Role for the Western European Union," *European Affairs*, 2:3, 31–38.
Hippel, Frank von & Roald Z. Sagdeev (eds.) 1990: *How to Achieve and Verify Deep Reductions in the Nuclear Arsenal* (New York: Gordon & Breach).
Hippler, Jochen 1993: "Krieg und Chaos: Irreguläre Kriegsführung und die Schwierigkeiten externer Intervention," in Matthies (ed.): 139–154.
——— 1993a: "Hoffnungsträger UNO? Die Vereinten Nationen zwischen Friedenseuphorie und realpolitischer Ernüchterung," in Matthies (ed.): 155–167.
Hjarvad, Stig 1985: "Alternativt forsvar—En genvej til nedrustning?" i VS' Antimilitaristiske Udvalg (ed.), *Hvordan kommer vi videre? Debatartikler om militarisme og fredspolitik* (Copenhagen: VS-Forlag): 81–91.
H. M. 1990: "Polen: Umgliederung der Streitkräfte und Verteidigungsdoktrin," *ÖMZ*, 28:3, 248–250.
Ho Chi Minh: See Minh, Ho Chi.
Hoffman, Mark 1991: "From Conformity to Confrontation: Arms Control," in Croft (ed.): 65–87.
——— (ed.) 1990: *UK Arms Control in the 1990s* (Manchester: Manchester U.P.).
Hoffmann, Stanley 1985: "Thoughts on the Concept of Common Security," in Väyrynen (ed.): 53–64.
——— 1990: "Reflections on the 'German Question'," *Survival*, 32:4, 291–298.

Hoffmans, D. W. 1985: "Developments in Precision Guided Munitions," *NSN*, May: 48–55.

Hofmann, Hans W. & Reinar K. Huber 1990: "On the Role of New Technologies for Conventional Stability in Europe," in Schmähling (ed.): 137–158.

Hofmann, Hans W., Reinar K. Huber & Karl Steiger 1986: "On Reactive Defense Options: A Comparative Systems Analysis of Alternatives for the Initial Defense Against the First Strategic Echelon of the Warsaw Pact in Central Europe," in Huber (ed.): 97–140.

———, Reiner K. Huber & Karl Steiger 1986a: "Some Remarks on the Costs of Reactive Defence Options," in Barnaby & Borg (eds.): 303–314.

Hogan, Michael J. (ed.) 1992: *The End of the Cold War: Its Meaning and Implications* (Cambridge: Cambridge U.P.).

Hogrefe, Bernd 1989: "Wider die sichere Niederlage: Zur Forderung nach struktureller Nichtangriffsfähigkeit," *Truppenpraxis*, 4: 387–388.

Holborn, Hajo 1941: "Moltke and Schlieffen: The Prussian-German School," in Earle (ed.): 172–205.

——— 1986: "The Prusso-German School: Moltke and the Rise of the General Staff," in Paret (ed.): 281–295.

Holbraad, Carsten 1986: "Denmark: Half-Hearted Partner," in Ørvik (ed.): 15–60.

Holcomb, James F. & Graham H. Turbiville 1988: "Soviet Desant Forces: Part 1, Soviet Airborne and Air-assault Capabilities" and "Soviet Desant Forces: Part 2, Broadening the Desant Concept," *IDR*, 9: 1077–1082; 10: 1259–1264.

Holden, Gerald 1989: *The WTO and Soviet Security Policy* (Oxford: Basil Blackwell).

——— 1990: "The End of an Alliance: Soviet Policy and the Warsaw Pact, 1989–90," *PRIF-R*, 16.

——— 1990a: "Alternative Defence and the Warsaw Treaty Organization," in Randle & Rogers (eds.): 71–89.

——— 1990b: "Soviet 'New Thinking' in Security Policy," in Kaldor, Holden & Kalk (eds.): 235–250.

——— 1991: *Soviet Military Reform: Conventional Disarmament and the Crisis of Militarized Socialism* (London: Pluto).

——— 1991a: "Der Zerfall der WVO," in Schwerdtfenger & al. (eds.): 151–162.

——— 1991b: "The Road to the Coup: Civil-Military Relations in the Soviet Crisis," *PRIF-R*, 23.

——— 1993: "Die russische Militärpolitik und die GUS," *HSFK-R*, 4.

——— (ed.) 1985: *The Second Superpower: The Arms Race and the Soviet Union* (London: CND).

Holdren, John 1989: "Interaction Between Nuclear and Conventional Arms-Control Measures (Report on a Pugwash Workshop, June 1988)," in Rotblat & Goldanskii (eds.): 93–96.

——— 1990: "Konventionelle Streitkräfte in Europa," in Weizsäcker (ed.): 218–229.

Holik, Josef 1989: "Conventional Stability," *Disarmament*, 12:2 (Summer): 26–33.

Hollins, Harry B. 1982: "A Defensive Weapons System," *BAS*, June/July: 63–65.

———, Averill L. Powers & Mark Sommer 1989: *The Conquest of War: Alternative Strategies for Global Security* (Boulder, Colo.: Westview).

Holloway, David 1984: *The Soviet Union and the Arms Race,* 2d ed. (New Haven: Yale U.P.).

——— 1989: "Gorbachev's New Thinking," *FA*, 68:1, 66–81.

——— 1989a: "State, Society and the Military under Gorbachev," *IS*, 14:3 (Winter): 524.

——— & Jane M.O. Sharp (eds.) 1984: *The Warsaw Pact: Alliance in Transition* (London: Macmillan).

Holm, Hans-Henrik & Nikolai Petersen (eds.) 1983: *The European Missile Crisis: Nuclear Weapons and Security Policy* (London: Pinter).

Holmes, Kim R. 1984: *The West German Peace Movement and the National Question* (Cambridge, Mass.: Institute for Foreign Policy Analysis).

Holmes, Leslie 1989: "Afghanistan and Sino-Soviet Relations," in Saikal & Maley (eds.): 122–141.

Holoboff, Elaine 1990: "The Soviet Concept of Reasonable Sufficiency: Conventional Arms Control in an Era of Transition," *Working Paper,* 29 (Ottawa: Canadian Institute for International Peace and Security).

——— 1993: "The Collapse of the Soviet Union and the Problems of the CIS," in Center for Defence Studies (ed.): 117–132.

Holst, Johan Jørgen 1983: "Moving Towards No First Use in Practice," in Steinbruner & Sigal (eds.): 173–196.

——— 1983a: "Confidence-Building Measures: A Conceptual Framework," *Survival,* 25:1 (Jananuary–February): 2–15.

——— 1985: "Norwegian Security Policy," in idem & al. (eds.): 93–123.

——— 1985a: "Confidence-Building and Nuclear Weapons in Europe," in Väyrynen (ed.): 193–206.

——— 1987: "Vertrauens- und Sicherheitsbildende Massnahmen in und für Europa—ein konzeptioneller Rahmen und eine längerfristige Perspektive," in Brauch (ed.): 360–382.

——— 1989: "Uncertainty and Opportunity in an Era of East-West Change," *AP,* 247: 3–15.

——— 1990: "Confidence and Security Building in Europe: Achievements and Lessons," *NUPIP,* 436.

——— 1990a: "The Changing European Environment: Political Trends and Prospects," *NUPIN,* 426.

——— 1990b: "The Reconstruction of Europe: A Norwegian Perspective," *NUPIN,* 423.

——— 1990c: "Civilian-Based Defence in a New Era: A Keynote Address," *NUPIN,* 422.

——— 1990d: "European Security: A View from the North," *NUPIP,* 438.

——— 1990e: "NATO and the Northern Region: Security and Arms Control," *NUPIP,* 437.

——— 1991: "Arms Control in the Nineties: A European Perspective," *Dædalus,* Winter: 83–110.

——— 1991a: "From Arctic to Baltic: The Strategic Significance of Norway," *NSN,* 36:3 (May–June): 23–35.

——— 1992: "Arms Control in the Nineties: A European Perspective," in Adler (ed.): 85–117.

———, Kenneth Hunt & Anders C. Sjaastad (eds.) 1985: *Deterrence and Defense in the North* (Oslo: Norwegian U.P.).

Holsti, Kalevi J. 1991: *Peace and War: Armed Conflicts and International Order, 1648–1989* (Cambridge: Cambridge U.P.).

Holsti, Ole R., Richard A. Brody & Robert C. North 1969: "The Management of International Crisis: Affect and Action in American-Soviet Relations," in Pruitt & Snyder (eds.): 62–79.

Holtzendorff, Friedrich 1990: "Das Risiko und die Chance der einstigen NVA-Soldaten," *EW,* 10: 562–564.

Holzweissig, Gunter 1990: "The National People's Army After the Upheaval in the GDR," *AuP,* 4: 399–404.

Honcharenko, A. N. & E. M. Lisitsyn 1993: "Armed Forces of Ukraine and Problems of National Security," in Center for Defence Studies (ed.): 154–168.

Honig, Jan Willem 1992: "The 'Renationalization' of Western European Defence," *SS*, 2:1 (Autumn): 122–138.

Hoover, Herbert & Hugh Gibson 1943: "The Problems of Lasting Peace," in Willkie & al.: 151–360.

Hoover, Robert 1980: *Arms Control: The Interwar Naval Agreements* (Denver: University of Denver Press).

Hopmann, P. Terrence 1982: "The 'No First Use' Proposal and Arms Control Initiatives in Europe," *BPP*, 13:3, 271–275.

——— 1987: "MBFR, CSCE, CDE: Negotiatiing Structures for Cooperative Security," in Windass & Grove (eds.) 1987:II (no pagination).

——— 1988: "Conventional Arms Control and the Nuclear Weapons Dilemma in Europe," in Hopmann & Barnaby (eds.): 322–337.

——— 1991: "Mutual Security and Arms Reductions in Europe," in Smoke & Kortunov (eds.): 269–287.

——— & Frank Barnaby (eds.) 1988: *Rethinking the Nuclear Weapons Dilemma in Europe* (New York: St. Martin's).

Hoppe, Thomas 1984: "Zur Konzeption defensiver Verteidigung in ethischer Sicht," *Wiener Blätter zur Friedensforschung*, 40/41: 34–49.

——— & Hans Joachim Schmidt 1990: "Konventionelle Stabilisierung. Militärstrategische und rüstungspolitische Fragen eines Kriegsverhütungskonzepts mit weniger Kernwaffen aus ethischer und politik-wissenschaftlicher Sicht," *HSFK-B*, 1.

Hopper, Thomas 1984: "A Call for Defensive Weapons," *BAS*, May: 54–56.

Horelik, Arnold L. (ed.) 1986: *U.S.-Soviet Relations: The Next Phase* (Ithaca: Cornell U.P.).

Høringsgruppen 1980: *Forsvar os vel: Hvidbog om dansk forsvars- og sikkerheds-politik i internationalt perspektiv* (Copenhagen: Information).

Horlohe, Thomas 1984: "Das AirLand Battle 2000-Konzept," in Müller (ed.): 23–52.

Horn, Erwin 1992: "Der atomwaffenfreie Korridor," in Lutz (ed.): 137–142.

——— & Heinrich Buch 1984: "Eine Auswertung der Anhörung 'Alternative Strategien'," *S+F*, 3: 22–26.

Horn, Gyula 1989: "Problems of Comparison," in UNIDIR (eds.): 30–34.

——— 1989a: "Reduction of Conventional Armed Forces and Armaments in Europe," *Disarmament,* 12:2 (Summer): 34–40.

Horne, Alastair 1977: *A Savage War of Peace: Algeria, 1954–1962* (Harmondsworth: Penguin Books).

Hornung, Volker 1970: "Amerikanische Bürgerrechtsbewegung und 'Black Power'-Revolte," in Ebert (ed.): 52–56.

Horowitz, Dan 1993: "The Israeli Concept of National Security," in Yaniv (ed.): 11–54.

Howard, Michael 1958: *Disengagement in Europe* (Harmondsworth: Penguin).

——— 1979: "Temperamente Belli: Can War Be Controlled?" in idem (ed.): 1–16.

——— 1982: "Reassurance and Deterrence: Western Defence in the 1980s," in idem 1984: 218–236.

——— 1984: *The Causes of Wars and Other Essays* (London: Unwin).

——— 1985: "Men Against Fire: Expectations of War in 1914" (*IS*, 9:1) in Miller (ed.): 41–57.

——— 1986: "The Causes of War" (including discussion with Kenneth Walz and Karl Deutsch), in Østerud (ed.): 17–43.

———— 1989: "1989: A Farewell to Arms?" *IA*, 65:3, 407–414.

———— 1990: "Kriegsverhütung, Militärstrategie und defensive Verteidigung," in Weizsäcker (ed.): 189–196.

———— 1990a: "Critical Remarks on Defensive Defence," in Cerutti & Ragionieri (eds.): 161–162.

———— 1991: "British Grand Strategy in World War I," in Kennedy (ed.): 31–42.

———— 1993: "The Historical Development of the UN's Role in International Security," in Roberts & Kingsbury (eds.): 63–80.

———— (ed.) 1979: *Restraints on War: Studies in the Limitation of Armed Conflict* (Oxford: Oxford U.P.).

Howard, Patrick 1990: "Naval Confidence-Building Measures: A CSCE Perspective," in Fieldhouse (ed.): 226–237.

Howlett, Darryl & John Simpson 1993: "Nuclearisation and Denuclearisation in South Africa," *Survival*, 35:3 (Autumn): 154–173.

Howorth, Jolyon 1984: *France: The Politics of Peace* (London: Merlin).

———— 1984a: "Defence and the Mitterand Government," in Howorth & Chilton (eds.): 94–134.

———— & Patricia Chilton (eds.) 1984: *Defence and Dissent in Contemporary France* (London: Croom Helm).

Hoyer-Boot, Bernd, Baron von 1992: "Erweiterte Luftverteidigung," *EuS*, 41:9 (September): 506–511.

HSFK (Hessische Stiftung für Friedens- und Konfliktforschung) (eds.) 1982: *Europa zwischen Konfrontation und Kooperation: Entspannungspolitik für die achtziger Jahren* (Frankfurt: Campus).

Hsü, Immanuel C. Y. 1975: *The Rise of Modern China*, 2d ed. (New York: Oxford U.P.).

Hubel, Helmut 1993: "Finland nach dem Ost-West-Konflikt," *EA*, 48:15 (10 August): 443–450.

Huber, Reiner K. 1986: "On Strategic Stability in Europe Without Nuclear Weapons," in Avenhaus & Kettele (eds.): 215–238.

———— 1986a: *Some Remarks on Structural Implications of Strategic Stability in Central Europe*. Working Paper 1 for the Fifth Workshop of the Pugwash Study Group on Conventional Forces in Europe, Castilionciello, Italy, 9–12 October 1986.

———— 1987: "Bemerkungen zur 'Strukturellen Nichtangriffsfähigkeit'," *S+F*, 1: 32–38.

———— 1988: "Crisis Stability in Europe Without Nuclear Weapons: Prerequisites and Implications for Security and Arms Control," *Fakultät für Informatik: Bericht*, S-8704 (München: Universität der Bundeswehr).

———— 1988a: "The Defence Efficiency Hypothesis and Conventional Stability in Europe: Implications for Arms Control," *Fakultät für Informatik: Bericht*, SS-8801 (München: Universität der Bundeswehr).

———— 1990: "An Analytical Approach to Cooperative Security: First Order Assessment of Force Posture Requirements in a Changing Security Environment," in idem (ed.): 165–184.

———— 1990a: "Multipolare Sicherheitssysteme für Europa: Systemanalytische Überlegungen zu einer militärischen Ausgestaltung," *ÖMZ*, 5: 412–418.

———— 1991: "Militärische Modelle für die künftige Stabilität," *EW*, 40:2 (February): 94–99.

———— 1991a: "Parity and Stability: Some Conclusions from Geometrical Models of Military Operations in Central Europe," *International Interactions*, 16:4, 225–238.

———— 1991b: "Stable Operational Minima for the Defence: Observations from the

Analysis of Simple Geometrical Models of Military Operations," in Avenhaus & al. (eds.): 108–136.

——— 1993: "Military Stability of Multipolar International Systems: An Analysis of Military Potentials in Post–Cold War Europe" (Neubiberg: Universität der Bundeswehr Institut für Angewandte Systemforschung und Operations Research, Fakultät für Informatik, München, June 1993).

——— 1993a: "Ethisch verantwortliche Sicherheitspolitik im neuen Europa," *EuS*, 42:11 (November): 566–568.

——— 1993b: "Über das Verhältnis von Offensiv- und Defensivfähigkeit von Streitkräften und ihre personellen Begrenzungen unter Stabilitätsgesichtspunkten," in Forndran & Pohlman (eds.): 285–314.

——— (ed.) 1986: *Modeling and Analysis of Conventional Defense in Europe: Assessment of Improvement Options* (New York: Plenum).

——— (ed.) 1990: *Military Stability: Prerequisites and Analysis of Requirements for Conventional Stability in Europe* (Baden-Baden: Nomos).

——— & Rudolf Avenaus 1993: "Problems of Multipolar International Stability," in Huber & Avenhaus (eds.): 11–20.

——— & Hans W. Hofmann 1984: "Gradual Defensivity: An Approach to a Stable Conventional Force Equlirium in Europe," in Brans (ed.): 197–211.

——— & Hans W. Hofmann 1984a: "Some Thoughts on Unilaterally Reducing the Conventional Imbalance in Central Europe: Gradual Defensivity as a Force Design Principle," *Fakultät für Informatik: Bericht*, 8402 (München: Universität der Bundeswehr).

——— & Hans Hofmann 1990: "The Defence Efficiency Hypothesis and Conventional Stability in Europe: Implications for Arms Control," in Boserup & Neild (eds.): 109–132.

——— & Hilmar Linnenkamp 1994: "Disarmament, Arms Control and NOD," in Møller & Wiberg (eds.): 25–36.

———, Hilmar Linnenkamp & Ingrid Schölch 1990: "Introduction," in Huber (ed.): 19–25.

——— & O. Schindler 1991: "Stability Implications of Different Views on the Effects of Air Support on the Force Ratio in Main-Thrust Sectors," *Bericht*, S–9101 (Neubiberg: Universität der Bundeswehr, München, Fakultät für Informatik).

——— & O. Schindler 1993: "Military Stability of Multipolar Power Systems: An Analytical Concept for Its Assessment, Exemplified for the Case of Poland, Byelarus, the Ukraine and Russia," in Huber & Avenhaus (eds.): 155–180.

——— & Rudolf Avenhaus (eds.) 1993: *International Stability in a Multipolar World: Issues and Models for Analysis* (Baden-Baden: Nomos).

Huck, Burkhardt J. 1990: "Bundeswehrplanung zwischen Umrüstung und Abrüstung," in Köllner & Huck (eds.): 39–58.

Hudak, Vasil 1992: "East-Central Europe and the Czech and Slovak Republics in a New Security Environment," *ES*, 1:4 (Winter): 118–145.

Hudson, George E. (ed.) 1990: *Soviet National Security Policy Under Perestroika* (Boston: Unwin Hyman).

Hufbauer, Gary Clyde, Jeffrey J. Scott & Kimberley Ann Elliott 1990: *Economic Sanctions Reconsidered*, 2d ed., vol. I: *History and Current Policy;* vol. II: *Supplemental Case Histories* (Washington, D.C.: Institute for International Economics).

Hug, Peter & Ruedi Meier 1992: *Rüstungskonversion: Die Umwandlung militärabhängiger Arbeitsplätze in zivile Beschäftigung* (Chur, Switzerland: Rüegger).

Huizer, Gerrit 1972: "Land Invasion as a Nonviolent Strategy of Peasant Rebellion," *JPR*, 9:2, 121–132.

Hunt, Kenneth 1985: "Comment on Chapters 2 and 3," in Roper (ed.): 61–77.

Hunter, F. Robert 1991: *The Palestinian Uprising: A War by Other Means* (Berkeley: University of California Press).

Huntington, Samuel 1983: "Conventional Deterrence and Conventional Retaliation in Europe" (*IS,* 8:3) in Miller (ed.) 1986: 251–275.

Hürlimann, Jacques & Kurt Spillmann 1991: "Der Bericht 90 zur schweizerischen Sicherheitspolitik im Urteil ausländischer Expertinnen und Experten," *Zürcher Beiträge zur Sicherheitspolitik und Konfliktforschung,* no. 15 (Zürich: Forschungsstelle für Sicherheitspolitik und Konfliktanalyse, Eidgenössische Technische Hochschule).

Husain, Roos Masood 1992: "Arms Control and the Proliferation of High-Technology Weapons in South Asia and the Middle East: A View from Pakistan," in Stahl & Kemp (eds.): 135–141.

Husbands, Joe & Anne Hessing Kahn 1988: "The Conventional Arms Transfer Talks: An Experiment in Mutual Arms Trade Restraint," in Ohlson (ed.): 110–125.

Huth, Paul K. 1988: *Extended Deterrence and the Prevention of War* (New Haven: Yale U.P.).

Huxley, Steven 1993: "Nonviolence Misconceived? A Critique of Civilian-Based Defence," in Balazs & Wiberg (eds.): 234–243.

Huzzard, Ron & Christopher Meredith (eds.) 1985: *World Disarmament: An Idea Whose Time Has Come* (London: Spokesman).

Hyde-Price, Adrian 1991: *European Security Beyond the Cold War: Four Scenarios for the Year 2010* (London: Sage, RUSI).

——— 1991a: "Ten Principles for a New European Security System." Paper presented to the conference "Sicherheit im neuen Europa," Evangelische Akademie Loccum and SAS, 27–29 September 1991.

IDDS (Institute for Defence and Disarmament Studies) (ed.) 1991: *A Chronological History of the CFE Negotiations* (Brookline, Mass.: IDDS).

IFSH (Institut für Friedens- und Sicherheitspolitik an der Universität Hamburg) 1985: "Angriff in die Tiefe: Ein Diskussionsbeitrag zum Rogers-Plan," *S+F Sonderdruck,* 1.

——— 1990: "Ein geeintes Deutschland in einem neuen Europa: Vom Blocksystem zur Sicherheitsgemeinschaft," *S+F,* 8:2, 74–85.

——— 1992: "Friedensforschung nach dem Ende des Ost-West Konflikts: Alte Probleme und neue Herausforderungen: Eine Studie aus dem IFSH," in Lutz (ed.): 523–553.

——— 1993: "Vom Recht des Stärkeren zur Stärke des Rechts: Die Europäische Sicherheitsgemeinschaft (ESG) als Garant von Sicherheit und Frieden—Eine Studie des IFSH," in Lutz 1993: 101–153.

Iivonen, Jyrki 1990: "Soviet Foreign Policy Doctrine in Transition," in Harle & Iivonen (eds.): 22–39.

Iklé, Fred Charles 1971: *Every War Must End* (New York: Columbia U.P.).

——— 1993: "The Case for a Russian-US Security Community," in Goodby & Morel (eds.): 9–22.

Ilnicki, Marek, Andrzej Karkoszka, Lech Kosciuk & Andrzej Makowski 1991: "Security and Politico-Military Stability in the Baltic Region," *PISM Occasional Papers,* no. 21 (Warsaw: Polish Institute of International Affairs).

IMEMO (Institute of World Economy and International Relations) (eds.) 1987: *Disarmament and Security: 1987 Yearbook* (Moscow: Novosti).

——— (eds.) 1989: *Disarmament and Security: 1988–1989 Yearbook* (Moscow: Novosti).

IMSF (Institut für Marxistische Studien) (eds.) 1989: *Rüstung, Abrüstung, Frieden:*

Tendenzen, Probleme, Perspektiven für eine neue Friedensordnung: Marxistische Studien: Jahrbuch des IMSF 15 (Frankfurt a.M.: IMSF).

Independent Commission on Disarmament and Security Issues: See Palme Commission.

Independent Defence Research Group 1988: "Saor-ollscoil ca h'Eireann: A Defence Policy for Ireland" (Dublin: Centre for Peace Research).

Informationsdienst Sicherheitspolitik 1984: "'Alternative Strategien': Machen sie den Frieden sicherer?" *ISP*, 1.

IPRA-DSG (International Peace Research Assocation, Disarmament Study Group, Workshop on Confidence-Building Measures) 1980: "Building Confidence in Europe: An Analytical and Action-Oriented Study," *BPP*, 11:2 (June): 1–17.

Ipsen, Knut 1992: "Sovereignty and Collective Security," in Lutz (ed.): 461–473.

Isaacs, John 1993: "Strengthening the World Policeman," *BAS*, 49:10 (December): 14–15.

Ischebeck, Otfried 1990: "Frankreich und die Abrüstung in Europa—eine deutsche Perspektive," in Weizsäcker (ed.): 282–292.

———— 1991: "Überlegungen zu einem zukünftigen europäischen Sicherheitssystem," in Lutz (ed.) 1991, 191–216.

———— 1991a: "Der Kampfpanzer der Zukunft," in Müller & Neuneck (eds.): 233–248.

———— 1992: "Evolution of Tanks and Anti-Tank Weapons: Assessment of Offence-Defence Dynamics and Arms Control Options," in Smit & al. (eds.): 177–196.

Istjagin, Leonid G. 1989: "Das neue Denken in der sowjetischen Aussenpolitik und alternative Sicherheitskonzepte," in Karl & Auffermann (eds.): 12–20.

Ivlev, Leonid 1994: "NOD in the USSR and Its Successors," in Møller & Wiberg (eds.): 138–142.

Izmalkov, Valeriy 1993: "Ukraine and Her Armed Forces: The Conditions and Process for Their Creation, Character, Structure and Military Doctrine," *ES*, 2:2 (Summer): 279–319.

Izyumov, Alexei 1991: "Conversion in the Soviet Union—and Possibilities for Cooperation in the Baltic Region," *BPP*, 22:3 (September): 271–279.

Jaberg, Sabine 1992: "KSZE 2001: Profil einer Europäischen Sicherheitsordnung: Bilanz und Perspektiven ihrer institutionellen Entwicklung," *HB*, 70.

———— 1993: "Frieden ist mehr als die Abwesenheit von Krieg: Deutschland als Friedens- und Zivilmacht in Europa—eine Perspektive," in Lutz (ed.): 159–182.

Jack, Hamer A. 1984: "Unilateralism to Break the Stalemate," *Disarmament*, 7:3, 88–94.

Jackson, Gabriel 1965: *The Spanish Republic and the Civil War, 1936–1939* (Princeton: Princeton U.P.).

Jackson, Robert H. 1992: "The Security Dilemma in Africa," in Job (ed.): 81–94.

Jacobsen, Carl 1992: "On the Search for a New World Security Order: 'The Inviolability of Borders': Prescription for Peace or War?" *ES*, 1:1 (Spring): 50–57.

———— (ed.) 1987: *The Uncertain Course: New Weapons, Strategies, and Mindsets* (Oxford: U.P., SIPRI).

Jacobsen, Hans-Adolf & Heinz-Georg Lemm (eds.) 1984: *Heere International* (Herford: Mittler).

Jaeschke, Paul 1986: "Politische Rahmenbedingungen einer neuen Sicherheitspolitik," in Gohl & Niesporek (eds.): 45–61.

Jahn, Egbert 1971: "Soziohistorische Voraussetzungen der sozialen Verteidigung," in Ebert (ed.): 28–35.

———— 1973: "Civilian Defense and Civilian Offense," *JPR,* 10:3, 285–294.

———— 1977: "Social and Political Conditions for the Expansion of the Non-Violent Movement: Historical Evolution and Future Prospects," *BPP,* 9:4, 335–338.

———— 1984: "Frieden durch Pazifismus: Das Konzept waffen- und gewaltfreier Sicherheitspolitik," in Vogt (ed.): 79–91.

———— 1984a: "Prospects and Impasses of the New Peace Movement: Alternative Horizons," *BPP,* 15:1, 47–56.

———— 1986: "Der Einfluss der Ideologie auf die sowjetische Aussen- und Rüstungspolitik," 3 parts, *Osteuropa,* no. 5: 356–374; no. 6: 447–461; no. 7: 509–521.

———— 1989: "Entscheidend ist die umfassende Friedenspolitik," in Gross & al. (eds.): 273–286.

———— 1990: "The Role of Governments, Social Organizations and Peace Movements in the New German and European Detente Process," in Kaldor, Holden & Kalk (eds.): 65–89.

———— & Yoshikazu Sakamoto (eds.) 1981: *Elements of World Instability: Armaments, Communication, Food, International Division of Labour.* Proceedings of the International Peace Research Association Eighth General Conference (Frankfurt: Campus).

Jaipal, Rikhi 1989: "Super Power Relations as a Key," in UNIDIR (eds.): 114–122.

Jalonen, Olli-Pekka 1990: "Non-Offensive Defence and the Security Dilemma in the Arctic Region," in Auffermann (ed.): 137–148.

———— 1990a: "Soviet Military Doctrine: Stability or Change," *CRPV,* 13:1, 17–28.

———— & Unto Vesa 1992: "Something Old, Something New, Something Borrowed, Something Blue: Finland's Defence Policy in a Changing Security Environment," *Cooperation and Conflict,* 27:4 (December): 377–396.

James, Alan 1993: "Internal Peace-Keeping: A Dead End for the UN?" *Security Dialogue,* 24:4 (December): 359–368.

Janning, Josef, Hans-Josef Legrand & Helmut Zander (eds.) 1987: *Friedensbewegungen: Entwicklung und Folgen in der Bundesrepublik Deutschland, Europa und den USA* (Köln: Wissenschaft & Politik).

Järvenpää, Pauli 1992: "Finland: Peace Treaty of 1947," in Tanner (ed.): 55–70.

Jasani, Bhupendra & Christopher Lee 1984: *Countdown to Space War* (London: Taylor & Francis, SIPRI).

Jaurés, Jean 1915: *L'Armée nouvelle* (Paris: l'Humanité).

Jean, Carlo 1990: "Airpower and Conventional Stability," in Boserup & Neild (eds.): 151–158.

Jegorow, Waleri Nikolajewitsch 1972: *Friedliche Koexistenz und revolutionärer Prozess* (Berlin: Staatsverlag der DDR).

Jencks, Harlan W. 1984: "'People's War Under Modern Conditions': Wishful Thinking, National Suicide, or Effective Deterrent?" *China Quarterly,* June, 305–319.

Jensen, Keld 1989: "Tidsgeografisk analyse af militær bevægelse og tilgænge-lighed: Indledende undersøgelse af forudsætningerne for opbygningen af en defensive militær infrastruktur," *Arbejdspapir,* 84 (Roskilde: RUC, Institute of Geography, Social Analysis and Datology).

Jervas, Gunnar 1983: "Sverige, Norden och kärnvapen," in Heurlin (ed.): 58–81.

———— 1987: "Sweden in a Less Benign Environment," in Sundelius (ed.): 57–74.

———— 1988: "Sveriges försvar i et framtidsperspektiv," *Internasjonal Politikk,* 1: 73–92.

Jervis, Robert 1970: *The Logic of Images in International Relations* (Princeton: Princeton U.P.).

———— 1976: *Perception and Misperception in International Politics* (Princeton: Princeton U.P.).

———— 1978: "Cooperation Under the Security Dilemma," *WP,* 30:2 (January): 167–214.

———— 1982: "Security Regimes," *International Organization,* 36:2, 357–378.

———— 1984: *The Illogic of American Nuclear Strategy* (Ithaca: Cornell U.P.).

———— 1985: "From Balance to Concert: A Study of International Security Cooperation," *WP,* 38:1 (October): 58–79.

———— 1987: "Strategic Theory: What's New and What's True," in Roman Kolkowicz (ed.), *The Logic of Nuclear Terror* (Boston: Allan & Unwin): 47–81.

———— 1988: "The Political Effects of Nuclear Weapons: Stability in the Postwar World," (*IS,* 13:2) in Lynn-Jones & al. (eds.) 1990: 28–38.

———— 1988a: "Realism, Game Theory, and Cooperation," *WP,* 40:3 (April): 317–349.

———— 1989: *The Meaning of the Nuclear Revolution: Statecraft and the Prospects of Armageddon* (Ithaca: Cornell U.P.).

———— 1989a: "Rational Deterrence: Theory and Evidence," *WP,* 41:2 (January): 183–207.

———— 1990: "Psychology and Crisis Stability," in Andrew Goldberg, Debra Van Opstal & James H. Barley (eds.), *Avoiding the Brink: Theory and Practice in Crisis Management* (London: Brassey's): 17–41.

———— 1991: "Domino Beliefs and Strategic Behavior," in idem & Snyder (eds.): 20–50.

———— 1991a: "Conclusion," in idem & Bialer (eds.): 302–313.

———— 1991b: "Will the New World Be Better?" in idem & Bialer (eds.): 7–19.

———— 1991c: "Arms Control, Stability, and Causes of War," *Dædalus,* Winter: 167–182.

———— 1991d: "The Future of World Politics: Will It Resemble the Past?" *IS,* 16:3 (Winter): 39–73.

———— 1992: "Arms Control, Stability, and Causes of War," in Adler (ed.): 175–193.

———— & Seweryn Bialer (eds.) 1991: *Soviet-American Relations After the Cold War* (Durham: Duke U.P.).

———— & Jack Snyder (eds.) 1991: *Dominoes and Bandwagons: Strategic Beliefs and Great Power Competition in the Eurasian Rimland* (New York: Oxford U.P.).

Jesse, Eckhard 1989: "Der 'dritte Weg' in der deutschen Frage: Über die Aktualität, Problematik und Randständigkeit einer deutschlandpolitischen Position," *DA,* 22:5, 543–559.

Jinbiao, Wang 1990: "Die Doktrin der Gemeinsamen Sicherheit aus der Sicht eines Chinesen," in Lutz & Schmähling (eds.): 203–216.

Job, Brian L. 1992: "The Insecurity Dilemma: National, Regime, and State Securities in the Third World," in idem (ed.): 11–35.

———— (ed.) 1992: *The Insecurity Dilemma: National Security of Third World States* (Boulder, Colo.: Lynne Rienner).

Joccheim, Gernot 1987: "Mohandas K. Gandhi (1873–1948): Die Einheit von Moral und Handeln," in Rajewsky & Riesenberger (eds.): 111–118.

Joenniemi, Pertti 1986: "Norden and the Development of Strategic Doctrines: From a Buffer Zone to a Grey Area," *CRPV,* 1–2: 54–73

———— 1988: "Debatten om säkerhetsutveckling i Östersjöområdet: Strukturer, riktlinjer och slutsatser," in Joenniemi & Vesa (eds.): 224–239.

—— 1989: "The Underlying Assumptions of Finnish Neutrality," in Kruzel & Haltzel (eds.): 49–60.

—— 1989a: "Finland and European Security: Traditions and Responses to Current Challenges," in Wæver & al. (eds.): 233–249.

—— (ed.) 1991: "Co-operation in the Baltic Sea Region: Needs and Prospects," *Research Report,* 42 (Tampere: TAPRI).

—— (ed.) 1993: *Cooperation in the Baltic Sea Region* (London: Taylor & Francis).

——— & Unto Vesa (eds.) 1988: "Säkerhetsutveckling i Östersjöområdet," *Research Report,* 35 (Tampere: TAPRI).

Joffe, Ellis 1987: *The Chinese Army After Mao* (London: Weidenfeld & Nicholson).

—— & Gerald Segal 1985: "The PLA Under Modern Conditions," *Survival,* 27:4 (July–August): 146–157.

Joffe, Josef 1984: "Squaring Many Circles: West German Security Policy Between Deterrence, Détente and Alliance," in Cooney & al. (eds.): 174–203.

—— 1985: "Nuclear Weapons, No First Use, and European Order," in Hardin & al. (eds.): 231–243.

—— 1988: *The Limited Partnership: Europe, the United States, and the Burdens of Alliance* (Cambridge, Mass.: Ballinger).

—— 1990: "Once More: The German Question," *Survival,* 32:2, 129–140.

—— 1990: "Failed Dreams: Collective Security and the Future of European Stability," in Clesse & Rühl (eds.): 168–174.

—— 1991: "The 'Revisionists': Germany and Russia in a Post-Bipolar World," in Clark & Serfaty (eds.): 95–125.

—— 1992: "Collective Security and the Future of Europe: Failed Dreams and Dead Ends," *Survival,* 34:1 (Spring): 36–50.

Johannsen, Margaret 1990: "Beyond Deterrence Through Common Security," in Cerutti & Ragionieri (eds.): 44–58.

Johansen, Robert C. 1990: "Toward Post-Nuclear Global Security: An Overview," in Weston (ed.): 220–260.

—— 1991: "Lessons for Collective Security," *WPJ,* 8:3 (Summer): 561–574.

—— 1991a: "Real Security Is Democratic Security," *Alternatives,* 16:2 (Spring): 209–242.

—— 1991b: "A Policy Framework for World Security," in Klare & Thomas (eds.): 401–424.

—— 1991c: "Do Preparations for War Increase International Peace and Security?" in Kegley (ed.): 224–244.

Johansson, Bertil & Lars O. Nordmark 1979: "Decentraliserat försvar," *Försvar i Nutid,* 1–2.

Johnsen, William T. 1990: "Is it Time for a New NATO Strategy?" in Clesse & Rühl (eds.): 220–234.

Johnson, Alfred Ross, Robert W. Dean & Alexander Alexiev 1982: *East European Military Establishments: The Warsaw Pact Northern Tier* (New York: Crane Russak).

Johnson, James Turner 1981: *Just War Tradition and the Restraint of War: A Moral and Political Inquiry* (Princeton: Princeton U.P.).

—— 1987: *The Quest for Peace: Three Moral Traditions in Western Cultural History* (Princeton: Princeton U.P.).

Johnson, Peter 1985: *Neutrality: A Policy for Britain* (London: Temple Smith).

Johnson, Robert H. 1989: "The Persian Gulf in U.S. Strategy: A Skeptical View," *IS,* 14:1 (Summer): 122–160.

Johnston, Alastair Iain 1990: "China and Arms Control in the Asia-Pacific Region," in Langdon & Ross (eds.): 173–204.

Johnston, Douglas M. 1991: "Naval Arms Control: The Burden of Proof," *Disarmament*, 14:1, 168–182.

Johnston, Wittle 1982: "Reflections on Wilson and the Problems of World Peace," in Arthur S. Link (ed.), *Woodrow Wilson and a Revolutionary World, 1913–1921* (Chapel Hill: University of North Carolina Press): 190–231.

Jomini, Antoine-Henri, de 1811: *Traite des grandes operations militaires*, 2d ed. (Paris: Magimel).

Jonah, James O.C. 1984: "Disarmament and the Strengthening of United Nations Machinery to Maintain International Peace and Security," *Disarmament*, 7:3, 95–99.

Jones, Archer 1988: *The Art of War in the Western World* (London: Harrap).

Jones, Christopher D. 1981: "Soviet Military Doctrine and Warsaw Pact Exercises," in Leebaert (ed.): 225–258.

———— 1981a: *Soviet Influence in Eastern Europe: Political Autonomy and the Warsaw Pact* (New York: Praeger).

———— 1984: "National Armies and National Sovereignity," in Holloway & Sharp (eds.): 87–110.

———— 1992: "Protection from One's Friends: The Disintegration of the Warsaw Pact," in Leebaert & Dickinson (eds.): 100–126.

Jones, Lynn F. 1990: "Conventional Arms Reductions: Need for Analysis," in Huber (ed.): 129–132.

Jones, Peter D. & Jo Vallentine 1990: "Transarmament: A Proposal to Widen the Defence Debate," in Cheeseman & Kettle (eds.): 175–190.

Jopp, Matthias (ed.) 1992: *Dimensionen des Friedens-Theorie, Praxis und Selbstverständnis der Friedensforschung* (Baden-Baden: Nomos).

Jordan, Gerhard 1989: "Das Bundesheer in Österreich und der Widerstand dagegen," in Gross & al. (eds.): 241–271.

Jordan, Robert 1990: "Introduction: The Changing Political Environment of Europe," in Jordan (ed.): 1–25.

———— (ed.) 1990: *Europe and the Superpowers: Essays on European International Politics* (London: Pinter).

Jost, Hans Ulrich 1990: "Switzerland's Atlantic Perspectives," in Miliovejic & Maurer (eds.): 110–121.

Jourkine (Zhurkin), Vitali, Sergui (Sergei) Karaganov & Andrei Kortounov 1989: "Un point de vue soviétique sur la suffisance raissonable," *Memento Defense— Desarmament 1989* (Brussels: GRIP): 71–76.

Joxe, Alain 1984: "Dissuassion infra-nucleaire: Principes de dissuassion civique," *CES*, 6.

———— 1990: *Le cycle de la dissuasion (1945–1990): Essai de stratégie critique* (Paris: La Découverte, Fondation pour les études de la défense nationale).

———— 1990a: "Nonoffensive Defence and French Strategic Thinking," in Cerutti & Ragionieri (eds.): 134–141.

———— 1991: *Vouage aux sources de la guerre* (Paris: Presses Universitaires de la France).

Joyner, Christopher C. 1992: "International Law," in Schraeder (ed.): 229–244.

Jukes, Geoffrey 1989: "The Soviet Armed Forces and the Afghan War," in Saikal & Maley (eds.): 82–100.

Jundzis, Tälavs 1991: "Defence Interests of Latvia Republic During the Period of Independence Resumption." Paper presented to the workshop "Soviet, Russian and Baltic Security," Svaneke (Denmark), 25–27 September 1991.

Jung, Ernst F. 1988: "Conventional Arms Control in Europe in Light of the MBFR Experience," *AuP*, 39:2, 150–168.

Jung, Kim Dae 1992: "The Once and Future Korea," *Foreign Policy,* 86 (Spring): 40–55.

Jung, Lothar 1987: *"Wir haben begonnen umzudenken": Michail Gorbatschows Reformkonzept für die UdSSR* (Köln: Bund).

———— 1989: "Die 'deutsche Frage' und die Abschreckungspolitik," *DA,* 22:9, 1011–1015.

———— 1990: "Der Warschauer Pakt und die konventionelle Abrüstung in Europa," in E. Müller & Neuneck (eds.): 45–62.

Jung, Werner 1988: "Zur konventionellen Stabilität in Europa," in Bahr & Lutz (eds.): 452–456.

———— 1988a: "Gemeinsame Sicherheit—Strukturelle Nichtangriffsfähigkeit— Defensive Verteidigung," in Bahr & Lutz (eds.): 239–254.

Juutilainen, Antti 1985: "Operational Decisions by the Defence Forces," in FCMH (eds.): 153–173.

Kade, Gerhard 1981: *Generale für den Frieden: Interviews von Gerhard Kade* (Köln: Pahl-Rugenstein).

Kaestner, Roland 1990: "Zur künftigen Wehrstruktur deutscher Streitkräfte in einem Europa kooperativer Sicherheit," *S+F,* 8:2, 86–93.

———— 1991: "Überlegungen zu einer künftigen Wehrstruktur deutscher Streitkräfte in einem Europa kooperativer und kollektiver Sicherheit," in Lutz (ed.) 1991: 233–272.

———— & Götz Neuneck 1991: "HighTech und der Krieg am Golf—die Kosten moderner Kriegsführung," *S+F,* 9:3, 127–133.

Kafkalas, Peter N. 1986: "The Light Division and Low-Intensity Conflict: Are They Losing Sight of Each Other?" *MR,* 66:1, 18–27.

Kahl, Suzanne 1990: "Cutting Tanks and ACVs in Europe: Centerpiece of the CFE Agreement," *Briefing Paper,* no. 1 (Brookline, Mass.: Institute for Defense and Disarmament Studies).

Kahn, Herman 1965: *On Escalation: Metaphors and Scenarios* (London: Pall Mall).

Kaiser, Karl 1989: "Wozu Atomwaffen in Zeiten der Abrüstung," in Wagemann (ed.): 189–202.

———— 1993: "Die ständige Mitgliedschaft im Sicherheitsrat: Ein berechtigtes Ziel der neuen deutschen Aussenpolitik," *EA,* 48:19 (10 October): 541–552.

————, Georg Leber, Alois Mertes & Franz-Josef Schulze 1982: "Nuclear Weapons and the Preservation of Peace" (*FA,* Summer) in W. Bundy (ed.) 1985: 77–100.

———— & Pierre Lellouche (eds.) 1986: *Deutsch-Französiche Sicherheitspolitik: Auf dem Wege zur Gemeinsamkeit?* (Bonn: Europa Union).

———— & John Roper (eds.) 1988: *British-German Defence Cooperation: Partners Within the Alliance* (London: Jane's).

Kaiser, Reinhard 1988: "Vor Beginn der Verhandlungen über konventionelle Rüstungskontrolle in Europa," *Blätter,* 9: 1098–1110.

Kaldor, Mary 1978: *The Disintegrating West* (New York: Hill & Wang).

———— 1981: *The Baroque Arsenal* (New York: Hill & Wang).

———— 1982: "The Role of Nuclear Weapons in Western Relations," in idem & Smith (eds.): 105–124.

———— 1983: "Beyond the Blocs: Defending Europe the Political Way," *WPJ,* 1:1, 121.

———— 1985: "The Concept of Common Security," in Väyrynen (ed.): 37–52.

———— 1986: "Disengaging Europe from the Superpowers," in Ackland & McGuire (eds.): 303–310.

———— 1987: "The Imaginary War," in Smith & Thompson (eds.): 72–99.

———— 1990: *The Imaginary War: Understanding the East-West Conflict* (Oxford: Basil Blackwell).

———— 1990a: "Avoiding a New Division of Europe," *WPJ,* 8:1, 181–193.

———— & Richard Falk 1987: Introduction to idem & Falk (eds.): 1–27.

———— & Richard Falk (eds.) 1987: *Dealignment: A New Foreign Policy Perspective* (Oxford: Basil Blackwell).

————, Gerard Holden & Richard Falk (eds.) 1989: *The New Detente: Rethinking East-West Relations* (London: Verso).

———— & Dan Smith (eds.) 1982: *Disarming Europe* (London: Merlin).

Kaldrack, Gerd & Paul Klein (eds.) 1992: *Die Zukunft der Streitkräfte angesichts weltweiter Abrüstungsbemühungen* (Baden-Baden: Nomos).

Kalicki, Jan & al. 1991: "Fundamental Deterrence and Mutual Security Beyond START," in Smoke & Kortunov (eds.): 249–268.

Kaltefleiter, Werner, William E. Griffith & al. 1989: *Security Perspectives of the West German Left: The SPD and the Greens in Opposition* (Cambridge, Mass.: Institute for Foreign Policy Analysis).

Kalyadin, Alexander N. 1988: "Non-Proliferation of Nuclear Weapons: Nuclear-Free Zones," in IMEMO (eds.): 511–524.

———— 1990: "Effect of Perestroika on Disarmament Strategy," *Disarmament,* 13:4, 1–15.

Kamal, Nazir 1992: "Defensive Security in Regions Other Than Europe," *Disarmament,* 15(4): 136–152.

Kamp, Karl-Heinz 1989: "Konventionelle Rüstungskontrolle vom Atlantik bis zum Ural: Sachstand und Probleme," *APZ,* B 8/89 (17 February).

———— 1993: "Das nukleare Erbe der Sowjetunion—eine Aufgabe westlicher Sicherheitspolitik," *EA,* 48:21 (10 November): 623–632.

Kanet, Roger E. & Edward A. Kolodziej (eds.) 1991: *The Cold War as Competition: Superpower Cooperation in Regional Conflict Management* (Baltimore: Johns Hopkins U.P.).

Kanet, Roger E., Tamara J. Resler & Deborah N. Miner (eds.) 1992: *Soviet Foreign Policy in Transition* (Cambridge: Cambridge U.P.).

Kanninen, Ermei & Vilho Tervasmäki 1985: "Development of Finland's National Defence After the Second World War," in FCMH (eds.): 267–293.

Kant, Immanuel 1788: *Kritik der praktischen Vernunft* (Stuttgart: Reclam, 1963).

———— 1795: *Zum ewigen Frieden: Ein philosophischer Entwurf* (Stuttgart: Reclam, 1963).

Kaplan, Morton A. 1987: "Planning for Peace," in Kaplan (ed.): 11–30.

———— (ed.) 1973: *SALT, Problems and Prospects* (Morristown, N.J.: General Learning).

———— (ed.) 1987: *Consolidating Peace in Europe: A Dialogue Between East and West* (New York: Paragon).

Karaganov, Sergei A. 1990: "Towards a New Security System in Europe," in Harle & Iivonen (eds.): 40–50.

———— 1990a: "The Problems of the USSR's European Policy," *IA(M),* 7: 72–80.

———— 1990b: "The Soviet Union and the New European Architecture," in Clesse & Rühl (eds.): 429–435.

———— 1992: "Implications of German Unification for the Former Soviet Union," in Stares (ed.): 331–364.

Karaosmanogulo, Ali 1988: "Turkey and the Southern Flank," in Chipman (ed.): 287–353.

Karas, Thomas 1984: *The New High Ground: Strategies and Weapons for Space Age War* (London: New English Library).

Karber, Philip A. 1976: "The Soviet Anti-Tank Debate," *Survival,* 18:3 (March–April): 105–111.

———— 1986: "NATO Doctrine and National Operational Priorities: The Central Front and the Flanks: Part I," *AP,* 207: 12–33.

———— 1986a: "Military Rationale and Operational Robustness: The Impact of Emerging Technology and Experimental Tactics on the Future of the Infantry," in Huber (ed.): 167–192.

———— 1988: "Conventional Arms Control Options, or Why 'Nunn' Is Better than None," in Nehrlich & Thomson (eds.): 158–181.

Karkoszka, Andrzej 1988: "Present Tendencies in the Arms Build-up: New Weapons Systems and European Security," in Rotfeld (ed.): 85–100.

———— 1988a: "The Ideological Aspect of Doctrine: A Polish View," in Kirk (ed.): 17–20.

———— 1988b: "Merits of the Jaruszelski Plan," *BAS,* 44:7 (September): 32–34.

———— 1989: "Transition of the East—A New Beginning for Europe?" *AP,* 247: 81–91.

———— 1989a: "Verifying Conventional Force Reductions and Limitations," in Blackwill & Larrabee (eds.): 306–341.

———— 1989b: "Links Between Nuclear and Conventional Arms Reductions," in Borg & Smit (eds.): 215–220.

———— 1989c: "The Relevance of Disengagement Zones," in Borg & Smit (eds.): 231–238.

———— 1990: "Prospects for Arms Control," in UNIDIR (eds.): 43–51.

———— 1992: "Transition to Defence-Oriented Configurations," *Disarmament,* 15(4): 95–111.

Karl, Wilfried 1990: "Vom Ost-West-Konflikt zur hegemonialen Integration?" in Calliess (ed.): 107–125.

———— 1991: "Militär und Herrschaft: Theoreme zum Militarismus," in Karl & Nielebock (eds.): 17–36.

———— & Burkhard Auffermann (eds.) 1989: "Alternative Sicherheitskonzepte für Europa—ein Ost-West-Dialog," *Dokumentation,* 71/90 (Berlin: Evangelisches Bildungswerk).

———— & Thomas Nielebock (eds.) 1991: *Die Zukunft des Militärs in Industriegesellschaften: Jahrbuch für Friedens- und Konfliktforschung,* Band XVIII (Baden-Baden: Nomos).

Karns, Margaret P. & Karen A. Mingst 1991: "Multilateral Institutions and International Security," in Klare & Thomas (eds.): 266–294.

Karp, Aaron 1991: "Ballistic Missile Proliferation," in *SIPRI,* 1991, 317–344.

———— 1991a: "Controlling Ballistic Missile Proliferation," *Survival,* 33(6): 517–530.

———— 1992: "Ballistic Missile Proliferation and the MTCR," in Neuneck & Ischebeck (eds.): 171–184.

Karp, Craig M. 1986: "The War in Afghanistan," *FA,* Summer: 1026–1047.

Karp, Regina Owen 1992: "The START Treaty and the Future of Strategic Nuclear Arms Control," *SIPRI:* 13–37.

———— (ed.) 1991: *Security with Nuclear Weapons? Different Perspectives on National Security* (London: Oxford U.P.).

———— (ed.) 1992: *Security Without Nuclear Weapons? Different Perspectives on Non-Nuclear Security* (Oxford: Oxford U.P.).

———— (ed.) 1993: *Central and Eastern Europe: The Challenge of Transition* (Oxford: Oxford U.P.).

Karpov, Victor 1989: "Reduction of Conventional Forces in Europe—Problems and Prospects," in UNIDIR (eds.): 3–6.

Karsh, Ephraim 1987: "The Iran-Iraq War: A Military Analysis," *AP,* 220.

———— 1988: *Neutrality and Small States* (London: Routledge).

———— 1991: "Neutralization: The Key to Arab-Israeli Peace," *BPP,* 22:1, 11–24.

———— 1991a: "Why Saddam Hussein Invaded Kuwait," *Survival,* 33:1 (January–February): 18–30.

———— & M. Zuhiar Diab 1993: "A Road to Arab-Israeli Peace?" in Centre for Defence Studies (ed.): 315–334.

Kashlev, Yuri 1989: "Conventional Disarmament in Europe," *Disarmament,* 12:2, 41–50.

Kastl, Jörg 1988: "Das neue Denken in der sowjetischen Aussenpolitik," *EA,* 20, 575–582.

Katzenstein, Peter J. & Nobuo Okawara 1993: "Japan's National Security: Structures, Norms, and Policies," *IS,* 17:4 (Spring): 84–118.

Kaufman, Daniel J., David S. Clark & Kevin P. Sheenan (eds.) 1991: *U.S. National Security Strategy for the 1990s* (Baltimore: Johns Hopkins U.P.).

Kaufman, Robert Gordon 1990: *Arms Control During the Pre-Nuclear Era: The United States and Naval Limitation Between the Two World Wars* (New York: Columbia U.P.).

———— 1992: "The United States and Naval Arms Control Between the Two World Wars: Implications for Contemporary and Future Arms Control," *Millennium,* 21:1 (Spring): 29–52.

Kaufmann, Richard P. 1988: "Economic Reform and the Soviet Military," *WQ,* 11:3, 201–211.

Kaufmann, William W. 1956: "The Requirements of Deterrence," in Kaufmann (ed.): *Military Policy and National Security* (Princeton: Princeton U.P.): 12–28.

———— 1983: "Nonnuclear Deterrence," in Steinbruner & Sigal (eds.): 43–90.

———— 1990: "A Plan to Cut Military Spending in Half," *BAS,* 46:2, 35–39.

———— 1990a: "Some Small Change for Defense," *Brookings Review,* Summer, 26–33.

———— 1992: *Assessing the Base Force: How Much Is Too Much?* (Washington, D.C.: Brookings).

Kaysen, Carl, Robert S. McNamara & George W. Rathjens 1991: "Nuclear Weapons After the Cold War," *FA,* 70:4 (Fall): 95–110.

Kearsley, Harold J. 1992: *Maritime Power and the Twenty-First Century* (Aldershot: Dartmouth).

Keatinge, Patrick 1987: "Ireland: A Case Apart," in Sundelius (ed.): 75–94.

———— 1989: "Neutrality and Regional Integration: Ireland's Experience in the European Community," in Kruzel & Haltzel (eds.): 61–80.

Keegan, John 1989: *The Second World War* (Harmondsworth: Penguin).

Kegley, Charles W. (ed.) 1991: *The Long Postwar Peace* (New York: HarperCollins).

Kelleher, Catherine McArdle 1983: "America Looks at Europe," in Freedman (ed.): 44–66.

———— 1987: "NATO Nuclear Operations," in Carter & al. (eds.): 445–469.

———— 1988: "Managing NATO's Tactical Nuclear Operations," *Survival,* 30:1, 59–78.

———— 1991: "Arms Control in a Revolutionary Europe," *Dædalus,* Winter 1991: 111–132.

———— 1992: "Arms Control in a Revolutionary Future: Europe," in Adler (ed.): 119–139.

———— 1993: "A Renewed Security Partnership? The United States and the European Community in the 1990s," *Brookings Review,* 11:4 (Fall): 30–35.

———— 1993a: "A New Definition of Security," in Rotblat (ed.): 24–27.

Kelley, P. X. 1986: "The Amphibious Warfare Strategy," *Proc,* 112:1 (January): 18–29.

Kelly, Andrew 1988: "The Conventional Balance and Conventional Deterrence in Europe," in Burt (ed.): 12–69.

Kelly, Petra K. & Gert Bastian 1989: "Kleinstaaten können andere Wege gehen— die Schweiz als mögliches Modell für ein entmilitarisiertes Europa," in Gross & al. (eds.): 193–203.

Kelstrup, Morten 1987: *Fælles Sikkerhed: Det tyske socialdemokratis opfattelse af sikkerhedspartnerskab* (Copenhagen: SNU).

———— 1990: "Fælles sikkerhed og ikke-offensivt forsvar som elementer i dansk sikkerhedspolitik," in Sørensen (ed.): 276–308.

———— 1990: "Die deutsche Einigung und die deutsche Frage," in Benjowski & al. (eds.): 30–38.

Kemp, Geoffrey 1990: "The Gulf Crisis: Diplomacy or Force?" *Survival,* 32:6 (November–December): 507–517.

———— (with Shelley A. Stahl) 1991: *The Control of the Middle East Arms Race* (New York: Carnegie Endowment for International Peace).

———— & Shelley A. Stahl 1992: "Arms Control in the Middle East and South Asia: Goals, Methods and Limitations," in Stahl & Kemp (eds.): 165–170.

Kemp, Ian (ed.) 1990: "JDW Country Survey: Finland," *Jane's Defence Weekly,* 14:8 (25 August 1990): 269–285.

Kende, Istvan 1989: "The History of the Peace Concept and Organizations from the Late Middle Ages to the 1870s," in Harle (ed.): 1–32.

Kennan, George F. 1958: *Russia, the Atom and the West: The BBC Reith Lectures, 1957* (London: Oxford U.P.).

———— 1982: *The Nuclear Delusion: Soviet-American Relations in the Atomic Age* (New York: Pantheon).

Kennedy, Paul 1988: *The Rise and Fall of the Great Powers: Economic Change and Conflict from 1500 to 2000* (London: Unwin Hymann).

Kennedy, Paul M. 1991: "Grand Strategy in War and Peace: Toward a Broader Definition," in idem (ed.): 1–7.

———— 1991a: "American Grand Strategy, Today and Tomorrow: Learning from the European Experience," in idem (ed.): 167–184.

———— (ed.) 1991: *Grand Strategies in War and Peace* (New Haven: Yale U.P.).

Kennedy, Robert 1979: "Precision ATGMs and NATO Defence," *Orbis,* 22:4, 897–927.

———— 1990: "The Soviet/Warsaw Pact Nonnuclear Threat and the Balance of Conventional Forces in Europe," in Brauch & Kennedy (eds.): 39–68.

Kennedy-Pipe, Caroline 1993: "The CIS: Sources of Stability and Instability," in Karp (ed.): 258–282.

Keohane, Dan 1993: *Labour Party Defence Policy Since 1945* (London: Leicester U.P.).

Keohane, Robert O. (ed.) 1986: *Neorealism and Its Critics* (New York: Columbia U.P.).

———— & Joseph S. Nye 1977: *Power and Interdependence: World Politics in Transition* (Boston: Little Brown).

————, Joseph S. Nye & Stanley Hoffmann (eds.) 1993: *After the Cold War: International Institutions and State Strategies in Europe, 1989–1991* (Cambridge: Harvard U.P.).

Kerby, William & Götz Neuneck 1990: "Überlegungen zur Modellierung von Abrüstung und Stabilität," in Weizsäcker (ed.): 84–103.

———— & Götz Neuneck 1992: "Die nukleare Abschreckung überwinden: Chancen und Risiken der Abrüstung nach Ende des Kalten Krieges," in Lutz (ed.): 383–393.

Kessler, Meryl A. & Thomas G. Weiss 1991: "The United Nations and Third World Security in the 1990s," in Weiss & Kessler (eds.): 105–124.

Kessler, Wolfgang 1987: "Leo Tolstoj (1828–1910), Sittlicher Anarchismus und Gewaltlosigkeit," in Rajewsky & Riesenberger (eds.): 96–102.

Kettle, St. John 1990: "Conclusion," in Cheeseman & Kettle (eds.): 191–205.

Kiessling, Günter 1989: "Zum Begriff der Neutralität in der deutschlandpolitischen Diskussion," *DA,* 22:4, 384–390.

———— 1989a: *Neutralität ist kein Verrat: Entwurf einer europäischen Friedensordnung* (Erlangen: Straube).

———— 1990: "Das Ende von Jalta," in Venohr (ed.): 136–159.

———— 1990a: *NATO oder Elbe: Modell für ein europäisches Sicherheitssystem* (Erlangen: Straube).

Kime, Steve F. 1990: "The Soviet Navy in a Changing World," in Gillette & Frank (eds.): 271–279.

King-Hall, Stephen 1959: *Defense in the Nuclear Age* (New York: Fellowship).

Kipp, Jacob W. 1987: "Conventional Force Modernization and the Assymmetries of Military Doctrine: Historical Reflections on Air/Land Battle and the Operational Manoeuvre Group," in Jacobsen (ed.): 137–166.

———— 1990: "Gorbachev's Gambit: Soviet Military Doctrine and Conventional Arms Control in an Era of Reform," in Sarkesian (ed.): 83–120.

———— 1991: "Alternative Politico-Military Futures," in Hemsley (ed.): 67–87.

———— 1991a: "Operational Art: The Evolution of Soviet Operational Art: The Significance of 'Strategic Defense' and 'Premeditated Defense' in the Conduct of Theater-Strategic Operations," *JSMS,* 4:4 (December): 621–648.

———— 1992: "The Uncertain Future of the Soviet Military, from Coup to Commonwealth: The Antecedents of National Armies," *ES,* 1:2 (Summer): 207–238.

———— 1992a: "Civil-Military Relations in Central and Eastern Europe," *MR,* 72:12 (December): 27–35.

———— 1993: "Toward Understanding: A Military-to-Military Conversation on Doctrine, Military Art, and Field Regulations," *Journal of Slavic Military Studies,* 6:2 (June): 203–207.

———— & Timothy Thomas 1992: "International Ramifications of Yugoslavia's Serial Wars: The Challenge of Ethno-National Conflicts for a Post–Cold War European Order," *ES,* 1:4 (Winter): 146–192.

Kirby, David 1991: "Demilitarising Europe," in Kirby & Hooper (eds.): 173–198.

———— & Nick Hooper 1991: "Peaceful Europe, Troubled World?" in idem & Hooper (eds.): 224–234.

———— & Nick Hooper (eds.) 1991: *The Costs of Peace: Assessing Europe's Security Options* (Chur, Switzerland: Harwood).

Kireyev, Alexei 1990: "Conversion in the Soviet Dimension," *IA(M),* 5: 92–102.

———— 1991: "The Price of the Peace Dividend," *IA(M),* 8: 8–17.

Kirk, Elizabeth J. (ed.) 1988: *The New Force Reduction Negotiations in Europe: Problems and Prospects.* Proceedings of an AAAS Annual Meeting symposium, 13 February 1988 (Washington, D.C.: AAAS).

Kissinger, Henry A. 1957: *A World Restored: The Politics of Conservatism in a Revolutionary Age* (Boston: Houghton Mifflin).

———— 1960: *The Necessity for Choice: Prospects of American Foreign Policy* (Westport, Conn.: Greenwood).

———— 1969: *Nuclear Weapons and Foreign Policy,* abridged ed. (New York: Norton).

———— 1979: "The Future of NATO," in Kenneth A. Myers (ed.), *NATO: The Next Thirty Years* (Boulder, Colo.: Westview): 5–8.

———— 1984: "A Plan to Reshape NATO," *Time* (5 March): 20–24.

———— (ed.) 1965: *Problems of National Strategy* (New York: Praeger).

Kittel, Gabriele, Manfred Makowittzki, Thomas Nielebock & Ralf Ott 1991: "Nuclear Weapon-Free Zones: Re-evaluating a Political Concept of Peacemaking," *BPP*, 22:2 (June): 217–226.

Klare, Michael T. 1981: *Beyond the "Vietnam Syndrome": U.S. Interventionism in the 1980s* (Washington, D.C.: IPS).

———— 1984: *American Arms Supermarket* (Austin: University of Texas Press).

———— 1992: "The Development of Low-Intensity-Conflict Doctrine," in Schraeder (ed.): 37–54.

———— 1993: "Application of Arms Control to Regional Conflicts in the Third World," in Rotblat (ed.): 107–116.

———— 1993a: "The Next Great Arms Race," *FA*, 72:3 (Summer): 136–152.

———— & Daniel C. Thomas (eds.) 1991: *World Security: Trends and Challenges at Century's End* (New York: St. Martin's).

Klein, Dieter 1988: "Politökonomische Grundlagen für einen friedensfähigen Kapitalismus," *IPW-B*, 2: 1–9.

———— 1988a: *Chancen für einen friedensfähigen Kapitalismus* (Berlin: Dietz).

Klein, Fritz & al. (eds.) 1987: *Friedliche Koexistenz: Erfahrungen-Chancen-Gefahren* (Berlin, DDR: Akademieverlag).

Klein, Jean 1988: "Conventional and Nuclear Stability: A French Approach," in Kirk (ed.): 21–24.

———— 1989: "The Links Between Nuclear and Conventional Forces," in Rotblat & Goldanskii (eds.): 72–78.

———— 1990: "Considérations sur le désarmement en Europe," *Arés*, 12: 13–31.

———— 1990a: "Frankreich und die Abrüstung in Europa," in Weizsäcker (ed.): 274–281.

Klein, Paul (ed.) 1991: *Wehrpflicht und Wehrpflichtige heute* (Baden-Baden: Nomos).

———— & Ekkehard Lippert 1987: "Funktionswandel des Militärs und Streitkräftestruktur: Oder: Wie viele Soldaten sind genug?" in Bald (ed.): 101–123.

———— & Rolf P. Zimmermann (eds.) 1993: *Beispielhaft? Eine Zwischenbilanz zur Eingliederung der Nationalen Volksarmee in die Bundeswehr* (Baden-Baden: Nomos).

Klein, Peter 1990: "A Structural Change in European Security Policy," in Clesse & Rühl (eds.): 514–526.

Kliot, Nurit & Stanley Waterman (eds.) 1991: *The Political Geography of Conflict and Peace* (London: Belhaven).

Klippenberg, Erik 1977: "New Weapons Technology and the Offence/Defence Balance," *AP*, 145: 26–32.

———— 1989: "Strategy: The Impact of Technology," *AP*, 237: 41–50.

Klitgaard, Robert E. 1971: "Gandhi's Non-Violence as a Tactic," *JPR*, 8:2, 143–154.

Klumper, Antonius Albertus 1983: "Sociale Verdediging en Nederlands Verzet '40–'45. Ideel concept getoets aan historische werklijkheid." Ph.D. diss., Rijksuniversiteit de Utrecht.

Kluss, Heinz 1993: "Die Abrüstung konventioneller Streitkräfte und ihre Kontrolle," *EA*, 46:6 (25 March): 167–178.

Kniest, N. 1992: "Export Controls and the Missile Technology Control Regime," in Neuneck & Ischebeck (eds.): 185–196.

Knight, Charles 1988: "New Realities. New Opportunities: A Proposal for Restructuring the U.S. Army in NATO," preliminary version (Cambridge, Mass.: Commonwealth Institute).

——, Lutz Unterseher & Carl Conetta 1992: "Military Research and Development After the Second Gulf War," in Smit & al. (eds.): 253–262.

Knippenberg, Angela 1992: "The Future of the United Nations," in Tehranian & Tehranian (eds.): 75–89.

Koch, Jutta 1989: "Die Zukunft militärischer Sicherung: Strukturwandel, Defensivorientierung, Zivilisierung?" *Das Parlament,* 7 (9 February): 12–13.

—— 1993: "Strikt defensive Verteidigungsstrukturen in Europa—Aufgabe der Zukunft oder Hirngespenst der Vergangenheit," in Forndran & Pohlman (eds.): 261–284.

Kochetkov, Gennedy & Victor Sergeev 1989: "Artificial Intelligence and Conventional Weapons Reduction," in Borg & Smit (eds.): 77–86.

Kodama, Katsuya & Unto Vesa (eds.) 1990: *Towards a Comparative Analysis of Peace Movements* (Aldershot: Dartmouth).

Koizumi, Maomi 1990: "Perestroika in the Soviet Military," in Hasegawa & Pravda (eds.): 155–177.

Kokoshin, Andrei A. 1988: "Strategic and Operational Contexts: A Soviet Approach," in Kirk (ed.): 9–12.

—— 1988a: "Restructure Forces, Enhance Security," *BAS,* 44:7 (September): 35–38.

—— 1989: "Conditions and Criteria for a Regime of Conventional Stability in Europe," in Clesse & Schelling (eds.): 322–331.

—— 1989a: "The Future of NATO and Warsaw Pact Strategy: Paper II," *AP,* 247: 60–65.

—— 1989b: "NATO/WTO Military Doctrine," in Blackwill & Larrabee (eds.): 212–230.

—— 1990: "The Europe We Need," *IA(M),* 12: 15–24.

—— 1991: "Arms Control: A View from Moscow," *Dædalus,* Winter: 133–144.

—— 1992: "Arms Control: A View from Moscow," in Adler (ed.): 141–151.

——, Andrey Afanasyevich & Valentin Veniaminovich Larionov 1992: "Confrontation of Conventional Forces in the Context of Ensuring Strategic Stability," in Brauch & Kennedy (eds.): 71–82.

—— & Valentin Larionov 1987: "The Battle of Kursk in the Light of Contemporary Defensive Doctrine," *IMEMO,* 5.

—— & Valentin Larionov 1989: "Four Models of WTO-NATO Strategic Interrelations," in Borg & Smit (eds.): 35–44.

—— & Valentin Larionov 1990: "The Confrontation of Conventional Forces in the Context of Ensuring Stability," in Boserup & Neild (eds.): 31–43.

Kokoski, Richard 1993: "The Treaty on Open Skies," *SIPRI:* 632–634.

—— & Sergey Koulik (eds.) 1990: *Verification of Conventional Arms Control: Technological Constraints and Opportunities* (Boulder, Colo.: Westview).

Kokoyev, Mikhail & Andrei Androsov 1990: *Verification: The Soviet Stance: Its Past, Present and Future* (New York: UNIDIR).

Kolar, Petr, Otto Pick & Miroslaw Polreich 1993: "The End of Czechoslovakia: Security Implications for Central Europe," *Perspectives: Review of Central European Affairs,* 1:1, 27–35.

Koli, Markku 1992: "Development of Military Technology and Its Impact on the Finnish Land War Doctrine," *Finnish Defence Studies,* 4 (Helsinki: War College).

Köllner, Lutz & Burkhardt J. Huck (eds.) 1990: *Abrüstung und Konversion:*

Politische Voraussetzungen und wirtschaftliche Folgen in der Bundesrepublik (Frankfurt a.M.: Campus).

Kolodziej, Edward A. 1987: "Modernization of British and French Nuclear Forces: Arms Control and Security Dimensions," in Jacobsen (ed.) 239–253.

——— 1989: "British-French Nuclearization and European Denuclearization: Implications for U.S. Policy," in Prestre (ed.): 105–146.

——— 1991: "The Cold War as Cooperation," in Kanet & Kolodziej (eds.): 3–30.

Komer, Robert W. 1984: "Strategy and Military Reform," in Clark & al. (eds.): 5–15.

Komitee für Grundrechte und Demokratie 1981: *Frieden mit anderen Waffen: Fünf Vorschläge zu einer alternativen Sicherheitspolitik* (Reinbek: Rowohlt).

König, Ernest 1990: "Auftrag und Reform: Unser Heer zwischen Doktrinismus, Bürokratismus und Ökonomie," *ÖMZ,* 28:4, 280–289.

König, Ernst 1994: "Österreichs Streitkräfte vor neuen Aufgaben," *ÖMZ,* 32:1 (January): 13–22.

Konovalov, Alexander A. 1989: "The Military Objectives of Conventional Arms Control," in Blackwill & Larrabee (eds.): 164–185.

——— 1990: "A Possible Stable Configuration of Warsaw Pact–NATO General Purpose Forces After Radical Reductions from the Atlantic to the Urals," in Boserup & Neild (eds.): 140–147.

——— 1990a: "Possible Ways to Stabilize the Balance in Tactical Aviation on the European Continent," in Boserup & Neild (eds.): 166–170.

——— 1991: "The Problem of Developing Stable Non-Offensive Structures for General-Purpose Forces and Conventional Arms in Europe," in Smoke & Kortunov (eds.): 288–316.

——— 1992: "Central European Security: A View from Moscow," *ES,* 1:4 (Winter): 65–79.

——— 1993: "Russia: Security in Transition," in Karp (ed.): 196–224.

——— & Valeri Mazing 1989: "Qualitative Elements of Conventional Stability and Technological Threats," in Borg & Smit (eds.): 49–60.

Konservative Folkeparti, det (eds.) 1986: "Debathæfte om: Defensivt Forsvar: Rapport fra en konservativ forsvarshøring" (Copenhagen: Det konservative Folkeparti).

Koole, Ruud & Paul Lucardie 1986: "The Netherlands: Social Democrats and Security Policy," in Ørvik (ed.): 108–147.

Kooten, Olaf van 1988: "ATBM Curtails the Implementation of Strategies Emphasizing Defence in Europe," in Baudisch (ed.): 36–40.

Koppe, Karl-Heinz 1985: "Auf der Suche nach einer politischen Alternative zur Sicherheitspolitik mit militärischen Mitteln," in Lutz (ed.) 1985a: 327–338.

——— 1992: "Der verdeckte Frieden: Zur Rationalität pazifistischer politik," in Lutz (ed.): 513–522.

——— 1992a: "Frieden ohne Militär," *S+F,* 10(3): 155–160.

Kostko, Yuri A. 1988: "Alternative Conceptions and Models of Security and Defence for Europe and the Soviet Military Doctrine," *CRPV,* 11:3, 87–94.

——— 1990: "'Non-Offensive Defence' and Security of Europe," *Peace and the Sciences,* December (Vienna: International Institute for Peace): 12–14.

Köszegevári, Tibor 1992: "The Defence Policy and Armed Forces of Hungary," *ES,* 1:3 (Autumn): 412–426.

Kowel, Joel 1983: *Against the State of Nuclear Terror* (London: Pan).

Kozak, Heinz 1988: "Überlegungen zum Konzept der Strukturellen Nichtangriffsfähigkeit," in Bahr & Lutz (eds.): 255–266.

Koziej, Stanislaw 1992: "Main Problems of Operational Art and Tactics of Poland's Ground Forces in the 1990s," *JSMS,* 5:4 (December): 567–574.

———— 1992a: "Pan-European Security System: Future Military Doctrine?" *MR*, 72:12 (December): 48–53.

———— 1993: "Polish Defense Policy," *Journal of Slavic Military Studies*, 6:2 (June): 149–165.

Kozyrev, Andrey 1990: "Promoting Restraint in International Arms Transfers: A Soviet Perspective," *Disarmament*, 13:4, 148–156.

Kraft, Michael 1983: "Forsvar for Fred—en skitse til et folkeligt, ikke-truende dansk totalforsvar" (Copenhagen: KT).

Krakau, Anton & Ole Diehl 1989: "'Annähernde Parität' der Streitkräfte un Europa? Eine kritische Analyse des östlichen Streitkräftevergleichs," *Berichte*, 32.

———— 1989a: "Die Militärexperten der sowjetischen Westforschungsinstitute und die innersowjetische Strategiediskussion," *Berichte*, 41.

Kramer, Mark 1990: "Beyond the Brezhnev Doctrine: A New Era in Soviet-East European Relations?" *IS*, 14:3, 25–67.

———— 1992: "The Global Arms Trade After the Persian Gulf War," *SS*, 2:2 (Winter): 260–309.

Krasner, Michael A. & Nikolaj Petersen 1986: "Peace and Politics: The Danish Peace Movement and Its Impact on National Security Policy," *BPP*, 23:2 (June): 155–174.

Krasner, Stephen D. (ed.) 1982: *International Regimes* (Ithaca: Cornell U.P.).

Krass, Allan & Dan Smith 1982: "Nuclear Strategy and Technology," in Kaldor & Smith (eds.): 3–34.

Krause, Christian 1983: "Theory and Conception of CBM in East and West," in Baudissin (ed.): 13–29.

———— 1984: "Kriterien für eine alternative Militärstrategie," in Bruns, Krause & Lübkemeier: 125–159.

———— 1986: "Can NATO Renounce the First Use of Nuclear Weapons? Comments on the Current Debate Over Security Policy in the United States and in the Federal Republic" (Bonn: Friedrich-Ebert-Stiftung).

———— 1987: "Muss Nichtangriffsfähigkeit strukturell sein? Kritische Gedanken zu einem unklaren Begriff," *S+F*, 1: 47–51.

———— 1987a: "Strukturelle Nichtangriffsfähigkeit: Eine Nachlese," *S+F*, 4: 272–274.

———— 1988: "'Strukturelle Nichtangriffsfähigkeit'—A Yardstick for Conventional Stability?" CRPV, 11:3, 121–129.

———— 1988a: "Strukturelle Nichtangriffsfähigkeit im Rahmen europäischer Entspannungspolitik," in Bahr & Lutz (eds.): 267–295.

———— 1988b: "'Strukturelle Nichtangriffsfähigkeit'—A Yardstick for Conventional Stability?" *CPRV*, 11:3, 121–129.

———— 1990: "Defensive Verteidigung in Europa—gestern und heute," *Studien*, 36.

———— 1990a: "Die Zukunft der konventionellen Abrüstung in Europa," *Studien*, 41.

———— 1990b: "Europe's Security 1990" (Bonn: Friedrich-Ebert-Stiftung).

———— 1992: "Wehrpflicht in alle Zukunft?" *Studien*, 51.

Krause, Joachim 1988: "Prospects for Conventional Arms Control in Europe," *Occasional Paper Series*, 8 (New York: Institute for East-West Security Studies).

———— & Charles K. Mallory 1992: *Chemical Weapons in Soviet Military Doctrine: Military and Historical Experience, 1915–1991* (Boulder, Colo.: Westview).

Krause-Brewer, Fides 1991: "'Friedensdividende': Wieviel und für wen?" *Europäische Sicherheit* (Europäische Wehrkunde), 6: 346–349.

Krauss, Melvyn 1986: *How NATO Weakens the West* (New York: Simon & Schuster).

Krell, Gert 1981: "The Development of the Concept of Security," in Jahn & Sakamoto (eds.): 238–254.

———— 1988: "Ostpolitik Dimensions of West German Security Policy," *PRIF-R*, 1.

———— 1989: "Die Ostpolitik der Bundesrepublik Deutschland und die deutsche Frage: Historische Entwicklungen und politische Optionen," *HSFK-R*, 7.

———— 1990: "The Federal Republic of Germany and Arms Control," *PRIF-R*, 10.

———— 1991: "West German Ostpolitik and the German Question," *JPR*, 28:3 (August): 311–324.

———— 1991a: "Aufbruch und Krise: Das Weltsystem nach dem Ost-West-Konflikt und die aktuelle Friedens-und Sicherheitsproblematik," *HFSK-B*, 6.

————, Thomas Risse-Kappen & Hans-Joachim Schmidt 1983: "The No-First-Use Question in Germany," in Steinbruner & Sigal (eds.): 147–172.

Kremenyuk, Victor A. 1991: "The Cold War as Cooperation: A Soviet Perspective," in Kanet & Kolodziej (eds.): 31–61.

Krepon, Michael 1988: "Verification of Conventional Arms Reductions," *Survival*, 30:6, 544–555.

———— 1990: "Open Skies, Production Monitoring and the CFE Negotiations: The Missing Link," in Brauch (ed.): 112–128.

———— 1992: "Less Offense, More Defense," *BAS*, 48:3 (April): 12–13.

Krieger, David & Frank Kelly (eds.) 1988: *Waging Peace in the Nuclear Age: Ideas for Action* (Santa Barbara: Capra).

Krippendorff, Ekkehardt 1983: "Einseitige Abrüstung," in Krippendorff & Stuckenbrock (eds.): 211–222.

———— 1984: *Staat und Krieg: Die historische Logik politischer Unvernunft* (Frankfurt: Suhrkamp).

———— 1987: "Die NATO-Strategie der Abschrekung: Auf dem Weg zur erhöhten Kriegsrisiko," in Heisenberg & Lutz (eds.): 674–681.

———— 1992: "Pazifismus—Bellizismus: Ein Essay," in Birkenback & al. (eds.): 28–40.

———— & R. Stuckenbrock (eds.) 1983: *Zur Kritik des Palme Berichts: Atomwaffenfrie Zonen in Europa* (Berlin: Europäische Perspektiven).

Krohn, Axel 1989: *Nuklearwaffenfreie Zone: Regionales Disengagement unter die Rahmenbedingung globaler Grossmachtinteressen: Das Fallbeispiel Nordeuropa* (Baden-Baden: Nomos).

———— 1991: "Sicherheitspolitik im Ostseeraum im Rahmen gesamteuropäischer Veränderungen," *S+F*, 9:2, 98–105.

———— 1991a: "Rolle und Struktur von Streitkräften in einem zukünftigen Europäischen Sicherheitssystem," in Lutz (ed.): 273–290.

———— 1991b: "The Vienna Military Doctrine Seminar," in *SIPRI*, 1991, 501–512.

———— 1991c: "A Nuclear Weapon-Free Zone in the Baltic: A Denuclearization Concept in a Changing Environment," in Joenniemi (ed.): 19–28.

———— 1992: "Arms Control, Disarmament, and Regional Security-Building: Prospects for the Baltic," in Wellmann (ed.): 210–218.

———— 1993: *Eine neue Sicherheitspolitik für den Ostseeraum* (Leverkusen: Leske & Budrich).

———— 1993a: "Naval Arms Control and Disarmament in the Baltic Sea Region," in Joenniemi (ed.): 99–123.

Krönig, Volker 1989: "Auf dem Wege zur gemeinsamen Sicherheit," in SAS (eds.): 25–33.

———— 1989a: "Vertrauensbildung, konventionelle Rüstungskontrolle und Strukturwandel der Verteidgigung in Europa," *Bremer Beiträge,* 1: 9–14.

Krüger, Klaus 1986: "Wege zu einer gesamteuropäischen Friedensordnung," in Gohl & Niesporek (eds.): 101–112.

Krummenacher, Heinz 1990: "Wenn Sicherheitspolitik verunsichert: Zur Notwendigkeit der Neugestaltung der schweizerischen Sicherheitskonzeption," in Vogt (ed.): 287–301.

Kruzel, Joseph 1988: "New Challenges for Swedish Security Policy," *Survival,* 30:6 (November–December): 529–543.

———— 1989: "The European Neutrals, National Defence, and International Security," in Kruzel & Haltzel (eds.): 133–160.

———— 1989a: "The Future of European Neutrality," in Kruzel & Haltzel (eds.): 295–311.

———— & Michael H. Haltzel (eds.) 1989: *Between the Blocs: Problems and Prospects for Europe's Neutrals and Non-Aligned States* (Cambridge: Cambridge U.P.).

Kubalkova, V., & A. A. Cruckshank 1980: *Marxism-Leninism and the Theory of International Relations* (London: Routledge & Kegan Paul).

Kubiak, Hieronim 1993: "Poland: National Security in a Changing Environment," in Karp (ed.): 69–100.

Kubbig, Bernd W. 1992: "Raketenabwehr als angemessene technologische Antwort auf das politische Proliferationsproblem?" *HSFK-R,* 9.

Kuglitsch, Franz Josef 1992: "Das KSZE-Forum für Sicherheitskooperation," *ÖMZ,* 30:6 (November–December): 485–490.

Kuhlmann, Jürgen, & Ekkehard Lippert 1992: "Wehrpflicht ade? Argumente wider und für die Wehrpflicht in Friedenszeiten," in Kaldrach & Klein (eds.): 41–76.

Kühne, Winrich 1993: "Ohne Soldaten geht es nicht! Rettung aus der Not durch 'robuste' Blauhelmeinsätze," in Matthies (ed.): 123–138.

Kühnrich, Heinz 1968: *Der Partisanenkrieg in Europa, 1939–1945* (Berlin, DDR: Dietz).

Kuklev, Vladimir A. 1989: "The Role of Armed Forces in a European Security System," in Wachter & Krohn (eds.): 71–78.

Kulasik, Karol K. 1993: "Die Landesverteidigung der Slowakei," ÖMZ, 31:3 (May–June): 224–228.

Kull, Steven 1993: "Co-operation or Competition: The Battle of Ideas in Russia and the USA," in Goodby & Morel (eds.): 209–223.

Kunzendorff, Volker 1989: "Verfahren für die Verifikation der Abrüstung konventioneller Streitkräfte in Europe durch VorOrtInspektionen," *S+F,* 7:2, 72–76.

———— 1989a: "Verification in Conventional Arms Control," *AP,* 245.

Kupchan, Charles A. 1989: "Setting Conventional Forces Requirements: Roughly Right or Precisely Wrong?" *WP,* 41:4, 536–578.

———— 1991: "Conventional Force Reductions: Assessment, Uncertainty, and the Search for Stability," in Danspeckgruber (ed.): 169–185.

———— & Clifford A. Kupchan 1991: "Concerts, Collective Security, and the Future of Europe," *IS,* 16:1 (Summer): 114–161.

Kwak, Tae-Hwan 1993: "Current Issues in Inter-Korean Arms Control and Disarmament Talks," *Korean Journal of National Unification,* 2: 177–206.

Labody, Laszlo 1989: "The Political Objectives of Conventional Arms Control," in Blackwill & Larrabee (eds.): 116–139.

Labour Party 1984: "Defence and Security for Britain: Statement to Annual Conference 1984 by the National Executive Committee" (Manchester: Labour Party).

Labour Party Defence Study Group 1977: *Sense About Defence* (London: Quartet Books).

Lachowski, Zdzislaw 1992: "The Second Vienna Seminar on Military Doctrine," *SIPRI:* 496–505.

———— 1993: "The Vienna Confidence- and Security-Building Measures in 1992," *SIPRI:* 618–631.

Lackey, Douglas P. 1984: *Moral Principles and Nuclear Weapons* (Totowa, N.J.: Rowman & Allanheld).

La Cour, Wilhelm (ed.) 1946: *Danmark under Besættelsen,* 3 vols. (Cobenhagen: Westermann).

Lacy, James L. 1990: "Thinking About Conventional Stability in Europe: A Commentary on von Müller's Approach," in Schmähling (ed.): 179–188.

Lafontaine, Oskar 1983: *Angst vor den Freunden* (Hamburg: Spiegel).

———— 1988: *Die Gesellschaft der Zukunft: Reformpolitik in einer veränderten Welt* (Hamburg: Hoffmann & Campe).

Laird, Robbin 1990: Introduction to Laird (ed.): 1–12.

———— (ed.) 1990: *West European Arms Control Policy* (Durham: Duke U.P.).

———— & David Robertson 1990: "British Arms Control Policy," in Laird (ed.): 13–70.

Lall, Betty G., & John Trepper Marlin 1992: *Building a Peace Economy: Opportunities and Problems of Post–Cold War Defense Cuts* (Boulder, Colo.: Westview).

Lamb, John 1993: "Defence Reform for a Democratic South Africa," *Communiqué,* July (Ottawa: Canadian Centre for Global Security).

Lambeth, Benjamin S. 1992: "A Generation Too Late: Civilian Analysis and Soviet Military Thinking," in Leebaert & Dickinson (eds.): 217–247.

Lamers, Karl 1988: "Konventionelle Abrüstung in Europa," *APZ,* B 18/88, 12–20.

———— 1990: "Modelle der Sicherheit," in Heisenberg & Lutz (eds.): II:340–351.

Lampe, John R., & Daniel N. Nelson (eds.) 1993: *East European Security Reconsidered* (Washington: Woodrow Wilson Center Press/Johns Hopkins U.P.).

Lanchester, F. W. 1916: *Aircraft in Warfare: The Dawn of the Fourth Arm* (London: Constable).

Landais-Stamp, Paul, & Paul Rogers 1989: *Rocking the Boat: New Zealand, the United States and the Nuclear-Free Zone Controversy in the 1980s* (Oxford: Berg).

Landgren, Signe 1992: "Post-Soviet Threats to Security," *SIPRI:* 531–562.

Langdon, Frank C., & Douglas A. Ross 1990: "Allliance Challenges of the Gorbachev Era," in Langdon & Ross (eds.): 269–287.

———— & Douglas A. Ross (eds.) 1990: *Superpower Maritime Strategy in the Pacific* (London: Routledge).

Lange, Peer H. 1989: "NATO and WTO 'Military Doctrines' in East-West Consultations," in Blackwill & Larrabee (eds.): 186–211.

Langille, Howard Peter 1990: *Changing the Guard: Canada's Defence in a World in Transition* (Toronto: University of Toronto Press).

Lapins, Wulf 1989: "Diskrepanzen und Kompromissmöglichkeiten bei den Wiener Verhandlungen über konventionelle Streitkräfte in Europa (VKSE) zwischen NATO und Warschauer Vertrag," *KAAD,* 32.

———— 1990: "Die Militärpolitik der UdSSR im urteil ziviler Experten: Gewogen und für kritikwürdig befunden," *Studien,* 38.

———— 1990a: "Durch Militzärreform zum Reformmilitär? Sowjetische Sicherheits- und Verteidigungspolitik im Wandel," *KAAD,* 44.

———— 1991: "Die Sowjetunion und die europäische Sicherheit: Von der kollektiv-en Sicherheit zur KSZE," *KAAD,* 46.

Laqueur, Walter (ed.) 1978: *The Guerilla Reader: A Historical Anthology* (London: Wildwood).

La Rocque, Gene R. 1988: "The Role of the Military," in Krieger & Kelly (eds.): 113–125.

——— 1989: "What Should We Defend? A New Military Strategy for the United States," *Defense Monitor*, 17:4.

Larrabee, F. Stephen 1984: "Soviet Crisis Management in Eastern Europe," in Holloway & Sharp (eds.): 111–138.

——— 1988: "Gorbachev and the Soviet Military," *FA*, Summer: 1002–1026.

——— 1990: "Long Memories and Short Fuses: Change and Instability in the Balkans," *IS*, 15:3 (Winter): 58–91.

——— 1993: "The Former Yugoslavia: Emerging Security Orientation," in Karp (ed.): 177–195.

——— & Allen Lynch 1987: "Vertrauensbildende Massnahmen und amerikanisch-sowjetische Beziehungen," in Brauch (ed.): 183–210.

Latham, Robert 1993: "Democracy and War-Making: Locating the International Liberal Context," *Millennium*, 22:2 (Summer): 139–164.

Lather, Karl-Heinz 1982: "Alternative Konzeptionen der Verteidigung, Part 3: 'Raumverteidigung' und 'Vorneverteidigung' als operative Konzeptionen," *Truppenpraksis*, 11: 787–794.

Laurence, Edward J. 1993: "Reducing the Negative Consequences of Arms Transfers Through Unilateral Arms Control," in Ramberg (ed.): 175–198.

———, Siemon T. Wezeman, & Herbert Wulf 1993: "Arms Watch: SIPRI Report on the First Year of the UN Register of Conventional Arms," *SIPRI Research Report*, 6 (Oxford: Oxford U.P.).

Laurent, Jacques, & Rene Ernould 1989: "URSS: vers une nouvelle 'Révolution dans les affaires militaires'," *Stratégique*, 2, 1989: 11–54.

Laurent, Paul 1991: "The Costs of Defence," in Croft (ed.): 88–103.

Laurenti, Jefferey 1990: "Reflections on a Nonoffensive Defence," in UNIDIR (eds.): 187–192.

Lawrence, Thomas Edward 1929: "The Lessons of Arabia," in Laqueur (ed.) 1978: 126–138.

——— 1935: *The Seven Pillars of Wisdom: A Triumph* (London: Cape).

Leaver, Richard, & Jim Richardson 1993: *Post Cold War Order* (St. Leonard's: Allen & Unwin).

Leavitt, Lloyd R., Jr., & Wallace H. Nutting 1990: "An Appraisal of Post–Cold War Military Postures for the US in NATO," in Gottfried & Bracken (eds.): 139–180.

Leavitt, Rob, & Conrad Miller 1990: "Doctrine Seminar Looks Beyond the Cold War," *DDA*, 3:1, 1–2.

Le Borgne, Claude 1991: "Guerre de huit ans, Guerre de cent heures," *Stratégique*, 51/52: 273–282.

Lebow, Richard Ned 1981: *Between Peace and War: The Nature of International Crisis* (Baltimore: Johns Hopkins U.P.).

——— 1983: "The Deterrence Deadlock: Is There a Way Out?" *Political Psychology*, 4:2, 333–354.

——— 1984: "Windows of Opportunity: Do States Jump Through Them?" *IS*, 9:1 (Summer) 147–186.

——— 1985: "Deterrence Reconsidered: The Challenge of Recent Research," *Survival*, January–February, 20–28.

——— 1985a: "The Soviet Offensive in Europe: The Schlieffen Plan Revisited?" in Lynn-Jones, Miller, & Evera (eds.) 1989: 312–346.

———— 1989: "Deterrence: A Political and Psychological Critique," in Stern & al. (eds.): 25–51.

Lech, Ann Mosley, & Mark Tessler (eds.) 1989: *Israel, Egypt and the Palestinians: From Camp David to Intifada* (Bloomington: Indiana U.P.).

Lederman, Itshak 1991: "Verification of Conventional Arms Control as a Stabilizing Tool of a New Security System in Europe," *BPP*, 22:3 (September): 291–302.

Lee, C. M. 1991: "The Future of Arms Control in the Korean Peninsula," *WQ*, 14:3 (Summer): 181–197.

Lee, John Marshall 1988: "No First Use of Nuclear Weapons," in Hopmann & Barnaby (eds.): 73–85.

Lee, Kent D. 1990: "Strategy and History: The Soviet Approach to Military History and Its Implications for Military Strategy," *JSMS*, 3:3, 409–445.

Leebaert, Derek (ed.) 1981: *Soviet Military Thinking* (London: Allen & Unwin).

———— & Timothy Dickinson (eds.) 1992: *Soviet Strategy and New Military Thinking* (New York: Cambridge U.P.).

Leech, John 1991: *Halt! Who Goes Where? The Future of NATO in the New Europe* (London: Brassey's).

Lefebvre, Stéphane 1992: *Military Doctrines in Flux: Defensive Defence: Bibliography and Compendium of Resources* (Ottawa: Department of National Defence, Canada).

Legwold, Robert 1989: "The Revolution in Soviet Foreign Policy," *FA*, 18:1, 82–98.

Leimbacher, Urs 1990: "Die Diskussion um die Abschaffung der Schweizer Armee," in Heisenberg & Lutz (eds.): II:260–272.

Leitenberg, Milton 1986: "Soviet Submarine Operations in Swedish Waters, 1980–1986," *Washington Papers*, 128.

Leites, Nathan 1981: "The Soviet Style of War," in Leebaert (ed.): 185–224.

Leitzel, Jim (ed.) 1993: *Economics and National Security*, Pew Studies in Economics and Security (Boulder, Colo.: Westview).

Lellouche, Pierre 1985: *L'Avenir de la Guerre* (Paris: Mazarine).

———— 1992: *Le nouveau monde: De l'ordre de Yalta au désorde des nations* (Paris: Grasset).

Lemaitre, Pierre 1989: "A Hungarian Concept of Security," in Waever & al. (eds.): 154–167.

———— 1990: "Hungarian Concepts and Policies of Security, 1945–1987," *WP-CPCR*, 20.

Lemke, Hans-Dieter 1989: "Militärische Stabilität durch Überlegenheit der Verteidigung," *SWP-AZ*, 2599 Fo.Pl. II 3c/89 (Ebenhausen: Stiftung Wissenschaft und Politik).

———— 1990: "Conventional Arms Control and Defence Planning," in Huber (ed.): 107–114.

———— 1990a: "Prospects for Conventional Arms Control in Europe," *Disarmament*, 13:1, 39–57.

———— 1991: "Models of Defence: Aspects of the West German Debate," in Reid & Dewar (eds.): 131–142.

———— 1992: "Zur Bedeutung militärischer Kräfteverhältnisse im künftigen Europa," in Heydrich & al. (eds.): 631–643.

Leonard, James, Martin Kaplan, & Benjamin Sanders 1993: "Verification and Enforcement in a NWFW," in Rotblat & al. (eds.): 132–144.

Lepingwell, John W. R. 1987: "The Laws of Combat? Lanchester Reexamined," *IS*, 12:1 (Summer): 89–139.

———— 1992: "Soviet Civil-Military Relations and the August Coup," *WP*, 44:4 (July): 539–572.
LeShan, Lawrence 1992: *The Psychology of War: Comprehending Its Mystique and Its Madness* (Chicago: Noble).
Leski, Christian 1988: "Strukturelle Angriffsunfähigkeit: Zur wissenschaftlichen Diskussion eines politischen Begriffes," *S+F*, 4: 253–259.
Levin, N. Gordon, Jr. 1972: *Woodrow Wilson and the Paris Conference* (Lexington, Mass.: Heath).
Levin, Norman D. 1993: "Prospects for U.S.-Japanese Security Cooperation," in Unger (ed.): 71–84.
Levine, Robert A. 1990: "The NATO Debate," in Schmähling (ed.): 51–62.
———— 1990a: *Still the Arms Debate: A RAND Corporation Research Study* (Aldershot: Dartmouth).
———— (ed.) 1992: *Transition and Turmoil in the Atlantic Alliance* (New York: Crane Russak).
Levite, Ariel 1990: *Offense and Defense in Israeli Military Doctrine* (Jerusalem: *Jerusalem Post;* Boulder, Colo.: Westview).
———— 1992: "Ballistic Missile Proliferation in the Middle East: Causes and Consequences," in Neuneck & Ischebeck (eds.): 221–227.
————, Bruce W. Jentleson, & Larry Berman (eds.) 1992: *Foreign Military Intervention: The Dynamics of Protracted Conflict* (New York: Columbia U.P.).
Levy, Jack S. 1984: "The Offensive/Defensive Balance of Military Technology: A Theoretical and Historical Analysis," *International Studies Quarterly* 28: 167–214.
———— 1987: "Declining Power and the Preventive Motivation for War," *WP*, 40:1 (October): 82–107.
———— 1990: "Preferences, Constraints, and Choices in July 1914," *IS*, 15:3, 151–186.
Levy, "Yank" (with Tom Wintringham) 1941: *Guerilla Warfare* (Harmondsworth: Penguin).
Lewis, Flora 1990: "Bringing in the East," *FA*, 69:4, 15–26.
Libiszewski, Stefano 1989: "Zum Verhältnis von Militärpolitik und europäischer Sicherheitsordnung: Überlegungen zu den politischen Aspekten alternartiver Verteidigungskonzepte," in Karl & Auffermann (eds.): 106–110.
Lichal, Robert 1989: "Staat und Sicherheit," *ÖMZ*, 6: 453–462.
Liddell Hart, Basil 1921: "The 'Man-in-the-Dark' Theory of Infantry Tactics and the 'Expanding Torrent' System of Attack," *Journal of the Royal United Services Institution,* 66 (February): 1–22.
———— 1932: "Aggression and the Problem of Weapons," *English Review,* 55 (July): 74–78.
———— 1934: *T. E. Lawrence in Arabia and After* (London: Cape).
———— 1937: *Europe in Arms* (London: Faber & Faber).
———— 1939: *The Defence of Britain* (London: Faber & Faber).
———— 1951: *The Other Side of the Hill: Germany's Generals, Their Rise and Fall, With Their Own Account of Military Events, 1939–1945,* rev. and enlarged ed. (London: Cassell).
———— 1960: *Deterrent or Defence* (London: Stevens).
———— 1967: "Lessons from Resistance Movements: Guerilla and Non-violent," in Roberts (ed.): 195–214.
———— 1967a: *Strategy: The Indirect Approach,* 2d, rev. ed. (New York: Signet, 1974).

Lider, Julian 1983: "Die sowjetische Militärwissenschaft: Beschreibung und kritische Bestandaufnahme," *ÖMZ*, 2: 143–153.

Life and Peace Institute: *Peace Policy for the 1980s* (Uppsala: Life and Peace Institute).

Lifton, Robert Jay, & Richard Falk 1982: *Indefensible Weapons: The Political and Psychological Case Against Nuclearism* (New York: Basic Books).

Liko, Karl 1992: "Das Wiener Dokument 1992: Vertrauens- und sicherheitsbildende Massnahmen für ein Europa im Wandel," *ÖMZ*, 30:4 (July–August): 293–299.

Lilly-Webber, Steve, & Klaus D. Remme 1990: "Bridging the Aircraft Gap," *Briefing Paper*, 2 (Brookline, Mass.: Institute for Defense and Disarmament Studies).

Lilow, Alexander 1989: "Europa vor der Wahl einer neuen Alternative: Ein bulgarischer Standpunkt," *Berichte*, 38. Also in Lutz & Schmähling (eds.) 1990: 182–202.

Lindberg, Steve 1987: "Submarine Incursions for Political Consumption," *Nordic Journal of Soviet & East European Studies*, 4(1): 61–77

Lindén, Johan 1991: *JAS 39 Gripen: Den havarerade stormaktsdrömmen* (Stockholm: Broderskaps Förlag).

Lindner, Philip R. 1984: "Considerations of a Conventional Defence of Central Europe," in Golden & al. (eds.): 109–120.

Lindsey, George 1989: "Strategic Stability in the Arctic," *AP*, 241.

Linkohr, Rolf 1984: "Perspektiven einer europäischen Friedensordnung," in Brauch (ed.): 263–274.

Linnenkamp, Hilmar 1990: "Harmel II. Überlegungen zur Fortschreibung der geschäftsgrundlage der NATO," in Vogt (ed.): 71–84.

———— 1992: "The Security Policy of the New Germany," in Stares (ed.): 93–125.

———— 1993: "The North Atlantic Cooperation Council: A Stabilizing Element of a New European Order?" in Ehrhart & al. (eds.): 219–228.

Lin Piao 1965: "Long Live the Victory in the People's War," in Laqueur (ed.) 1978: 197–202.

Lindsay, Franklin A. 1962: "Unconventional Warfare," in Kissinger (ed.) 1965: 344–356.

Lippert, Ekkehard 1992: "Friedenserhaltung als Aufgabe der Bundeswehr," *EuS*, 41:11 (November): 607–612.

————, Mathias Schönborn, & Günther Wachtler 1984: "Gesellschaftliche und politische Konsequenzen alternativer Verteidigungskonzepte," in Weizsäcker (ed.): 179–198.

———— & Günther Wachtler (eds.) 1988: *Frieden: Ein Handwörterbuch* (Opladen: Westdeutscher).

Lippert, Günter 1987: "GSFG. Spearhead of the Red Army," *IDR*, 20:5 (May): 553–562.

Liska, George 1962: *Nations in Alliance: The Limits of Interdependence* (Baltimore: Johns Hopkins U.P.).

———— 1991: "Between East and West: East-Central Europe's Future Past," in Clark & Serfaty (eds.): 161–189.

Litwak, Robert S. 1990: "Soviet Policies in the Third World: Objectives, Instruments and Constraints," in Allison & Williams (eds.): 29–48.

Lobow, Wladimir Nikolajewitsch 1993: "Russland—Strategische Lage und Militärdoktrin," *ÖMZ*, 31:1 (January–February): 6–11.

Lock, Peter 1990: "Militärische Sicherheitspolitik in Deutschland—Kernproblem für die Konstruktion eines 'Gemeinsamen Europäischen Hauses'," in Calliess (ed.): 571–578.

Lockhart, Greg 1990: "Revolutionary War," in Ball & Downes (ed.): 129–145.
Lockman, Zachary, & Jopel Beinin (eds.) 1989: *Intifada: The Palestinian Uprising Against Israeli Occupation* (London: Tauris).
Lockwood, Dunbar 1993: "Nuclear Arms Control," *SIPRI:* 549–573.
Lodgaard, Sverre 1982: "Nordic Initiatives for a Nuclear Weapon-Free Zone in Europe," in *SIPRI,* 1982, 75–93.
——— 1983: "Nuclear Disengagement," in idem & Thee (eds.): 3–52.
——— 1984: "Policies of Common Security in Europe," *CRPV,* 7:4, 200–208.
——— 1985: "Policies of Common Security in Europe," in Väyrynen (ed.): 157–176.
——— 1985a: "Approaches to Nuclear Arms Restraint: Between Arms Control and Disarmament," *BPP,* 16:4, 335–348.
——— 1986: "The Building of Confidence and Security at the Negotiations in Stockholm and Vienna," in *SIPRI,* 1986, 423–446.
——— 1987: "Threats to European Security: The Main Elements," in idem & Birnbaum (eds.): 3–36.
——— 1987a: "A New Deal for Confidence and Security," in idem & Birnbaum (eds.): 177–211.
——— 1988: "Soneforslagene og de nordiske landenes Østersjø-politik," in Joenniemi & Vesa (eds.): 196–206.
——— 1989: "Confidence Building in the Baltic Region," in Westing (ed.): 99–112.
——— 1989a: "Naval Arms Control in the North—Some Specific Proposals," *BPP,* 20:2, 147–155.
——— 1989b: "Naval Forces: Arms Restraint and Confidence Building," in Borg & Smit (eds.): 207–213.
——— 1989c: "The Nuclear Arms Race and Deterrence in Europe," in Tromp (ed.): 101–118.
——— 1989d: "Europe in the Aftermath of INF: Political Order and Military Deterrence," in Nakarada & Øberg (eds.): 95–115.
——— 1991: "CSBMs: A Critical Appraisal," *PRIO Report,* 2 (Oslo: PRIO).
——— 1991a: "Vertical and Horizontal Proliferation in the Middle East/Persian Gulf," *BPP,* 22:1, 3–10.
——— 1992: "Competing Schemes for Europe: The CSCE, NATO and the European Union," *Security Dialogue,* 23:3 (September): 57–68.
——— 1992a: "Redefining Norden," in Øberg (ed.): 281–300.
——— 1993: "In Defence of International Peace and Security: New Missions for the United Nations," *Peace and the Sciences,* September (International Institute for Peace, Vienna): 18–24 .
——— & Per Berg 1982: "Long-Range Theatre Nuclear Forces in Europe," in *SIPRI,* 1982: 3–50.
——— & Per Berg 1983: "Disengagement and Nuclear Weapon-Free Zones: Raising the Nuclear Threshold," in Lodgaard & Thee (eds.): 101–114.
——— & Per Berg 1985: "Disengagement in Central Europe," in Rotblat & Hellman (eds.): 242–259.
——— & John P. Holdren 1990: "Naval Arms Control," in Lodgaard (ed.): 1–41.
——— (ed.) 1990: *Naval Arms Control* (London: Sage).
——— & Karl Birnbaum (eds.) 1987: *Overcoming Threats to Europe: A New Deal for Confidence and Security* (Oxford: Oxford U.P.).
——— & Marek Thee (eds.) 1983: *Nuclear Disengagement in Europe* (London: Taylor & Francis, SIPRI).
Lohs, Karlheinz 1986: "Pros and Cons of a Chemical Weapons-Free Zone in Europe: The Genesis of a Concept," in Rotblat & d'Ambrosio (eds.): 197–119.

Lombardy, David 1986: "Surface-to-Air Missiles: The Future Medium SAM," *NSN*, August: 64–67.

Longerich, Peter (ed.) 1990: "Was ist des Deutschen Vaterland?" *Dokumente zur Frage der deutschen Einheit, 1800–1990* (München: Piper).

Loquai, Heinz 1982: "Alternative Konzeptionen der Verteidigung, II Teil: Beurteilung aus sicherheitspolitischer und strategischer Sicht," *Truppenpraxis*, 10: 703–712.

———— 1987: "Alternative Verteidigungskonzeptionen," in Buchbender, Bühl, & Quaden (eds.): 261–269.

Lorentz, Joseph P. 1993: "Collective Security After the Cold War," in Brown & Schraub (eds.): 211–238.

Löser, Jochen (ed.) 1981: *Weder rot noch tot: Überleben ohne Atomkrieg: Eine Sicherheitspolitische Alternative* (München: Olzog).

———— 1985: "Defensive Optionen der NATO in den neunziger Jahren," *ÖMZ*, 6: 525–532.

———— 1985a: "Defensive Optionen Mitteleuropas: Ein kollektives Sicherheitssystem der kleinen Schritte," in Lutz (ed.): 294–309.

———— 1986: "Gedanken zur Weiterentwicklung von Strategie, Operation und Taktik," *Allgemeine Schweizerische Militärzeitschrift*, 9: 529–532.

———— 1986a: "Modern Defence Technologies for Non-Nuclear Border Defence" in Tromp (ed.): 195–200.

———— 1987: "Europäische Friedenszone, Gemeinsame Sicherheit und Defensive Optionen," *Dialog*, 10: 111–114.

———— 1987a: "Strukturelle Nichtangriffsfähigkeit," *S+F*, 1: 51–60.

———— 1988: "Diskussionsbeiträge zur konventionellen Stabilität," in Bahr & Lutz (eds.): 457–460.

———— 1990: "Gegenseitige Sicherheit durch defensive Stabilität in Mitteleuropa," in Weizsäcker (ed.): 133–142.

———— & al. 1982: *Gegen den dritten Weltkrieg: Strategie der Freien* (Herford: Mittler).

———— & Harald Anderson 1984: *Antwort auf Genf: Sicherheit für West und Ost* (München: Olzog).

———— & Lothar Penz 1988: "Verstärkung der nicht-atomaren Verteidigung, aber wie?" *Allgemeine Schweitzerische Militärzeitschrift*, 1: 10–14.

———— & Ulrike Schilling 1984: *Neutralität für Mitteleuropa: Das Ende der Blöcke* (München: Bertelsmann).

Loth, Wilfried 1989: *Ost-West-Konflikt und deutsche Frage* (München: dtv).

———— 1990: "Welche Einheit soll es sein? Beobachtungen zur Lage der Nation," *Blätter*, 3: 301–309.

Löwe, Bernd P. 1990: "Deutschland in Europa-zwischen Utopie und Pragmatismus," *S+F*, 8:2, 65–69.

Lowe, Karl, & Thomas Durell-Young 1991: "Multinational Corps in NATO," *Survival*, 33:1 (January–February): 66–77.

Luard, Evan 1987: "Western Europe and the Reagan Doctrine," *IA*, 4: 563–574.

Lübkemeier, Eckhard 1988: "Abschreckung und Gemeinsame Sicherheit," *NGFH*, 1: 82–85.

———— 1988a: "Current NATO Strategy and No First Use: What Can They Accomplish?" in Hopmann & Barnaby (eds.): 115–131.

———— 1990: "NATO's Identity Crisis," *BAS*, 46:8, 30–33.

———— 1990a: "Sicherheit jenseits der Blockkonfrontation: Die strategischen Inplikationen der Revolutionen in Osteuropa," *KAAD*, 40.

———— 1990b: "Europas Sicherheit nach dem Kalten Krieg," *Reihe Eurokolleg*, 5 (Bonn: Friedrich-Ebert-Stiftung).

——— 1990c: "Security and Peace in Post–Cold War Europe," in Clesse & Rühl (eds.): 183–201.

——— 1991: "The Political Upheaval in Europe and the Reform of NATO Strategy," *NATO Review,* 39:3 (June): 17–21.

——— 1993: "The United Germany in the Post-Bipolar World," *Studien,* 56.

——— 1993a: "Konzeptionelle Überlegungen zur militärischen und politischen Stabilität in Europa," in Forndran & Pohlman (eds.): 115–138.

Lucas, Michael 1988: "The United States and Post-INF Europe: The Need for a Pan-European Policy," *WPJ,* 5:2, 183–234.

——— 1990: *The Western Alliance After INF: Redefining U.S. Policy Toward Europe and the Soviet Union* (Boulder, Colo.: Lynne Rienner).

——— (ed.) 1993: *The CSCE in the 1990s: Constructing European Security and Cooperation* (Baden-Baden: Nomos).

Ludlow, Peter 1991: "The German-German Negotiations and the 'Two-Plus-Four' Talks," in Heisenberg (ed.): 15–27.

Luft, Christa 1991: "The Network of FRG-GDR Agreements and the Future of Europe," in Rotfeld & Stützle (eds.): 9–19.

Luif, Paul 1986: "Neutralität und Frieden: Grundlegende Bemerkungen zu Geschichte und Gegenwart," *Dialog,* 6: 17–76.

Lukic, Reneo 1984: "La dissuassion populaire yougoslave," *CES,* 5.

Lummer, Heinrich 1988: "Zwei Friedensverträge mit Deutschland? Zu einem Vorschlag von Egon Bahr," *DA,* 21:4, 370–374.

Lumsden, Malvern 1978: *Anti-personnel Weapons* (London: Taylor & Francis, SIPRI).

Lunn, Simon 1991: "The Military Balance, Arms Control, and the Central Region," in Reid & Dewar (eds.): 79–92.

Luschev, Pyotr 1989: "The Soviet Military Doctrine: Significance, Aims and Realization," *NSN,* December, 26–28.

Lütgemeier-Davin, Reinhold 1990: "'Wiederwehrhaftmachung' oder 'Abrüstung': Die militärische Sicherheitspolitik der Weimarer Republik im Licht pazifistischer Öffentlichkeit," in Steinweg (ed.): 186–231.

Lütken, Carsten A. 1990: "Confidence and Security Building—A Naval Perspective," in Lodgaard (ed.): 138–145.

Luttwak, Edward N. 1976: *The Grand Strategy of the Roman Empire: From the First Century A.D. to the Third* (Baltimore: Johns Hopkins U.P.).

——— 1980: "The Operational Level of War" (*IS,* 5:3) in Miller (ed.) 1986: 211–229.

——— 1983: *The Grand Strategy of the Soviet Union* (New York: St. Martin's).

——— 1984: *The Pentagon and the Art of War* (New York: Simon & Schuster).

——— 1986: *On the Meaning of Victory: Essays on Strategy* (New York: Simon & Schuster).

——— 1987: *Strategy: The Logic of War and Peace* (Cambridge: Harvard U.P.).

——— 1991: "The Traditional Approaches to Peace," in Thompson & Jensen (eds.): 3–12.

Lutz, Dieter S. 1981: *Weltkrieg wider Willen: Eine Kräftevergleichsanalyse der Nuklearwaffen in und für Europa* (Reinbek: Rowohlt).

——— 1982: "Neutralität: (K)eine sicherheitspolitische Alternative für die Bundesrepublik Deutschland?" in idem & Grosse-Jütte (eds.): 742.

——— 1982a: "Stabilität durch Vorhersehbarkeit und Vertrauensbildung: Kriegsbilder, unkalkulierbare Realitat und Worst-Case Orientierung," in idem & Müller (eds.): 23–52.

——— 1983: "Auf dem Weg zu einer Neuen Europäischen Friedensordnung.

Programmatische Überlegungen zur Utopie und Machbarkeit eines Systems Kollektiver Sicherheit in und für Europa," in DGFK (eds.): 505–532.

———— 1984: "Eröffnungsreferat zur Anhörung 'Alternative Strategien'," *S+F*, 3: 65–68.

———— 1984a: "Arms Control and Arms Renunciation: A Critical Review and Proposals," *Disarmament*, 7:3, 103–108.

———— 1985: "Auf dem Weg zu einem System Kollektiver Sicherheit in und für Europa," in idem (ed.): 22–44.

———— 1985a: "Für eine neue Friedens- und Sicherheitspolitik der SPD," *NGFH*, 9: 836–842.

———— 1986: "Common Security: The New Concept: Distinctive Features and Structural Elements of Common Security Compared with Other Security Policy Methods and Strategies," *HB*, 8.

———— 1986a: "Die SPD auf der Suche nach einer neuen friedens- und sicherheitspolitischen Konzeption, oder: Was ist Gemeinsame Sicherheit?" *S+F*, 2: 109–114.

———— 1986b: "Gemeinsame Sicherheit: Das neue Konzept: Definitionsmerkmale und Strukturelemente im Vergleich mit anderen sicherheitspolitischen Modellen und Strategien," in Bahr & Lutz (eds.): 45–82.

———— 1986c: "Ist eine rot-grüne Sicherheitspolitik möglich?" *S+F*, 4: 228–231.

———— 1986d: "Security Partnership and/or Common Security," *HB*, 8.

———— 1986e: "Seemacht und Marinedoktrin: Eine Einführung in die theoretische Grundlagen der Rüstung zur See," in Lutz, Müller, & Pott: 7–54.

———— 1986f: "Sicherheitspartnerschaft und/oder Gemeinsame Sicherheit? Zur Entstehung und Entwicklung der Begriffe und ihrer Inhalte," in Bahr & Lutz (eds.): 29–42.

———— 1987: "Der Status Quo in Westeuropa: Die NATO," in Heisenberg & Lutz (eds.): 251–259.

———— 1987a: "Politische und militärische Modelle der Sicherheit," in Heisenberg & Lutz (eds.): 561–591.

———— 1987b: "Stabilität und Abrüstung durch Umrüstung: Zur Abrüstungs- und Stabilitätsorientierung Struktureller Nichtangriffsfähigkeit im Rahmen Gemeinsamer Sicherheit," *S+F*, 3: 167–169.

———— 1987c: "Was heisst 'Strukturelle Nichtangriffsfähigkeit' (i. e. S.)? Versuch einer Annäherung," *S+F*, 1: 2–3.

———— 1987d: "Zu den verfassungsrechtlichen Rahmenbedingungen Gemeinsamer Sicherheit nach dem Grundgesetz der Bundesrepublik Deutschland," in Bahr & Lutz (eds.): 85–104.

———— 1987e: "Towards a New European Peace Order (NEPO)," in Kaplan (ed.): 124–151.

———— 1988: "On the Theory of Structural Inability to Launch an Attack," *HB*, 25.

———— 1988a: "SIA and Defensive Zones," *BPP*, 20:1, 71–80.

———— 1988b: "Teilelement einer politischen Konzeption: Funktionen, Zielk.iterien, konzeptioneller Rahmen der strukturellen Nichtangriffsfähigkeit," *NGFH*, 6: 547–553.

———— 1988c: "Vom atomwaffenfreien Korridor zur panzerfreien Zone," in Bahr & Lutz (eds.): 493–500.

———— 1988d: *Strukturelle Angriffsunfähigkeit im Rahmen Defensiver Abhaltung und einer Konzeption Gemeinsamen Friedens: Disussionsvorschläge, verabschiedet vom August-Bebel-Kreis* (Bonn: SPD).

———— 1988e: "Zur Genesis Struktureller Nichtangriffsfähigkeit," in Lutz & Bahr (eds.): 26–35.

———— 1989: *Gemeinsame Sicherheit: Defensive Abhaltung und strukturelle*

Angriffsunfähigkeit, Bd. IV: *Zur Genesis und Theorie Struktureller Angriffsunfähigkeit im Rahmen einer Strategie Defensiver Abhaltung und ihrer konzeptionellen Einbettung in die Gemeinsame Sicherheit* (Baden-Baden: Nomos).

———— 1989a: "Basic Law. Security and Peace: Armament and Disarmament," *HB*, 38.

———— 1989b: "Gemeinsamer Frieden und Europäische Sicherheit," in Krohn, Axel, & Lutz, "Europäische Sicherheit II," *HB*, 40: 1–39.

———— 1989c: "Luftstreitkräfte, Strukturelle Angriffsunfähigkeit und Abrüstung," *HB*, 41.

———— 1989d: "SIA and Defensive Zones," *BPP*, 20:1, 71–80.

———— 1989e: "'Alles was fliegt, muss weg'? Luftstreitkräfte, Strukturelle Angriffsunfähigkeit und Abrüstung," *S+F*, 7:2, 145–149. Also in E. Müller & Neuneck (eds.) 1990: 81–92.

———— 1989f: "In Zukunft keine nationale Sicherheit mehr," in Gross & al. (eds.): 287–293.

———— 1990: "Deutsche Einheit—Europäische Sicherheit oder Brauchen wir noch (deutsche) Streitkräfte?" *HB*, 43.

———— 1990a: "Die Golfkriese, das Grundgesetz, die Gemeinsame Sicherheit," *S+F*, 8:4, 233–237.

———— 1990b: "Ein Europa des Gemeinsamen Friedens mit der Sowjetunion?" in Lutz & Schmähling (eds.): 13–22.

———— 1990c: "Gemeinsamer Frieden: Zur Bewältigung grenzüberschreitender Gefahren durch ein System Kollektiver Sicherheit in und für Europa," in Vogt (ed.): 41–51.

———— 1990d: "Strukturelle Angriffsunfähigkeit und Abrüstung: Das Beispiel 'Luftstreitkräfte'," in Lutz & Schmähling (eds.): 253–268.

———— 1990e: "Towards a European Peace Order and a System of Collective Security," *BPP*, 21:1, 71–76.

———— 1990f: "Wieviel (deutsche) Soldaten braucht Europa?" *S+F*, 1: 18–22.

———— 1990g: *Brauchen wir nich deutsche Streitkräfte: Plädoyer des August-Bebel-Kreises für ein System Kollektiver Sicherheit in und für Europa* (Bonn: August-Bebel-Kreis).

———— 1991: "Deutsche Einheit—Europäische Sicherheit," in idem (ed.): 49–90.

———— 1991a: "Friedensforschung und Völkerrecht: Von einem Recht der Staaten zu einem Recht der Völker und Menschen," in idem (ed.), "Völkerrecht und Friedensordnung. Diskussionsbeiträge," *HB*, 59: 7–12.

———— 1991b: *Sicherheit 2000: Gemeinsame Sicherheit im Übergang vom Abschreckungsregime zu einbem System Kollektiver Sicherheit in und für Europa* (Baden-Baden: Nomos).

———— 1992: "Völkerrecht und Friedensordnung aus der Sicht des Grundgesetzes," *S+F*, 10:1, 32–36.

———— 1992a: "Egon Bahr—'Vater' der Gemeinsamen Sicherheit," in idem (ed.): 165–182.

———— 1992b: "Europe on the Way to a Regional System of Collective Security," *Disarmament*, 15:4, 13–26.

———— 1993: "Krieg als ultima ratio? Zum Einsatz der Bundeswehr ausserhalb des Territoriums der Bundesrepublik Deutschland," *HB*, 72.

———— 1993a: *Deutschland und die Kollektive Sicherheit: Politische, rechtliche und programmatische Aspekte* (Leverkusen: Leske & Budreich).

———— 1993b: "Das Grundgesetz fordert Friedenspolitik: Zum Streit um unsere sicherheitspolitische Zukunft," in idem (ed.): 27–52.

———— (ed.) 1985: *Kollektive Sicherheit in und für Europa: Eine Alternative?*

Beiträge zur Utopie und Umsetzung einer neuen Friedens- und Sicherheitsprogrammatik: Pro und Contra (Baden-Baden: Nomos).

—— (ed.) 1985a: *Im Dienst fur Frieden und Sicherheit: Festschrift für Wolf Graf von Baudissin* (Baden-Baden: Nomos).

—— (ed.) 1987: *Lexikon Rüstung, Frieden, Sicherheit* (München: Beck).

—— (ed.) 1991: *Gemeinsame Sicherheit, Kollektive Sicherheit, Gemeinsamer Frieden*, Bd. VI: *Auf dem Weg zu einer Neuen Europäischen Friedensordnung* (Baden-Baden: Nomos).

—— (ed.) 1992: *Das Undenkbare denken: Festschrift für Egon Bahr zum siebzigsten Geburtstag* (Baden-Baden: Nomos).

—— (ed.) 1993: *Deutsche Soldaten weltweit?* (Reinbek: Rowohlt).

—— & Alexander Theilmann 1986: "Gemeinsame Sicherheit: Auf der Suche nach Inhalten: Auf dem Weg zur Akzeptanz? Zu den Kritiken und Reaktionen in der Bundesrepublik und in der DDR," in Bahr & Lutz (eds.): 223–251.

——, Erwin Müller, & Andreas Pott 1986: *Seemacht und Sicherheit: Beiträge zur Diskussion maritimer Rüstung und Rüstungskontrolle* (Baden-Baden: Nomos).

—— & Annemarie Grosse-Jütte (eds.) 1982: *Neutralität: Eine Alternative?* (Baden-Baden: Nomos).

—— & Erwin Müller (eds.) 1982: *Vertrauensbildende Massnahmen: Zur Theorie und Praxis einer sicherheitspolitischen Strategie* (Baden-Baden: Nomos).

—— & Elmar Schmähling (eds.) 1990: *Gemeinsame Sicherheit: Internationale Diskussion*, Bd. V: *Beiträge und Dokumente aus Ost und West* (Baden-Baden: Nomos).

Lutz, Ernst 1990: "Soziale Verteidigung: Zur Konzeption und Problematik eines alternativen Modells," in Heisenberg & Lutz (eds.): II:225–233.

Lynch, Allen (ed.): *Building Security in Europe: Confidence-Building Measures and the CSCE* (New York: Institute for East-West Security Studies).

—— 1992: *The Cold War Is Over—Again* (Boulder, Colo.: Westview).

Lynch, Thomas F., III 1990: "The Military and Alternative Security: New 'Missions' for Stable Conventional Security," in Weston (ed.): 7–42.

Lynn, John A. 1984: *The Bayonets of the Republic: Motivation and Tactics in the Army of Revolutionary France, 1791–94* (Urbana: University of Illinois Press).

Lynn-Jones, Sean 1988: "The Incidents at Sea Agreement," in George, Farley, & Dallin (eds.): 482–509.

—— 1990: "Applying and Extending the USA-USSR Incidents at Sea Agreement," in Fieldhouse (ed.): 203–219.

—— 1992: "The Future of International Security Studies," in Ball & Horner (eds.): 71–107.

—— (ed.) 1991: *The Cold War and After: Prospects for Peace: An International Security Reader* (Cambridge: MIT).

—— & Stephen R. Rock 1988: "From Confrontation to Cooperation: Transforming the U.S.-Soviet Relationship," in Nye, Allison, & Carnesale (eds.): 111–131.

——, Steven E. Miller, & Stephen Van Evera (eds.) 1989: *Soviet Military Policy. An* International Security *Reader* (Cambridge: MIT).

——, Steven E. Miller, & Stephen Van Evera (eds.) 1990: *Nuclear Diplomacy and Crisis Management: An* International Security *Reader* (Cambridge: MIT).

MacDonald, Douglas J. 1992: "Anti-Interventionism and the Study of American Foreign Policy in the Third World," *SS*, 2:2 (Winter): 225–246.

MacFarlane, S. Neil 1993: "Russia, the West and European Security," *Survival*, 35:3 (Autumn): 3–25.

MacGregor, Douglas 1986: "Uncertain Allies? East European Forces in the Warsaw Pact," *Soviet Studies,* 38(2): 227–247.

——— 1989: *The Soviet-East German Military Alliance* (Cambridge: Cambridge U.P.).

Mack, Andrew 1984: "Arms Control, Disarmament and the Concept of Defensive Defence," *Disarmament,* 7:3, 109–119.

——— 1986: "Defence versus Offence: The Dibb Report and Its Critics," *Working Paper,* 14 (Canberra: Research School of Pacific Studies).

——— 1989: "Alternative Defence Concepts: The European Debate," *Working Paper,* 68 (Canberra: Australian National University, Peace Research Centre).

——— 1990: "Arms Control in the Pacific: The Naval Dimension," *Working Paper,* 88 (Canberra: Australian National University, Peace Research Centre).

——— 1990a: "CSBMs and Military Security," *Working Paper,* 83 (Canberra: Australian National University, Peace Research Centre).

——— 1990b: "What Is Wrong with Deterrence Theory?" *Working Papers,* 76 (Canberra: Australian National University, Peace Research Centre).

——— 1990c: "Political and Strategic Limits to 'Denuclearization' in Australia and New Zealand," in Langdon & Ross (eds.): 239–265.

——— 1990d: "A New Australian Militarism?" *Working Paper,* 89 (Canberra: Australian National University, Peace Research Centre).

——— 1990e: "The Strategy of Non-Provocative Defence: The European Debate," in Ball & Downes (ed.): 163–190.

——— 1991: "Nuclear Issues and Arms Control on the Korean Peninsula," *Working Paper,* 96 (Canberra: Australian National University, Peace Research Centre).

——— 1991a: "Missile Proliferation in the Asia/Pacific Region," in Findlay (ed.): 37–52.

——— 1991b: "Beyond MTCR: Additional Responses to the Missile Proliferation Problem," in Findlay (ed.): 123–131.

——— 1992: "Prospects for Security in the Asia-Pacific," Peace Review, 4:2 (Summer): 7–12.

——— 1993: "Arms Control at Sea," in Smith & Bergin (eds.): 83–98.

Mackay, Louis, & David Fernbach (eds.) 1983: *Nuclear-Free Defence* (London: Heretic Books).

Mackenzie, J. J. G. 1991: "The Counter-offensive," in Reid & Dewar (eds.): 161–180.

Mackinlay, John 1993: "The Requirement for a Multinational Enforcement Capability," in Weiss (ed.): 139–152.

MacLear, Michael 1981: *Vietnam: The Ten Thousand Day War* (London: Thames Methuen).

MacShane, Dennis 1981: *Soliadarity: Poland's Independent Trade Union* (London: Spokesman).

Magenheimer, Heinz 1992: "Erfolge und Grenzen der Rüstungskontrolle," *ÖMZ,* 30:3 (May–June): 218–229.

——— 1993: "Die Streitkräfte in den Staaten Ostmitteleuropas," *ÖMZ,* 31:5 (September–October): 453–455.

——— 1994: "Zum deutsch-sowjetischen Krieg 1941: Neue Quellen und Erkenntnisse," *ÖMZ,* 32:1 (January): 51–60.

Mahan, Alfred T. 1894: *The Influence of Sea Power upon History, 1660–1783,* 5th ed. (Boston: Little Brown; reprint, New York: Dover, 1987).

Mahncke, Dieter 1988: "Are We Talking About the Right Issue?" in Hopmann & Barnaby (eds.): 183–204.

Maizière, Ulrich de 1989: "Wehrdienst mit gutem Gewissen," in Wagemann (ed.): 211–217.

Makarevsky, Vadim 1988: "Military Detente in Europe," in Fedorenko (ed.): 113–136.

——— 1989: "Non-Offensive Defence: the Tactical Aspect," in IMEMO (eds.): 235–242.

Maley, William 1989: "The Geneva Accords of April 1988," in Saikal & Maley (eds.): 12–28.

Malkov, B. 1989: "The Defensive Doctrine: Military-Technological Aspects," *SMR*, 7: 3–7.

Mallin, Maurice A. 1990: *Tanks, Fighters and Ships: U.S. Conventional Force Planning Since WWII* (Washington: Brassey's).

Mamula, Bramko 1989: "The Strategic Model of Defence of Small Countries," in Singh & Vekaric (eds.): 147–159.

Mandela, Nelson 1993: "South Africa's Future Foreign Policy," *FA*, 72:5 (November–December): 86–97.

Mandelbaum, Michael 1989: "Ending the Cold War," *FA*, 68:2 (Spring): 16–36.

Mao Tse-tung: See Mao Zedong.

Manfrass Sirjacques, Françoise 1991: "La politique de défense en Allemange," in Buffotot (ed.): 15–29.

Mao Zedong 1936: "Problems of Strategy in China's Revolutionary War," in idem 1975: I:179–254.

——— 1938: "On Protracted War," in idem 1975: II:113–194.

——— 1938a: "Problems of Strategy in Guerilla War Against Japan," in idem 1975: II:79–112.

——— 1938b: "Problems of War and Strategy," in idem 1975: II:219–235.

——— 1975: *Selected Works of Mao Zedong,* 4 vols. (Peking: Foreign Languages Press).

Maresca, John J. 1988: "Helsinki Accord, 1975," in George, Farley, & Dallin (eds.): 106–122.

Marighella, Carlos 1978: "Minimanual of the Urban Guerilla," in Laqueur (ed.): 219–228.

Marks, John 1992: "A Helsinki-Type Process for the Middle East," in Spiegel (ed.): 71–77.

Markusen, Ann 1992: "Dismantling the Cold War Economy," *WPJ*, 9:3 (Summer): 389–400.

——— & Joel Yudken 1992: *Dismantling the Cold War Economy* (New York: Basic Books).

Marolz, Josef 1987: "Neutralität—Nützlichkeit aus österreichischer Sicht," in Buchbender, Bühl, & Quaden (eds.): 105–110.

——— 1987a: "Raumverteidigung in der Praxis—das österreichische Modell," in Bühl (ed.): 185–196.

Martin, Brian 1991: "Revolutionary Social Defence," *BPP*, 22:1, 97–106.

Martin, Laurence 1979: "Limited Nuclear War," in Howard (ed.): 103–122.

——— 1989: "The Influence of Soviet Military Doctrine on Western Strategy," in Flynn (ed.): 383–410.

Martin, Matthias 1991: "Die Haltung der Sowjetunion gegenüber einem System Kollektiver Sicherheit," in Lutz (ed.) 1991: 161–190.

——— & Paul Schäfer 1993: "Militärische Dimensionen der neuen deutschen Aussenpolitik," *Blätter*, 38:10 (October): 1185–1199.

Martinez-Pujalte, Antonio-Luis 1991: "Three Essential Principles of Non-Offensive Defence, and Some Timely Considerations," in Møller (ed.): 9–29.

Maruh, Jürgen, & Manfred Wilke (eds.) 1984: *Wohin treibt die SPD? Wende oder Kontinuität sozialdemokratischer Sicherheitspolitik* (München: Olzog).

Mason, R. A. 1989: "Airpower in Conventional Arms Control," *Survival,* September/October, 397–414.

Mason, Tony 1987: "The Air Defence of Europe: Ways Forward," in Windass & Grove (eds.) 1987: I (no pagination).

———— 1988: "Airpower Related Confidence Building Measures," in Windass & Grove (eds.): no pagination.

Mastny, Vojtech 1992: *The Helsinki Process and Reintegration of Europe: Analysis and Documentation, 1986–1990* (London: Pinter).

Matthews, Lloyd J., & Dale E. Brown (eds.) 1989: *The Parameters of Military Ethics* (Washington, D.C.: Pergamon-Brassey's).

Matthies, Volker 1993: "Die Schrecken des Krieges und die (Ohn)macht der internationalen Gemeinschaft," in idem (ed.): 7–23.

———— (ed.) 1993: *Frieden durch Einmischung? Die Schrecken des Krieges und die (Ohn)Macht der internationalen Gemeinschaft* (Bonn: Dietz).

May, Michael, George F. Bing, & John D. Steinbruner 1988: "Strategic Arsenals After Start: The Implications of Deep Cuts," *IS,* 13:1, 90–133.

————, George F. Bing, & John D. Steinbrumer 1989: "Deep Cuts and Strategic Targeting: A Clarification," *IS,* 14:3, 194–197.

May, Michael M., & John R. Harvey 1987: "Nuclear Operations and Arms Control," in Carter, Steinbruner, & Zraket (eds.): 704–735.

Mayer, Walter 1992: "Spannocchi und die Raumverteidigung," *ÖMZ,* 30:6 (November–December): 481–485.

Mazarr, Michael J. 1990: *Light Forces and the Future of U.S. Military Strategy* (Washington: Brassey's).

———— 1993: "The Military Dilemmas of Humanitarian Intervention," *Security Dialogue,* 24:2 (June): 151–162.

Mazing, Valeri 1989: "On the Evolution of Soviet Military Doctrine," in Borg & Smit (eds.): 115–122.

McAdams, A. James 1986: "Inter-German Détente: A New Balance," *FA,* Fall, 136–153.

———— 1990: "An Obituary for the Berlin Wall," *WPJ,* 7:2, 357–375.

McCall, Malcolm, & Oliver Ramsbotham (eds.) 1990: *Just Deterrence: Morality and Deterrence into the Twenty-First Century* (London: Brassey's).

McCausland, Jeff 1986: "The East German Army: Spear Point or Weakness?" *Defense Analysis,* 2: 137–153.

———— 1993: "The Gulf Conflict: A Military Analysis," *AP,* 282.

MccGwire, Michael 1979: "Naval Power and Soviet Global Strategy" in Miller & Evera (eds.) 1988: 115–170.

———— 1981: "Soviet Naval Doctrine and Strategy," in Leebaert (ed.): 125–184.

———— 1984: "The Dilemmas and Delusions of Deterrence," in Prins (ed.): 75–97.

———— 1986: "Deterrence: The Problem, Not the Solution," *IA,* 1: 55–70.

———— 1987: "Update: Soviet Military Objectives," *WPJ,* Fall, 723–731.

———— 1987a: "Why the Soviets Want Arms Control," *Technology Review,* 90:2, 37–45.

———— 1987b: *Military Objectives in Soviet Foreign Policy* (Washington, D.C.: Brookings).

———— 1988: "A Mutual Security Regime for Europe," *IS,* 12:3, 361–379.

———— 1988a: "New Directions in Soviet Arms-Control Policy," *WQ,* 11:3, 185–200.

————— 1988b: "Rethinking War: The Soviets and European Security," *Brookings Review,* Spring, 3–12.

————— 1989: "The New Challenge of Europe," in Clesse & Schelling (eds.): 283–303.

————— 1991: *Perestroika and Soviet National Security* (Washington, D.C.: Brookings).

————— 1991a: "Evolving Soviet Political Doctrine," in Hemsley (ed.): 13–30.

McCormick, Kip 1987: "The Evolution of Soviet Military Doctrine. Afghanistan," *MR,* July 1987, 61–72.

McCoubrey, H., & N. D. White 1992: *International Law and Armed Conflict* (Aldershot: Dartmouth).

McCoy, James 1990: "Naval Arms Control-Ante Up," *Proc,* 116:9 (September): 34–39.

McDowall, David 1989: *Palestine and Israel: The Uprising and Beyond* (London: Tauris).

McFarland, Stephen L. 1986: "Preparing for What Never Came: Chemical and Biological Warfare in World War II," *Defense Analysis,* 2:2, 107–121.

McInnes, Colin 1991: "NATO Strategy and Conventional Defence," in Booth (ed.): 165–188.

————— 1991a: "Conventional Forces," in Croft (ed.): 29–46.

————— 1991b: "Trident," in Byrd (ed.): 67–87.

————— 1993: "The Military Security Agenda," in Rees (ed.): 71–86.

McIntosh, M. 1986: *Japan Rearmed* (New York: St. Martin's).

McLean, Scilla (ed.) 1986: *How Nuclear Weapons Decisions Are Made* (London: Macmillan).

McMillan, Stuart 1987: *Neither Confirm Nor Deny: The Nuclear Ships Dispute between New Zealand and the United States* (New York: Praeger).

McNamara, Robert S. 1986: *Blundering into Disaster: Surviving the First Century of the Nuclear Age* (New York: Pantheon).

————— 1989: *Out of the Cold: New American Foreign and Defense Policy in the 21st Century* (New York: Simon & Schuster).

—————, George F. Kennan, McGeorge Bundy, & Gerard Smith 1985: "The President's Dream: Star Wars or Arms Control," in Bundy (ed.): 277–291.

McNeill, Terry P. 1991: "European Security," in Kirby & Hooper (eds.): 11–33.

McNeill, William H. 1982: *The Pursuit of Power: Technology, Armed Force, and Society since A.D. 1000* (Oxford: Blackwell).

McSweeney, Bill 1990: "Constructive Ambiguity: Options for Neutrals in the 1990s," in Randle & Rogers (eds.): 147–158.

Mead, Walter Russell 1990: "The Once and Future Reich," *WPJ,* 7:4 (Winter): 593–638.

————— 1993: "An American Grand Strategy. The Quest for Order in a Disordered World," *WPJ,* 10:1 (Spring): 9–38.

Mearsheimer, John J. 1979: "Precision-Guided Munitions and Conventional Deterrence," *Survival,* March–April, 68–77.

————— 1981: "Maneuver, Mobile Defense, and the NATO Central Front" (*IS,* 6:3) in Miller (ed.) 1986: 231–249.

————— 1982: "Why the Soviets Can't Win Quickly in Central Europe" (*IS,* 7:1) in Miller (ed.) 1986: 121–157.

————— 1983: *Conventional Deterrence* (Ithaca: Cornell U.P.).

————— 1983a: "The Military Reform Movement: A Critical Assessment," *Orbis,* Summer: 285–300.

————— 1986: "A Strategic Misstep: The Maritime Strategy and Deterrence in Europe" (*IS,* 11:2) in Miller & Evera (eds.) 1988: 47–101.

———— 1986a: "Prospects for Conventional Deterrence in Europe," in Ackland & McGuire (eds.): 335–344.

———— 1986b: "Nuclear Weapons and Deterrence in Europe," in Tromp (ed.): 3–30.

———— 1988: "Numbers, Strategy, and the European Balance," *IS*, 12:4 (Spring): 174–185.

———— 1988a: *Liddell Hart and the Weight of History* (London: Brassey's).

———— 1989: "Assessing the Conventional Balance: The 3:1 Rule and Its Critics," *IS*, 13:4 (Spring): 54–89.

———— 1989a: "Nuclear Weapons and Deterrence in Europe," in Tromp (ed.): 71–100.

———— 1990: "Back to the Future: Instability in Europe After the Cold War," *IS*, 15:1 (Summer): 5–52.

———— 1993: "The Case for a Ukrainian Nuclear Deterrent," *FA*, 72:3 (Summer): 50–66.

————, Barry R. Posen, & Eliot A. Cohen 1989: "Correspondence: Reassessing Net Assessment," *IS*, 13:4 (Spring): 128–179.

Mechtersheimer, Alfred 1984: *Rüstung und Frieden: Argumente für eine neue Sicherheitspolitik*, 2d ed. (Reinbek: Rowohlt).

———— 1986: "The 1987 Peace Plan: 13 Demands for a New Peace and Security Policy," in Størner & al. (eds.): 213–230.

———— 1987: "Der Stellenwert alternativer Verteidigungsstrategien fur das Ost-West-Verhaltnis," in Bühl (ed.): 149–154.

———— & Peter Barth 1986: *Militarisierungsatlas der Bundesrepublik: Streitkräfte, Waffen und Standorte: Kosten und Risiken* (Darmstadt: Sammlung Luchterhand).

———— & Peter Barth (eds.) 1983: *Den Atomkrieg führbar und gewinnbar machen?* (Reinbek: Rowohlt).

Mehl, Regine 1993: "Auf dem Prüfstand: Die KSZE zwischen Zivillokig und Militärlogik," in Birkenback & al. (eds.): 61–71.

Meisner, Boris 1991: "Die Wechselbeziehungen zwischen der Innen- und Aussenpolitik Gorbatschows," in Brahm & al.: 87–114.

Mellon, Christian 1984: "Peace Organisations in France Today," in Howorth & Chilton (eds.): 202–216.

————, Jean-Marie Muller & Jaques Semelin 1985: *La Dissuasion Civile* (Paris: Fondation pour les Études de défense nationale).

Melman, Seymour 1970: *Pentagon Capitalism* (New York: McGraw-Hill).

———— 1985: *The Permanent War Economy* (New York: Simon & Schuster).

———— 1987: *Conversion from Military to Civilian Economy: An Economic Alternative* (Washington, D.C.: SANE).

———— 1988: *The Demilitarized Society: Disarmament and Conversion* (London: Spokesman).

Melville, Andrei Y. 1988: "Nuclear Revolution and the New Way of Thinking," in Gromyko & Hellman (eds.): 176–184.

Mendl, Wolf 1984: *Western Europe and Japan Between the Superpowers* (London: Croom Helm).

Mendlowitz, Saul H. (ed.) 1975: *On the Creation of a Just World Order* (New York: Free Press).

———— & R. B. J. Walker (eds.) 1987: *Towards a Just World Peace* (London: Butterworths).

Menning, Bruce W. 1992: "Redefined Soviet Military Doctrine in Perspective: Military-Political Science, Forecasting, and Offensive-Defensive Paradigms," in Brauch & Kennedy (eds.): 83–99.

Menon, Anand, Anthony Forster, & William Wallace 1992: "A Common European Defence?" *Survival,* 34:3 (Autumn): 98–118.

Merkl, Peter H. 1993: *German Unification in the European Context* (University Park: Pennsylvania State U.P.).

Mertes, Alois 1985: "Common Security and Defensive Security," in Väyrynen (ed.): 187–192.

Messenger, Charles 1987: "Counter-Mobility, A Conventional NATO Force Multiplier," in Windass & Grove (eds.): no pagination.

Meulman, F. H. 1991: "Defensive Defense: Shadow or Substance of the 'New Elan', and a Reflection on a Conceptual Framework for Aerospace Power under a Defensive Defense Regime" (Air Power Research Institute, Center for Aerospace Doctrine, Research and Education, Air University, Maxwell AFB, Alabama, manuscript).

Meurant, Jacques 1987: "Evolution and Nature of Humanitarian Law," *JPR,* 24:3 (September): 237–250.

Mey, Holger H. 1992: *NATO-Strategie vor der Wende: Die Entwicklung des Verständnisses nuklearer Macht im Bündnis zwischen 1967 und 1990* (Baden-Baden: Nomos).

———— & Michael Rühle 1991: "German Security Interests and NATO's Nuclear Strategy," *AuP,* 42:1, 20–30.

Meyer, Berthold 1986: "Atomwaffenfreie Zonen in Europa—ein langer Weg der kleinen Schritte," in Albrecht (ed.): 16–56.

———— 1987: "Die Parteien in der Bundesrepublik Deutschland und die sicherheits-politische Zusammenarbeit in Westeuropa," *HSFK-R,* 2.

———— 1989: "Common Security versus Western Security Cooperation? The Debate on European Security in the Federal Republic of Germany," in Wæver & al. (eds.): 168–185.

———— 1990: "Vertrauensbildung im Wandel der Ost-West-Beziehungen," in Schweizerische Friedensstiftung (eds.): 7–25.

———— 1990a: "Auf Vertrauen bauen? Über die Erfolge und Effekte des KSZE/KVAE-Prozesses für die Friedensentwicklung," in Vogt (ed.): 85–96.

———— 1991: "Parteipositionen zur europäischen Sicherheit im Elften Deutschen Bundestag und ihr Wandel aufgrund der DDR-Revolution," in Lutz (ed.): 13–32.

———— 1993: "Das Hamburger Modell und die Realität gesamteuropäischer Sicherheitskooperation," *S+F,* 11:3, 147–152.

———— 1993: "'Wehret den Anfängen!' Konzepte und Wege der Kriegsverhütung," in Matthies (ed.): 43–56.

Meyer, Dierk, & Allen Sens 1992: "Multinational Formations in NATO: From Forward Defence to Rapid Reaction," in Haglund & Mager (eds.): 91–112.

Meyer, Stephen M. 1984: "Verification and Risk in Arms Control," *IS,* 8:4, 111–126.

———— 1985: "Soviet Perspectives on the Paths to Nuclear War," in Allison, Carnesale, & Nye (eds.): 167–205.

———— 1988: "The Sources and Prospects of Gorbachev's New Political Thinking on Security" (*IS,* 13:2) in Lynn-Jones, Miller, & Evera (eds.) 1989: 110–149.

———— 1991: "How the Threat (and the Coup) Collapsed: The Politicization of the Soviet Military," *IS,* 16:3 (Winter): 5–38.

Meyer, Thomas 1987: "Ein neuer Rahmen für den Ost-West Dialog; Das gemein-same Papier von SED und SPD: Kein nationales Memorandum," *NGFH,* 34:10, 870–877.

Meyer, Wolfgang 1992: "Neue Konzepte für Rüstungskontrolle," *EuS,* 41:7 (July): 370–373.

Meyer-Plath, Christian 1989: "The Role of Military Forces within a ESS 2020: Some Thoughts on the System, the Military Doctrines and the Force Structure," in Wachter & Krohn (eds.): 51–70.

Meyers, Reinhard 1989: "Zur Einleitung: Grundzüge der Debatte um alternative Verteidigungskonzepte," in Gerber: 1–18.

Michejew, Viktor 1987: "Vertrauensbildende Massnahmen. Der sowjetische Standpunkt," in Brauch (ed.): 211–224.

Michta, Andrew A. 1992: *East-Central Europe After the Warsaw Pact: Security Dilemmas in the 1990s* (New York: Greenwood).

Midlarsky, Manus I. (ed.) 1992: *The Internationalization of Communal Strife* (London: Routledge).

Miettinen, Jorma 1986: "Non-provocative Conventional Defence—a Provision for Peace in Europe," in Tromp (ed.): 239–244.

Miggiano, Paolo 1989: "Potenzialita e limiti della difesa non militare," in Ragionieri (ed.): 229–246.

Mihalisko, Kathleen 1993: "Security Issues in Ukraine and Belarus," in Karp (ed.): 225–257.

Miksche, Ferdinand Otto 1952: *Unconditional Surrender* (London: Faber & Faber).

———— 1955: *Atomic Weapons and Armies* (London: Faber & Faber).

———— 1959: *The Failure of Atomic Strategy* (London: Faber & Faber).

———— 1977: "Les armes guidées avec précision changent les formes de combat," *Défence Nationale*, 33:12, 59–70.

Miliovejic, Marko 1990: "The Swiss Armed Forces," in Miliovejic & Maurer (eds.): 3–48.

———— & Pierre Maurer (eds.) 1990: *Swiss Neutrality and Security: Armed Forces, National Defence and Foreign Policy* (New York: Berg).

Militärgeschichtliches Forschungsamt 1975: *Verteidigung im Bündnis: Planung, Aufgaben und Bewährung, der Bundeswehr 1950–1972* (München: Bernard & Graefe).

Millar, T. B. 1993: *Asian-Pacific Security after the Cold War* (St. Leonards: Allen & Unwin).

Miller, Benjamin 1992: "Explaining Great Power Cooperation in Conflict Management," *WP*, 45:1 (October): 1–46.

Miller, David 1991: "Multinationality: Implications of NATO's Evolving Strategy," *IDR*, 3: 211–213.

Miller, James N. 1988: "Zero and Minimal Nuclear Weapons," in Nye, Allison, & Carnesale (eds.): 11–32.

Miller, Robert F. 1989: "Afghanistan and Soviet Alliances," in Saikal & Maley (eds.): 101–212.

Miller, Steven E. 1990: "Distinctive Features of Maritime Capabilities: Do They Warrant New Approaches to Confidence-Building, Arms Control and Disarmament at Sea?" in Lodgaard (ed.): 84–117.

———— 1990a: "US Withdrawal and NATO's Northern Flank: Impact and Implications," in Sharp (ed.): 303–348.

———— 1991: "Is Arms Control a Path to Peace?" in Thompson & Jensen (eds.): 46–63.

———— 1992: "Europe and Naval Arms Control in the Gorbachev Era," in Fürst & al. (eds.): 3–12.

———— 1993: "Russian-US Security Co-Operation on the High Seas," in Goodby & Morel (eds.): 249–271.

———— 1993a: "The Case Against a Ukrainian Nuclear Deterrent," *FA*, 72:3 (Summer): 66–80.

———— (ed.) 1984: *Strategy and Nuclear Deterrence: An International Security Reader* (Princeton: Princeton U.P.).

———— (ed.) 1985: *Military Strategy and the Origins of the First World War: An* International Security *Reader* (Princeton: Princeton U.P.).

———— (ed.) 1986: *Conventional Forces and American Defense Policy: An* International Security *Reader* (Princeton: Princeton U.P.).

———— & Stephen van Evera (eds.) 1986: *The Star Wars Controversy. An* International Security *Reader* (Princeton: Princeton U.P.).

———— & Stephen Van Evera (eds.) 1988: *Naval Strategy and National Security: An* International Security *Reader* (Princeton: Princeton U.P.).

———— & Sean Lynn-Jones (eds.) 1989: *Conventional Forces and American Defence Policy: An* International Security *Reader,* rev. ed. (Cambridge: MIT).

Mills, Greg 1993: "Armed Forces in Post-Apartheid South Africa," *Survival,* 35:3 (Autumn): 78–96.

Milosavljevic, Gordana 1989: "Security and Changes in the Contemporary World," in Singh & Vekaric (eds.): 83–89.

———— (ed.) 1991: "Conversion of Military Potential for Civilian Purposes: Significance for the Development Strategy of Yugoslavia," *CSS Papers,* 5 (Belgrade: Center for Strategic Studies).

Milshstein, Mikhail: See Milstein, Mikhail.

Milstein [Milshstein], Mikhail 1988: "New Concept of a Safe World," in Fedorenko (ed.): 40–60.

———— 1989: "Nuclear Forces and Their Relations to Conventional Armaments," in Rotblat & Goldanskii (eds.): 79–85.

Minh, Ho Chi 1973: *Selected Writings (1920–1969)* (Hanoi: Foreign Languages Publishing House).

Minni, Sergio, & Alberto Salvato 1989: "Una zona denuclearizzata nel Trivenetto," in Ragionieri (ed.): 163–182.

Mintz, Alex (ed.) 1992: *The Political Economy of Military Spending in the United States* (London: Routledge).

———— & Nehemia Geva 1993: "Why Don't Democracies Fight Each Other? An Experimental Study," *Journal of Conflict Resolution,* 37:3 (September): 484–503.

Mische, Patricia M. 1989: "Ecological Security in an Interdependent World," *Breakthrough* (New York: Global Education Associates), 10:4–11:1, 7–17.

Miszczak, Krzyszof 1993: "Die Sicherheitspolitik der Republik Polen," *EuS,* 42:10 (October): 515–519.

Moeller, Kay, & Markus Tidten 1994: "North Korea and the Bomb: Radicalisation in Isolation," *AuP,* 45:1 (1st Quarter): 99–109.

Mohan, C. Raja 1989: "Restructuring the Strategic Environment," in Singh & Vekaric (eds.): 26–62.

Moissejev, Michail 1989: "Wir müssen alle lernen, auf neue Art zu denken," (interview) *Truppenpraxis,* 5: 487–493.

Molander, Johan 1992: "The United Nations and the Elimination of Iraq's Weapons of Mass Destruction: The Implementation of a Cease-Fire Condition," in Tanner (ed.): 137–158.

Møller, Bjørn 1985: "Atomvåbenfri zone: Hvordan? Hvorfor?" in idem (ed.): 7–40.

———— 1986: "Defensivt forsvar," in idem (ed.): 31–57.

———— 1986a: "The Strategy Crisis of NATO and the Prospects of Non-Offensive Defence," in Størner & al. (eds.): 231–248.

———— 1987: "A Non-offensive Maritime Strategy for the Nordic Area: Some Preliminary Ideas," *WP-CPCR,* 3.

———— 1987a: "The Need for an Alternative NATO Strategy," *JPR,* 24(1): 61–74.

———— 1987b: "Ikkeførstebrug af atomvåben og ikkeoffensivt forsvar," in Henning Sørensen (ed.), *Sådan skal Danmark forsvares* (Copenhagen: Nyt fra Samfundsvidenskaberne): 163–207.

———— 1987c: "Disengagement and Non-Offensive Defence," *WP-CPCR,* 3.

———— 1988: "Common Security and Military Posture," in Baudisch (ed.): 41–54.

———— 1988a: "No-First-Use of Nuclear Weapons and Non-Offensive Defence," in Alger & Stohl (eds.): 436–454.

———— 1988b: "Perspectives of Disarmament in Europe—the Significance of Non-Offensive Defence Strategies," *CRPV,* 11:3, 104–114.

———— 1989: "Air Power and Non-Offensive Defence: A Preliminary Analysis," *WP-CPCR,* 2.

———— 1989a: "Common Security: Some Military and Non-Military Aspects," *WP-CPCR,* 8.

———— 1989b: "Crisis Stability and Non-Offensive Defence: A Nordic Perspective," in Chalmers, Møller, & Stevenson: 1–65.

———— 1989c: "Is War Impossible in Europe? A Critique of the Hypothesis of the Totally Destructive Conventional War (TDCW)," *WP-CPCR,* 11.

———— 1989d: "Ist Krieg in Europa unmöglich geworden? Eine kritik der Hypothese vom totalverheerenden konventionellen Krieg," *S+F,* 4: 232–237.

———— 1989e: "Restructuring the Naval Forces Towards Non-Offensive Defence," in Borg & Smit (eds.): 189–206.

———— 1989f: "The Nordic Regional Security Complex: A Preliminary Analysis," in Leif Ohlson (ed.), *Case Studies on Regional Conflicts and Regional Conflict Resolution* (Gotenburg: Padrigu Peace Studies): 45–82.

———— 1989g: "Nicht-offensive Verteidigung in Ost- und Westeuropa," in Karl & Auffermann (eds.): 62–79.

———— 1990: "A Common Security System for Europe?" *WP-CPCR,* 6.

———— 1990a: "Conversion: A Comprehensive Agenda," *WP-CPCR,* 19.

———— 1990b: "Current and Alternative Trends in the Military Confrontation in the Baltic Area—with a Special Emphasis on Naval Forces," *Acta Politica,* 2 (Szczecin: Uniwersytet Szczecinski): 89–118.

———— 1990c: "Gemeinsame Sicherheit und strukturelle Angriffsunfähigkeit aus dänischer Sicht," in Lutz & Schmähling (eds.): 217–243.

———— 1990d: "Nicht-offensive Verteidigung (NOV) und die deutsche Frage," in Benjowski, Heurlin, Putensen, & Wiberg (eds.): 54–71.

———— 1990e: "NOD und die deutsche Frage," *Arbeitshefte: Zeitschrift der Juso-Hochschulgruppen,* 89 (November): 25–34.

———— 1990f: "Non-Offensive Defence, the Armaments Dynamics, Arms Control and Disarmament," in Auffermann (ed.): 43–102.

———— 1990g: "Ikke-offensivt forsvar og europæisk våbenkontrol," in Sørensen (ed.): 237–269.

———— 1990h: "Teorien om ikke-offensivt forsvar," in Sørensen (ed.): 13–65.

———— 1990i: "Teknologi og forsvar," in Henning Sørensen (ed.), *Forsvar i Forandring* (Copenhagen: Samfundslitteratur, 1991): 63–96.

———— 1991: *Resolving the Security Dilemma in Europe: The German Debate on Non-Offensive Defence* (London: Brassey's).

———— 1991a: "Civilian-Based and Non-Offensive Defence," *WP-CPCR,* 3.

———— 1991b: "European Security Structures for the Nineties and Beyond: A Nordic Perspective," *AFR,* 40.

———— 1991c: "Guerilla Strategies and Non-Offensive Defence," *WP-CPCR,* 4.

———— 1991d: "Transarmament and Conversion: A Comprehensive Agenda," in idem (ed.) 1991: 45–89.

———— 1991e: "Ikke-offensivt forsvar," in Henning Gotttlieb, Bertel Heurlin, & Jørgen Teglers (eds.), *Fred og konflikt* (Copenhagen: SNU): 98–113.

———— 1991f: "A New Security System for Europe?" in Danspeckgruber (ed.): 251–264.

———— 1992: *Common Security and Non-Offensive Defense: A Neorealist Perspective* (Boulder, Colo.: Lynne Rienner).

———— 1992a: "What Is Defensive Security? Non-Offensive Defence and Stability in a Post-Bipolar World," *WP-CPCR*, 10.

———— 1992b: "Conversion Through the Ages: Outline of a Historical Analysis," *WP-CPCR*, 11.

———— 1992c: "Den globale Våbensmedje," *WP-CPCR*, 17.

———— 1992d: "A Common Security System for Europe," in Hanna Newcombe (ed.), *Hopes and Fears: The Human Future* (Toronto: Science for Peace): 1–26.

———— 1993: "Security Concepts: New Challenges and Risks," *WP-CPCR*, 18.

———— 1993a: "Multinationality, Defensivity, Collective Security," *WP-CPCR*, 20.

———— 1993b: "Defence Restructuring and Conversion in Post–Cold War Europe," *WP-CPCR*, 21.

———— 1993c: "Forsvar uten trussel," *Ikkevold*, 158: 22–24.

———— 1994: "Germany and NOD," in Møller & Wiberg (eds.): 153–165.

———— (ed.) 1985: *Norden Atomvåbenfri Zone* (Egtved: Fredsbevægelsens Forlag).

———— (ed.) 1986: *Forsvar—det bedste forsvar* (Egtved: Fredsbevægelsens Forlag).

———— (ed.) 1991: "Non-Offensive Defense in Europe: IPRA Defense and Disarmaments Study Group Paper 2," *AFR*, 42.

———— & Håkan Wiberg 1989: "Non-Offensive Defence and Armaments Dynamics: A Theoretical and Empirical Assessment," *WP-CPCR*, 3.

———— & Håkan Wiberg 1990: "Nicht-offensive Verteidigung als Vertrauensbildende Massnahme? Probleme und Konzepte," in Schweizerische Friedensstiftung (ed.): 39–79.

———— & Håkan Wiberg (eds.) 1994: *Non-Offensive Defence for the Twenty-First Century* (Boulder, Colo.: Westview).

Möller-Gulland, Niels 1993: "The Forum for Security Cooperation and Related Security Issues," in Lucas (ed.): 31–60.

Montgailhard, Jean Desazars de 1989: "Conventional Disarmament in Europe," *Disarmament*, 12:2, 51–59.

Moon, John Ellis van Courtland 1984: "Chemical Weapons and Deterrence: The World War II Experience," *IS*, 8:4 (Spring): 3–35.

Moraczewski, Marian, & Wojciech Multan 1990: "Elimination of Capability for Surprise Attack and Large-Scale Offensive Operation," in Huber (ed.): 97–196.

Moreillon, Jacques 1987: "Contemporary Challenges to Humanitarian Law," *JPR*, 24:3 (September): 215–218.

Moreton, Edwina 1987: "The German Question in the 1980s," in Moreton (ed.): 320.

———— (ed.) 1987: *Germany Between East and West* (Cambridge: Cambridge U.P.).

Morgenthau, Hans 1960: *Politics Among Nations: The Struggle for Power and Peace*, 3d ed. (New York: Knopf).

Morphet, Sally 1993: "UN Peacekeeping and Election-Monitoring," in Roberts & Kingsbury (eds.): 183–239.

Morrison, John 1993: "A Homeland Defence Option," in Clarke & Sabin (eds.): 50–69.

Morse, John H. 1975: "New Weapons Technologies: Implications for NATO," *Orbis*, 19:2, 497–513.

—— 1976: "Advanced Technology in Modern War," *RJ*, 121:2, 8–16.

Mortimer, Edward 1992: "European Security After the Cold War," *AP*, 271.

Möttölä, Kari 1984: "Neutralitets og forsvarspolitik: Finsk sikkerhedspolitik siden begyndelsen af 1970erne," in Heurlin (ed.): 139–173.

—— 1988: "Zonprocessen i ett brydningsskede," in Joenniemi & Vesa (eds.): 207–211.

—— 1990: "Maritime CSBMs and Naval Arms Control in the North of Europe: Obstacles and Opportunities," in Lodgaard (ed.): 146–158.

—— 1993: "Prospects for Cooperative Security in Europe: The Role of the CSCE," in Lucas (ed.): 1–30.

—— (ed.) 1988: *The Arctic Challenge: Nordic and Canadian Approaches to Security and Cooperation in an Emerging International Region* (Boulder, Colo.: Westview).

Mouritzen, Hans 1988: *Finlandization: Towards a General Theory of Adaptive Politics* (Aldershot: Gower).

Mueller, John 1988: "The Essential Irrelevance of Nuclear Weapons: Stability in the Postwar World" (*IS*, 13:2) in Lynn-Jones, Miller & Evera (eds.) 1990: 3–27.

—— 1989: *Retreat from Doomsday: The Obsolescence of Major War* (New York: Basic Books).

—— 1990: "A New Concert of Europe," *Foreign Policy*, 77 (Winter 1989–1990): 3–16.

—— 1991: "Taking Peace Seriously: Two Proposals," in Jervis & Bialer (eds.): 262–275.

—— 1991a: "Pearl Harbour: Military Inconvenience, Political Disaster," *IS*, 16:3 (Winter): 172–203.

Muijen, Marie-Louise van 1990: "NPD en de internationale politiek," in *Jaarboek Vrede en Veiligheid 1989: Bewapening, Vredesbeweging en het Nederlandse Veiligheidsbelied* (Nijmegen: Studiecentrum voor Vredesvraagstukken): 183–197.

—— 1993: *Better Safe Than Provocative: A Policy Science Perspective on the West European Non-Provocative Defence Debate in the 1980s* (Amsterdam: VU U.P.).

Müller, Albrecht A. C. von 1984: *Die Kunst des Friedens: Grundzüge einer europäischen Sicherheitspolitik für die 80er und 90er Jahre* (München: Hanser).

—— 1985: "Confidence-Building by Hardware Measures," in Rotblat & Hellman (eds.): 275–286.

—— 1986: "Integrated Forward Defence: Outline of a Modified Conventional Defence for Central Europe," in Tromp (ed.): 201–224.

—— 1986a: "Neue Chancen für den Rüstungskontrollprozess," *EW*, 35:7, 380–382.

—— 1986b: "Structural Stability at the Central Front," in Boserup, Christensen, & Nathan (eds.): 239–256.

—— 1987: "Stellungsnahme zum Thema 'Strukturelle Nichtangriffsfähigkeit'," *S+F*, 1: 60–62.

—— 1987a: "Technischer Fortschritt und Rüstungskontrolle: Überlegungen zur Integration von Sicherheits- und Verteidigungspolitik am Beispiel der Landstreitkräfte der Bundeswehr," in Heisenberg & Lutz (eds.): 498–522.

—— 1987b: "Das Konzept der strukturellen Nichtangriffsfähigkeit," *Dialog*, 10: 53–100.

———— 1988: "Beantwortung des Fragebogens für die Anhörung zur 'Konventionellen Stabilität'," in Bahr & Lutz (eds.): 461–463.

———— 1988a: "Die Schaffung konventioneller Stabilität: Ein modifiziertes Rahmenkonzept für die Rüstungskontrollpolitik und die Streikräfte-Modernisierung der NATO," in Bahr & Lutz (eds.): 81–92.

———— 1988b: "Die Transformation des Ost-West-Konflikts: Ein Rahmenkonzept für eine zweite Entspannungspolitik," in Bülow, Funk, & Müller: 47–68.

———— 1988c: "Konventionelle Stabilität: Plädoyer für eine verstärkte Integration von Sicherheits- und Verteidigungspolitik," in Bülow, Funk, & Müller: 69–94.

———— 1989: "Conventional Stability 2000: How Could NATO Regain the Initiative?" in Clesse & Schelling (eds.): 332–341.

———— 1989a: "Towards Conventional Stability in Europe," in Rotblat & Goldanski (eds.): 118–135.

———— 1989b: "Conventional Stability in Europe," in Tromp (ed.): 237–258.

———— 1990: "Conventional Stability and Defense Dominance," in UNIDIR (eds.): 3–22.

———— 1990a: "Conventional Stability and Arms Control," in Boserup & Neild (eds.): 47–54.

———— 1990b: "Conventional Stability in Europe: Outlines of the Military Hardware for a Second Détente," in Schmähling (ed.): 159–178.

———— 1990c: "Conventional Arms Control: The Agenda and Its Dangers," in Boserup & Neild (eds.): 19–30.

———— 1990d: "Umrisse einer neuen Sicherheitsarchitektur für Europa," in Heisenberg & Lutz (eds.): II:147–159.

———— 1990e: "A Tripolar Security Structure for Europe," in Clesse & Rühl (eds.): 264–273.

———— & Andrzej Karkoszka 1988: "A Modified Approach to Conventional Arms Control," in Bülow, Funk, & Müller: 220–224; and *BAS*, 44:7, 39–41.

———— & Andrzej Karkoszka 1990: "An East-West Negotiation Proposal," in Boserup & Neild (eds.): 135–139.

Müller, Barbara (ed.) 1991: "Keine Bundeswehr in Krisengebiete," with an enclosure of Knut Ipsen, "Der Einsatz der Bundeswehr zur Verteidigung, im Spannungs- und Verteidigungsfall sowie im internen bewaffneten Konflikt," *F&A*, special issue 1.

Müller, Erwin 1982: "Vertrauensbildende Massnahmen: Chancen und Probleme eines sicherheitspolitischen Konzepts," in Lutz & Müller (eds.): 67–128.

———— 1984: "Dilemma Sicherheitspolitik: Tradierte Muster westdeutscher Sicherheitspolitik und Alternativoptionen: Ein Problem und Leistungsvergleich," in idem (ed.): 53–170.

———— 1985: "Die Supermächte und ein System Kollektiver Sicherheit in Europa," in Lutz (ed.): 151–171.

———— 1985a: "Flexible Response oder konventionelles Alternativkonzept: Ein sicherheitspolitischer Problem-und Leistungsvergleich," in Lutz (ed.): 359–374.

———— 1986: "Gemeinsame Sicherheit: Profil eines Konzepts alternativer Sicherheitspolitik," in Bahr & Lutz (eds.): 159–198.

———— 1986a: "Rüstungsdynamik und Rüstungssteuerung im historischen Vergleich: Erfolgsbedingungen und Misserfolgsursachen von Rüstungskontrollpolitik," in Lutz, Müller, & Pott: 165–197.

———— 1987: "Zur Logik politischer Bedrohungsanalysen," in Heisenberg & Lutz (eds.): 79–97.

———— 1988: "Überlegungen zum Problemfeld 'Konventionelle Stabilität'," in Bahr & Lutz (eds.): 44–61.

———— 1988a: "Vertrauensbildung in Europa: Die KVAE und ihre Folgen," in Moltmann (ed.): 131–135.

———— 1989: "Konventionelle Stabilität durch Strukturelle Angriffsunfähigkeit," *S+F,* 7:2, 139–145, also in Müller & Neuneck (eds.) 1990: 75–80.

———— 1991: "Grundsatzprobleme kollektiver Sicherheit: Historische Erfahrungen und Lehren für die Zukunft," in Lutz (ed.): 93–114.

———— 1993: "Angst vor den Deutschen?" in Lutz (ed.): 63–82.

———— (ed.) 1984: *Dilemma Sicherheit: Beiträge zur Diskussion über militärische Alternativkonzepte* (Baden-Baden: Nomos).

———— & Götz Neuneck (eds.) 1990: *Abrüstung und Konventionelle Stabilität in Europa* (Baden-Baden: Nomos).

———— & (eds.) 1991: *Rüstungsmodernisierung und Rüstungskontrolle: Neue Technologien, Rüstungsdynamik und Stabilität* (Baden-Baden: Nomos).

Müller, Harald 1989: "Transforming the East-West Conflict: The Crucial Role in Verification," in Altmann & Rotblat (eds.): 2–15.

———— 1991: "Maintaining Non-Nuclear Weapon Status," in Karp (ed.): 301–339.

———— 1992: "Arms Control in the Middle East: A European Perspective," in Spiegel (ed.): 173–179.

———— 1992a: "The Role of Hegemonies and Alliances," in Karp (ed.): 226–249.

———— 1993: "Nukleare Nichtverbreitung: Ein umfassender Strategieentwurf," *HSFK-R,* 7.

———— 1993a: *Die Chance der Kooperation: Regime in den internationalen Beziehungen* (Darmstadt: Wissenschaftliche Buchgesellschaft).

Muller, Jean-Marie 1984: *Vous avez dit "pacifisme"? De la ménace nucleaire a la défense civile non-violente* (Paris: Cerf).

Müller, Klaus, & Christian Wellmann 1993: "Verfassungskonsens adé: Weltweite Einsätze der Bundeswehr," in Birkenback & al. (eds.): 139–151.

Müller, Manfred 1988: "European Security and Non-Offensive Defence," *Scientific World,* 32:1, 11–13.

———— 1988a: "Opening Address," in Baudisch (ed.): 5–10.

———— 1989: "Constraints," in Blackwill & Larrabee (eds.): 405–421.

———— 1990: "Deutschlands Rolle in einem neugestalteten europäischen Sicherheitssystem," *EA,* 6: 221–224.

Müller, Martin 1988: "Thesen zur Kritik der Theorie struktureller Angriffsunfähigkeit," *S+F,* 6:1, 41–45.

———— 1988a: *Politik und Bürokratie: Die MBFR-Politik der Bundesrepublik Deutschland zwischen 1967 und 1973* (Baden-Baden: Nomos).

———— 1990: "Deutsche Einheit und europäische Sicherheit: Eine Kritik der Studie der IFSH," *S+F,* 8:3, 166–170.

———— & Volker Matthies 1990: "Die Strukturwandel des internationalen Systems und die Sicherheit der Bundesrepublik Deutschland," in Calliess (ed.): 484–524.

Mumenthaler, Hans 1990: "Civil Defence: Means for Disaster Relief," in Miliovejic & Maurer (eds.): 49–64.

Murray, Douglas J., & Paul R. Viotti (eds.) 1982: *The Defense Policies of Nations: A Comparative Study* (Baltimore: Johns Hopkins U.P.).

Mushaben, Joyce Marie 1986: "Grassroots and Gewaltfreie Aktionen: A Study of Mass Mobilization Strategies in the West German Peace Movement," *JPR,* 23:2, 141–154.

Mutz, Reinhard 1986: "Gemeinsame Sicherheit: Grundzüge einer Alternative zum Abschreckungsfrieden," in Bahr & Lutz (eds.): 83–157.

———— 1987: "Militärische Aspekte Gemeinsamer Sicherheit: Anmerkungen zur

Frage des konzeptkonformen Sicherheitsbegriffs," in Bahr & Lutz (eds.): 267–276.

——— 1988: "Die Kontrolle konventioneller Rüstung in Europa," in Moltmann (ed.): 121–130.

——— 1988a: "Konventionelle Stabilität: Eine Einführung," in Bahr & Lutz (eds.): 36–43.

——— 1990: "Europäische Sicherheit und deutsche Einheit: Zur Kritik Martin Müllers an der Stellungsnahme des IFSH," *S+F*, 8:3, 171–176.

——— 1991: "Die NATO auf der Suche nach ihrer neuen Rolle," in Schwerdtfenger & al. (eds.): 129–139.

——— 1993: "Schiessen wie die Andren: Eine Armee sucht ihren Zweck," in Lutz (ed.): 11–26.

——— & Götz Neuneck 1990: "Konventionelle Stabilität und qualitative Rüstungsdynamik," in E. Müller & Neuneck (eds.): 119–134.

Myers, David B. 1990: *New Soviet Thinking and U.S. Nuclear Policy* (Philadelphia: Temple U.P.).

Myrdal, Alva 1976: *The Game of Disarmament: How the United States and Russia Run the Arms Race* (New York: Pantheon).

——— 1981: "Dynamics of European Nuclear Disarmament," in Myrdal & al.: 209–276.

——— & al. 1981: *Dynamics of European Nuclear Disarmament* (Nottingham: Spokesman).

Nacht, Michael 1985: *The Age of Vulnerability: Threats to the Nuclear Stalemate* (Washington, D.C.: Brookings).

Nagler, Michael N. 1991: "Ideas of World Order and the Map of Peace," in Thompson & Jensen (eds.): 373–392.

Nakarada, Radmilla, & Jan Øberg (eds.) 1989: *Surviving Together: The Olof Palme Lectures on Common Security, 1988* (London: Dartmouth).

Nardin, Terry, & David R. Mapel (eds.) 1993: *Traditions in International Ethics* (Cambridge: Cambridge U.P.).

Nardulli, Bruce R. 1988: "The View From Berlin: A Conference Report," in Nehrlich & Thomson (eds.): 211–236.

Nardulli, Giuseppe 1989: "La difesa non provocatoria e il Mediterraneo: la dimensione navale," in Ragionieri (ed.): 69–88.

——— 1990: "Nuclear Weapons and Defensive Defense in the Mediterranean Sea," in Cerutti & Ragionieri (eds.): 151–160.

Narveson, Jan 1985: "Getting on the Road to Peace: A Modest Proposal," in Hardin & al. (eds.): 213–229.

Nassauer, Otfried 1992: "Verteidigungsplanung für das geeinte Deutschland," in Butterwege & Senghaas-Knobloch (eds.): 99–120.

Natalegawa, Marty 1993: "De-Nuking Southeast Asia," *Pacific Research*, 6:1, 8–10.

NATO Information Service 1989: *The North Atlantic Treaty Organisation: Facts and Figures*, 11th ed. (Brussels: NATO).

Naumann, Klaus 1989: "Defensive Doktrinen und Streitkräftestrukturen," *EA*, 22: 667–676.

——— 1989a: "The Restructured Bundeswehr," *NSN*, October, 31–35.

Navias, Martin 1993: *Going Ballistic: The Build-up of Missiles in the Middle East* (London: Brassey's).

Nazarenko, Vladimir 1989: "Europe: From the Balance of Terror to Equal Security," *SMR*, 6: 52–54.

Nedelmann, Carl, Ingrid Angermann, Christian Buhrmeister, Klaus Hennel, & Jörg

Scharff 1987: "Psychologische Aspekte Gemeinsamer Sicherheit," in Bahr & Lutz (eds.) 1987: 201–222.

Nehrlich, Uwe 1986: "Der Bedeutungswandel der französischen Nuklearstreitmacht," in Kaiser & Lellouche (eds.): 165–179.

———— 1988: "Conventional Arms Control in Europe: The Objectives," in Nehrlich & Thomson (eds.): 182–202.

———— 1993: "Neue Sicherheitsfunktionen der NATO," *EA,* 48:23 (10 December): 663–672.

———— (ed.) 1982: *Die Einhegung sowjettischer Macht: Kontrolliertes militärisches Gleichgewicht als Bedingung europäischer Sicherheit* (Baden-Baden: Nomos).

———— & James A. Thomson (eds.) 1988: *Conventional Arms Control and the Security of Europe* (Boulder, Colo.: Westview).

Neild, Robert 1981: *How to Make Up Your Mind About the Bomb* (London: André Deutsch).

———— 1984: "What Political Signals Should Our Armed Forces Send?" in Prins (ed.): 210–217.

———— 1986: "The Implications of the Increasing Accuracy of Non-Nuclear Weapons," in Rotblat & d'Ambrosio (eds.): 93–106; abridged version in Boserup & Neilds (eds.) 1990: 55–62.

———— 1986a: "The Case Against Arms Negotiations and for a Reconsideration of Strategy," *Arms Control,* 7:2, 133–155.

———— 1989: "The Alternative Paradigm," in Barfoed & al. (eds.): 34–39.

———— 1989a: "Non-Offensive Defence: The Way to Achieve Common Security in Europe," *Background Paper,* January (Ottawa: Canadian Institute for International Peace and Security).

———— 1990: *An Essay on Strategy as It Affects the Achievement of Peace in a Nuclear Setting* (London: Macmillan).

———— 1992: "Military Aspects of Defensive Security," *Disarmament,* 15:4, 126–135.

———— & Anders Boserup 1987: "Beyond INF: A New Approach to Nonnuclear Forces," *WPJ,* 4:4 (Fall): 605–620.

Nelson, Daniel 1991: "The Soviet Union and Europe," in Jordan (ed.): 48–65.

———— 1991a: "Security After Hegemony," *BPP,* 22:3 (September): 335–345.

———— 1991b: "The Costs of Demilitarization in the USSR and Eastern Europe," *Survival,* 33:4 (July–August): 312–326.

———— 1993: "Creating Security in the Balkans," in Karp (ed.): 155–176.

Neuhold, Hanspeter 1989: "Challenges to Neutrality in an Interdependent World," in Kruzel & Haltzel (eds.): 83–97.

Neumann, Hans Dieter 1988: "The Special Case of Germany," in Kirk (ed.): 13–16.

Neumann, Leo D. 1988: "World Government," in Nye, Allison, & Carnesale (eds.): 197–214.

Neuneck, Götz 1990: "Das Problem der Rüstungsmodernisierung bei konventionellen Streitkräften," *S+F,* 8:4, 221–230.

———— 1991: "Die Modernisierung konventioneller Streitkräfte," in Müller & Neuneck (eds.): 265–285.

———— 1993: "Wohin marschiert die Bundeswehr? Der Bundeswehrplan '94: Die Hardthöhe und die neue Unübersichtlichkeit," in Lutz (ed.): 83–114.

———— & Otfried Ischebeck (eds.) 1992: *Missile Proliferation, Missile Defence, and Arms Control* (Baden-Baden: Nomos).

New Democrats 1988: "Canada's Stake in Common Security." Report by the

International Affairs Committee of the New Democratic Party of Canada (Ontario: New Democrats).

Newhouse, John 1973: *Cold Dawn: The Story of SALT* (New York: Reinhart & Winston).

—————— 1989: *The Nuclear Age: From Hiroshima to Star Wars* (London: Michael Joseph).

Newman, William W., & Judyth L. Twigg 1993: "Building a Eurasian-Atlantic Security Community: Co-Operative Management of the Military Transition," in Goodby & Morel (eds.): 224–248.

Nichols, Thomas M. 1993: *The Sacred Cause: Civil-Military Conflict over Soviet National Security, 1917–1992* (Ithaca: Cornell U.P.).

—————— & Theodore Krasik 1990: "Civil-Military Relations Under Gorbachev: The Struggle over National Security," in Green & Karasik (eds.): 29–61.

Nielsen, Peter Michael 1990: *CSCE: Sikkerhed og samarbejde i Europa* (Copenhagen: SNU).

Niesporek, Heinrich 1986: "Verteidigung statt Selbstvernichtung," in Gohl & Niesporek (eds.): 165–173.

Niezing, Johan 1982: "Rethinking Utopia: Some Sceptical Remarks on the Methodology of Civilian-Based Defense Research," *BPP*, 13:4, 329–334.

—————— 1987: *Sociale Verdediging als Logisch Alternatief: Van Utopie Naar Optie* (Antwerpen/Assen: EPO/Von Gorchum).

—————— 1989: "Die besondere friedenspolitische Herausforderung des Atomzeitalters," in Gross & al. (eds.): 351–362.

—————— 1990: "Non-Offensive Defense (NOD) and Social Defense (SD): Their Convergence in Time and Space," in Czempiel & al. (eds.): 45–49.

Nikitin, Alexander I. 1988: "The Concept of Universal Security: A Revolution of Thinking and Policy in the Nuclear Age," in Gromyko & Hellman (eds.): 168–175.

Nikolayev, Konstantin 1989: *Nuclear Deterrence: Past and Future* (Moscow: Novosti).

Nincic, Miroslav 1985: *How War Might Spread to Europe* (London: Taylor & Francis).

Nishihara, Masashi 1993: "Northeast Asia and Japanese Security," in Unger (ed.): 85–98.

Nitze, Paul 1986: "On the Road to a More Stable Peace," in Snyder (ed.): 221–227.

Nobel, Jaap (ed.) 1991: *The Coming of Age of Peace Research: Studies in the Development of a Discipline* (Groningen: Styx).

Noel-Baker, Philip J. 1979: *The First World Disarmament Conference, 1932–33* (Oxford: Pergamon).

Nokkala, Arto 1990: "The Problem of Military Credibility in Non-Offensive Defence: Some Outlines for a Framework," in Auffermann (ed.): 13–42.

—————— 1991: "Non-Offensive Defence: A Criteria Model of Military Credibility," *Finnish Defence Studies* (Helsinki: War College): 3.

Nolan, Janne E. 1988: "The U.S.-Soviet Conventional Arms Transfer Negotiations," in George, Farley, & Dallin (eds.): 510–523.

—————— 1991: *Trappings of Power: Ballistic Missiles in the Third World* (Washington, D.C.: Brookings).

—————— 1992: "Stemming the Proliferation of Ballistic Missiles: An Assessment of Arms Control Options," in Stahl & Kemp (eds.): 171–186.

Nolte, Hans-Heinrich 1990: "Perestrojka und Internationales System: Zur Rolle der Rüstung," *Das Argument*, 183: 759–768.

—————— & Wilhelm Nolte 1984: *Ziviler Widerstand und Autonome Abwehr* (Baden-Baden: Nomos).

Nolte, Wilhelm 1984: "Die Militärstrategien von NATO und Warschauer Pakt und die Veränderungen in den Nukleardoktrinen der USA und der UdSSR," in Brauch (ed.): 113–127.

———— 1985: "Soziale Verteidigung im System Kollektiver Sicherheit?" in Lutz (ed.): 310–320.

———— 1986: "Deutschlandargumente für Autonome Abwehr," in Gohl & Niesporek (eds.): 174–186.

———— 1986a: "Verteidigung in Deutschland: Wiedervereinigung mit militärischen Mitteln?" S+F, 3: 154–158.

———— 1986b: "Autonomous Protection," in Tromp (ed.): 171–178.

———— 1987: "Autonome Abwehr: Alternative zu den Alternativen!" in Bühl (ed.): 219–233.

———— 1987a: "Militärische Aspekte Gemeinsamer Sicherheit: Können Soldaten Beiträge zu Gemeinsamer Sicherheit erbringen?" in Bahr & Lutz (eds.): 301–307.

———— 1987b: "Zum Thema 'Strukturelle Nichtangriffsfähigkeit'," S+F, 1: 63–72.

———— 1988: "Anmerkungen und Kritik zur Konventionellen Stabilität," in Bahr & Lutz (eds.): 464–471.

———— 1988a: "Dominanz der Defensive: Streitbare Anmerkungen zur Krause-Studie," S+F, 1: 37–40.

———— 1988b: "Neun aufbauende Thesen, eine streitbare Anmerkung und eine graphische Skizze des Alternativkonzeptes 'Autonome Abwehr'," in Calliess (ed.): 174–179.

———— 1989: "Civilian-Military Security Co-Function: Autonomous Protection, Alternative Defence from a West German Perspective," in Tromp (ed.): 259–280.

———— 1989: "Common Security and Autonomous Protection: Focus on Civilian-Military Defence in Europe," in Nakarada & Øberg (eds.): 131–143.

———— 1990: "Zur Transpositionsfähigkeit zu mehr nichtmilitärischen Mitteln," in Weizsäcker (ed.): 158–170.

———— 1990a: "'Wer wagt . . .' Plädoyer für das Zusammenwirken Gewaltfreien Widerstandes mit Konventioneller Verteidigung," in Vogt (ed.): 263–276.

———— 1990b: "Zivilisierung von Verteidigungsstrukturen: Annäherung an die Auflösung des Verteidigungs-Vernichtungs-Paradoxons," in Calliess (ed.): 663–686.

———— 1992: "Die Bundeswehr—Zwischen Landesverteidigung und Machtprojektion," Blätter, 37:10 (October): 1206–1215.

———— 1993: "Zivilisierung von Verteidigung in ihren Strukturen: Annäherungen an die Auflösung des Verteidigungs-Vernichtungs-Paradoxons," in Klaus Dieter Wolf (ed.), Ordnung zwischen Gewaltproduktion und Friedensstuftung (Baden-Baden: Nomos): 175–184.

Nordic Study Group on Port Call Policies 1990: "The Port Call Issue: Nordic Considerations," BPP, 21:3, 337–352.

Nordlinger, Eric A. 1991: "America's Strategic Immunity: The Basis of a National Security Strategy," in Jervis & Bialer (eds.): 239–261.

Nørgaard, Ole 1990: "New Political Thinking in East and West: A Comparative Perspective," in Harle & Iivonen (eds.): 51–83.

North, Robert C. 1990: War, Peace, Survival: Global Politics and Conceptual Synthesis (Boulder, Colo.: Westview).

Nötzold, Jürgen 1991: "Wirtschaft und Aussenpolitik in der Perestroika," in Hamann & Matthies (eds.): 81–98.

Nowak, Jerzy M. 1990: "Rethinking the European Security System: A Polish Perspective," in Clesse & Rühl (eds.): 359–371.

Nozko, Kazimierz 1992: "Directions for Improving the Defense System of the Polish Republic (With an Introduction by Peter Podbielski)," *ES*, 1:2 (Summer): 144–164.

Nye, Joseph S., Jr., Graham T. Allison, & Albert Carnesale (eds.) 1988: *Fateful Visions: Avoiding Nuclear Catastrophe* (Cambridge, Mass.: Ballinger).

Nygren, Bertil 1984: "Fredelig samexistens: Klaskamp, fred och samarbete: Sovjetunions detentedoktrin," *Research Report*, 12 (Stockholm: Swedish Institute of International Affairs).

O'Ballance, Edgar 1972: *The Third Arab-Israeli War* (London: Faber).

———— 1978: *No Victor, No Vanquished: The Yom Kippur War* (Novato, Calif.: Presidio Press).

Øberg, Jan 1980: *Myter om vor sikkerhed: En kritik af dansk forsvarspolitik i et udviklingsperspektiv* (Copenhagen: Mellemfolkeligt Samvirke).

———— 1983: *At udvikle sikkerhed og sikre udvikling* (Copenhagen: Vindrose).

———— 1983a: "Samfund og forsvarsform," in Thoft & Jacobsen (eds.): 58–68.

———— 1984: "Three Minutes to a New Day?" *JPR*, 21:2, 95–99.

———— 1989: "Conventional Warfare and Alternative Defence Models: Some Theoretical Notes," in Tromp (ed.): 157–163.

———— 1989a: "Avskräckning som bestraffning och antropocentrism," in Wiberg (ed.): 147–186.

———— 1990: "Coping With the Loss of a Close Enemy: Perestroika as a Challenge to the West," *BPP*, 21:3, 287–298.

———— (ed.) 1992: *Nordic Security in the 1990s: Options in the Changing Europe* (London: Pinter).

Ochmanek, David A., & Edward L. Warner III 1989: "Conventional Arms Reduction Approaches," in Blackwill & Larrabee (eds.): 231–257.

Ocokoljic, Stanislaw 1989: "Implications of Changing US-Soviet Military Relations," in Singh & Vekaric (eds.): 90–110.

Odom, William E. 1990: "Gorbachev's Strategy and Western Security: Illusions vs. Reality," *WQ*, 13:1 (Winter): 145–168.

———— 1991: "Unreasonable Sufficiency? Assessing the New Soviet Strategy," in Frost (ed.): 91–120.

Oelrich, Ivan 1990: "Monitoring Production of Conventional Weapons," in Kokoski & Koulik (eds.): 155–163.

Oeser, Edith 1990: "Vier plus Zwei gleich Fünf? Vier Mächte und zwei deutsche Staaten," *Blätter*, 4: 426–433.

Office of Technology Assessment: See OTA.

Ofstad, Harald, & Marlene Öhberg (eds.) 1982: *Vår röst, en makt: Nordiska inlägg i fredsdebatten* (Stockholm: Gidlunds).

Ohlson, Thomas (ed.) 1988: *Arms Transfer Limitations and Third World Security* (Oxford: Oxford U.P.).

———— & Stephen John Stedman 1993: "Security in Post-Apartheid Southern Africa," *Security Dialogue*, 24:4 (December): 415–428.

Okawara, Yoshio 1993: "Japan's Global Responsibilities," in Unger (ed.): 55–67.

Oldberg, Ingmar 1987: "Incidents in the Baltic: A Dark Side in Soviet-Swedish Relations Since 1945," *Nordic Journal of Soviet & East European Studies*, 4:1, 15–48.

Oldenburg, Fred 1990: "Sowjetische Deutschland-Politik nach der Oktober-Revolution in der DDR," *DA*, 23:1, 68–76.

Olson, Thomas 1985: "Social Defence and Deterrence: Their Interrelationsship," *BPP*, 16:1, 33–40.

Olson, William J. 1985: "The Light Force Initiative," *MR*, 65:6, 217.

Ondarza, Henning von 1990: "Das Heer in einer Phase des Übergangs: Historische Chancen der Neugestaltung," *EW*, 7: 406–413.

—— 1990a: "The German Army on Its Way to the Year 2000," *NSN,* February–March, 31–37.

—— 1993: "Pillar of the New European Security Architecture—The North Atlantic Cooperation Council," *NSN,* 38(5/6): 41–43.

Orbons, Sjef 1986: "Verteidigungsoptionen für Mitteleuropa: Defensivität, Technologie, Konsequenzen" (Delft/Bonn: SAS).

—— 1989: "A Classification and Evaluation of Concepts for the Defense of NATO's Central Sector," *AFR,* 33.

Orden, Geoffrey Van 1991: "The Bundeswehr in Transition," *Survival,* 33:4 (July–August): 352–370.

Orenstein, Harold S. 1989: "Warsaw Pact Views on Trends in Ground Forces Tactics," *IDR,* 9: 1147–1154.

Ørvik, Niels 1986: Introduction to Ørvik (ed.): 1–14.

—— 1986a: "Norway: Deterrence Versus Nonprovocation," in Ørvik (ed.): 186–247.

—— (ed.) 1986: *Semialignment and Western Security* (London: Croom Helm).

Oseth, John M. 1984: "An Overview of the Reform Debate," in Clark & al. (eds.): 44–61.

Osgood, Charles E. 1959: "Suggestions for Winning the Real War with Communism," *Journal of Conflict Resolution,* 3:4 (December): 295–325.

—— 1962: *An Alternative to War or Surrender* (Urbana: University of Illinois Press).

—— 1986: "The Way GRIT Works," in Carlson & Cromstock (eds.) 1986a: 24–30.

Osherenko, Gail, & Oran R. Young 1989: *The Age of the Arctic: Hot Conflicts and Cold Realities* (Cambridge: Cambridge U.P.).

Østerud, Øyvind (ed.) 1986: *Studies of War and Peace* (Oslo: Norwegian U.P.).

OTA (Office of Technology Assessment) 1985: *Strategic Defenses: Ballistic Missile Defense Technologies; Anti-Satellite Weapons: Countermeasures and Arms Control. Two Reports by the Office of Technology Assessment* (Princeton: Princeton U.P.).

—— 1987: *New Technology for NATO: Implementing Follow-On Forces Attack* (Washington, D.C.: GPO).

—— 1992: *Building Future Security: Strategies for Restructuring the Defense Technology and Industrial Base* (Washington, D.C.: GPO).

Ouagrham, Sonia Ben 1993: "Le désarmement et la conversion de l'industrie militaire en Russie," *Travaux de Récherche,* 24 (New York: UN; Geneva: UNIDIR).

Oudnaren, John Van 1988: "Conventional Arms Control in Europe: Soviet Policy and Objectives," in Nehrlich & Thomson (eds.): 40–64.

—— 1991: "Gorbachev and His Predecessors: Two Faces of New Thinking," in Clark & Serfaty (eds.): 3–28.

—— 1991a: *Détente in Europe: The Soviet Union and the West Since 1953* (Durham: Duke U.P.).

Outze, Børge 1946: "Sabotage," in La Cour (ed.): III:135–210.

Owen, Henry, & Edward C. Meyer 1989: "Central European Security," *FA,* 68:3 (Summer): 22–40.

Oxford Research Group 1990: "New Threats and New Responses: Proposals for Future Security Decision-Making in Europe," *Current Decisions Report,* 4 (Oxford: Oxford Research Group).

Oye, Kenneth A. 1985: "Explaining Cooperation Under Anarchy: Hypotheses and Strategies," *WP,* 38:1 (October): 1–24.

Oznobitschev, Sergei, & Viktor Moskvin 1992: "Global Defense: What Is the Russian Interest?" in Neuneck & Ischebeck (eds.): 163–170.

Paffenholz, Thania 1993: "'Die Waffen nieder!': Konzepte und Wege der Kriegsbeendigung," in Matthies (ed.): 57–68.

Palme Commission (Independent Commission on Disarmament and Security Issues) 1982: *Common Security: A Blueprint for Survival. With a Prologue by Cyrus Vance* (New York: Simon & Schuster).

———— 1989: *A World at Peace: Common Security in the Twenty-first Century* (Stockholm: Palme Commission).

Palmer, John 1988: *Europe Without America? The Crisis in Atlantic Relations* (London: Oxford U.P.).

Palmisona 1993: "Lithauen und seine Landesverteidigung," *ÖMZ*, 31:5 (September–October): 456–458.

Pant, K. C. 1989: "The Philosophy of Indian Defence," in Singh & Vekaric (eds.): 206–214.

Pantev, Plamen 1988: "The Offensive/Defensive Reliant Strategic Posture and European Security: Conceptual, Political and International Legal Aspects," in Baudisch (ed.): 55–61.

Papp, Daniel S. 1984: "Potential Soviet Responses to a NATO Retaliatory Offensive Strategy," in Keith A. Dunn & William O. Staudemeier (eds.), *Military Strategy in Transition: Defense and Deterrence in the 1980s* (Boulder, Colo.: Westview): 147–165.

———— 1990: "The Changing Face of U.S.-Soviet Relations: An American Perspective," in Brauch & Kennedy (eds.): 3–19.

Paret, Peter (ed.) 1986: *Makers of Modern Strategy from Machiavelli to the Nuclear Age* (Princeton: Princeton U.P.).

Paskins, Barrie (ed.) 1986: *Ethics and European Security* (London: Croom Helm).

Paukert, Liba, & Peter Richards (eds.) 1991: *Defence Expenditure, Industrial Conversion and Local Employment* (Geneva: ILO).

Paul, Derek (ed.) 1985: *Defending Europe: Options for Security* (London: Taylor & Francis).

Paul, T. V. 1993: "The Politics of Unilateral Nuclear-Free Zones: The Case of the South Pacific," in Ramberg (ed.): 159–174.

Paulson, Dennis (ed.) 1986: *Voices of Survival in the Nuclear Age* (London: Wisdom).

Pay, John, & Geoffrey Till 1990: "Introduction," Pay & Till (eds.): 1–9.

———— & Geoffrey Jill (eds.) 1990: *East-West Relations in the 1990s: The Naval Dimension* (London: Pinter).

Payne, Wilbour B. 1990: "Conventional Force Stability," in Huber (ed.): 133–142.

Paznyak, Vyacheslav E. 1993: "'Security Aspects of Belarus' Policies Towards the CIS and Western Europe," in Ehrhart & al. (eds.): 91–100.

Peattie, Lisa R. 1989: "Economic Conversion as a Set of Organizing Ideas," in Dumas & Thee (eds.): 23–36.

Pedersen, Knud P. 1981: *Alternativ til Ragnarok* (Århus: Grevas).

Pedersen, Ole Karup 1989: "The Debate on European Security in Denmark," in Wæver & al. (eds.): 269–282.

Pedersen, Robert 1982: *Fra neutralitet til engagement: Socialdemokratiet og forsvaret gennem 110 år* (Copenhagen: Chr. Erichsen).

Pedroncini, Guy 1991: "Remarque sur les grandes décisions stratégiques françaises de 1914 a 1940," *Stratégique*, 50 (1991/2): 289–305.

Perkins, Ken (ed.) 1987: *Weapons and Warfare: Conventional Weapons and Their Role in Battle* (London: Brassey's).

Perrett, Bryan 1985: *Lightning War: A History of Blitzkrieg* (London: Panther).

Perry, William J. 1991: "Desert Storm and Deterrence," *FA,* 70:4 (Fall): 64–82.

Peter, Hans 1989: "Offensivkonzepte der NATO gefährden die Sicherheit Europas," *IPW-B,* 2: 16–22.

Peters, Ingo 1993: "Sicherheitspolitische Vertrauensbildung im neuen Europa: Herausforderunbgen und Entwicklungen in der VSBM-Politik," in Forndran & Pohlman (eds.): 363–389.

Peters, John 1992: "International Relations Theory and European Conventional Arms Control: Practical Linkages," *ES,* 1:2 (Summer): 109–125.

Petersen, Gert 1981: *Om fredens nødvendighed* (Copenhagen: Vindrose).

———— 1982: *Om nødvendigheden af en dansk fredspolitik* (Copenhagen: Vindrose).

Petersen, Nikolai 1984: "Das Scandilux-Experiment: Auf dem Wege zu einer transnationalen sozialdemokratischen Sicherheitsperspektive," *EA,* 16: 493–500.

———— 1988: "Denmark, Greenland, and Arctic Security," in Möttölä (ed.): 39–74.

———— 1988a: "Dansk sikkerhedspolitik i Østersøområdet," in Joenniemi & Vesa (eds.): 54–67.

Petersen, Phillip A. 1987: "Soviet Offensive Operations in Central Europe," *IDR,* 8: 26–32.

———— 1990: "Soviet Military Strategy in an Age of New Thinking on International Security," in Pay & Till (eds.): 46–69.

———— 1992: "Security Policy in the Post-Baltic States," *ES,* 1:1 (Spring): 13–49.

———— 1992a: "Security Policy in the Post-Soviet Heartland and Moldova," *ES,* 1:3 (Autumn): 295–351.

Petraeus, David H. 1984: "Light Infantry in Europe: Strategic Flexibility and Conventional Deterrence," *MR,* 64:12, 35–55.

Petrovsky, Vladimir 1990: "New Technologies and Security: Prospects After the Cold War," *Disarmament,* 13:4, 45–57.

Peuthert, Joachim 1989: "Konventionelle Stabilität, Strukturelle Angriffsunfähigkeit: Militärische Konsequenzen sicherheitspolitischer Absichten," *S+F,* 7:2, 110–113.

Pfaltzgraff, Robert L., Jr., Kim R. Holmes, Claus Clemens, & Werner Kaltefleiter 1983: *The Greens of West Germany: Origins, Strategies, and Transatlantic Implications* (Cambridge, Mass.: Institute for Foreign Policy Analysis).

Pfeiler, Wolfgang 1986: "Zum Verhältnis von Abschreckung und Entspannung in der Sicherheitspolitik," in Gohl & Niesporek (eds.): 127–148.

———— 1987: "Hat das sowjetische 'neue politische Denken' auch zu einem neuen militärischen Denken geführt?" *APZ,* B 44/87 (31 October): 28–38.

———— 1991: "Gorbatschows Deutschlandpolitik," in Brahm & al.: 115–130.

Philips, Sheena 1990: "Waffenentwicklung und neue Optionen für die konventionelle Verteidigung der NATO," in Heisenberg & Lutz (eds.): II:79–93.

Phillips, Ann L. 1990: "The German Democratic Republic and the New European Political Order," in Jordan (ed.): 87–114.

Phillips, Thomas R. (ed.) 1985: *Roots of Strategy. The 5 Greatest Military Classics of All Time* (Harrisburg: Stackpole).

Pickus, Robert 1991: "New Approaches," in Thompson & Jensen (eds.): 228–252.

Piepmeier, Rainer 1987: "Immanuel Kant (1724–1804): Friede als Ziel der Geschichte," in Rajewski & Riesenberger (eds.): 17–25.

Pierre, Andrew J. 1982: *The Global Politics of Arms Sales* (Princeton: Princeton U.P.).

———— 1986: "Enhancing Conventional Defense: A Question of Priorities," in Pierre (ed.): 9–39.

———— (ed.) 1986: *The Conventional Defense of Europe: New Technologies and New Strategies* (New York: Council on Foreign Relations).

Pike, John 1985: "Assessing the Soviet ABM Programme," in E. P. Thompson (ed.), *Star Wars* (Harmondsworth: Penguin): 50–67.

———— & Christopher Bolkcom 1992: "Prospects for an International Control Regime for Attack Aircraft," in Brauch & al. (eds.): 313–328.

———— & Eric Stambler 1992: "Constraints on the R&D and Transfer of Ballistic Missiles Defence Technology," in Brauch & al (eds.): 61–74.

————, Eric Stambler, & Chris Bolkcom 1992: "The Role of Land-Based Missile Defense," in Neuneck & Ischebeck (eds.): 125–145.

Pilat, Joseph 1993: "Technology Deployment and Denial: A Unilateral Approach to Arms Control?" in Ramberg (ed.): 113–139.

Piotrowskii, Harry 1992: "The Structure of the International System," in Schraeder (ed.): 209–228.

Piziali, Stefano 1989: "Il sistema difensivo jugoslavo," in Ragionieri (ed.): 247–271.

Plat, Joseph F., & Robert E. Pendley 1990: "NATO Strategy for the Nineties: The Federal Republic, the United States, and Continued Alliance Consensus," in Schmähling (cd.): 87–102.

Platias, Athanassios G., & R. J. Rydell 1982: "International Security Regimes: The Case of a Balkan Nuclear-Free Zone," in Carlton & Schaerf (eds.): 273–298.

Platt, Alan 1992: "Arms Control in the Middle East," in Spiegel (ed.): 139–150.

Plechanov, Sergeij 1989: "Disarmament and European Security," in Calliess (ed.): 19–26.

Pleiner, Horst 1990: "Aktuelle militärstrategische Entwicklungen und mögliche Auswirkungen auf das Bundesheer der neunziger Jahre," *ÖMZ*, 5: 369–379.

Plügge, Matthias 1990: "The East German Army: High-speed Perestroika," *IDR*, 4: 361–362.

Plummer, R. C. 1990: "Gorbachev and Soviet Defence Policy," in Pugh & Williams (eds.): 161–175.

Poch, Ulrich 1970: "Anpassungspolitik ohne Kollaboration: Der dänische Widerstand von 1940–45," in Ebert (ed.): 237–293.

Podlesnyi, Pavel 1989: "Non-military Aspects of Security: The View of a Soviet Scholar," in Wæver & al. (eds.): 73–84.

Poeppel, Hans 1990: "Wenn die direkte Drohung schwindet . . . Operative Planbung für die Zukunft," *EW*, 9: 532–535.

Poggiolini, Illaria, & Leopoldo Nuti 1992: "The Italian Peace Treaty of 1947: The Enemy/Ally Dilemma and Military Limitations," in Tanner (ed.): 27–38.

Pohlman, Hartmut 1993: "Wirkungen Rüstungstechnischer Veränderungen auf die sicherheitspolitische Stabilität," in Forndran & Pohlman (eds.): 221–236.

Poirier, Lucien 1991: "La Guerre de Golfe dans la généalogie de la stratégie," *Stratégique*, 51/52: 29–70.

Polish Institute of International Affairs 1990: "Towards a New Security Structure in Europe," *Peace and the Sciences*, December (Vienna: International Institute for Peace): 7–11.

Pollard, A. J. G. 1991: "Future Defence of the Central Region," in Reid & Dewar (eds.): 111–129.

Pond, Elizabeth 1990: *After the Wall: American Policy Toward Germany* (New York: Priority Press).

Pontara, Giuliano 1965: "The Rejection of Violence in Gandhian Ethics of Conflict Resolution," *JPR*, 2:3, 197–215.

Poole, J. B. & R. Guthrie (eds.) 1993: *Verification 1993: Peacekeeping, Arms Control and the Environment* (London: Brassey's).

Porch, Douglas 1991: "Arms and Alliances: French Grand Strategy and Policy in 1914 and 1940," in Kennedy (ed.): 125–143.

Posen, Barry R. 1982: "Inadvertant Nuclear War? Escalation and NATO's Northern Flank" (*IS*, Fall 1982) in Miller (ed.) 1984: 85–111.

——— 1984: "Measuring the European Conventional Balance: Coping with Complexity in Threat Assessment" (*IS*, 9:3) in Miller (ed.) 1986: 79–120.

——— 1984a: *The Sources of Military Doctrine: France, Britain, and Germany Between the World Wars* (Ithaca, N.Y.: Cornell U.P.).

——— 1988: "Is NATO Decisively Outnumbered?" *IS*, 12:4 (Summer): 186–202.

——— 1991: "Crisis Stability and Conventional Arms Control," *Dædalus*, Winter 1991: 217–232.

——— 1991: *Inadvertent Escalation: Conventional War and Nuclear Risks* (Ithaca, N.Y.: Cornell U.P.).

——— 1992: "Crisis Stability and Conventional Arms Control," in Adler (ed.): 231–246.

——— 1993: "The Security Dilemma of Ethnic Conflict," *Survival*, 35:1 (Spring): 27–47.

——— & Stephen Van Evera 1983: "Reagan Administration Defense Policy: Departure from Containment" (*IS* 8:1) in Miller (ed.) 1986: 19–61.

Postol, Theodore A. 1991: "Lessons of the Gulf War Experience with Patriot," *IS*, 16:3 (Winter): 119–171.

Pott, Andreas 1985: "Dic Scc und cin Systcm Kollektiver Sicherheit in Europa," in Lutz (ed.): 344–357.

——— 1986: "Maritime Rüstungssteuerung als Sonderfall der kooperativen Rüstungssteuerung," in Lutz, Müller, & Pott: 137–163.

——— 1987: "Die Strategie der Abschreckung: Mythenbild der Sicherheitspolitik," in Heisenberg & Lutz (eds.): 153–165.

Powers, John R., & Joseph E. Muckerman 1994: "Policy Brief: Rethinking the Nuclear Threat," *Orbis*, 38:1 (Winter): 99–108.

Prados, John 1982: *The Soviet Estimate: U. S. Intelligence Analysis and Russian Military Strength* (New York: Dial).

——— 1986: *Presidents' Secret Wars: CIA and Pentagon Covert Operations Since World War II* (New York: Morrow).

Pravda, Alex 1989: "Linkages Between Domestic and Foreign Policy in the Soviet Union," in Clesse & Schelling (eds.): 92–107.

——— 1990: "The Politics of Foreign and Security Policy," in Stephen White, Alex Pravda, & Zvi Gitelman (eds.), *Developments in Soviet Politics* (London: Macmillan): 207–227.

——— (ed.) 1991: *Yearbook of Soviet Foreign Relations* (London: Tauris).

——— (ed.) 1992: *The End of the Outer Empire: Soviet-East European Relations in Transition* (London: Sage).

Prawitz, Jan 1983: "A Nordic Nuclear Weapon-Free Zone: Model and Procedure," in Lodgaard & Thee (eds.): 191–198.

——— 1988: "Rustningskontroll och förtroendeskapande åtgärder i Östersjööm-rådet," in Joenniemi & Vesa (eds.): 216–223.

——— 1990: "Naval Arms Control: History and Observations," in Fieldhouse (ed.): 43–63.

Prestre, Philippe G. Le (ed.) 1989: *French Security Policy in a Disarming World: Domestic Challenges and International Constraints* (Boulder, Colo.: Lynne Rienner).

PRIM (Peace Research Institute, USSR Academy of Sciences) (ed.) 1991: *Ways to Security*, 2 vols. (Moscow: PRIM).

Primakov, Eugeny, Vladen Martynov, & Herman Diligensky 1990: "New Thinking," *IMEMO Discussion Papers* (Moscow: IMEMO).

Prins, Gwyn 1984: "Introduction: The Paradox of Security," in Prins (ed.): ix–xvii.

———— 1986: "Arms Control: Lessons Learned and the Future," in Avenhaus & al. (eds.): 51–96.

———— 1988: "Naval Requirements: A Note on Missiles and Mines," *Pugwash Newsletter,* 25:3 (January): 110–116.

———— 1988a: "Perverse Paradoxes in the Application of the Paradoxical Logic of Strategy," *Millennium,* 17:3, 539–551.

———— 1990: "Implications of the Changing Consensus on Nuclear Deterrence," in Schmähling (ed.): 17–36.

———— 1990a: "Military Restructuring and the Challenge of Europe," in UNIDIR (eds.): 52–105.

———— 1990b: "Politics and the Environment," *IA,* 66:4, 711–730.

———— 1990c: "Bonfire of the Certainties," in Huber (ed.): 115–126.

———— 1991: "The United Nations and Naval Power in the Post–Cold War World," *Disarmament,* 16:1, 183–215.

———— 1991a: "The United Nations and Peace-Keeping in the Post–Cold War World: The Case of Naval Power," *BPP,* 22:2 (June): 135–156.

———— 1992: "A New Focus for Security Studies," in Ball & Horner (eds.): 178–222.

———— (ed.) 1983: *Defended to Death: A Study of the Nuclear Arms Race* (Harmondsworth: Penguin).

———— (ed.) 1984: *The Nuclear Crisis Reader* (New York: Vintage; published in the U.K. as *The Choice*).

Proektor, Daniel M. 1988: "Clausewitz und die Gegenwart," *ÖMZ,* 2: 139–143.

———— 1989: "Über die sowjetische Militärdoktrin," in Wagemann (ed.): 177–188.

Pruitt, Dean G., & Richard C. Snyder (eds.) 1969: *Theory and Research on the Causes of War* (Englewood Cliffs, N.J.: Prentice-Hall).

Prystrom, Janusz 1988: "The Problem of Non-Offensive Defence—the Case of Poland," *CRPV,* 11:3, 115–120.

———— 1988a: "Stabilität—Vertrauen—Sicherheit: Wege zu einer neuen europäischen Friedensordnung," in Calliess (ed.): 251–256.

———— 1989: "The East-European Discussion About Alternative Defence Concepts," in Karl & Auffermann (eds.): 56–61.

———— 1992: "The Military-Strategic Emancipation of Poland," *PRIF-R,* 25.

———— 1994: "Eastern Europe and NOD," in Møller & Wiberg (eds.): 143–152.

Pugh, Michael C. 1990: "Prospects for International Order," in Pugh & Williams (eds.): 176–196.

———— 1992: "Maritime Peacekeeping: Scope for Deep Blue Berets?" *Working Paper,* 119 (Canberra: Australian National University, Peace Research Centre).

———— & Phil Williams (eds.) 1990: *Superpower Politics: Change in the United States and the Soviet Union* (Manchester: Manchester U.P.).

Pülöp, Mihàly 1992: "The Military Clauses of the Paris Peace Treaties with Rumania, Bulgaria, and Hungary," in Tanner (ed.): 39–54.

Purkrabek, Miroslav 1993: "Social Problems of the Democratic Transformation of the Human Potential of the Czechoslovak Army," *Journal of Slavic Military Studies,* 6:2 (June): 221–233.

Pursell, Carroll W., Jr. (ed.) 1972: *The Military Industrial Complex* (New York: Harper & Row).

Purver, Ronald G. 1988: "Arms Control Proposals for the Arctic: A Survey and Critique," in Möttölä (ed.): 183–221.

————— 1988a: "Arctic Arms Control: Constraints and Opportunities," *Occasional Papers,* 3 (Ottawa: Canadian Institute for International Peace and Security).

————— 1990: "SSBN Sanctuaries for Submarine Stand-Off Zones: A Possible Naval Arms Control Trade-Off," in Lodgaard (ed.): 224–239.

Putensen, Gregor, & Klaus Benjowski 1990: "Der deutsch-deutsche Einigungsprozess und seine Wirkungen für Europas Sicherheit," in Benjowski & al. (eds.): 72–86.

PvdA (Partij van der Arbeid) 1984: "Vrede en veiligheid: Een nota van de Tweede Kammerfractie Partij van der Arbeid" (Den Haag: PvdA).

————— 1987: "Vrede en Veiligheid in Europe: Defensienota van de Tweede Kamerfractie Partij van der Arbeid" (Den Haag: PvdA).

Pyle, Kenneth B. 1993: "Japan and the Future of Collective Security," in Unger (ed.): 99–117.

Pyskir, Bohdan 1993: "The Silent Coup: The Building of Ukraine's Military," *ES,* 2:1 (Spring): 139–161.

Quester, George 1966: *Deterrence Before Hiroshima* (New York: Wiley).

————— 1977: *Offense and Defense in the International System* (New York: Wiley).

————— 1985: "Substituting Conventional for Nuclear Weapons: Some Problems and Some Possibilities," in Hardin & al. (eds.): 245–266.

————— 1986: "Avoiding Offensive Weapons and Strengthening the Defense," in idem 1986a: 229–250.

————— 1986a: *The Future of Nuclear Deterrence* (New York: Wiley).

————— 1986b: "War Termination and Nuclear Targeting Stratcgy," in Ball & Richelson (eds.): 285–305.

————— 1988: *Offense and Defense in the International System,* 2d ed. (New Brunswick and Oxford: Transaction Books).

————— 1989: "The Necessary Moral Hypocrisy of the Slide into Mutual Assured Destruction," in Shue (ed.): 227–270.

————— 1989a: "Alternative Explanations of Soviet Policies on Conventional Defence," in Borg & Smit (eds.): 123–132.

————— 1989b: "Offence and Defence in the American Tradition," in Borg & Smit (eds.): 149–159.

————— 1989c: "Some Thoughts on 'Deterrence Failures'," in Stern & al. (eds.): 52–65.

————— 1990: "Finlandization as a Problem or an Opportunity," in Heisler (ed.): 33–45.

————— 1990a: "Some Barriers to Thinking About Conventional Defence," in Schmähling (ed.): 105–116.

————— 1991: "The Soviet Opening to Nonprovocative Defence," in Jervis & Bialer (eds.): 133–147.

————— 1991a: "Predicting the Future of Deterrence," in Nobel (ed.): 121–132.

————— 1992: "The Future of Nuclear Deterrence," *Survival,* 34:1 (Spring): 74–88.

————— 1992a: "Nuclear Proliferation and the Elimination of Nuclear Weapons," in Karp (ed.): 202–226.

————— 1993: "Unilateral Self-Restraint on Nuclear Proliferation: Canada, Sweden, Switzerland, and Germany," in Ramberg (ed.): 141–157.

"R." 1992: "Die Gemeinschaft Unabhängiger Staaten, GUS," *ÖMZ,* 30:6 (November–December): 563–565.

"R." 1993: "Ukraine: Die Frage der Atomwaffen," *ÖMZ,* 31:4 (July–August): 358–359.

"R." 1993a: "Letland—Die Landesverteidigung," *ÖMZ,* 31:3 (May–June): 363–365.

Rabert, Bernhard 1993: "South Africa's Nuclear Weapons—a Defused Time Bomb?" *AuP,* 44:3 (Autumn): 232–242.

Radinovic, Radovan 1989: "Yugoslavia's Doctrine of Security," in Singh & Vekaric (eds.): 181–205.

Raevsky, Andrei 1993: "Development of Russian National Security Policies: Military Reform," *Research Paper,* 25 (New York: UN; Geneva: UNIDIR).

Ragionieri, Rodolfo 1989: *Sicurezza comune: Une nuova strategia di pace oltra la deterrenza* (Firenze: Edizioni Cultura della Pace).

———— 1989a: "Introduction," in idem (ed.): 15–31.

———— 1989b: "L'Italia e il Mediterraneo: minaccia, scenari di crisi e problemi 'out-of-area'," in Ragionieri (ed.): 35–68.

———— 1990: "Italia: Fuerzas y estrategia," *Papeles para la Paz,* 39–40 (Madrid: Centro de Investigation para la Paz): 87–110.

———— 1990a: "Political and Military Aspects of Security in the Mediterranean Region," in Cerutti & Ragionieri (eds.): 142–150.

———— 1990b: "Reflections on Conflict Resolution in the Mediterranean," in Czempiel & al. (eds.): 141–160.

———— 1991: "Italy, NATO and the Mediterranean: Cooperative Security Versus Interventionism," in Møller (ed.): 31–43.

———— (ed.) 1989: *La sicurezza dell'Italia: Problemi e alternative* (Torino: Marietti).

Rajewsky, Christiane 1987: "Zur Geschichte des Pazifismus seit 1933," in Rajewsky & Riesenberger (eds.): 412–446.

———— & Dieter Riesenberger (eds.) 1987: *Wider den Krieg: Grosse Pazifisten von Immanuel Kant bis Heinrich Böll* (München: Beck).

Ramberg, Bennett 1986: "Targeting Nuclear Energy," in Ball & Richelson (eds.): 57–83.

———— 1993: "Arms Control Without Negotiation: A Concepual Overview," in idem (ed): 1–17.

———— (ed.) 1993: *Arms Control Without Negotiation: From the Cold War to the New World Order* (Boulder, Colo.: Lynne Rienner).

Ramet, Sabrina P. 1992: *Nationalism and Federalism in Yugoslavia, 1962–1991,* 2d ed. (Bloomington: Indiana U.P.).

Ramsbotham, Oliver 1989: *Modernizing NATO's Nuclear Weapons: "No Decisions Have Been Made"* (London: Macmillan).

———— 1991: "Election 1991/92: The Missing Defence Debate," *Current Decisions Report,* 6 (Oxford: Oxford Research Group).

———— & Hugh Miall 1992: *Beyond Deterrence: Britain, Germany and the New European Security Debate* (London: Macmillan).

Randle, Michael 1986: "Alternative Defence and Deterrence in the Nuclear Age," in Tromp (ed.): 245–256.

———— 1990: "Dissolving the Blocs in Europe," in Randle & Rogers (eds.): 91–107.

———— & Paul Rogers (eds.) 1990: *Alternatives in European Security* (Aldershot: Dartmouth).

Ranelagh, James 1987: *The Agency: The Rise and Decline of the CIA* (New York: Simon & Schuster).

Ranft, Bryan 1979: "Restraints on War at Sea before 1945," in Howard (ed.): 39–56.

Range, Clemens 1993: "Die sicherheitspolitischen Verhältnisse zwischen den GUS-Republiken," *EuS,* 42:1 (January): 45–48.

Rania, Peter 1978: *Political Opposition in Poland 1954–1977* (London: Thames Methuen).

Rapoport, Anatoli 1968: "Editor's Introduction," in Clausewitz, *On War* (Harmondsworth: Penguin, 1982).

—————— 1970: "Prisoner's Dilemma: Recollections and Observations," in Rapoport (ed.): 17–34.

—————— 1970a: "Critique of Strategic Thinking," in Naomi Rosenbaum (ed.), *Readings on the International Political System* (Englewoood Cliffs, N.J.: Prentice-Hall): 201–227.

—————— 1974: *Fights, Games and Debates* (Ann Arbor: University of Michigan Press).

—————— 1989: *The Origins of Violence: Approaches to the Study of Conflict* (New York: Paragon).

—————— 1992: *Peace: An Idea Whose Time Has Come* (Ann Arbor: University of Michigan Press).

—————— (ed.) 1970: *Game Theory as a Theory of Conflict Resolution* (Dordrecht: Reidel).

—————— & Albert M. Chammah 1970: *Prisoner's Dilemma: A Study in Conflict and Cooperation* (Ann Arbor: University of Michigan Press).

Rathjens, George 1973: "The Dynamics of the Arms Race," in Herbert York (ed.), *Arms Control: Readings from the Scientific American* (San Francisco: Freeman): 177–187.

Rattinger, Hans 1985: "The Federal Republic of Germany: Much Ado About (Almost) Nothing," in Flynn & Rattinger (eds.): 101–174.

Rauchensteiner, Manfred 1989: "Kriterien einer defensiven Verteidigung," in Wagemann (ed.): 65–76.

Rautenberg, Hans-Jürgen, & Norbert Wiggershaus 1977: *Die Himmeroder Denkschrift v. Oktober 1950: Politische und militärische Überlegungen für einen Beitrag der Bundesrepublik Deutschland zur westeuropäischen Verteidigung* (Karlsruhe: Militärgeschichtliches Forschungsamt).

Rauter, Gerhard 1991: "'Bundesdienstpflicht' statt Wehrpflicht? Gedanken zu Konzeptionen einer Verfassungsänderung," *ÖMZ* 29:5 (September–October): 377–384.

Ravenal, Earl C. 1978: *Never Again: Learning from America's Foreign Policy Failures* (Philadelphia: Temple U.P.).

—————— 1983: "No First Use: A View from the United States," *BAS,* April, 11–16.

—————— 1988: "Coupling and Decoupling: The Prospects for Extended Deterrence," in Hopmann & Barnaby (eds.): 59–70.

Record, Jeffrey 1993: *Hollow Victory: A Contrary View of the Gulf War* (Washington, D.C.: Brassey's).

Reeb, Hans Joachim 1990: "Eine Botschaft der Bundeswehr: Innere Führung auch für die NVA," *EW,* 7: 388–392.

Rees, Wyn 1991: "Britain and NATO," in Byrd (ed.): 88–104.

—————— (ed.) 1993: *International Politics in Europe: The New Agenda* (London: Routledge).

Reich, Utz-Peter 1990: "Die Kunst der Dialektik als Methode der Friedensforschung," in Weizsäcker (ed.): 104–109.

Reichart, John F., & Steven R. Sturm (eds.) 1983: *American Defense Policy,* 5th ed. (Baltimore: Johns Hopkins U.P.).

Reichel, Daniel 1988: "Jomini, ein 'AntiClausewitz'?" *ÖMZ,* 3: 241–247.

Reid, Brian Holden 1987: *J. F. C. Fuller: Military Thinker* (New York: St. Martin's).

—————— 1991: "The Counter-offensive: A Theoretical and Historical Perspective," in Reid & Dewar (eds.): 143–160.

———— & Michael Dewar (eds.) 1991: *Military Strategy in a Changing Europe* (London: Brassey's).

Reinfried, Hubert, & Ludwig Schulte 1987: *Ausstrieg aus der Nuklearstrategie? Chancen und Risiken für die Sicherheit Europas* (Herford: Mittler).

Reinhold, Otto 1988: "Wichtiger Beitrag und viele Diskussionsfragen: Zu Egon Bahrs Buch 'Zum europäischen Frieden'," *NGFH,* 35:10, 915–919.

Reiter, Erich 1990: "Die Bundesheerdebatte zwischen Anspruch und Wirklichkeit," *ÖMZ,* 28:3, 181–187.

Remacle, Eric 1990: "Project for European Security in the Year 2000," in Calliess (ed.): 419–427.

———— 1990a: "Les négotiations CFE et les bouleversements a l'Est dessinnent le nouveau paysage de l'Europe," *Memento Defese-Desarmement, 1990* (Bruxelles: GRIP): 25–34.

———— 1993: "The Yugoslav Crisis as a Test Case for CSCE's Role in Conflict Prevention and Crisis Mamagement," in Lucas (ed.): 109–123.

———— 1994: "Conventional Stability and NOD," in Møller & Wiberg (eds.): 51–59.

Renner, Michael G. 1989: "Conversion to a Peaceful Economy: Criteria, Objectives and Constituencies," in Dumas & Thee (eds.): 37–54.

———— 1989: "National Security: The Economic and Environmental Dimensions," *Worldwatch Paper,* 89 (Washington, D.C.: Worldwatch Institute).

———— 1990: "Swords Into Plowshares: Converting to a Peace Economy," *Worldwatch Paper,* 96 (Washington, D.C.: Worldwatch Institute).

———— 1992: *Economic Adjustments After the Cold War: Strategies for Conversion* (Aldershot: Dartmouth).

———— 1993: "Critical Juncture: The Future of Peacekeeping," *Worldwatch Paper,* 114 (Washington, D.C.: Worldwatch Institute).

Rentsch, Hellmuth 1962: *Partisanenkampf: Erfahrungen und Lehren* (Frankfurt: Bernard & Graefe).

Resor, Stanley, John D. Steinbruner, Paul Warnke, & Jack Mendelsohn 1989: "Special Briefing: European Arms Control After the NATO Summit," *Arms Control Today,* June/July, 38.

Reusch, Jürgen 1988: *Neue Sicherheitspolitik im Nuklearzeitalter* (Köln: Pahl-Rugenstein).

———— 1989: "Neues Denken und die marxistische Wissenschaft vom Frieden," in IMSF (eds.): 11–54.

Rice, Condolezza 1984: "Defense Burden-Sharing," in Holloway & Sharp (eds.): 59–86.

———— 1987: "The Party, the Military, and Decision Authority in the Soviet Union," *WP,* 1: 55–81.

———— 1988: "SALT and the Search for a Security Regime," in George, Farley, & Dallin (eds.): 293–306.

———— 1989: "Gorbachev and the Military: A Revolution in Security Policy, Too?" *Harriman Institute: Forum,* 2:4.

———— 1991: "The Evolution of Soviet Grand Strategy," in Kennedy (ed.): 145–164.

Richardson, Dick 1989: "Process and Progress in Disarmament: Some Lessons of History," in Harle & Sivonen (eds.): 26–44.

Richardson, William R. 1986: "FM 100-5: The AirLand Battle in 1986," *MR,* 66:3, 4–11.

Richtenwald, Marion 1988: "Perestrojka in den sowjetische Streitkräften," *Berichte,* 10.

Richter, Brigitta 1991: "Von der Westeuropäischen Union zu einer Europäischen Sicherheitsgemeinschaft," in Lutz (ed.): 303–316.

Richter, Claus 1987: "Die internationale Strategiediskussion," in Bühl (ed.): 29–54.

Richter, Frank (ed.) 1991: "Sowjetunion: Wieviele Armeen gibt es im Land?" *Osteuropa,* 41:10 (October): A556–A563.

———— & Jobst Echterling 1991: "Military Reform Develops in the Soviet Union," *AuP,* 42:1, 48–57.

Richter, Horst Eberhard 1987: "Das Problem Gemeinsame Sicherheit-Psychologische Aspekte," in Bahr & Lutz (eds.): 195–200.

Ridgway, Rozanne L. 1989: "The United States and Europe's Neutrals," in Kruzel & Haltzel (eds.): 286–291.

Ries, Tomas 1989: "Swedish and Finnish Defence Policies: A Comparative Study," in Kruzel & Haltzel (eds.): 200–221.

———— 1989a: *Cold Will: The Defence of Finland* (London: Brassey's).

———— 1992: "Russia's Military Inheritance," *IDR,* 25:3 (March): 223–226.

Riesenberger, Dieter 1987: "Zur Geschichte des Pazifismus von 1800 bis 1933," in Rajewsky & Riesenberger (eds.): 212–228.

———— 1987a: "Alfred Hermann Fried (1864–1921): Die Überwindung des Krieges durch zwischenstaatliche Organisation," in Rajewsky & Riesenberger (eds.): 54–60.

Rikhye, Indar Jit, & Kjell Skjelsbæk (eds.) 1990: *The United Nations and Peacekeeping: Results, Limitations and Prospects: The Lessons of 40 Years of Experience* (London: Macmillan).

Risse-Kappen, Thomas 1988: *Die Krise der Sicherheitspolitik: Neuorientierungen und Entscheidungsprozesse im politischen System der Bundesrepublik Deutschland, 1977–1984* (München: Kaiser).

———— 1990: "Predicting the New Europe," *BAS,* 46:8, 25–29.

———— 1991: "From Mutual Containment to Common Security: Europe During and After the Cold War," in Klare & Thomas (eds.): 123–143.

———— 1991a: "Did 'Peace Through Strength' End the Cold War? Lessons from INF," *IS,* 16:1 (Summer): 162–188.

———— 1991b: "Toward a Nuclear-Free Central Europe: The Future of Nuclear Weapons in a Post–Cold War Europe," in Danspeckgruber (ed.): 77–90.

Rittberger, Volker, Manfred Efinger, & Martin Mendler 1990: "Toward an East-West Security Regime: The Case of Confidence- and Security-Building Measures," *JPR,* 27:1, 55–74.

Rittstieg, Helmut 1987: "Gemeinsame Sicherheit als Maxime der Verteidigungspolitik beider deutschen Staaten?" in Bahr & Lutz (eds.): 47–54.

Rix, Christiane 1987: "Plädoyer für ein europäisches System Kollektiver Sicherheit," in Heisenberg & Lutz (eds.): 697–709.

———— & Peter Wilke 1986: "Gemeinsame Sicherheit zwischen Entspannungspolitik, Rüstungskontrolle und einseitiger Abrüstung," in Bahr & Lutz (eds.): 197–219.

Robert, Denis, & Cynthia Robert 1985: *How to Secure Peace in Europe* (London: Harney & Jones).

Roberts, Adam 1967: "Civilian Defence Strategy," in Roberts (ed.): 215–254.

———— 1967a: "Transarmament to Civilian Defence," in Roberts (ed.): 291–301.

———— 1969: "Civilian Resistance in Defence: The Defeat of the Generals' Revolt," in Anon (ed.): 146–154.

———— 1970: "Eine Schlacht gewonnen, den Krieg verloren? Zwischenbilanz des Widerstandes in der CSSR," in Ebert (ed.): 307–317.

———— 1976: *Nations in Arms: The Theory and Practice of Territorial Defence* (New York: Praeger).

———— 1977: "Civil Resistance and Swedish Defence Policy," in Geeraerts (ed.): 121–152.

———— 1978: "Civilian Defence Twenty Years On," *BPP*, 9:4, 293–300.

———— 1983: "Britain and Non-Nuclear Defence," in Baylis (ed.): 191–218

———— 1983a: "The Trouble With Unilateralism: The UK, the 1983 General Election, and Non-Nuclear Defence," *BPP*, 14:4, 305–312.

———— 1989: "The Possible Role of Territorial Defence in NATO Strategy," in Tromp (ed.): 213–222.

———— 1991: "New Peace Research, Old International Relations," in Nobel (ed.): 1–23.

———— 1991a: "Civil Resistance in the East European and Soviet Revolutions," *Monograph Series*, 4 (Cambridge, Mass.: Albert Einstein Institution).

———— 1992: "International Law and the Use of Force: Paper I," *AP*, 266: 53–63.

———— 1993: "Humanitarian War: Military Intervention and Human Rights," *IA*, 69:3 (July): 429–450.

———— 1993a: "The United Nations and International Security," *Survival*, 35:2 (Summer): 3–30.

———— 1993b: "The Changing Face of the United Nations," in Center for Defence Studies (ed.): 279–294.

———— (ed.) 1967: *The Strategy of Civilian Defence: Non-Violent Resistance to Aggression* (London: Faber & Faber).

———— & Benedict Kingsbury 1993: "The UN's Roles in International Society Since 1945," in Roberts & Kingsbury (eds.): 1–62.

———— & Benedict Kingsbury (eds.) 1993: *United Nations, Divided World: The UN's Role in International Relations*, new expanded ed. (Oxford: Oxford U.P.).

Roberts, Brad 1993: "From Nonproliferation to Antiproliferation," *IS*, 18:1 (Summer): 139–173.

Robinson, Julian Perry 1982: "The Changing Status of Chemical and Biological Warfare: Recent Technical, Military and Political Developments," in *SIPRI*, 1982: 317–362.

———— 1983: "Quasinuclear Weapons," in William Gutteridge & Trevor Taylor (eds.), *The Danger of New Weapons Systems* (London: Macmillan): 151–165.

———— 1986: "NATO Chemical Weapons Policy and Posture," *ADIU Occasional Paper*, 4.

———— 1992: "The Australia Group: A Description and Assessment," in Brauch & al. (eds.): 157–176.

————, Thomas Stock & Ronald G. Sutherland 1993: "The Chemical Weapons Convention: The Success of Chemical Disarmament Negotiations," *SIPRI*: 705–734.

Robles, Alfonso Garcia 1969: "The Treaty for the Prohibition of Nuclear Weapons in Latin America (Treaty of Tlatelolco)," in *SIPRI*, 1969/1970: 218–236.

Rochester, J. Martin 1993: *Waiting for the Millennium: The United Nations and the Future of World Order* (Columbia: University of South Carolina Press).

Rochon, Thomas 1989: *Mobilizing for Peace: The Antinuclear Movements in Western Europe* (Princeton, N.J.: Princeton U.P.).

Rodley, Nigel S. 1992: "Collective Intervention to Protect Human Rights and Civilian Populations: The Legal Framework," in Rodley (ed.): 14–42.

———— (ed.) 1992: *To Loose the Bands of Wickedness: International Intervention in Defence of Human Rights* (London: Brassey's).

Rogers, Bernard 1982: "The Atlantic Alliance: Prescriptions for a Difficult Decade," *FA*, 60:3 (Summer): 1145–1156.

———— 1983: "Sword and Shield: ACE Attack of Warsaw Pact Follow-On Forces," *NSN*, 28:1, 16–26.

———— 1983a: "Greater Flexibility for NATO's Flexible Response," *Atlantic Community Quarterly*, 21:3 (Fall): 233–243.

———— 1985: "Follow-On Forces Attack," *NATO Review*, 1: 1–9.

Rogers, Paul 1988: "The Nuclear Connection," *BAS*, 44:7 (September): 20–23.

———— 1990: "Alternative Military Options in Europe," in Randle & Rogers (eds.): 31–43.

———— 1991: "Alternative Strategies," in Byrd (ed.): 126–141.

———— & Malcolm Dando 1992: *A Violent Peace: Global Security After the Cold War* (London: Brassey's).

————, Malcolm Dando, & Peter Van Den Dungen 1981: *As Lambs to the Slaughter* (London: Arrow).

Roggemann, Herwig 1993: *Krieg und Frieden auf dem Balkan* (Berlin: Arno Spitz).

Rogov, Sergey M. 1988: "Abrüstung und europäische Sicherheit," in Calliess (ed.): 37–44.

———— 1990: "The Changing Face of Soviet-American Relations: A Soviet Perspective," in Brauch & Kennedy (eds.): 20–35.

———— 1990a: "The End of the Cold War and Soviet Military Spending," *Disarmament*, 13:3, 9–23.

———— 1992: "Military Reform: Now or Never," *ES*, 1:1 (Spring): 5–12.

———— 1993: "A National Security Policy for Russia," in Goodby & Morel (eds.): 75–80.

Röhling, Bert V. A. 1978: "Feasibility of Inoffensive Deterrence," *BPP*, 9:4, 339–347.

———— 1981: "Arms Control, Disarmament and Small Countries," *Impact of Science on Society*, 31:1, 97–112.

———— & Olge Sukovic 1976: *The Law of War and Dubious Weapons* (Stockholm: Almqvist & Wiksell, SIPRI).

Rohwer, Jürgen 1990: "Alternating Russian and Soviet Naval Strategies," in Gillette & Frank (eds.): 95–120.

———— 1990a: "Russian and Soviet Naval Strategy," in Skogan & Brundtland (eds.): 3–17.

Roick, Michael, & Fritz Vilmar 1988: "Strikt defensive Verteidigung—für eine sicherheitspolitische Alternative nach der Null-Lösung," in Calliess (ed.): 161–173.

Romanov, V. 1991: "The Murmansk Initiative: Path to Reduction of Level of Military Confrontation in the Arctic," in PRIM (ed.): I:85–95.

Romberg, Walter 1986: *Krisenstabile militärische Sicherheit in Mitteleuropa: Kriterien, Modelle und ethische Aspekte* (Berlin: Bund der Evangelischen Kirchen in der DDR).

———— 1988: "Towards Non-Offensive Defence Through Unilateral, Limited and Reciprocated Reductions: On a Gradualistic Approach to Military Crisis Stability in Central Europe," in Baudisch (ed.): 62–67.

———— 1989: "Ein gradualistischer Zugang zu militärischer Krisenstabilität," in SAS (eds.): 215–222.

———— 1989a: "The Relation Between Unilateral and Bilateral Steps," in Borg & Smit (eds.): 239–244.

———— 1990: "Thesen zu sicherheitspolitischen Alternativen für Osteuropa," *Bremer Beiträge*, 3: 6–8.

Romer, Jean-Christophe 1992: "Une armée russe: quelle armée?" *Politique étrangère*, 57:1 (Spring): 63–73.

Roose, Frank de 1990: "Military Responses to Civilian-Based Defense," *BPP*, 21:4, 421–429.

Roper, John (ed.) 1985: *The Future of British Defence Policy* (Aldershot: Gower).

————, Masashi Nishihara, Olara A. Ottunu, & Enid C.B. Schoettle 1993: *Keeping the Peace in the Post–Cold War Era: Strengthening Multilateral Peacekeeping: A Report to the Trilateral Commission* (New York: Trilateral Commission).

Ropp, Theodore 1941: "Continental Doctrines of Sea Power," in Earle (ed.): 446–456.

Rosas, Alan 1983: "Comments on a Draft Declaration," in Lodgaard & Thee (eds.): 233–237.

———— 1983a: "Non-Use of Nuclear Weapons and NWFZs," in Lodgaard & Thee (eds.): 221–230.

———— & Pär Stenbäck 1987: "Frontiers of Humanitarian Law," *JPR*, 24:3 (September): 219–236.

Rosenkranz, Erhard 1977: "Die Weizsäckersche Herausforderung," *EW*, 26:4, 178–182.

Rosolowsky, Diane 1987: *West Germany's Foreign Policy: The Impact of the Social Democrats and the Greens* (New York: Greenwood).

Rossbach, Anton 1989: "Konventionelle Abrüstung und die Umstrukturierung des Verteidigungskonzepte: Anmerkungen zu den Fragen," in Calliess (ed.): 130–138.

Rotblat, Joseph 1967: *Pugwash: A History of the Conferences on Science and World Affairs* (Prague: Czechoslovak Academy of Sciences).

———— 1981: *Nuclear Radiation in Warfare* (London: Taylor & Francis, SIPRI).

———— 1992: "Towards a Nuclear-Weapons-Free World: Societal Verification," *Security Dialogue*, 23:4 (December): 51–61.

———— 1993: "Past Attempts to Abolish Nuclear Weapons," in Roblat & al. (eds.): 17–32.

———— 1993a: "Societal Verification," in Rotblat & al. (eds.): 103–118.

———— (ed.) 1993: *Striving for Peace, Security and Development in the World: Annals of Pugwash, 1991* (London: World Scientific).

———— & Ubiratan d'Ambrosio (eds.) 1986: *World Peace and the Developing Countries: Annals of Pugwash, 1985* (London: Macmillan).

———— & V. I. Goldanski (eds.) 1989: *Global Problems of Common Security: Annals of Pugwash, 1988* (Berlin: Springer).

———— & Sven Hellman (eds.) 1985: *Nuclear Strategy and World Security: Annals of Pugwash, 1984* (London: Macmillan).

———— & J. P. Holdren (eds.) 1990: *Building Global Security Through Cooperation: Annals of Pugwash, 1989* (London: Macmillan).

————, Jack Steinberger, & Bhalachandra Udgaonkar (eds.) 1993: *A Nuclear-Weapon-Free World: Desirable? Feasible? Pugwash Monograph* (Boulder, Colo.: Westview).

Rotfeld, Adam Daniel 1983: "CBM in and for Europe," in Baudissin (ed.): 73–82.

———— 1986: "Developing a Confidence-Building System in East-West Relations: Europe and the CSCE," in Lynch (ed.): 69–127.

———— 1987: "Vertrauens- und Sicherheitsbildende Massnahmen zwischen Helsinki und Stockholm: Theorie und Erfahrungen aus polnischer Sicht," in Brauch (ed.): 105–151.

———— 1988: "The Formation of a System of Mutual Confidence and the Limitation of Armaments," in idem (ed.): 219–264.

———— 1989: "Confidence- and Security-Building Measures," in Blackwill & Larrabee (eds.): 359–378.

———— 1989a: "Confidence- and Security-Building Measures in Europe," in Rotblat & Goldanskii (eds.): 136–142.

———— 1990: "Changes in the European Security System," in Clesse & Rühl (eds.): 103–123.

———— 1991: "New Security Structures in Europe: Concepts, Proposals and Decisions," in *SIPRI*, 1991, 585–616 (with appendices).

———— 1991a: "The Future of Europe and Germany: Conference Summary," in Rotfeld & Stützle (eds.): 76–89.

———— 1991b: "New Security Structures in Europe: Concepts, Proposals and Decisions," in *SIPRI*: 585–616.

———— 1992: "Transformation of the European Security System: The Agenda Ahead," in Lutz (ed.): 409–422.

———— 1992a: "European Security Structures in Transition," *SIPRI*: 563–594.

———— 1993: "The CSCE: Towards a Security Organization," *SIPRI*: 171–221.

———— (ed.) 1986: *From Confidence to Disarmament* (Warsaw: PWN).

———— & Walther Stützle (eds.) 1991: *Germany and Europe in Transition* (Oxford: Oxford U.P., SIPRI).

Roth, Günter 1987: "Die Trias strategischer Herausforderungen der Zukunft," in Bald (ed.): 83–100.

———— 1989: "Deutsche Rüstungspolitik im Zeitalter der Weltkriege," in Wagemann (ed.): 47–56.

Rothenberg, Günther E. 1986: "Moltke, Schlieffen, and the Doctrine of Strategic Envelopment," in Paret (ed.): 296–325.

Rothschild, Emma 1985: "Common Security and Deterrence," in Väyrynen (ed.): 85–196.

———— 1992: "Egon Bahr and Common Security," in Lutz (ed.): 153–164.

Rotter, Manfred 1984: "Bewaffnete Neutralität: Das Beispiel Österreich," *Militärpolitik Dokumentation*, 8: 38.

Rougeron, Camille 1982: "The Historical Dimension of Guerilla Warfare," in Chaliand (ed.): 35–51.

Roy, Denny 1993: "Neorealism and Kant: No Pacific Union," *JPR*, 30:4 (November): 451–453.

Rozemond, Sam, & Jan Geert Siccima 1989: "Déja vu: Europe in Conventional Arms," in Tromp (ed.): 283–292.

Rubinstein, Alvin Z. 1991: "New World Order or Hollow Victory?" *FA*, 70:4 (Fall): 53–65.

Rücker, Brigitte, & Fritz Vilmar 1986: "Grundzüge und Hauptprobleme der 'Defensiv-Verteidigung'," *Mediatus*, 4: 3–7.

Rudney, Robert, & Luc Reuchler (eds.) 1988: *European Security Beyond the Year 2000* (New York: Praeger).

Ruehl, Lothar: see Rühl, Lothar.

Ruggie, John Gerard 1993: "Wandering in the Void: Charting the U.N.'s New Strategic Role," *FA*, 72:5 (November–December): 26–31.

Rühl, Lothar 1982: "MBFR: Lessons and Problems," *AP*, 176.

———— 1988: "An Increased Emphasis on Conventional Defence: A European View," in Nehrlich & Thomson (eds.): 35–39.

———— 1990: "Konventionelle Abrüstung im Sturmwind der Ereignisse," *EA*, 8: 264–274.

———— 1991: "Der Vertrag über Konventionelle Streitkräfte inm Europa und die neue Konsellation der europäischen Sicherheit," *EA*, 46:3, 81–90.

———— 1991a: "The Agreement on Conventional Forces in Europe: Culmination and End of European Arms Control?" *AuP*, 42:2, 116–127.

———— 1991b: "Sicherheit in Europa: Zu mehr Stabilität ohne Instabilität," in Wellershoff (ed.): 198–204.

———— 1991c: "Offensive Defence in the Warsaw Pact," *Survival,* 33:5 (September–October): 442–450.

———— 1991d: "Die 'Vorwärtsverteidigung' der NVA und der sowjetischen Streitkräfte in Deutschland bis 1990," *ÖMZ,* 29:6, 501–508.

Rühle, Hans 1984: "Wunschträume statt Alpträume," *EW,* 11.

Rühle, Michael 1991: "How Much Conventional Deterrence for NATO?" in Danspeckgruber (ed.): 107–122.

Ruloff, Dieter 1988: *Weltstaat oder Staatenwelt: Über die Chancen globaler Zusammenarbeit* (München: Beck).

Rummel, R. J. 1979: *Understanding Conflict and War,* vol. IV: *War, Power, and Peace* (Beverly Hills, Calif.: Sage).

———— 1985: "Libertarian Propositions on Violence Within and Between Nations: A Test Against Published Research Results," *Journal of Conflict Resolution,* 29:3 (September): 419–455.

———— 1991: "Political Systems, Violence, and War," in Thompson & Jensen (eds.): 350–370.

Rusman, Paul 1992: "A Conventional Arms Transfer Regime in the European Community," in Brauch & al. (eds.): 271–281.

Russett, Bruce M. 1990: "Politics and Alternative Security: Toward a More Democratic, Therefore More Peaceful, World," in Weston (ed.): 107–136.

———— 1993: *Grasping the Democratic Peace: Principles for a Post–Cold War World* (Princeton: Princeton U.P.).

———— & William Antholis 1992: "Do Democracies Fight Each Other? Evidence from the Peloponnesian War," *JPR,* 29:4 (November): 415–434.

————, Thomas Risse-Kappen, & John J. Mearsheimer 1990: "Correspondence: Back to the Future, Part III: Realism and the Realities of European Security," *IS,* 15:3, 216–222.

———— & James S. Sutterlin 1991: "The U.N. in a New World Order," *FA,* 70:2, 69–83.

Sabin, Philip A. G. 1986: "Reassurance, Consensus and Controversy: The Domestic Dilemmas of European Defence," in Flanagan & Hampson (eds.): 137–162.

Sadowski, Yahya 1992: "Sandstorm with a Silver Lining? Prospects for Arms Control in the Arab World," *Brookings Review,* 10:3 (Summer): 7–11.

Saeter, Martin 1983: "Norsk Atomvåpenpolitikk," in Heurlin (ed.): 114–139.

———— 1983a: "Nuclear Disengagement Efforts, 1955–80: Politics of *Status Quo* or Political Change," in Lodgaard & Thee (eds.): 53–70.

Saferworld Foundation 1990: "Security After the Cold War: Redirecting Global Resources" (Bristol: Saferworld).

———— 1990a: "Diversification of the Defence Industry" (Bristol: Saferworld).

———— 1990b: "Security in the New Europe" (Bristol: Saferworld).

———— 1990c: "UK Defence Requirements, 1990–2000" (Bristol: Saferworld).

———— 1990d: "The Gulf Crisis: Test Case for a New World Order" (Bristol: Saferworld).

———— 1990e: "European Security: The New Agenda" (Bristol: Saferworld).

———— 1991: "Western Security and Soviet Reform: A Programme for Action" (Bristol: Saferworld).

Sagan, Carl 1983: "Nuclear War and Climatic Catastrophe" (*FA,* Winter 1983–1984) in Bundy (ed.) 1985: 117–152.

Sagdeev, Roald Z., & Andrey S. Kokoshin 1990: "Stability of the Nuclear Balance After Deep Reductions I," in Hippel & Sagdeev (eds.): 9–22.

Saikal, Amin 1989: "The Regional Politics of the Afghan Crisis," in Saikal & Maley (eds.): 52–66.

—— & William Maley (eds.) 1989: *The Soviet Withdrawal from Afghanistan* (Cambridge: Cambridge U.P.).

Sailer, Michael 1989: "Die Bedeutung des kerntechnischen Risikopotentials für die Kriegsfähigkeit einer Industriegesellschaft," *S+F,* 7:4, 211–219.

Sakanaka, Tomohisa 1990: "International Relations in Asia," in UNIDIR (eds.): 119–131.

Saksena, K. P. 1993: *Reforming the United Nations: The Challenge of Relevance* (New Delhi: Sage).

Salicath, Carl P., & Harald W. Støren 1988: "Sveriges forhold til Sovietunionen 1981–1986: Normalisering eller tilpasning?" *Internasjonal Politikk,* 1: 93–121.

Salminen, Pertti 1992: "The Impact of Arms Technology on Military Doctrines: Documentation," *Finnish Defence Studies,* 5 (Helsinki: War College).

Salmon, Jack D. 1988: "Can Nonviolence Be Combined with Military Means for National Defence?" *JPR,* 25:1, 69–80.

SALSS (Sozialwissenschaftliche Analysen zur Leistungssteigerung Sozialer Systeme) 1984: "Konventionelle Landstreitkräfte für Mitteleuropa: Eine Militärische Bedrohungsanalyse" (Bonn: SALSS).

—— 1984a: "Spezialisierung auf die Defensive: Einführung und Quellensammlung: Ein Übersichtsgutachten" (Bonn: SALSS).

—— 1985: "Die Zukunft der Bundeswehr: Eine Durchleuchtung der amtlichen Planung: Gutachten" (Bonn: SALSS).

Salter, Stephen H. 1986: "I Cut, You Choose," in Carlson & Cromstock (eds.) 1986a: 293–304.

Sampson, Anthony 1977: *The Arms Bazaar* (Sevenoaks, Kent: Hodder & Stoughton).

Sandole, Dennis J. D., & Hugo van der Merwe (eds.) 1993: *Conflict Resolution Theory and Practice: Integration and Application* (Manchester: Manchester U.P.).

Saner, Hans 1989: "Wäre es nicht besser, die Schweiz würde konsequent auf Frieden setzen?" in Gross & al. (eds.): 133–146.

Sanguinetti, Antoine 1984: "French Defence: A Military Critique," in Howorth & Chilton (eds.): 173–189.

—— 1988: "Thoughts on the Possibility of Restoring European Balance: Prerequisites for the Current Situation," in Fedorenko (ed.): 137–161.

Saperstein, Alvin M. 1985: "New Technology Prospers Best with New Organization: Appropriate Use of Local Personnel in the Defense of Europe," *BPP,* 16:2, 121–127.

—— 1986: "Predictability, Chaos and the Transition to War," *BPP,* 17:1, 87–93.

—— 1986a: "Fluid Dynamic and Kinetic Theory Models in a Non-Provocative Land Defence of Central Europe," *American Journal of Physics,* 54:7, 607–611.

—— 1987: "An Enhanced Non-Provocative Defence in Europe: Attrition of Aggressive Armoured Forces by Local Militias," *JPR,* 24:1, 47–60.

—— 1988: "Provoking a Discussion on Non-Provocative Defence," *JPR,* 25:1, 91–93.

Sapir, Jacques, René Ernould, Daniel Pineye, & al. 1992: "La Décomposition de l'Armée Sovietique," *Dossier,* 45 (Paris: Fondation pour les études de défense nationale).

Sarkesian, Sam C. 1986: *The New Battlefield: The United States and Unconventional Conflicts* (New York: Greenwood Press).

———— (ed.) 1972: *The Military-Industrial Complex: A Reassessment* (Beverly Hills and London: Sage).

———— & John Allen Williams (eds.) 1990: *The U.S. Army in a New Security Era* (Boulder, Colo.: Lynne Rienner).

SAS 1984: "Landstreitkräfte zur Verteidigung der Bundesrepublik Deutschland.: Ein Modell für die 90er Jahre," in SAS (eds.): 156–173.

———— 1984a: "Das Personal alternativer Landstreitkräfte," in SAS (eds.): 176–180.

———— 1984b: "Über die Luftstreitkräfte einer defensiven Verteidigung," in SAS (eds.): 181–182.

———— 1987: "Das SAS-Konzept: Strukturwandel für die 90er Jahre (Stand Herbst 1987)" (Bonn: SAS).

———— 1989: "Prüfsteine für die Entwicklung einer stabilitätsgerechten konventionellen Verteidigung in Mitteleuropa," in SAS (eds.): 235–239.

———— 1989a: "'Verteidigerüberlegenheit' (SWP) und 'Vertrauensbildende Verteidigung' (SAS)" (Bonn: SAS).

———— (eds.) 1984: *Strukturwandel der Verteidigung: Entwürfe für eine konsequente Defensive* (Opladen: Westdeutscher).

———— (eds.) 1989: *Vertrauensbildende Verteidigung: Reform deutscher Sicherheitspolitik* (Gerlingen: Bleicher).

Sauerwein, Brigitte 1990: "David: A Non-offensive Defense System for Germany," *IDR,* 23:12 (December): 1405–1406.

———— 1993: "Rich in Arms, Poor in Tradition: The Ukrainian Armed Forces," *IDR,* 26:4 (April): 317–318.

Savelyev, Alexander 1993: "The ABM Treaty: Should We Keep It?" *Comparative Strategy,* 12:1 (January–March): 103–107.

———— & Yuri Streltsov 1989: "Problems of NATO Military Policy: Discussions in the West," in IMEMO (eds.): 225–235.

Savkin, V. Ye. 1972: *The Basic Principles of Operational Art and Tactics: A Soviet View* (Moscow), translated and published by the U.S. Air Force (Washington, D.C.: GPO).

Sawyer, Ralph D. (ed. and trans.; with Mei-chün Sawyer) 1993: *The Seven Military Classics of Ancient China* (Boulder, Colo.: Westview).

Schachter, Oscar 1991: "The Role of International Law in Maintaining Peace," in Thompson & Jensen (eds.): 65–127.

———— 1993: "Sovereignty and Threats to Peace," in Weiss (ed.): 19–44.

Schäfer, Helmut 1988: "Aufgaben konventioneller Rüstungskontrolle in Europa und die Perspektiven des KSZE-Proxesses," in Calliess (ed.): 63–72.

Scheer, Hermann 1984: "Der Ost-West Konflikt und die Antwort der NATO: Eine kritische Auseinandersetzung mit dem AirLand Battle-Konzept," in Brauch (ed.): 204–213.

———— 1985: "Chancen und Klippen einer deutsch-französischen Sicherkeitskooperation," *NGFH,* 8: 732–736.

———— 1986: "Mit Schlagworten gegen eine Denkschule," *EW,* 1: 5–6.

———— 1986a: *Die Befreiung von der Bombe: Weltfrieden, europäischer Weg und die Zukunft der Deutschen* (Köln: Bund).

———— 1987: "Miliz—nüchtern betrachtet," in Bald (ed.): 125–133.

———— 1989: "Sind Atomwaffen zur Kriegsverhütung erforderlich?" in SAS (eds.): 48–54.

Scheer, Robert 1982: *With Enough Shovels: Reagan, Bush and Nuclear War* (New York: Random House).

Scheffran, Jürgen 1991: "Die Modernisierung von C3I-Systemen in Europa," in Müller & Neuneck (eds.): 287–309.

Scheler, Wolfgang, & al. 1989: *Frieden. Krieg. Streitkräfte: Historisch-materialistischer Abriss* (Berlin, GDR: Militärverlag der DDR).

Schell, Jonathan 1984: *The Abolition* (London: Picador).

Schelling, Thomas C. 1960: *The Strategy of Conflict* (Cambridge: Harvard U.P.).

——— 1966: *Arms and Influence* (New Haven: Yale U.P.).

——— 1967: "Some Questions on Civilian Defence," in Roberts (ed.): 302–308.

——— 1986: "What Went Wrong with Arms Control," in Østerud (ed.): 90–109.

——— 1992: "The Thirtieth Year," in Adler (ed.): 21–31.

——— & Morton H. Halperin 1961: *Strategy and Arms Control* (New York: Pergamon, 1985).

Scherbak, Igor 1991: *Confidence-building Measures and International Security: The Political and Military Aspects: A Soviet Approach* (Geneva: UNIDIR).

Scheven, Werner von 1992: "The Merger of Two Formerly Hostile German Armies," *AuP*, 43:2, 164–173.

Schiemann, Catherine 1991: *Neutralität in Krieg und Frieden: Die Aussenpolitik der Vereinigten Staaten gegenüber der Schweiz 1941–1949: Eine Diplomatiegeschichtliche Untersuchung* (Zürich: Rüegger).

Schierholz, H. 1986: "Scheitert ein rot-grünes Bündnis an der NATO-Frage? Anmerkungen zur Friedenspolitik der Grünen," *Blätter*, 7: 1214–1225.

Schimmelfennig, Frank 1993: "The CSCE as a Model for the Third World? The Middle East and African Cases," in Lucas (ed.): 319–334.

——— 1993a: "The Dissolution of the Soviet Union and the Western Strategy of Security Through Reintegration," in Ehrhart & al. (eds.): 205–214.

Schirmeister, Helga, 1989: *Sicherheit durch Vertrauen* (Berlin, GDR: Staatsverlag).

Schirmer, Klaus 1989: "Bemerkungen zu Grundlagen und Kriterien des Streitkräftevergleichs," in Calliess (ed.): 67–78.

Schlichting, Ursel, & Jörg Wallner 1992: "Die Gemeinschaft Unabhängiger Staaten und die ehemals sowjetischen Streitkräfte: Ein schwieriges Erbe," *HB*, 66.

Schlotter, Peter 1982: "Die Ost-West Beziehungen als pluralistisches Sicherheitssystem," in HSFK (eds.): 42–61.

——— 1992: "Frieden durch Integration? Europäische Institutionen und die künftige Friedensordnung," in Wasmuht (ed.): 255–270.

Schmähling, Elmar 1987: "Bemerkungen zum Problemfeld 'Strukturelle Nichtangriffsfahigkeit'," *S+F*, 1: 72–77.

——— 1988: "Auch wir brauchen ein 'Neues Denken': Abschied von der Militarisierung der äusseren Sicherheit," *NGFH*, 6: 542–547.

——— 1988a: "Das Ende unserer Sicherheit oder Abschied von einer Illusion? Warum die Falken gegen das INF-Abkommen sind," *S+F*, 1: 34–37.

——— 1988b: "Gedanken zu den gestellten Fragen über konventionelle Stabilität," in Bahr & Lutz (eds.): 472–476.

——— 1988c: "Von der 'Strukturellen Nichtangriffsfähigkeit' zur 'Verteidigerüberlegenheit': Eine militärische Würdigung," in Bahr & Lutz, (eds.): 62–80.

——— 1989: "Armee als Wagenburg? Über Tendenzen der Abschrottung in der Bundeswehr und falsche Reaktionen auf den gesellschaftlichen Wandel-ein Plädoyer für demokratisierte Streitkräfte," in Vogt (ed.): 252–264.

——— 1989a: "German Security Policy Beyond American Hegemony," *WPJ*, 6:2, 371–384.

——— 1989b: "Rüstungskontrolle und Abrüstung bei Seestreitkräften," *S+F*, 7:2, 150–152; also in E. Müller & Neuneck (eds.) 1990: 93–100.

——— 1989c: "Seestreitkräfte und Rüstungsbeschränkung: Möglichkeiten und Probleme der Einbeziehung von Seestreitkräften in den Prozess von

Abrüstung, Rüstungskontrolle und Vertrauensbildung," *APZ*, B 8/89 (February): 17.

——— 1990: "Conclusion: A New Way of Thinking for a Future Security Regime," in Schmähling (ed.): 189–199.

——— 1990a: "Geht die Nachkriegszeit für die Bundesrepublik Deutschland zu Ende?" in Lutz & Schmähling (eds.): 23–34.

——— 1990b: "NATO's Nuclear Strategy in a Dilemma," in Schmähling (ed.): 3–10.

——— 1990c: "The Vulnerability of Surface Forces to Modern Stand-Off Weapons," in Boserup & Neild (eds.): 173–178.

——— 1990d: "Vom nationalen Kriegsdienst zum europäischen Friedensdienst— die Zukunft der Bundeswehr," in H-E. Bahr (ed.): 119–126.

——— 1990e: *Der unmögliche Krieg: Sicherheit und Verteidigung vor der Jahrtausendwende* (Düsseldorf: ECON).

——— 1990f: "Naval Forces in the Process of Disarmament, Arms Control and Confidence-Building," in Lodgaard (ed.): 118–129.

——— (ed.) 1990: *Life Beyond the Bomb: Global Stability Without Nuclear Deterrence* (München: Berg).

——— & Dieter Mahncke 1990: "Hoffnung ohne Garantie," *NGFH*, 37:2, 150–156.

Schmid, A. P., & al. 1985: *Social Defence and Soviet Military Power: An Inquiry into the Relevance of an Alternative Defence Concept* (Leiden: Center for the Sudy of Social Conflict, State University of Leiden).

Schmid, Günther 1983: "Positionen in der sicherheitspolitischen Diskussion und ihre Vertreter in der Bundesrepublik Deutschland," *ÖMZ*, 6: 504–512.

——— 1987: "Alternative sicherheitspolitische Konzepte—Inhalte, Möglichkeiten und Grenzen," in Bühl (ed.): 235–254.

Schmidt, Hans-Joachim 1988: "Conventional Arms Control in Europe," *PRIF-R*, 6.

——— 1990: "Die Anfangsphase der Wiener Verhandlungen über konventionelle Streitkräfte in Europa," in E. Müller & Neuneck (eds.): 11–36.

——— 1991: "Konventionelle Rüstungskontrolle in Europa (KSE)," in Schwerdtfenger & al. (eds.): 278–293.

Schmidt, Helmut 1984: "Zur Lage der Sicherheitspolitik: Zweiter Verhandlungstag des ausserordentlichen Parteitages der SPD in Köln vom 18. bis 19.11 1983," in Maruh & Wilke (eds.): 129–164.

——— 1986: *Eine Strategie für den Westen* (Berlin: Siedler).

——— 1987: "Deutsch-französische Zusammenarbeit in der Sicherheitspolitik," *EA*, 11: 303–307.

——— 1993: "Lessons of the German Reunification for Korea," *Security Dialogue*, 24:4 (December): 397–408.

Schmidt, Max 1989: "Zur Kriegsuntauglichkeit hochindustrialisierter Gesellschaften," *S+F*, 2: 106–109

——— 1991: "German Unity and Security in Europe," in Rotfeld & Stützle (eds.): 69–73.

——— & al. 1989: *Sicherheit und Friedliche Koexistenz: Umfassende internationale Sicherheit Umsetzung freidlicher Koexistenzbeziehungen heute* (Berlin: Staatsverlag der DDR).

——— & Wolfgang Schwarz 1988: "Das Gemeinsame Haus Europa: Realitäten, Herausforderungen, Perspektiven," 2 parts, *IPW-B*, 9: 1–10; 10: 1–11.

——— & Wolfgang Schwarz 1989: "DDR und BRD: Sicherheitspartner im gemeinsamen Haus Europa," *IPW-B*, 8: 1–10.

——— & Wolfgang Schwarz 1989a: "Frieden als Überlebungsbedingung," *S+F*, 7:4, 206–211.

———— & Wolfgang Schwarz 1989b: "From East Germany: It's NATO's Move Now," *BAS*, 45:7, 5–7.

———— & Wolfgang Schwarz 1989c: "Herausforderungen und Perspektiven für Gemeinsame Sicherheit und das gemeinsame Haus Europa," in IMSF (eds.): 248–267.

———— & Wolfgang Schwarz 1989d: "Strukturelle Angriffsunfähigkeit in Europa: Vernünftige Alternative zur Abschreckung," *Horizont*, 3: 3–4.

———— & Wolfgang Schwarz 1989e: "Umfassende internationale Sicherheit Erfordernis unserer Zeit," in Schmidt & al.: 10–61.

———— & Wolfgang Schwarz 1990: "German Unity and European Security," *Peace and the Sciences* (Vienna: International Institute for Peace): 1–9.

———— & Wolfgang Schwarz 1990a: "Mutual Inability to Launch Attacks, and Arms Reductions in Europe," *Disarmament*, 13:1, 69–82

———— & Wolfgang Schwarz 1990b: "Von der nuklearen Abschreckung zur Gemeinsamen Sicherheit," in Lutz & Schmähling (eds.): 85–112.

Schmidt, Peter 1992: "Partners or Rivals: NATO, WEU, EC and the Reorganization of European Security Policy—Taking Stock," in idem (ed.): 187–228.

———— (ed.) 1992: *In the Midst of Change: On the Development of West European Security and Defence Cooperation* (Baden-Baden: Nomos).

Schmidt-Eenboom, Erich 1989: "Strukturelle Verteidigungsuntauglichkeit urbaner Zentren und Regionen," *S+F*, 7:4, 226–231.

Schmige, Georg 1988: "Staatenvertrag statt Friedensvertrag," *DA*, 21:11, 1167–1170.

Schmitz, Walter, Otto Reidelhuber, & Klaus Niemeyer 1986: "Some Long-Term Trends in Force Structuring," in Huber (ed.): 141–166.

Schmückle, Gerd, & Albrecht A. C. von Müller 1988: "Das Konzept der 'Stabilen Abhaltung': Umrisse eines Gesamtkonzepts zur Um- und Afrüstung in Europa" (Starnberg: Max Planck Society).

Schnappertz, Jürgen, & Wolfgang Bruckmann 1990: "Die Westintegration als Modell für Gesamteuropa," in Calliess (ed.): 271–280.

Schneider, Harald 1990: "Konzeption und Realität der Verteidigung im Bündnis," in Heisenberg & Lutz (eds.): I:239–258.

Schneider, William, Jr. 1993: "The Future of the ABM Treaty," *Comparative Strategy*, 12:1 (January–March): 53–56,

Schneider, Wolfgang 1987: "Unser Konzept der militärischen Landesverteidigung: Statt auf Risikostrategien wird auf klare Defensive gesetzt," *Dialog*, 10: 227–234.

Schneiter, George 1984: "The ABM Treaty Today," in Carter & Schwartz (eds.): 221–250.

Schnell, Harry H. 1993: "Multinationale Streitkräftestrukturen," *EuS*, 42:8 (August): 417–420.

Schölch, Ingrid 1990: *Abschreckung, Sicherheit und Stabilität: Grundsatzprobleme der sicherheitspolitischen Situation in Europa* (Baden-Baden: Nomos).

———— 1990a: "Political and Military Determinants of Military Stability in Europe," in Huber (ed.): 65–70.

Scholz, Rupert 1989: "Alternative Verteidigung? Die Strategie der 'Spinne im Netz'," *EW*, 11: 643–646.

———— 1989a: "Sicherheit trotz Abrüstung?" in Wagemann (ed.): 107–114.

———— 1990: "Deutsche Frage und europäische Sicherheit: Sicherheitspolitik in einem sich einigenden Deutschland und Europa," *EA*, 7: 239–246.

Schönbohm, Jörg 1992: *Zwei Armeen und ein Vaterland: Das Ende der Nationalen Volksarmee* (Berlin: Siedler).

Schraeder, Peter J. 1992: "Studying U.S. Intervention in the Third World," in idem (ed.): 1–19.

———— 1992a: "U.S. Intervention in Perspective," in idem (ed.): 385–405.

———— (ed.) 1992: *Intervention into the 1990s: U.S. Foreign Policy in the Third World* (Boulder, Colo.: Lynne Rienner).

Schramm, Friedrich-Karl, Wolfram-Georg Riggert, & Alois Friedel (eds.) 1972: *Sicherheitskonferenz in Europa: Dokumentation 1954–1972: Die Bemühungen um Entspannung und Annäherung im politischen, militärischen, wirtschaftlichen, wissenscahftlich-technologischen und kulturellen Bereich* (Frankfurt: Metzner).

Schreiber, Wilfried 1988: "Zur Militärdoktrin der Staaten des Warschauer Vertrages," in Calliess (ed.): 140–145.

Schreiner, Karl H. 1987: "'Strukturelle Nichtangriffsfähigheit': Eine Auseinandersetzung mit den Thesen von Albrecht A. C. von Müller," in Bühl (ed.): 173–183.

Schröder, Diethelm (ed.) 1984: *Krieg oder was sonst? NATO: Strategie des Unsicherheit* (Reinbek: Rowohlt).

Schröder, Hans-Henning 1989: "Die Verteidigungspolitik der UdSSR 1987–1989," *Berichte*, 14.

———— 1991: "Die Abrüstungs- und Konversionsprozesse in der Sowjetunion: Erkenntnisse aus Erfahrungen," *Dialog*, 20: 175–188.

———— 1991a: "Vormalige Säulen der Macht: Streitkräfte und KGB," *Blätter* 36:10 (October): 1179–1183.

———— 1991b: "Militär und Politik in der Sowjetunion: Eine Analyse mit aktuellem Nachwort," in Birkenbach & al. (eds.): 123–136.

———— 1992: "Vereinigte Streitkräfte und nationale Armeen," *OE*, 42:8 (August): 669–679.

Schrogl, Kai-Uwe 1990: "Die Begrenzung konventioneller Rüstung in Europa: Ein regimeanalytisches Konfliktmodell," *Tübinger Arbeitspapiere zur internationalen Politik und Friedensforschung* (Tübingen: Universität Tübingen, Institut für Politikwissenschaft).

Schubert, Klaus von 1970: *Wiederbewaffnung und Westintegration: Die innere Auseinandersetzung um die militärische und aussenpolitische Orientierung der Bundesrepublik, 1950–1952* (Stuttgart: Deutsche Verlags-Anstalt).

———— 1988: "Strategische Voraussetzungen einer zukünftigen Wehrstruktur," in Bald & Klein (eds.): 11–22.

———— 1989: "Gemeinsame Sicherheit—das neue Paradigma: Ein Beitrag aus der Sicht der Bundesrepublik Deutschland," *NGFH*, 36:11, 978–982.

———— 1990: "Mutual Responsibility: A Strategy for West and East," in Schmähling (ed.): 73–86.

———— 1992: *Von der Abschreckung zur gemeinsamen Sicherheit* (Baden-Baden: Nomos).

————, Egon Bahr, & Gert Krell 1988: "Stellungnahme: Zur Entwicklung der Ost-West-Beziehungen nach dem INF-Vertrag," in Moltmann (ed.): 9–14.

Schulte, Gregory L. 1993: "NATO's Nuclear Forces in a Changing World," *NATO Review*, 41:1 (February): 17–22.

Schulte, Ludwig 1985: "Das Schlagwort von der 'defensiven Umrüstung'," *EW*, 9: 474–479.

———— 1990: *Trumpf der Verteidigung: Mehr Stabilität durch moderne Technologien* (Baden-Baden: Nomos).

Schultz, Eberhard 1988: "Militärische und politische Sicherheit," *NGFH*, 6: 525–531.

———— 1989: "Berlin-Politik: Über das Provisorium hinausgedacht: Ein Ansatz zu einer Antwort auf die deutsche Frage in Selbstbestimmung," *DA*, 22:10, 1094–1106.

———— 1989a: "Europäische Sicherheit als ein Aspekt der deutschen Frage," *DA*, 22:5, 524–532.

Schulz, Erich (ed.) 1992: "Neuorganization der polnischen Streitkräfte: Neue Westvorstellungen, Motivationen, Zielsetzungen und Aufgaben," *OE*, 42:8 (August): A430–A447.

———— (ed.) 1993: "Sicherheit und Rückbesinnung auf Tradition: Zur Entwicklung der ungarischen Streitkräften," *OE*, 43:3 (April): A202–A211.

———— (ed.) 1993a: "Eine Armee geht auseinander: Probleme der tschechoslowakischen Streitkräfte bei der Trennung der Föderation," *OE*, 43:5 (May): A257–A269.

———— (ed.) 1993b: "Reformprozesse und Demokratisierungsbestrebungen in der Bulgarischen Armee," *OE*, 43:3 (April): A212–A224.

———— (ed.) 1993c: "Verteidigung der Sicherheit des Landes und der Demokratie: Zur neuesten Entwicklung in den rumänischen Streitkräften," *OE*, 43:11 (November): A662–A642.

———— (ed.) 1993c: "Neuaufbau und Umstrukturierung der tschechoslowakischen Streitkräfte vor der Teilung," *OE*, 43:1 (January): A22–A40.

Schulze, Franz-Joseph 1986: "Die Notwendigkeit unverzüglicher gemeinsamer Abwehrreaktionen," in Kaiser & Lellouche (eds.): 150–159.

Schulze-Marmeling, Dieter 1987: *Die NATO: Anatomie eines Militärpaktes* (Göttingen: Werkstatt).

Schuster, Dieter, & Ulrike C. Wasmuht 1984: "'Alternative Strategien': Eine Auswertung der öffentlichen Anhörung im Verteidigungsausschuss des Deutschen Bundestages," *S+F*, 3: 2–16.

Schütze, Walter 1983: "European Security Policy in the 1980s: Rethinking Western Strategy," in Lodgaard & Thee (eds.): 81–91.

———— 1987: "Vorschläge für die Ausgestaltung der Vertrauens- und Sicherheitsbildenden Massnahmen," in Brauch (ed.): 227–243.

———— 1988: "What Kind of Military Posture Would Corresspond to the Idea of Common Security?" in Baudisch (ed.): 68–75.

———— 1989: "Die Auswirkungen der deutsch-französischen Zusammenarbeit auf die alternative europäische Sicherheit," in Karl & Auffermann (eds.): 98–105.

———— 1990: "Bauplan für eine gesamteuropäische Architektur," *Blätter*, 5: 585–591.

Schwartz, David 1984: "Past and Present: The Historical Legacy," in Carter & Schwartz (eds.): 330–349.

Schwartz, Klaus-Dieter (ed.) 1977: *Sicherheitspolitik: Analysen zur politischen und militärischen Sicherheit*, 2d ed. (München: Osang).

———— (ed.) 1978: *Sicherheitspolitik: Analysen zur politischen und militärischen Sicherheit*, 3d ed. (München: Osang).

Schwarz, Wolfgang 1988: "Non-offensive Defence: Goals and Issues of Disarmament Between NATO and WTO in Europe," *IPW-P*, November.

———— 1989: "Strukturelle Angriffsunfähigkeit in Europa: Erfordernisse, Konzepte, Hemnisse," *IPW-F*, 2.

———— 1989a: "Strukturelle Angriffsunfähigkeit in Europa: NATO und Warschauer Vertragsorganisation," *Beiträge zur Konfliktforschung*, 2.

Schweizerische Friedensstiftung (ed.) 1990: *Blocküberwindende Vertrauensbildung nach dem europäischen Herbst '89* (Bern: Schweizerische Friedensstiftung).

Schwerdtfenger, Johannes, Egon Bahr, & Gert Krell (eds.) 1991: *Friedensgutachten 1991* (Münster: LIT).

Schwok, René 1990: "The European Community and Switzerland: Fewer Frontiers but Continuing Neutrality," in Miliovejic & Maurer (eds.): 217–231.

Scott, William F. 1990: "Soviet Military Doctrine: Continuity and Change?" in Green & Karasik (eds.): 1–11.

Scowcroft Commission 1983: "The Future of US Strategic Forces: Report of the President's Commission on Strategic Forces, 11 April 1983 (Excerpts)," *Survival,* 25:4, 177–186.

SDP & Liberal Party 1986: "Defence and Disarmament: Report of the Joint SDP-Liberal Alliance Commission" (Hebden Bridge: Hebden Royd).

Seethaler, Frank A. 1988: "Switzerland's Current Defence Posture," *IDR,* 6: 625–632.

Segal, Gerald 1985: *Defending China* (New York: Oxford U.P.).

———— 1991: "North-East Asia: Common Secuirity or á la carte?" *IA,* 67(4): 755–767.

———— 1991a: "China," in Karp (ed.): 189–205.

———— & William T. Tow (eds.) 1984: *Chinese Defence Policy* (London: Macmillan).

Segbers, Klaus 1991: "Sowjetische Aussenpolitik im Wandel," in Hamann & Matthies (eds.): 99–127.

Seidelmann, Reimund 1987: "SPD und SED zur Schaffung eines atomwaffenfreien Korridors," *DA,* 2: 134–148.

———— 1989: "Zu den europapolitischen Aussagen des neuen SPD-Programms," *NGFH,* 36:9, 1035–1037.

Seiffert, Wolfgang 1986: *Das ganze Deutschland: Perspektiven der Wiedervereinigung* (München: Piper).

———— 1989: *Die Deutschen und Gorbatschow: Chancen für einen Interessenausgleich* (Erlangen: Straube).

———— 1990: "Das Selbstbestimmungsrecht der Deutschen," in Venohr (ed.): 91–102.

Sella, Amnon 1991: "Soviet Collective Security Perception," in Hemsley (ed.): 51–66.

Semlitsch, Karl 1987: "Zum Abrüsten durch Umrüsten auf defensive Wehrstrukturen—eine Idee der neutralen und blockfreien Staaten Europas," in Brauch (ed.): 390–400.

Senghaas, Dieter 1969: *Abschreckung und Frieden: Studien zur Kritik der organisierten Friedlosigkeit* (Frankfurt: Suhrkamp).

———— 1972: *Rüstung und Militarismus* (Frankfurt: Suhrkamp).

———— 1985: "Die Ostproblematik eines europäischen Systems Kollektiver Sicherheit," in Lutz (ed.): 294–309.

———— 1986: *Die Zukunft Europas: Probleme der Friedensgestaltung* (Frankfurt: Suhrkamp).

———— 1987: "Conventional Forces in Europe: Dismantle Offense, Strengthen Defense," *BAS,* 43:10 (December): 911.

———— 1988: *Konfliktformationen im Internationalen System* (Frankfurt: Suhrkamp).

———— 1988a: "Die Entmilitarisierung des Systemkonflikts: Europas Change," in Calliess (ed.): 45–57.

———— 1990: "Die Zivilisierung des Ost-West-Konfliktes," in Weizsäcker (ed.): 171–188.

———— 1990a: "Europe 2000: A Peace Plan," *Alternatives,* 15:4, 467–478.

———— 1990b: "Requirements for a Timely Peace Policy," *BPP,* 21:4, 441–447.

———— 1990c: "Was friedenspolitisch jetzt zu tun wäre," in H-E. Bahr (ed.): 169–188.

——— 1990d: *Europa 2000: Ein Friedensplan* (Frankfurt: Suhrkamp).

——— 1990e: "Frieden in einem Europa demokratischer Rechtsstaaten: Ausgangslage, Perspektiven, Probleme," *Dialog,* 18: 179–197.

——— 1990f: "Arms Race Dynamics and Arms Control," in Gleditsch & Njølstad (eds.): 15–30.

——— 1990g: "Auch für die 'Soziale Verteidigung' schlägt jetzt die Stunde der Wahrheit," in Calliess (ed.): 627–647.

——— 1991: "Die politische Konversion Europas: Friedenspolitische Überlegungen an der Schwell der neunziger Jahre," *Dialog,* 20: 203–213.

——— 1991a: "Friedliche Streitbeilegung und kollektive Sicherheit im neuen Europa," *EA,* 46:10, 311–317.

——— 1991b: "Peace Theory and the Restructuring of Europe," *Alternatives,* 16:3 (Summer): 353–366.

——— 1992: "Weltinnenpolitik—Ansätze für ein Konzept," *EA,* 47:22 (25 November): 643–652.

——— 1992a: "Friedensforschung an der Schwelle der neunziger Jahre," in Bächler (ed.) 1992a: 3–16.

——— 1992b: "Die Welt als Schrecken: Internationale Politik zwischen Zivilisierung und Regression," in Lutz (ed.): 347–355.

——— 1992c: "Europe: Die Teilung überwinden," in Butterwege & Senghaas-Knobloch (eds.): 25–33.

——— 1993: "Was sind der Deutschen Interessen?" *Blätter,* 38:6 (June): 673–687.

——— 1993a: "Frieden und Krieg in dieser Zeit: Sechs Thesen," *S+F,* 11:3, 159–165.

——— & Eva Senghaas-Knobloch 1992: "Si vis pacem, para pacem: Überlegungen zu einem zeitgemässen Friedenskonzept," *Leviathan: Zeitschrift für Sozialwissenschaft,* 20:2 (June): 230–251.

——— (ed.) 1971: *Kritische Friedensforschung* (Frankfurt: Suhrkamp).

Serebryannikov, Vladimir 1989: "More on the Defence Doctrine Dilemma," *SMR,* 9: 23–24.

Setar, John 1989: "Quantitative Modelling of Combat and Non-Provocative Defence," in Borg & Smit (eds.): 87–98.

Shahi, Aga 1990: "Defence, Disarmament and Collective Security," in UNIDIR (eds.): 172–186.

Shambaugh, David 1992: "China's Security Policy in the Post–Cold War Era," *Survival,* 34:2 (Summer): 88–106.

Sharfman, Peter 1991: "The Future of FOFA," in Reid & Dewar (eds.): 5–110.

Sharp, Gene 1967: "The Technique of Nonviolent Action," in Roberts (ed.): 87–105.

——— 1973: *The Politics of Non-violent Action,* 3 vols. (Boston: Porter Sargent).

——— 1978: "Nonviolent Action as a Substitute for International War," *BPP,* 9:4, 321–328.

——— 1979: *Gandhi as a Political Strategist—With Essays on Ethics and Politics* (Boston: Porter Sargent).

——— 1985: *Making Europe Unconquerable: The Potential of Civilian-Based Deterrence and Defence* (London: Taylor & Francis).

——— 1985a: *National Security Through Civilian-Based Defense* (Omaha, Nebr.: Association for Transarmament Studies).

——— 1990: "The Role of Power in Nonviolent Struggle," *Monograph Series,* 3 (Cambridge, Mass.: Albert Einstein Institution).

——— 1992 (with Bruce Jenkins): *Self-Reliant Defense Without Bankruptcy or War: Considerations for the Baltics, East Central Europe, and Members of the*

Commonwealth of Independent States (Cambridge, Mass.: Albert Einstein Institution).

——— 1993: "L'Abolition de la guerre, un but réaliste," *Cahiers de la Non-Violence,* 4 (Quebec: Le Centre de ressources sur la Non-violence).

——— & Bruce Jenkins 1990: *Civilian-Based Defense: A Post-Military Weapons System* (Princeton: Princeton U.P.).

Sharp, Jane M. O. 1984: "Are the Soviets Still Interested in Arms Control?" *WPJ,* 1:4, 814–849.

——— 1985: "Arms Control and Alliance Commitments," *Political Science Quarterly,* 100:4, 649–667.

——— 1987: "After Reykjavik: Arms Control and the Allies," *IA,* 63:2, 239–257.

——— 1987a: "NATO's Security Dilemma," *BAS,* 43:3 (March): 42–44.

——— 1988: "Conventional Arms Control in Europe: Problems and Prospects," in *SIPRI,* 1988: 315–337.

——— 1989: "Conventional Arms Control in Europe," in *SIPRI:* 369–402.

——— 1989a: "Belief Systems and Arms Control Possibilities," in Carlton & Schaerf (eds.): 114–134.

——— 1990: "Continuity and Change in Soviet Arms Control Policy," *JSMS,* 3:1, 1–31.

——— 1990a: "Why Discuss US Withdrawal," in Sharp (ed.): 3–33.

——— 1990b: "Conventional Arms Control in Europe," in *SIPRI:* 459–507.

——— 1991: "Conventional Arms Control in Europe," in *SIPRI:* 407–474 (with appendices).

——— 1991a: "Disarmament and Arms Control: A New Beginning?" in Booth (ed.): 110–139.

——— 1992: "Conventional Arms Control in Europe: Developments and Prospects in 1991," in *SIPRI:* 459–479.

——— 1992a: "Arms Control and the Atlantic Alliance," in Hampson & al. (eds.): 13–44.

——— 1993: "Conventional Arms Control in Europe," in *SIPRI:* 591–617.

——— (ed.) 1990: *Europe After an American Withdrawal: Economic and Military Issues* (Oxford: Oxford U.P.).

Shaskolsky, Nikolai 1989: "The Problem of Limiting and Reducing Naval Activities and Naval Armaments," in IMEMO (eds.): 319–340.

Shaver, David E. 1990: "Force Structures: The United States and Europe in the Coming Decade," in Sarkesian (ed.): 152–173.

Shaw, Martin 1992: *Post-Military Society: Militarism, Demilitarization and War at the End of the Twentieth Century* (London: Polity).

Shawky, A. H. 1991: "A Secure Middle East," *European Affairs,* 5:4 (August–September): 45–48.

Shenfield, Steven 1984: "The Soviet Undertaking Not to Use Nuclear Weapons First and Its Significance," *Detente,* October, 10–11.

——— 1985: "Assertive and Reactive Threats," in Windass (ed.): 63–76.

——— 1987: "The Nuclear Predicament," *Chatham House Papers,* 37 (London: Routledge & Kegan Paul).

——— 1988: "Recent Soviet Thinking on Common Security," in Windass & Grove (eds.): no pagination.

Sherwood, Elizabeth D. 1990: *Allies in Crisis: Meeting Global Challenges to Western Security* (New Haven: Yale U.P.).

Shevchenko, Alexander 1988: "Disarmament, Confidence and Co-operation," in Fedorenko (ed.): 10–19.

Shevtsova, Lilia 1992: "The August Coup and the Soviet Collapse," *Survival,* 34:1 (Spring): 5–18.

Shields, Henry S. 1985: "Why the OMG?" *MR*, 65:11 (November): 4–13.

Shim, Jae-Kwon 1991: "A Korean Nuclear-Free Zone: A Perspective," *Working Paper*, 110 (Canberra: Australian National University, Peace Research Centre).

Shimshoni, Jonathan 1990: "Technology, Military Advantage, and World War I: A Case for Military Entrepreneurship," *IS*, 15:3, 187–215.

Showalter, Dennis E. 1991: "Total War for Limited Objectives: An Interpretation of German Grand Strategy," in Kennedy (ed.): 105–124.

Shue, Henry (ed.) 1989: *Nuclear Deterrence and Moral Restraint, Critical Choices for American Strategy* (Cambridge: Cambridge U.P.).

Shuey, Robert 1992: "Assessment of the Missile Technology Control Regime," in Brauch & al. (eds.): 177–190.

Shustov, Vladimir 1990: "From Interbloc Confrontation Towards a New Peaceful Order in Europe," in Clesse & Rühl (eds.): 436–445.

Shy, John 1986: "Jomini," in Paret (ed.): 143–185.

——— & Thomas W. Collier 1986: "Revolutionary War," in Paret (ed.): 815–862.

Siccima, Jan Geert 1985: "The Netherlands Depillarized: Security Policy in a New Domestic Context," in Flynn (ed.): 113–170.

——— 1988: "Toward a European Defence Entity?" in Alford & Hunt (eds.): 13–40.

——— 1988a: "The Netherlands: 'Hollandities' in Remission?" in Rudney & Reuchler (eds.): 103–114.

Sidorenko, A. A. 1970: *The Offensive* (Moscow 1970); translated and published by the U.S. Air Force (Washington, D.C.: GPO).

Siedschlag, Horst 1989: "Grundlagen und Kriterien für das Zählen und Vergleichen," in Calliess (ed.): 42–66.

Siegelberg, Jens (ed.) 1991: *Die Kriege 1985 bis 1990: Analyse ihrer Ursachen* (Münster: LIT).

Siegmann, Heinrich 1989: "Enhancing Conventional Stability in Europe," in Altmann & Rotblat (eds.): 131–139.

——— 1989: "'Alternative' Konzepte auf dem Weg nach Wien," in Karl & Auffermann (eds.): 80–88.

Sigal, Leon V. 1983: "No First Use and NATO's Nuclear Posture," in Steinbruner & Sigal (eds.): 106–133.

——— 1983a: "Political Prospects for No First Use," in Steinbruner & Sigel (eds.): 134–146.

——— 1983b: *Nuclear Forces in Europe: Enduring Dilemmas, Present Prospects* (Washington, D.C.: Brookings).

——— 1987: "Conventional Forces in Europe: Signs of a Soviet Shift," *BAS*, 43:10 (December): 16–21.

Sigmund, Hans 1985: "Die schwedische Rüstungsindustrie," *ÖMZ*, 5: 416–426

——— 1986: "Die Rüstungswirtschaft der Schweiz," *ÖMZ*, 1: 37–48.

Sikkerheds- og Nedrustningspolitiske Udvalg, det: See SNU.

Simon, Jeffrey 1993: "Does Eastern Europe Belong in NATO?" *Orbis*, 37:1 (Winter): 21–35.

Simoni, Arnold 1990: "Regional Security Associations: A New Security Architecture for Non-Nuclear Powers: Overview," in Calliess (ed.): 687–712.

Simons Thomas W., Jr. 1993: *Eastern Europe in the Postwar World*, 2d ed. (New York: St. Martin's).

Simpkin, Richard E. 1983: *Red Armour: An Examination of the Soviet Mobile Force Concept* (London: Brassey's).

——— 1986: *Race to the Swift: Thoughts on 21st Century Warfare* (London: Brassey's).

———— 1987: "Tank Warfare. The Last Decades of the Dinosaurs," in Perkins (ed.): 165–192.
Singer, E. 1963: "A Bargaining Model for Disarmament Negotiations," *Journal of Conflict Resolution,* 7: 21–25.
Singer, Max, & Aaron Wildawsky 1993: *The Real World Order: Zones of Peace/Zones of Turmoil* (Chatham, N.J.: Chatham House).
Singh, Hari 1993: "Prospects for Regional Stability in Southeast Asia in the Post–Cold War Era," *Millennium,* 22:2 (Summer): 279–300.
Singh, Jasjit 1989: "Evolution of Politico-Military Doctrines," in Singh & Vekaric (eds.): 1–25.
———— 1989a: "The Indian Experience," in Singh & Vekaric (eds.): 215–233.
———— 1989b: "Towards Equal Security," in Singh & Vekaric (eds.): 234–256.
———— 1990: "Defence Strategies," in UNIDIR (eds.): 23–39.
———— 1992: "Regional and Global Stability and the Role Played by Missiles," in Neuneck & Ischebeck (eds.): 241–255.
———— 1992a: "Arms Control and Proliferation of High-Technology Weapons in South Asia and the Middle East: A View from India," in Stahl & Kemp (eds.): 123–134.
———— 1992b: "Defensive Security: The Conceptual Challenges," *Disarmament,* 15:4, 112–125.
———— 1993: "Nuclear Weapons Threat," in Singh & Bernauer (eds.): 57–79.
———— 1993a: "Security in a Period of Change," in Singh & Bernauer (eds.): 151–161.
———— 1994: "NOD and Southern Asia," in Møller & Wiberg (eds.): 195–207.
———— & Thomas Bernauer (eds.) 1993: *Security of Third World Countries* (Aldershot: Dartmouth).
———— & Vatroslav Vekaric (eds.) 1989: *Non-Provocative Defence: The Search for Equal Security* (New Delhi: Lancer).
Singh, Ravinder Pal 1992: "Prospects of a Conventional Arms Export Control Regime: A Perspective from the South," in Brauch & al. (eds.): 331–341.
Singham, A. W. 1993: "The National Security State and the End of the Cold War: Security Dilemmas for the Third World," in Singh & Bernauer (eds.): 5–24.
Sinn, Norbert 1991: "Der Golfkrieg II—Militärische Erfahrungen: Ableitungen für die Landesverteidigung Österreichs," *ÖMZ,* 29(6): 518–525.
SIPRI 1973: *CBW and the Law of War The Problem of Chemical and Biological Warfare: A Study of the Historical, Technical, Military, Legal and Political Aspects of CBW, and Possible Disarmament Measures,* vol. IIII (Stockholm: Almqvist & Wicksell).
———— 1980: "The Prohibition of Radiological Warfare," in *SIPRI,* 1980: 381–388.
———— 1981: "The Prohibition of Inhumane and Indiscriminate Weapons," in *SIPRI,* 1981: 445–467.
———— (eds.) 1978: *Tactical Nuclear Weapons: European Perspectives* (London: Taylor & Francis).
Sivonen, Pekka 1989: "NATO and the Political Consequences of Nuclear Disarmament in Europe," in Harle & Sivonen (eds.): 68–92.
———— 1990: "European Security: New, Old and Borrowed," *JPR,* 27:4, 385–397.
———— 1991: "The Changing Context of Nuclear Security Issues in Europe," in Harle & Sivonen (eds.): 19–48.
Skingsley, Anthony 1991: "Interdiction and Follow-on Forces Attack," in Reid & Dewar (eds.): 207–218.
Skjelsbæk, Kjell 1990: "United Nations Peacekeeping: Expectations, Limitations and Results: Forty Years of Mixed Experience," in Czempiel & al. (eds.): 77–90.

Skjold, Anne Marit 1989: "Eliminering av ballistiske raketter: Et bridrag til styrket krisestabilitet og rustningsstabilitet i Øst/Vest-forholdet?," *NUPIR,* 125.

Skodvin, Magne 1967: "Norwegian Nonviolent Resistance During the German Occupation," in Roberts (ed.): 136–153.

Skogan, John Kristen 1990: "The Evolution of the Four Soviet Fleets, 1968–87," in Skvgan & Brundtland (eds.): 18–33.

———— 1991: "Rustningskontroll til sjøs," *Internasjonal Politikk,* 49:2, 157–170.

———— (ed.) 1993: "Civil-Military Relations in the Post-Communist States in Eastern and Central Europe," *Research Report,* 167 (Oslo: NUPI).

———— & Arne Brundtland (eds.) 1990: *Soviet Sea Power in Northern Waters: Facts, Motivations, Impact and Responses* (London: Pinter).

Skuhra, Anselm 1986: "Die Rüstungspolitik Österreichs," *Dialog,* 7: 173–267.

———— 1986a: "Friedensbewegungen, Neutralisierungsbestrebungen und ihre Rahmenbedingungen im Europa der 80er Jahre," *Dialog,* 6: 77–180.

———— 1987: "Austria and the New Cold War," in Sundelius (ed.): 117–147.

Sloan, Stanley 1986: *NATO's Future: Towards a New Transatlantic Bargain* (London: Macmillan).

Slocombe, Walter 1987: "Preplanned Operations," in Carter & al. (eds.): 121–141.

Sloss, Leon 1984: "The Strategist's Perspective," in Carter & Schwartz (eds.): 24–48.

Smart, Ian 1992: "Untangling the Priorities: Weapons, Vehicles, and the Objectives of Arms Control," in Stahl & Kemp (eds.): 145–164.

Smit, Wim A. 1986: "The Patriot Missile: An Arms Control Impact Analysis," in Barnaby & Borg (eds.): 156–176.

———— 1989: "Defence Technology Assessment and the Control of Emerging Technologies," in Borg & Smit (eds.): 61–76.

———— 1989a: "Conventional Attack on Nuclear Facilities," in Tromp (ed.): 293–318.

———— 1990: "Controlling Military Technological Innovation—the Role of Verification," in Grin & Graaf (eds.): 53–70.

————, Hans van Gool, Dik van Houwelinger, & Eert Schooten 1987: "ATBM: Its Implications for Arms Control and European Security," in Borg & Smit (eds.): 113–125.

————, John Grin, & Lev Voronkov (eds.) 1992: *Military Technological Innovation and Stability in a Changing World: Politically Assessing and Influencing Weapon Innovation and Military Research and Development* (Amsterdam: VU U.P.).

Smith, Chris 1988: "Third World Arms Control, Military Technology and Alternative Security," in Ohlson (ed.): 58–72.

Smith, Dan 1980: *The Defence of the Realm in the 1980s* (London: Croom Helm).

———— 1981: "Non-Nuclear Military Options for Britain," *Peace Studies Paper,* 6 (Bradford: School of Peace Studies).

———— 1983: "The Crisis of Atlanticism," in Baylis (ed.): 219–238.

———— 1987: "The Cold War," in Smith & Thompson (eds.): 44–71.

———— 1989: *Pressure: How America Runs NATO* (London: Bloomsbury).

———— 1990: "Themes in the Non-Nuclear Defence Debate," in Randle & Rogers (eds.): 7–20.

———— & E. P. Thompson (eds.) 1987: *Prospectus for a Habitable Planet* (Harmondsworth: Penguin).

Smith, D. L., & A. L. Meier 1987: "Ogarkov's Revolution: Soviet Military Doctrine for the 1990s," *IDR,* 7: 869–873.

Smith, Gary 1990: "Two Rhetorics of Region," in Cheeseman & Kettle (eds.): 111–132.

———— 1991: "The State and the Armed Forces: Defence as Militarism," *Working Paper,* 100 (Canberra: Australian National University, Peace Research Centre).

Smith, Hugh, & Anthony Bergin (eds.) 1993: *Naval Power in the Pacific Toward the Year 2000* (Boulder, Colo.: Lynne Rienner).

Smith, Michael Joseph 1986: *Realist Thought from Weber to Kissinger* (Baton Rouge: Louisiana State U.P.).

Smith, Peter 1990: "Beyond the Blocs: New Security Principles for the New Europe," *Discussion Paper* (Appleby, U.K.: Just Defence).

———— 1991: "East West Peace: A Common Concern," *Chartrist for Democratic Socialis,* Autumn: 16–18.

———— & Mike Gapes 1984: "Real Defence: Britain Without the Bomb" (Leeds: Independent Labour Publications).

Smith, William D. 1993: "Theatre Ballistic Missile Defence for Europe," *NSN,* 38(5/6): 45–51.

Smock, David R. (ed.) 1992: *Religious Perspectives on War: Christian, Muslim, and Jewish Attitudes Toward Force After the Gulf War* (Washington, D.C.: United States Institute of Peace).

Smoke, Richard 1988: "For a NATO Defensive Deterrent," in Hopmann & Barnaby (eds.): 205–220.

———— 1991: "A Theory of Mutual Security," in Smoke & Kortunov (eds.): 59–111.

———— & Andrei Kortunov (eds.) 1991: *Mutual Security: A New Approach to Soviet-American Relations* (London: Macmillan).

Smyser, W. R. 1991: "U.S.S.R.-Germany: A Link Restored," *Foreign Policy,* 84 (Fall): 125–141.

Snow, Edgar 1972: *Red Star over China,* rev. ed. (Harmondsworth: Penguin Books).

SNU (Det Sikkerheds- og Nedrustningspolitiske Udvalg) 1982: *Dansk sikkerheds-politik og forslagene om Norden som kernevåbenfri zone* (Copenhagen: SNU).

———— (ed.) 1988: *CST: Konventional nedrustning i Europa* (Copenhagen: SNU).

———— (ed.) 1989: *"Fornuftig tilstrækkelighed" og "Ikkeoffensivt forsvar": Den sovjetiske debat* (Copenhagen: SNU).

Snyder, Craig (ed.) 1986: *The Strategic Defense Debate: Can "Star Wars" Make Us Safe?* (Philadelphia: University of Pennsylvania Press).

Snyder, Glenn H. 1961: *Deterrence and Defence: Toward a Theory of National Security* (Princeton: Princeton U.P.).

———— 1984: "The Security Dilemma in Alliance Politics," *WP,* 36:4, 461–495.

———— & Diesing 1977: *Conflict Among Nations: Bargaining, Decision-Making and System Structure* (Princeton: Princeton U.P.).

Snyder, Jack 1984: *The Ideology of the Offensive: Military Decision Making and the Disasters of 1914* (Ithaca: Cornell U.P.).

———— 1985: "Civil-Military Relations and the Cult of the Offensive, 1914 and 1984," in Miller (ed.): 108–146.

———— 1986: "The Gorbachev Revolution: A Waning of Soviet Expansionism?" (*IS,* 12:3) in Lynn-Jones, Miller, & Evera (eds.): 71–109.

———— 1987: "Limiting Offensive Conventional Forces: Soviet Proposals and Western Options" (*IS,* 12:4) in Miller & Lynn-Jones (eds.) 1989: 292–321.

———— 1990: "Averting Anarchy in the New Europe," *IS,* 14:4 (Spring): 5–41.

———— 1990a: "Controlling Nationalism in the New Europe," in Clesse & Rühl (eds.): 49–62.

———— 1991: "Introduction," in idem (eds.): 3–19.

—— 1991a: *Myths of Empire: Domestic Politics and International Ambition* (Ithaca: Cornell U.P.).

Socialdemokratiet 1986: "Socialdemokratiets forslag til modernisering af forsvaret" (Copenhagen: Socialdemokratiet).

Solo, Pam 1988: *From Protest to Policy: Beyond the Freeze to Common Security* (Cambridge, Mass.: Ballinger).

Sommer, Mark 1983: "Forging a Preservative Defense," *BAS,* August–September, 5–7.

—— 1984: "Beating Our Swords into Shields" (Miranda, Calif.: Center for Preservative Defense).

—— 1985: *Beyond the Bomb: Living Without Nuclear Weapons* (New York: EXPRO, Talman).

—— 1986: "Alternative Security Systems," in Carlson & Cromstock (eds.) 1986a: 318–338.

—— 1988: "An Emerging Consensus: Common Security Through Qualitative Disarmament" (New York: Alternative Defense Project).

—— 1989: "Reuniting Europe: Shared Prosperity as a Path to Peace" (Berkeley: Pacific Institute for Studies in Development, Environment and Security).

Sonntag, Philip 1990: "Quantität und Qualität in der Rüstungssteuerung," in Weizsäcker (ed.): 77–83.

Sørensen, Bent 1985: *Fred og frihed: Om kapnedrustning i vor tid* (Copenhagen: Borgen).

—— 1985a: "Security Implications of Alternative Defense Options for Western Europe," *JPR,* 22:3, 197–209.

—— 1985b: "An Alternative Defense Plan for Western Europe," *Peace Research Series,* 3 (Roskilde: RUC/IMFUFA).

—— 1986: "Steps to Security: An Alternative Defence Plan for Western European and Pacific Countries," *Peace Dossier,* 14 (Melbourne: Victorian Association for Peace Studies).

Sørensen, Georg 1992: "Utopianism in Peace Research: The Gandhian Heritage," *JPR,* 29:2 (May): 135–144.

—— 1992a: "Kant and Processes of Democratization: Consequences for Neorealist Thought," *JPR,* 29:4 (November): 397–414.

—— 1993: "A Rejoinder," *JPR,* 30:4 (November): 453–454.

Sørensen, Henning 1990: "Den danske befolknings holdning til Ikke-Offensivt Forsvar," in idem (ed.): 177–189.

—— 1990a: "Ikke-offensivt forsvar i den danske dagspresse," in idem (ed.): 202–226.

—— & Lutz Unterseher 1989: "NOD Principles Implemented in Danish Defence," *WP-CPCR,* 14.

—— & Lutz Unterseher 1990: "Ikke-offensive elementer indpasset i dansk forsvar," in Sørensen (ed.): 95–115.

—— (ed.) 1990: *Ikke-Offensivt Forsvar: En Introduktion* (Copenhagen: Samfundslitteratur).

Sorokin, Konstantin E. 1993: "The Creation of Russian Armed Forces," in Center for Defence Studies (ed.): 133–153.

—— 1994: "The Nuclear Strategy Debate," *Orbis,* 38:1 (Winter): 19–40.

Soroos, Marvin S. 1990: "Global Policy Studies and Peace Research," *JPR,* 27:2, 117–126.

—— (ed.) 1990: "Special Issue on the Challenge of Global Policy," *JPR,* 27:2.

Souchon, Lennart 1990: *Neue deutsche Sicherheitspolitik* (Herford: Mittler).

—— 1992: *Die Renaissance Europas: Europäische Sicherheitspolitik—Ein internationales Modell* (Herford: Mittler).

Sozialdemokratische Partei Deutschlands: See SPD.

Southwood, Peter 1991: *Disarming Military Industries: Turning an Outbreak of Peace into an Enduring Legacy* (London: Macmillan).

Spanger, Hans-Joachim 1989: "The GDR in East-West Relations," *AP*, 240.

Spannocchi, Emilio 1976: "Verteidigung ohne Selbstzerstörung," in Weizsäcker (ed.): 15–91.

———— 1990: "Sicherheit ohne militärische Operation," in Weizsäcker (ed.): 197–203.

SPD (Sozialdemokratische Partei Deutschlands) 1986: "Leitantrag 'Friedens- und Sicherheitspolitik' in der vom Bundesparteitag der SPD in Nürnberg am 27. August 1986 beschlossenen Fassung," *Blätter*, 10: 1268–1278.

———— 1988: "Frieden und Abrüstung in Europa: Beschluss zur Friedens- und Abrüstungspolitik des SPD-Bundesparteitages in Münster 30.8–2.9.1988," *Politik: Informationsdienst der SPD*, 10 September.

———— & PVAP 1985: "Gemeinsame Erklärung von SPD und PVAP zur Sicherheit und Zusammenarbeit in Europa durch Massnahmen der gegenseitigen Vertrauensbildung," *Blätter*, 12: 1515–1519.

———— & PVAP 1988: "Kriterien und Schritte zu vertrauensschaffende Sicherheitsstrukturen in Europa," *Blätter*, 4: 508–510.

———— & SED 1985: *Für eine chemiewaffenfreie Zone in Europe: Gemeinsame politische Initiative der Sozialistischen Einheitspartei Deutschlands und der Sozialdemokratischen Partei Deutschlands* (Dresden: Zeit im Bild).

———— & SED 1987: "'Der Streit der Ideologien und die gemeinsame Sicherheit': Gemeinsame Erklärung der Grundwertekommission der SPD und der Akademie für Gesellschaftswissenschaften beim ZK der SED vom 27. August 1987," *Blätter*, 10: 1365–1371.

———— & SED 1987a: "Initiative von SPD und SED für eine atomwaffenfreie Zone in Mitteleuropa," *DA*, 2: 211–215.

Spear, Joanna 1990: "Britain and Conventional Arms Transfer Restraint," in Hoffman (ed.): 170–189.

———— & Stuart Croft 1990: "Superpowers and Arms Transfers to the Third World," in Allison & Williams (eds.): 89–106.

Spector, Leonard S. 1984: *Nuclear Proliferation Today* (New York: Vintage).

———— (with Jacqueline R. Smith) 1990: *Nuclear Ambitions: The Spread of Nuclear Weapons, 1989–1990* (Boulder, Colo.: Westview).

Spero, Joshua 1992: "*Déja Vu* All Over Again? Poland's Attempt to Avoid Entrapment Between Two Belligerents," *ES*, 1:4 (Winter): 92–117.

———— 1992a: "The Budapest-Prague-Warsaw Triangle: Central European Security after the Visegrad Summit," *ES*, 1:1 (Spring): 58–83.

Spescha, Mark 1989: "Das Völkerrecht zwingt die Schweiz nicht zur Bewaffnung," in Gross & al. (eds.): 93–108.

Spiegel, Steven L. (ed.) 1992: *The Arab-Israeli Search for Peace* (Boulder, Colo.: Lynne Rienner).

Spillmann, Kurt R. 1989: "Beyond Soldiers and Arms: The Swiss Model of Comprehensive Security Policy," in Kruzel & Haltzel (eds.): 161–174.

———— 1990: "Swiss Security Policy in a Changing Environment," in Miliovejic & Maurer (eds.): 207–216.

Spitzer, Hartwig 1990: "Überwachung der Abrüstung konventioneller Streitkräfte in Europa aus der Luft und aus dem Weltraum," in E. Müller & Neuneck (eds.): 161–174.

Spreafico, M. Clelia, & Paolo Farinella 1989: "Difesa territoriale e difesa civile in Italia: I precedenti di un dibattito da riaprire," in Ragionieri (ed.): 213–228.

Sprout, Margaret T. 1941: "Mahan: Evangelist of Sea Power," in Earle (ed.): 415–456.

Stahl, Shelley A., & Geoffrey Kemp (eds.) 1992: *Arms Control and Weapons Proliferation in the Middle East and South Asia* (New York: St. Martin's).

Staley, Robert Stephen, II 1992: *The Wave of the Future: The United Nations and Naval Peacekeeping* (Boulder, Colo.: Lynne Rienner).

Stares, Paul B. 1991: *Command Performance: The Neglected Dimension of European Security* (Washington, D.C.: Brookings).

—— (ed.) 1992: *The New Germany and the New Europe* (Washington, D.C.: Brookings).

—— & John D. Steinbruner 1992: "Cooperative Security in the New Europe," in Stares (ed.): 218–248.

Starke, Bernd 1991: "Die politische und rechtliche Aktualität eines regionalen Systems der kollektiven Sicherheit (SKS) in Europa vor dem Hintergrund der Integration der Deutschen Demokratischen Republik in die Bundesrepublik Deutschland," in Lutz (ed.) 1991: 133–146.

Starr, Harvey 1992: "Democracy and War: Choice, Learning and Security Communities," *JPR*, 29:2 (May): 207–213.

Statz, Albert 1988: "Unilateral Steps to Disarmament and Deliberate Self-Containment: Elements of a Green Foreign Policy Strategy for West Germany," *Green Peace Policy*, 13: 2–4.

—— 1990: "Zur Rolle Deutschlands und Berlins in einem gesamteuropäischen Friedensprozess," in Schweizerische Friedensstiftung (eds.): 105–120.

Steele, Jonathan 1985: *The Limits of Soviet Power: The Kremlin's Foreign Policy—Brezhnev to Chernenko* (Harmondsworth: Penguin).

—— 1987: "East-West and North-South," in Smith & Thompson (eds.): 124–145.

Stefanic, David 1988: "The Rapacki Plan: A Case Study of East European Diplomacy," *East European Quarterly*, 21:4, 401–412.

Stefanick, Tom 1987: *Strategic Antisubmarine Warfare and Naval Strategy* (Lexington, Mass.: Lexington Books).

Stein, Arthur A. 1990: *Why Nations Cooperate: Circumstance and Choice in International Relations* (Ithaca: Cornell U.P.).

Stein, George 1993: "The Euro-Corps and Future European Security Architecture," *ES*, 2:2 (Summer): 200–226.

Stein, Mark, & Kathleen Fahey 1987: "Bibliography of Alternative Defense Research" (Brookline, Mass.: IDDS).

Steinberger, Jack, Essam Galal, & Mikhail Milstein 1993: "A Nuclear-Weapon-Free World: Is It Desirable? Is It Necessary?" in Rotblat & al. (eds.): 52–62.

Steinbruner, John D. 1983: "Alliance Security," in Steinbruner & Sigal (eds.): 197–207.

—— 1988: "The Prospects of Cooperative Security," *Brookings Review*, 6:12 (Winter 1988/1989): 53–62.

—— 1990: "Revolution in Foreign Policy," in Aaron (ed.): 65–110.

—— 1991: "The Consequences of the Gulf War: Forging the New World Order Under a Forced Schedule," *Brookings Review*, 9:2 (Spring): 6–13.

—— 1992: "Business as Usual? The Redesign of National Security," *Brookings Review*, 10:3 (Summer): 12–15.

—— & Leon V. Sigal (eds.) 1983: *Alliance Security: NATO and the No-First-Use Question* (Washington, D.C.: Brookings).

Steiner, Kurt 1988: "Negotiations for an Austrian State Treaty," in George, Farley, & Dallin (eds.): 46–82.

Steininger, Rolf 1989: *Wiederbewaffnung: Die Entscheidung für einen west-deutschen Verteidigungsbeitrag: Adenauer und die Westmächte, 1950* (Erlangen: Straube).

———— 1990: "Deutsche Frage und Berliner Konferenz 1954," in Venohr (ed.): 37–90.

Steinke, Rudolph, & Michel Vale (eds.) 1983: *Germany Debates Defense: The NATO Alliance at the Crossroad* (New York: Sharpe).

Steinsaecker, HeinzEberhard Freiherr von 1982: "Zur 'Raumverteidigung'," *Wehrwissenschaftliche Rundschau,* 1: 1–11.

Steinweg, Reiner (ed.) 1985: *Rüstung und Soziale Sicherheit,* Friedensanalysen, vol. 20 (Frankfurt: Suhrkamp).

———— (ed.) 1986: *Unsere Bundeswehr? Zum 25jährigen Bestehen einer umstritte-nen Institution* (Franfurt: Suhrkamp).

———— (ed.) 1987: *Kriegsursachen,* Friedensanalysen, vol. XXI (Frankfurt: Suhrkamp).

———— (ed.) 1989: *Militärregime und Entwicklungspolitik,* Friedensanalysen, vol. XXII (Frankfurt: Suhrkamp).

———— (ed.) 1990: *Lehren aus der Geschichte? Historische Friedensforschung,* Friedensanalysen, vol. XXII (Frankfurt: Suhrkamp).

———— & Christian Wellmann (eds.) 1990: *Die vergessene Dimension interna-tionaler Konflikte: Subjektivität* (Frankfurt: Suhrkamp).

Stembera, Milan 1992: "SFR-Streitkräfte und Rüstung," *ÖMZ,* 30:5 (September–October): 390–396.

———— 1993: "Die sicherheitspolitische Konsequenzen der Teilung der CSFR," *ÖMZ,* 31:3 (May–June): 216–222.

Stenseth, Dagfinn 1992: "The New Russia, CIS and the Future," *Security Dialogue,* 23:3 (September): 19–26.

Stepashin, Sergey Vadimovich 1992: "On the Future Defence Policy of Russia," *RJ,* 137:2 (April): 35–37.

———— 1993: "Russia and NATO: A Vital Partnership for European Security," *RJ,* 138:4 (August): 11–17.

Stephenson, Carolyn 1988: "The Need for Alternative Forms of Security: Crises and Opportunities," *Alternatives,* 13:1, 55–76.

———— 1992: "New Conceptions of Security and Their Implications for Means and Methods," in Tehranian & Tehranian (eds.): 47–61.

Stern, Maria 1991: "Security Policy in Transition: Sweden After the Cold War," *Padrigu Thesis Series* (Gotherburg: Padrigu): 1.

Stern, Paul C., Robert Axelrod, Robert Jervis, & Roy Radner (eds.) 1989: *Perspectives on Deterrence* (New York: Oxford U.P.).

Sternstein, Wolfgang 1967: "The Ruhrkampf of 1923: Economic Problems of Civilian Defence," in Roberts (ed.): 106–135.

———— 1970: "Die Befreiung der Dritten Welt: Ghandis Konzeption einer gewalt-freien Revolution," in Ebert (ed.): 131–160.

———— 1977: "Through Non-violent Action in Intrasocietal Conflicts to Civilian Defence: Some Remarks on the Past and Future of the Research on Civilian Defence," in Geeraerts (ed.): 54–65.

———— 1988: "Soziale Verteidigung," in Lippert & Wachtler (eds.): 390–399.

Stettler, Edwin 1990: "Relations Between National Defence and the Economy in Switzerland," in Miliovejic & Maurer (eds.): 81–93.

Stevens, Sayre 1984: "The Soviet BMD Program," in Carter & Schwartz (eds.): 182–220.

Stewart, Philip D., & Margaret G. Hermann 1990: "The Soviet Debate over 'New

Thinking' and the Restructuring of U.S.-Soviet Relations," in Hudson (ed.): 47–70.

Stinnes, Manfred 1977: "Theoretical Perspectives in Civilian Defence: An Issue in West German Peace Research," in Geeraerts (ed.): 77–102.

Stock, Wolfgang, & Klaus Krottmayer 1986: "Soziale Verteidigung im dauernd neutralen Staat: Verfassungsrechtliche Aspekte der Sozialen Verteidigung Österreichs," *Dialog,* 7: 117–172.

Stockton, Paul 1986: "Strategic Stability Between the Superpowers," *AP,* 213.

Stoecker, Sally W. 1990: "The Historical Roots of the Current Debates on Soviet Military Doctrine and Defense," *JSMS,* 3:3, 363–389.

Stoltenberg, Gerhard 1990: "Die transatlantische Allianz: Lebenswichtiger Faktor der Zukunft," *EW,* 3: 136–144.

Støren, Harald W. 1987: "Sjømakt og Politisk Innflytelse: Begreper, Modeller og Relevans," *NUPIR,* 113.

Størner, Torben, Jørgen Kjær, Henning Høgh Laursen, Gert Østergård, & Wolfgang Kahlig (eds.) 1986: *Peace and the Future: Proceedings of the II International Peace University, Aarhus, October 1985* (Århus: Aarhus U.P.).

Strachan, Hew 1984: "Conventional Defence in Europe," *IA,* 61:1 (Winter): 27–45.

Stratmann, K.-Peter 1986: "The Nuclear Relationship: Conventional Defence and Nuclear Posture in the Strategy of Deterrence," in Huber (ed.): 5–14.

———— 1988: "Arms Control and the Military Balance: The West German Debate," in Kaiser & Roper (eds.): 90–112.

———— 1989: "The Military Objectives of Conventional Arms Control," in Blackwill & Larrabee (eds.): 140–164.

Strauss, Barry S., & Josiah Ober 1990: *The Anatomy of Error: Ancient Military Disasters and Their Lessons for Modern Strategists* (New York: St. Martin's).

Strekal, Oleg 1993: "Die Ukraine und die europäische Sicherheit," *EuS,* 42:7 (July): 344–346.

Streltsov, Yuri 1989: "New-Generation Confidence- and Security-Building Measures," in IMEMO (eds.): 385–392.

Stromseth, Jane E. 1988: *The Origins of Flexible Response: NATO's Debate over Strategy in the 1960's* (New York: St. Martin's).

Strübel, Manfred (ed.) 1985: *Friedens- und Sicherheitspolitische Alternativen: Konzepte: Kontroversen, Perspektiven* (Giessen: Schmitz).

Strübel, Michael 1984: "Friedensbewegungen gestern und heute: Herausforderungen für die SPD," in Brauch (ed.): 71–86.

Studiengruppe Alternative Sicherheitspolitik: See SAS.

Stülpnagel, Joachim von 1924: "Gedanken über den Krieg der Zukunft," in Bald (ed.) 1987: 52–66.

———— 1924a: "Den Krieg unter anderen Voraussetzungen," in Bald (ed.) 1987: 67–70.

Stutz, Alfred: Raumverteidigung—Utopie oder Alternative (Zürich: Verlag Neue Züricher Zeitung).

Subrahmanyam, K. 1989: "Non-provocative Defence and International Environment," in Singh & Vekaric (eds.): 160–180.

Suganami, Hidemi 1989: *The Domestic Analogy and World Order Proposals* (Cambridge: Cambridge U.P.).

———— 1990: "Bringing Order to the Causes of War Debates," *Millennium,* 19:1, 19–35.

Suh, Mark B. M. 1992: "Normalization and Unification Prospects in Korea," *AuP,* 43:3 (Autumn): 256–266.

Summerscale, Peter 1981: "The Continuing Validity of the Brezhnev Doctrine," in Dawisha & Hanson (eds.): 26–40.

Sundelius, Bengt 1987: "Dilemmas and Security Strategies for the Neutral Democracies," in Sundelius (ed.): 11–32.

———— 1989: "National Security Dilemmas and Strategies for the European Neutrals," in Kruzel & Haltzel (eds.): 98–121.

———— (ed.) 1987: *The Neutral Democracies and the New Cold War* (Boulder, Colo.: Westview).

Sun Tzu 500 B.C.: "The Art of War," in Sawyer (ed.) 1993: 157–186.

Sur, Serge (ed.) 1991: *La vérification des accords sur le désarmament et la limitation des armements: Moyens, méthode et pratique* (New York: United Nations; Geneva: UNIDIR).

———— (ed.) 1992: *Disarmament and Limitation of Armaments: Unilateral Measures and Policies* (New York: United Nations, UNIDIR).

———— (ed.) 1993: "Nuclear Deterrence: Problems and Perspectives in the 1990's" (New York: United Nations; Geneva: UNIDIR).

Surotchak, Jan Erik 1989: "One Step Forward, Two Steps Back? Assessing the Status and Influence of the West German Greens," in Griffith & al.: 66–82.

Sutton, Boyd D., John R. Landry, Malcolm B. Armstrong, Howell M. Esles III, & Wesley K. Clark 1984: "Deep Attack Concepts and the Defence of Central Europe," *Survival*, 26:2, 50–78.

Suvorov, Viktor 1983: "Speznaz: The Soviet Union's Special Forces," *IDR*, 9: 1209–1216.

Svechin, Aleksandr 1992: *Strategy*, ed. by Kent D. Lee; introductory essays by Andrei A. Kokoshin, Valentin V. Larionov, Vladimir N. Lobov, and Jacob W. Kipp (Minneapolis, Minn.: East View).

Sveics, V. V. 1969: *Small Nation Survival: Political Defense in Unequal Conflicts* (New York: Exposition).

Svensson, Per 1980: "Vort sårbare samfund—og det totale forsvar," *Forsvarets langtidsplanlægningsserie,* 8 (Copenhagen: Ministry of Defence).

Swita, Bogdan: "The OMG in the Defense," *MR*, 72:7 (July): 35–45.

Symonides, Janusz 1989: "The General State of European Security," in Blackwill & Larrabee (eds.): 22–43.

Szabo, Stephen F. 1990: *The Changing Politics of German Security* (London: Pinter).

———— 1990a: Introduction to Szabo (ed.): 1–9.

———— 1993: "The New Germany and Central European Security," in Lampe & Nelson (eds.): 35–54.

———— (ed.) 1990: *The Bundeswehr and Western Security* (London: Macmillan).

Szentesi, György, Laszlo Tolnay, & Peter Vajda 1987: *Military Balance and Doctrines* (Prague: International Organization of Journalists).

Szilard, Leo 1964: "'Minimal Deterrent' vs. Saturation Parity," in Kissinger (ed.): 376–391.

Talbot, Strobe 1984: *Deadly Gambits: The Reagan Administration and the Stalemate in Nuclear Arms Control* (London: Picador).

Tamamoto, Masaru 1990: "Japan's Search for a World Role," *WPJ*, 7:3 (Summer): 493–520.

———— 1990a: "Trial of an Ideal: Japan's Debate over the Gulf Crisis," *WPJ*, 8:1 (Winter): 89–106.

———— 1993: "The Japan That Wants to Be Liked," in Unger (ed.): 37–54.

Tammes, Rolf (ed.) 1990: "Soviet 'Reasonable Sufficiency' and Norwegian Security," *Forsvarsstudier,* 5 (Oslo: Institutt for Forsvarsstudier).

Tanner, Fred 1990: "Switzerland and Arms Control: Constraints and Opportunities," in Miliovejic & Maurer (eds.): 136–161.

———— 1991: "Die Schweiz und die Rüstungskontrolle: Grenzen und Möglichkeiten eines Kleinstaates," *Zürcher Beiträge zur Sicherheitspolitik und Konfliktforschung,* 14 (Zürich: Forschungsstelle für Sicherheitspolitik und Konfliktanalyse, Eidgenössische Technische Hochschule).

———— 1992: "Versailles: German Disarmament After World War I," in Ganner (ed.): 5–26.

———— (ed.) 1992: *From Versailles to Baghdad: Post-War Armament Control of Defeated States* (New York: United Nations; Geneva: UNIDIR).

Tanner, Jakob 1989: "Angenommen, die Schweiz hätte 1945 die Armee abgeschafft," in Gross & al. (eds.): 71–83.

Tarleton, Gael Donelan 1992: "Red Star of the Sea: The Soviet Navy and Strategic Policy," in Leebaert & Dickinson (eds.): 127–153.

Tarr, David W. 1988: "NATO Strategy: Can Conventional Forces Deter a Conventional Attack?" in Hopmann & Barnaby (eds.): 221–235.

Tatchell, Peter 1985: *Democratic Defence: A Non-Nuclear Alternative* (London: GMP).

Tauschitz, Othmar 1990: "Die Militärdoktrin Österreichs," *ÖMZ,* 28:2, 89–91.

Taylor, A. J. P. 1969: *War by Time-Table: How the First World War Began* (London: MacDonald).

Taylor, Richard, & Nigel Young (eds.) 1987: *Campaigns for Peace: British Peace Movements in the Twentieth Century* (Manchester: Manchester U.P.).

Taylor, Trevor 1988: "A European Defence Entity: European Institutions and Defence," in Alford & Hunt (eds.): 177–220.

———— 1991: "NATO and Central Europe," *NATO Review,* 39:5 (October): 17–22.

Tegnelia, James A. 1985: "Emerging Technology for Conventional Deterrence," *IDR,* 5: 643–652.

Tehranian, Katharine, & Majid Tehranian (eds.) 1992: *Restructuring for World Peace: On the Threshold of the Twenty-First Century* (Creskill, N.J.: Hampton).

Tessler, Mark 1989: "Israeli Politics and the Palestinian Problem after Camp David," in Lesch & Tessler (eds.): 140–174.

———— 1989a: "Epilogue: Thinking About Territorial Compromise in Israel," in Lesch & Tessler (eds.): 272–284.

———— & Ann Mosley Lesch 1989: "Israel's Drive into the West Bank and Gaza," in Lesch & Tessler (eds.): 194–222.

Tetzlaff, Martin 1988: "Überlegungen zur Konventionellen Stabilität," in Bahr & Lutz (eds.): 477–480.

Thee, Marek 1983: "The NWFZ as a Popular Move: Informed, Rational and Imaginative Politics," in Lodgaard & Thee (eds.): 255–264.

———— 1983a: "Conceptual Issues Related to European Security, Arms Control and Confidence-Building Measures," *PRIO Report* (Oslo: PRIO): 9.

———— 1989: "Swords into Ploughshares: The Quest for Peace and Human Development," in Dumas & Thee (eds.): 55–69.

———— 1990: "Science-Based Military Technology as a Driving Force Behind the Arms Race," in Gleditsch & Njølstad (eds.): 105–120.

Theilmann, Alexander 1986: "Ausländische Stellungnahmen zum Thema Gemeinsame Sicherheitein erster Überblick," in Bahr & Lutz (eds.): 253–275.

———— 1986a: "Die Diskussion über militärische Defensivkonzepte. Entwicklungen, Inhalte, Perspektiven," *HB,* 9.

Thimann, Christian 1989: "Das Personalmodell einer alternativen Verteidigung," in SAS (eds.): 204–211.

————— 1989a: "Drei militärische Modelle auf einen Blick," in SAS (eds.): 17–22.

————— & al. 1988: "Zum Personalmodell von H. Funk: 'Die Integrierte Vorneverteidigung Mitteleuropa'" (Bonn: SAS).

Thoft, Jens 1983: "Elementer til en ny sikkerhedspolitik," in Thoft & Jacobsen (eds.): 112–123.

————— (ed.) 1974: *Ikkevold: Strategi i klassekampen* (Copenhagen: GMT).

————— & Hanne Jacobsen (eds.) 1983: *Oprustning, omrustning, nedrustning—fred* (Århus: SP Forlag).

Thomas, Caroline 1992: *The Environment in International Relations* (London: RIIA).

Thomas, Timothy L. 1991: "Military Doctrine: The Soviet Military on 'Desert Storm': Redefining Doctrine," *JSMS,* 4:4 (December): 594–620.

Thompson, E. P. & Dan Smith (eds.) 1980: *Protest and Survive* (Harmondsworth: Penguin).

Thompson, Gordon, & Pam Solo (eds.) 1988: "New Directions for NATO: Adapting the Atlantic Alliance to the Needs of the 1990s" (Cambridge, Mass.: Institute for Resource and Security Studies & Institute for Peace and International Security).

Thompson, Julian 1991: *The Lifeblood of War: Logistics in Armed Conflict* (London: Brassey's).

Thompson, W. Scott, & Kenneth M. Jensen (eds.) 1991: *Approaches to Peace: An Intellectual Map* (Washington, D.C.: United States Institute for Peace).

Thomson, James A., & Nanette C. Ganz 1988: "Conventional Arms Control Revisited: Objectives in the New Phase," in Nehrlich & Thomson (eds.): 108–120.

Thorburn, Hugh 1986: "The New Democratic Party and National Defence," in Ørvik (ed.) 169–185.

Thoreau, Henry David 1849: "Civil Disobedience" (excerpt from "Resistance to Civil Government") in George Woodcock (ed.), *The Anarchist Reader* (London: Fontana, 1977): 197–204.

Thorsson, Inga 1985: *Med sikte på nedrustning: Omställning från militär til civil produktion i Sverige, Betänkande av Utredningen om samband mellan nedrustning och utveckling. Del 2: Särskilda rapporter* (Stockholm: Utrikesdepartementet, Statens offentlige udredninger 1985): 43.

Thränert, Oliver 1991: "Vom Rüstungswettlauf zum Anrüstungswettlauf? US-Präsident Bushs Abrüstungsinitiative vom 27. September 1991 und die Antwort des sowjetischen Präsidenten Gorbatschow," *Studie,* 47.

Tiedtke, Stephan 1984: "Alternative Military Defence Strategies as a Component of Detente and Ostpolitik," *BPP,* 15:1, 13–24.

————— 1986: "Abschreckung und ihre Alternativen: Die sowjettische Sicht einer westlichen Debatte," *Texte und Materialien der Forschungsstätte der Evangelischen Studiengemeinschaft,* Reihe A, 20 (Heidelberg: F. E. St.).

————— 1987: "Die sowjetische Militärdoktrin und sicherheitspolitische Stabilität," in Bühl (ed.): 79–90.

Till, Geoffrey 1982: *Maritime Strategy and the Nuclear Age* (London: Macmillan).

Tilson, John 1991: "Considering New Forces for NATO." Paper presented to the conference "Sicherheit im neuen Europa," Evangelische Akademie Loccum and SAS, 27–29 September 1991.

Tirman, John (ed.) 1984: *The Fallacy of Star Wars* (New York: Vintage).

————— (ed.) 1986: *Empty Promise: The Growing Case Against Star Wars* (Boston: Beacon).

Tito, Josip Broz 1966: *Selected Military Works* (Belgrade: Vojnoizdavacki Zavod).

Titus, Keith 1991: "The Future Shape of Warfare," in Hemsley (ed.): 91–103.

Tobias, Sheila, Peter Goudinoff, Stefan Leader, & Shelah Leader 1982: *What Kinds of Guns Are They Buying for Your Butter? A Beginner's Guide to Defense, Weaponry, and Military Spending* (New York: Morrow).

Tolnay, Laszlo 1987: "Military Doctrines and Armament," in Szentesi & al.: 133–154.

Tolstoy, Leo 1894: *The Kingdom of God Is Within You* (London: Heinemann).

Törnudd, K. 1983: "Possible Rules Regulating the Transit of Nuclear Weapons Inside and Their Deployment Outside a Nordic NWFZ," in Lodgaard & Thee (eds.): 199–208.

Tow, William T. 1993: "Contending Security Approaches in the Asia-Pacific Region," *SS*, 3:1 (Autumn): 75–116.

Trachtenberg, Marc 1990: "The Meaning of Mobilization in 1914," *IS*, 15:3 (Winter 1990–1991): 120–150.

―――― 1991: "The Past and Future of Arms Control," *Dædalus*, Winter 1991: 203–216.

―――― 1991a: *History and Strategy* (Princeton: Princeton U.P.).

Tranholm-Mikkelsen, Jeppe 1991: "Neo-Functionalism: Obstinate or Obsolete? A Reappraisal in the Light of the New Dynamism of the EC," *Millennium*, 29:1, 1–22.

Trapp, Ralf (ed.) 1987: "Chemical Weapon Free Zones?" *SIPRI Chemical and Biological Warfare Studies*, 7 (Oxford: Oxford U.P., SIPRI).

Trempnau, Thomas 1983: "Horst Afhcldt's Defensive Response Model," in Steinke & Vale (eds.): 187–196.

Trenin, Dmitri 1994: "NOD in the USSR and Successor States," in Møller & Wiberg (eds.): 127–137.

Trevorton, Gregory F. 1991: "Deterrence and Collective Security," in Thompson & Jensen (eds.): 15–27.

Trgo, Fabijan 1966: "Survey of the People's Liberation War," in Tito 1966: 307–336.

Tritten, James J. 1990: "'Naval' Arms Control: A Poor Choice of Words and an Idea Whose Time Has Yet to Come," *CRPV*, 13:2, 65–86.

Trofiemenko, Henry 1989: "The Emergence of Mutual Security: Its Objective Basis," in Allison & Ury (eds.): 167–180.

Trommer, Aage 1971: *Jernbanesabotagen i Danmark under 2: Verdenskrig: En krigshistorisk undersøgelse* (Odense: Odense Universitetsforlag).

Tromp, Hylke W. 1978: "The Dutch Research Project on Civilian Defence, 1974–1978," *BPP*, 9:4, 301–307.

―――― 1984: "Friedenssicherung durch Alternatives der Sicherheitspolitik?" in Vogt (ed.): 133–144.

―――― 1985: "Vorschläge zur Veränderung der gegenwärtigen Sicherheitspolitik," in Lutz (ed.) 1985a: 339–358.

―――― 1988: "Interdependence and Security: The Dilemma of the Peace Research Agenda," *BPP*, 19:2, 151–158.

―――― 1989: "Syllabus errorum: Security Policy in the Nuclear Age," in idem (ed.): 331–350.

―――― 1990: "Nonoffensive Defense and Conventional Stability in Europe: The Wrong Problem, The Wrong Solution," in Cerutti & Ragionieri (eds.): 169–176.

―――― 1990a: "Nonoffensive Defence, Conventional Stability, and the Military Balance," in Schmähling (ed.): 37–48.

―――― 1991: "Peace Research and the Inter-Paradigm Debate," in Nobel (ed.): 97–109.

———— (ed.) 1986: *Non-Nuclear War in Europe: Alternatives for Nuclear Defence* (Groningen: Polemological Institute).

———— (ed.) 1989: *War in Europe: Nuclear and Conventional Perspectives* (Aldershot: Gower).

Tronnberg, Stefan 1985: *Nedrustning under mellankrigstiden: Sverige och nedrustningskonferensen i Geneve 1932* (Stockholm: Seelig).

Truger, Arno 1986: "Neutralitätspolitische Konzeptionen der Friedensbewegung," *Dialog,* 6: 243–292.

Tschernyschew, Wladimir 1989: "Strukturelle Angriffsunfähigkeit: Die Diskussion in der Sowjetunion über das Prinzip des 'hinlänglichen Verteidigungsmasses'," *Truppenpraxis,* 4: 388–394.

———— 1989a: "Strukturelle Angriffsunfähigkeit: Ansicht der UdSSR," *ÖMZ,* 2: 115–119.

Tsipis, Kosta 1985: "Third-Generation Nuclear Weapons," in *SIPRI,* 1985: 98–101.

———— 1992: "Military Technological Innovation and Stability," in Smit & al. (eds.): 167–176.

———— & Mark Andersen 1992: "New Technologies and Defense Policy for the European Theater," in Brauch & Kennedy (eds.): 173–189.

Tsypkin, Mikhail 1992: "Will the Military Rule Russia?," *SS,* 2:1 (Autumn): 38–73.

Tuchman, Barbara 1962: *The Guns of August* (New York: Bantam, 1976).

Tucker, Robert W. 1993: "The Triumph of Wilsonianism?" *WPJ,* 10:4 (Winter): 83–99.

Tunander, Ola 1989: *Cold Water Politics: The Maritime Strategy and Geopolitics of the Northern Front* (London: Sage).

———— 1990: "Naval Hierarchies, European 'Neutralism' and Regional Restraint at Sea," in Lodgaard (ed.): 159–186.

Turner, Stansfield, & G. Thibault 1982: "Preparing for the Unexpected: The Need for a New Military Strategy," *FA,* Fall: 122–135.

Türpe, André 1989: "Von der Philosophie der Abschreckung zu einem gesichertem Frieden für Europa," in Wagemann (ed.): 133–142.

Twigg, Judyth L. 1993: "Defence Planning: The Potential for Transparency and Co-Operation," in Goodby & Morel (eds.): 272–288.

Twining, David T. 1986: "An East-West Center for Military Cooperation," in Borawski (ed.): 173–183.

UCS (Union of Concerned Scientists) 1983: *No-First-Use* (Washington, D.C.: UCS).

Udgaonkar, Bhalchandra, Raja Mohan, & Maj Britt Theorin 1993: "Approaches Towards a Nuclear-Weapon-Free World," in Rotblat & al. (eds.): 201–220.

Ueta, Takato 1992: "Japan: A Case of a Non-Control Regime," in Tanner (ed.): 101–114.

Uhle-Wettler, Franz 1966: *Leichte Infanterie im Atomzeitalter* (Darmstadt: Wehr & Wissen).

———— 1980: *Gefechtsfeld Mitteleuropa: Gefahr der Übertechnisierung von Streitkräften* (Gütersloh: Bernard & Graefe).

———— 1990: "Geographische Faktoren in der Verteidigung," in Weizsäcker (ed.): 123–132.

Ullman, Richard H. 1985: "Denuclearizing International Politics," in Hardin & al. (eds.): 191–212.

———— 1990: "Enlarging the Zone of Peace," *Foreign Policy,* 80 (Fall): 102–120.

———— 1991: *Securing Europe.* A Twentieth Century Fund Book (Princeton: Princeton U.P.).

Umbach, Frank 1994: "Nuklearmacht sein oder nicht sein: Hintergründe zu nuklearen Ambitionen der Ukraine, Weissrusslands und Kasachstans," part 1, *ÖMZ,* 32:1 (January): 23–32.

Unger, Danny 1993: "The Problem of Global Leadership: Waiting for Japan," in Unger (ed.): 3–20.

———— & Paul Blackburn (eds.) 1993: *Japan's Emerging Global Role* (Boulder, Colo.: Lynne Rienner).

UNIDIR (ed.) 1989: *Problems and Perspectives of Conventional Disarmament in Europe* (New York: Taylor & Francis).

———— (ed.) 1990: "Transparency in International Arms Transfers," *Disarmament Topical Papers,* 3 (New York: United Nations).

———— (ed.) 1990a: *Nonoffensive Defense: A Global Perspective* (New York: Taylor & Francis).

———— (ed.) 1991: "Conversion: Economic Adjustments in an Era of Arms Reduction," vol. 1, *Disarmament Topical Papers,* 5 (New York: UNIDIR).

Union of Concerned Scientists: See UCS.

Unterseher, Lutz 1984: "Friedenssicherung durch Vermeidung von Provokation? Ein pragmatischer Vorschlag für eine alternative Landesverteidigung," in Vogt (ed.): 95–103

———— 1984a: "Für eine tragfähige Verteidigung der Bundesrepublik: Grundgedanken und Orientierungen," in SAS (eds.): 108–130.

———— 1984b: "Konventionelle Verteidigung Mitteleuropas: Etablierte Struktur und Alternativen im Test," in Brauch (ed.): 214–222.

———— 1986: "Emphasizing Defence: An Ongoing Non-Debate in the Federal Republic of Germany," in Barnaby & Borg (eds.): 116–124.

———— 1987: "Bewegung, Bewegung! Zur Kritik eingefahrener Vorstellungen vom Krieg," *S+F,* 2: 90–97.

———— 1987a: "Fünf Thesen zur Vermeidung von Missverständnissen um eine alternative Verteidigung," in Bahr & Lutz (eds.): 283–300.

———— 1987b: "Stellungnahme der Studiengruppe Alternative Sicherheitspolitik (SAS) zum Thema 'Vertrauensbildende Strukturen'," *S+F,* 1: 77–83.

———— 1987c: "The Standard Battalion of Line Infantry: An Exercise in Force Design with Robust Technologies" (Bonn: SAS).

———— 1987d: "Zur Kritik westlicher Militärstrategien für Europa: Die Notwendigkeit einer Alternative," *Dialog,* 10: 33–52.

———— 1988: "Defusing the Military Confrontation in Europe: Some Guidelines for Bilateral Talks," in Baudisch (ed.): 83–85.

———— 1988a: "Some Naughty Thoughts on Italy's Defence." Paper prepared for the international conference "Rethinking European Security," at the Istiuto Universitario Europeo, University of Florence, 23–24 September.

———— 1988b: "Spider and Web: The Case for a Pragmatic Defence Alternative" (Bonn: SAS).

———— 1988c: "Tactical Air Forces: For a Disarmament Initiative of the West" (Bonn: SAS).

———— 1988d: "Vertrauensbildende Verteidigung: Technik—Struktur—Doktrin," in Calliess (ed.): 279–282.

———— 1989: ". . . Der Ost macht ernst: Über die militärische und politische Bedeutung einseitiger Truppenreduzierungen der UdSSR und ihrer Verbündeten," *S+F,* 7:4, 248–251.

———— 1989a: "'Sicher ist Sicher': Militärisches Rückversicherungsdenken gegen strikte Defensivorientierung," in Vogt (ed.): 82–98.

———— 1989b: "Bauprinzipien alternativer Landstreitkräfte," in SAS (eds.): 149–164.

———— 1989c: "Defending Europe: Toward a Stable Conventional Deterrent," in Shue (ed.): 293–342.

———— 1989d: "Ein anderes Heer: Wesentliche Einzelheiten," in SAS (eds.): 240–270.

———— 1989e: "MP(S)G: The Mobile Protected (Shot) Gun: An Exercise in Alternative Weapons Design" (Bonn: SAS).

———— 1989f: "Spinne im Netz: Ein tragfähiger Weg zur Vertrauensbildenden Verteidigung," *Bremer Beiträge,* 1: 15–26.

———— 1989g: "Umrisse einer stabilen Lutfverteidigung," in SAS (eds.): 188–203.

———— 1989h: "Urban Warfare," for Trevor Dupuy (ed.), *Military Encyclopedia* (manuscript).

———— 1989i: "Seven Theses on Defusing Europe as a Potential Source and Theatre of Military Conflict," in Tromp (ed.): 165–172.

———— 1990: "A Note on the Intricacies of Military Force Assessment," in Brauch & Kennedy (eds.): 69–78.

———— 1990a: "Der Abrüstungsprozess ist kein Selbstlaufer," *Bremer Beiträge,* 3: 40–44.

———— 1990b: "The Case for Alternative Defense: A Detailed Proposal and Its Implications," in Cerutti & Ragionieri (eds.): 105–117.

———— 1990c: "Vertrauensbildende Verteidigung," in Weizsäcker (ed.): 143–149.

———— 1991a: "Bundeswehr: Wohin (damit)?" Paper presented to the conference "Sicherheit im neuen Europa," Evangelische Akademie Loccum and SAS, 27–29 September 1991.

———— 1991b (with Hartmut Bebermeyer & Malcolm Chalmers): "Stabile Sicherheitsstrukturen in Europa—Programm für die Jahrtausendwende." Paper for the Österreichisches Institut für Friedensforscung und Friedenserziehung, Stadtschlaining (Bonn: SAS).

———— 1994: "NOD and the Land Forces: Confidence-Building Defence for the New Europe," in Møller & Wiberg (eds.): 60–73.

Urban, Jan 1993: "The Czech and Slovak Republics: Security Consequences of the Breakup of the CSFR," in Karp (ed.): 101–121.

Urban, mark 1988: *War in Afghanistan* (New York: St. Martin's Press).

Urquhart, Brian 1991: "The U.N.'s Crucial Choice," *Foreign Policy,* 84 (Fall): 157–165.

———— 1993: "For a UN Volunteer Military Force," *New York Review of Books,* 40:11 (10 June): 3–4.

———— 1993a: "The UN and International Security After the Cold War," in Roberts & Kingsbury (eds.): 81–103.

Ury, William J. 1985: *Beyond the Hotline: How Crisis Control Can Prevent Nuclear War* (Boston: Houghton Mifflin).

Vajda, Peter 1987: "The Rogers Plan and the Real Correlation of Conventional Forces in Europe," in Szentesi & al. (eds.): 84–132.

Valiliev, Alexei A. 1989: "Nuclear Weapons and Conventional Arms Control," in Blackwill & Larrabee (eds.): 442–461.

Valki, Laszlo 1985: "The Concept of Defensive Defence," *Külpolitika,* 1: 99–112.

———— 1986: "Arguments and Counter-arguments Concerning Defensive Defence," in Barnaby & Borg (eds.): 110–115.

———— 1988: "Uncertainties About WTO and NATO Military Doctrine," in Alger & Stohl (eds.): 419–435.

———— 1990: "Structural Defensiveness: Debates and Views," in Cerutti & Ragionieri (eds.): 118–133.

———— 1990a: "Die Antworten Osteuropas auf das Konzept der defensiven Verteidigung," in Weizsäcker (ed.): 264–273.

———— 1992: "Security Concerns in Central Europe," *ES,* 1:4 (Winter): 1–14.
———— 1993: "A Future Security Architecture for Europe," in "International Security and East-Central Europe," *Occasional Papers* (Budapest: Centre for Security Studies, Hungarian Academy of Sciences): 103–142.
Van Creveld, Martin: see Creveld, Martin Van.
Vanecek, Vaclav 1964: "Historical Significance of the Peace Project of King George of Bohemia and the Research Problems Involved," in Vanecek (ed.): 9–66.
———— (ed.) 1964: *The Universal Peace Organization of King George of Bohemia: A Fifteenth Century Plan for World Peace, 1462/1464* (Prague: Publishing House of the Czechoslovak Academy of Sciences).
Van Evera, Stephen W.: see Evera, Stephen W.
Van Orden, Geoffrey: See Orden, Geoffrey Van.
Vare, Raivo 1991: "Ensuring the National Security of the Republic of Estonia." Paper presented to the workshop "Soviet, Russian and Baltic Security," Svaneke (Denmark) 25–27 September 1991.
Vares, Peter, & Mare Haab 1993: "The Baltic States: *Quo vadis?*" in Karp (ed.): 283–306.
Väyrynen, Paavo 1989: "Neutrality in the European Context: The Finnish Perspective," in Kruzel & Haltzel (eds.): 277–282.
Väyrynen, Raimo 1982: "Die finnische Verteidigungspolitik und ihre militärische Infrastruktur," in Lutz & Grosse-Jütte (eds.): 141–166.
———— 1983: "Kernevåben og Finlands udenrigspolitik," in Heurlin (cd.): 18–57.
———— 1985: "Introduction: Towards a Strategy of Common Security," in Väyrynen (ed.): 1–15.
———— 1985a: "The European Cooperation and Security Process: Security Dilemmas and Confidence-Building Measures," *BPP,* 16:4, 349–361.
———— 1987: "Adaptation of a Small Power to International Tensions: The Case of Finland," in Sundelius (ed.): 33–56.
———— 1987a: "Neutrality, Dealignment and Political Order in Europe," in Kaldor & Falk (eds.): 163–184.
———— 1988: "Disarmament, Deterrence and Common Security," in Hellman (ed.): 65–70.
———— 1989: "Arms Race: Theory and Action," in Harle & Sivonen (eds.): 6–25.
———— 1989a: "Common Security and the State System," in Nakarada & Øberg (eds.): 39–57.
———— 1990: "Conflict Tranformation and Cooperation in Europe," *BPP,* 21:3, 299–306.
———— (ed.) 1985: *Policies for Common Security* (London: Taylor & Francis, SIPRI).
———— (ed.) 1987: *The Quest for Peace: Transcending Collective Violence and War Among Societies, Cultures and States* (London: Sage).
———— (ed.) 1991: *New Directions in Conflict Theory: Conflict Resolution and Conflict Transformation* (London: Sage).
Vekaric, Vatroslav 1989: "Current Trends in International Affairs," in Singh & Vekaric (eds.): 63–82.
Venohr, Wolfgang (ed.) 1990: *Deutschland wird es sein* (Erlangen: Straube).
Veremis, Thanos 1988: "Greece and NATO: Continuity and Change," in Chipman (ed.): 236–286.
———— 1990: "The Future of European Security: The South East," in Clesse & Rühl (eds.): 342–350.
Vesa, Unto 1988: "Översikt över de säkerhetspolitiska tendenserna och rust-

ningskontrollens möjligheter i Östersjöområdet," in Joenniemi & Vesa (eds.): 1–40.

Vetschera, Heinz 1985: "Der Weg zu Staatsvertrag und Neutralität," *ÖMZ,* 3: 223–228.

———— 1985a: "Die Rüstungsbeschränkungen des Österreichischen Staatsvertrags aus rechtlicher und politischer Sicht," *ÖMZ,* 6: 500–505.

———— 1985b: "Neutrality and Defence: Legal Theory and Military Practice in the European Neutrals' Defence Policies," *Defense Analysis,* 1:1 (March): 51–64.

———— 1989: "Reduction of Conventional Armed Forces in Europe: Problems and Prospects," *Disarmament,* 12:2, 70–80.

———— 1991: "The Future of Arms Control for European Security: From Arms Control to Force Control," in Danspeckgruber (ed.): 147–168.

———— 1992: "Armament Limitations of the Vienna State Treaty," in Tanner (ed.): 115–135.

———— & James V. Rocca 1985: "Österreich in den Ost-West-Beziehungen," *ÖMZ,* 3: 229–305.

Victor, Jean Christophe 1985: "The Conflict in Afghanistan," in *SIPRI,* 1985: 577–610.

Vigor, Peter H. 1982: "Soviet Echeloning," *MR,* 62:8 (August): 69–74.

———— 1983: *Soviet Blitzkrieg Theory* (New York: St. Martin's).

Villiers, J. W. de, Roger Jardine, & Michell Reiss 1993: "Why South Africa Gave Up the Bomb," *FA,* 72:5 (November–December): 98–109.

Viñas, Angel 1988: "Spain and NATO: Internal Debate and External Challenges," in Chipman (ed.): 140–194.

Visuri, Pekka 1985: "Die Entwicklung der finnischen Verteidigungsdoktrin nach dem Zweiten Weltkrieg," *ÖMZ,* 1: 9–16.

———— 1989: "Alternative Defence in Central European Debate," in Harle & Sivonen (eds.): 93–109.

———— 1989a: *Totaalisesta sodasta kriisinhallintaan: Puolustusperiaatteiden kehitys läntisessä Keski-Euroopassa ja Suomessa vuosina, 1945–1985* (Keuruu: Otava).

———— 1990: "Evolution of the Finnish Military Doctrine, 1945–1985," *Finnish Defence Studies,* 1 (Helsinki: War College).

———— 1990a: "Political Change in Europe and the Idea of Non-Offensive Defence: In Search for New Perspectives," in Auffermann (ed.): 115–136.

———— 1991: "Die militärischen Artikel des Pariser Friedensvertrages: Vorgeschichte der Beschränkungen für die finnischen Streitkräfte und Probleme der Auslegung," *ÖMZ,* 29:1 (January–February): 14–17.

———— 1991a: "The Changing Strategic Status of the Baltic Region." Paper presented to the workshop "Soviet, Russian and Baltic Security," Svaneke (Denmark) 25–27 September 1991.

Vladimirov, A. I. 1992: "Some Military Aspects of the Security Strategy of our Civilisation," *ES,* 1:2 (Summer): 184–206.

Vogel, Hans 1987: "Switzerland and the New Cold War: International and Domestic Determinants of Swiss Security Policy," in Sundelius (ed.): 95–116.

Vogt, Roland 1993: "Zivile Weltordnung oder militärischer Interventionismus," *Blätter,* 38:8 (August): 960–971.

Vogt, Wolfgang R. 1989: "Legitimitätsverfall der Sicherheitspolitik," in SAS (ed.): 67–101.

———— 1989a: "Militär und Risikogesellschaft," *S+F,* 7:4, 198–205.

———— 1990: "The Crisis of Acceptance of the Security Policy with Military Means in the Federal Republic of Germany," in Brauch & Kennedy (eds.): 165–188.

——— 1990a: "Positiven Frieden wagen: Zur Transformation militarisierter Friedenssicherung in zivilizierte Friedensgestaltung," in Vogt (ed.): 1–37.

——— 1990b: "Wozu noch Militär? Die Debatte über alte und neue Funktionen der Bundeswehr," in Birkenbach & al. (eds.): 142–155.

——— 1991: "Militärlogik gegen Zivillogik: Sind Kriege abschaffbar?" in Birkenbach & al. (eds.): 39–53.

——— 1992: "Militär—Eine Institution auf der Suche nach Legitimation: Zur Debatte über die zukünftigen Funktionen militärischer Macht im Prozess der Friedenssicherung und -gestaltung," in Kaldrach & Klein (eds.): 9–40.

——— & Ingeborg Rubber-Vogt 1989: "Angst vorm Frieden: Von den Schwierigkeiten, tradiertes Sicherheitsdenken zu überwinden und positiven Frieden zu wagen," in Vogt (ed.): 1–32.

——— (ed.) 1984: *Streitfall Frieden: Positionen und Analysen zur Sicherheitspolitik und Friedensbewegung* (Heidelberg: Müller).

——— (ed.) 1989: *Angst vorm Frieden: Über die Schwierigkeiten der Friedensentwicklung für das Jahr 2000* (Darmstadt: Wissenschaftliche Buchgesellschaft).

——— (ed.) 1990: *Mut zum Frieden: Über die Möglichkeiten einer Friedensentwicklung für das Jahr 2000* (Darmstadt: Wissenschaftliche Buchgesellschaft).

Voigt, Karsten D. 1984: "Interim Report of the Sub-Commitee on Conventional Defence in Europe" (Bruxelles: Military Committee of the North Atlantic Assembly).

——— 1985: "Gemeinsame Wege zur chemischen Abrüstung," *NGFH*, 8: 736–741.

——— 1985a: "Sozialdemokratische Sicherheitspolitik: Ursprung und Perspektiven," in Lutz (ed.) 1985a: 69–98.

——— 1986: "Die Grünen nach Hannover, oder Warum es zwischen ihnen und der SPD keine Koalition geben kann," *S+F*, 4: 220–222.

——— 1987: "Der Vorschlag für einen nuklear-waffenfreien Korridor: Ein deutsch-deutsches Pilotprojekt für eine Sicherheitspartnerschaft im Ost-West Konflikt," *DA*, 2.

——— 1987a: "New Strategies and Operational Concepts." Sub-Committee on Conventional Defence: *Report* (Bruxelles: Military Committee of the North Atlantic Assembly).

——— 1987b: "Von der nuklearen zur konventionellen Abrüstung in Europa: Kriterien konventioneller Stabilität und Möglichkeiten der Rüstungskontrolle," *EA*, 14: 409–418.

——— 1988: "Konventionelle Stabilisierung und Strukturelle Nichtangriffsfähigkeit: Ein systematischer Vergleicherschiedener Konzepte," *APZ*, B 18/88 (29 April): 21–34. (English version: "The Prospects of Conventional Stabilization in Europe," in Kaldor, Holden, & Falk (eds.) 1990: 207–234.)

——— 1989: "Defense Alliances in the Future: West European Integration and All-European Co-operation," *BPP*, 20:4, 355–361.

——— 1989a: "Durch 'Neues Denken' zur Neuen Friedenspolitik," *NGFH*, 4: 305–311.

——— 1989b: "Durchbruch bei der konventionellen Abrüstung möglich: Analyse und Bewertung des bisherigen Verlaufs der Wiener Verhandlungen zur konventionellen Abrüstung," *S+F*, 7:2, 136–139; also in E. Müller & Neuneck (eds.) 1990: 37–44.

——— 1989c: "Eine Sicherheitspolitik für Europa," in Wagemann (ed.): 115–122.

———— 1990: "Deutsche Einheit und gesamteuropäische Ordnung des Friedens und der Freiheit," *DA,* 4: 562–568.

———— 1990a: "Eine europäische Friedensordnung ohne den Ungeist des Nationalismus," *EW,* 3: 146–154.

———— 1990b: "KSZE und europäische Gemeinschaft als wichtigste Grundlagen einer europäischen Friedensordnung," in H. E. Bahr (ed.): 207–211.

———— 1990c: "Overcoming the Division of Europe," *BPP,* 21:1, 3–14.

———— 1990d: "Schritte zur Demilitarisierung Europas," in Weizsäcker (ed.): 317–331.

———— 1990e: "Europäische Friedensordnung und deutsche Einheit," in Heisenberg & Lutz (eds.): II:417–425.

———— 1990f: "Die demokratische Revolution in Osteuropa und gesamteuropäische Friedensordnung," in Calliess (ed.): 232–241.

Volger, Helmut 1991: "Braucht die UNO neue Strukturen? Konzepte und Chancen einer Reform der Weltorganization," *Blätter,* 36:9 (September): 1087–1098.

Volgyes, Ivan 1982: "Regional Differences Within the Warsaw Pact," *Orbis,* Fall: 665–678.

———— 1992: "Military Security in the Post-Communist Age: Reflections on Myths and Misperceptions," *ES,* 1:4 (Winter): 56–64.

Volle, Angelika 1988: "The Political Debate on Security Policy in the Federal Republic," in Kaiser & Roper (eds.): 41–59.

Vo Nguyen Giap: See Giap, Vo Nguyen.

Vukadinovic, Radovan 1989: "The Various Conceptions of European Neutrality," in Kruzel & Haltzel (eds.): 29–46.

Vuono, Carl E. 1991: "Desert Storm and the Future of Conventional Forces," *FA,* 70:2, 49–68.

Vuorenmaa, Anssi 1985: "Defensive Strategy and Basic Operational Decisions in the Finland-Soviet Winter War, 1939–1940," in FCMH (eds.): 74–96.

———— 1985a: "Finland's Defence Forces: The Years of Construction," in FCMH (eds.): 39–54.

Wachter, Gerhard, & Axel Krohn 1989: "The Role of Military Forces within a European Security System," in Wachter & Krohn (eds.): 1–26.

———— & Axel Krohn (eds.) 1989: *Stability and Arms Control in Europe: The Role of Military Forces Within a European Security System* (Stockholm: SIPRI).

Wæver, Ole 1989: "Conceptions of Détente and Change: Some Non-Military Aspects of Security Thinking in the FRG," in idem al. (eds.): 186–224.

———— 1989a: "Ideologies of Stabilization, Stabilization of Ideologies: Reading German Social Democrats," in Harle & Sivonen (eds.): 110–139.

———— 1989b: "Conflicts of Vision: Visions of Conflict," in idem & al. (eds.): 283–325.

———— 1990: "Three Competing Europes: German, French, Russian," *IA,* 66:3, 477–493.

————, Pierre Lemaitre, & Elzbieta Tromer (eds.) 1989: *European Polyphony: Perspectives Beyond East-West Confrontation* (London: Macmillan).

————, Barry Buzan, Morten Kelstrup, & Pierre Lemaitre 1993: *Identity, Migration and the New Security Agenda in Europe* (London: Pinter).

Wagemann, Eberhard (ed.) 1989: *Frieden ohne Rüstung?* (Herford: Mittler).

Wagenlehner, Günther 1987: "Friedliche Koexistenz und ideologischer Kampf," in Buchbender, Bühl, & Quaden (eds.): 129–139.

———— 1991: "Die Militärpolitik Gorbatschows und der Warschauer Pakt," in Brahm & al.: 131–150.

Wagner, Richard L. 1990: "A Path to Enduring Improvements, Over the Long Term, in the East/West Security Relationship," in Huber (ed.): 29–50.

Wagner, Wolfgang 1989: "Auf der Suche nach einer neuen Ordnung in Europa," *EA*, 44:23, 693–702.

———— 1993: "Der ständige Sitz im Scherhetsrat: Wer braucht wen: Die Deutschen diesen Sitz? Der Sicherheitsrat die Deutschen?" *EA*, 48:19 (10 October): 533–540.

Walendowski, Edmund 1988: *Combat Motivation of the Polish Forces* (London: Macmillan).

Walker, Jeronne 1991: "New Thinking About Conventional Arms Control," *Survival*, 33:1, 53–65.

———— 1993: "Regional Organizations and Ethnic Conflict," in Karp (ed.): 45–68.

Walker, J. R. 1987: "Air-to-Surface Weapons," in Perkins (ed.): 1–56.

Walker, Paul F. 1981: "Precision-Guided Weapons," *Scientific American*, 25:2, 21–29.

———— 1985: "Trends in Military Technology," in Windass (ed.): 19–42.

———— 1986: "Emerging Technologies and Conventional Defence," in Barnaby & Borg (eds.): 27–43.

———— 1992: "From Cold War to Common Security: The Need for a Peace Perestroika," in Bächler (ed.) 1992a: 249–260.

Walker, Rachel 1990: "'New Thinking' and Soviet Foreign Policy," in Pugh & Williams (eds.): 142–160.

Walker, R. B. J. 1988: *One World, Many Worlds: Struggles for a Just World Peace* (Boulder, Colo.: Lynne Rienner).

Wallace, William 1985: "Shifts in British Defence Policy: How Would the European Allies React?" in Roper (ed.): 103–115.

Wallach, Jehuda L. 1967: *Das Dogma der Vernichtungsschlacht: Die Lehren von Clausewitz und Schlieffen und ihre Wirkungen in zwei Weltkriegen* (Frankfurt: Bernard & Graefe).

———— 1972: *Kriegstheorien: Ihre Entwicklung im 19. und 20. Jahrhundert* (Frankfurt: Bernard & Graefe).

———— 1986: "'Misperceptions of Clausewitz' *On War* by the German Military," *JSS*, 9:23, 213–239.

Wallander, Celeste A. 1992: "International Institutions and Modern Security Strategies," *POC*, 41:1–2 (January–April): 44–62.

Wallensteen, Peter 1988: "Crisis Management by Crisis Prevention: Reducing Major Power Involvement in Third World Conflicts," in Hopmann & Barnaby (eds.): 267–286.

Wallin, Lars B. 1981: "Översikt av den internationalla debatten kring det framtida konventionella Europakrigets karaktär," *FOA Rapport*, C–10118-M3 (Stockholm: Försvarets Forskningsanstalt).

———— 1991: "Sweden," in Karp (ed.): 360–381.

Walt, Stephen M. 1987: *The Origins of Alliances* (Ithaca: Cornell U.P.).

———— 1989: "The Case for Finite Containment: Analyzing U.S. Grand Strategy," *IS*, 14:1, 5–49.

Walter, Franz 1993: "Russlands 'neue' Streitkräfte," *OE*, 43:5 (May): 413–428.

Waltz, Kenneth N. 1979: *Theory of International Politics* (Reading, Mass.: Addison-Wesley).

———— 1981: "The Spread of Nuclear Weapons: More May Be Better," *AP*, 171.

Walzer, Michael 1977: *Just and Unjust Wars: A Moral Argument with Historical Illustrations* (New York: Harper Torchbooks).

Warnke, Paul C. 1984: "Should the United States Commit Itself Not to Be the First to Use Nuclear Weapons?" in Blackaby & al. (eds.): 121–127.

Wasmuht, Ulrike 1987: *Friedensbewegungen der 80er Jahre. Zur Analyse ihrer strukturellen und aktuellen Entstehungsbedingungen in der Bundesrepublik*

Deutschland und den Vereinigten Staaten von Amerika nach 1945: Ein Vergleich (Giessen: Focus).

———— (ed.) 1992: *Ist Wissen Macht? Zur aktuellen Funktion von Friedensforschung* (Baden-Baden: Nomos).

Watkins, James D. 1986: "The Maritime Strategy," *Proc,* 112:1 (January): 2–17.

Watson, Adam 1992: *The Evolution of International Society* (London: Routledge).

Webber, Philip 1990: *New Defence Strategies for the 1990s. From Confrontation to Coexistence* (London: Macmillan).

Weber, Bernd 1993: "Die sicherheitspolitische Lage und das Militärpotential in der ehemaligen Sowjetunion," *OE,* 43:7 (July): 641–656.

Weber, Steve 1991: "Cooperation and Interdependence," *Dædalus,* (Winter): 183–202.

Wechselmann, Maj 1980: *Till kamp mot vapnen!* (Stockholm: Gidlunds).

Weck, Hervé de, & Pierre Maurer 1990: "Swiss National Defence Policy Revisited," in Miliovejic & Maurer (eds.): 65–80.

Weede, Erich 1984: "Democracy and War Involvement," *Journal of Conflict Resolution,* 28:4, 649–664.

———— 1990: "Beyond Pax Atomica: Is Conventional Stability Conceivable? Does Tension-Reduction Matter?" in Huber (ed.): 51–64.

———— 1992: "Some Simple Calculations on Democracy and War Involvement," *JPR,* 29:4 (November): 377–384.

Weidenbaum, Murray 1992: *Small Wars, Big Defense: Paying for the Military After the Cold War* (Oxford: Oxford U.P.).

Weinberg, Gerhard L. 1964: "Airpower and Partisan Warfare," in Armstrong (ed.): 361–387.

Weiner, Klaus-Peter 1989: "Die 'Europäisierung der Sicherheitspolitik': Tendenzen, Probleme, Perspektiven," in IMSF (eds.): 212–223.

Weiner, Milton G. 1986: "Analyzing Alternative Concepts for the Defence of NATO," in Huber (ed.): 83–96.

Weiss, Thomas G. (ed.) 1993: *Collective Security in a Changing World* (Boulder, Colo.: Lynne Rienner).

———— & Maryl A. Kessler (eds.) 1991: *Third World Security in the Post–Cold War Era* (Boulder, Colo.: Lynne Rienner).

Weizsäcker, Carl Friedrich von 1979: *Wege in der Gefahr: Eine Studie über Wirtschaft, Gesellschaft und Kriegsverhütung* (München: dtv).

———— 1983: *Der bedrohte Friede: Politische Aufsätze, 1945–1981* (München: dtv).

———— 1990: "Bedingungen des Friedens in Europa," in idem (ed.): 17–22.

———— (ed.) 1971 : *Kriegsfolgen und Kriegsverhütung,* 2d ed. (München: Hanser).

———— (ed.) 1976: *Verteidigung ohne Schlacht* (München: Deutscher Taschenbuch).

———— (ed.) 1984: *Die Praxis der defensiven Verteidigung* (Hameln: Sponholz).

———— (ed.) 1990: *Die Zukunft des Friedens in Europa: Politische und militärische Voraussetzungen* (München: Hanser).

Weizsäcker, Ernst U., von 1992: *Erdpolitik: Ökologische Realpolitik an der Schwelle zum Jahrhundert der Umwelt,* 3d, updated ed. (Darmstadt: Wissenschaftliche Buchgesellschaft).

Welch, David 1988: "Internationalism," in Nye, Allison, & Carnesale (eds.): 171–196.

Weller, Jac 1974: "Middle East Tank Killers," *RJ,* 119:4, 28–35.

Weller, Lawrence D. 1983: "No First Use: A History," *BAS,* February, 28–34.

Weller, M. (ed.) 1993: *Iraq and Kuwait: The Hostilities and Their Aftermath. Cambridge International Documents,* vol. III (Cambridge: Grotius).

Wellershoff, Dieter 1991: "Von der Konfrontation über den Dialog zur Kooperation: Der militärische Beitrag zur Absicherung des Wandels," in Wellershoff (ed.): 191–197.

———— (ed.) 1991: *Frieden ohne Macht? Sicherheitspolitik und Streitkräfte im Wandel* (Bonn: Bouvier).

Welles, Sumner 1943: "Blue-print for Peace," in Willkie & al.: 417–437.

Wellmann, Christian 1990: "Konversion in der DDR: Werden die Chancen genutzt?" in Birkenbach & al. (eds.): 184–193.

———— (ed.) 1992: *The Baltic Sea Region: Conflict or Cooperation? Region-Making, Security, Disarmament and Conversion* (Münster: LIT).

Welsh, Ian 1987: "Deterrence or Defence" (Lancaster: Richardson Institute).

Wendt, James C. 1992: "Conventional Arms Control for Korea: A Proposed Approach," *Survival,* 34:4 (Winter): 108–124.

Werbik, Hans 1985: "'Gemeinsame Sicherheit' als erkenntnisleitendes Prinzip für kulturpsychologische Forschungen," *Dialog,* 2–3: 187–196.

———— 1987: "Kulturpsychologische Aspekte der 'Gemeinsamen Sicherheit'," in Bahr & Lutz (eds.): 223–234.

Wernicke, Joachim, & Ingrid Schöll 1985: *Verteidigen statt vernichten: Wege aus der atomaren Falle* (München: Kösel).

Westander, Henrik 1988: *Bofors svindlande affärer* (Stockholm: Ordfront).

Westing, Arthur 1978: "The Military Impact on Human Environment," in *SIPRI,* 1978: 43–68.

———— (ed.) 1989: *Comprehensive Security for the Baltic: An Environmental Approach* (London: Sage).

Weston, Burns H. 1990: "Law and Alternative Security: Toward a Just World Peace," in Weston (ed.): 78–106.

———— (ed.) 1990: *Alternative Security: Living Without Nuclear Deterrence* (Boulder, Colo.: Westview).

Wette, Wolfram 1984: "Das Verhältnis von SPD und bewaffneter Macht (1863–1945)," in Brauch (ed.): 25–36.

———— 1989: "Die Deutschen zwischen Militarismus und Pazifismus," *Blätter,* August, 974–980.

———— 1990: "Reichstag und 'Kriegsgewinnlerei' (1916–1918): Die Anfänge parlamentarischer Rüstungskontrolle in Deutschland," in Steinweg (ed.): 117–159.

———— 1993: "Von neuer 'militärischer Normalität' und 'gewachsener Verantwortung' Deutschlands: Ein Essay," in Birkenback & al. (eds.): 21–33.

Wettig, Gerhard 1967: *Entmilitarisierung und Wiederbewaffnung in Deutschland, 1943–1955: Internationale Auseinandersetzungen um die Rolle der Deutschen in Europa* (München: Oldenburg).

———— 1988: "Sowjetische Sicherheitspolitik im Zeichen des 'neuen Denkens'," *ÖMZ,* 1: 6–11.

———— 1990: "Zur Philosophie, Konzeption und Fragwürdigkeit eines Alternativmodells: Kollektive Sicherheit in und für Europa," in Heisenberg & Lutz (eds.): II:315–327.

———— 1991: "Soviet Foreign Policy Under Gorbachev: New Political Thinking and Its Impact," in Frost (ed.): 35–64.

———— 1991a: *Changes in Soviet Policy Towards the West* (London: Pinter).

Wheeler, Nicholas J. 1990: "The Dual Imperative of Britain's Nuclear Deterrent: The Soviet Threat, Alliance Politics and Arms Control," in Hoffman (ed.): 32–52.

———— 1991: "Perceptions of the Soviet Threat," in Croft (ed.): 161–178.

———— 1991a: "Europe's Future Security Commitments," in Kirby & Hooper (eds.): 81–106.

———— 1992: "Minimum Deterrence and Nuclear Abolition," in Karp (ed.): 250–280.

———— & Ken Booth 1987: "Beyond the Security Dilemma: Technology, Strategy and International Security," in Jacobsen (ed.): 313–337.

White, Paul C., Robert E. Pendley, & Patrick J. Garrity 1992: "Thinking About No Nuclear Forces: Technical and Strategic Constraints on Transitions and End-Points," in Karp (ed.): 103–127.

White, Ralph K. 1990: "Psychology and Alternative Security: Needs, Perceptions, and Misperceptions," in Weston (ed.): 176–205.

Wiberg, Håkan 1977: "The Conditions of Nonviolent Action," in Geeraerts (ed.): 103–120.

———— 1981: "Détente in Europe," in Jahn & Sakamoto (eds.): 303–312.

———— 1983: "Terrorbalancen och det offensiva försvaret," in Jan Øberg (ed.), *Försvar för en kärnvapenfri värld* (Stockholm: Wahlström & Widstrand): 21–33.

———— 1986: "The Nordic Countries: A Special Kind of System," *CRPV*, 12: 2–12.

———— 1986a: "Nuclear Weapon Free Zones as a Process: The Nordic Case," in Størner & al. (eds.): 191–208.

———— 1989: "Avskräckningens Grundproblem," in idem (ed.): 27–54.

———— 1989a: "Tillitsskapande åtgärder, gemensam säkerhet, en ny världsordning: Realism hos altervativ som en fråga om logik och som en fråga om makt," in idem (ed.): 187–209.

———— 1989b: "Balance of Power and Nuclear Deterrence," *WP-CPCR*, 15.

———— 1989c: "Non-Offensive Defence and the Korean Problem," *WP-CPCR*, 22.

———— 1989d: "The Nukes Are Ill," in Barfoed & al. (eds.): 157–164.

———— 1989e: "Alternatives to Deterrence," in Nakarada & Øberg (eds.): 59–69.

———— 1990: "Maritime tillidsskabende foranstaltninger," *WP-CPCR*, 3.

———— 1990a: "The East/West Heritage and European Interdependence," *WP-CPCR*, 11.

———— 1990b: "New Perspectives on Non-Offensive Defence," *WP-CPCR*, 22.

———— 1990c: "Nonoffensive Defence and the Korean Peninsula," in UNIDIR (eds.): 132–142.

———— 1990d: "Ikke-offensivt forsvar som konfliktløsning og konfliktobjekt," in Sørensen (ed.): 71–85.

———— 1990e: "Armament Dynamics, Why Worry?" in Gleditsch & Njølstad (eds.): 352–375.

———— 1990f: *Konfliktteori och fredsforskning,* 2d ed. (Stockholm: Almquist & Wicksell).

———— 1991: "Security and Arms Control Policies of the Nordic Countries," *WP-CPCR*, 6.

———— 1991a: "The Dynamics of Disarmament in the Middle East." Background paper for the Peace Building in the Middle East Commission of IPRA, 1991.

———— 1992: "Divided States and Divided Nations as a Security Problem: The Case of Yugoslavia," *WP-CPCR*, 14.

———— 1992a: "Post–1989 Challenges for Peace Research," in Bächler (ed.) 1992a: 39–54.

———— 1993: "The Dynamics of Disarmament in the Middle East," in Balazs & Wiberg (eds.): 181–195.

———— 1994: "Neutral and Non-Aligned States in Europe and NOD," in Møller & Wiberg (eds.): 176–194

———— (ed.) 1989: *Farväl til Avskräckningen?* (Lund: Lund U.P.).

Wickham, John A. 1985: "Light Infantry Divisions in Defense of Europe," *NSN*, Special Issue 1: 100–107.

Widmer, Sigmund 1989: "Forms of Neutrality," in Kruzel & Haltzel (eds.): 17–28.

Wieck, Hans Georg 1989: "Sicherheitspolitisches Denken in West und Ost," in Wagemann (ed.): 123–132.

Wieczorek, Pawel 1990: "Non-Offensive Defence and Conversion," *Peace and the Sciences,* December (Vienna: International Institute for Peace): 15–16.

Wieczorek-Zeul, Heidemarie, & Hermann Scheer 1990: "Deutschland und die NATO: Denkschrift zur Frage der Mitgliedschaft des vereinigten Deutschland in der NATO," *F&A,* 1: 11–16.

———— & Hermann Scheer 1990a: "Politischer Strukturwandel in Europa und das Ende der Abschreckung," *F&A,* 1: 5–10.

Wiejacz, J. 1983: "A Central European View on a Nuclear Weapons-Free Zone in the North," in Lodgaard & Thee (eds.): 71–79.

Wieland, Lothat 1987: "Heinrich Ströbel (1869–1944): Eutwurf einer deutschen Friedenspolitik in der Zwischenkriegszeit," in Rajewsky & Riesenberger (eds.): 139–146.

———— 1990: "Sozialdemokratie und Abrüstung in der Weimarer Republik," in Steinweg (ed.): 160–185.

Wiesendahl, Elmar 1989: "Militärische Funktionskriese durch Frieden? Zum Verschwinden von Verteidigungswillen und Kampfbereitschaft in der Bundesrepublik Deutschland," in Vogt (ed.): 229–238.

Wiesner, Jerome B. 1985: "Unilateral Actions as Confidence-Building Measures," in Väyrynen (ed.): 135–142.

Wijk, Rob de 1986: "Deep Strike," in Barnaby & Borg (eds.): 73–88.

———— 1986a: "Follow-on Forces Attack: A European Perspective," in Tromp (ed.): 165–170.

———— 1989: "Trends in NATO Strategy: 'Improving' Conventional Defence," in Borg & Smit (eds.): 133–148.

Wild, Adolf 1987: "Jean Jaurés (1859–1914): Der sozialistische Internationalismus gegen den Krieg," in Rajewsky & Riesenberger (eds.): 69–81.

Wilde, Jaap de 1991: *Saved from Oblivion: Interdependence Theory in the First Half of the 20th Century: A Study on the Causality Between War and Complex Interdependence* (Aldershot: Dartmouth).

Wilenski, Peter 1993: "The Structure of the UN in the Post–Cold War Period," in Roberts & Kingsbury (eds.): 437–467.

Wilke, Peter: "Ökonomische Aspekte Gemeinsamer Sicherheit: Anmerkungen zum Stand der Debatte," in Bahr & Lutz (eds.) 1987: 151–168.

Williams, Phil 1985: "Meeting Alliance and National Needs," in Roper (ed.): 9–27.

———— 1987: "Crisis Management in Europe: Problems and Opportunities," in Windass & Grove (eds.): II (no pagination).

———— 1990: "British Security and Arms Control Policy: The Changing Context," in Hoffman (ed.): 11–31.

———— & William Wallace 1984: "Emerging Technologies and European Security," *Survival,* 26:2 (March–April): 70–78.

Willkie, Wendell L., & al. 1943: *Prefaces to Peace* (New York: Simon & Schuster).

Wilson, Dick 1977: *The Long March 1935: The Epic of Chinese Communism's Survival* (Harmondsworth: Penguin Books).

Wilson, Peter A. 1992: "Emerging New Technologies for the Defense of Central Europe and Geostrategic Implications," in Brauch & Kennedy (eds.): 149–171.

Windass, Stan 1985: "Essentials of Defensive Deterrence," in Windass (ed.): 43–62.

———— 1985a: "Nuclear Guidelines: Next Steps," in Windass (ed.): 99–121.

———— 1985b: "Problems of NATO Defence," in Windass (ed.): 1–18.

———— 1987: "Cooperative Security: A Conceptual Framework," in Windass & Grove (eds.) 1987: II (no pagination).

———— 1987a: "The Defence of Europe: A Conceptual Framework," in Windass & Grove (eds.) 1987: I (no pagination).

———— (ed.) 1985: *Avoiding Nuclear War: Common Security as a Strategy for the Defence of the West* (London: Brassey's).

———— & Eric Grove 1987 (with the assistance of Manfred Bertele and Reiner Huber): "The Land Defence of Europe: Ways Forward," in Windass & Grove (eds.) 1987: I (no pagination).

———— & Eric Grove (eds.) 1987: *Common Security in Europe,* vol. I: *The Defence of Europe; Common Security in Europe,* vol. II: *Cooperative Security* (Adderbury, Oxfordshire: Foundation for International Security).

———— & Eric Grove (eds.) 1988: *Common Security in Europe, 1988* (Adderbury, Oxfordshire: Foundation for International Security).

Windsor, Philip 1985: "The Role of Germany," in Windass (ed.): 87–98.

———— 1989: "Neutral States in Historical Perspective," in Kruzel & Haltzel (eds.): 3–9.

Winkler, Franz 1986: *Partisanenkampf in Jugoslawien: Winteroffensive deutscher Verbände im Gorski Kotar und im kroatischen Küstengebiet 1944* (Frauenfeld: Huber).

Winkler, Theodor 1992: "Central Europe and the Post–Cold War European Security Order," *ES,* 1:4 (Winter): 15–40.

Winnefeld, James A. 1989: "Avoiding the Conventional Arms Control Bottle," *Proc,* 115:4 (April): 30–36.

Wintringham, Tom 1940: *Armies of Freemen* (London: Routledge).

———— 1940a: *New Ways of War* (Harmondsworth: Penguin).

———— 1941: *Freedom Is Our Weapon: A Policy for Army Reform* (London: Kegan Paul).

———— 1941a: *The Politics of Victory* (London: Routledge).

———— 1942: *People's War* (Harmondsworth: Penguin).

Wirtz, James J. 1993: "Strategic Conventional Deterrence: Lessons from the Maritime Strategy," *SS,* 3:1 (Autumn): 117–151.

Wiseman, Geoffrey 1989: *Common Security and Non-Provocative Defence: Alternative Approaches to the Security Dilemma* (Canberra: Australian National University Research School of Pacific Studies, Peace Research Centre).

———— 1994: "NOD and the Asia-Pacific," in Møller & Wiberg (eds.): 208–223.

Wissdorf, Jörg 1992: *Doktrin und Struktur: Eine Untersuchung über den Zusammenhang von sicherheitspolitischen Konzepten und Streitkräftestrukturen unter besonderer Berücksichtigung von defensiven Einsatzoptionen für Luftstreitkräfte* (Baden-Baden: Nomos).

Wittmann, Klaus 1989: "Challenges of Conventional Arms Control," *AP,* 239.

———— 1992: "Zur Entwicklung der künftigen NATO-Strategie," in Kaldrach & Klein (eds.): 117–130.

Woddis, Jack 1977: *Armies and Politics* (London: Lawrence & Wishart).

Wohlstetter, Albert 1959: "The Delicate Balance of Terror" (*FA,* 37:2) in Kissinger (ed.) 1965: 34–58.

———— & Richard Brody 1987: "Continuing Control as a Requirement for Deterring," in Carter & al. (eds.): 142–196.

Wohlstetter, Roberta 1962: *Pearl Harbor: Warning and Decision* (Stanford: Stanford U.P.).

Wolf, Reinhard 1992: "Kollektive Sicherheit und das neue Europa," *EA*, 47:13 (10 July): 365–374.

Wolfers, Arnold 1965: *Discord and Collaboration: Essays on International Politics* (Baltimore: Johns Hopkins U.P.).

Woodward, Bob 1987: *Veil: The Secret Wars of the CIA, 1981–1987* (New York: Simon & Schuster).

Wörister, Karl 1986: "Landesverteidigung und Neutralität im Spannungsfeld zwischen Anspruch und Wirklichkeit," *Dialog*, 7: 41–116.

World Commission on Environment and Development: See Brundtland Commission.

Wörner, Manfred 1985: "Kritik an den Entwürfen zur Alternativen Verteidigung," in Lutz (ed.): 366–378.

———— 1986: "A Missile Defence for NATO Europe," *Strategic Review*, Winter: 13–20.

Woyke, Wichard 1993: "NATO Faces New Challenges," *AuP*, 44:2 (2d Quarter): 120–126.

Wright, David 1992: "SDI After the Cold War: The Reorientation of Strategic Defense," in Neuneck & Ischebeck (eds.): 147–162.

Wulf, Herbert 1990: "Ökonomische Umrüstung: Erfahrungen und Ergebnisse der Konversion von Rüstungsproduktion," in Vogt (ed.): 191–203.

———— 1991: "Rüstungsindustrie und Konversion," in Schwerdtfenger & al. (eds.): 294–307.

———— 1991a: "Arms Production," in *SIPRI*, 1991: 281–310.

———— 1991b: "United Nations Deliberations on the Arms Trade," in Anthony (ed.): 228–237.

———— 1992: "Waffenexport zu Beginn der neunziger Jahre," *EA*, 47:11 (10 July): 313–322.

———— 1993: "The United Nations Register of Conventional Arms," *SIPRI*: 533–544.

———— 1993a: "Arms Industry Limited: The Turning Point in the 1990s," in Wulf (ed.): 3–26.

———— (ed.) 1993: *Arms Industry Limited* (Oxford: Oxford U.P.).

Yaniv, Avner 1993: "A Question of Survival: The Military and Politics Under Siege," in Yaniv (ed.): 81–104.

———— (ed.) 1993: *National Security and Democracy in Israel* (Boulder, Colo.: Lynne Rienner).

Yasov, Dmitri 1987: "Warsaw Treaty Military Doctrine: For Defence of Peace and Socialism," *IA(M)*, 10 (October): 38.

Young, Oran R. 1986: "International Regimes: Toward a New Theory of Institutions," *WP*, 39:1 (October): 104–122.

Yost, David 1982: "Ballistic Missile Defense and the Atlantic Alliance," in Miller & Evera (eds.) 1986: 131–162.

———— 1985: "France's Deterrent Posture and Security in Europe" ("Part I: Capabilities and Doctrine" and "Part II: Strategic and Arms Control Implications"), *AP*: 194–195.

Zagladin, Vadim V. 1985: "The Soviet Concept of Security," in Väyrynen (ed.): 65–72.

———— 1986: *Our Aim: Universal International Security* (Moscow: Novosti).

———— 1988: "Entideologisierung und Entmilitarisierung der Ost-West-Beziehungen: Vom 'Gleichgewicht der Kräfte' zum 'Ausgleich der Interessen'," *Blätter*, 12: 1426–1435.

———— 1989: "Perestroika and Soviet Foreign Policy," in Clesse & Schelling (eds.): 193–211.

Zagoria, Donald S. 1991: "Soviet Policy in East Asia: The Quest for Constructive Engagement," in Jervis & Bialer (eds.): 164–182.

Zagorski, Andrei 1992: "Implications of the Disintegration of the Soviet Union for the Evolution of the Nuclear Environment of International Politics," in Neuneck & Ischebeck (eds.): 31–38.

———— 1992a: "Post-Soviet Nuclear Proliferation Risks," *Security Dialogue*, 23:3 (September): 27–39.

———— 1993: "Developments in the CIS: Challenges for Russia," *AuP*, 44:2 (2d Quarter): 144–152.

Zarski, Zdzislaw 1989: "The Role of Military Forces Within a European Security System," in Wachter & Krohn (eds.): 27–50.

Zawodny, J.K. 1962: "Unconventional Warfare," in Kissinger (ed.) 1965: 333–343.

Zehrer, Hartmut (ed.) 1992: *Der Golfkonflikt: Dokumentation, Analyze und Auswertung aus militärischer Sicht* (Herford: Mittler).

Zeidler, Manfred 1993: *Reichswehr und Rote Armee, 1920–1933: Wege und Stationen einer ungewöhnlichen Zusammenarbeit* (München: Oldenburg).

Zellner, Wolfgang 1989: "Abrüstung: Das Mandat von Wien," *Blätter*, 3: 323–338.

———— 1991: "Der Vertrag über konventionelle Abrüstung in Europa," *Blätter*, 1: 70–77.

Zhong, Yang 1992: "The Transformation of the Soviet Military and the August Coup," *Armed Forces and Society*, 19:1 (Fall): 47–70.

Zhurkin, V., S. Karaganov, & A. Kortunov 1989: *Reasonable Sufficiency and New Political Thinking* (Moscow: Nauka Publishers). See also Jourkine, Karaganov, & Kortounov.

Zielonka, Jan 1992: "Security in Central Europe," *AP*, 272.

Ziemke, Earl 1964: "Composition and Morale of the Partisan Movement," in Armstrong (ed.): 135–196.

Zimmerman, Peter D. 1992: "Bronze Medal Technology and Demand-Side Controls on Missile Proliferation," in Neuneck & Ischebeck (eds.): 19–29.

Zinner, Paul E.: "German and U.S. Perceptions of Arms Control," in Wolfram E. Hanrieder (ed.), *Arms Control, the FRG, and the Future of East-West Relations* (Boulder, Colo.: Westview): 11–28.

Zlenko, Anatoly 1992: "Independent Ukraine: Risk of Stability?" *RJ*, 137:2 (April): 38–42.

Zubok, Vladislav, & Andrei Kokoshin 1989: "Opportunities Missed in 1932?" *IA(M)* 2: 112–121.

"Zur Militärdoktrin der Russischen Föderation" 1994, *ÖMZ*, 32:1 (January): 75–82.

About the Book
and the Author

This comprehensive guide to the increasingly important subject of alternative defense is written by one of the most well-respected researchers in the field. The main section of the book—a dictionary of more than 550 detailed entries and an equal number of cross-references—is preceded by an introduction explaining the purposes of alternative defense, the history of the alternative defense debate, and the status of current ideas and concepts. The alphabetically arranged dictionary includes entries from the following broad categories:

- Individuals
- Groups and organizations
- Countries and regions
- Events
- Technologies
- Concepts
- Theories

A major feature of this important reference tool is its bibliography of more than two thousand books, journals, and articles, useful to the librarian, the specialist, and the general reader.

Bjørn Møller (Ph.D. and MA) is senior research fellow at the Centre for Peace and Conflict Research, Copenhagen, and part-time lecturer at the University of Copenhagen. He is the author and editor of several books on alternative defense, and the founder and director of the Global Non-Offensive Defence Network.